OF THE ELEMENTS

	IB	IIB	IIIA	IVA	VA	VIA	VIIA	0
								2 **He** Helium 4.00260
			5 **B** Boron 10.81	6 **C** Carbon 12.011	7 **N** Nitrogen 14.0067	8 **O** Oxygen 15.9994	9 **F** Fluorine 18.99840	10 **Ne** Neon 20.179
			13 **Al** Aluminum 26.98154	14 **Si** Silicon 28.086	15 **P** Phosphorus 30.97376	16 **S** Sulfur 32.06	17 **Cl** Chlorine 35.453	18 **Ar** Argon 39.948
28 **Ni** Nickel 58.71	29 **Cu** Copper 63.546	30 **Zn** Zinc 65.38	31 **Ga** Gallium 69.72	32 **Ge** Germanium 72.59	33 **As** Arsenic 74.9216	34 **Se** Selenium 78.96	35 **Br** Bromine 79.904	36 **Kr** Krypton 83.80
46 **Pd** Palladium 106.4	47 **Ag** Silver 107.868	48 **Cd** Cadmium 112.40	49 **In** Indium 114.82	50 **Sn** Tin 118.69	51 **Sb** Antimony 121.75	52 **Te** Tellurium 127.60	53 **I** Iodine 126.9045	54 **Xe** Xenon 131.30
78 **Pt** Platinum 195.09	79 **Au** Gold 196.9665	80 **Hg** Mercury 200.59	81 **Tl** Thallium 204.37	82 **Pb** Lead 207.2	83 **Bi** Bismuth 208.9804	84 **Po** Polonium (210)a	85 **At** Astatine (210)a	86 **Rn** Radon (222)a

metals ← → nonmetals

63 **Eu** Europium 151.96	64 **Gd** Gadolinium 157.25	65 **Tb** Terbium 158.9254	66 **Dy** Dysprosium 162.50	67 **Ho** Holmium 164.9304	68 **Er** Erbium 167.26	69 **Tm** Thulium 168.9342	70 **Yb** Ytterbium 173.04	71 **Lu** Lutetium 174.97
95 **Am** Americium (243)a	96 **Cm** Curium (247)a	97 **Bk** Berkelium (249)a	98 **Cf** Californium (251)a	99 **Es** Einsteinium (254)a	100 **Fm** Fermium (253)a	101 **Md** Mendelevium (256)a	102 **No** Nobelium (254)a	103 **Lr** Lawrencium (257)a

Introduction to Chemistry

Introduction to Chemistry

Charles E. Mortimer
Muhlenberg College

D. VAN NOSTRAND COMPANY
New York Cincinnati Toronto London Melbourne

TO L.R.M. AND E.H.M.

D. Van Nostrand Company Regional Offices:
New York Cincinnati

D. Van Nostrand Company International Offices:
London Toronto Melbourne

Library of Congress Catalog Card Number: 76-58670
ISBN: 0-442-25569-1

Published by **D. Van Nostrand Company**
135 West 50th Street, New York, N.Y. 10020

10 9 8 7 6 5 4 3

Preface

Designing an introductory chemistry course for college students is difficult. The subject is large and growing. Today's students have widely different interests, academic backgrounds, and abilities. The course must take into account the requirements of diverse career objectives. A text should help instructors respond to their particular problems. INTRODUCTION TO CHEMISTRY was written as an alternative to *CHEMISTRY: A CONCEPTUAL APPROACH* and in response to some instructors' suggestions as to how their needs could be met, especially in the organization of topics.

The beginning chapters offer a gradual introduction to the subject. Atomic structure and chemical bonding, which some students find abstract and difficult, are postponed until the students are better able to handle them. The principles of stoichiometry, which are central to an understanding of all chemical concepts, are introduced in Chapter 2. The early introduction of this topic permits it to be expanded and reinforced throughout the entire course. The basic principles of thermochemistry, which also are of fundamental importance, are introduced in Chapter 3. The early and systematic discussion of thermochemistry prepares the way for the use of energy concepts (such as lattice energy, ionization energy, and bond energy) in the development of later topics that focus in other directions. The topical order of INTRODUCTION TO CHEMISTRY should facilitate the design of a good laboratory program closely correlated to classroom work.

At the same time, the format of this book is intended to permit relatively wide latitude in course organization. Instructors should not feel constrained to present all the material in this book in the exact order in which it is given.

In recent years, the career goals as well as the mathematical and scientific backgrounds of students who enroll in introductory chemistry have become more diverse than ever. Instructors, therefore, can assume no uniform level of attainment on the part of entering students. In INTRODUCTION TO CHEMISTRY, each concept is developed slowly and thoroughly in a way that should make it understandable to a student who has had no prior exposure to chemistry. Important terms are set in boldface type the first time they are used. Numerous problems are solved as examples. Step-by-step directions for the solution of basic types of problems are set off in boxes. Students will find this boxed material useful for initial assignments and also for reference in later work. The use of calculus is avoided.

Approximately one thousand chapter-end problems of varying difficulty are included. Similar problems are grouped together, and the problems of a chapter are arranged in the order in which the topics are discussed in that chapter. Answers to selected numerical problems are given in Appendix G. A *Solutions Manual* containing worked-out solutions for mathematical problems and an *Answer Booklet* containing answers only are available. The *Student Self-Study Guide,* by Michaeleen P. Lee, is keyed to the text and designed to be used independently by students.

Historical anecdotes appear throughout the text to enliven the material. They illustrate that chemistry is a human endeavor, that the science grows and changes, that creative imagination plays a role in the development of chemical thought, and that a full understanding of any phenomenon is never obtained—truth is only approached.

The growing acceptance of the International System of Units (SI) makes it important that students be introduced to this official metric system. With the exception of the United States, all the major industrial powers of the world have adopted SI units exclusively. In the interest of world-wide scientific communication and international trade, sooner or later the United States must convert. Many scientific groups and agencies (including the National Bureau of Standards) have recommended or officially adopted SI units. Some upper level chemistry and physics books use the units and the secondary schools of some states have begun to use SI units.

The rules for the use of the International System appear in *Le Système International d'Unités,* published by the International Bureau of Weights and Measures, which is available in an English translation prepared jointly by the National Bureau of Standards, of the United States, and the National Physical Laboratory, of the United Kingdom.* In INTRODUCTION TO CHEMISTRY, the joule, rather than the calorie is used for energy measurements. The ångstrom unit is not employed. Nanometers ($1 \text{ nm} = 10^{-9} \text{ m}$) are used in place of ångstrom units for the wavelengths of spectral lines ($4500 \text{ Å} = 450 \text{ nm}$), and picometers ($1 \text{ pm} = 10^{-12} \text{ m}$) are used for bond distances ($1.98 \text{ Å} = 198 \text{ pm}$). The standard atmosphere, which may be used with SI units for the present, is employed instead of the pascal, which is the SI unit of pressure. Since the International Committee of Weights and Measures considers it preferable to avoid the use of the torr, this unit is not employed to express the results of pressure measurements.

I sincerely thank the following for their suggestions and comments: David Herlocker, Western Maryland College, George B. Kauffman, California State University, Fresno, Michaeleen P. Lee, Bucks County Community College, Karl Marhenke, Cabrillo College, and Louise S. Shive, Muhlenberg College.

Suggestions for the improvement of this book will be welcomed.

Charles E. Mortimer

*Page, Chester H., and Vigoureux, Paul, ed. *The International System of Units (SI)*, National Bureau of Standards Special Publication 330, U.S. Government Printing Office, Washington, D.C. (1972).

Contents

Chapter 1

Introduction

Until the early years of the nineteenth century, the tendency to specialize in a particular branch of the sciences was not strong. Scientists commonly followed their interests no matter where they led. John Dalton (1766–1844) is best known for his atomic theory. But the scientific interests of Dalton ranged widely and included topics in botany, meteorology, chemistry, physiology, and physics.

As scientific knowledge grew, scientific specialization increased. In the first half of the nineteenth century, such specialization became the norm. Some older scientists resisted the change. Michael Faraday (1791–1867), whose major accomplishments are in the chemistry of electrolysis and the physics of electricity and magnetism, refused to be called a physicist, a chemist, or any other kind of -ist; he regarded himself as a natural philosopher.

By the beginning of the twentieth century, science had become highly organized. The interests of the various branches of science, however, frequently overlap. Ernest Rutherford (1871–1937), who is noted for his investigations into the nature of radioactive decay and the nuclear structure of the atom, considered himself to be a physicist. Rutherford, a man with a keen sense of humor, once wrote to a chemist friend that the authors of some articles that were critical of Rutherford's work were "damned fools, whom I think must once have been chemists."

In 1908, Rutherford, who was then a professor of physics at the University of Manchester, was awarded a Nobel prize. The prize was awarded, however, in chemistry, not physics! When the awards were presented, a banquet was given in Stockholm in honor of the prize winners. At the banquet, Rutherford made a speech in which he said that he had observed many transformations in his scientific work, but not one was as rapid as his own transformation from physicist to chemist.

The boundaries of the various branches of science are indistinct. The interests of the branches coincide at times, and they share common scientific concepts and methods. A precise definition of chemistry, therefore, is impossible.

Nevertheless, there is a common, if somewhat vague, understanding of what chemistry includes. **Chemistry** is usually defined as the science that deals with the composition, properties, and transformations of matter. Since matter is anything that occupies space and has mass, the scope of chemistry is enormous. This book should provide an introduction to chemistry and its unifying principles.

1.1 THE DEVELOPMENT OF MODERN CHEMISTRY

Modern chemistry, which emerged late in the eighteenth century, took hundreds of years to develop. The story of its development can be divided roughly into five periods.

1. **Practical arts (--- to 600 BC)** The production of metals from ores, the manufacture of pottery, brewing, baking, and the preparation of medicines, dyes, and drugs are ancient arts. Archaeological evidence proves that the inhabitants of ancient Egypt and Mesopotamia were skilled in these crafts, but how and when they developed is not known.

   These arts, which are chemical processes, became highly developed during this period. The development, however, was *empirical,* that is, based on practical experience alone without reference to underlying chemical principles. The Egyptian craftsman knew how to obtain copper by heating malachite ore with charcoal. He did not know, nor did he seek to know, why his process worked and what actually occurred in the fire.

2. **Greek (600 BC to 300 BC)** The philosophical aspect (or theoretical aspect) of chemistry began in classical Greece about 600 BC. The foundation of Greek science was the search for principles through which an understanding of nature could be obtained. Two theories of the Greeks became very important in the centuries that followed.
   a. A concept that all substances are composed of four elements (earth, air, fire, and water) in various proportions originated with Greek philosophers of this period.
   b. A theory that matter consists of separate and distinct units called **atoms** was proposed by Leucippus in the fifth century BC.

   Plato proposed that the atoms of one element differ in shape from the atoms of another. Furthermore, he believed that atoms of one element could be changed (or **transmuted**) into atoms of another by changing the shape of the atoms.

   The concept of transmutation is also found in Aristotle's theories. Aristotle (who did not believe in the existence of atoms) proposed that the elements, and therefore all substances, are composed of the same primary matter and differ only in the forms that this primary matter assumes. To Aristotle, the form included not only the shape but also the qualities (such as color and hardness) that distinguish one substance from others. He proposed that changes in form constantly occur in nature and that all material things (animate and inanimate) grow and develop from immature forms to adult forms. (Throughout the middle ages, men believed that minerals could grow and that mines would be replenished after minerals were removed from them.)

3. **Alchemy (300 BC to 1650 AD)** The philosophical tradition of ancient Greece and the craft tradition of ancient Egypt met in Alexandria, Egypt (the city founded by Alexander the Great in 331 BC), and *alchemy* was the result of the union. The early alchemists used Egyptian techniques for the handling of materials to investigate theories concerned with the nature of matter. Books written in Alexandria (the oldest known works on chemical topics) contain diagrams of chemical apparatus and descriptions of many laboratory operations, for example, distillation, crystallization, and sublimation.

   The philosophical content of alchemy incorporated elements of astrology and mysticism into the theories of the earlier Greeks. A dominant interest of the alchemists was the transmutation of base metals, such as iron or lead, into the noble metal, gold. They believed that a metal could be changed by changing its qualities (notably its color) and that such changes occur in nature—that metals strive for the perfection represented by gold. Furthermore, the alchemists believed that these changes could be brought about by means of a very small amount of a powerful transmuting agent (later called the *philosopher's stone*).

In the seventh century AD, the Arabs conquered the centers of Hellenistic civilization (including Egypt in 640 AD), and alchemy passed into their hands. Greek texts were translated into Arabic and served as the foundation for the work of Arab alchemists. The Arabs called the philosopher's stone *aliksir* (which was later corrupted into *elixir*). Arab alchemists believed that this substance could not only ennoble metals by transmuting them into gold but also could ennoble life by curing all diseases. For centuries afterward, the two principal goals of alchemy were the transmutation of base metals into gold and the discovery of an *elixir of life* that could make man immortal by preventing death.

In the twelfth and thirteenth centuries, alchemy was gradually introduced into Europe by means of the translation of Arabic works into Latin. Most of the translations were made in Spain where, after the Islamic conquest in the eighth century, a rich Moorish culture was established and flourished.

European alchemists added little that was new to alchemical theory. Their work is important because they preserved the large body of chemical data that they received from the past, added to it, and passed it on to later chemists.

Alchemy lasted until the seventeenth century. Gradually the theories and attitudes of the alchemists began to be questioned. The work of Robert Boyle, who published *The Sceptical Chymist* in 1661, is noteworthy. Although Boyle believed that the transmutation of base metals into gold might be possible, he severely criticized alchemical thought. Boyle emphasized that chemical theory should be derived from experimental evidence.

4. Phlogiston (1650 to 1790) Throughout most of the eighteenth century, the phlogiston theory dominated chemistry. This theory, which was later shown to be erroneous, was principally the work of Georg Ernst Stahl. *Phlogiston* (a "fire principle") was assumed to be a constituent of any substance that could undergo combustion.

Upon combustion, a substance was thought to lose its phlogiston and be reduced to a simpler form. Air was believed to function in a combustion by carrying off the phlogiston as it was released. Whereas we would think of the combustion of wood in the following terms

$$\text{wood} + \text{oxygen gas (from air)} \rightarrow \text{ashes} + \text{oxygen-containing gases}$$

according to the phlogiston theory,

$$\text{wood} \rightarrow \text{ashes} + \text{phlogiston (removed by air)}$$

Wood, therefore, was believed to be a compound composed of ashes and phlogiston. Readily combustible materials were thought to be rich in phlogiston.

The phlogiston theory interpreted **calcination** in a similar way. The formation of a metal oxide (called a calx) by heating a metal in air is called a calcination.

metal + oxygen gas (from air) → calx (metal oxide)

According to the phlogiston theory, a metal is assumed to be a compound composed of a calx and phlogiston. Calcination, therefore, was thought to be the loss of phlogiston by a metal:

metal → calx + phlogiston (removed by air)

The phlogiston theory was extended to explain many other chemical phenomena. The preparation of certain metals, for example, can be accomplished by heating the metal oxide with carbon:

calx (metal oxide) + carbon → metal + carbon monoxide gas

In a process of this type, the carbon (supposedly rich in phlogiston) was thought to replace the phlogiston lost by calcination:

calx + phlogiston (from carbon) → metal

One difficulty inherent in the phlogiston theory was never adequately explained. When wood burns, it supposedly *loses* phlogiston and the resulting ashes weigh *less* than the original piece of wood. On the other hand, in calcination, the *loss* of phlogiston is accompanied by an increase in weight since the calx weighs *more* than the original metal. The adherents of the phlogiston theory recognized this problem, but throughout most of the eighteenth century, the importance of weighing and measuring was not realized.

5. **Modern chemistry (1790 ---)** The work of Antoine Lavoisier in the late eighteenth century is generally regarded as the beginning of modern chemistry. Lavoisier deliberately set out to overthrow the phlogiston theory and revolutionize chemistry. He relied on the results of quantitative experimentation (he used the chemical balance extensively) to arrive at his explanations of a number of chemical phenomena.

The **law of conservation of mass** states that there is no detectable change in mass during the course of a chemical reaction. In other words, the total mass of all materials entering into a chemical reaction equals the total mass of all the products of the reaction. This law was implicit in the work of earlier men, but Lavoisier stated it explicitly and used it as the cornerstone of his science. To Lavoisier, therefore, the phlogiston theory was impossible.

The roles that gases play in reactions proved to be a stumbling block to the development of chemical theory. When the law of conservation of mass is applied to a combustion or to a calcination, the masses of the gases used or produced in these reactions must be taken into account. The correct interpretation of these processes, therefore, had to wait until chemists identified the gases involved and developed methods to handle and measure gases. Lavoisier drew upon the results of other scientists' work with gases to explain these reactions.

In interpreting chemical phenomena, Lavoisier used the modern definitions of elements and compounds (Section 1.2). The phlogiston theory regarded a metal as a *compound* composed of a calx and phlogiston. Lavoisier showed that a metal is an *element* and that the corresponding calx is a *compound* composed of the metal and oxygen from the air.

In his book *Traité Elémentaire de Chimie* (*Elementary Treatise on Chemistry*), published in 1789, Lavoisier used essentially modern terminology. The present-day language of chemistry is based on the system of nomenclature that Lavoisier devised.

The achievements of scientists since the 1790's are described throughout this book. More has been learned about chemistry in the two centuries following Lavoisier than in the twenty centuries preceding him. Chemistry has gradually developed five principal branches; these divisions, however, are arbitrary and subject to criticism.

a. **Organic chemistry**—the chemistry of most of the compounds of carbon. At one time it was assumed that these compounds could only be obtained from plant or animal life or derived from other compounds that had been obtained from living material.
b. **Inorganic chemistry**—the chemistry of all the elements except carbon. Some simple carbon compounds (for example, carbon dioxide and the carbonates) are traditionally classified as inorganic compounds.
c. **Analytical chemistry**—the identification of the composition, both qualitative and quantitative, of substances.
d. **Physical chemistry**—the study of the principles that underlie the structure of matter and chemical transformations.
e. **Biochemistry**—the chemistry of living systems, both plant and animal.

1.2 ELEMENTS, COMPOUNDS, AND MIXTURES

The ancient Greeks originated the concept that all matter is composed of a limited number of simple substances called **elements.** The Greeks assumed that all terrestrial matter is derived from four elements: *earth, air, fire,* and *water.* Since heavenly bodies were thought to be perfect and unchangeable, celestial matter was assumed to be composed of a different element, which later came to be known as *quintessence* (from Latin, meaning *fifth element*). This Greek theory dominated scientific thought for centuries.

In 1661, Robert Boyle proposed an essentially modern definition of an element in his book *The Sceptical Chymist:*

> I now mean by Elements . . . certain Primitive and Simple, or perfectly unmingled bodies; which not being made of any other bodies, or of one another, are the Ingredients of which all those call'd perfectly mixt Bodies are immediately compounded, and into which they are ultimately resolved.

Boyle made no attempt to identify specific substances as elements. He did, however, emphasize that the proof of the existence of elements, as well as the identification of them, rested on chemical experimentation.

Boyle's concept of a chemical element was firmly established by Lavoisier in the following century. Lavoisier accepted a substance as an element if it could not be decomposed into simpler substances. Furthermore, he showed that a compound is produced by the union of elements. Lavoisier correctly identified 23 elements (although he incorrectly included light, heat, and several simple compounds in his list).

At the present time, 106 elements are known. Of these, 85 have been isolated from natural sources; the remainder have been prepared by nuclear reactions (Section 23.5).

Each element is assigned a one or two letter symbol that has been decided upon by international agreement. Whereas the name of an element may differ from one language to another, the symbol does not. Nitrogen, for example, is called *azoto* in Italian and *stickstoff* in German, but the symbol for nitrogen is N in any language. These symbols are listed in the table of the elements that appears inside the back cover of this book.

Most of the symbols correspond closely to the English names of the elements. Some of them, however, do not. The symbols for some of the elements have been assigned on the basis of their Latin names; these elements are listed in Table 1.1. The symbol for tungsten, W, is derived from the German name for the element, wolfram.

Table 1.1 Symbols of elements derived from Latin names

ENGLISH NAME	LATIN NAME	SYMBOL
antimony	stibium	Sb
copper	cuprum	Cu
gold	aurum	Au
iron	ferrum	Fe
lead	plumbum	Pb
mercury	hydrargyrum	Hg
potassium	kalium	K
silver	argentum	Ag
sodium	natrium	Na
tin	stannum	Sn

The 15 most abundant elements in the earth's crust, bodies of water, and atmosphere are listed in Table 1.2. This classification includes those parts of the universe from which we can obtain the elements. The earth consists of a core (which is probably composed of iron and nickel) surrounded successively by a mantle and a thin crust. The crust is about 20 to 40 miles thick and constitutes only about 1% of the earth's mass.

Table 1.2 Abundance of the elements (earth's crust, bodies of water, and atmosphere)

RANK	ELEMENT	SYMBOL	PERCENT BY WEIGHT
1	oxygen	O	49.2
2	silicon	Si	25.7
3	aluminum	Al	7.5
4	iron	Fe	4.7
5	calcium	Ca	3.4
6	sodium	Na	2.6
7	potassium	K	2.4
8	magnesium	Mg	1.9
9	hydrogen	H	0.9
10	titanium	Ti	0.6
11	chlorine	Cl	0.2
12	phosphorus	P	0.1
13	manganese	Mn	0.1
14	carbon	C	0.09
15	sulfur	S	0.05
	all others		0.56

If the entire earth were considered, a list different from that of Table 1.2 would result, and the most abundant element would be iron. On the other hand, the most abundant element in the universe as a whole is hydrogen, which is thought to constitute about 75% of the total mass of the universe.

Whether an element finds wide commercial use depends not only upon its abundance but also upon its accessibility. Some familiar elements (such as copper, tin, and lead) are not particularly abundant but are found in nature in deposits of ores from which they can be obtained readily. Other elements that are more abundant (such as titanium, rubidium, and zirconium) are not widely used because either their ores are widespread in nature or the extraction of the elements from their ores is difficult or expensive.

Compounds are substances that are composed of two or more elements in fixed proportions. The **law of definite proportions** (first proposed by Joseph Proust in 1799) states that a pure compound always consists of the same elements combined in the same proportion by mass.* The compound *water,* for

* Mass is a measure of the *quantity* of matter. The mass of a body does not change. Weight is the gravitational *force* of attraction exerted by the earth on a body. The weight of a body varies with the distance of that body from the center of the earth.

example, is always formed from the elements *hydrogen* and *oxygen* in the proportion 11.19% hydrogen to 88.81% oxygen. Over ten thousand inorganic compounds are known, and over one million organic compounds have been either synthesized or isolated from natural sources. Compounds have properties that are different from the properties of the elements of which they are composed.

An element or a compound is called a **pure substance.** All other kinds of matter are mixtures. **Mixtures** consist of two or more pure substances and have variable compositions. The properties of a mixture depend upon the composition of the mixture and the properties of the pure substances that form the mixture. There are two types of mixtures. A **heterogeneous mixture** is not uniform throughout but consists of parts that are physically distinct. A sample containing iron and sand, for example, is a heterogeneous mixture. A **homogeneous mixture** appears uniform throughout and is usually called a **solution.** Air, salt dissolved in water, and a silver–gold alloy are examples of a gaseous, a liquid, and a solid solution, respectively. The classification of matter is summarized in Figure 1.1.

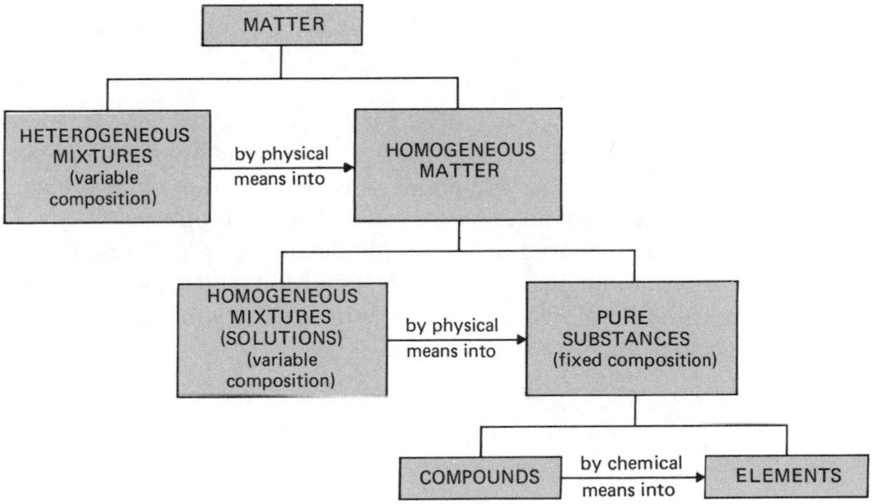

Figure 1.1 Classification of matter.

1.3 THE METRIC SYSTEM

The metric system of measurement is used in all scientific studies. As a result of a treaty signed in 1875, metric conventions are established and modified when necessary by international agreement. From time to time, an international group, the General Conference of Weights and Measures, meets to ratify improvements in the metric system. The currently approved **International System of Units** (*Le Système International d'Unités,* officially abbreviated **SI**) is a modernization and simplification of an old system which developed from one proposed by the French Academy of Science in 1790. Lavoisier was a member of the committee that formulated the original system.

The International System is founded on seven **base units** and two **supplementary units** (see Table 1.3 and Appendix A). The selection of primary standards for the base units is arbitrary. For example, the primary standard of mass, the kilogram, is defined as the mass of a cylinder of platinum–irridium alloy that is kept at the International Bureau of Weights and Measures at Sèvres, France. Throughout the years, the primary standards for some base units have been changed when new standards appeared to be superior to old ones.

Multiples or fractions of base units are indicated by the use of prefixes (see Table 1.4). The base unit of length, the meter (m), is usually not used to record the distances between cities. A larger unit, the kilometer (km), is more convenient. The kilometer is equal to 1000 meters, and the name for this unit is obtained by adding the prefix *kilo-* (which means 1000 ×) to the name of the base unit:

$$1 \text{ km} = 1000 \text{ m}$$

The centimeter (cm) is a smaller unit than the meter. The prefix *centi-* means 0.01 ×, and 1 centimeter is 0.01 meter:

$$1 \text{ cm} = 0.01 \text{ m}$$

Note that the name for the base unit for mass, the kilogram, contains a prefix. The names of other units of mass are obtained by substituting other prefixes for the prefix *kilo*. The names of no other base units contain prefixes.

Other SI units, called **derived units,** are obtained from the base units by algebraic combination. Examples are the SI unit for volume, which is the cubic meter (abbreviated m^3) and the SI unit for velocity, which is the meter per second (abbreviated m/s or $m \cdot s^{-1}$).

Table 1.3 Base units and supplementary units of the International System of units

	MEASUREMENT	UNIT	SYMBOL
Base units	length	meter	m
	mass	kilogram	kg
	time	second	s
	electric current	ampere	amp
	temperature	kelvin	K
	amount of substance	mole	mol
	luminous intensity	candela	cd
Supplementary units	plane angle	radian	rad
	solid angle	steradian	sr

Table 1.4 Prefixes used to modify unit terms in the metric system

PREFIX	ABBREVIATION	FACTOR	
tetra-	T-	1 000 000 000 000 ×	(10^{12})
giga-	G-	1 000 000 000 ×	(10^{9})
mega-	M-	1 000 000 ×	(10^{6})
kilo-	k-	1 000 ×	(10^{3})
hecto-	h-	100 ×	(10^{2})
deka-	da-	10 ×	(10)
deci-	d-	0.1 ×	(10^{-1})
centi-	c-	0.01 ×	(10^{-2})
milli-	m-	0.001 ×	(10^{-3})
micro-	μ-	0.000 001 ×	(10^{-6})
nano-	n-	0.000 000 001 ×	(10^{-9})
pico-	p-	0.000 000 000 001 ×	(10^{-12})
femto-	f-	0.000 000 000 000 001 ×	(10^{-15})
atto-	a-	0.000 000 000 000 000 001 ×	(10^{-18})

Some derived units are given special names. The SI unit for force, for example, is the newton, N. This unit is derived from the base units for mass (the kilogram, kg), length (the meter, m), and time (the second, s). The **newton** is the force that gives a mass of 1 kg an acceleration of 1 m/s^2.

$$1\ \text{N} = 1\ \text{kg}\cdot\text{m/s}^2$$

Rather than derive all the units that are employed in this book here, each unit, except for the simpler ones, is described when the need for it arises. Note, also, that a discussion of metric units appears in the Appendix.

The current terminology of the International System has been developed since 1960. Some units that were defined prior to this time do not conform to SI rules and are not SI units. The use of some of these units is, however, permitted. The liter, for example, which is defined as 1 cubic decimeter (and hence is 1000 cm^3), may be used in addition to the official SI unit of volume, the cubic meter. Certain other units that are not a part of SI are to be retained for a limited period of time. The standard atmosphere (atm, a unit of pressure) falls into this category. The use of other units that are outside the International System is discouraged. For example, the International Committee of Weights and Measures considers it preferable to avoid the use of the calorie as an energy unit.

Not all scientists have adopted SI units, but use of the system appears to be growing. Strict adherence to the International System, however, poses a problem since it eliminates some units that previously have been used widely. Since much of the data found in the chemical literature has been recorded in units that are not SI units, one must be familiar with both the old and the new units.

1.4 SCIENTIFIC NOTATION

Very large and very small numbers are frequently encountered in scientific studies. The velocity of light in a vacuum, for example, is

29,979,000,000 cm/s

and the distance between the centers of two hydrogen atoms in an H_2 molecule is

0.0000000075 cm

Scientific notation is used to simplify the handling of cumbersome values such as these. When scientific notation is employed, the value is expressed in the form

$$a \times 10^n$$

where *a*, the decimal part, is a number with one digit to the left of the decimal point and all others to the right, and *n*, the exponent of 10, is a positive or negative integer or zero.

A number can be converted into this form by moving the decimal point until there is only one nonzero digit to the left of it. For each place the decimal point is moved to the *left*, *n* is *increased* by one. For each place the decimal point is moved to the *right*, *n* is *decreased* by one. For example,

$$29{,}979{,}000{,}000 \text{ cm/sec} = 2.9979 \times 10^{10} \text{ cm/s}$$
$$0.0000000075 \text{ cm} = 7.5 \times 10^{-9} \text{ cm}$$

Mathematical operations involving numbers expressed in this manner are carried out in the following ways.

1. **Multiplication** The decimal parts are multiplied and the exponents of 10 are added algebraically.

$$\begin{aligned}(3.0 \times 10^5)(2.0 \times 10^2) &= (3.0 \times 2.0) \times 10^{5+2} \\ &= 6.0 \times 10^7 \\ (4.0 \times 10^7)(5.0 \times 10^{-3}) &= (4.0 \times 5.0) \times 10^{7+(-3)} \\ &= 20 \times 10^4 \\ &= 2.0 \times 10^5\end{aligned}$$

2. **Division** The decimal parts are divided, and the exponent of 10 in the denominator is algebraically subtracted from the exponent of 10 in the numerator.

$$\begin{aligned}\frac{6.89 \times 10^{-7}}{3.36 \times 10^3} &= \left(\frac{6.89}{3.36}\right) \times 10^{(-7)-(+3)} \\ &= 2.05 \times 10^{-10}\end{aligned}$$

3. **Addition and subtraction** The numbers must all be expressed with the same power of 10. The answer, which has this same power of 10, is found by adding or subtracting the decimal parts.

$$(6.25 \times 10^3) + (3.0 \times 10^2) = (6.25 \times 10^3) + (0.30 \times 10^3)$$
$$= 6.55 \times 10^3$$

4. **Taking a root** When a square root is taken, the number is written in such a way that the exponent of 10 is perfectly divisible by 2. The answer is obtained by taking the square root of the decimal part and dividing the power of 10 by 2.

$$\sqrt{2.21 \times 10^{-7}} = \sqrt{22.1 \times 10^{-8}}$$
$$= 4.70 \times 10^{-4}$$

When a cube root is taken, the cube root of the decimal part is obtained and the exponent of 10 is divided by 3.

$$\sqrt[3]{1.86 \times 10^8} = \sqrt[3]{186 \times 10^6}$$
$$= 5.71 \times 10^2$$

5. **Raising to a power** When a number is squared, the decimal part is squared and the exponent of 10 is multiplied by 2.

$$(1.36 \times 10^4)^2 = (1.36)^2 \times 10^{2(+4)}$$
$$= 1.85 \times 10^8$$

When a number is cubed, the decimal is cubed and the exponent of 10 is multiplied by 3.

$$(2.06 \times 10^{-5})^3 = (2.06)^3 \times 10^{3(-5)}$$
$$= 8.74 \times 10^{-15}$$

In general,

$$(a \times 10^n)^p = a^p \times 10^{p(n)}$$

1.5 SIGNIFICANT FIGURES

Every measurement is uncertain to some extent. Suppose, for example, that we wish to measure the mass of an object. If we use a platform balance, we can determine the mass to the nearest 0.1 g. An analytical balance, on the other hand, is capable of giving results correct to the nearest 0.0001 g. The exactness, or precision, of the measurement depends upon the limitations of the measuring device and the skill with which it is used.

The precision of a measurement is indicated by the number of figures used to record it. The digits in a properly recorded measurement are **significant figures.** These figures include all those that are known with certainty plus one more which is an estimate.

Suppose that a platform balance is used, and the mass of an object is determined to be 12.3 g. The chances are slight that the actual mass of the object is exactly 12.3 g, no more nor less. We are sure of the first two figures (the 1 and the 2); we know that the mass is greater than 12 g. The third figure (the 3), however, is somewhat inexact. At best, it tells us that the true mass lies closer to 12.3 g than to either 12.2 g or 12.4 g. If, for example, the actual mass were 12.28 . . . g or 12.33 . . . g, the value would be correctly recorded in either case as 12.3 g *to three significant figures*.

If, in our example, we add a zero to the measurement, we indicate a value containing *four significant figures* (12.30 g) which is incorrect and misleading. This value indicates that the actual mass is between 12.29 g and 12.31 g. We have, however, no idea of the magnitude of the integer of the second decimal place since we have determined the value only to the nearest 0.1 g. The zero does not indicate that the second decimal place is unknown or undetermined. Rather, it should be interpreted in the same way that any other figure is (see, however, rule 1 that follows). Since the uncertainty in the measurement lies in the 3, this digit should be the last significant figure reported.

The following rules can be used to determine the proper number of significant figures to be recorded for a measurement.

1. Zeros used to locate the decimal point are not significant. Suppose that the distance between two points is measured as 3 cm. This measurement could also be expressed as 0.03 m since 1 cm is 0.01 m.

$$3 \text{ cm} = 0.03 \text{ m}$$

Both values, however, contain only *one* significant figure. The zeros in the second value, since they merely serve to locate the decimal point, are not significant. The precision of a measurement cannot be increased by changing units.

Zeros that arise as a part of a measurement are significant. The number 0.0005030 has four significant figures. The zeros after 5 are significant. Those preceding the numeral 5 are not significant since they have been added only to locate the decimal point.

Occasionally, it is difficult to interpret the number of significant figures in a value that contains zeros, such as 600. Are the zeros significant, or do they merely serve to locate the decimal point? This type of problem can be avoided by using scientific notation. The decimal point is located by the power of 10 employed; the first part of the term contains only significant figures. The value 600, therefore, can be expressed in any of the following ways depending upon how precisely the measurement has been made.

6.00×10^2	(three significant figures)
6.0×10^2	(two significant figures)
6×10^2	(one significant figure)

2. Certain values, such as those that arise from the definition of terms, are exact. For example, by definition, there are *exactly* 1000 ml in 1 liter. The value 1000 may be considered to have an infinite number of significant figures (zeros) following the decimal point.

   Values obtained by counting may also be exact. The H_2 molecule, for example, contains exactly 2 atoms, not 2.1 or 2.3. Other counts, however, are inexact. The population of the world, for example, is estimated and is not derived from an actual count.
3. At times, the answer to a calculation contains more figures than are significant. The following rules should be used to round off such a value to the correct number of digits.
   a. If the figure following the last number to be retained is less than 5, all the unwanted figures are discarded and the last number is left unchanged.

3.6247	is	3.62	to three significant figures

   b. If the figure following the last number to be retained is greater than 5, or 5 with other digits following it, the last figure is increased by 1 and the unwanted figures are discarded.

7.5647	is	7.565	to four significant figures
6.2501	is	6.3	to two significant figures

   c. If the figure following the last figure to be retained is 5 and there are only zeros following the 5, the 5 is discarded and the last figure is increased by 1 if it is an odd number or left unchanged if it is an even number. In a case of this type, the last figure of the rounded-off value is always an even number. Zero is considered to be an even number.

3.250	is	3.2	to two significant figures
7.635	is	7.64	to three significant figures
8.105	is	8.10	to three significant figures

The number of significant figures in the answer to a calculation depends upon the numbers of significant figures in the values used in the calculation. Consider the following problem. If we place 2.38 g of salt in a container that has a mass of 52.2 g, what will be the mass of the container plus salt? Simple addition gives 54.58 g. But we cannot know the mass of the two together any more precisely than we know the mass of one alone. The result must be rounded off to the nearest 0.1 g, which gives 54.6 g.

4. The result of an *addition* or *subtraction* should be reported to the same number of decimal places as that of the term with the least number of decimal places. The answer for the addition

$$\begin{array}{r} 161.032 \\ 5.6 \\ \underline{32.4524} \\ 199.0844 \end{array}$$

should be reported as 199.1 since the number 5.6 has only one digit following the decimal point.

5. The answer to a *multiplication* or *division* is rounded off to the same number of significant figures as is possessed by the least precise term used in the calculation. The result of the multiplication

$$152.06 \times 0.24 = 36.4944$$

should be reported as 36, since the least precise term in the calculation is 0.24 (two significant figures).

1.6 CHEMICAL CALCULATIONS

Units should be indicated as an integral part of all measurements. It makes little sense to say that the length of an object is 5.0. What does this value mean: 5.0 cm, 5.0 m, 5.0 ft? Careful use of units will simplify problem solving and reduce the probability of making errors.

The unit labels that are included in the terms employed in a calculation should undergo the same mathematical operations as the numbers. In any calculation, the units that appear in both numerator and denominator are canceled, and those remaining appear as a part of the answer. If the answer does not have the unit sought, a mistake has been made in the way that the calculation has been set up.

Many problems may be solved by the use of one or more **conversion factors.** A factor of this type is derived from an equality and is designed to convert a measurement from one unit into another. Suppose, for example, that we wish to calculate the number of centimeters in 5.00 inches (in.). The conversion factor that we need to solve the problem is derived from the exact relation

$$2.54 \text{ cm} = 1.00 \text{ in.}$$

If we divide both sides of this equation by 1.00 in., we obtain

$$\frac{2.54 \text{ cm}}{1.00 \text{ in.}} = 1$$

The factor (2.54 cm/1.00 in.) is equal to 1 since the numerator and the denominator are equivalent.

Our problem can be stated in the following way.

$$? \text{ cm} = 5.00 \text{ in.}$$

Multiplication by the conversion factor that we derived solves the problem

$$? \text{ cm} = 5.00 \cancel{\text{in.}} \left(\frac{2.54 \text{ cm}}{1.00 \cancel{\text{in.}}}\right) = 12.7 \text{ cm}$$

Since the factor is equal to 1, this operation does not change the value of the given quantity. Notice that the *inch* labels cancel, which leaves the answer in the desired unit, *centimeters*.

A second conversion factor can be derived from the equation

$$2.54 \text{ cm} = 1.00 \text{ in.}$$

by dividing both sides of the equation by 2.54 cm

$$1 = \frac{1.00 \text{ in.}}{2.54 \text{ cm}}$$

This factor, which is also equal to 1, is the reciprocal of the factor previously derived and can be used to convert centimeters to inches. For example, the number of inches that equals 20.0 cm can be found in the following way.

$$? \text{ in.} = 20.0 \cancel{\text{cm}} \left(\frac{1.00 \text{ in.}}{2.54 \cancel{\text{cm}}}\right) = 7.87 \text{ in.}$$

A single equality that relates two units, therefore, can be used to derive two conversion factors. The factors are reciprocals of one another. In the solution to a problem, the correct factor to use is the one that will lead to the cancellation of the unit, or units, that must be eliminated. Notice that this unit should be found in the denominator of the factor.

The solution to some problems requires the use of several factors. If we wish to find the number of centimeters in 0.750 ft, we can state the problem in the following way.

$$? \text{ cm} = 0.750 \text{ ft}$$

Since 1.00 ft = 12.0 in., we derive the conversion factor (12.0 in./1.00 ft), which is of course equal to 1. Multiplication by this factor converts feet into inches but does not complete the solution.

$$? \text{ cm} = 0.750 \cancel{\text{ft}} \left(\frac{12.0 \text{ in.}}{1.00 \cancel{\text{ft}}}\right)$$

The factor needed to convert inches into centimeters is (2.54 cm/1.00 in.), and thus

$$? \text{ cm} = 0.750 \cancel{\text{ft}} \left(\frac{12.0 \cancel{\text{in.}}}{1.00 \cancel{\text{ft}}}\right)\left(\frac{2.54 \text{ cm}}{1.00 \cancel{\text{in.}}}\right) = 22.9 \text{ cm}$$

Factors can also be derived from percentages, as shown in Example 1.1.

Example 1.1 Sterling silver is an alloy consisting of 92.5% Ag and 7.5% Cu. How many kilograms of sterling silver can be made from 3.00 kg of pure silver?

Solution The problem is stated in the following way.

$$? \text{ kg sterling} = 3.00 \text{ kg Ag}$$

It is clear that we need a factor in which the unit in the denominator is *kg Ag*. We can derive the desired factor from the percentage of silver in sterling silver. Since sterling silver is 92.5% Ag by mass, 100.0 kg of sterling silver contains 92.5 kg of silver.

100.0 kg sterling = 92.5 kg Ag

The factor we need, therefore, is (100.0 kg sterling/92.5 kg Ag). Notice that the label *kg Ag* appears in the denominator of this factor and will cancel the unit of the given quantity.

$$? \text{ kg sterling} = 3.00 \cancel{\text{kg Ag}} \left(\frac{100.0 \text{ kg sterling}}{92.5 \cancel{\text{kg Ag}}} \right) = 3.24 \text{ kg sterling}$$

Frequently, information is given in the form of ratios. The cost per unit item, the distance traveled per unit time, and the number of items per unit mass are examples. The word *per* implies division, and the number in the denominator is 1 (exactly) unless specified otherwise. A speed of 50 kilometers per hour is 50 km/1 hr.

The numerator and denominator of such a ratio are equivalent.

50 km = 1 hr

These ratios may be used, therefore, as conversion factors—either in the form in which they are given (50 km/1 hr) or in the inverted form (1 hr/50 km).

If a ratio is desired as the answer to a problem, the calculation is set up by using the numerator of the ratio as the quantity desired and the denominator of the ratio as the quantity given. If, for example, we wish to find a rate of travel in *kilometers per hour,* we state the problem in the following way.

? km = 1 hr

Example 1.2 What is the speed of a car (in km/hr) if it travels 16 km in 13 min?

Solution The problem is stated by writing

? km = 1 hr

The only information that is given in the problem and that can be used to derive a conversion factor is

16 km = 13 min

Since the answer is to be expressed in *km,* the factor that has this unit in the numerator is the one that should be used, (16 km/13 min).

The unit in the denominator of the factor (min), however, will not cancel the unit of the given quantity (hr). This cancellation can be brought about by using another factor, one derived from

$$60 \text{ min} = 1 \text{ hr}$$

The factor derived from this equality is 60 min/1 hr since the unit in the denominator (hr) will cancel the unit of the given quantity. Use of both factors gives

$$? \text{ km} = 1 \cancel{\text{hr}}\left(\frac{60 \cancel{\text{min}}}{1 \cancel{\text{hr}}}\right)\left(\frac{16 \text{ km}}{13 \cancel{\text{min}}}\right) = 74 \text{ km}$$

The car traveled at the rate of 74 km/hr.*

THE CONVERSION FACTOR METHOD OF PROBLEM SOLVING

1. State the problem. Write the unit in which the answer should be expressed, an equal sign, and that quantity given in the problem that will lead to a solution.
2. Derive a conversion factor in which the unit in the denominator is the same as the unit of the given quantity. The factor may be derived from information given in the problem or from the definition of a unit.
3. Write the conversion factor after the given quantity (written in step 1) to indicate multiplication. Cancel units. When this multiplication is performed, the answer will be expressed in the unit in the numerator of the factor.
4. If this unit is not the one sought, additional conversion factors must be employed. The unit in the denominator of each factor should cancel the unit in the numerator of the preceding factor.
5. Continue the process until the only uncanceled unit is the desired unit.
6. Perform the mathematical operations indicated and obtain the answer.

* Cancellation will not be indicated in future examples.

PROBLEMS

1.1 Give the names of the elements for which the symbols are: (a) Na, (b) Au, (c) Cu, (d) Cr, (e) Ag, (f) K, (g) Sn, (h) Fe, (i) Pb, (j) Hg.

1.2 What are the symbols for the following elements? (a) phosphorus, (b) carbon, (c) calcium, (d) chlorine, (e) sodium, (f) silicon, (g) sulfur, (h) aluminum, (i) magnesium, (j) manganese.

1.3 How do mixtures and compounds differ?

1.4 The reaction of 16 g of oxygen and 2 g of hydrogen produces 18 g of water. What law is illustrated by this fact?

1.5 What law is demonstrated by the fact that ammonia is always 82.2% nitrogen and 17.8% hydrogen?

1.6 (a) How many centimeters are in 1 kilometer? (b) How many kilograms are in 1 milligram? (c) How many nanoseconds are in 10 milliseconds? (d) How many tetrameters are in 100 micrometers?

1.7 The ångstrom unit (Å), which is defined as 10^{-8} cm, is not an SI unit. (a) How many nanometers and how many picometers are in 500 Å? (b) How many nanometers and how many picometers are in 0.50 Å?

1.8 How many liters are in 1 cubic meter? The liter is defined as 1 cubic decimeter.

1.9 The elevation of Denver, Colorado, is 5,470 feet. What is it in kilometers? One inch is exactly 2.54 cm.

1.10 A magnum is equal to $\frac{2}{5}$ of a gallon. How many liters is this? There are 473.2 ml in 1 pint, 1000 ml in 1 liter, and 8 pints in 1 gallon.

1.11 Write the following in scientific notation. (a) 35, (b) 7963, (c) 0.08205, (d) 2030.0, (e) 97,000,000, (f) 0.000510.

1.12 Write the following in scientific notation. (a) 4, (b) 0.000060008, (c) 219×10^3, (d) 63×10^{-2}, (e) 0.00295×10^5, (f) 0.0224×10^{-3}.

1.13 Determine the number of significant figures in each of the following. (a) 25, (b) 38.16, (c) 30.09, (d) 20.0, (e) 200, (f) 0.78596, (g) 0.07083, (h) 97,000,000, (i) 0.0002.

1.14 Perform the following calculations, and report the answer using the proper number of significant figures. (a) $6.973 + 21.52 + 300.1$, (b) $35.782 - 22.682$, (c) $822.43 - 90.5$, (d) $32 \times 587 \times 256$, (e) $7.8 + 0.002$, (f) $(1.61 \times 10^{15}) + (1.61 \times 10^{16})$.

1.15 Perform the following calculations and report the answer using the proper number of significant figures. (a) $(5.8 \times 10^{30}) - (3.2 \times 10^{29})$, (b) $4.3/8.71$, (c) $(10.5 \times 1.9)/6.82$, (d) $\sqrt{4 \times 10^3}$, (e) $(5.6 \times 10^{-2})^2$.

1.16 According to the U.S. Department of Agriculture, the average weight of a six-foot tall man is 167 pounds. If a six-foot tall man weighs 90.8 kg, is he overweight or underweight? One pound is 453.6 g.

1.17 Which is deeper, the Red Sea at 7,254 feet or the Bering Sea at 4.091 km? One inch is exactly 2.54 cm.

1.18 The human body is 0.004% iron. How many grams of iron does a

150 pound person contain? One pound is 453.6 g.

1.19 The speed limit on a highway is 50 miles per hour. Convert this value to kilometers per hour. The following relations are exact: 1 inch = 2.54 cm, 12 inches = 1 foot, 5,280 feet = 1 mile.

1.20 Convert the measurements of the following recipe (for biscuits) into milliliters. $1\frac{3}{4}$ cup flour, 1 teaspoon salt, $2\frac{1}{2}$ teaspoons baking powder, 4 tablespoons butter, $\frac{2}{3}$ cup milk. The following relations are given: 1 pint = 473.2 ml, 1 pint = 2 cups, 1 cup = 16 tablespoons, 1 tablespoon = 3 teaspoons. (This recipe is continued in Problem 3.1.)

1.21 The metric tonne (symbol, t) is defined as 1000 kg. Although this unit is not an SI unit, its use has been authorized. The pound and short ton (2000 pounds) are units of the English system. Since 1 pound is 453.6 g, how many metric tonnes are equivalent to one short ton? How many short tons are in 1.00 t?

1.22 If the metric system is adopted for everyday use, fabric will be sold in meters (not yards), milk in liters (not quarts), and meat in kilograms (not pounds). Given the following: 1 inch = 2.540 cm, 1 pint = 473.2 ml, and 1 pound = 453.6 g, what percentage increase does each of the following changes represent? (a) fabric: 1.00 meter instead of 1.00 yard, (b) milk: 1.00 liter instead of 1.00 quart, (c) meat: 1.00 kilogram instead of 2.00 pound.

Chapter 2

Stoichiometry

Alfred North Whitehead, the philosopher and mathematician, wrote, "All science as it grows towards perfection becomes mathematical in its ideas." Modern chemistry began when Lavoisier and chemists of his time recognized the importance of careful measurement and began to ask questions that could be answered quantitatively. **Stoichiometry** (derived from the Greek *stoicheion,* meaning "element," and *metron,* meaning "to measure") is the branch of chemistry that deals with the quantitative relationships between elements and compounds in chemical reactions. The atomic theory of matter is basic to this study.

2.1 DALTON'S ATOMIC THEORY

Credit for the first atomic theory is usually given to the ancient Greeks, but this concept may have had its origins in even earlier civilizations. Two theories prevailed among the Greeks. Aristotle (fourth century BC) believed that matter is continuous and hence, hypothetically, can be divided endlessly into smaller and smaller particles. The atomic theory of Leucippus and Democritus (fifth century BC) held that the subdivision of matter would ultimately yield atoms, which could not be further divided. The word *atom* is derived from the Greek word *atomos,* which means "uncut" or "indivisible."

The theories of the ancient Greeks were based on abstract thought, not on planned experimentation. For approximately two thousand years the atomic theory remained as mere speculation. The existence of atoms was accepted by Robert Boyle in his book *The Sceptical Chymist* (1661) and by Isaac Newton in his books *Principia* (1687) and *Opticks* (1704). John Dalton, however, proposed an atomic theory, which he developed in the years 1803 to 1808, that is a landmark in the history of chemistry.

Many scientists of the time believed that all matter consists of atoms, but Dalton went further. Dalton made the atomic theory quantitative by showing that it is possible to determine the relative masses of the atoms of different elements. The principal postulates of Dalton's theory are:

1. Elements are composed of extremely small particles called atoms. All atoms of the same element are alike, and atoms of different elements are different.
2. The separation of atoms and the union of atoms occur in chemical reactions. In these reactions, no atom is created or destroyed, and no atom of one element is converted into an atom of another element.
3. A chemical compound is the result of the combination of atoms of two or more elements in a simple numerical ratio.

Dalton believed that all atoms of a given element have equal atomic masses. Today we know that many elements consist of several types of atoms that differ in mass (see the discussion of isotopes in Section 5.6). All atoms of the same element, however, react chemically in the same way. We can, therefore, work with Dalton's theory by using an average mass for the atoms of each element. In most calculations, no mistake is made by assuming that an element consists of only one type of atom with an average mass, which is called the **atomic weight.**

Dalton derived the quantitative aspects of his theory from the laws of chemical change. His second postulate accounts for the **law of conservation of mass,** which states that there is no detectable change in mass during the course of a chemical reaction. Since chemical reactions consist of the separating and joining of atoms and since atoms are neither created nor destroyed in these processes, the total mass of all the materials entering into a chemical reaction must equal the total mass of all the products of the reaction.

The third postulate of Dalton's theory explains the **law of definite proportions,** which states that a pure compound always contains the same elements combined in the same proportions by mass. Since a given compound is the result of the combination of atoms of two or more elements in a fixed ratio, the proportions by mass of the elements present in the compound are also fixed.

On the basis of his theory, Dalton proposed a third law of chemical combination, the **law of multiple proportions.** This law states that when two elements, A and B, form more than one compound, the amounts of A that are combined in these compounds with a fixed amount of B are in a small whole-number ratio. This law follows from Dalton's view that the atoms in a compound are combined in a fixed proportion. If two elements form two compounds, the atom ratio of the compounds must be different. Since these atomic ratios are simple whole-number values, the relationship between the atom ratios of the two compounds is also a simple whole-number ratio.

For example, carbon and oxygen form two compounds: carbon dioxide and carbon monoxide. In carbon dioxide, *two* atoms of oxygen are combined with one atom of carbon, and in carbon monoxide, *one* atom of oxygen is combined with one atom of carbon. When the two compounds are compared, therefore, the masses of oxygen that combine with a fixed mass of carbon stand in the ratio of 2:1. The experimental verification of the law of multiple proportions was strong support for Dalton's theory.

A very important aspect of Dalton's work is that he showed that it is possible to determine the relative masses of atoms. Water is a compound that consists of 88.8% oxygen and 11.2% hydrogen. Dalton incorrectly assumed that one atom of oxygen is combined with one atom of hydrogen in water. On this basis, the mass of a single oxygen atom would be 88.8/11.2 (which is approximately 8/1) times the mass of a single hydrogen atom. The assignment of a mass of 1 to the hydrogen atom would give a mass of approximately 8 to the oxygen atom on this scale.*

The formulation that Dalton used for water is incorrect. Actually, one atom of oxygen is combined with *two* atoms of hydrogen. The oxygen atom, therefore, has a mass that is approximately 8 times the mass of *two* hydrogen atoms, or 16 times the mass of one hydrogen atom. Thus oxygen should have been assigned a relative mass of approximately 16.

The fact that Dalton made errors in assigning relative atomic masses does not detract from his achievement. He clearly showed that it is possible to derive relative atomic weights provided accurate analyses of compounds and the numbers of atoms combined in each compound are known.

Any relative atomic weight scale must be based on the arbitrary assignment of a value to one atom, which is chosen as a standard. Dalton used the hydrogen atom as his standard and assigned it a value of 1. In later years, chemists used naturally occurring oxygen as the standard and set its atomic weight equal to exactly 16. Today, a particular type of carbon atom (assigned a value of exactly 12) is employed as the standard (see Section 5.7). Carbon occurs in nature as a mixture of several types of atoms; the particular one

* Since the figures that Dalton used for the percent composition of water were very inaccurate, Dalton actually assigned a relative mass of 7 to the oxygen atom.

employed as the standard for atomic weights is the most abundant one. When the percentages and masses of all the types of carbon atoms are taken into account, the average atomic mass for naturally occurring carbon is 12.011, which is recorded as the atomic weight of carbon. A table of modern atomic weights appears inside the back cover of this book.

2.2 FORMULAS

The chemical symbols that are assigned to the elements (Section 1.2) are used to write formulas that describe the atomic composition of compounds. The formula of water is H_2O, which indicates that there are two atoms of hydrogen for every one atom of oxygen in the compound. The subscripts of the formula indicate the relative number of atoms of each type that are combined. If a symbol carries no subscript, the number one is assumed. The formula of sulfuric acid is H_2SO_4, which indicates that the combining ratio of this compound is two atoms of hydrogen to one atom of sulfur to four atoms of oxygen.

A **molecule** is a particle formed from two or more atoms. Some (but not all) compounds occur in molecular form. In these cases, the formula gives the number of atoms of each type in a single molecule of the compound. Formulas of this type are sometimes called **molecular formulas.** Both water and sulfuric acid are molecular in nature, and both H_2O and H_2SO_4 are molecular formulas.

The molecular formula of hydrogen peroxide, H_2O_2, indicates that there are two atoms of hydrogen and two atoms of oxygen in a molecule of hydrogen peroxide. Notice that the ratio of hydrogen atoms to oxygen atoms (2 to 2) is not the *simplest* whole-number ratio (which is 1 to 1). A formula that is written using the simplest whole-number ratio is called the **simplest formula,** or **empirical formula.** The *molecular* formula of hydrogen peroxide is H_2O_2; the *empirical* formula is HO. At times the data at hand are sufficient to derive only an empirical formula.

For some molecular compounds, the molecular and empirical formulas are identical; examples are H_2O, H_2SO_4, CO_2, and NH_3. For many molecular compounds, however, the molecular and empirical formulas are different. The molecular formulas

$$N_2H_4 \qquad B_3N_3H_6 \qquad C_6H_6$$

correspond to the empirical formulas

$$NH_2 \qquad BNH_2 \qquad CH$$

Notice that the atomic ratio for an empirical formula can be obtained by reducing the atomic ratio of the molecular formula to the lowest possible set of whole numbers.

Some compounds are not composed of molecules. Sodium chloride, for example, is made up of **ions**—particles that bear positive or negative charges (Chapter 6). The sodium and chloride ions, which are derived from sodium and chlorine atoms, form a crystal. A sodium chloride crystal is composed of large numbers of these ions, which are held together by the attractions between the dissimilar charges of the ions (see Figure 2.1).

In the crystal, there is one sodium ion (Na^+) for every one chloride ion (Cl^-). The formula of the compound is NaCl. The formula does not describe a molecule nor does it indicate that the ions are paired, since no ion in the crystal can be considered as belonging exclusively to another. Rather, the formula gives the *simplest* ratio of atoms of each type required to produce the compound. The formula NaCl, therefore, is an empirical formula.

The formulas of ionic compounds are derived from the formula of their ions. Since a crystal is electrically neutral, the total charge of the positive ions must equal the total charge of the negative ions. Consider barium chloride. The formula of the barium ion is Ba^{2+}, and the formula of the chloride ion is Cl^-. The two ions

Ba^{2+} and $2Cl^-$ form $BaCl_2$

The arrangement of ions in a $BaCl_2$ crystal is different from that in the NaCl crystal. The ions in each of these crystals are present in the ratio indicated by their formulas (*one* Ba^{2+} to *two* Cl^- in $BaCl_2$, and *one* Na^+ to *one* Cl^- in NaCl).

The formulas of ionic compounds such as $BaCl_2$ and NaCl are empirical formulas. The formulas of a few ionic compounds, however, can be reduced to simpler terms. Sodium peroxide is such a compound. In sodium peroxide, two sodium ions (Na^+) are present for every one peroxide ion (O_2^{2-}).

$2Na^+$ and O_2^{2-} form Na_2O_2

This formula can be reduced to NaO which is the *empirical* formula of sodium peroxide. What do we call the formula Na_2O_2? It seems odd to call Na_2O_2 a molecular formula since Na_2O_2 is an ionic compound and molecules of Na_2O_2 do not exist. For this reason, some chemists prefer to call all molecular formulas **true formulas.**

The problem encountered with the formula of sodium peroxide is not com-

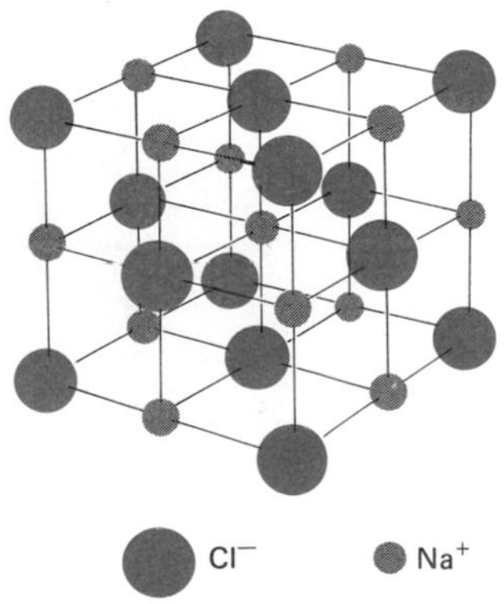

Figure 2.1 Sodium chloride crystal lattice.

mon. The formulas of most ionic compounds are empirical formulas, and the atomic ratios they indicate cannot be reduced. The problem arises for a few (but not all) ionic compounds in which one of the ions consists of more than one atom. Sodium peroxydisulfate, for example, contains

$2Na^+$ and $S_2O_8^{2-}$

and the *true* formula and *empirical* formula are

$Na_2S_2O_8$ and $NaSO_4$

On the other hand, sodium phosphate contains

$3Na^+$ and PO_4^{3-}

and the *true* formula is the same as the *empirical* formula

Na_3PO_4

Formulas are also used to designate the atomic composition of the molecules of which some elements are composed. A number of elements occur in nature as **diatomic molecules**—molecules that contain two atoms joined together. These elements, together with their molecular formulas, are listed in Table 2.1. Some elements are composed of molecules that are formed from more than two atoms. Sulfur molecules, for example, consist of eight atoms and have the molecular formula S_8. The molecular formula of the phosphorus molecules is P_4. The *empirical formulas* of *all* elements, of course, consist of only the symbols for the elements.

Table 2.1 Elements that occur in nature as diatomic molecules

ELEMENT	FORMULA
hydrogen	H_2
nitrogen	N_2
oxygen	O_2
fluorine	F_2
chlorine	Cl_2
bromine	Br_2
iodine	I_2

2.3 THE MOLE

The atomic weight of fluorine is 19.0 and of hydrogen is 1.0, which means that an atom of fluorine is 19 times heavier than an atom of hydrogen.* If we take 100 fluorine atoms and 100 hydrogen atoms, the mass of the collection of fluorine atoms will be 19 times that of the collection of hydrogen atoms. The masses of any two samples of fluorine and hydrogen that contain the same num-

* For most problem work, atomic weights will be rounded off to the first figure following the decimal point.

ber of atoms will stand in the ratio of 19.0 to 1.0, which is the ratio of their atomic weights.

Now suppose that we take 19.0 g of fluorine and 1.0 g of hydrogen, which are values in grams numerically equal to the atomic weight of the elements. Since the masses of the samples stand in the ratio of 19.0 to 1.0, the samples must contain the same number of atoms. In fact, a sample of any element that has a mass in grams numerically equal to the atomic weight of the element will contain this same number of atoms.

This number is called **Avogadro's number,** named for Amadeo Avogadro who first interpreted the behavior of gases in terms of the number of reacting atoms (Section 4.6). The value of Avogadro's number has been experimentally determined to be 6.022×10^{23}. The amount of a substance that contains Avogadro's number of elementary units is called a **mole** (abbreviated mol), which is an SI base unit.

Thus, a sample of an element that has a mass in grams numerically equal to the atomic weight of the element is a mole of atoms of the element and contains Avogadro's number of atoms. The atomic weight of beryllium, for example, is 9.0. A 9.0 g sample of beryllium is a mole of beryllium atoms and contains 6.022×10^{23} beryllium atoms (Avogadro's number).

Example 2.1 How many moles of aluminum are there in 100 g of Al?
Solution First we state the problem in the following way.

$$? \text{ mol Al} = 100 \text{ g Al}$$

Next, we derive a conversion factor to solve the problem. Since the atomic weight of Al is 27.0,

$$1 \text{ mol Al} = 27.0 \text{ g Al}$$

We use the conversion factor that has the unit *g Al* in the denominator since this is the unit that must be eliminated.

$$? \text{ mol Al} = 100 \text{ g Al} \left(\frac{1 \text{ mol Al}}{27.0 \text{ g Al}}\right) = 3.70 \text{ mol Al}$$

Example 2.2 How many grams of gold constitute 0.2500 mol of Au?
Solution We state the problem.

$$? \text{ g Au} = 0.2500 \text{ mol Au}$$

The atomic weight of Au is 197.0, therefore,

$$1 \text{ mol Au} = 197.0 \text{ g Au}$$

The unit that must be eliminated in this problem is *mol Au,* and the conversion factor must have this unit in the denominator.

$$? \text{ g Au} = 0.2500 \text{ mol Au} \left(\frac{197.0 \text{ g Au}}{1 \text{ mol Au}}\right) = 49.25 \text{ g Au}$$

Example 2.3 How many carbon atoms are there in a 1.00 carat diamond? Diamond is pure carbon, and one carat is 0.200 g.

Solution The problem is

$$? \text{ atoms C} = 0.200 \text{ g C}$$

We derive a conversion factor from

$$1 \text{ mol C} = 12.0 \text{ g C}$$

with the unit *g C* in the denominator so that this unit will cancel.

$$? \text{ atoms C} = 0.200 \text{ g C} \left(\frac{1 \text{ mol C}}{12.0 \text{ g C}}\right)$$

At this point the calculation would give an answer expressed in *mol C*. A second conversion factor is derived from

$$1 \text{ mol C} = 6.02 \times 10^{23} \text{ atoms C}$$

with the unit *mol C* in the denominator. Multiplication by this factor completes the solution.

$$? \text{ atoms C} = 0.200 \text{ g C} \left(\frac{1 \text{ mol C}}{12.0 \text{ g C}}\right)\left(\frac{6.02 \times 10^{23} \text{ atoms C}}{1 \text{ mol C}}\right) = 1.00 \times 10^{22} \text{ atoms C}$$

A mole contains Avogadro's number of entities. A mole of *molecules* consists of Avogadro's number of *molecules*.

The **molecular weight** of a molecular substance is the sum of the atomic weights of the atoms that constitute the molecule. For molecular materials, a sample that has a mass in grams numerically equal to the molecular weight of the substance is a mole of *molecules* and contains Avogadro's number of *molecules*.

The molecular weight of H_2O, for example, is 18.0 (twice the atomic weight of H plus the atomic weight of O). A mole of H_2O has a mass of 18.0 g and contains 6.02×10^{23} H_2O molecules. Since there are two atoms of H and one atom of O in a molecule of water, a mole of H_2O molecules (18.0 g) contains two moles of H atoms (2.0 g) and one mole of O atoms (16.0 g).

When the mole designation is used, the type of entity being measured *must* be specified. A mole of H *atoms* contains 6.02×10^{23} *H atoms* and has a mass of 1.0 g; a mole of H_2 *molecules* contains 6.02×10^{23} H_2 *molecules* and has a mass of 2.0 g. In the case of fluorine

$1 \text{ mol F} = 6.02 \times 10^{23} \text{ F atoms} = 19.0 \text{ g fluorine}$
$1 \text{ mol } F_2 = 6.02 \times 10^{23} \text{ } F_2 \text{ molecules} = 38.0 \text{ g fluorine}$

What about ionic substances? The designation "1 mol NaCl" means that a sample of NaCl contains 6.02×10^{23} NaCl formula units. Since the atomic weight of Na is 23.0 and the atomic weight of Cl is 35.5, one mole of NaCl has a mass of 58.5 g. In reality, one mole of NaCl (6.02×10^{23} NaCl units) consists of one mole of Na^+ ions (6.02×10^{23} Na^+ ions) and one mole of Cl^- ions (6.02×10^{23} Cl^- ions). Barium chloride, $BaCl_2$, contains

$1 \text{ mol } Ba^{2+} = 6.022 \times 10^{23} \text{ } Ba^{2+} \text{ ions} = 137.3 \text{ g barium*}$
$2 \text{ mol } Cl^- = 2(6.022 \times 10^{23}) \text{ } Cl^- \text{ ions} = 2(35.5) \text{ g } Cl^- = 71.0 \text{ g chlorine}$

which together make up

$1 \text{ mol } BaCl_2 = 6.022 \times 10^{23} \text{ } BaCl_2 \text{ units} = 208.2 \text{ g } BaCl_2$

2.4 DERIVATION OF FORMULAS

Data from the chemical analysis of a compound are used to derive the empirical formula of the compound. The analysis gives the proportions by mass of the elements that make up the compound. The simplest or empirical formula indicates the atomic proportions of the compound—the relative numbers of atoms of various types that make up the compound.

Since a mole of atoms of one element contains the same number of atoms as a mole of atoms of any other element, the ratio by moles is the same as the ratio by atoms. The number of moles of each element present in a sample of the compound is readily obtained from the mass of each element present. The simplest whole-number ratio by moles (which is the same as the ratio by atoms) is used to write the empirical formula. The procedure is illustrated in the examples that follow.

Example 2.4 What is the empirical formula of a compound that contains 43.6% P and 56.4% O?

Solution We assume, for convenience, that we have a sample that has a mass of 100.0 g. On the basis of the percentage composition, this sample contains 43.6 g P and 56.4 g O.

Next, we use the method illustrated in Example 2.1 to find the number of moles of P atoms and O atoms in these quantities. The atomic weight of P is 31.0 and of O is 16.0.

$$? \text{ mol P} = 43.6 \text{ g P} \left(\frac{1 \text{ mol P}}{31.0 \text{ g P}}\right) = 1.41 \text{ mol P}$$

* For all practical purposes, an ion has the same mass as the atom from which it is derived.

$$? \text{ mol O} = 56.4 \text{ g O} \left(\frac{1 \text{ mol O}}{16.0 \text{ g O}}\right) = 3.53 \text{ mol O}$$

The ratio by atoms is the same as the ratio by moles of atoms. There are, therefore, 1.41 atoms of P for every 3.53 atoms of O in the compound. We need, however, the simplest *whole-number* ratio in order to write the formula. By dividing the two values by the smaller value, we get

$$\text{for P, } \frac{1.41}{1.41} = 1.00 \qquad \text{for O, } \frac{3.53}{1.41} = 2.50$$

We still do not have a whole-number ratio, but we can get one by multiplying each of these values by 2. Hence, the simplest whole-number ratio is 2 to 5, and the empirical formula is P_2O_5.

Example 2.5 Caffeine, which occurs in coffee, tea, and kola nuts, is a stimulant for the central nervous system. A 1.261 g sample of pure caffeine contains 0.624 g C, 0.065 g H, 0.364 g N, and 0.208 g O. What is the empirical formula of caffeine?

Solution The results of a chemical analysis are usually reported in terms of percentages. However, any mass ratio can be converted into a mole ratio and, in this form, used to derive an empirical formula. There is no need to convert the data given in this example into percentages. We calculate the number of moles of each element present in the sample.

$$? \text{ mol C} = 0.624 \text{ g C} \left(\frac{1 \text{ mol C}}{12.0 \text{ g C}}\right) = 0.0520 \text{ mol C}$$

$$? \text{ mol H} = 0.065 \text{ g H} \left(\frac{1 \text{ mol H}}{1.0 \text{ g H}}\right) = 0.065 \text{ mol H}$$

$$? \text{ mol N} = 0.364 \text{ g N} \left(\frac{1 \text{ mol N}}{14.0 \text{ g N}}\right) = 0.0260 \text{ mol N}$$

$$? \text{ mol O} = 0.208 \text{ g O} \left(\frac{1 \text{ mol O}}{16.0 \text{ g O}}\right) = 0.0130 \text{ mol O}$$

Division of each of these values by the smallest value (0.0130) gives the ratio:

4 mol C: 5 mol H: 2 mol N: 1 mol O

and the empirical formula of caffeine, therefore, is $C_4H_5N_2O$.

The molecular formula of a compound can be derived from the empirical formula if the molecular weight of the compound is known.

Example 2.6 What is the molecular formula of the oxide of phosphorus that has the empirical formula P_2O_5 (derived in Example 2.4) if the molecular weight of this compound is 284?

Solution The value obtained by adding the atomic weights indicated by the empirical formula P_2O_5 is 142. If we divide this formula weight into the actual molecular weight, we get

$$\frac{284}{142} = 2$$

There are, therefore, twice as many atoms of each kind present in a molecule as is indicated by the empirical formula. The molecular formula is P_4O_{10}.

Example 2.7 The molecular weight of caffeine is 194 and the empirical formula of caffeine is $C_4H_5N_2O$. What is the molecular formula of caffeine?

Solution The formula weight indicated by $C_4H_5N_2O$ is 97. Since the molecular weight is twice this value, the molecular formula of caffeine is $C_8H_{10}N_4O_2$.

Example 2.8 Glucose, a simple sugar, is a constituent of human blood and tissue fluids and is utilized by cells as a principal source of energy. The compound contains 40.0% C, 6.73% H, and 53.3% O and has a molecular weight of 180.2. What is the molecular formula of glucose?

Solution The most convenient way to solve the problem is to calculate the number of moles of each element present in a mole of glucose. We first determine the number of grams of each element in a mole of glucose (180.2 g). Since the compound contains 40.0% C, there are 40.0 g of C in 100.0 g of glucose, and we use the factor (40.0 g C/100.0 g glucose).

$$? \text{ g C} = 1 \text{ mol glucose} \left(\frac{180.2 \text{ g glucose}}{1 \text{ mol glucose}}\right)\left(\frac{40.0 \text{ g C}}{100.0 \text{ g glucose}}\right) = 72.1 \text{ g C}$$

In like manner, the number of grams of H and of O can be found.

$$? \text{ g H} = 1 \text{ mol glucose} \left(\frac{180.2 \text{ g glucose}}{1 \text{ mol glucose}}\right)\left(\frac{6.73 \text{ g H}}{100.0 \text{ g glucose}}\right) = 12.1 \text{ g H}$$

$$? \text{ g O} = 1 \text{ mole glucose} \left(\frac{180.2 \text{ g glucose}}{1 \text{ mol glucose}}\right)\left(\frac{53.3 \text{ g O}}{100.0 \text{ g glucose}}\right) = 96.0 \text{ g O}$$

Next, we determine the number of moles of atoms that each of these values represents.

$$? \text{ mol C} = 72.1 \text{ g C} \left(\frac{1 \text{ mol C}}{12.0 \text{ g C}}\right) = 6.00 \text{ mol C}$$

$$? \text{ mol H} = 12.1 \text{ g H} \left(\frac{1 \text{ mol H}}{1.01 \text{ g H}}\right) = 12.0 \text{ mol H}$$

$$? \text{ mol O} = 96.0 \text{ g O} \left(\frac{1 \text{ mol O}}{16.0 \text{ g O}}\right) = 6.00 \text{ mol O}$$

These values are the number of moles of atoms of each element present in a mole of glucose molecules. They are also the number of atoms of each type present in a molecule of glucose. The molecular formula, therefore, is $C_6H_{12}O_6$.

The problem may also be solved by first determining the empirical formula from the analytical data (it is CH_2O) and then using the molecular weight to derive the molecular formula.

DERIVATION OF EMPIRICAL FORMULAS

1. If the data are given in terms of percentage composition, base the calculation on a 100.0 g sample of the compound. In this instance, the number of grams of each element present in the sample will be numerically equal to the percentage of that element present in the compound.

   There is no need to find percentages if the data are given in terms of the number of grams of each element present in a sample.
2. Convert the number of grams of each element present in the sample to the number of moles of *atoms* of each element. The conversion factors needed are derived from the fact that 1 mol of *atoms* of an element (numerator) is an *atomic* weight in grams (denominator).
3. Divide each of the values obtained in step 2 by the smallest value. If every number obtained in this way is not a whole number multiply *each number* by the same simple integer in such a way that whole numbers will result.
4. A ratio by moles of atoms is the same as a ratio by atoms. The whole numbers obtained in step 3 are the subscripts of the empirical formula.

2.5 PERCENTAGE COMPOSITION OF COMPOUNDS

The percentage composition of a compound is readily calculated from the formula of the compound. The subscripts of the formula give the number of moles of each element in a mole of the compound. From this information, and the atomic weights of the elements, we can obtain the number of grams of each element contained in a mole of the compound. The percentage of a given element is 100 times the mass of the element divided by the mass of a mole of the compound. The process is illustrated in Example 2.9.

Example 2.9 What is the percentage of Fe in Fe_2O_3?
Solution One mole of Fe_2O_3 contains

$$\begin{aligned} 2 \text{ mol Fe} &= 2(55.8) \text{ g Fe} = 111.6 \text{ g Fe} \\ 3 \text{ mol O} &= 3(16.0) \text{ g O} = \underline{\ 48.0 \text{ g O}\ } \\ & \qquad\qquad\qquad\quad 159.6 \text{ g} \end{aligned}$$

The sum of the masses, 159.6 g, is the mass of one mole of Fe_2O_3. The percentage of Fe in Fe_2O_3 is

$$\frac{111.6 \text{ g Fe}}{159.6 \text{ g } Fe_2O_3} \times 100\% = 69.92\% \text{ Fe in } Fe_2O_3$$

The percentage composition of a compound is frequently determined by chemical analysis. This data can then be used to find the empirical formula of the compound. Example 2.10 is an illustration of a method used for the analysis of organic compounds.

Example 2.10 Nicotine is a compound that contains carbon, hydrogen, and nitrogen. If a 1.215 g sample of nicotine is burned in oxygen, 3.300 g of CO_2, 0.945 g of H_2O, and 0.210 g of N_2 are the products of the combustion. What is the percentage composition of nicotine?

Solution We first calculate the quantity of each element present in the 1.215 g sample of nicotine. The carbon of the nicotine formed 3.300 g of CO_2. We ask, therefore,

$$? \text{ g C} = 3.300 \text{ g } CO_2$$

The conversion factor that we use to solve this problem is the fraction that we would use to find the percentage of C in CO_2. Since 1 mol CO_2 (44.0 g CO_2) contains 1 mol C (12.0 g C),

$$12.0 \text{ g C} = 44.0 \text{ g } CO_2$$

We derive the conversion factor (12.0 g C/44.0 g CO_2).

$$? \text{ g C} = 3.300 \text{ g } CO_2 \left(\frac{12.0 \text{ g C}}{44.0 \text{ g } CO_2}\right) = 0.900 \text{ g C}$$

The same procedure is used to find the number of grams of hydrogen in the nicotine sample. The hydrogen of the nicotine formed 0.945 g H_2O. In 1 mol of H_2O (18.0 g H_2O) there are 2 mol of H atoms (2.0 g H). Therefore,

$$? \text{ g H} = 0.945 \text{ g } H_2O \left(\frac{2.0 \text{ g H}}{18.0 \text{ g } H_2O}\right) = 0.105 \text{ g H}$$

In a combustion such as the one described, the nitrogen does not combine with oxygen but is evolved as N_2. Hence, the sample contained 0.210 g N.

The quantity of each element present in the 1.215 g sample of nicotine is used to determine the percentage composition of nicotine.

$$\frac{0.900 \text{ g C}}{1.215 \text{ g nicotine}} \times 100\% = 74.1\% \text{ C in nicotine}$$

$$\frac{0.105 \text{ g H}}{1.215 \text{ g nicotine}} \times 100\% = 8.6\% \text{ H in nicotine}$$

$$\frac{0.210 \text{ g N}}{1.215 \text{ g nicotine}} \times 100\% = 17.3\% \text{ N in nicotine}$$

These data can be used to find the empirical formula of nicotine which is $C_{10}H_{14}N_2$.

Some simple stoichiometric problems can be solved by the use of proportions derived from formulas.

Example 2.11 Silver sulfide, Ag_2S, occurs as the mineral argentite, which is an ore of silver. How many grams of silver are theoretically obtainable from 250 g of an impure ore that is 70.0% Ag_2S?
Solution The problem can be stated as follows.

$$? \text{ g Ag} = 250 \text{ g ore}$$

If we take 100.0 g of the ore, we will get 70.0 g of Ag_2S since the ore is 70.0% Ag_2S. Therefore,

$$70.0 \text{ g } Ag_2S = 100.0 \text{ g ore}$$

and the factor (70.0 g Ag_2S/100.0 g ore) can be derived.

$$? \text{ g Ag} = 250 \text{ g ore} \left(\frac{70.0 \text{ g } Ag_2S}{100.0 \text{ g ore}}\right)$$

The *g ore* labels cancel, and at this point we have an answer in *g* Ag_2S.

The problem can be solved by use of the same factor that would be used to find the percentage of Ag in Ag_2S. From the formula Ag_2S, we derive

$$2 \text{ mol Ag} = 1 \text{ mol } Ag_2S$$
$$2(107.9) \text{ g Ag} = 247.9 \text{ g } Ag_2S$$
$$215.8 \text{ g Ag} = 247.9 \text{ g } Ag_2S$$

Therefore,

$$? \text{ g Ag} = 250 \text{ g ore} \left(\frac{70.0 \text{ g } Ag_2S}{100.0 \text{ g ore}}\right)\left(\frac{215.8 \text{ g Ag}}{247.9 \text{ g } Ag_2S}\right) = 152 \text{ g Ag}$$

2.6 CHEMICAL EQUATIONS

Chemical equations are representations of reactions in terms of the symbols and formulas of the elements and compounds involved. The reactants are indicated on the left and the products on the right. An arrow is used instead of the customary equal sign of the algebraic equation; it may be considered as an abbreviation for the word "yields."

Chemical equations report the results of experimentation. One of the goals of chemistry is the discovery and development of principles that make it possible to predict the products of chemical reactions; careful attention will be given to any such generalizations. All too often, however, the products of a particular set of reactants must be memorized, and any prediction is subject to modification if experiment dictates. What may appear reasonable on paper is not necessarily what occurs in the laboratory.

The first step in writing a chemical equation is to ascertain the products of the reaction in question. Carbon disulfide, CS_2, reacts with chlorine, Cl_2, to produce carbon tetrachloride, CCl_4, and disulfur dichloride, S_2Cl_2. To represent this, we write

$$CS_2 + Cl_2 \rightarrow CCl_4 + S_2Cl_2$$

This equation is not quantitatively correct because it violates the law of conservation of mass. There must be as many atoms of each element, combined or uncombined, indicated on the left side of an equation as there are on the right side. Although one carbon atom and two sulfur atoms are indicated on both the left side and the right side of the equation, two chlorine atoms (one Cl_2 molecule) appear on the left and six chlorine atoms appear on the right. The equation can be balanced* by indicating that three molecules of chlorine should be used for the reaction. Thus,

$$CS_2 + 3Cl_2 \rightarrow CCl_4 + S_2Cl_2$$

The simplest types of chemical equations are balanced by trial and error as the following examples will illustrate. When steam is passed over hot iron, hydrogen gas and an oxide of iron that has the formula Fe_3O_4 are produced. Thus,

$$Fe + H_2O \rightarrow Fe_3O_4 + H_2$$

It is tempting to substitute another oxide of iron, FeO, for Fe_3O_4, since this would immediately produce a balanced equation. Such an equation, however, would be without value; experiment indicates that Fe_3O_4, not FeO, is a product of the reaction. Balancing an equation is never accomplished by altering the formulas of the products of the reaction. In the equation for the reaction of iron and steam, three atoms of Fe and four molecules of H_2O are needed to provide the iron and oxygen atoms required for the formation of Fe_3O_4.

* Strictly speaking, the expression is not an "equation" until it is balanced.

$$3Fe + 4H_2O \rightarrow Fe_3O_4 + H_2$$

The equation is now balanced except for hydrogen which may be balanced as follows:

$$3Fe + 4H_2O \rightarrow Fe_3O_4 + 4H_2$$

In much the same manner we can balance an equation for the complete combustion of ethane (C_2H_6) in oxygen. The products of this reaction are carbon dioxide and water.

$$C_2H_6 + O_2 \rightarrow CO_2 + H_2O$$

To balance the two carbon atoms of C_2H_6, the production of two molecules of CO_2 must be indicated; and the six hydrogen atoms of C_2H_6 require that three molecules of H_2O be produced.

$$C_2H_6 + O_2 \rightarrow 2CO_2 + 3H_2O$$

Only the oxygen remains unbalanced; there are seven atoms of oxygen on the right and only two on the left. In order to get seven atoms of oxygen on the left, we would have to take $3\frac{1}{2}$, or $\frac{7}{2}$, molecules of O_2.

$$C_2H_6 + \tfrac{7}{2}O_2 \rightarrow 2CO_2 + 3H_2O$$

Customarily, equations are written with whole number coefficients. By multiplying the entire equation by two, we get

$$2C_2H_6 + 7O_2 \rightarrow 4CO_2 + 6H_2O$$

2.7 PROBLEMS BASED ON CHEMICAL EQUATIONS

A chemical equation can be interpreted in several different ways. Consider, for example, the equation

$$2H_2 + O_2 \rightarrow 2H_2O$$

On the simplest level this equation shows that hydrogen reacts with oxygen to produce water. On the atomic–molecular level, it states that

$$2 \text{ molecules } H_2 + 1 \text{ molecule } O_2 \rightarrow 2 \text{ molecules } H_2O$$

It can also be read as

$$2 \text{ mol } H_2 + 1 \text{ mol } O_2 \rightarrow 2 \text{ mol } H_2O$$

since 1 mol of H_2, 1 mol of O_2, and 1 mol of H_2O all contain the same number of molecules (Avogadro's number).

The last interpretation is the one that enables us to solve stoichiometric problems. The coefficients of the chemical equation give the ratios, by moles, in which the substances react and are produced. Since 2 mol of H_2 react with 1 mol of O_2, 10 mol of H_2 would require 5 mol of O_2 for reaction. Since 2 mol of H_2 produce 2 mol of H_2O, the reaction of 10 mol of H_2 would produce 10 mol of H_2O.

Example 2.12 Chlorine can be prepared by the reaction

$$MnO_2 + 4HCl \rightarrow MnCl_2 + Cl_2 + 2H_2O$$

(a) How many grams of HCl are required to react with 25.0 g of MnO_2? (b) How many grams of Cl_2 are produced by the reaction?

Solution (a) The problem is

$$? \text{ g HCl} = 25.0 \text{ g } MnO_2$$

The stoichiometric ratio to be derived from the chemical equation will be expressed in moles. Therefore, we convert *g MnO_2* into *mol MnO_2*. The molecular weight of MnO_2 is 86.9.

$$? \text{ g HCl} = 25.0 \text{ g } MnO_2 \left(\frac{1 \text{ mol } MnO_2}{86.9 \text{ g } MnO_2}\right)$$

The chemical equation gives the relation

$$1 \text{ mol } MnO_2 = 4 \text{ mol HCl}$$

from which we derive the conversion factor (4 mol HCl/1 mol MnO_2).

$$? \text{ g HCl} = 25.0 \text{ g } MnO_2 \left(\frac{1 \text{ mol } MnO_2}{86.9 \text{ g } MnO_2}\right)\left(\frac{4 \text{ mol HCl}}{1 \text{ mol } MnO_2}\right)$$

At this point, the calculation would give the number of moles of HCl required. We must, therefore, convert *mol HCl* into *g HCl* to get our answer. The molecular weight of HCl is 36.5.

$$? \text{ g HCl} = 25.0 \text{ g } MnO_2 \left(\frac{1 \text{ mol } MnO_2}{86.9 \text{ g } MnO_2}\right)\left(\frac{4 \text{ mol HCl}}{1 \text{ mol } MnO_2}\right)\left(\frac{36.5 \text{ g HCl}}{1 \text{ mol HCl}}\right) = 42.0 \text{ g HCl}$$

(b) The same procedure is used to solve this problem. Grams of MnO_2 is converted into moles of MnO_2. The mole relation from the chemical equation

$$1 \text{ mol } MnO_2 = 1 \text{ mol } Cl_2$$

is used to find the number of moles of Cl_2 produced. Finally, moles of Cl_2 is converted into grams of Cl_2.

$$? \text{ g } Cl_2 = 25.0 \text{ g } MnO_2 \left(\frac{1 \text{ mol } MnO_2}{86.9 \text{ g } MnO_2}\right)\left(\frac{1 \text{ mol } Cl_2}{1 \text{ mol } MnO_2}\right)\left(\frac{71.0 \text{ g } Cl_2}{1 \text{ mol } Cl_2}\right) = 20.4 \text{ g } Cl_2$$

Example 2.13 The amount of carbon monoxide in a sample can be determined by the reaction

$$I_2O_5 + 5CO \rightarrow I_2 + 5CO_2$$

If a gas sample liberates 0.192 g of I_2, how many grams of CO were present in the sample?

Solution The relation, by moles, of the two substances of interest is obtained from the chemical equation. It is

$$5 \text{ mol CO} = 1 \text{ mol } I_2$$

We also need to know

$$1 \text{ mol CO} = 28.0 \text{ g CO}$$
$$1 \text{ mol } I_2 = 253.8 \text{ g } I_2$$

Conversion factors derived from these three relations are needed to solve the problem. The solution is

$$? \text{ g CO} = 0.192 \text{ g } I_2 \left(\frac{1 \text{ mol } I_2}{253.8 \text{ g } I_2}\right)\left(\frac{5 \text{ mol CO}}{1 \text{ mol } I_2}\right)\left(\frac{28.0 \text{ g CO}}{1 \text{ mol CO}}\right) = 0.106 \text{ g CO}$$

In some problems, quantities are given for two or more reactants. Suppose, for example, that we are asked how much H_2O can be prepared from 2 mol of H_2 and 2 mol of O_2. The chemical equation

$$2 H_2 + O_2 \rightarrow 2H_2O$$

states that 2 mol of H_2 will react with only 1 mol of O_2. In the problem, however, 2 mol of H_2 and 2 mol of O_2 are given. More O_2 has been supplied than can be used. When all the H_2 has been consumed, the reaction will stop. At this point, 1 mol of O_2 will have been used and 1 mol of O_2 will remain unreacted. The amount of H_2 supplied limits the reaction and determines how much H_2O will be formed. Hydrogen, therefore, is called the **limiting reactant.**

Whenever the quantities of two or more reactants are given in a problem, we must determine which one limits the reaction before the problem can be solved.

Example 2.14 How many grams of N_2F_4 can theoretically be prepared from 4.00 g of NH_3 and 14.0 g of F_2. The chemical equation for the reaction is

$$2NH_3 + 5F_2 \rightarrow N_2F_4 + 6HF$$

Solution The first step is to determine which reactant limits the reaction. We find the number of moles of each reactant present before reaction. The molecular weight of NH_3 is 17.0 and of F_2 is 38.0.

$$? \text{ mol } NH_3 = 4.00 \text{ g } NH_3 \left(\frac{1 \text{ mol } NH_3}{17.0 \text{ g } NH_3}\right) = 0.235 \text{ mol } NH_3$$

$$? \text{ mol } F_2 = 14.0 \text{ g } F_2 \left(\frac{1 \text{ mol } F_2}{38.0 \text{ g } F_2}\right) = 0.368 \text{ mol } F_2$$

If we read the chemical equation in terms of moles, we note that

$$2 \text{ mol } NH_3 = 5 \text{ mol } F_2$$

We compare the number of moles supplied with these quantities. The problem specified 0.235 mol of NH_3 which is

$$\frac{0.235 \text{ mol } NH_3}{2 \text{ mol } NH_3} = 0.118$$

of the amount stated in the relation derived from the chemical equation. The 0.368 mol F_2 is

$$\frac{0.368 \text{ mol } F_2}{5 \text{ mol } F_2} = 0.0736$$

of the amount given in the relation from the chemical equation. The F_2, therefore, is the limiting reactant since a smaller *proportionate* amount has been supplied (0.0736 is a smaller number than 0.118). The problem is solved on the basis of the amount of F_2 supplied.

$$? \text{ g } N_2F_4 = 0.368 \text{ mol } F_2$$

The relation between the quantity of F_2 employed and the quantity of N_2F_4 produced is derived from the chemical equation. It is

$$5 \text{ mol } F_2 = 1 \text{ mol } N_2F_4$$

The molecular weight of N_2F_4 is 104.0.

$$1 \text{ mol } N_2F_4 = 104.0 \text{ g } N_2F_4$$

The solution to the problem is

$$? \text{ g } N_2F_4 = 0.368 \text{ mol } F_2 \left(\frac{1 \text{ mol } N_2F_4}{5 \text{ mol } F_2}\right)\left(\frac{104.0 \text{ g } N_2F_4}{1 \text{ mol } N_2F_4}\right) = 7.65 \text{ g } N_2F_4$$

Frequently, the quantity of a product actually obtained from a reaction is less than the amount calculated. It may be that part of the reactants do not react, or that part of the reactants react in a way different from that desired (side reactions), or that not all of the product is recovered. The **percent yield** relates the amount of product that is actually obtained (the actual yield) to the amount that theory would predict (the theoretical yield).

$$\text{percent yield} = \frac{\text{actual yield}}{\text{theoretical yield}} \times 100\%$$

Example 2.15 If 4.80 g of N_2F_4 is obtained from the experiment described in Example 2.14, what is the percent yield?

Solution The theoretical yield is the result of the calculation of Example 2.14, 7.65 g N_2F_4. The actual yield is 4.80 g N_2F_4. Therefore, the percent yield is

$$\frac{4.80 \text{ g } N_2F_4}{7.65 \text{ g } N_2F_4} \times 100\% = 62.7\%$$

CALCULATIONS BASED ON CHEMICAL EQUATIONS

1. State the problem. Indicate the substance sought (using grams as the desired unit), an equal sign, and the mass of the substance given (in grams).
2. Enter the factor that will convert the mass of the substance given into moles of substance given. The conversion factor is derived from the fact that 1 mol of a substance (numerator) is a molecular weight in grams (denominator).
3. Enter the conversion factor that is derived from the coefficients of the chemical equation and that relates the number of moles of substance sought (numerator) to the number of moles of substance given (denominator).
4. Enter the factor that will convert the number of moles of substance sought to grams of substance sought. The molecular weight of the substance sought in grams (numerator) is 1 mol of the substance sought (denominator).
5. Carry out the mathematical operations indicated to obtain the answer. All units should cancel except grams of the substance sought.

If more than one quantity is given in the problem:

1. Calculate the number of moles of each of the given reactants (see preceding step 2).
2. Divide each of these values by the coefficient of the chemical equation that pertains to the substance being considered.
3. The smallest number obtained in step 2 pertains to the reactant that limits the extent of the reaction. Use the quantity of this reactant to solve the problem in the way previously outlined.

THE SYMBOLS OF CHEMISTRY

Chemical symbols have been used since the earliest days of alchemy. The symbols that the alchemists used for the seven known metals (see Figure 2.2) were derived from the symbols for the seven known planets (which, to them, included the sun and the moon). The noble metals, gold and silver, were represented by the symbols for the sun and the moon—the rulers of the day and the night. The assignment was partially based on color; to this day we speak of the *golden sun* and the *silvery moon.*

Since the implements of war were made from iron, this metal was assigned the symbol for the god of war, Mars. The symbol itself represents a shield with a spear behind it and is used as the biological symbol for *male.* Copper, on the other hand, was regarded as a feminine metal and assigned the symbol for Venus, goddess of love. This symbol, said to depict the looking glass of the goddess, is the biological symbol for *female.*

The alchemists accumulated extensive lists of symbols, which included representations for apparatus and procedures as well as chemical substances. The use of these symbols, however, only served to make the subject more mystical and obscure; in fact, they may have been used for this purpose.

When John Dalton developed his atomic theory (in the years 1803 to 1808), he designed a system of symbols to go with it (see Figure 2.3). Dalton represented a single atom of a given element by a circle with a distinguishing mark. Molecules were represented by clusters of the appropriate circles. Dalton was a teacher who also gave numerous public lectures on scientific topics. He designed his symbols to serve not only as an aid to chemists but also as an aid in instructing beginners.

Jöns Berzelius, a Swedish chemist who accepted the atomic theory, rejected Dalton's symbolism. In 1813, Berzelius proposed a system of chemical symbols that employed letters to represent atoms and from which the modern system of symbols and formulas has developed. The scheme of Berzelius rapidly

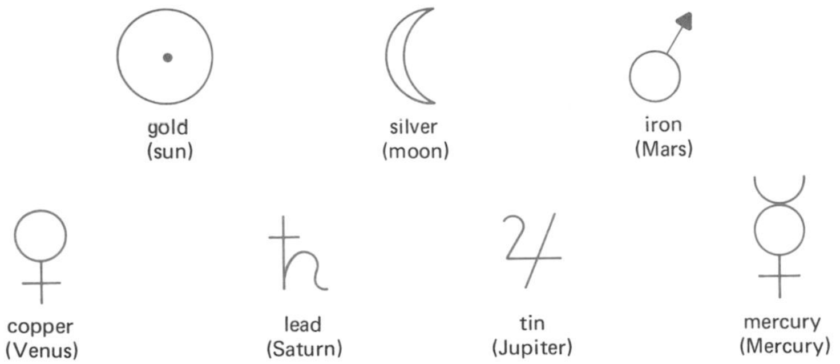

Figure 2.2 Alchemical symbols for the seven known metals, derived from the symbols for the planets.

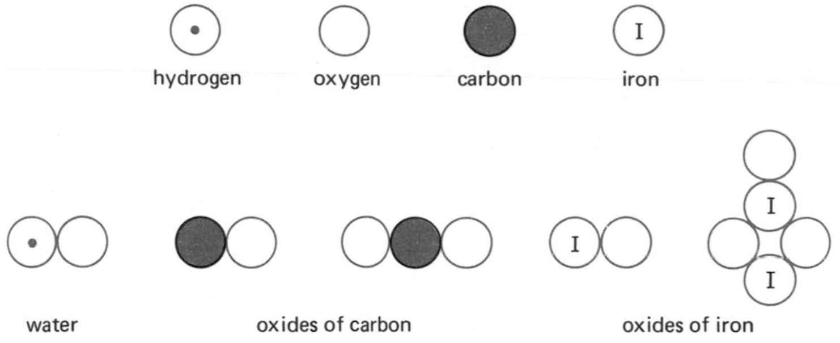

Figure 2.3 Some of Dalton's symbols. Note that the formula for water is incorrect.

found wide acceptance in continental Europe, while in Britain, Dalton's system was largely employed.

The British Association for the Advancement of Science formed a committee to study the question of chemical notation. In 1835, the committee recommended that Berzelius' system be used, not Dalton's. The committee believed that British adoption of the system that was in use on the continent would help scientific communication. Furthermore, Dalton's symbols for molecules depicted the arrangement of atoms in space, a matter thought to be purely theoretical and of which there was then no practical knowledge. Berzelius' formulas, on the other hand, implied only the numbers and types of atoms present in compounds, matters open to experimental proof.

Dalton thought that a system that made no attempt to answer the question of how atoms were arranged in molecules was only half a system and that his system was "the only one representing nature." Dalton's position was subsequently vindicated. In later years, organic chemists found it absolutely essential that their formulas indicate the arrangement of atoms in molecules, and various notation systems have been developed to indicate this aspect of molecular structure.

Dalton felt deeply that his system was superior. In 1837, an argument concerning the question is said to have caused Dalton to have a serious, but not fatal, stroke. Berzelius, too, was not without strong convictions concerning his system of symbolism. He quarreled with Justis von Liebig because Liebig printed (in a journal that he edited) an article by Berzelius in which the numbers in the formulas were placed below the atomic symbols as they are today (for example, Fe_2O_3) instead of above the symbols as Berzelius thought they should be (for example, Fe^2O^3).

PROBLEMS

2.1 (a) What is the mass of a single platinum atom? (b) How many atoms are present in a 10.00 g sample of platinum?

2.2 If one atom of an element has a mass of 3.330×10^{-22} g, what is that element's atomic weight?

2.3 How many moles and how many molecules are contained in 10.0 g of each of the following? (a) O_2, (b) H_2O, (c) OF_2.

2.4 Calcium is present in the blood at a concentration of about 10 mg per 100 ml. How many moles of calcium are present in the body's 5 liter of blood?

2.5 In which of the following compounds is the percent oxygen the highest? the lowest? SO_3, SO_2, Cl_2O, P_4O_{10}.

2.6 In which of the following compounds is the percent zinc the highest? the lowest? ZnO, $ZnCl_2$, ZnS, $ZnSO_4$.

2.7 Potassium cyanide, KCN, is an extremely toxic substance. A lethal dose is approximately 5 mg per kg of body weight. How many milligrams of KCN is a lethal dose for a 70 kg person? How many moles of KCN? How many molecules of KCN?

2.8 What mass of silicon is combined with 6.40 g of oxygen in the compound SiO_2?

2.9 Pewter is 85% tin and 15% antimony. How many moles of antimony are contained in 50 g of pewter?

2.10 Plaster of Paris, a hydrate of calcium sulfate, has the formula $(CaSO_4)_2 \cdot H_2O$. What percent of plaster of Paris is water?

2.11 What percentage by mass of nickel carbonyl, $Ni(CO)_4$, is nickel?

2.12 A sample of thiogenal, a barbiturate, contains 36.03 g C, 5.04 g H, 7.00 g N, 8.00 g O, and 16.03 g S. What is its simplest formula?

2.13 The molecular weight of thiogenal is 288.4. Using the simplest formula found in Problem 12, determine its molecular formula.

2.14 White gold is an alloy of gold and palladium. If a sample of white gold is prepared from 0.019 mol gold and 0.20 mol palladium, what percentage of each metal is present?

2.15 What percentage by mass of borax, $Na_2B_4O_5(OH)_4 \cdot 8H_2O$, is boron? sodium?

2.16 Cocaine is 67.30% C, 6.98% H, 21.10% O, and 4.62% N. What is its simplest formula?

2.17 A sample of the mineral crysoberyl contains 2.25 g Be, 13.49 g Al, and 4.00 g O. What is its empirical formula?

2.18 What is the empirical formula of the mineral ilmenite if it is 36.8% Fe, 31.6% Ti, and 31.6% O?

2.19 How many grams of uranium can theoretically be extracted from a 10.0 kg sample of pitchblende which is 92.0% UO_2?

2.20 A sample of strychnine contains 84.08 g C, 7.392 g H, 9.338 g N, and

10.67 g O. What is its empirical formula?

2.21 What is the empirical formula of a compound that is 18.3% Li, 71.1% Al, and 10.6% H?

2.22 What is the empirical formula of a compound that is 47.7% K, 13.2% B, and 39.1% O?

2.23 If the compound in Problem 2.22 has a molecular weight of 246, what is its molecular formula?

2.24 Determine the molecular formulas of the compounds for which the following empirical formulas and molecular weights pertain. (a) P_2O_3, 220; (b) $Ca_3(PO_4)_2$, 310; (c) $C_{21}H_{27}O_3N$, 682; (d) CH_2O, 180; (e) BNH_2, 80.4.

2.25 Most iron is obtained from hematite, Fe_2O_3. What is the percentage of iron in Fe_2O_3?

2.26 What mass of chromium is theoretically obtainable from 100 kg of chromite that is 72.0% pure $FeCr_2O_4$?

2.27 What percent by mass of CoAsS is arsenic?

2.28 What is the empirical formula of a compound that is 75.92% C, 17.71% N, and 6.37% H?

2.29 What is the percent composition by mass of each element in Al_4C_3?

2.30 A 0.7234 g sample of hemoglobin is isolated and analyzed for iron. If the hemoglobin is found to contain 0.342% Fe, and it is assumed that four ferrous ions, Fe^{2+}, are incorporated in each hemoglobin molecule, calculate the molecular weight of hemoglobin.

2.31 Balance the following equations.
(a) $NaN_3 \rightarrow Na + N_2$
(b) $PbS + PbSO_4 \rightarrow Pb + SO_2$
(c) $P_4O_{10} + H_2O \rightarrow H_3PO_4$
(d) $P_4O_6 + I_2 \rightarrow P_2I_4 + P_4O_{10}$
(e) $C_2H_6 + O_2 \rightarrow CO_2 + H_2O$

2.32 Balance the following equations.
(a) $CaSi_2 + SbCl_3 \rightarrow Si + Sb + CaCl_2$
(b) $NH_3 + O_2 \rightarrow NO + H_2O$
(c) $C_{10}H_8 + O_2 \rightarrow CO_2 + H_2O$
(d) $NbBr_5 + Nb \rightarrow NbBr_4$
(e) $Na + SF_6 \rightarrow Na_2S + NaF$

2.33 How many grams of beryllium fluoride, BeF_2, can be prepared from 20.0 g of beryllium and 20.0 g of fluorine?

2.34 Aspirin, acetylsalicylic acid, is synthesized industrially by the following reaction.

$$\underset{\text{salicylic acid}}{C_7H_6O_3} + \underset{\text{acetic anhydride}}{C_4H_6O_3} \rightarrow \underset{\text{aspirin}}{C_9H_8O_4} + \underset{\text{acetic acid}}{C_2H_4O_2}$$

How many grams of salicylic acid must be used to prepare 25.0 kg of aspirin?

2.35 Commercially, iron is obtained from the reduction of hematite, Fe_2O_3.

$$Fe_2O_3 + 3CO \rightarrow 2Fe + 3CO_2$$

How many grams of iron can be obtained from 10.50 g Fe_2O_3?

2.36 What percentage of hematite, Fe_2O_3, is present in 2000 kg of ore if it yields 1184 kg of iron?

2.37 Sodium carbonate (Na_2CO_3) and niobium(V) oxide (Nb_2O_5) react to form sodium orthoniobate (Na_3NbO_4)

and carbon dioxide. (a) Write the equation for the reaction. (b) What mass of Nb_2O_5 is needed to prepare 1.00 g Na_3NbO_4? (c) What mass of Na_2CO_3 is required?

2.38 The Edison or nickel–iron battery is based on the reaction

$$Fe + 2NiO(OH) + 2H_2O \rightarrow Fe(OH)_2 + 2Ni(OH)_2$$

What mass of NiO(OH) is required for the reaction with 5.00 g Fe?

2.39 Upon drying at 150°C, a 7.80 g sample of hydrated copper sulfate, $CuSO_4 \cdot xH_2O$, weighed 4.98 g after all of the water was driven off. What is the value of *x*? (That is, how many moles of water are present per mole of $CuSO_4$?)

2.40 In the Van Arkel process, impure zirconium is heated with iodine, forming ZrI_4.

$$Zr + 2I_2 \rightarrow ZrI_4$$

The ZrI_4 is decomposed, and pure zirconium is obtained.

$$ZrI_4 \rightarrow Zr + 2I_2$$

If 20.0 g of impure zirconium yields 17.6 g of pure zirconium, what percent zirconium was present in the original metal?

2.41 Calcium cyanamide, CaNCN, is used as a fertilizer. It is obtained by a high-temperature reaction between calcium carbide, CaC_2, and nitrogen.

$$CaC_2 + N_2 \rightarrow CaNCN + C$$

What mass of calcium cyanamide is formed from 500 kg calcium carbide?

2.42 How many grams of $NaNO_3$ will react with 3.50 g $NaNH_2$ in the following reaction?

$$3NaNH_2 + NaNO_3 \rightarrow NaN_3 + 3NaOH + NH_3$$

2.43 Ethyl alcohol, C_2H_5OH, is a product of the fermentation of sugars. In the presence of enzymes, sucrose, $C_{12}H_{22}O_{11}$, is converted into ethyl alcohol and carbon dioxide.

$$C_{12}H_{22}O_{11} + H_2O \xrightarrow{\text{enzymes}} 4C_2H_5OH + 4CO_2$$

Usually only half of the sucrose is converted into alcohol, a 50% yield. How much ethyl alcohol can be obtained from 342 g of sucrose?

2.44 A sample of a compound containing only carbon and hydrogen was burned in oxygen, and 0.88 g CO_2 and 0.54 g H_2O were obtained. (a) How many moles of C atoms and moles of H atoms did the sample contain? (b) What is the empirical formula of the compound? (c) What was the mass of the sample that was burned? (d) If 0.0100 mol of the compound was burned, what is the molecular formula of the compound?

2.45 Aluminum carbide, Al_4C_3, reacts rapidly with water to produce methane, CH_4, and aluminum hydroxide, $Al(OH)_3$. (a) Write a balanced equation for this reaction. (b) What mass of $Al(OH)_3$ can be produced from 1.00 g Al_4C_3?

2.46 How many grams of $CsXeF_7$ can be produced by the reaction of 15.0 g of CsF and 20.0 g of XeF_6?

2.47 Carborundum, SiC, is used as an abrasive. It is formed by the combination of SiO_2 and carbon.

$$SiO_2 + 3C \rightarrow SiC + 2CO$$

What mass of carborundem is formed from 5.00 g SiO_2 and 2.00 g C?

2.48 What mass of MoF_5 is obtained from 25.0 g MoF_6 and 5.00 g $Mo(CO)_6$ by the reaction

$$5MoF_6 + Mo(CO)_6 \rightarrow 6\,MoF_5 + 6CO$$

2.49 Diphosphorus tetraiodide may be prepared by the reaction (unbalanced)

$$P_4O_6 + I_2 \rightarrow P_2I_4 + P_4O_{10}$$

What mass of P_2I_4 can be prepared from 2.50 g P_4O_6 and 4.00 g I_2?

2.50 Phosphorus is industrially prepared by heating a mixture of phosphate rock, sand, and coke in an electric furnace. The chemical changes are represented by the following equations.

$$Ca_3(PO_4)_2 + SiO_2 \rightarrow CaSiO_3 + P_4O_{10}$$

$$P_4O_{10} + C \rightarrow P_4 + CO$$

(a) Balance the equations. (b) If 202 kg of phosphate rock yields 25.0 kg of condensed P_4, what percent $Ca_3(PO_4)_2$ was initially present in the rock?

2.51 Denitrification is a process that occurs in the natural nitrogen cycle. During this process, glucose ($C_6H_{12}O_6$) plus potassium nitrate (KNO_3) form carbon dioxide, water, KOH, and N_2O. (a) Write a balanced equation for the reaction. (b) What mass of N_2O is formed from 29.8 g glucose and 19.9 g potassium nitrate?

2.52 Serum calcium can be determined by precipitation with oxalate, dissolution of the precipitate in sulfuric acid, and titration with potassium permanganate.

$$Ca^{2+} + C_2O_4^{2-} \rightarrow CaC_2O_4$$

$$CaC_2O_4 + H_2SO_4 \rightarrow H_2C_2O_4 + CaSO_4$$

$$2KMnO_4 + 5H_2C_2O_4 + 3H_2SO_4 \rightarrow K_2SO_4 + 2MnSO_4 + 10CO_2 + 8H_2O$$

If a serum sample contains 0.20 mg Ca^{2+}, what mass of potassium permanganate will be needed for completion of the titration reaction?

2.53 What mass of chromyl chloride, CrO_2Cl_2, can be produced by 10.0g $K_2Cr_2O_7$ and 10.0g KCl in the following reaction?

$$K_2Cr_2O_7 + 4KCl + 3H_2SO_4 \rightarrow 2CrO_2Cl_2 + 3K_2SO_4 + 3H_2O$$

Chapter 3

Thermochemistry

The last chapter served as an introduction to calculations concerned with the masses of substances involved in chemical changes. Calculations involving the heat effects that accompany physical and chemical changes are also of central importance to the chemist. In succeeding chapters, calculations involving energy effects will be frequently encountered. In this chapter, this type of calculation will be introduced.

3.1 ENERGY MEASUREMENT

It is common to think of force as thc application of physical strength—as pushing. If the effects of friction are neglected, a body in motion remains in motion at a constant velocity, and a body at rest stays at rest (its velocity is zero). If these bodies are pushed, there will be a change in their velocities. The increase in velocity per unit time is called the **acceleration.**

Suppose, for example, that we have a body moving at a velocity of 1 m/s. Assume that this body is acted on by a constant force, that is, given a sustained push. The body will move faster and faster. At the end of 1 second(s) it may be moving at the rate of 2 m/s. At the end of 2 s, its speed may be 3 m/s. If the body picks up speed at the rate of one meter per second in a second, its acceleration is said to be 1 m/s^2.

A force that gives a *one gram* body an acceleration of 1 m/s^2 is not so large as a force that gives a *one kilogram* body the same acceleration. The magnitude of a **force** (F), therefore, is proportional to the mass of the body (m) as well as to the acceleration (a) that the force produces.

$$F = ma$$

The SI unit of force is called the newton (symbol, N) and is derived from the base units of mass (the kilogram), length (the meter), and time (the second).

$$\begin{aligned} F &= ma \\ 1\ \text{N} &= (1\ \text{kg})(1\ \text{m/s}^2) \\ 1\ \text{N} &= 1\ \text{kg}\cdot\text{m/s}^2 \end{aligned}$$

Work (W) is defined as the force times the distance through which the force acts (d).

$$W = Fd$$

In the International System, the unit of work is the joule (symbol, J). The **joule** is defined as the work done when a force of one newton acts through a distance of one meter.

$$\begin{aligned} W &= Fd \\ 1\ \text{J} &= (1\ \text{N})(1\ \text{m}) \\ &= 1\ \text{N}\cdot\text{m} \\ &= 1\ \text{kg}\cdot\text{m}^2/\text{s}^2 \end{aligned}$$

Energy may be defined as the capacity to do work. There are many forms of energy, such as heat energy, electrical energy, and chemical energy. When one form of energy is converted into another form, energy is neither created nor destroyed. The SI unit of work, the joule, is the unit used for all energy measurements, including heat measurements. The unit is named in honor of James Joule (1818–1889), a student of John Dalton, who demonstrated that a given quantity of work always produces the same quantity of heat.

3.2 TEMPERATURE AND HEAT

Most liquids expand as the temperature increases. The mercury thermometer is designed to use the expansion of mercury to measure temperature. The thermometer consists of a small bulb sealed to a tube that has a narrow bore (called a capillary tube). The bulb and part of the tube contain mercury, the space above the mercury is evacuated, and the upper end of the tube is sealed. When the temperature increases, the mercury expands and rises in the capillary tube.

The **Celsius temperature scale,** named for Anders Celsius, a Swedish astronomer, is employed in scientific studies and is part of the International System. The scale is based on the assignment of 0°C to the normal freezing point of water and 100°C to the normal boiling point of water. When a thermometer is placed in a mixture of ice and water, the mercury will stand at a height that is marked on the tube as 0°C. When the thermometer is placed in boiling water, the mercury will rise to a position that is marked 100°C. The tube is marked between these two fixed points to indicate 100 equal divisions, each of which represents one degree. The thermometer is calibrated below 0°C and above 100°C by marking off degrees of the same size. The Celsius scale was formerly called the **centigrade scale,** derived from the Latin words *centum* (a hundred) and *gradus* (a degree).

On the **Fahrenheit temperature scale** (named for G. Daniel Fahrenheit, a German instrument maker) the normal freezing point of water is 32°F and the normal boiling point of water is 212°F. Since there are 100 Celsius degrees and 180 (212 − 32) Fahrenheit degrees between these two fixed points, 5 Celsius degrees equal 9 Fahrenheit degrees (5°C = 9°F).

The Fahrenheit temperature scale is not used in scientific work. Conversion of a temperature from the Fahrenheit scale to the Celsius scale can be accomplished in the following way.

1. Subtract 32° from the Fahrenheit reading. The value obtained tells how many Fahrenheit degrees the temperature is above the freezing point of water.
2. Since 5 Celsius degrees equal 9 Fahrenheit degrees, $\frac{5}{9}$ of the value obtained is the number of Celsius degrees above the freezing point of water, which is 0°C.

Hence,

$$°C = \frac{5}{9}(°F - 32°)$$

In Figure 3.1, the two temperature scales are compared. The thermodynamic temperature scale, called the Kelvin scale, is described in Section 4.3.

Temperature is defined as the degree of hotness. Heat, on the other hand, is a form of energy. In the past, chemists have customarily measured heat in calories. The **specific heat** of a substance is defined as the amount of heat required to raise the temperature of 1 g of the substance by 1°C. The calorie was originally defined in terms of the specific heat of water. The one-degree temperature interval had to be specified, however, since the specific heat of water changes slightly as the temperature changes. For many years, the **calorie** was defined as the amount of heat required to raise the temperature of 1 g of water from 14.5°C to 15.5°C.

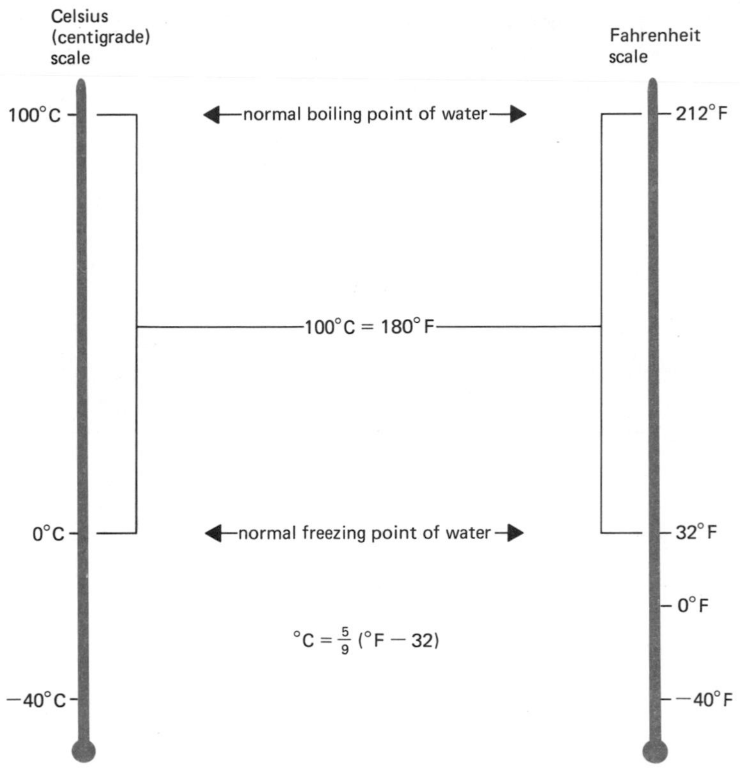

Figure 3.1 Comparison of the Celsius (centigrade) and Fahrenheit temperature scales.

Very precise determinations of heat energy, in joules, can be made by electrical measurements. The joule, therefore, is a better primary standard of heat than the specific heat of water. The calorie is now defined by its joule equivalent.

1 cal = 4.184 J (exactly)

Several points should be noted.

1. The joule and the calorie are relatively small units for measurement of thermochemical values. Such values are frequently reported in kilojoules (a kJ is 1000 J) and kilocalories (a kcal is 1000 cal).
2. The International Committee of Weights and Measures recommends that all energy measurements be based on the joule and that the calorie no longer be used. In the past, however, thermochemical values have customarily been recorded in calories and kilocalories.
   a. To convert a value given in *calories to joules,* multiply by (4.184 J/1 cal).
   b. To convert a value given in *kilocalories to kilojoules,* multiply by (4.184 kJ/1 kcal).

3. For our purposes, the specific heat of water can be considered to be a constant (4.18 J/g°C or 1.00 cal/g°C) over any temperature interval between the freezing point and boiling point of water.

3.3 CALORIMETRY

The **heat capacity** (C) of a given mass of a substance is the amount of heat required to raise the temperature of the mass by 1°C. Specific heat is the heat capacity of *one gram* of a substance—the amount of heat required to raise the temperature of 1 g of the substance by 1°C. Therefore,

$$C = (\text{mass})(\text{specific heat})$$

Since the specific heat of water is 4.18 J/g °C, the heat capacity of 500 g of water is

$$\begin{aligned} C &= (500\ \text{g})(4.18\ \text{J/g °C}) \\ &= 2090\ \text{J/°C} = 2.09\ \text{kJ/°C} \end{aligned}$$

This sample absorbs 2.09 kJ of heat for each degree that the temperature increases. Twice this amount of heat (4.18 kJ) would be required to raise the temperature by 2°C. In general,

$$q = C(t_2 - t_1)$$

where q is the heat absorbed by the sample, C is the heat capacity of the sample, t_2 is the final temperature, and t_1 is the initial temperature. The heat absorbed by a 500 g sample of water when it is heated from 20.00°C to 25.00°C can be calculated in the following way.

$$\begin{aligned} q &= C(t_2 - t_1) \\ &= (2.09\ \text{kJ/°C})(25.00\text{°C} - 20.00\text{°C}) \\ &= (2.09\ \text{kJ/°C})(5.00\text{°C}) \\ &= 10.4\ \text{kJ} \end{aligned}$$

A device called a **calorimeter** is used to measure the heat effects that accompany chemical changes. A calorimeter consists of a vessel, in which the reaction is conducted, submerged in a weighed quantity of water in a well-insulated container. The reaction is run using known quantities of reactants, and the heat evolved by the reaction increases the temperature of the water and the calorimeter. The amount of heat liberated by the reaction can be calculated from the increase in temperature if the total heat capacity of the calorimeter and its contents is known.

Example 3.1 A bomb type of calorimeter (Figure 3.2) is used to measure the heat evolved by the combustion of glucose, $C_6H_{12}O_6$.

$$C_6H_{12}O_6(s) + 6O_2(g) \rightarrow 6CO_2(g) + 6H_2O(l)$$

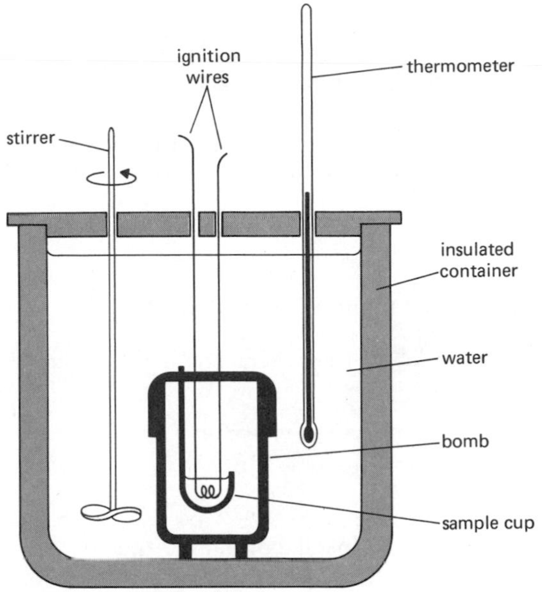

Figure 3.2 Bomb calorimeter

A 3.00 g sample of glucose is placed in the bomb, which is then filled with oxygen gas under pressure. The bomb is placed in a well-insulated calorimeter vessel that is filled with 1.20 kg of water. The initial temperature of the assembly is 19.00°C. The reaction mixture is ignited by the electrical heating of a wire within the bomb. The reaction causes the temperature of the calorimeter and its contents to increase to 25.50°C. The heat capacity of the calorimeter is 2.21 kJ/°C. The molecular weight of glucose is 180. How much heat is evolved by the combustion of 1 mol of glucose?

Solution Since 1200 g of water is employed and the specific heat of water is 4.18 J/g °C, the heat capacity of the water in the calorimeter, C_{H_2O}, is

$$\begin{aligned} C &= \text{(mass)(specific heat)} \\ C_{H_2O} &= (1200\text{ g})(4.18\text{ J/g °C}) \\ &= 5016\text{ J/°C} = 5.02\text{ kJ/°C} \end{aligned}$$

The heat capacity of the calorimeter, C_{cal}, is 2.21 kJ/°C. The total heat capacity, C_{total}, is

$$\begin{aligned} C_{total} &= C_{H_2O} + C_{cal} \\ &= 5.02\text{ kJ/°C} + 2.21\text{ kJ/°C} \\ &= 7.23\text{ kJ/°C} \end{aligned}$$

Thus, 7.23 kJ of heat is needed to raise the temperature of the assembly by 1°C.

The amount of heat *absorbed* by the calorimeter and the water is

$$\begin{aligned} q &= C(t_2 - t_1) \\ &= (7.23\text{ kJ/°C})(25.50\text{°C} - 19.00\text{°C}) \end{aligned}$$

$= (7.23 \text{ kJ/°C})(6.50°\text{C})$
$= 47.0 \text{ kJ}$

This quantity (47.0 kJ) is also the amount of heat *evolved* by the combustion of 3.00 g of glucose.

When a system absorbs heat, the quantity added is given a positive sign. When a system evolves heat, the quantity lost is given a negative sign. Therefore,

$$-47.0 \text{ kJ} = 3.00 \text{ g } C_6H_{12}O_6$$

For 1 mol of glucose (180 g of glucose), the quantity of heat evolved is

$$?\text{kJ} = 180 \text{ g } C_6H_{12}O_6 \left(\frac{-47.0 \text{ kJ}}{3.00 \text{ g } C_6H_{12}O_6}\right) = -2.82 \times 10^3 \text{ kJ}$$

3.4 THERMOCHEMICAL EQUATIONS

If a reaction that produces a gas is run in a closed container, the pressure inside the container will increase. Most reactions, however, are run in containers that are open to the atmosphere. For these reactions, the pressure is constant whether gases are produced or not.*

The heat effects of reactions that are conducted under constant pressure can be related to a property that is called **enthalpy** and is given the symbol **H.** Every pure substance is assumed to have an enthalpy (which is also called a heat content). A given set of reactants, therefore, has a definite total enthalpy, $H_{\text{reactants}}$. The corresponding set of products also has a definite total enthalpy, H_{products}. The heat of reaction is the difference between these enthalpies and is therefore given the symbol ΔH. The upper case Greek delta, Δ, is used to indicate a difference.

$$\Delta H = H_{\text{products}} - H_{\text{reactants}}$$

* Reactions run in a calorimeter bomb may or may not cause the pressure inside the bomb to change. The equation for the reaction described in Example 3.1 is

$$C_6H_{12}O_6(s) + 6O_2(g) \rightarrow 6CO_2(g) + 6H_2O(l)$$

Notice that six moles of gas (O_2 gas) are consumed and six moles of gas (CO_2 gas) are produced. The pressure inside the bomb, therefore, does not change when the reaction occurs.

If a reaction produces more moles of gas than it uses, the pressure inside a calorimeter bomb builds up. If such a reaction occurs in a container that is open to the atmosphere, the gases that are produced escape. The pressure in this case remains constant and equal to the pressure of the atmosphere. For such a reaction, the heat effect measured when the pressure is allowed to change is slightly different from the heat effect measured at constant pressure. In a case of this type a correction is applied to the value obtained by use of a bomb calorimeter (see Section 17.2).

There are no appreciable pressure effects for many reactions, including those that do not involve gases (for example, reactions run in solution) and those in which the number of moles of gas used equals the number of moles of gas produced.

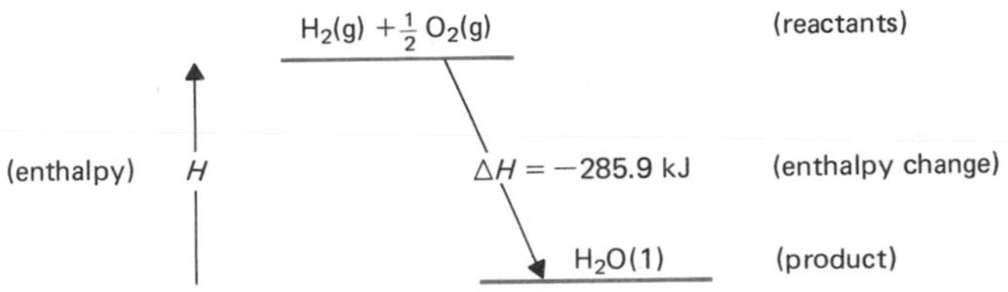

Figure 3.3 Enthalpy diagram for an exothermic reaction.

Reactions that liberate heat are called **exothermic reactions.** For reactions of this type, the products have a lower enthalpy than the reactants; ΔH is a negative value. When the reaction occurs, the products replace the reactants in the system. The enthalpy of the reaction system declines (ΔH is negative), and the difference is given off as heat (see Figure 3.3).

Reactions that absorb heat are called **endothermic reactions.** For reactions of this type, the enthalpy of the products is higher than the enthalpy of the reactants, and ΔH is positive. When the reaction occurs, heat must be supplied in order to raise the enthalpy of the system (see Figure 3.4).

The enthalpies of chemical substances depend upon temperature and pressure. By convention, ΔH values are usually reported for reactions run at 25°C and standard atmospheric pressure (Section 4.1). If any other conditions are employed, they are noted.

Thermochemical data may be given by writing a chemical equation for the reaction under consideration and listing beside it the ΔH value for the reaction *as it is written.* The appropriate value of ΔH is the one required when the equation is read in molar quantities. Contrary to usual practice, fractional coefficients may be used to balance the chemical equation. A fractional coefficient simply indicates a fraction of a mole of a substance. Thus,

$$H_2(g) + \tfrac{1}{2}O_2(g) \rightarrow H_2O(l) \qquad \Delta H = -286 \text{ kJ}$$

When 1 mol of hydrogen gas reacts with $\frac{1}{2}$ mol of oxygen gas to produce 1 mol of liquid water, 286 kJ of heat is evolved.

The state of each substance in the reaction must be indicated in the equation. A designation, such as (g) for gas, (s) for solid, (l) for liquid, or (aq) for "in aqueous solution," is placed after each formula. The need for this conven-

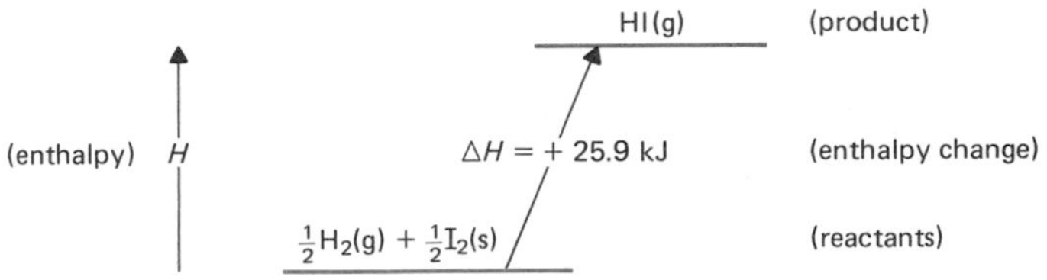

Figure 3.4 Enthalpy diagram for an endothermic reaction.

tion can be demonstrated by comparing the following equation with the preceding one.

$$H_2(g) + \tfrac{1}{2}O_2(g) \rightarrow H_2O(g) \qquad \Delta H = -242 \text{ kJ}$$

Notice that 44 kJ less heat is liberated by the second reaction than by the first reaction. This quantity of heat is used to convert 1 mol of $H_2O(l)$ to 1 mol of $H_2O(g)$.

When an equation is reversed, the sign of ΔH is changed. A reaction that is endothermic in one direction is exothermic in the opposite direction.

$$\tfrac{1}{2}H_2(g) + \tfrac{1}{2}I_2(s) \rightarrow HI(g) \qquad \Delta H = +25.9 \text{ kJ}$$
$$HI(g) \rightarrow \tfrac{1}{2}H_2(g) + \tfrac{1}{2}I_2(s) \qquad \Delta H = -25.9 \text{ kJ}$$

If the coefficients of the substances in a chemical equation are multiplied by a factor, the ΔH value must be multiplied by the same factor. For example, if the last equation is multiplied through by 2, the ΔH value must also be multiplied by 2.

$$2HI(g) \rightarrow H_2(g) + I_2(s) \qquad \Delta H = 2(-25.9 \text{ kJ}) = -51.8 \text{ kJ}$$

In like manner, the coefficients of an equation and the ΔH value can be divided by the same number.

The conventions for writing thermochemical equations can be summarized as follows.

1. For exothermic reactions (the reaction system loses heat), ΔH is negative. For endothermic reactions (the reaction system absorbs heat), ΔH is positive.
2. Unless otherwise noted, ΔH values are given for reactions run at 25°C and standard atmospheric pressure.
3. Designations such as (g), (s), (l), and (aq) are placed behind the formulas in the equation to indicate the physical state of each substance.
4. The coefficients of the substances of the chemical equation indicate the number of moles of each substance involved in the reaction (fractions may be used), and the ΔH value given corresponds to these quantities of materials.
5. If the coefficients in the chemical equation are multiplied or divided by a factor, the ΔH value must be multiplied or divided by the same factor.
6. When a chemical equation is reversed, the sign but not the magnitude of the ΔH value is changed.

3.5 THE LAW OF HESS

The basis of many thermochemical calculations is the **law of constant heat summation,** which was established experimentally by G. H. Hess in 1840. This **law of Hess** states that the change in enthalpy for any chemical reaction is constant, whether the reaction occurs in one step or in several steps. Thermochemical data, therefore, may be treated algebraically.

Consider, for example, the reaction of graphite with oxygen that produces carbon dioxide gas.

$$C(\text{graphite}) + O_2(g) \rightarrow CO_2(g) \qquad \Delta H = -393.5 \text{ kJ}$$

This transformation also can be brought about in two steps: the reaction of graphite with O_2 that forms CO followed by the reaction of the CO with O_2 that forms CO_2. Addition of the equations for the steps gives a result that is identical with the equation for the direct reaction (See Figure 3.5).

$$\begin{array}{rlr} C(\text{graphite}) + \frac{1}{2}O_2(g) \rightarrow CO(g) & & \Delta H = -110.5 \text{ kJ} \\ CO(g) + \frac{1}{2}O_2(g) \rightarrow CO_2(g) & & \underline{\Delta H = -283.0 \text{ kJ}} \\ \hline C(\text{graphite}) + \quad O_2(g) \rightarrow CO_2(g) & & \Delta H = -393.5 \text{ kJ} \end{array}$$

Since thermochemical data can be treated algebraically, it is possible to derive an enthalpy of reaction from measurements made on other reactions. Suppose, for example, that the following thermochemical equations are given.

$$C(\text{graphite}) + O_2(g) \rightarrow CO_2(g) \qquad \Delta H = -393.5 \text{ kJ} \qquad (1)$$
$$H_2(g) + \tfrac{1}{2}O_2(g) \rightarrow H_2O(l) \qquad \Delta H = -285.9 \text{ kJ} \qquad (2)$$
$$CH_4(g) + 2O_2(g) \rightarrow CO_2(g) + 2H_2O(l) \qquad \Delta H = -890.4 \text{ kJ} \qquad (3)$$

These relations can be used to find the ΔH for the reaction in which methane, CH_4, is prepared from carbon and hydrogen. This enthalpy change cannot be measured directly.

$$C(\text{graphite}) + 2H_2(g) \rightarrow CH_4(g) \qquad \Delta H = ?$$

Since 1 mol of C(graphite) appears on the left of equation (1) and also on the left of the desired equation, equation (1) is written as given.

$$C(\text{graphite}) + O_2(g) \rightarrow CO_2(g) \qquad \Delta H = -393.5 \text{ kJ} \qquad (1)$$

Two moles of $H_2(g)$ appear on the left of the desired equation, and only 1 mol of $H_2(g)$ appears on the left of equation (2). Equation (2), therefore, is multiplied through by 2, and the ΔH value is multiplied by 2.

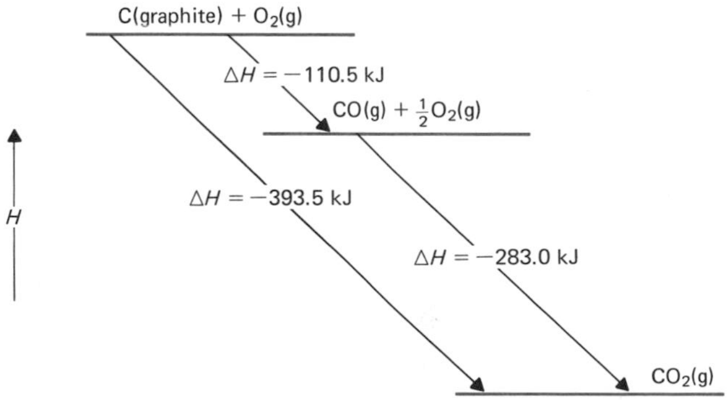

Figure 3.5 Enthalpy diagram to illustrate the law of Hess.

$$2H_2(g) + O_2(g) \rightarrow 2H_2O(l) \qquad \Delta H = -571.8 \text{ kJ} \qquad (4)$$

One mole of $CH_4(g)$ appears on the *right* side of the desired equation. Therefore, equation (3) is reversed, and the sign of the ΔH value is changed.

$$CO_2(g) + 2H_2O(l) \rightarrow CH_4(g) + 2O_2(g) \qquad \Delta H = +890.4 \text{ kJ} \qquad (5)$$

Equations (1), (4), and (5) are added. Terms common to both sides of the final equation ($2O_2$, CO_2, and $2H_2O$) are canceled.

$$C(\text{graphite}) + 2H_2(g) \rightarrow CH_4(g) \qquad \Delta H = -74.9 \text{ kJ}$$

The resulting ΔH value is the enthalpy of reaction that was sought.

3.6 ENTHALPIES OF FORMATION

The **enthalpy of formation,** ΔH_f, of a compound is the enthalpy change for the reaction in which 1 mol of the compound is made from its elements in their most stable forms at 25°C and standard atmospheric pressure. Enthalpy of formation, therefore, is a specific type of enthalpy change. The ΔH values for the reactions shown in Figures 3.3 and 3.4 are in reality the ΔH_f values of $H_2O(l)$ and $HI(g)$, respectively,

$$H_2(g) + \tfrac{1}{2}O_2(g) \rightarrow H_2O(l) \qquad \Delta H = -285.9 \text{ kJ}$$
$$\tfrac{1}{2}H_2(g) + \tfrac{1}{2}I_2(s) \rightarrow HI(g) \qquad \Delta H = +25.9 \text{ kJ}$$

Enthalpies of formation are either measured directly or calculated from other thermochemical data by applying the law of Hess. The result of the calculation of this type given in Section 3.5,

$$C(\text{graphite}) + 2H_2(g) \rightarrow CH_4(g) \qquad \Delta H = -74.9 \text{ kJ}$$

is the enthalpy of formation of $CH_4(g)$. Some enthalpies of formation are listed in Table 3.1.

The enthalpy change for a reaction can be calculated from the enthalpies of formation of the compounds involved in the reaction. For example, the enthalpy change for the reaction

$$C_2H_4(g) + H_2(g) \rightarrow C_2H_6(g) \qquad \Delta H = ?$$

can be calculated from the enthalpies of formation of ethylene, $C_2H_4(g)$, and ethane, $C_2H_6(g)$.

$$2C(\text{graphite}) + 2H_2(g) \rightarrow C_2H_4(g) \qquad \Delta H_f = +52.30 \text{ kJ} \qquad (6)$$
$$2C(\text{graphite}) + 3H_2(g) \rightarrow C_2H_6(g) \qquad \Delta H_f = -84.68 \text{ kJ} \qquad (7)$$

The reverse of equation (6) indicates a transformation in which $C_2H_4(g)$ breaks down into its elements. The appropriate value of the enthalpy change for the *reversed* reaction is $-\Delta H_f(C_2H_4)$ or -52.30 kJ. The elements from the decomposition of $C_2H_4(g)$ plus an additional mole of $H_2(g)$ can be imagined to form $C_2H_6(g)$; equation (7) is written as shown. When these two equations are added, the desired thermochemical expression is obtained.

Table 3.1 Enthalpies of formation at 25°C and 1 atm

COMPOUND	ΔH_f (kJ/mol)	COMPOUND	ΔH_f (kJ/mol)
$H_2O(g)$	−241.8	$COCl_2(g)$	−223.
$H_2O(l)$	−285.9	$SO_2(g)$	−296.9
$HF(g)$	−269.	$CO(g)$	−110.5
$HCl(g)$	−92.30	$CO_2(g)$	−393.5
$HBr(g)$	−36.2	$NO(g)$	+90.37
$HI(g)$	+25.9	$NO_2(g)$	+33.8
$H_2S(g)$	−20.2	$HNO_3(l)$	−173.2
$HCN(g)$	+130.5	$NH_4NO_3(s)$	−365.1
$NH_3(g)$	−46.19	$NaCl(s)$	−411.0
$PH_3(g)$	+9.25	$MgO(s)$	−601.83
$CH_4(g)$	−74.85	$CaO(s)$	−635.5
$C_2H_6(g)$	−84.68	$Ca(OH)_2(s)$	−986.59
$C_2H_4(g)$	+52.30	$CaCO_3(s)$	−1206.9
$C_2H_2(g)$	+226.7	$Ca_3P_2(s)$	−504.17
$C_6H_6(l)$	+49.04	$BaO(s)$	−588.1
$CH_3OH(g)$	−201.2	$BaCO_3(s)$	−1218.
$CH_3OH(l)$	−238.6	$Al_2O_3(s)$	−1669.8
$CH_3NH_2(g)$	−28.	$Fe_2O_3(s)$	−822.2
$NF_3(g)$	−113.	$AgCl(s)$	−127.0
$CF_4(g)$	−913.4	$HgBr_2(s)$	−169.
$CHCl_3(l)$	−132.	$ZnO(s)$	−348.0

$$\begin{array}{lr}
C_2H_4(g) \rightarrow 2C(\text{graphite}) + 2H_2(g) & \Delta H = -\Delta H_f = -52.30 \text{ kJ} \\
2C(\text{graphite}) + 3H_2(g) \rightarrow C_2H_6(g) & \Delta H_f = -84.68 \text{ kJ} \\
\hline
C_2H_4(g) + H_2(g) \rightarrow C_2H_6(g) & \Delta H = -136.98 \text{ kJ}
\end{array}$$

The ΔH of the reaction is, therefore, $\Delta H_f(C_2H_6) - \Delta H_f(C_2H_4)$.

In general, a ΔH value for a reaction may be obtained by subtracting the sum of the enthalpies of formation of the reactants from the sum of the enthalpies of formation of the products.

$$\Delta H = \Sigma\Delta H_f(\text{products}) - \Sigma\Delta H_f(\text{reactants})$$

The uppercase Greek sigma, Σ, indicates a sum. By reversing the sign of $\Sigma\Delta H_f$(reactants), we indicate a process in which the reactants are broken down into the elements. The formation of the products from these elements is indicated by the term $\Sigma\Delta H_f$(products).

Two factors frequently cause trouble in the use of this approach for the calculation of ΔH values.

1. Heats of formation are given in kilojoules per *mole*. The chemical equation that corresponds to a value listed in Table 3.1 pertains to the formation of only *one* mole of the compound. If more than one mole

(or less than one mole) of the compound is involved in the reaction being studied, the ΔH_f value must be multiplied by the number of moles involved.

2. The heat of formation of an *element* is zero (the heat effect when an element is prepared from itself). No terms for elements are added into the sums $\Sigma\Delta H_f$(products) and $\Sigma\Delta H_f$(reactants).

Consider the reaction

$$2NH_3(g) + 3Cl_2(g) \rightarrow N_2(g) + 6HCl(g) \qquad \Delta H = ?$$

The enthalpy change for the reaction can be calculated in the following way.

$$\begin{aligned}\Delta H &= \Sigma\Delta H_f(\text{products}) - \Sigma\Delta H_f(\text{reactants})\\ &= 6\Delta H_f(HCl) - 2\Delta H_f(NH_3)\\ &= 6(-92.30\text{ kJ}) - 2(-46.19\text{ kJ})\\ &= -553.80\text{ kJ} + 92.38\text{ kJ} = -461.42\text{ kJ}\end{aligned}$$

This calculation may be checked by the addition of the appropriate thermochemical equations. Since six moles of HCl(g) are prepared, the equation for the formation of HCl(g) is multiplied through by 6 and the enthalpy of formation of HCl(g) is multiplied by 6. Two moles of NH_3(g) are *consumed* in the reaction. The equation for the *formation* of NH_3(g) is multiplied through by 2 and *reversed;* the value of ΔH_f is multiplied by 2 and the sign changed. These two thermochemical expressions are added. No separate equations are introduced for the elements that are involved in the reaction (Cl_2 and N_2).

$$\begin{array}{rl}3H_2(g) + 3Cl_2(g) \rightarrow 6HCl(g) & \Delta H = 6\Delta H_f = -553.80\text{ kJ}\\ 2NH_3(g) \rightarrow N_2(g) + 3H_2(g) & \Delta H = -2\Delta H_f = +92.38\text{ kJ}\\ \hline 2NH_3(g) + 3Cl_2(g) \rightarrow N_2(g) + 6HCl(g) & \Delta H = -461.42\text{ kJ}\end{array}$$

The terms "$3H_2(g)$" cancel in the addition. Notice that "$3Cl_2(g)$" and "$N_2(g)$" appear in the final equation even though no special provision was made for their introduction.

Example 3.2 Use enthalpies of formation to calculate the ΔH of the reaction

$$Fe_2O_3(s) + 3CO(g) \rightarrow 2Fe(s) + 3CO_2(g)$$

Solution The values needed can be found in Table 3.1

$$\begin{aligned}\Delta H &= \Sigma\Delta H_f(\text{products}) - \Sigma\Delta H_f(\text{reactants})\\ &= 3\Delta H_f(CO_2) - [\Delta H_f(Fe_2O_3) + 3\Delta H_f(CO)]\\ &= 3(-393.5\text{ kJ}) - [(-822.2\text{ kJ}) + 3(-110.5\text{ kJ})]\\ &= -1180.5\text{ kJ} + 1153.7\text{ kJ} = -26.8\text{ kJ}\end{aligned}$$

Example 3.3 Given the following data:

$$CH_4(g) + 2O_2(g) \rightarrow CO_2(g) + 2H_2O(l) \qquad \Delta H = -890.4\text{ kJ}$$

the enthalpy of formation of $CO_2(g)$ is −393.5 kJ, the enthalpy of formation of $H_2O(l)$ is −285.9 kJ. Calculate the enthalpy of formation of $CH_4(g)$.

Solution In this case, the value of ΔH for a reaction is known and a ΔH_f value for one of the reactants is sought.

$$\Delta H = \Sigma\Delta H_f(\text{products}) - \Sigma\Delta H_f(\text{reactants})$$
$$\Delta H = \Delta H_f(CO_2) + 2\Delta H_f(H_2O) - \Delta H_f(CH_4)$$
$$-890.4\text{ kJ} = (-393.5\text{ kJ}) + 2(-285.9\text{ kJ}) - \Delta H_f(CH_4)$$
$$-890.4\text{ kJ} = -965.3\text{ kJ} - \Delta H_f(CH_4)$$
$$\Delta H_f(CH_4) = -74.9\text{ kJ}$$

USING ENTHALPIES OF FORMATION TO DERIVE ΔH VALUES

1. Write the chemical equation for the reaction.
2. Substitute in the equation:

$$\Delta H = \Sigma\Delta H_f(\text{products}) - \Sigma\Delta H_f(\text{reactants})$$

(a) The term $\Sigma\Delta H_f$(products) is the sum of the enthalpies of formation of the *compounds* that appear on the *right* side of the chemical equation.

(b) The term $\Sigma\Delta H_f$(reactants) is the sum of the enthalpies of formation of the *compounds* that appear on the *left* side of the chemical equation.

To obtain either of these sums, multiply the enthalpy of formation of each compound by the number of moles of that compound involved in the reaction (given by the appropriate coefficient of the chemical equation). No enthalpy terms are introduced for *elements* that appear in the chemical equation since their enthalpies of formation are zero.

3.7 BOND ENERGIES

Atoms are held together in molecules by chemical bonds (see Chapter 7). The energy required to *break* the bond that holds two atoms together in a diatomic molecule is called the **bond dissociation energy.** These values are reported in kilojoules per mole of bonds. In the following equations, which illustrate this process, dashes are used to represent the bonds between atoms; H_2, for example, appears as H—H.

$$\text{H—H(g)} \rightarrow \text{2H(g)} \qquad \Delta H = +435\text{ kJ}$$
$$\text{Cl—Cl(g)} \rightarrow \text{2Cl(g)} \qquad \Delta H = +243\text{ kJ}$$
$$\text{H—Cl(g)} \rightarrow \text{H(g)} + \text{Cl(g)} \qquad \Delta H = +431\text{ kJ}$$

Each of these ΔH values is *positive,* which indicates that energy is *absorbed* in each process. The bond in the H_2 molecule is the strongest of the three. It takes the most energy to pull the atoms of the H_2 molecule apart.

If one of these equations is reversed, the sign of the ΔH value must be changed.

$$\text{H(g)} + \text{Cl(g)} \rightarrow \text{H—Cl(g)} \qquad \Delta H = -431 \text{ kJ}$$

When a bond forms, energy is *released*—the same amount that is *required* to break the bond.

Bond energies may be used to determine ΔH values. Consider the reaction

$$\text{H}_2\text{(g)} + \text{Cl}_2\text{(g)} \rightarrow 2\text{HCl(g)} \qquad \Delta H = 2\Delta H_f = -184.6 \text{ kJ}$$

The ΔH for this reaction is twice the enthalpy of formation of HCl(g) since the equation indicates the formation of two moles of HCl(g). We can derive this ΔH value from bond energies in the following way. The enthalpy change is the sum of the energy *required* to break 1 mol of H—H bonds, the energy *required* to break 1 mol of Cl—Cl bonds, and the energy *evolved* by the formation of 2 mol of H—Cl bonds.

$$\begin{aligned} \text{H—H(g)} &\rightarrow 2\text{H(g)} & \Delta H &= +435 \text{ kJ} \\ \text{Cl—Cl(g)} &\rightarrow 2\text{Cl(g)} & \Delta H &= +243 \text{ kJ} \\ 2\text{H(g)} + 2\text{Cl(g)} &\rightarrow 2\text{H—Cl(g)} & \Delta H &= 2(-431)\text{kJ} = -862 \text{ kJ} \end{aligned}$$

The sum of these equations is

$$\text{H—H(g)} + \text{Cl—Cl(g)} \rightarrow 2\text{H—Cl(g)} \qquad \Delta H = -184 \text{ kJ}$$

A molecule that contains more than two atoms, such as H_2O, is called a **polyatomic molecule.** Molecules of this type contain more than one bond. There are, for example, two H—O bonds in the H_2O molecule. The ΔH for

$$\text{H—O—H(g)} \rightarrow 2\text{H(g)} + \text{O(g)} \qquad \Delta H = +926 \text{ kJ}$$

refers to a process in which *two* moles of H—O bonds are broken. The **average bond energy** of the H—O bond, therefore, is +926 kJ/2 mol, or +463 kJ/mol.

In the H_2O molecule, the H—O bonds are equivalent. If the bonds were broken one at a time, however, the ΔH values would not be the same.

$$\begin{aligned} \text{H—O—H(g)} &\rightarrow \text{H(g)} + \text{O—H(g)} & \Delta H &= +501 \text{ kJ} \\ \text{O—H(g)} &\rightarrow \text{O(g)} + \text{H(g)} & \Delta H &= +425 \text{ kJ} \end{aligned}$$

The average of the ΔH values for the steps is +463 kJ/mol, which is the average bond energy. In general, the second bond of a molecule such as H_2O is easier to break than the first. The fragment remaining after one H has been removed (O—H) is not as stable as the original molecule (H—O—H).

The bond energy of a given type of bond is not the same in all molecules containing that bond. The H—O bond energy in H—O—H is different from the H—O bond energy in H—O—Cl. The values listed in Table 3.2 for di-

Table 3.2 Average bond energies

BOND	AVERAGE BOND ENERGY (kJ/mol)	BOND	AVERAGE BOND ENERGY (kJ/mol)
H—H	435	P—H	318
H—F	565	N—Cl	201
H—Cl	431	P—Cl	326
H—Br	364	C—C	347
H—I	297	C=C	619
F—F	155	C≡C	812
Cl—Cl	243	C—H	414
Br—Br	193	C—O	335
I—I	151	C=O	707
O—O	138	C—F	485
O_2^*	494	C—Cl	326
O—H	463	C—N	293
O—F	184	C=N	616
O—Cl	205	C≡N	879
N—N	159	S—H	339
N=N	418	S—S	264
N≡N	941	S—Cl	276
N—H	389		

* Double bond of molecular oxygen.

atomic molecules are bond dissociation energies. The other values listed are average bond energies, and each of these values is an average derived from a large number of cases. Since average bond energies are approximations, a ΔH value obtained by use of these values must be regarded as an estimate.

In some molecules, two atoms are bonded together by more than one bond. Two nitrogen atoms, for example, can be joined by a single bond (N—N), a double bond (N=N), or a triple bond (N≡N) depending on the molecule. Multiple bonds are indicated in Table 3.2.

Example 3.4 Use average bond energies to calculate the value of ΔH for the reaction

$$2\text{H—}\overset{\displaystyle\text{H}\atop|}{\text{N}}\text{—H(g)} + 3\text{Cl—Cl(g)} \rightarrow \text{N}\equiv\text{N(g)} + 6\text{H—Cl(g)}$$

Solution We can imagine the reaction to take place by a series of steps. Energy is absorbed (ΔH is positive) when a bond is broken, and energy is evolved (ΔH is negative) when a bond is formed.

Six N—H bonds are broken:

$$2\text{H—}\overset{\displaystyle\text{H}\atop|}{\text{N}}\text{—H(g)} \rightarrow 2\text{N(g)} + 6\text{H(g)} \qquad \Delta H = 6(+389)\text{kJ} = +2334\text{ kJ}$$

Three Cl—Cl bonds are broken:

$$3Cl{-}Cl(g) \rightarrow 6Cl(g) \qquad \Delta H = 3(+243)\text{kJ} = +729\ \text{kJ}$$

One N≡N bond is formed:

$$2N(g) \rightarrow N{\equiv}N(g) \qquad \Delta H = -941\ \text{kJ}$$

Six H—Cl bonds are formed:

$$6H(g) + 6Cl(g) \rightarrow 6H{-}Cl(g) \qquad \Delta H = 6(-431)\ \text{kJ} = -2586\ \text{kJ}$$

The sum of these steps is the answer to the problem.

$$2NH_3(g) + 3Cl_2(g) \rightarrow N_2(g) + 6HCl(g) \qquad \Delta H = -464\ \text{kJ}$$

In Section 3.6, enthalpies of formation were used to calculate the value of ΔH for this reaction. The value obtained in this way (−461 kJ) is a more reliable value than the one derived from bond energies (−464 kJ).

COUNT RUMFORD AND THE NATURE OF HEAT

Probably the best known scientific work of Count Rumford is his research on the nature of heat. The theory that prevailed at the end of the eighteenth century regarded heat as a material fluid called caloric, which could flow from one body to another. Lavoisier included caloric in his list of elements. Some of Rumford's experiments and numerous publications were undertaken to disprove the caloric theory of heat.

In 1797, Rumford was engaged in supervising the boring of cannon at the Munich arsenal. The process, in which the bore of a cannon is drilled into a solid metal cannon barrel, liberates a large quantity of heat. If this heat were originally contained in the metal as caloric and released when the metal was ground up in the drilling operation, then the metal shavings produced should contain a smaller quantity of caloric than the solid metal.

Rumford found that the shavings had the same heat properties as the solid metal. To Rumford, the heat was produced by mechanical motion. His experiments convinced him that "heat is nothing but a vibratory motion taking place among the particles of bodies," which is close to the modern theory of heat.

Count Rumford was born Benjamin Thompson in Woburn, Massachusetts, in 1753. At the age of nineteen, he married a wealthy widow (who was 14 years older than he) from the town of Concord, New Hampshire (which had previously belonged to Massachusetts and been called Rumford).

At the time of the American revolution, Thompson was twice formally charged with "being unfriendly to the cause of Liberty." He strenuously denied these accusations, and both times the charges were dropped for lack of evidence. Documents discovered in the twentieth century (one of them in invisible ink), however, have proved that Thompson was a secret agent for the British. In 1776 he deserted his wife and child and left America for England.

In 1784, Thompson entered the service of the ruler of Bavaria, Karl Theodor, an Elector of the Holy Roman Empire. Thompson who was given free rein by the Elector, reorganized the army, founded industrial schools, instituted social reforms, established work houses for the employment of the poor, and supervised the beautification of the city of Munich. In 1792, the Elector was instrumental in having Thompson made a Count of the Holy Roman Empire. For his new name he chose Rumford, the original name of the American town in which he had been married. In the same year that he was made a count, his wife died.

In 1805, Rumford married the widow of Lavoisier, the great French chemist who had been guillotined during the reign of terror of the French revolution. The marriage between Rumford and his new wife (who insisted that she be called Comtesse Lavoisier de Rumford) was not a success. Their incompatibility was a subject of gossip. When he locked the gate to the grounds of their house so that guests that she had invited could not enter, she poured boiling water on his rose bushes. He called her a "female dragon" and left her in 1808; they were divorced in 1809. He moved to Anteul, France, and lived there until his death in 1814.

PROBLEMS

3.1 The recipe for biscuits given in Problem 1.20 states that the biscuits should be baked for 12 to 15 minutes in an oven heated to 450°F. What is this temperature in degrees Celsius?

3.2 Normal body temperature is 98.6°F. What is normal body temperature in degrees Celsius?

3.3 A thermostat is set at 68.0°F. What is the setting in degrees Celsius?

3.4 What temperature in degrees Celsius corresponds to −40°F?

3.5 What is the heat capacity of 150 g of water?

3.6 What amount of heat is required to raise the temperature of 150 g of water from 20.00°C to 24.00°C?

3.7 How many kilojoules of heat are produced by a combustion if the heat liberated is sufficient to raise the temperature of 2.00 kg of water from 25.00°C to 43.00°C?

3.8 What is the difference between heat and temperature?

3.9 What is the heat capacity of 1500 g of chloroform if the specific heat of chloroform is 0.962 J/g°C?

3.10 The heat capacity of 500 g of mercury is 69.0 J/°C. What is the specific heat of mercury?

3.11 A 10.0 g piece of silver requires 6.99 J of heat to increase its temperature from 19.00°C to 22.00°C. What is the specific heat of silver?

3.12 The specific heat of brass is 0.384 J/g°C. How much heat is required to raise the temperature of 113 g of brass from 25.2°C to 50.4°C?

3.13 The specific heat of gold is 0.132 J/g°C. A 10.0 g sample of gold requires 19.8 J of heat to increase its temperature to 35.0°C. What was the initial temperature of the sample?

3.14 If 66 kJ of heat raises the temperature of a sample of marble from 22.0°C to 52.0°C, what is the mass of the sample? The specific heat of marble is 0.88 J/g°C.

3.15 A calorimeter that has a heat capacity of 2.25 kJ/°C is used to measure the heat evolved by the combustion of 8.13 g of nicotine, $C_{10}H_{14}N_2$ (molecular weight, 162). The calorimeter contains 2.00 kg of water, and the initial temperature of the calorimeter and its contents is 20.00°C. The combustion liberates 299 kJ of heat. (a) What is the final temperature of the calorimeter and its contents? (b) What quantity of heat would be evolved by the combustion of 1 mol of nicotine?

3.16 A 1.60 g sample of methyl alcohol, CH_3OH, was burned in excess oxygen in a calorimeter that contained 1200 g of water. The heat capacity of the calorimeter was 1450 J/°C. The temperature of the calorimeter and its contents increased from 21.36°C to 31.95°C. What is the enthalpy of combustion per mole of methyl alcohol?

3.17 Glycine, $C_2H_5O_2N$, is a simple amino acid that occurs in proteins. A 2.50 g sample of glycine was burned in excess oxygen in a calorimeter that contained 1620 g of water. The temperature of the calorimeter and contents increased from 24.46°C to 28.07°C. The heat capacity of the calorimeter was 2.22 kJ/°C. What is the molar enthalpy of combustion of glycine?

3.18 The reaction of 12.45 g of P_4O_{10} with an appropriate quantity of water is run in a reaction vessel of a calorimeter. The vessel is surrounded by 950.0 g of water. The temperature of the calorimeter and its contents increases from 22.815°C to 26.885°C. The heat capacity of the calorimeter is 0.6904 kJ/°C. Calculate the enthalpy change for the reaction

$$P_4O_{10}(s) + 6H_2O(l) \rightarrow 4H_3PO_4(l)$$

3.19 The enthalpy of combustion of lactose, $C_{12}H_{22}O_{11}$, is −5652 kJ/mol. A 2.50 g sample of lactose was burned in a calorimeter that contained 1350 g of water. The heat capacity of the calorimeter is 1630 J/°C, and the initial temperature was 24.58°C. What was the final temperature?

3.20 The heat liberated by the combustion of glutaric acid, $H_2C_5H_6O_4$, is 2154 kJ/mol. The heat from the combustion of 1.50 g of glutaric acid is evolved in a calorimeter that contains 1250 g of water and that has a heat capacity of 9267 J/°C. What is the final temperature of the calorimeter and its contents if the initial temperature was 20.00°C?

3.21 The enthalpy of combustion of camphor, $C_{10}H_{16}O$, is −5898 kJ/mol. A 2.00 g sample of camphor is burned in a calorimeter that contains 1500 g of water. The temperature increases from 21.00°C to 29.85°C. What is the heat capacity of the calorimeter?

3.22 The enthalpy of combustion of oxalic acid, $H_2C_2O_4$, is −246.1 kJ/mol. A 5.00 g sample of oxalic acid is burned in a calorimeter that contains 1250 g of water. The temperature of the calorimeter and its contents increases from 24.00°C to 26.04°C. What is the heat capacity of the calorimeter?

3.23 The thermite reaction is highly exothermic.

$$2Al(s) + Fe_2O_3(s) \rightarrow 2Fe(s) + Al_2O_3(s) \qquad \Delta H = -848 \text{ kJ}$$

How much heat is liberated when 36.0 g of Al reacts with excess Fe_2O_3?

3.24 The combustion of 132 g of propane, $C_3H_8(g)$, liberates 6600 kJ of heat. What is the enthalpy change for the reaction

$$C_3H_8(g) + 5O_2(g) \rightarrow 3CO_2(g) + 4H_2O(l)$$

3.25 The combustion of 1.00 mol of the amino acid alanine, $C_3H_7O_2N$, liberates 1622 kJ of heat. What quantity of heat is liberated by 37.2 g of alanine?

3.26 Given the thermochemical equation

$$2NiS(s) + 3O_2(g) \rightarrow 2NiO + 2SO_2(g) \qquad \Delta H = -936.2 \text{ kJ}$$

(a) What quantity of heat is liberated by the reaction of 50.0 g of NiS with excess O_2? (b) What quantity of heat is liberated by a reaction that produces 50.0 g of SO_2?

3.27 It is sometimes said that the standard enthalpy of all elements, H, is set equal to zero in determining ΔH_f values for compounds. Criticize this statement.

3.28 Write the thermochemical equations that correspond to the following enthalpies of formation measured at 25°C and 1 atm. (a) $H_2O(l)$, −285.9 kJ/mol; (b) HBr(g), −36.2 kJ/mol; (c) $CHCl_3(l)$, −132 kJ/mol.

3.29 Write the thermochemical equations that correspond to the following enthalpies of formation measured at 25°C and 1 atm. (a) $NO_2(g)$, +33.8 kJ/mol; (b) MgO(s), −601.83 kJ/mol; (c) $Al_2O_3(s)$, −1669.8 kJ/mol.

3.30 Use enthalpies of formation (Table 3.1) to calculate the enthalpy change at 25°C for the following reaction.

$$3NO_2(g) + H_2O(l) \rightarrow 2HNO_3(l) + NO(g)$$

3.31 Use enthalpies of formation (Table 3.1) to calculate the enthalpy change at 25°C for the following reaction.

$$4NH_3(g) + 5O_2(g) \rightarrow 4NO(g) + 6H_2O(l)$$

3.32 Calculate the enthalpy change for the reaction

$$BaO(s) + CO_2(g) \rightarrow BaCO_3(s)$$

from the enthalpies of formation at 25°C (Table 3.1).

3.33 If limestone ($CaCO_3$) is roasted, quicklime (CaO) is produced. The addition of water to quicklime produces slaked lime, $Ca(OH)_2$.

$$CaCO_3(s) \rightarrow CaO(s) + CO_2(g)$$
$$CaO(s) + H_2O(l) \rightarrow Ca(OH)_2(s)$$

Use enthalpies of formation (Table 3.1) to calculate the enthalpy changes for the reactions.

3.34 Phosphine gas can be made by the reaction of calcium phosphide with water.

$$Ca_3P_2(s) + 6H_2O(l) \rightarrow 3Ca(OH)_2 + 2PH_3(g)$$

Use enthalpies of formation (Table 3.1) to calculate the enthalpy change for this reaction.

3.35 (a) Write the chemical equation for the combustion of one mole of methylamine, $CH_3NH_2(g)$, in $O_2(g)$. The products of the reaction are $CO_2(g)$, $H_2O(l)$, and $N_2(g)$. (b) Use enthalpies of formation (Table 3.1) to calculate the enthalpy change for the reaction.

3.36 (a) Write chemical equations for the combustion of acetylene, $C_2H_2(g)$, and of ethane, $C_2H_6(g)$, in $O_2(g)$. The products of both combustions are $CO_2(g)$ and $H_2O(l)$. (b) Use enthalpies of formation (Table 3.1) to calculate the enthalpy change for each reaction. (c) Compare the heating value of acetylene and ethane per gram of fuel.

3.37 The combustion of $CS_2(l)$ in $O_2(g)$ produces $CO_2(g)$ and $SO_2(g)$. The enthalpy change for the reaction in which *one mole* of $CS_2(l)$ is burned is -1075.7 kJ. (a) Write the thermochemical equation for the reaction. (b) Use enthalpies of formation (Table 3.1) to calculate the enthalpy of formation of $CS_2(l)$.

3.38 The enthalpy change for the combustion of *one mole* of pentane, $C_5H_{12}(l)$, to $CO_2(g)$ and $H_2O(l)$ is -3526.3 kJ. Calculate the enthalpy of formation of pentane.

3.39 The enthalpy change for the combustion of *one mole* of cyclohexane, $C_6H_{12}(l)$, to $CO_2(g)$ and $H_2O(l)$ is -3920 kJ. Calculate the enthalpy of formation of cyclohexane.

3.40 For the following reaction, $\Delta H = -401.7$ kJ.

$$CH_4(g) + 4Cl_2(g) \rightarrow CCl_4(l) + 4HCl(g)$$

Calculate the enthalpy of formation of $CCl_4(l)$.

3.41 For the following reaction, $\Delta H = -1010.0$ kJ.

$$2NH_3(g) + 3N_2O(g) \rightarrow 4N_2(g) + 3H_2O(l)$$

Calculate the enthalpy of formation of $N_2O(g)$.

3.42 Calculate the enthalpy change for the reaction

$$C_6H_6(l) + 3H_2(g) \rightarrow C_6H_{12}(g)$$

For the combustion of *one mole* of $C_6H_6(l)$ to $CO_2(g)$ and $H_2O(l)$, $\Delta H = -3268$ kJ. For the combustion of *one mole* of $C_6H_{12}(l)$ to $CO_2(g)$ and

$H_2O(l)$, $\Delta H = -3920$ kJ. The enthalpy of formation of $H_2O(l)$ is -285.9 kJ/mol.

3.43 For the reaction of the monatomic elements

$$H(g) + Cl(g) \rightarrow HCl(g)$$

$\Delta H = -431.62$ kJ. The enthalpy of formation of HCl(g) is -92.30 kJ/mol. The enthalpy of dissociation of $H_2(g)$,

$$H_2(g) \rightarrow 2H(g)$$

is $+435.89$ kJ/mol H_2. What is the enthalpy of dissociation of $Cl_2(g)$?

3.44 Use average bond energies (Table 3.2) to calculate the enthalpy of formation of HF(g). Compare the value to the value found in Table 3.1.

3.45 Use average bond energies (Table 3.2) to calculate the enthalpy of formation of $NH_3(g)$. Compare the value to the value listed in Table 3.1.

3.46 Use average bond energies (Table 3.2) to calculate ΔH for the reaction

$$\text{F—O—F(g)} + \text{H—O—H(g)} \rightarrow \text{O}_2\text{(g)} + 2\text{H—F(g)}$$

3.47 (a) Use average bond energies (Table 3.2) to calculate ΔH for the reaction

$$4\ \text{H—}\underset{}{\overset{\overset{\displaystyle\text{H}}{|}}{\text{N}}}\text{—H (g)} + 3\text{O}_2\text{(g)} \rightarrow 2\ \text{N}{\equiv}\text{N (g)} + 6\ \text{H—O—H (g)}$$

(b) Use enthalpies of formation (Table 3.1) to calculate ΔH for the reaction. How well do the values agree?

Chapter 4

Gases

Gases are believed to consist of widely separated molecules in rapid motion. Any two (or more) gases can be mixed in any proportion to prepare a perfectly uniform mixture; no such generalization can be made for liquids. Since the molecules of a gas are separated by comparatively large distances, the molecules of one gas can easily fit between the molecules of another gas. This molecular model can also be used to explain the fact that gases are readily compressed. Compression consists of forcing gas molecules closer together.

A gas expands to fill any container into which it is introduced. When a gas that has an odor is released into a room, it can soon be detected in all parts of the room. Gases diffuse because the gas molecules are in constant, rapid motion. Furthermore, in the course of their random motion, gas molecules strike the walls of the container. These impacts explain the fact that gases exert pressure.

4.1 PRESSURE

Pressure is defined as force per unit area. The pressure of a gas is equal to the force that the gas exerts on the walls of the container divided by the surface area of the container.

$$\text{pressure} = \frac{\text{force}}{\text{area}}$$

The SI unit of pressure is the **pascal** (abbreviated Pa), which is defined as the pressure equivalent to a force of one newton ($1\ \text{N} = 1\ \text{kg·m/s}^2$) acting on one square meter.

$$1\ \text{Pa} = \frac{1\ \text{N}}{1\ \text{m}^2}$$

$$= \frac{1\ \text{kg·m/s}^2}{1\ \text{m}^2} = 1\ \text{kg/m·s}^2$$

The chemist, however, usually measures gas pressures by relating them to the pressure of the atmosphere.

A **barometer** is used to measure the pressure that the atmosphere exerts on the surface of the earth. This instrument was devised in the seventeenth century by Evangelista Torricelli, a pupil of Galileo. A tube, approximately 850 mm in length and sealed at one end, is filled with mercury and inverted in an open container of mercury (Figure 4.1). The mercury falls in the tube but does not completely run out. The pressure of the atmosphere on the surface of the mercury in the dish supports the column of mercury in the tube.

The space above the mercury inside the tube is a nearly perfect vacuum. Since mercury is not very volatile at room temperature, only a negligible

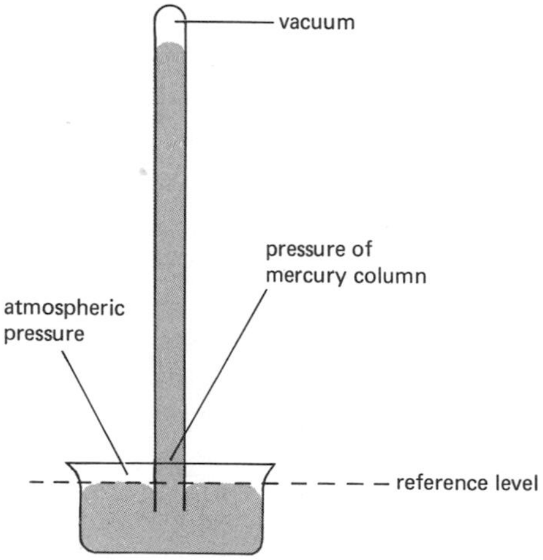

Figure 4.1 Barometer

amount of mercury vapor occupies this space. Therefore, practically no pressure is exerted on the upper surface of the mercury in the column. The pressure inside the tube and above the reference level shown in Figure 4.1 results from the weight of the mercury column alone. This pressure is equal to the atmospheric pressure outside the tube and above the reference level.

The height of mercury in the tube serves as a measure of the atmospheric pressure. When the atmospheric pressure rises, it pushes the mercury higher in the tube. Remember that pressure is force *per unit area*. Whether the tube has a relatively large or small cross-sectional area, a given atmospheric pressure will support the mercury in the tube to the same height.

The pressure of the atmosphere changes from day to day and from place to place. The average pressure at sea level supports a column of mercury to a height of 760 mm; this value is called 1 atmosphere (abbreviated atm). The definition of the **standard atmosphere,** however, is given in terms of the pascal.

$$1 \text{ atm} = 101{,}325 \text{ Pa} = 101.325 \text{ kPa}$$

The pressure equivalent to a height of 1 mm of mercury is called 1 **torr** (named for Torricelli); therefore,

$$1 \text{ atm} = 760 \text{ torr}$$

The International Committee of Weights and Measures recommends that pressures not be recorded in torrs. The preceding relationship can be used to convert readings made in terms of the height of a mercury column (in torrs) into atmospheres.

A **manometer,** a device that is used to measure the pressure of a sample of a gas, is patterned after the barometer. The type of manometer shown in Figure 4.2 consists of a U tube containing mercury. One arm of the U tube is open

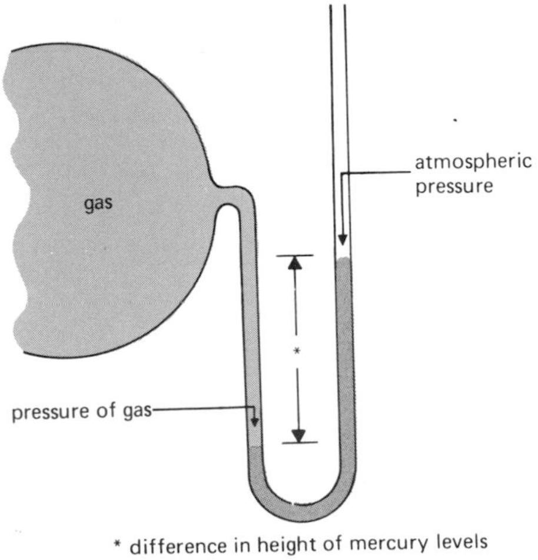

Figure 4.2 A type of manometer.

to the atmosphere; atmospheric pressure is exerted on the mercury in this arm. The other arm is connected to a container of a gas in such a way that the gas exerts pressure on the mercury in this arm.

If the gas sample were under a pressure *equal* to atmospheric pressure, the mercury would stand at the same level in both arms of the U tube. In the experiment illustrated in Figure 4.2, the gas pressure is *greater* than atmospheric pressure. The difference in height between the two mercury levels (in mm of mercury) must be added to the barometric pressure (in mm of mercury or torr) to obtain the pressure of the gas (in mm of mercury or in torr). If the pressure of the gas were *less* than atmospheric, the mercury in the left arm of the manometer shown in Figure 4.2 would stand at a higher level than the mercury in the right arm, and the difference in height would have to be subtracted from atmospheric pressure.

4.2 BOYLE'S LAW

The relationship between the volume and the pressure of a sample of a gas was studied by Robert Boyle in 1662. Boyle found that increasing the pressure on a sample of a gas causes the volume of the gas to decrease proportionately. If the pressure is doubled, the volume is cut in half. If the pressure is increased threefold, the volume is decreased to one-third its original value. **Boyle's law** states that at constant temperature the volume of a sample of gas varies inversely with the pressure.

$$V \propto \frac{1}{P}$$

The proportionality can be changed into an equality by the introduction of a constant, k.

$$V = \frac{k}{P} \qquad \text{or} \qquad PV = k$$

The value of the constant depends upon the size of the sample and the temperature. Pressure–volume data for an ideal gas (see Section 4.5) are plotted in Figure 4.3.

The volume of a gas is customarily measured in liters. The liter is defined as 1 cubic decimeter (1 dm^3 = 1000 cm^3). Since there are 1000 ml in 1 liter, 1 ml is 1 cm^3.

Example 4.1 A sample of a gas occupies 360 ml under a pressure of 0.750 atm. If the temperature is held constant, what volume will the sample occupy under a pressure of 1.000 atm?

Solution First, we tabulate the data given in the problem.

Initial conditions:	$V = 360$ ml	$P = 0.750$ atm
Final conditions:	$V = ?$ ml	$P = 1.000$ atm

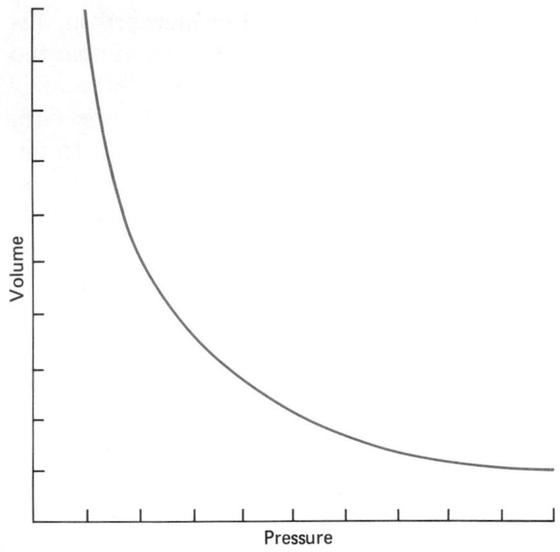

Figure 4.3 Pressure–volume curve for an ideal gas (Boyle's law).

The final volume can be obtained by correcting the initial volume for the change in pressure. The units sought are the same as the units of the quantity given.

? ml = 360 ml (pressure correction)

Correction factors are *not* the same as *conversion* factors. A correction factor is not equal to 1. We can derive two pressure-correction factors from the data given.

(1.000 atm/0.750 atm) (0.750 atm/1.000 atm)

Which one should be used?

Since the pressure *increases* from 0.750 atm to 1.000 atm, the volume must *decrease*. The correction factor must be a fraction less than 1. Hence,

$$? \text{ ml} = 360 \text{ ml} \left(\frac{0.750 \text{ atm}}{1.000 \text{ atm}}\right) = 270 \text{ ml}$$

Example 4.2 At 0°C and 5.00 atm, a given sample of a gas occupies 75.0 liter. The gas is compressed to a final volume of 30.0 liter at 0°C. What is the final pressure?

Solution

Initial conditions:	$V = 75.0$ liter	$P = 5.00$ atm	$t = 0$°C
Final conditions:	$V = 30.0$ liter	$P = ?$ atm	$t = 0$°C

The initial pressure must be corrected for the volume change. The temperature is constant, so no temperature correction is necessary.

? atm = 5.00 atm (volume correction)

Volume and pressure are inversely related. Since the volume decreases, the pressure must increase. The volume-correction factor must be a fraction greater than 1. We put the larger volume measurement in the numerator of the factor.

$$? \text{ atm} = 5.00 \text{ atm} \left(\frac{75.0 \text{ liter}}{30.0 \text{ liter}}\right) = 12.5 \text{ atm}$$

4.3 CHARLES' LAW

The relationship between the volume and the temperature of a gas sample was studied by Jacques Charles in 1787. His work was considerably extended by Joseph Gay-Lussac in 1802.

A gas expands when it is heated at constant pressure. Experimental data show that for each Celsius degree rise in temperature, the volume of a gas increases 1/273 of its value at 0°C if the pressure is held constant. A sample of a gas that has a volume of 273 ml at 0°C would expand 1/273 of 273 ml, or 1 ml, for each degree rise in temperature. At 1°C, the volume of the sample would be 274 ml. At 10°C, the volume would have increased by 10 ml to a value of 283 ml. At 273°C, the sample would have expanded 273 ml to a volume of 546 ml, which is double the original volume. These data are recorded in Table 4.1.

Although the volume increases in a regular manner with increase in temperature, the volume is *not* directly proportional to the Celsius temperature. An increase in temperature from 1°C to 10°C, for example, does not increase the volume tenfold but only from 274 ml to 283 ml. An absolute temperature scale, with temperatures measured in kelvins, is defined in such a way that volume *is* directly proportional to kelvin temperature.

A **kelvin reading** (denoted by T) is obtained by adding 273 to the Celsius temperature (denoted by t).

$$T = t + 273$$

Note that absolute temperatures are given in kelvins (abbreviated K), not

Table 4.1 Variation of the volume of a sample of gas with temperature

VOLUME (ml)	TEMPERATURE (°C)	(K)
273	0	273
274	1	274
283	10	283
546	273	546

degrees kelvin, and that the degree sign is not used in the abbreviation. Absolute temperatures are listed in the last column in Table 4.1.

The fact that volume is directly proportional to absolute temperature is readily recognized from the data in Table 4.1 since the volume was selected to point up this relationship. For example, when the absolute temperature is doubled (273 K to 546 K), the volume doubles (273 ml to 546 ml). The volume of any sample of a gas varies directly with *absolute* temperature if the pressure is held constant. This generalization is known as **Charles' law.**

$$V \propto T$$
$$V = kT$$

The numerical value of the proportionality constant k depends upon the pressure and the size of the gas sample.

The absolute temperature scale was first proposed by William Thomson, Lord Kelvin, in 1848; the unit is named in his honor. Any absolute measurement scale must be based on a zero point that represents the complete absence of the property being measured. On scales of this type, negative values are impossible. A length given in centimeters is an absolute measurement since 0 cm represents the complete absence of length. One can say that 10 cm is twice 5 cm because these are absolute measurements.

The Celsius temperature scale is not an absolute scale. The zero point, 0°C, is the freezing point of water, not the lowest possible temperature. Negative Celsius temperatures are possible, and doubling the Celsius temperature of a sample of a gas does not double the volume of the gas. On the other hand, the kelvin scale is absolute: 0 K is the lowest possible temperature, and negative kelvin temperatures are as impossible as negative lengths or negative volumes. Doubling the kelvin temperature of a sample of gas doubles the volume.

If volume versus temperature is plotted for a sample of a gas, a straight line results (see Figure 4.4). Since volume is directly proportional to absolute temperature, the volume of the gas *theoretically* should be zero at absolute zero. Upon cooling, gases liquefy and then solidify before temperatures this low are reached. No substance exists as a gas at a temperature near absolute zero. The straight line temperature–volume curve, however, can be extended to a volume of zero.

The temperature that corresponds to zero volume is −273.15°C. The kelvin is the same size as the Celsius degree, but the zero point of the kelvin scale is moved to −273.15°C. Conversion of a Celsius temperature into kelvins can be accomplished, therefore, by adding 273.15 to the Celsius reading.

$$T = t + 273.15$$

For most problem work, this value can be rounded off to 273 without introducing significant error.

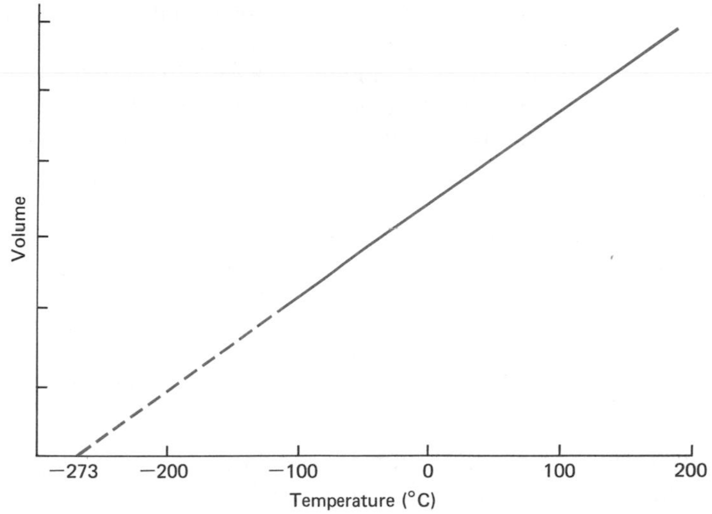

Figure 4.4 Temperature–volume curve for an ideal gas (Charle's law).

Example 4.3 A sample of a gas has a volume of 79.5 ml at 45°C. What volume will the sample occupy at 0°C when the pressure is held constant?

Solution We tabulate the data given in the problem. The Celsius temperatures (t) are converted into absolute temperatures (T): $T = t + 273$.

Initial conditions:	$V = 79.5$ ml	$t = 45°C$	$T = 318$ K
Final conditions:	$V =$? ml	$t = 0°C$	$T = 273$ K

Therefore,

$$? \text{ ml} = 79.5 \text{ ml (temperature correction)}$$

Since the temperature decreases from 318 K to 273 K, the volume must decrease. A correction factor with a value less than 1 must be used.

$$? \text{ ml} = 79.5 \text{ ml} \left(\frac{273 \text{ K}}{318 \text{ K}}\right) = 68.2 \text{ ml}$$

4.4 AMONTONS' LAW

The pressure of a gas confined in a container increases when the gas is heated. The mathematical relationship between pressure and temperature is similar to that between volume and temperature. The pressure of a gas varies directly with *absolute* temperature when the volume is constant.

$$P \propto T$$
$$P = kT$$

In this instance the value of k depends upon the amount of gas considered and its volume.

This generalization is sometimes called **Amontons' law.** In 1703, Guillaume Amontons constructed an air thermometer based on the principle that the pressure of a gas is a measure of the temperature of the gas.

Example 4.4 A 10.0-liter container is filled with a gas to a pressure of 2.00 atm at 0°C. At what temperature will the pressure inside the container be 2.50 atm?

Solution

Initial conditions:	$V = 10.0$ liter	$P = 2.00$ atm	$T = 273$ K
Final conditions:	$V = 10.0$ liter	$P = 2.50$ atm	$T = ?$ K

No volume correction is needed since the volume is constant. All temperatures must be expressed in kelvins in any gas problem. Therefore,

$$? \text{ K} = 273 \text{ K (pressure correction)}$$

Pressure varies directly with absolute temperature, The temperature must increase to produce the increase in pressure observed. A factor greater than 1 must be used.

$$? \text{ K} = 273 \text{ K} \left(\frac{2.50 \text{ atm}}{2.00 \text{ atm}}\right) = 341 \text{ K}$$

The answer can then be converted to the Celsius scale.

$$t = T - 273$$
$$= 341 \text{ K} - 273 \text{ K} = 68°\text{C}$$

4.5 IDEAL GAS LAW

The volume of a gas, at fixed temperature and pressure, varies directly with the number of moles of gas considered. Obviously, 1 mol of gas (a gram molecular weight) occupies half the volume that 2 mol occupy when the temperature and pressure of both samples are the same. Furthermore, the volume of 1 mol of a given gas is the same as the volume of 1 mol of any other gas if the volumes are measured at the same temperature and pressure (Avogadro's principle, described in Section 4.6). If n is the number of moles of gas,

$$V \propto n$$

or

$$V = kn$$

The numerical value of the proportionality constant, k, depends upon the temperature and pressure of the gas.

THE CORRECTION FACTOR METHOD FOR SOLVING GAS LAW PROBLEMS

1. Tabulate the data given in the problem. List the *initial conditions* (V, P, and t) and the *final conditions* (V, P, and t).
2. Convert temperature readings that are given in degrees Celsius (t) into kelvins (T). $T = t + 273.15$. For most problem work 273 is acceptable.
3. The solution consists of finding the *final* value of one of the three variables (V, P, or T) by correcting the *initial* value of this variable. Multiply the initial value by correction factors to correct for changes in the other two variables.
4. Consider each correction separately. A correction factor consists of a fraction derived from the initial and final values of the same variable (V, P, or T). One of these values is placed in the numerator of the factor, the other in the denominator. Two fractions, therefore, can be derived—one numerically greater than 1, the other, less than 1. Decide whether the change being considered should result in an increase or a decrease in the value being corrected. On this basis, select the fraction to be employed as the correction factor.
5. Since the units in the numerator and denominator of a correction factor are the same, they cancel. The answer has the same units as the value being corrected.
6. If a temperature is sought, it will be obtained in kelvins. The Celsius equivalent may be found at the end of the process. $t = T - 273.15$

This relationship can be combined with expressions for Boyle's law and Charles' law to give a general equation relating volume, temperature, pressure, and number of moles. Volume is inversely proportional to pressure and directly proportional to absolute temperature and to number of moles.

$$V \propto \frac{1}{P} \qquad V \propto T \qquad V \propto n$$

Therefore,

$$V \propto \left(\frac{1}{P}\right)(T)(n)$$

The proportionality can be changed into an equality by the use of a constant. In this case, the constant is given the designation R.

$$V = R\left(\frac{1}{P}\right)(T)(n)$$

Rearrangement gives

$$PV = nRT$$

Under ordinary conditions of temperature and pressure, most gases conform well to the behavior described by this equation. Deviations occur, however, under extreme conditions (low temperature and high pressure; Section 4.13). A hypothetical gas that follows the behavior described by the equation exactly, under all conditions, is called an **ideal gas.** The equation is known, therefore, as the **equation of state for an ideal gas.**

By convention, **standard temperature and pressure** (**STP**) are defined as 0°C (which is 273.15 K) and exactly 1 atm pressure. The volume of 1 mol of an ideal gas at STP, derived from experimental measurements, is 22.4136 liter. These data can be used to evaluate the ideal gas constant, R. Solution of the equation of state for R yields

$$R = \frac{PV}{nT}$$

Substitution of the data for the STP molar volume of an ideal gas gives

$$R = \frac{(1 \text{ atm})(22.4136 \text{ liter})}{(1 \text{ mol})(273.15 \text{ K})} = 0.082056 \text{ liter·atm/K·mol}$$

When this value of R is employed, volume must be expressed in liters, pressure in atmospheres, and temperature in kelvins. Values of R in other units appear in Table 4.2.

Table 4.2 Values of the ideal gas constant, *R*, in various units

R	UNITS
8.2056×10^{-2}	liter·atm/K·mol
8.3143×10^{3}	liter·Pa/K·mol
8.3143	J/K·mol*

* Since the joule is a newton meter (1 J = 1 N·m), this value of R is used when P is expressed in pascals (1 Pa = 1 N/m^2), V in cubic meters (m^3), n in moles (mol), and T in kelvins (K). This label is the equivalent, therefore, to m^3·Pa/K·mol.

The number of moles of gas in a sample, n, is equal to the mass of the sample, g, divided by the molecular weight of the gas, M.

$$n = \frac{g}{M}$$

Substitution of (g/M) for n in $PV = nRT$ gives

$$PV = \left(\frac{g}{M}\right) RT$$

Many problems can be solved by use of this form of the equation of state.

Example 4.5 The volume of a sample of a gas is 462 ml at 35°C and 1.15 atm. Calculate the volume of the sample at STP.

Solution

Initial conditions:	$V = 462$ ml	$T = 308$ K	$P = 1.15$ atm
Final conditions:	$V = ?$ ml	$T = 273$ K	$P = 1.00$ atm

The correction factor approach can be used to solve the problem.

? m = 462 ml (temperature correction)(pressure correction)

Each of these corrections is considered separately. First, the decrease in temperature, from 308 K to 273 K, causes the volume to decrease by a factor of (273 K/308 K). Second, the decrease in pressure, from 1.15 atm to 1.00 atm, causes the volume to increase by a factor of (1.15 atm/1.00 atm) since pressure and volume are inversely related.

$$? \text{ ml} = 462 \text{ ml} \left(\frac{273 \text{ K}}{308 \text{ K}}\right)\left(\frac{1.15 \text{ atm}}{1.00 \text{ atm}}\right) = 471 \text{ ml}$$

Example 4.6 At what pressure will 0.250 mol of $N_2(g)$ occupy 10.0 liter at 100°C?

Solution

$P = ?$ atm $\quad V = 10.0$ liter $\quad n = 0.250$ mol $\quad T = 373$ K

Problems in which only one set of conditions are stated are readily solved by substitution in the equation of state.

$$PV = nRT$$
$$P(10.0 \text{ liter}) = (0.250 \text{ mol})(0.0821 \text{ liter·atm/K·mol})(373 \text{ K})$$
$$P = 0.766 \text{ atm}$$

Example 4.7 How many moles of CO are present in a 500 ml sample of CO(g) collected at 50°C and 1.50 atm?

Solution The units of the values substituted into $PV = nRT$ must correspond to the units in which R is expressed. Therefore, we express the volume in liters and the temperature in kelvins.

$$PV = nRT$$
$$(1.50 \text{ atm})(0.500 \text{ liter}) = n(0.0821 \text{ liter·atm/K·mol})(323 \text{ K})$$
$$n = 0.0283 \text{ mol}$$

Example 4.8 What volume will 10.0 g of $CO_2(g)$ occupy at 27°C and 2.00 atm?

Solution The problem can be solved in one step if (g/M) is substituted for n in the equation of state. The molecular weight of CO_2 is 44.0.

$$PV = \left(\frac{g}{M}\right) RT$$

$$(2.00 \text{ atm})V = \left(\frac{10.0 \text{ g}}{44.0 \text{ g/mol}}\right) (0.0821 \text{ liter·atm/K·mol})(300 \text{ K})$$

$$V = 2.80 \text{ liter}$$

Example 4.9 What is the density of $NH_3(g)$ at 100°C and 1.15 atm?

Solution The density of a gas is the number of grams of gas in 1 liter. We can find the density, therefore, by setting V equal to 1.00 liter and solving for g. The molecular weight of NH_3 is 17.0.

$$PV = \left(\frac{g}{M}\right) RT$$

$$(1.15 \text{ atm})(1.00 \text{ liter}) = \left(\frac{g}{17.0 \text{ g/mol}}\right) (0.0821 \text{ liter·atm/K·mol})(373 \text{ K})$$

$$g = 0.638 \text{ g}$$

The density is 0.638 g/liter.

Example 4.10 (a) Cyclopropane is a gas that is used as a general anesthetic. The gas has a density of 1.50 g/liter at 50°C and 0.948 atm. What is the molecular weight of cyclopropane? (b) The empirical formula of cyclopropane is CH_2. What is the molecular formula of the compound?

Solution (a) Since the density is 1.50 g/liter, 1.50 g of the gas will occupy 1.00 liter under the conditions specified.

$$PV = \left(\frac{g}{M}\right) RT$$

$$(0.948 \text{ atm})(1.00 \text{ liter}) = \left(\frac{1.50 \text{ g}}{M}\right) (0.0821 \text{ liter·atm/K·mol})(323 \text{ K})$$

$$M = 42.0 \text{ g/mol}$$

(b) The formula weight of the empirical formula, CH_2, is 14.0. If we divide this formula weight into the molecular weight, we get $(42.0/14.0) = 3$. There are, therefore, three times as many atoms in the molecule as are indicated by the empirical formula. The molecular formula of cyclopropane is C_3H_6.

4.6 GAY-LUSSAC'S LAW OF COMBINING VOLUMES AND AVOGADRO'S PRINCIPLE

In 1808, Joseph Gay-Lussac reported the results of his experiments with reacting gases. When measured at constant temperature and pressure, the volumes of gases that are used or produced in a chemical reaction can be expressed in ratios of small whole numbers. This statement is called **Gay-Lussac's law of combining volumes.**

One of the reactions that Gay-Lussac studied is the reaction in which hydrogen chloride *gas* is produced from hydrogen *gas* and chlorine *gas*. If the volumes of all the gases are measured at the same temperature and pressure,

$$1 \text{ volume hydrogen} + 1 \text{ volume chlorine} \rightarrow 2 \text{ volumes hydrogen chloride}$$

For example,

$$10 \text{ liter hydrogen} + 10 \text{ liter chlorine} \rightarrow 20 \text{ liter hydrogen chloride}$$

Gay-Lussac did not know the formulas for these substances and did not write chemical equations. The volume ratio, however, is given by the coefficients of the gases in the chemical equation.

$$H_2(g) + Cl_2(g) \rightarrow 2HCl(g)$$

The law is applicable only to gases. The volumes of solids and liquids cannot be treated in this way.

The explanation for the law of combining volumes was proposed in 1811 by Amadeo Avogadro. **Avogadro's principle** states: equal volumes of all gases at the same temperature and pressure contain the same number of molecules.

Gay-Lussac's data show that equal volumes of $H_2(g)$ and $Cl_2(g)$ react. Since the reaction requires equal numbers of H_2 and Cl_2 molecules, a given volume of either gas must contain the same number of molecules. According to Gay-Lussac, the volume of HCl(g) produced is twice the volume of $H_2(g)$ used. The equation shows that the number of HCl molecules produced is twice the number of H_2 molecules used. We conclude that the number of HCl molecules in a "volume" is the same as the number of H_2 molecules in a "volume." The same comparison can be made between HCl and Cl_2.

The total volume of the reacting gases need not equal the volume of the gases produced. This fact is illustrated by another of Gay-Lussac's examples.

$$2CO(g) + O_2(g) \rightarrow 2CO_2(g)$$

The volume ratio for this reaction is

$$2 \text{ volumes } CO(g) + 1 \text{ volume } O_2(g) \rightarrow 2 \text{ volumes } CO_2(g)$$

The relative numbers of molecules involved in the reaction is given by the chemical equation. If we take $2x$ CO molecules, then

$$2x \text{ molecules } CO(g) + x \text{ molecules } O_2(g) \rightarrow 2x \text{ molecules } CO_2(g)$$

By comparing these two statements, we see that one "volume" of any of the gases contains the same number of molecules, x.

At the time that Gay-Lussac discovered the law of combining volumes, molecular structure was not well understood. Dalton believed that gaseous elements consisted of uncombined atoms. He interpreted the reaction between hydrogen and chlorine as a reaction between atoms.

$$H + Cl \rightarrow HCl$$

Consequently, Dalton believed that if equal volumes of gases contained equal numbers of *particles,* only one volume of hydrogen chloride (not two) should result from the reaction of one volume of hydrogen and one volume of chlorine.

Avogadro explained the difficulty by proposing that hydrogen and chlorine consisted of diatomic molecules.*

$$H_2(g) + Cl_2(g) \rightarrow 2HCl(g)$$

To Dalton, however, atoms combined to form molecules of compounds (such as HCl) because the atoms were different from one another and had a chemical affinity for one another. Dalton refused to accept the idea that two atoms that are alike, such as two hydrogen atoms, could combine to form a molecule. In subsequent years, a variety of experiments proved that some elements exist as diatomic molecules.

Example 4.11 (a) What volume of oxygen is required for the complete combustion of 15.0 liter of ethane, $C_2H_6(g)$, if all gases are measured at the same temperature and pressure?

$$2C_2H_6(g) + 7O_2(g) \rightarrow 4CO_2(g) + 6H_2O(g)$$

(b) What volume of carbon dioxide gas is produced?

Solution (a) The relationship between the volume of $C_2H_6(g)$ and the volume of $O_2(g)$ is given by the coefficients of the chemical equation.

$$2 \text{ liter } C_2H_6 = 7 \text{ liter } O_2$$

This relationship is used to derive a conversion factor.

$$? \text{ liter } O_2 = 15.0 \text{ liter } C_2H_6 \left(\frac{7 \text{ liter } O_2}{2 \text{ liter } C_2H_6}\right) = 52.5 \text{ liter } O_2$$

(b) In this case, the relationship is

$$2 \text{ liter } C_2H_6 = 4 \text{ liter } CO_2$$

Therefore,

$$? \text{ liter } CO_2 = 15.0 \text{ liter } C_2H_6 \left(\frac{4 \text{ liter } CO_2}{2 \text{ liter } C_2H_6}\right) = 30.0 \text{ liter } CO_2$$

* Although Avogadro proposed that hydrogen, chlorine, and hydrogen chloride molecules were diatomic, other formulas would work as well. For example, $H_4 + Cl_4 \rightarrow 2H_2Cl_2$. History has shown that Avogadro's guess was correct.

4.7 AVOGADRO'S PRINCIPLE AS A BASIS FOR ATOMIC AND MOLECULAR WEIGHTS

Avogadro's principle was almost entirely ignored after its presentation in 1811. Stanislao Cannizzaro was the first to see the full significance of this principle and was largely responsible for its universal acceptance (1858–1864) after the death of Avogadro.

Cannizzaro was the first to use gas densities to assign atomic and molecular weights. He based his atomic weight scale on hydrogen and assigned the hydrogen atom an atomic weight of 1, which is close to the atomic weight in current use. The molecular weight of H_2, therefore, was set equal to 2.

At any fixed temperature and pressure, the density of carbon dioxide gas is approximately 22 times greater than the density of hydrogen gas. In other words, the mass of 1 liter of carbon dioxide is 22 times the mass of 1 liter of hydrogen gas. According to Avogadro's principle, the number of molecules in 1 liter of carbon dioxide is the same as the number of molecules in 1 liter of hydrogen. A carbon dioxide molecule, therefore, must have a mass that is 22 times larger than the mass of a hydrogen molecule. Since the molecular weight of H_2 was set equal to 2, carbon dioxide was assigned a molecular weight of 22×2, or 44.

Avogadro's principle may also be used to assign molecular weights in a slightly different way. A mole of a substance (a molecular weight in grams) contains 6.022×10^{23} molecules (Avogadro's number, Section 2.3). According to Avogadro's principle, equal numbers of molecules of any two gases under the same conditions of temperature and pressure occupy equal volumes. A mole of a gas, therefore, should occupy the same volume as a mole of any other gas if the temperature and pressure are fixed. At STP, the volume of a mole of any gas is 22.4 liter. The molecular weight of a gas, therefore, is the mass, in grams, of 22.4 liter of the gas at STP. For most gases, the deviation from this ideal value is less than 1%.

Cannizzaro also showed that it is possible to use data of this type to assign atomic weights. Since carbon dioxide contains 27.3% carbon, one molecular weight of carbon dioxide (44) must contain 0.273×44, or 12 units of carbon. This value must represent either the atomic weight of carbon or a whole-number multiple of the atomic weight. The number of carbon atoms in a carbon compound must be a whole number.

Similar data for other gaseous compounds of carbon are summarized in Table 4.3. Notice that the smallest amount of carbon present in a molecular weight of any compound listed is 12 and that all other values are multiples of 12. Carbon, therefore, is assigned an atomic weight of 12. The compounds listed in Table 4.3 are assumed to contain 1, 2, 2, 1, 1, and 3 atoms of carbon, respectively.

The STP molar volume of a gas, 22.4 liter, can be used to solve some problems. All such calculations, however, can also be made by using the equation of state, $PV = nRT$.

Table 4.3 Composition of some carbon compounds

COMPOUND	MOLECULAR WEIGHT	% CARBON	MASS OF CARBON PER MOLE (g)
carbon dioxide	44	27.3	12
cyanogen	52	46.2	24
ethane	30	80.0	24
hydrogen cyanide	27	44.4	12
methane	16	75.0	12
propane	44	81.8	36

Example 4.12 What is the density of fluorine gas, $F_2(g)$, at STP?

Solution The molecular weight of F_2 is 38.0.

$$1 \text{ mol } F_2 = 38.0 \text{ g } F_2$$

At STP the volume of a mole of any gas is 22.4 liter.

$$1 \text{ mol } F_2 = 22.4 \text{ liter } F_2$$

Therefore,

$$? \text{ g } F_2 = 1 \text{ liter } F_2 \left(\frac{1 \text{ mol } F_2}{22.4 \text{ liter } F_2}\right)\left(\frac{38.0 \text{ g } F_2}{1 \text{ mol } F_2}\right) = 1.70 \text{ g } F_2$$

The density is 1.70 g/liter.

Example 4.13 How many moles of $CH_4(g)$ are contained in 1.00 liter of $CH_4(g)$ which is measured at STP?

Solution

$$? \text{ mol } CH_4 = 1.00 \text{ liter } CH_4 \left(\frac{1 \text{ mol } CH_4}{22.4 \text{ liter } CH_4}\right) = 0.0446 \text{ mol } CH_4$$

4.8 STOICHIOMETRY AND GAS VOLUMES

Stoichiometric problems may be based on the volumes of gases that are involved in a chemical reaction. Gay-Lussac's law of combining volumes is used to solve problems that deal with the volumes of two gases (see Section 4.6). Some problems concern the relationship between the *volume* of a gas and the *mass* of another substance. Examples of this type of problem follow. The clue to their solution is, as usual, the mole.

Example 4.14 A 0.400 g sample of sodium azide, $NaN_3(s)$, is heated and decomposes.

$$2NaN_3(s) \rightarrow 2Na(s) + 3N_2(g)$$

What volume of $N_2(g)$, measured at 25°C and 0.980 atm, is obtained?

Solution We find the number of moles of NaN_3 in the sample. The molecular weight of NaN_3 is 65.0.

$$? \text{ mol } NaN_3 = 0.400 \text{ g } NaN_3 \left(\frac{1 \text{ mol } NaN_3}{65.0 \text{ g } NaN_3}\right) = 0.00615 \text{ mol } NaN_3$$

From the chemical equation, we derive

$$2 \text{ mol } NaN_3 = 3 \text{ mol } N_2$$

Therefore,

$$? \text{ mol } N_2 = 0.00615 \text{ mol } NaN_3 \left(\frac{3 \text{ mol } N_2}{2 \text{ mol } NaN_3}\right) = 0.00923 \text{ mol } N_2$$

We find the volume of $N_2(g)$ by using the equation of state.

$$PV = nRT$$
$$(0.980 \text{ atm})V = (0.00923 \text{ mol})(0.0821 \text{ liter·atm/K·mol})(298 \text{ K})$$
$$V = 0.230 \text{ liter}$$

Example 4.15 How many liters of CO(g), measured at STP, are needed to reduce 1.00 kg of $Fe_2O_3(s)$? The chemical equation is

$$Fe_2O_3(s) + 3CO(g) \rightarrow 2Fe(s) + 3CO_2(g)$$

Solution We first find the number of moles of Fe_2O_3 in 1000 g of Fe_2O_3. The molecular weight of Fe_2O_3 is 159.6.

$$? \text{ mol } Fe_2O_3 = 1000 \text{ g } Fe_2O_3 \left(\frac{1 \text{ mol } Fe_2O_3}{159.6 \text{ g } Fe_2O_3}\right) = 6.27 \text{ mol } Fe_2O_3$$

From the equation, we see that

$$1 \text{ mol } Fe_2O_3 = 3 \text{ mol CO}$$

Therefore,

$$? \text{ mol CO} = 6.27 \text{ mol } Fe_2O_3 \left(\frac{3 \text{ mol CO}}{1 \text{ mol } Fe_2O_3}\right) = 18.8 \text{ mol CO}$$

At STP,

$$1 \text{ mol CO} = 22.4 \text{ liter CO}$$

Therefore,

$$? \text{ liter CO} = 18.8 \text{ mol CO} \left(\frac{22.4 \text{ liter CO}}{1 \text{ mol CO}}\right) = 421 \text{ liter CO}$$

This last step could have been solved by using the equation of state.

$$PV = nRT$$
$$(1 \text{ atm})V = (18.8 \text{ mol})(0.0821 \text{ liter·atm/K·mole})(273 \text{ K})$$
$$V = 421 \text{ liter}$$

Example 4.16 How many grams of Fe are needed to produce 100 liter of $H_2(g)$, measured at STP?

$$3Fe(s) + 4H_2O \rightarrow Fe_3O_4(s) + 4H_2(g)$$

Solution First, the number of moles of H_2 is found. At STP,

$$1 \text{ mol } H_2 = 22.4 \text{ liter } H_2$$

$$? \text{ mol } H_2 = 100 \text{ liter } H_2 \left(\frac{1 \text{ mol } H_2}{22.4 \text{ liter } H_2}\right) = 4.46 \text{ mol } H_2$$

[This step could also be accomplished by substituting in $PV = nRT$.] From the equation,

$$3 \text{ mol Fe} = 4 \text{ mol } H_2$$

we find,

$$? \text{ mol Fe} = 4.46 \text{ mol } H_2 \left(\frac{3 \text{ mol Fe}}{4 \text{ mol } H_2}\right) = 3.35 \text{ mol Fe}$$

The atomic weight of Fe is 55.8.

$$? \text{ g Fe} = 3.35 \text{ mol Fe} \left(\frac{55.8 \text{ g Fe}}{1 \text{ mol Fe}}\right) = 187 \text{ g Fe}$$

4.9 KINETIC THEORY OF GASES

The kinetic theory of gases provides a model to explain the regularity that is observed in the behavior of all gases. In 1738, Daniel Bernoulli explained Boyle's law by assuming that the pressure of a gas results from the collisions of gas molecules with the walls of the container. Bernoulli's explanation constitutes an early, and simple, expression of key aspects of the kinetic theory. The theory was enlarged and developed in the middle of the nineteenth century by many scientists, notably Krönig, Clausius, Maxwell, and Boltzmann.

The kinetic theory of gases includes the following postulates.

1. Gases consist of molecules widely separated in space. The actual volume of the individual molecules is negligible in comparison to the total volume of the gas as a whole. The word *molecule* is used here to designate the smallest particle of any gas; some gases (for example, the noble gases) consist of uncombined atoms.
2. Gas molecules are in constant, rapid, straight-line motion; they collide with each other and with the walls of the container. Although energy may be transferred from one molecule to another in these collisions, no kinetic energy (which is energy of motion) is lost.

CALCULATIONS THAT ARE BASED ON CHEMICAL EQUATIONS AND THAT INVOLVE GAS VOLUMES

The following types of problems may be encountered.

1. A *volume* of gas A is given and the *volume* of gas B is sought.
   a. *The volumes of both gases are measured under the same conditions of temperature and pressure.* Use Gay-Lussac's law of combining volumes. (See Example 4.11.)
   b. *The volumes of the two gases are measured under different conditions.* Use Gay-Lussac's law of combining volumes to find the volume of gas B under the conditions given for gas A. Use correction factors to correct this volume of gas B so that it conforms to the final conditions given in the problem.
2. A *mass* of substance A is given and the *volume* of gas B is sought.
   a. Find the number of moles of A.
   b. Use this number of moles of A to find the number of moles of B. The mole relationship of B to A is given by the chemical equation.
   c. Find the volume of gas B by substituting in $PV = nRT$; n is the number of moles found in step (b); P and T are the conditions under which the volume of B is measured. (See Examples 4.14 and 4.15.)
3. A *volume* of gas A is given and the *mass* of substance B is sought.
   a. Determine the number of moles of gas A by using $PV = nRT$.
   b. Use this number of moles of A to find the number of moles of B. The mole relationship of B to A is given by the chemical equation.
   c. Find the mass of B from the number of moles of B determined in step (b). One mole of B is the molecular weight of B in grams. (See Example 4.16.)

Solutions to problems of types 2 and 3 may be simplified if the gas volumes are measured at STP. The volume of 1 mol of any gas at STP is 22.4 liters. This relationship can be used in steps 2(c) and 3(a) in place of $PV = nRT$. (See Examples 4.15 and 4.16.)

3. The average kinetic energy of the molecules of a gas depends upon the temperature, and it increases as the temperature increases. At a given temperature, the molecules of all gases have the same average kinetic energy.
4. The attractive forces between gas molecules are negligible.

The gas laws can be explained by the kinetic theory. Consider *Boyle's law*. According to the theory, gas pressure is caused by molecular collisions with the walls of the container. If the number of molecules per unit volume (the molecular concentration) is increased, a higher pressure will result because of the larger number of collisions per unit time. Reducing the volume of a gas crowds the molecules into a smaller space, which produces a higher molecular concentration and a proportionally higher pressure.

Charles' law and *Amontons' law* relate the properties of gases to changes in temperature. The average kinetic energy of the molecules of a gas is proportional to the absolute temperature. At absolute zero, the kinetic energy of the molecules is theoretically zero; the molecules are at rest. Since the volume of the molecules of an ideal gas is negligible, the volume of an ideal gas at absolute zero is theoretically zero.

As the temperature is increased, the molecules move at increasing speeds. The collisions of the gas molecules with the walls of the container become increasingly more vigorous and more frequent. As a result, the pressure increases in the manner described by Amontons' law.

The pressure of a gas that is being heated can be held constant if the gas is allowed to expand. The increasing volume keeps the pressure constant by reducing the number of collisions that the molecules make with the container walls in a given time. In this way, the decreasing frequency of the collisions compensates for the increasing intensity of the collisions. Charles' law describes this situation.

4.10 DALTON'S LAW OF PARTIAL PRESSURES

The behavior of a mixture of gases that do not react with one another is frequently of interest. The pressure that a component of such a mixture would exert if it were the only gas present in the volume under consideration is the **partial pressure** of that component. **Dalton's law of partial pressures** (1801) states that the total pressure of a mixture of gases that do not react is equal to the sum of the partial pressures of all the gases present. If the total pressure is P_{total} and the partial pressures are $p_A, p_B, p_C, \ldots,$

$$P_{total} = p_A + p_B + p_C + \ldots$$

Supposc that 1 liter of gas A at 0.2 atm pressure and 1 liter of gas B at 0.4 atm pressure are mixed. If the *final volume is 1 liter* and the temperature is constant, the pressure of the mixture will be 0.6 atm.

According to the kinetic theory, the molecules of gas A have the same average kinetic energy as the molecules of gas B since the two gases are at the same temperature. Furthermore, the kinetic theory assumes that the gas molecules do not attract one another if the gases do not react chemically. Consequently, the act of mixing two or more gases does not change the average kinetic energy of any of the gases. Each gas exerts the same pressure that it would exert if it were the only gas present in the container.

If n_A mol of gas A and n_B mol of gas B are mixed, the total number of moles of gas in the mixture is $(n_A + n_B)$. The ratio of the number of moles of A to the total number of moles present is called the **mole fraction** of A, X_A.

$$X_A = \frac{n_A}{n_A + n_B} = \frac{n_A}{n_{total}}$$

The fraction of the total pressure that is due to gas A is given by the mole fraction of A. The partial pressure of A, therefore, is

$$p_A = \left(\frac{n_A}{n_A + n_B}\right) P_{total} = X_A P_{total}$$

The partial pressure of B is equal to the mole fraction of B times the total pressure.

$$p_B = \left(\frac{n_B}{n_A + n_B}\right) P_{total} = X_B P_{total}$$

Notice that the sum of the mole fractions is one.

$$X_A + X_B = 1$$

$$\frac{n_A}{n_A + n_B} + \frac{n_B}{n_A + n_B} = \frac{n_A + n_B}{n_A + n_B} = 1$$

Suppose that a mixture contains 1 mol of A and 4 mol of B. Then, the total number of moles is five, the mole fraction of A is one-fifth, and the mole fraction of B is four-fifths. The partial pressure of A, therefore, is one-fifth of the total pressure, and the partial pressure of B is four-fifths of the total pressure.

A gas evolved in the course of a laboratory experiment is frequently collected over water if it is not very soluble in water. The gas is conducted into an inverted bottle that has been filled with water. The gas displaces the water, and the collected gas is mixed with water vapor. The total pressure of the mixture is the sum of the partial pressure of the gas and the partial pressure of the water vapor. In Figure 4.5, the total pressure is equal to the barometric

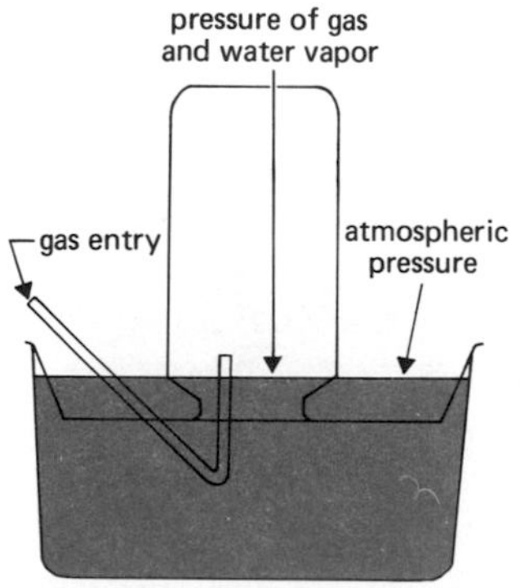

Figure 4.5 Schematic diagram of the collection of gas over water.

Table 4.4 Vapor pressure of water

TEMPERATURE (°C)	PRESSURE (atm)	(torr)	TEMPERATURE (°C)	PRESSURE (atm)	(torr)
0	0.0060	4.6	25	0.0313	23.8
1	0.0065	4.9	26	0.0332	25.2
2	0.0070	5.3	27	0.0352	26.7
3	0.0075	5.7	28	0.0373	28.3
4	0.0080	6.1	29	0.0395	30.0
5	0.0086	6.5	30	0.0419	31.8
6	0.0092	7.0	31	0.0443	33.7
7	0.0099	7.5	32	0.0470	35.7
8	0.0106	8.0	33	0.0496	37.7
9	0.0113	8.6	34	0.0525	39.9
10	0.0121	9.2	35	0.0555	42.2
11	0.0130	9.8	40	0.0728	55.3
12	0.0138	10.5	45	0.0946	71.9
13	0.0148	11.2	50	0.122	92.5
14	0.0158	12.0	55	0.155	118.0
15	0.0168	12.8	60	0.197	149.4
16	0.0179	13.6	65	0.247	187.5
17	0.0191	14.5	70	0.308	233.7
18	0.0204	15.5	75	0.380	289.1
19	0.0217	16.5	80	0.467	355.1
20	0.0231	17.5	85	0.571	433.6
21	0.0245	18.7	90	0.692	525.8
22	0.0261	19.8	95	0.834	633.9
23	0.0277	21.1	100	1.000	760.0
24	0.0294	22.4	105	1.192	906.1

pressure since the water stands at the same level inside the bottle as outside it. The pressure of the dry gas is found by subtracting the vapor pressure of water at the temperature of the experiment (Table 4.4) from the barometric pressure.

Example 4.17 A 370 ml sample of oxygen is collected over water at 23°C and a barometric pressure of 0.992 atm. What volume would this sample occupy dry and at STP?

Solution The vapor pressure of water at 23°C is 0.0277 atm. Consequently, the initial pressure of the oxygen is

$$0.992 \text{ atm} - 0.028 \text{ atm} = 0.964 \text{ atm}$$

Therefore,

Initial conditions:	$V = 370$ ml	$P = 0.964$ atm	$T = 296$ K
Final conditions:	$V = ?$ ml	$P = 1.000$ atm	$T = 273$ K

$$? \text{ ml} = 370 \text{ ml} \left(\frac{0.964 \text{ atm}}{1.000 \text{ atm}}\right)\left(\frac{273 \text{ K}}{296 \text{ K}}\right) = 329 \text{ ml}$$

Example 4.18 A mixture of 40.0 g of oxygen and 40.0 g of helium has a total pressure of 0.900 atm. What is the partial pressure of oxygen?

Solution The molecular weight of O_2 is 32.0. Forty grams of O_2 is 40.0/32.0, or 1.25, mol of O_2. Helium is a monatomic gas. Its atomic weight is 4.00. Therefore, 40.0/4.0, or 10.0, mol of He is present in the mixture.

Therefore,

$$X_{O_2} = \frac{n_{O_2}}{n_{O_2} + n_{He}}$$

$$X_{O_2} = \frac{1.25 \text{ mol}}{(1.25 + 10.0) \text{ mol}} = \frac{1.25 \text{ mol}}{11.2 \text{ mol}} = 0.112$$

The partial pressure of O_2 is

$$\begin{aligned} P_{O_2} &= X_{O_2} P_{\text{total}} \\ &= (0.112)(0.900 \text{ atm}) \\ &= 0.101 \text{ atm} \end{aligned}$$

4.11 GRAHAM'S LAW OF EFFUSION

Suppose that samples of two gases, A and B, are confined separately in identical containers under the same conditions of temperature and pressure. The kinetic theory of gases states that gases at the same temperature have the same average kinetic energy. The average kinetic energy of the molecules of gas A (KE_A), therefore, is the same as the average kinetic energy of the molecules of gas B (KE_B).

$$KE_A = KE_B$$

The kinetic energy of a body with a mass m moving in a straight line at a constant speed u is

$$KE = \tfrac{1}{2}mu^2$$

Therefore,

$$KE_A = \tfrac{1}{2}m_A u_A^2 \qquad \text{and} \qquad KE_B = \tfrac{1}{2}m_B u_B^2$$

The molecules of gas A (or gas B) do not all move at the same speed. The symbol u_A (as well as u_B) stands for the speed of a molecule that has the average kinetic energy. Since

$$\begin{aligned} KE_A &= KE_B \\ \tfrac{1}{2}m_A u_A^2 &= \tfrac{1}{2}m_B u_B^2 \end{aligned}$$

or

$$m_A u_A^2 = m_B u_B^2$$

By rearranging the equation, we get

$$\frac{u_A^2}{u_B^2} = \frac{m_B}{m_A}$$

Extracting the square root of both sides of the equation, gives us

$$\frac{u_A}{u_B} = \sqrt{\frac{m_B}{m_A}}$$

The ratio of the molecular masses, m_B/m_A, is the same as the ratio of the molecular weights, M_B/M_A. Therefore,

$$\frac{u_A}{u_B} = \sqrt{\frac{M_B}{M_A}}$$

Now let us suppose that each container has an identical, extremely small opening (called an orifice) in it. Gas molecules will escape through these orifices; the process is called **molecular effusion.** The rate of effusion, *r*, is equal to the rate at which molecules strike the orifice, which in turn is proportional to molecular speed, *u*. Molecules that move rapidly will effuse at a faster rate than slower moving molecules. The ratio u_A/u_B, therefore, is the same as the ratio of the effusion rates, r_A/r_B.

$$\frac{r_A}{r_B} = \sqrt{\frac{M_B}{M_A}}$$

This equation is an expression of **Graham's law of effusion,** which Thomas Graham experimentally derived during the period 1828 to 1833.

The relationship may also be expressed in terms of *gas densities*. Density is mass per unit volume. A gas density is usually expressed in grams per liter. We can find the density of gas A (d_A) by dividing the mass of a mole of gas A (a gram molecular weight, M_A) by the volume that a mole of gas A occupies (which we will indicate as V_m).

$$d_A = \frac{M_A}{V_m}$$

One mole of a gas occupies the same volume as one mole of any other gas if the temperature and pressure are fixed. Therefore,

$$d_B = \frac{M_B}{V_m}$$

If we divide both numerator and denominator of the ratio M_B/M_A by V_m, we get

$$\frac{M_B}{M_A} = \frac{M_B/V_m}{M_A/V_m} = \frac{d_B}{d_A}$$

The fraction d_B/d_A may be substituted for M_B/M_A in the equation for Graham's law.

$$\frac{r_A}{r_B} = \sqrt{\frac{d_B}{d_A}}$$

It is not surprising that the lighter of two molecules with the same kinetic energy will effuse more rapidly than the heavier one. (Notice the *inverse* relationship.) The molecular weight of O_2 is 32, and the molecular weight of H_2 is 2. Since

$$\frac{r_{H_2}}{r_{O_2}} = \sqrt{\frac{M_{O2}}{M_{H2}}}$$

$$\frac{r_{H_2}}{r_{O_2}} = \sqrt{\frac{32}{2}} = \sqrt{16} = 4$$

Hydrogen will effuse four times more rapidly than oxygen.

Example 4.19 What is the molecular weight of gas X if it effuses 0.876 times as rapidly as $N_2(g)$?
Solution The ratio of the rate of effusion of gas X to the rate of effusion of $N_2(g)$ is

$$\frac{r_X}{r_{N_2}} = 0.876$$

The molecular weight of N_2 is 28.0. Therefore,

$$\sqrt{\frac{M_{N_2}}{M_X}} = \frac{r_X}{r_{N_2}}$$

$$\sqrt{\frac{28.0}{M_X}} = 0.876$$

We square both sides of the equation and solve for M_X.

$$\frac{28.0}{M_X} = 0.767$$

$$M_X = \frac{28.0}{0.767} = 36.5$$

4.12 MOLECULAR SPEEDS

Gas molecules are in constant, rapid motion. The speed of a hydrogen molecule, for example, that has the average kinetic energy at 0°C is 1.84×10^3 m/s. Since the H_2 molecule is the lightest of all molecules, its speed is comparatively high.

The direction in which a H_2 molecule moves is constantly being changed through collisions with other H_2 molecules. At 1 atm pressure and 0°C, a H_2 molecule, on the average, undergoes about 1.4×10^{10} collisions in 1 s. The average distance traveled between collisions is only 1.3×10^{-5} cm.

Not all of the molecules of a gas have the same kinetic energy and travel at the same speed. Since energy can be exchanged in these collisions, the speed as well as the direction of a molecule change continually. In any sample of a gas, however, there is a large number of molecules, so that the molecular speeds are distributed over a range in a definite manner.

Distributions of the molecular speeds of a gas, called Maxwell–Boltzmann distributions, at two temperatures are shown in Figure 4.6. The fraction of the total number of molecules that has a particular speed is plotted against molecular speed. Each curve has a maximum, and the speed corresponding to the maximum is the most probable speed for that distribution. In other words, more molecules move at this speed than at any other. Comparatively few molecules possess very high or very low speeds.

When the temperature of a gas is increased, the curve broadens and shifts toward higher speeds. Fewer molecules than previously move at the lower speeds and more molecules move at the higher speeds. The addition of heat has caused the molecules, on the average, to move faster. The speed of a hydrogen molecule that has the average kinetic energy at 0°C is 1.84×10^3 m/s; at 100°C, it is 2.15×10^3 m/s.

A typical curve for the distribution of molecular energies is shown in Figure 4.7. Distribution curves of this type can be drawn for liquids and solids as well as gases.

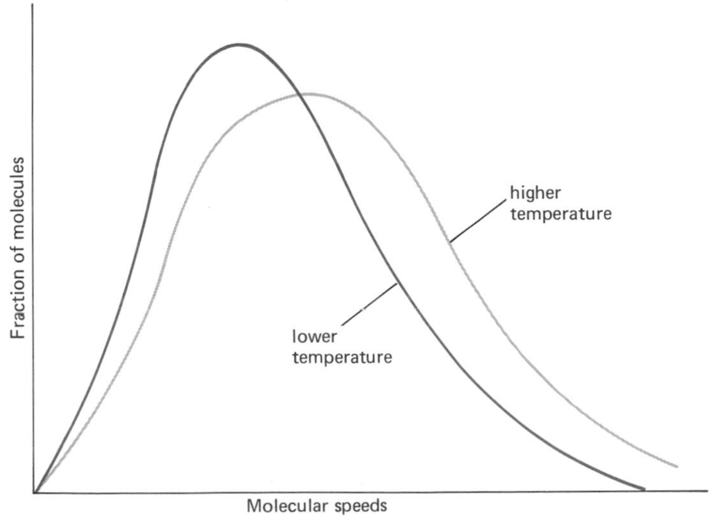

Figure 4.6 Distribution of molecular speeds.

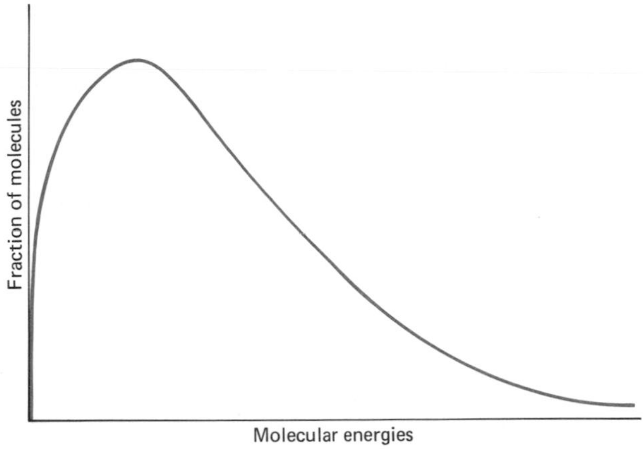

Figure 4.7 Distribution of molecular energies.

4.13 REAL GASES

The gas laws describe the behavior of an ideal or perfect gas—a gas defined by the kinetic theory. Under ordinary conditions of temperature and pressure, real gases follow the ideal gas laws fairly closely. At low temperatures and/or high pressures, however, they do not. The behavior of real gases deviates from the behavior of ideal gases for two reasons.

1. The kinetic theory assumes that there are no attractive forces between gas molecules. Such attractions must exist, however, because all gases can be liquefied. Intermolecular attractions hold the molecules together in the liquid state.

   At high temperatures, gas molecules move so rapidly that the attractive forces between molecules have little effect. On the other hand, at low temperatures the molecules move more slowly. The attractive forces pull the molecules together so that the volume observed at low temperatures is less than that predicted by the ideal gas law.
2. The kinetic theory assumes that gas molecules are points in space and that the actual volume of the gas molecules is not significant. At absolute zero, therefore, the temperature at which molecular motion stops, the volume of an ideal gas is zero. Real gases, of course, do not have zero molecular volumes. When the pressure is increased, the space between the molecules is reduced, but the volume of the molecules themselves cannot be changed. A given gas molecule cannot move through the entire volume of the container since other molecules are present. The volume through which they can move can be obtained by subtracting an amount called the **excluded volume** from the total volume.

   At pressures up to a few atmospheres, gas molecules are far apart, and the free space between the molecules is relatively large. Under such conditions, neglecting the excluded volume does not introduce a large error. At high pressures, however, the molecules are closer together. The excluded volume is no longer negligible in comparison to the total volume.

Johannes van der Waals in 1873 modified the equation of state for an ideal gas to take into account these two effects. The **van der Waals equation** is

$$\left(P + \frac{n^2a}{V^2}\right)(V - nb) = nRT$$

The numerical values of the constants a and b for each gas are determined by experiment. Typical values are listed in Table 4.5.

Table 4.5 van der Waals constants

	a (liter^2 atm/mol^2)	b (liter/mol)
H_2	0.244	0.0266
He	0.0341	0.0237
N_2	1.39	0.0391
O_2	1.36	0.0318
Cl_2	6.49	0.0562
NH_3	4.17	0.0371
CO	1.49	0.0399
CO_2	3.59	0.0427

The term n^2a/V^2 is added to the measured pressure, P, to correct for the intermolecular attractive forces. Pressure is caused by the collisions of the gas molecules with the walls of the container. The effect of a given collision would be greater if the molecule were not held back by the attraction of other molecules. Consequently, the pressure that is measured is *less* than it would be if these attractive forces did not exist. The term n^2a/V^2 is added to P so that $(P + n^2a/V^2)$ represents the pressure of an ideal gas, one in which there are no molecular forces.

The constant b is the excluded volume per mole of a given gas. For n moles of the gas, the term nb is subtracted from the measured volume, V. The factor $(V - nb)$ represents the volume of an ideal gas, one in which the molecules themselves have no volume.

4.14 LIQUEFACTION OF GASES

Liquefaction of a gas occurs under conditions that permit the intermolecular attractive forces to bind the gas molecules together in the liquid form. If the pressure is high, the molecules are close together, and the effect of the attractive forces is appreciable. The attractive forces are opposed by the motion

Table 4.6 Pressures needed to liquefy carbon dioxide (vapor pressures of liquid CO_2)

TEMPERATURE (°C)	PRESSURE (atm)
−50	6.7
−30	14.1
−10	26.1
10	44.4
20	56.5
30	71.2
31	72.8

of the gas molecules; thus, liquefaction is favored by low temperatures where the average kinetic energy of the molecules is low. The behavior of a gas deviates more and more from ideality as the temperature is lowered and the pressure is raised. At extremes of these conditions, gases liquefy.

The higher the temperature of a gas, the more difficult it is to liquefy and the higher the pressure that must be employed (Table 4.6). For each gas there is a temperature above which it is impossible to liquefy the gas no matter how high the applied pressure. This temperature is called the **critical temperature** of the gas under consideration. The **critical pressure** is the minimum pressure needed to liquefy a gas at its critical temperature. The critical constants of some common gases are listed in Table 4.7.

The critical temperature of a gas gives an indication of the strength of the intermolecular attractive forces of that gas. A substance with weak attractive forces would have a low critical temperature; above this temperature, the molecular motion is too violent to permit the relatively weak forces to hold the molecules in the liquid state. The substances of Table 4.7 are listed in order of increasing critical temperature; the magnitude of the intermolecular

Table 4.7 Critical point data

GAS	CRITICAL TEMPERATURE (K)	CRITICAL PRESSURE (atm)
He	5.3	2.26
H_2	33.3	12.8
N_2	126.1	33.5
CO	134.0	35.0
O_2	154.4	49.7
CH_4	190.2	45.6
CO_2	304.2	72.8
NH_3	405.6	111.5
H_2O	647.2	217.7

attractive forces (related to the a of Table 4.5) increases in this same order. Helium, which has weak attractive forces, can exist as a liquid below 5.3 K only; the strong attractive forces of water permit it to be liquefied up to a temperature of 647.2 K. The critical constants have been used to evaluate the constants of the van der Waals equation.

THE DISCOVERY OF GASES

To this day the four-element theory of the Greeks is remembered in common phrases. A house offers "protection from the elements." When we go out into a storm, we "brave the elements." The elements we talk about are air, water, fire, and earth. In these phrases, weather conditions, especially violent ones, are regarded as the activity of these elements.

For centuries, scientists thought in terms of the elements of the Greeks. They believed all gases were air or mixtures of air with various impurities. In the early seventeenth century, Johann van Helmont recognized that there are many distinct kinds of gases. He coined the word *gas* from the Greek word *chaos* (a term applied to formless matter). The importance of van Helmont's distinction, however, was not recognized for over one hundred years.

In the early years of the eighteenth century, the method for collecting gases over water was devised (see Section 4.10 and Figure 4.5). For the first time, chemists could trap and study pure samples of a gas. Stephen Hales, who used this method, actually isolated several different gases, but he thought that all of them were forms of air.

In 1754, Joseph Black, in an important series of experiments, studied "fixed air" (CO_2), which had been discovered earlier by van Helmont. In 1766, Henry Cavendish rediscovered "inflammable air" (H_2), which had been described by Boyle earlier. During the next twenty years, many "airs" (most of them hitherto unknown) were carefully investigated. Two important chemists of this period who are known for their work with gases, are Carl Scheele, a Swede, and Joseph Priestley, an Englishman.

Priestley became interested in gases because he lived next door to a brewery! He noticed that "fixed air" (CO_2) is a product of the fermentation process, and he obtained samples of the gas to study. His first chemical publication was *Directions for Impregnating Water with Fixed Air,* published in 1772, in which he described how to make soda water. He believed the drink that he had discovered had medicinal properties.

All in all, Priestley discovered nine gases, including oxygen in 1774. (Scheele had discovered oxygen in 1772, but his work was not published until

1777 and Priestley did not know of it.) Priestley called the gas "dephlogisticated air" because he thought that it was air from which all the phlogiston had been removed. A substance burns more brightly in pure oxygen than in air. Priestley believed that "dephlogisticated air" would carry off the phlogiston produced by a combustion better than ordinary air, which he thought already contained some phlogiston.

When he visited Paris in 1774, Priestley met Lavoisier, whom he told about his experiments with oxygen. Lavoisier also received a letter from Scheele in which the preparation of the gas was described. It was Lavoisier who proposed the correct interpretation of the role of oxygen (as he named it) in a combustion. In his explanation he used van Helmont's word *gas* (which Lavoisier spelled *gaz*). He also claimed (falsely) that Priestley, Scheele, *and he* had discovered oxygen "about the same time"!

Priestley, who refused to accept Lavoisier's system of chemistry, became the last important chemist to support the phlogiston theory. His last chemical publication was *Doctrine of Phlogiston Established,* published in 1800.

Priestley's liberal views on religion and politics—especially his sympathy for the French revolution—caused him trouble. His house was ransacked and eventually he was ostracized by English scientists. In 1794, he left England for America, where he settled in Northumberland, Pennsylvania. In the same year, Lavoisier was guillotined because of his past associations with the French monarchy.

PROBLEMS

4.1 The volume of a sample of gas is 3.00 liter at 3.00 atm. If the temperature is held constant, what is the volume of the sample at (a) 1.00 atm, (b) 0.30 atm, (c) 10.00 atm?

4.2 The volume of a sample of a gas is 2.50 liter at 2.50 atm. If the temperature is held constant, what will be the volume of the sample at (a) 1.30 atm, (b) 0.30 atm, (c) 1.00 atm?

4.3 The volume of a sample of gas is 250 ml at 1.00 atm. If the temperature is held constant, what is the pressure of the sample at a volume of (a) 300 ml, (b) 150 ml, (c) 1.00 liter?

4.4 The volume of a sample of gas is 30.0 liter at 2.00 atm. If the temperature is held constant, what is the pressure of the sample at a volume of (a) 15.0 liter, (b) 35.0 liter, (c) 3.00 liter?

4.5 Convert the following temperatures to the Kelvin temperature scale. (a) 0°C, (b) 100°C, (c) −63°C, (d) 547°C.

4.6 Convert the following temperatures to the Celsius temperature scale. (a) 0 K, (b) 100 K, (c) 273 K, (d) 450 K.

4.7 The volume of a gas sample is 100 ml at 100°C. If the pressure is held constant, at what temperature will the sample have a volume of 200 ml?

4.8 The volume of gas in a gas thermometer is 100.0 ml at 0°C. What is the temperature when the volume is 125.0 ml? Assume that the pressure remains constant.

4.9 A container is filled with a gas to a pressure of 3.00 atm at 0°C. (a) What pressure will develop inside the sealed container if it is warmed to 100°C? (b) At what temperature would the pressure be 2.00 atm? (c) At what temperature would the pressure be 4.00 atm?

4.10 A gas in a 1.00 liter container expands to a volume of 5.00 liter. The initial pressure is 0.980 atm. What is the final pressure if the temperature does not change?

4.11 A vessel containing argon at 0.10 atm is heated from 0°C to 100°C. If the volume does not change, what is the final pressure?

4.12 A 75 ml sample of gas is heated at constant pressure from 0°C to 33°C. What is the volume at 33°C?

4.13 A 50 ml gas sample is cooled from 100°C to 12°C while the pressure is held constant. What is the final volume?

4.15 How many moles of gas at 1.00 atm and 25°C are contained in a 5.00 liter vessel?

4.16 A 1.0 liter flask contains 3.5 moles of gas at 27°C. What is the pressure of the gas?

4.17 What volume would 2.10 g of CH_4(g) occupy at 77°C and 0.750 atm?

4.18 Calculate the number of molecules in 1.00 liter of O_2 gas at 56°C and 1.08 atm.

4.19 A balloon containing 1.00 liter of He is purchased in an air-conditioned store, in which the temperature is 25°C, and is carried outside where it heats to 45°C. If there is no pressure change, what is the final volume?

4.20 If the pressure on the balloon described in Problem 4.19 does change from 0.980 atm to 0.996 atm during the temperature change, what is the final volume?

4.21 A gas sample is heated from −10°C to 87°C, and the volume is increased from 1.00 liter to 3.30 liter. If the initial pressure is 0.750 atm, what is the final pressure?

4.14 Complete the following table for an ideal gas.

Pressure P	Volume V	Moles n	Temperature T
1.00 atm	______	1.00 mol	273 K
0.500 atm	______	1.00 mol	0°C
202 atm	10.0 liter	______	100°C
0.452 atm	5.00 ml	1.00×10^{-3}mol	______
______	5.00 liter	1.25 mol	104°C

4.22 What is the density of $PH_3(g)$, in grams per liter, at 22°C and 0.950 atm?

4.23 If the temperature is held constant at 0°C, at what pressure will the density of He(g) be 1.00 g/liter?

4.24 A gas has a density of 2.04 g/liter at 32°C and a pressure of 0.850 atm. What is the molecular weight of the gas?

4.25 An evacuated 1.00 liter flask weighs 104.35 g. A quantity of nitrogen gas is added to the flask. After the nitrogen is added, the flask and contents weigh 105.68 g, and the pressure exerted by the gas is 2.00 atm. What is the temperature inside the flask?

4.26 Hydrogen cyanide, a highly poisonous compound, is commercially prepared by the following reaction run at a high temperature in the presence of a catalyst.

$$2CH_4(g) + 3O_2(g) + 2NH_3(g) \rightarrow 2HCN(g) + 6H_2O(g)$$

How many liters of $CH_4(g)$, $O_2(g)$, and $NH_3(g)$ are required and how many liters of $H_2O(g)$ are produced in the preparation of 15.0 liter of HCN(g)? Assume that all gas volumes are measured under the same conditions of temperature and pressure.

4.27 Cyanogen, $(CN)_2(g)$, is a flammable, extremely poisonous gas. It can be prepared by a catalyzed, gas-phase reaction between HCN(g) and $NO_2(g)$. The products of the reaction are $(CN)_2(g)$, NO(g), and $H_2O(g)$. Assume that all gas volumes are measured under the same conditions of temperature and pressure. (a) Write the equation for the reaction. (b) How many liters of $(CN)_2(g)$ can be prepared by the reaction of 10.0 liter of HCN(g) with excess NO_2? (c) How many liters of NO(g) and $H_2O(g)$ are prepared in the reaction described in part (b)?

4.28 What is the density in g/ml of oxygen gas at STP?

4.29 What is the density of phosgene gas, $COCl_2$, at STP?

4.30 What is the molecular weight of a gas with a density of 3.71 g/liter at STP?

4.31 What is the molecular weight of a gas with a density of 0.178 g/liter at STP?

4.32 What volume would 1 billion (10^9) molecules of $H_2(g)$ occupy at STP?

4.33 W. W. Ruby estimated that the total mass of atmospheric oxygen is 15×10^{20} g. If all this oxygen were at STP, what volume in liters would it occupy?

4.34 A room which is 4.0 m wide, 5.0 m long, and 2.5 m high is at STP. How many moles of ideal gas are in the room? How many molecules?

4.35 Approximately 20% of the volume occupied by air is oxygen. How many moles of oxygen are in the room described in problem 4.34? How many grams of oxygen are in the room?

4.36 About 1.5×10^8 metric tonnes (1 metric tonne = 1000 kg) of CO are released into the atmosphere each year. What is the volume of this quantity of CO at STP?

4.37 How many liters of oxygen measured at STP are needed to burn 10.0 g of magnesium?

$$2Mg(s) + O_2(g) \rightarrow 2MgO(s)$$

4.38 Copper (II) oxide can be reduced to metallic copper with hydrogen at 200°C.

$$CuO(s) + H_2 \rightarrow Cu(s) + H_2O(g)$$

What volume of H_2O vapor is produced by the reduction of 7.65 g CuO?

4.39 Five liters of methane measured at STP are burned.

$$CH_4(g) + 2O_2(g) \rightarrow CO_2(g) + 2H_2O(g)$$

What is the total volume of the gaseous products measured at 1.00 atm and 505°C?

4.40 The complete combustion of octane yields carbon dioxide and water.

$$2C_8H_{18}(g) + 25O_2(g) \rightarrow 16CO_2(g) + 18H_2O(g)$$

What volume of gas is produced from the complete combustion of 0.670 g octane if the temperature is 400°C and the pressure is 1.20 atm?

4.41 Write the equation for the reaction of lanthanum carbide, $La_2(C_2)_3(s)$, with water to yield lanthanum hydroxide, $La(OH)_3(s)$, and acetylene, $C_2H_2(g)$. How many grams of $La_2(C_2)_3$ would be needed to prepare 1.000 liter of C_2H_2 (measured at STP)?

4.42 Calcium hydride, $CaH_2(s)$, reacts with water to yield hydrogen gas and calcium hydroxide. Write the equation for the reaction. What volume of $H_2(g)$, measured at STP, would be obtained from the reaction of 1.000 g of CaH_2 with excess H_2O?

4.43 A sample of sodium azide, NaN_3, is decomposed by heating, and 462 ml of N_2 gas, measured at 24°C and 0.980 atm is obtained. What did the sample of NaN_3 weigh?

4.44 In the absence of a catalyst, ammonia, $NH_3(g)$, reacts with oxygen gas to yield nitrogen gas and water vapor. Write an equation for the reaction. What volume of $N_2(g)$ can be obtained from 10.0 liter of $NH_3(g)$ and 10.0 liter of $O_2(g)$? The volumes of all gases are measured under the same conditions.

4.45 A mixture is prepared from 15.0 liter of ammonia and 15.0 liter of chlorine. These substances react according to the equation

$$2NH_3(g) + 3Cl_2(g) \rightarrow N_2(g) + 6HCl(g)$$

If the volumes of all of the gases are measured at the same temperature and pressure, list the volumes of all of the substances present at the conclusion of the reaction.

4.46 Magnesium and aluminum react with aqueous acids to give hydrogen gas.

$$Mg(s) + 2H^+(aq) \rightarrow H_2(g) + Mg^{2+}(aq)$$

$$2Al(s) + 6H^+(aq) \rightarrow 3H_2(g) + 2Al^{3+}(aq)$$

What volume of $H_2(g)$ measured at 20°C and a pressure of 1.01 atm can be obtained from 4.00 g of an alloy that is 70.0% Mg and 30.0% Al?

4.47 The complete combustion of 11.2 liter of a gaseous compound that contains only carbon and hydrogen requires 50.4 liter of $O_2(g)$ and produces 33.6 liter of $CO_2(g)$ and 27.0 g of $H_2O(l)$. All gas measurements are made at STP. (a) For each substance involved in the reaction, calculate the number of moles used or produced. (b) Use your answers from part (a) to derive the whole-number coefficients of the balanced chemical equation for the reaction. (c) Determine a formula for the hydrocarbon and write the equation for the reaction.

4.48 Assume that 10 liter of $NH_3(g)$ and 10 liter of $O_2(g)$ are mixed and react according to the equation

$$4NH_3(g) + 5O_2(g) \rightarrow 4NO(g) + 6H_2O(g)$$

If the conditions under which the gases are measured are constant and such that all materials are gaseous, list the volumes of all materials at the conclusion of the reaction.

4.49 Acetylene is formed by the reaction of water and calcium carbide.

$$CaC_2(s) + 2H_2O(l) \rightarrow C_2H_2(g) + Ca(OH)_2(s)$$

What volume of acetylene can be produced at STP from 4.00 g CaC_2 and 2.75 g H_2O?

4.50 Traces of oxygen can be removed from streams of gas by reaction with heated copper.

$$2Cu(s) + O_2(g) \rightarrow 2CuO(s)$$

If the original copper weighs 0.3750 g and the oxidized copper weighs 0.3880 g, what volume of O_2 at STP was removed from the gas stream?

4.51 A sample of $N_2(g)$, collected over water at 25°C and a pressure of 1.050 atm, occupies 150 ml. What volume would the N_2 occupy if dry and at STP? The vapor pressure of water at 25°C is 0.031 atm.

4.52 A 1.00 liter sample of neon is collected over water at 27°C and 0.983 atm pressure. If the gas is dried and placed in a 1.50 liter container at 72°C, what pressure will it exert?

4.53 A sample of an ideal gas is passed through water and 500 ml collected at 22°C. The wet gas exerts a pressure of 0.993 atm. The sample of gas, when dried and confined in the same volume, exerts the same pressure, 0.993 atm, at 30°C. From these data, calculate the vapor pressure of water at 22°C.

4.54 A 250 ml sample of gas is collected over water at 20°C and at a barometric pressure of 1.00 atm. What volume would the gas occupy dry and at 25°C?

4.55 A gas, X, diffuses 3.1 times faster than fluorine gas. What is the molecular weight of gas X?

4.56 Under certain conditions of temperature and pressure, the density of oxygen gas is 1.30 g/liter. A volume of 21.0 ml of $O_2(g)$ effuses through an apparatus in 1.00 sec. The rate of effusion of gas X through the same apparatus is 15.0 ml/sec. Calculate the density of gas X under the experimental conditions.

4.57 At 27°C and 1 atm, the density of $CO_2(g)$ is 1.79 g/liter. The rate of effusion of $CO_2(g)$ through an apparatus is 7.50 ml/sec. What is the density of an unknown gas if it effuses at a rate of 6.90 ml/sec through the same apparatus and under the same conditions? (b) What is the approximate molecular weight of the gas?

4.58 Calculate the density of a gas at STP if a given volume of the gas effuses through an apparatus in 5.00 min and the same volume of oxygen, at the same temperature and pressure, effuses through this apparatus in 6.30 min.

4.59 Use Graham's law to calculate the molecular weight of a gas if a given volume of the gas effuses through an apparatus in 282 sec and the same volume of nitrogen, under the same conditions of temperature and pressure, effuses through the apparatus in 215 sec.

4.60 Calculate the pressure exerted by 1.00 mole of NH_3 gas confined to a volume of 1.00 liter at 27°C by (a) the ideal gas law and (b) van der Waals equation.

4.61 Calculate the pressure exerted by 1.00 mole of O_2 gas confined to a volume of 1.00 liter at 0°C by (a) the ideal gas law and (b) the van der Waals equation.

Chapter 5

Atomic Structure

Modern atomic theory has provided a foundation upon which modern chemistry has been built. An understanding of atomic structure and the ways in which atoms interact is central to an understanding of chemistry.

Individual atoms cannot be weighed, measured, or examined directly. Indirect evidence, therefore, has been used to develop the atomic theory. Some nineteenth-century scientists (Wilhelm Ostwald, for example) did not believe that atoms exist. Others (Michael Faraday, for example) thought of the atomic theory as a convenient concept that may or may not represent reality. Still others (Ludwig Boltzmann, for example) believed in the physical existence of atoms.

Whether atoms are real or imaginary was an open question as late as 1904. In that year a debate, attended by noted international scientists, was held at the St. Louis World Fair. Ostwald argued against the existence of atoms, and Jacobus van't Hoff and Boltzmann defended the other side.

5.1 THE ELECTRON

In Dalton's theory, as in the theories of the Greeks, atoms were regarded as the smallest possible components of matter. Toward the end of the nineteenth century, it began to appear that the atom itself might be composed of even smaller particles. This change in viewpoint was brought about largely by evidence from experiments with electricity.

In 1807–8, the English chemist Humphry Davy discovered five elements (potassium, sodium, calcium, strontium, and barium) by using electricity to decompose compounds. This work led Davy to propose that elements are held together in compounds by attractions that are electrical in nature.

In 1832–33, Michael Faraday ran an important series of experiments on chemical electrolysis, in which compounds are decomposed by electricity. Faraday determined the quantitative relationships between the amount of electricity used in an electrolysis and its chemical effect. He found, for example, that the same quantity of electricity is required to produce 1 mol of sodium by electrolysis as is required to produce 1 mol of potassium. On the other hand, exactly twice this quantity of electricity is required to produce 1 mol of calcium. On the basis of Faraday's work, George Johnstone Stoney proposed in 1874 that units of electrical charge were associated with atoms. In 1891, Stoney suggested that these units be called **electrons.**

Attempts to pass an electric current through a vacuum led to the discovery of **cathode rays** by Julius Plücker in 1859 (Figure 5.1). Two electrodes are sealed in a glass tube from which air is almost completely removed. When a high voltage is impressed across these electrodes, rays stream from the negative electrode, called the cathode. These rays are negatively charged, travel in straight lines, and cause the walls opposite the cathode to glow.

In the latter part of the nineteenth century, cathode rays were extensively studied. The results of the experiments of many scientists led to the conclusion that the rays are streams of fast-moving, negatively charged particles. These particles were eventually called electrons as suggested by Stoney. The

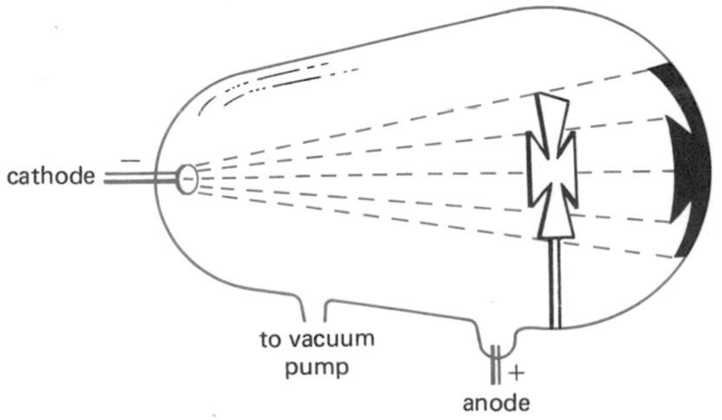

Figure 5.1 Cathode-ray tube.

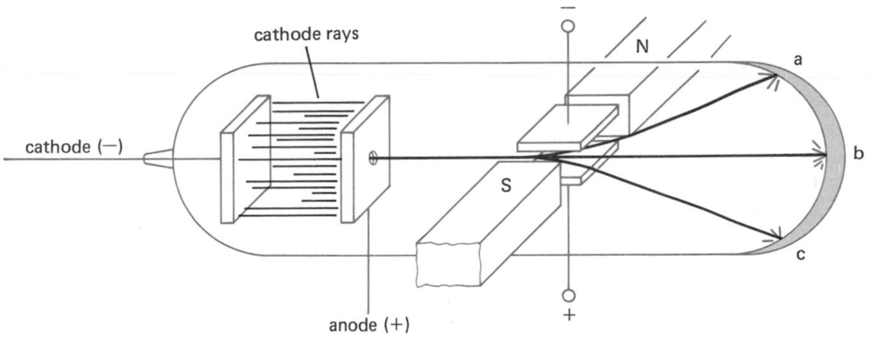

Figure 5.2 Deflections of cathode rays. (a) In a magnetic field. (b) With magnetic and electric fields balanced; no net deflection. (c) In an electric field.

electrons, which originate from the metal of which the cathode is composed, are the same no matter what material is employed for the electrode.

Since unlike charges attract, the streams of electrons that constitute a cathode ray are attracted toward the positive plate when two oppositely charged plates are placed on either side of them (Figure 5.2, path c). The rays, therefore, are deflected from their usual straight-line path in an electric field. The degree of deflection varies

1. *Directly* with the size of the charge of the particle, e. A particle with a high charge is deflected more than one with a low charge.
2. *Inversely* with the mass of the particle, m. A particle with a large mass is deflected less than one with a small mass.

The ratio of charge to mass, e/m, therefore, determines the extent to which the electrons are deflected from a straight-line path in an electric field. Electrons are deflected in a magnetic field also. In this case, however, the deflection is at right angles to the applied field (Figure 5.2, path a).

In 1897, Joseph J. Thomson determined the value of e/m for the electron by studying the deflections of cathode rays in electric and magnetic fields. The value is

$$e/m = -1.7588 \times 10^8 \text{ coul/g}$$

The *coulomb* (coul)* is the SI unit of electric charge. One coulomb is the quantity of charge that passes a given point in an electric circuit in one second when the current is one ampere.

The first precise measurement of the charge of the electron was made by Robert A. Millikan in 1909. In Millikan's experiment (Figure 5.3), electrons are produced by the action of X rays on the molecules of which air is composed. Very small drops of oil pick up electrons and acquire electric charges.

* The symbol *coul* for coulomb is used in this book, rather than the SI symbol, C, to avoid confusion since the chemical symbol for carbon is C and the symbol for degrees Celsius is °C.

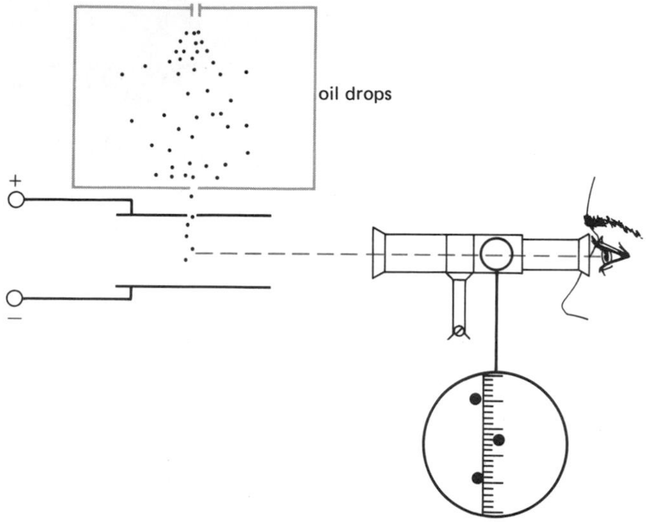

Figure 5.3 Millikan's determination of the charge on the electron.

The oil drops are allowed to settle between two horizontal plates, and the mass of a single drop is determined by measuring its rate of fall.

When the plates are charged, the rate of fall of the drop is altered because the negatively charged drop is attracted to the positive plate. Measurement of the rate of fall in these circumstances enables calculation of the charge on the drop. Since a given drop can pick up one or more electrons, the charges calculated in this way are not identical. They are, however, all simple multiples of the same value, which is assumed to be the charge on a single electron.

$$e = -1.6022 \times 10^{-19} \text{ coul}$$

The mass of the electron can be calculated from the value of e/m and the value of e.

$$m = \frac{e}{e/m} = \frac{-1.6022 \times 10^{-19} \text{ coul}}{-1.7588 \times 10^{8} \text{ coul/g}} = 9.1096 \times 10^{-28} \text{ g}$$

5.2 THE PROTON

If one or more electrons are removed from a neutral atom or molecule, the residue has a positive charge equal to the sum of the negative charges of the electrons removed. If one electron is removed from a Ne atom, a Ne^{+} ion results; if two are removed, a Ne^{2+} ion results; and so forth. Positive particles of this type (positive ions) are formed in an electric discharge tube when cathode rays rip electrons from the atoms or molecules of the gas present in the tube. These positive ions move *toward* the negative electrode (the cathode). If holes have been bored in this electrode, the positive ions pass through them

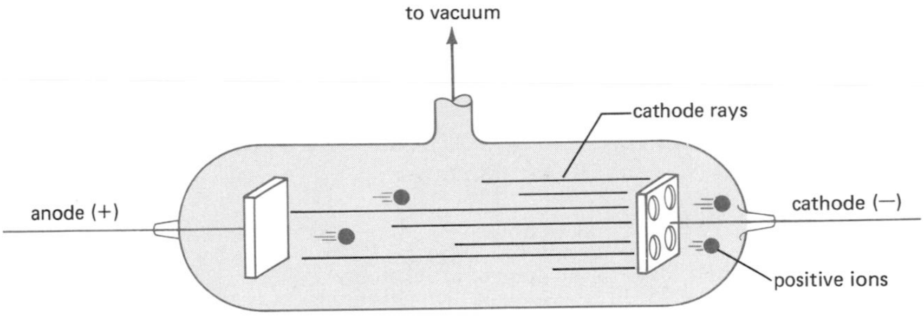

Figure 5.4 Positive rays.

(see Figure 5.4). The electrons of the cathode rays, since they are negatively charged, move in the opposite direction.

These streams of positive ions, called **positive rays,** were first observed by Eugen Goldstein in 1886. The deflections of positive rays in electric and magnetic fields were studied by Wilhelm Wien (1898) and J. J. Thomson (1906). Values of e/m were determined for the positive ions by using essentially the same method employed in the study of cathode rays.

When different gases are used in the discharge tube, different types of positive ions are produced. When hydrogen gas is used, a positive particle results that has the smallest mass (and hence the largest e/m value) of any positive ion observed.

$$e/m = +9.5791 \times 10^4 \text{ coul/g}$$

These particles, now called **protons,** are assumed to be a component of all atoms. The proton has a charge equal in magnitude to that of the electron but opposite in sign.

$$e = +1.6022 \times 10^{-19} \text{ coul}$$

This charge is called the **unit electrical charge.** The proton is said to have a unit positive charge and the electron, a unit negative charge.

The mass of the proton, which is 1836 times the mass of the electron, can be calculated from these data.

$$m = \frac{e}{e/m} = \frac{+1.6022 \times 10^{-19} \text{ coul}}{+9.5791 \times 10^4 \text{ coul/g}} = 1.6726 \times 10^{-24} \text{ g}$$

5.3 THE NEUTRON

Since atoms are electrically neutral, a given atom must contain as many electrons as protons. To account for the total masses of atoms, Ernest Rutherford in 1920 postulated the existence of an uncharged particle. Since this particle is uncharged, it is difficult to detect and characterize. In 1932, however, James Chadwick published the results of his work which established the existence of the **neutron.** He was able to calculate the mass of the neutron

Table 5.1 Subatomic particles

PARTICLE	MASS GRAMS	MASS UNIFIED ATOMIC MASS UNITS[a]	CHARGE[b]
electron	9.1096×10^{-28}	0.00054859	1 −
proton	1.6726×10^{-24}	1.007277	1 +
neutron	1.6749×10^{-24}	1.008665	0

[a] The unified atomic mass unit (u) is $\frac{1}{12}$ the mass of a ^{12}C atom (Section 5.7).
[b] The unit charge is 1.6022×10^{-19} coul.

from data on certain nuclear reactions (see Chapter 23) in which neutrons are produced. By taking into account the masses and energies of all particles used and produced in these reactions, Chadwick determined the mass of the neutron, which is very slightly larger than that of the proton. The neutron has a mass of 1.6749×10^{-24} g and the proton, 1.6726×10^{-24} g.

The properties of the electron, proton, and neutron are summarized in Table 5.1. At the present time, many other subatomic particles have been identified. Their relationship to atomic structure, however, is not completely understood. For the study of chemistry, atomic structure is adequately explained on the basis of the electron, proton, and neutron.

5.4 THE NUCLEAR ATOM

Certain atoms are unstable combinations of the subatomic particles. These atoms spontaneously emit rays and in this way change into atoms with a different chemical identity. This process, called **radioactivity,** was discovered by Henri Becquerel in 1896. In subsequent years, Ernest Rutherford explained the nature of the three types of rays emitted by radioactive substances that occur in nature (see Table 5.2). Other types of rays have now been identified, but these result from the disintegration of man-made atoms.

Alpha rays consist of particles that carry a 2+ charge and have a mass approximately four times that of the proton. These α particles are ejected from the radioactive substance at speeds around 16,000 km/s. When α particles were first studied, the neutron had not yet been discovered. We now know that

Table 5.2 Types of radioactive emissions

RAY	SYMBOL	COMPOSITION	CHARGE OF COMPONENT
alpha	α	particles containing 2 protons and 2 neutrons	2+
beta	β	electrons	1−
gamma	γ	very short wavelength electromagnetic radiation	0

an α particle consists of two protons and two neutrons. **Beta rays** are streams of electrons that travel at approximately 130,000 km/s. **Gamma radiation** is essentially a highly energetic form of light. Gamma rays are uncharged and are similar to X rays.

In 1911, Rutherford reported the results of experiments in which α particles were used to investigate the structure of the atom. A beam of α particles was directed against a very thin (about 0.0004 cm thick) foil of gold, platinum, silver, or copper. The large majority of the α particles went directly through the foil in use. Some, however, were deflected from their straight-line path, and a few recoiled back towards their source (Figure 5.5).

Rutherford explained the results of these experiments by proposing that a **nucleus** exists in the center of the atom. Most of the mass and all of the positive charge of the atom are concentrated in the nucleus. The electrons (which occupy most of the total volume of the atom) are outside the nucleus (**extranuclear**) and in rapid motion around it. Since an atom is electrically neutral, the positive charge of the nucleus equals the total negative charge of all the electrons. The nucleus is now believed to contain protons and neutrons. The number of protons is equal to the number of electrons present in the atom. The neutrons and protons together account for the mass of the nucleus.

It is important to grasp the scale of this model of the atom. If a nucleus of

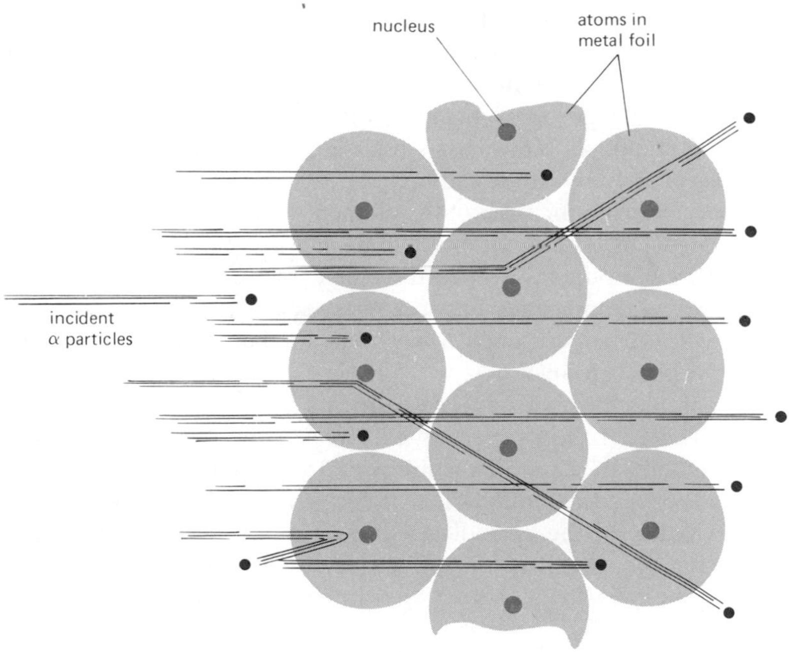

Figure 5.5 Deflections and recoil of α particles by nuclei of metal foil in Rutherford's experiment (not to scale).

an atom were the size of a tennis ball, the atom would have a diameter of over one mile. Since most of the volume of an atom is empty space, most α particles pass directly through the target foils. The comparatively light electrons do not deflect the heavier, fast-moving α particles. A close approach of an α particle (which is positively charged) to a nucleus (which is also positively charged) results in the repulsion of the α particle and deflection from its straight-line path. In a very few instances, when an α particle scores a direct hit on a nucleus, the particle is reflected back toward its source.

The composition and stability of the nucleus is discussed in Chapter 23.

5.5 ATOMIC NUMBER AND THE PERIODIC LAW

Early in the nineteenth century, chemists became interested in the chemical and physical similarities that exist between elements. In 1817 and 1829, Johann W. Döbereiner published articles in which he examined the properties of sets of elements that he called triads (Ca, Sr, Ba; Li, Na, K; Cl, Br, I; and S, Se, Te). The elements of each set have similar properties, and the atomic weight of the second element of a set is approximately equal to the average of the atomic weights of the other two elements of the set.

In following years, many chemists made attempts to find regularities in the properties of the elements, and several charts designed to feature similarities between elements were published. In the years 1863–66, John A. R. Newlands proposed and developed his "law of octaves." Newlands stated that when the elements are listed by increasing atomic weight, the eighth element is similar to the first, the ninth to the second, and so forth. He compared this relationship to octaves of musical notes. Unfortunately, the actual relationship is not so simple as Newlands supposed. His work seemed forced and was not taken seriously by other chemists.

The modern periodic classification of the elements stems from the works of Julius Lothar Meyer (1869) and, in particular, Dmitri Mendeleev (1869). Mendeleev proposed a periodic law: when the elements are studied in order of increasing atomic weight, similarities in properties recur periodically. Mendeleev's table listed the elements in such a way that similar elements appeared in vertical columns, called groups.

Mendeleev left blanks for undiscovered elements in his table. On the basis of his system, he predicted the properties of three of the missing elements. The subsequent discovery of scandium, gallium, and germanium, each of which was found to have properties much like those predicted by Mendeleev, demonstrated the validity of the periodic system. The existence of the noble gases (He, Ne, Ar, Kr, Xe, and Rn) was unforeseen by Mendeleev. Nevertheless, after their discovery in the years 1892–98, these elements readily fitted into the periodic table.

The plan of the periodic table required that three elements (K, Ni, and I) be placed out of the order determined by increasing atomic weight. Iodine, for example, should be element number 52 on the basis of atomic weight. Instead, iodine was arbitrarily made element number 53 so that it would fall in a group of the table with chemically similar elements (F, Cl, and Br). Subsequent study of the periodic classification convinced many that some fundamental property other than atomic weight is the cause of the observed periodicity. It was proposed that this fundamental property is in some way related to atomic number, which at that time was only a serial number derived from the periodic system.

The work of Henry G. J. Moseley in the years 1913 and 1914 solved the problem. When high-energy cathode rays are focused on a target, X rays are produced (Figure 5.6); the nature of the X rays depends upon the substance used as the target. Moseley studied the X rays produced by the cathode-ray bombardment of 38 elements with atomic numbers between 13 (aluminum) and 79 (gold).

X Rays are transmitted through space in a wave motion (see Figure 5.8 in Section 5.8). The wavelengths of X rays are, in general, short in comparison to the wavelengths of other forms of radiant energy (such as visible light, Section 5.8). Moseley found that there is a simple mathematical relationship between the wavelengths of the X rays produced and the atomic numbers of the elements used as targets (but not the atomic weights). He was able to assign the correct atomic number to an element on the basis of the wavelengths of the X rays that result when the element is used as a target.

In this way, Moseley settled the problem involving the classification of elements that have atomic weights out of line with those of their neighbors (K, Ni, and I). He determined that four elements before number 79 (gold) remained to be discovered (numbers 43, 61, 72, and 75). He also showed how the lanthanides (elements number 58 to 71) fit into the periodic classification. On the basis of Moseley's work, the **periodic law** was redefined: the chemical and physical properties of the elements are periodic functions of *atomic number*.

Moseley's atomic numbers agreed roughly with the nuclear charges that

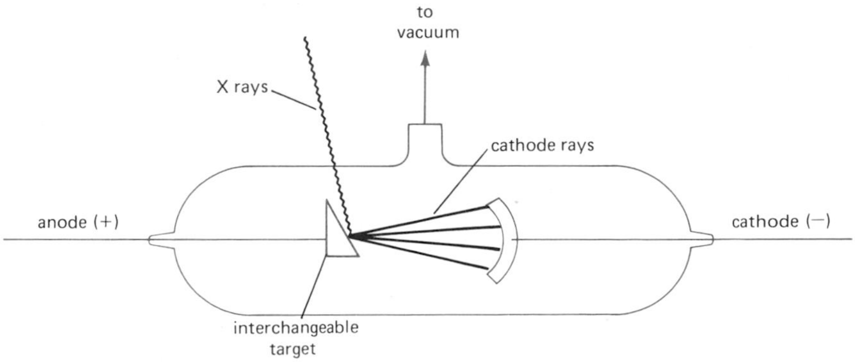

Figure 5.6 X-ray tube.

Rutherford had calculated on the basis of the α-particle scattering experiments. Moseley proposed, therefore, that the atomic number, which is given the symbol Z, is the number of units of positive charge of the atomic nucleus. He said, "There is in the atom a fundamental quantity, which increases by regular steps as we pass from one element to the next. This quantity can only be the charge on the central positive nucleus."

The most popular form of the periodic table is the long form that is found inside the front cover of this book. The **periods** consist of those elements that are arranged in horizontal rows in the table. The organization is such that elements of similar chemical and physical properties (called **groups** or **families**) appear in vertical columns. In this table, the set of elements (the lanthanides) that appears at the bottom should actually appear in the body of the chart in proper order by atomic number so that the sixth period has its requisite 32 elements. For convenience in reproduction this arrangement is not used. The chart should be vertically cut, the sections separated, and the lanthanides inserted in their proper positions. For the actinides, which appear below the lanthanides at the bottom of the periodic table, the same considerations pertain. They should be inserted in the seventh period (as yet incomplete).

The first period consists of only two elements, hydrogen and helium. Subsequent periods have 8, 8, 18, 18, and 32 elements. With the exception of the first period, each complete period begins with an **alkali metal** (group I A)—a highly reactive, light, silvery metal—and ends with a **noble gas** (group 0)—a colorless gas of low reactivity. The element before the noble gas of each complete period (except for the first period) is a **halogen** (group VII A)—a very reactive nonmetal.

A general pattern exists for each period after the first. Starting with an alkali metal, the properties change from element to element—the metallic properties fade and are gradually replaced by nonmetallic characteristics. After a highly reactive nonmetal (a halogen) is reached, each complete period ends with a noble gas.

5.6 ATOMIC SYMBOLS AND ISOTOPES

An atom is identified by two numbers: the atomic number and the mass number. Thc **atomic number,** Z, is the number of unit positive charges on the nucleus. Since the proton has a 1+ charge, the atomic number is equal to the number of protons in the nucleus of the atom. An atom is electrically neutral. It must have as many electrons (at 1− each) as it has protons (at 1+ each). The atomic number, therefore, also indicates the number of extranuclear electrons in the atom.

The **mass number,** A, of an atom is the total number of protons and neutrons (collectively called **nucleons**) in the nucleus of the atom. The number of neutrons can be calculated by subtracting the atomic number (the number of protons) from the mass number (the number of protons and neutrons taken together).

number of neutrons $= A - Z$

The mass *number* is the total *number* of nucleons in a nucleus and not the mass of the nucleus. The mass number, however, is a whole-number approximation of the atomic mass in atomic mass units (u) since the mass of the proton as well as the mass of the neutron is approximately equal to 1 u and the mass of the electron is negligible.

An atom of an element is designated by the **chemical symbol** for the element with the atomic number of the element placed at the lower left and the mass number placed at the upper left.*

$${}^{A}_{Z}\text{Symbol}$$

The symbol ${}^{35}_{17}Cl$ designates an atom of chlorine that has 17 protons (Z) and 18 neutrons ($A - Z$) in the nucleus and 17 extranuclear electrons (Z). An atom of sodium with the symbol ${}^{23}_{11}Na$ has 11 protons and 12 neutrons in the nucleus and 11 electrons in motion around this nucleus.

All atoms of a given element have the same atomic number. Some elements, however, consist of several types of atoms that differ from one another in mass number. Atoms that have the same atomic number but different mass numbers are called **isotopes.**

Two isotopes of chlorine occur in nature: ${}^{35}_{17}Cl$ and ${}^{37}_{17}Cl$. The atomic compositions of these isotopes are:

${}^{35}_{17}Cl$	17 protons	18 neutrons	17 electrons
${}^{37}_{17}Cl$	17 protons	20 neutrons	17 electrons

Isotopes, therefore, differ in the number of neutrons in the nucleus, which means that they differ in atomic mass.

The chemical properties of an atom depend principally upon the numbers of protons and electrons that the atom contains, which are determined by the atomic number. Isotopes of the same element, therefore, have very similar (in most cases, indistinguishable) chemical properties. Some elements exist in nature in only one isotopic form (for example, sodium, beryllium, and fluorine). Most elements, however, have more than one natural isotope—tin has 10.

The **mass spectrometer** is used to determine the types of isotopes present in an element, the exact atomic masses of these isotopes, and the relative amount of each isotope present. The essential features of the instrument are shown in Figure 5.7. Positive ions are produced from vaporized material by bombardment with electrons. These ions are attracted toward a negatively

* The other two corners are reserved for other designations: the upper right for the charge if the atom has lost or gained electrons and become an ion, the lower right corner for the number of atoms present in a molecule or a formula unit.

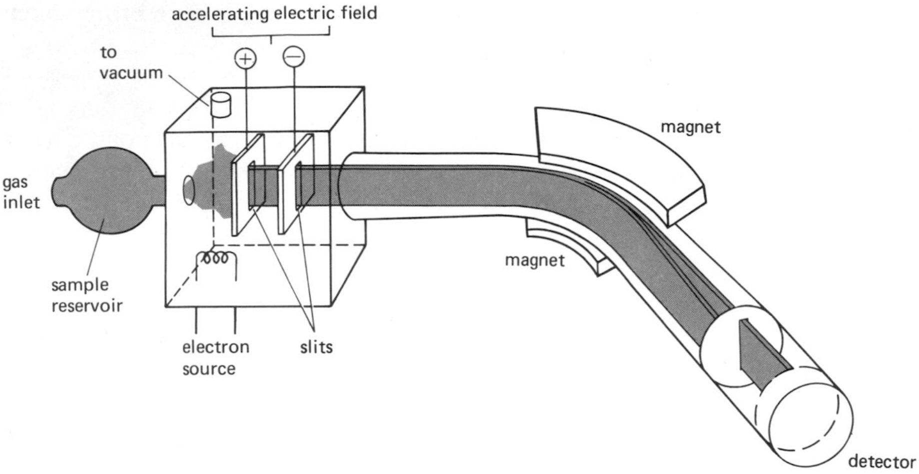

Figure 5.7 Essential features of a mass spectrometer.

charged slit. They are accelerated by the attraction and pass through the slit with a high velocity.

The beam of ions is then passed through a magnetic field. Charged particles follow a circular path in a magnetic field. The higher the charge on the particle (e), the more it is deflected from a straight-line path. The mass of a particle (m), on the other hand, is inversely related to the degree of deflection. The larger the mass of a particle, the less it is deflected from a straight-line path. The radius of the path of a positive ion in the magnetic field, therefore, depends upon the value of e/m.

All the ions that pass through the final slit have the same value of e/m. Ions with other values of e/m can be made to pass through this slit by adjusting the magnetic field strength or the voltage used to accelerate the ions. Consequently, each type of ion present can be made to pass through the slit separately. The detector measures the intensity of each ion beam, which depends upon the relative amount of each isotope present in the sample.

5.7 ATOMIC WEIGHTS

The **atomic mass unit** (u) is defined as one-twelfth the mass of the ${}^{12}_{6}C$ atom. The assignment of exactly 12 u to the mass of this isotope of carbon is arbitrary. Other standards have been used in the past.*

* Prior to 1961, two atomic mass scales were employed. Chemists based their scale on the assignment of exactly 16 to the *average* mass of oxygen atoms as they occur in nature. Physicists, however, based their scale on the mass of the *single* isotope ${}^{16}_{8}O$. Since naturally occurring oxygen consists of three isotopes, the two scales were different. The unified scale, based on ${}^{12}_{6}C$, was developed to eliminate this duality. The SI symbol u was selected so that it would indicate measurements made on the *unified* scale.

On this scale, the mass of the proton is 1.007277 u, the mass of the neutron is 1.008665 u, and the mass of the electron 0.0005486 u (Table 5.1). The mass of an atom, however, cannot be calculated from these values. With the exception of ^1_1H, the sum of the masses of the particles that make up a nucleus is always larger than the actual mass of the nucleus.

Einstein has shown that mass and energy are equivalent:

$$E = mc^2$$

where E is energy (in joules), m is mass (in kilograms), and c is the speed of light (3.00×10^8 m/s). If the required nucleons were brought together to form a nucleus, some of their mass would be converted into energy. This quantity of energy, called the **binding energy,** is also the amount of energy required to pull the nucleus apart. Binding energy is discussed more fully in Section 23.6.

Most elements occur in nature as mixtures of isotopes. With very few exceptions, however, these mixtures have constant compositions. Data obtained by use of a mass spectrometer show that the element chlorine consists of 75.53% $^{35}_{17}Cl$ atoms (mass, 34.97 u) and 24.47% $^{37}_{17}Cl$ atoms (mass, 36.95 u). Any sample of chlorine from a natural source consists of these two isotopes in this proportion.

The **atomic weight** of the element chlorine is the weighted average of the atomic masses of the natural isotopes. We can *not* find the average mass by adding the masses of the isotopes and dividing by 2. The value obtained in this way would be correct only if the element consisted of equal numbers of atoms of the two isotopes. Instead, the *weighted* average is found by multiplying the atomic mass of each isotope by its fractional abundance and adding the values obtained. A fractional abundance is the decimal equivalent of the percent abundance.

	(fractional abundance)(mass)
$^{35}_{17}Cl$	(0.7553)(34.97 u) = 26.41 u
$^{37}_{17}Cl$	(0.2447)(36.95 u) – 9.04 u
	atomic weight Cl = 35.45 u

The accepted value for chlorine is 35.453 u.

There is no atom of chlorine with a mass of 35.453 u, but it is convenient to think in terms of one. In most calculations, no mistake is made by assuming that a sample of an element consists of only one type of atom with the average mass, the atomic weight. This assumption is valid because there is a huge number of atoms, in even a small sample of matter. There are for example more atoms in a drop of water than there are people on earth.

Example 5.1 What is the atomic weight of magnesium? The element consists of 78.70% $^{24}_{12}Mg$ atoms (mass, 23.99 u), 10.13% $^{25}_{12}Mg$ atoms (mass, 24.99 u), and 11.17% $^{26}_{12}Mg$ atoms (mass, 25.98 u).

Solution

	(abundance)(mass)	
$^{24}_{12}Mg$	(0.7870)(23.99 u) =	18.88 u
$^{25}_{12}Mg$	(0.1013)(24.99 u) =	2.53 u
$^{26}_{12}Mg$	(0.1117)(25.98 u) =	2.90 u
	atomic weight Mg =	24.31 u

Example 5.2 Carbon occurs in nature as a mixture of $^{12}_{6}C$ and $^{13}_{6}C$. The atomic mass of $^{12}_{6}C$ is exactly 12 u, by definition, and the atomic mass of $^{13}_{6}C$ is 13.003 u. The atomic weight of carbon is 12.011 u. What is the atom percent of $^{12}_{6}C$ in natural carbon?

Solution The equation to determine the atomic weight of carbon is

$$(\text{abundance } ^{12}_{6}\text{C})(\text{mass } ^{12}_{6}\text{C}) + (\text{abundance } ^{13}_{6}\text{C})(\text{mass } ^{13}_{6}\text{C}) = \text{atomic weight C}$$

If we let x equal the abundance of $^{12}_{6}C$, then $(1-x)$ is the abundance of $^{13}_{6}C$. Therefore,

$$\begin{aligned}(x)(12.000) + (1-x)(13.003) &= 12.011\\ 12.000x + 13.003 - 13.003x &= 12.011\\ -1.003x &= -0.992\\ x &= 0.989\end{aligned}$$

Atoms of $^{12}_{6}C$ constitute 98.9% of the total number.

Very small amounts of $^{14}_{6}C$ also occur in nature. The amount is so small, however, that the atomic weight of carbon can be calculated by neglecting this isotope.

5.8 ELECTROMAGNETIC RADIATION

Electromagnetic radiation travels through space in a wave motion (Figure 5.8). The following terms describe these waves.

1. The **wavelength,** λ (lambda), is the distance from crest to crest (or trough to trough).
2. The **amplitude,** a, of a wave is its height. The **intensity** (or brightness) of the radiation is proportional to the square of the amplitude, a^2.

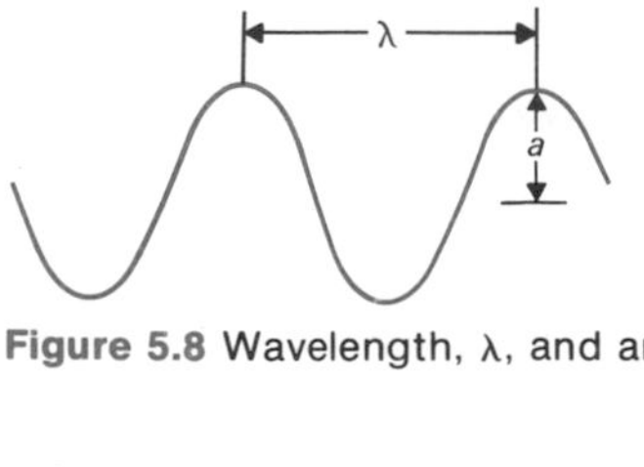

Figure 5.8 Wavelength, λ, and amplitude, *a*, of a wave.

3. In a vacuum, all waves, regardless of wavelength, travel at the same speed, 2.9979×10^8 m/s. This speed is called the **speed of light** and is given the symbol c.
4. The **frequency** of the radiation, ν (nu), is the number of waves that pass a given spot in a second. For a given type of radiation, the wavelength times the number of waves per second (the frequency) equals the distance traveled per second (the speed of light).

$$\lambda\nu = c$$

and therefore,

$$\nu = \frac{c}{\lambda}$$

The electromagnetic spectrum is shown in Figure 5.9. Radio waves have very long wavelengths, infrared waves (radiant heat) have moderate wavelengths, and γ rays (from radioactive decay) have extremely short wavelengths. White light (visible light) consists of rays with wavelengths in the approximate range of 4×10^{-5} cm to 7.5×10^{-5} cm (which is 400 nm to 750 nm).*

The wave theory successfully interprets many properties of electromagnetic radiation. Other properties, however, require that such radiation be considered as consisting of particles. In 1900, Max Planck proposed the

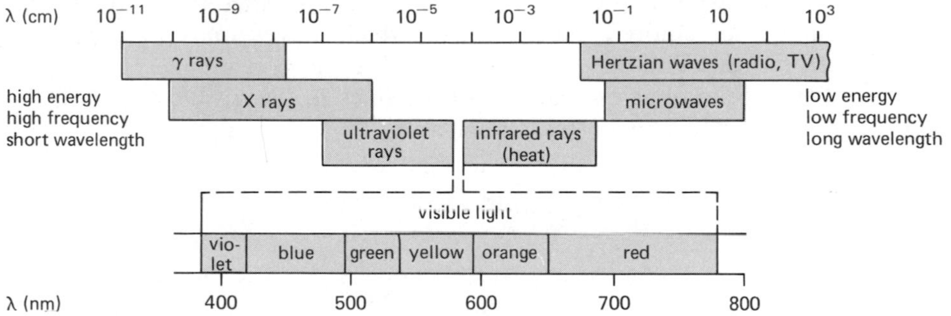

Figure 5.9 Electromagnetic radiation. Note that the approximate ranges of electromagnetic radiations are plotted on a logarithmic scale in the upper part of the diagram. The spectrum of visible light is not plotted in this way.

* In the past, wavelengths have been measured in ångstrom units (Å), 10^{-8} cm. This unit is not a part of the International System. The International Committee of Weights and Measures recommends that the nanometer (nm), 10^{-7} cm, be used instead.

$$1\ \text{Å} = 10^{-8}\ \text{cm} = 10^{-10}\ \text{m}$$
$$1\ \text{nm} = 10^{-7}\ \text{cm} = 10^{-9}\ \text{m}$$

Therefore, $1\ \text{Å} = 0.1$ nm and $1\ \text{nm} = 10\ \text{Å}$

The range 400 nm to 750 nm corresponds to 4000 Å to 7500 Å.

quantum theory of radiant energy. Planck suggested that radiant energy could be absorbed or given off only in definite quantities, called **quanta.** The energy of a quantum, E, is proportional to the frequency of the radiation, ν.

$$E = h\nu$$

The proportionality constant, h, is **Planck's constant,** 6.6262×10^{-34} J·s.

High energy radiation, therefore, has a high frequency. A high frequency means that a large number of waves pass a spot in 1s. The wavelength, therefore, must be short. In 1905, Albert Einstein proposed that Planck's quanta are discontinuous bits of energy, which were later named **photons.**

5.9 ATOMIC SPECTRA

When a ray of light is passed through a prism, the ray is bent, or refracted. The amount that a wave is refracted depends upon its wavelength. A wave with a short wavelength is bent more than one with a long wavelength. Since ordinary white light consists of waves with all the wavelengths in the visible range, a ray of white light is spread out into a wide band called a **continuous spectrum.** This spectrum is a rainbow of colors with no blank spots—violet merges into blue, blue into green, and so on.

When gases or vapors of a chemical substance are heated in an electric arc or a Bunsen flame, light is emitted. If a ray of this light is passed through a prism, a **line spectrum** is produced (Figure 5.10). This spectrum consists of a limited number of colored lines, each of which corresponds to a different wavelength of light. The line spectrum of each element is unique.

The frequencies that correspond to the lines in the visible region of the hydrogen spectrum are given by the equation:

$$\nu = \frac{c}{\lambda} = (3.290 \times 10^{15}/\text{s})\left(\frac{1}{2^2} - \frac{1}{n^2}\right) \qquad \text{n} = 3, 4, 5, \ldots$$

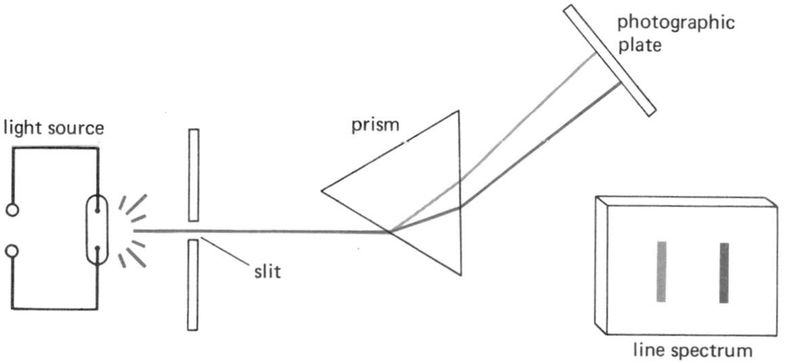

Figure 5.10 The spectroscope.

where n is an integer equal to, or greater than, 3. This relationship was discovered in 1885 by J. J. Balmer, and the series of lines in the visible region is called the **Balmer series.**

In 1913, Niels Bohr proposed a theory for the electronic structure of the hydrogen atom that explained the line spectrum of this element. The hydrogen atom contains one electron and a nucleus that consists of a single proton. Bohr's theory includes the following points.

1. The electron of the hydrogen atom can exist only in certain spherical **orbits** (which are also called **energy levels** or **shells**). These shells are arranged concentrically around the nucleus. Each shell is designated by a letter (K, L, M, N, O, . . .) or a value of n (1, 2, 3, 4, 5, . . .).
2. The electron has a definite energy characteristic of the orbit in which it is moving. The K level ($n = 1$), the shell closest to the nucleus, has the smallest radius. An electron in the K level has the lowest possible energy. With increasing distance from the nucleus (K, L, M, N, O; $n = 1, 2, 3, 4, 5$), the radius of the shell and the energy of an electron in the shell increase. The electron cannot have an energy that would place it between the permissible shells.
3. When the electrons of an atom are as close to the nucleus as possible (for hydrogen, one electron in the K shell), they are in the condition of lowest possible energy, called the **ground state.** When the atoms are heated in an electric arc or Bunsen flame, electrons absorb energy and jump to outer levels, which are higher energy states. The atoms are said to be in **excited states.**
4. When an electron falls back to a lower level, it emits a definite amount of energy. The energy difference between the high energy state and low energy state is emitted in the form of a quantum of light. The light quantum has a characteristic frequency (and wavelength) and produces a characteristic spectral line. In spectral studies, many atoms are absorbing energy at the same time that many others are emitting it. Each spectral line corresponds to a different electron transition.

Bohr derived an equation for the energy that an electron would have in each orbit, E_{orbit}. This equation can be simplified to

$$E_{\text{orbit}} = -\frac{(2.180 \times 10^{-18}\,\text{J})}{n^2} \qquad n = 1, 2, 3, \ldots$$

We shall let the energy of the electron in an outer level (n_o) be indicated by E_o and the energy of the electron in an inner level (n_i) be indicated by E_i. When the electron falls from the outer level to the inner level, $(E_o - E_i)$ is given off as a photon of light. According to Planck's equation, the energy of a photon is equal to $h\nu$. Therefore,

$$h\nu = E_o - E_i \qquad \text{(for } E_o \rightarrow E_i\text{)}$$

$$h\nu = \frac{(-2.180 \times 10^{-18}\text{J})}{n_o^2} - \frac{(-2.180 \times 10^{-18}\text{J})}{n_i^2}$$

$$h\nu = (2.180 \times 10^{-18}\text{J})\left(\frac{1}{n_i^2} - \frac{1}{n_o^2}\right)$$

Since $h = 6.626 \times 10^{-34}\text{J·s}$,

$$\nu = \left(\frac{2.180 \times 10^{-18}\text{J}}{6.626 \times 10^{-34}\text{J·s}}\right)\left(\frac{1}{n_i^2} - \frac{1}{n_o^2}\right)$$

$$\nu = (3.290 \times 10^{15}/\text{s})\left(\frac{1}{n_i^2} - \frac{1}{n_o^2}\right)$$

The lines produced by electron transitions to the $n = 2$ level from higher levels are described by the equation

$$\nu = (3.290 \times 10^{15}/\text{s})\left(\frac{1}{2^2} - \frac{1}{n_o^2}\right) \qquad n = 3, 4, 5, \ldots$$

This equation is the same as the equation derived by Balmer from experimental data.

The relationship between some of the electronic transitions of the hydrogen atom and the spectral lines is illustrated in Figure 5.11. Since electron transitions to the $n = 1$ level **(Lyman series)** release more energy than those to the $n = 2$ level (Balmer series), the wavelengths of the lines of the Lyman series are *shorter* than those of the Balmer series. The lines of the Lyman series occur in the ultraviolet region. On the other hand, the lines of the **Paschen series,**

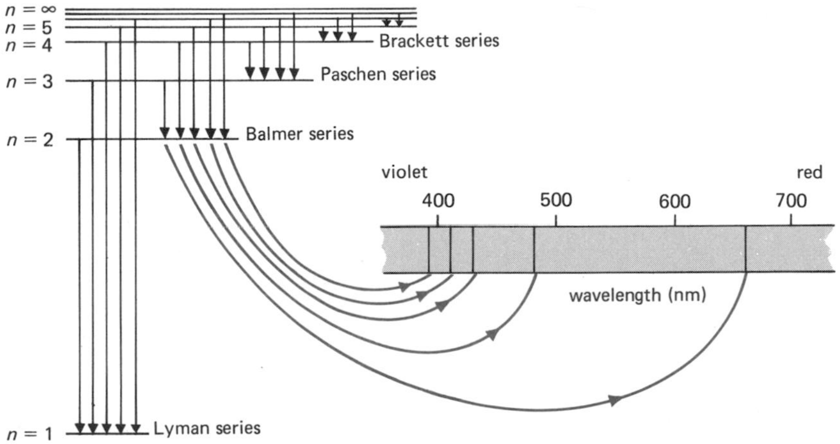

Figure 5.11 The relation between some electron transitions of the hydrogen atom and the spectral lines of the visible region.

which represent transitions to the $n = 3$ level, occur at wavelengths *longer* than the Balmer series. The Paschen lines appear in the infrared region.

The Bohr theory is highly successful in interpreting the spectrum of hydrogen. It fails, however, to explain the spectra of atoms that contain more than one electron.

Example 5.3 What are the frequency and wavelength of the line in the hydrogen spectrum that corresponds to an electron transition from the $n = 3$ level to the $n = 2$ level?

Solution

$$\nu = (3.290 \times 10^{15}/\text{s})\left(\frac{1}{n_i^2} - \frac{1}{n_o^2}\right)$$

$$= (3.290 \times 10^{15}/\text{s})\left(\frac{1}{2^2} - \frac{1}{3^2}\right)$$

$$= (3.290 \times 10^{15}/\text{s})\left(\frac{1}{4} - \frac{1}{9}\right)$$

$$= 0.4569 \times 10^{15}/\text{s} = 4.569 \times 10^{14}/\text{s}$$

Since, $\nu\lambda = c$

$$\lambda = \frac{c}{\nu}$$

$$= \frac{2.998 \times 10^{8}\ \text{m/s}}{4.569 \times 10^{14}/\text{s}}$$

$$= 6.562 \times 10^{-7}\ \text{m} = 656.2\ \text{nm}$$

Note that the conversion from meters to nanometers can be accomplished in the following way.

$$?\text{nm} = 6.562 \times 10^{-7}\ \text{m}\left(\frac{1\ \text{nm}}{10^{-9}\ \text{m}}\right) = 6.562 \times 10^{2}\ \text{nm} = 656.2\ \text{nm}$$

5.10 WAVE MECHANICS

Bohr regarded the electron as a charged particle in motion, and he assumed that the electron in a hydrogen atom can possess only certain definite quantities of energy. Other aspects of the theory, however, he derived from the laws of classical physics that pertain to the behavior of charged particles. It soon became evident that this approach was inadequate and that a new one was needed.

In order to predict the path of a moving body, we must know both its position and velocity at the same time. Werner Heisenberg's **uncertainty principle** (1926) states that it is impossible to determine simultaneously the exact *position* and exact *momentum* of a body as small as the electron. The more precisely we try to determine one of these values, the more uncertain we are of the other.

We see objects by noting interference with the light rays used to illuminate them. Radiation with an extremely short wavelength would be needed to locate an object as small as the electron. Such radiation has a high frequency ($\nu = c/\lambda$) and is therefore very energetic ($E = h\nu$). When it strikes the electron, the impact causes the direction of motion and speed of the electron to change. The attempt to locate the electron changes the momentum of the electron drastically.

Photons with longer wavelengths are less energetic and would have a smaller effect on the momentum of the electron. Because of their longer wavelengths, such photons, however, would not indicate the position of the electron very precisely. It appears therefore, that an exact description of the path of an electron in a Bohr orbit is not possible.

In the same way that light has both a wave and a particle-like character, matter has a dual nature. In 1924, Louis de Broglie proposed that electrons and other particles have wave properties. He showed that it is possible to assign a wavelength to a moving electron. In 1926, Erwin Schrödinger used de Broglie's relation to develop an equation that describes the electron in terms of its wave character.

The **Schrödinger equation** is the basis of **wave mechanics.** The equation is written in terms of a **wave function, ψ** (psi), for an electron. When the equation is solved for the electron in the hydrogen atom, a series of wave functions is obtained. Each wave function corresponds to a definite energy state for the electron and pertains to a region in which the electron may be found. The wave function of an electron is called an **orbital** to distinguish it from the orbit of Bohr.

The intensity of a wave is proportional to the square of its amplitude. The wave function, ψ, is an amplitude function. At any position in space, the value of ψ^2 for a very small volume is proportional to the electron charge density. The charge of the electron can be assumed to be spread out into a charge cloud by the rapid motion of the electron. The cloud is denser in some regions than others. The probability of finding the electron in a given region is proportional to the density of the charge cloud at that spot. The probability is high in a region

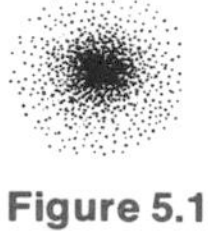

Figure 5.12 Cross section of the charge cloud of an electron in the $n = 1$ state of the hydrogen atom.

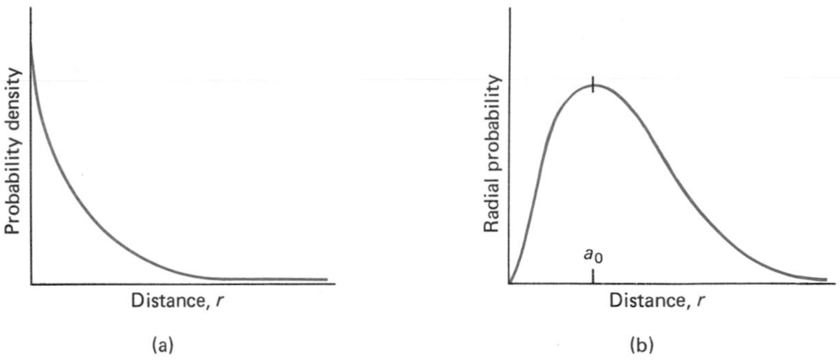

Figure 5.13 Probability curves for an electron in the $n = 1$ state of the hydrogen atom. (a) Probability of finding the electron per unit volume versus distance from the nucleus. (b) Probability of finding the electron at a given distance versus distance from the nucleus.

where the cloud is dense. This interpretation does not attempt to describe the path of the electron, it merely predicts where an electron is likely to be found.

For an electron in the $n = 1$ state of the hydrogen atom, the charge cloud has the greatest density near the nucleus and becomes thinner as the distance from the nucleus increases (Figure 5.12). More information about this probability distribution can be obtained from the curves of Figure 5.13. In curve (a), ψ^2 is plotted against distance from the nucleus. The probability of finding the electron in a small volume segment is greatest near the nucleus and approaches zero as the distance from the nucleus increases.

Curve (b) is a radial probability curve. The *total* probability of finding the electron at a given distance from the nucleus is plotted against distance. Imagine a group of very thin spherical shells arranged one after the other with the center of each shell at the center of the nucleus. What is the probability of finding the electron in each of these shells? The probability of finding the electron in a small volume segment is greatest near the nucleus. A shell close to the nucleus, however, contains fewer of these volume segments than one farther out. The radial probability takes both factors into account.

The curve shows a maximum at a distance a_0. The total probability of finding the electron at all points of distance r from the nucleus is greatest when r is equal to a_0. This value is the same as the value determined by the Bohr theory for the radius of the $n = 1$ shell. In the Bohr theory, a_0 is the distance at which the electron is *always* found in the $n = 1$ shell. In wave mechanics, a_0 is the distance at which the electron is likely to be found most often.

It is not possible to draw a shape that bounds a region in which the probability of finding the electron is 100%. A surface can be drawn, however, that connects points of equal probability and that encloses a volume in which the probability of finding the electron is high (for example, 90%). Such a representation, called a **boundary surface diagram,** is shown in Figure 5.14 for the electron in $n = 1$ state of the hydrogen atom.

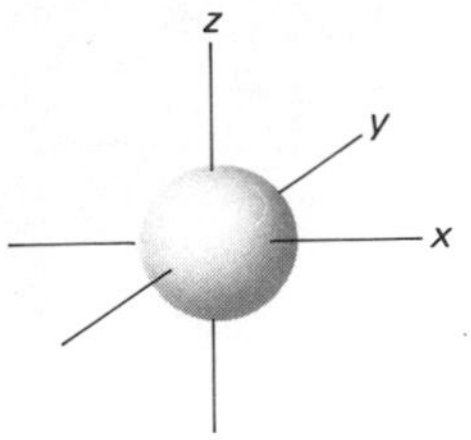

Figure 5.14 Boundary surface representation of an electron in the $n = 1$ state of the hydrogen atom. Volume encloses 90% of the electron density. Nucleus is at the origin.

5.11 QUANTUM NUMBERS

In wave mechanics, the electron distribution of an atom containing a number of electrons is divided into *shells*. The shells, in turn, are thought to consist of one or more *subshells,* and the subshells are assumed to be composed of one or more *orbitals,* which the *electrons* occupy. Each electron of an atom is identified by a combination of four quantum numbers, which loosely indicate shell, subshell, orbital, and electron.

The **principal quantum number,** n, corresponds approximately to the n introduced by Bohr. It identifies the shell, or level, to which the electron belongs. These shells are regions where the probability of finding an electron is high. The value of n is a positive integer.

$$n = 1, 2, 3, \ldots$$

The larger the value of n, the farther the shell is from the nucleus.

Each shell consists of one or more subshells, or sublevels. The number of subshells in a principal shell is equal to the value of n. There is only one subshell in the $n = 1$ shell, there are two in the $n = 2$ shell, three in the $n = 3$ shell, and so on. Each subshell in a shell is assigned a **subsidiary quantum number,** l. The values of l for the sublevels of a shell are determined by the shell's value of n. There is one value of l for each term in the series

$$l = 0, 1, 2, 3, \ldots, (n - 1)$$

When $n = 1$, the only value of l is 0, and there is only one subshell. When $n = 2$, there are two subshells that have l values of 0 and 1, respectively. When $n = 3$, the three subshells have l values of 0, 1, and 2.

Other symbols are used at times to denote subshells. A letter is used to represent each value of l in the following way.

$$l = 0, 1, 2, 3, 4, 5, \ldots$$
$$\text{notation} = s, p, d, f, g, h, \ldots$$

The first four notations are the initial letters of adjectives formerly used to identify spectral lines: sharp, principal, diffuse, and fundamental. For l values higher than 3, the letters proceed alphabetically—*g, h, i,* and so on. Combining the principal quantum number with one of these letters gives a convenient way to designate a subshell. The subshell with $n = 2$ and $l = 0$ is called the 2*s* subshell. The subshell with $n = 2$ and $l = 1$ is called the 2*p* subshell. Table 5.3 contains a summary of subshell notation for the first four shells.

Each subshell consists of one or more orbitals. The number of orbitals in a subshell is given by the equation

$$\text{number of orbitals} = 2l + 1$$

Therefore,

$$\begin{aligned} \text{notation} &= s, p, d, f, g, \ldots \\ l &= 0, 1, 2, 3, 4, \ldots \\ \text{number of orbitals} &= 1, 3, 5, 7, 9, \ldots \end{aligned}$$

An *s* subshell consists of one orbital, a *p* subshell consists of three orbitals, a *d* subshell consists of five orbitals, and so on.

Each orbital within a given subshell is identified by a **magnetic orbital quantum number,** m_l. For any subshell, the values of m_l are given by the terms in the series

$$m_l = +l, +(l - 1), \ldots, 0, \ldots, -(l - 1), -l$$

Thus, for $l = 0$, the only permitted value of m_l is 0 (one *s* orbital). For $l = 1$, m_l can be +1, 0, and −1 (three *p* orbitals). For $l = 2$, m_l can be +2, +1, 0, −1, and −2 (five *d* orbitals). Notice that the values of m_l are derived from *l*, and that the values of *l* are derived from *n*.

Each orbital in an atom, therefore, is identified by a set of values for *n*, *l*, and m_l. An orbital described by the quantum numbers $n = 2$, $l = 1$, and $m_l = 0$, is an orbital in the *p* subshell of the second shell—a 2*p* orbital. The quantum numbers for the orbitals of the first four shells are given in Table 5.4.

Table 5.3 Subshell notations

n	*l*	SPECTROSCOPIC NOTATIONS
1	0	1*s*
2	0	2*s*
2	1	2*p*
3	0	3*s*
3	1	3*p*
3	2	3*d*
4	0	4*s*
4	1	4*p*
4	2	4*d*
4	3	4*f*

Table 5.4 The orbitals of the first four shells

SHELL n	SUBSHELL l	ORBITAL m_l	SUBSHELL NOTATION	NUMBER OF ORBITALS PER SUBSHELL
1	0	0	1*s*	1
2	0	0	2*s*	1
	1	+1, 0, −1	2*p*	3
3	0	0	3*s*	1
	1	+1, 0, −1	3*p*	3
	2	+2, +1, 0, −1, −2	3*d*	5
4	0	0	4*s*	1
	1	+1, 0, −1	4*p*	3
	2	+2, +1, 0, −1, −2	4*d*	5
	3	+3, +2, +1, 0, −1, −2, −3	4*f*	7

In Section 5.10, the electron charge cloud of the 1*s* orbital was discussed (see Figures 5.12, 5.13, and 5.14). Each of the three parts of Figure 5.15 pertains to the charge cloud of the 2*s* orbital. Figure 5.15(a) is a radial probability curve for the 2*s* orbital. The curve shows two places where the probability of finding an electron is relatively high: one is close to the nucleus and the other is farther out. In the charge cloud of the 2*s* orbital, therefore, there are two regions where the electron density is relatively high. They appear in Figure 5.15(b), which is a cross section of the electron density of the orbital. The boundary surface diagram of the 2*s* orbital, however, appears the same as that for the 1*s* orbital, except for size. Compare Figure 5.15(c) with Figure 5.14. All *s* orbitals are spherical.

Boundary surface diagrams for the three 2*p* orbitals are shown in Figure 5.16. In these diagrams, the nucleus is at the origin. The electron density of a *p* orbital is not spherical. Instead, each *p* orbital consists of two sections, called lobes, that are on either side of a plane that passes through the nucleus. The shapes of the three orbitals are identical. They differ, however, in the way

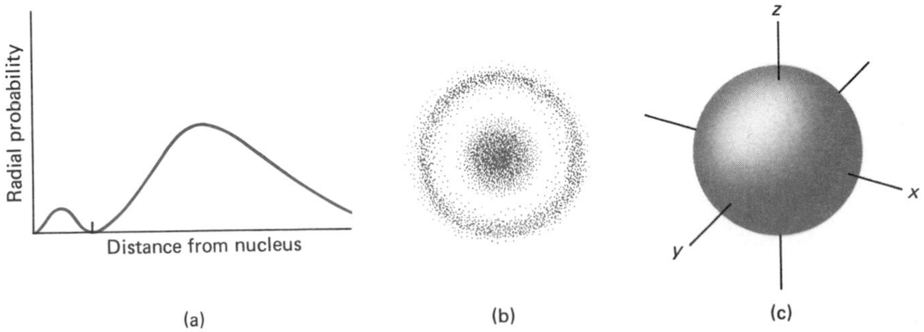

Figure 5.15 Diagrams for the 2*s* orbital. (a) Radial probability distribution; (b) cross section of electron charge cloud; (c) boundary surface diagram.

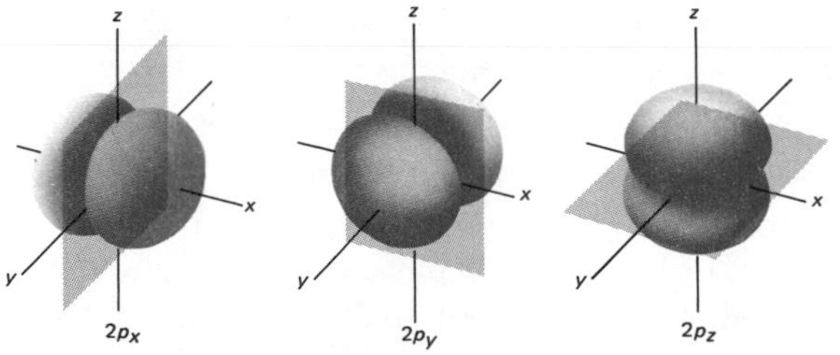

Figure 5.16 Boundary surface diagrams for the 2*p* orbitals.

that the lobes are directed. Since the lobes may be considered to lie along the *x*, *y*, or *z* axis, they are given the designations $2p_x$, $2p_y$, and $2p_z$.

A magnetic field has no effect on the energy of an *s* electron since an *s* orbital is spherical. No matter how we turn a sphere, it always looks the same with regard to a fixed frame of reference. A spherical orbital (an *s* orbital) in a magnetic field presents only one aspect toward the lines of force.

The *p* orbitals are not spherical. Each *p* sublevel consists of three orbitals that differ in their orientation. Each *p* orbital presents a different aspect toward the lines of force of a magnetic field. These *p* orbitals are identical in terms of energy. In the absence of the magnetic field, no distinction between electrons that occupy different *p* orbitals is noted. If spectral studies are run using a magnetic field, however, some of the lines of the spectrum are split into several lines. This effect, called the **Zeeman effect,** disappears when the magnetic field is removed. The m_l value of an orbital is associated with the orientation of the orbital in relation to a direction fixed by a magnetic field.

Boundary surface diagrams of the five 3*d* orbitals are shown in Figure 5.17. The shape of the d_{z^2} orbital is different from the others, but they are all equivalent in terms of energy.

The fourth quantum number is the **magnetic spin quantum number,** m_s. An electron in an orbital has properties that can be explained by assuming that it spins on its own axis. Since only two directions of spin are possible, there are only two values of m_s: $+\frac{1}{2}$ and $-\frac{1}{2}$. At times it is convenient to indicate electron spins by arrows; $\uparrow$ is $+\frac{1}{2}$ and $\downarrow$ is $-\frac{1}{2}$. A spinning charge generates a magnetic field. Each electron, therefore, has a magnetic moment associated with it because of its spin. The magnetic moments of two electrons that spin in opposite directions cancel each other. Each orbital can hold two electrons with opposed spins.

Each electron, therefore, may be described by a set of four quantum numbers.

1. *n* gives the shell and the relative average distance of the electron from the nucleus.
2. *l* gives the subshell and the shape of the orbital for the electron. Each orbital of a given subshell is equivalent in energy.

3. m_l designates the orientation of the orbital.
4. m_s refers to the spin of the electron.

The **exclusion principle** of Wolfgang Pauli states that no two electrons in the same atom may have identical sets of all four quantum numbers. Even if two electrons have the same values for n, l, and m_l they will differ in their m_s values. This situation indicates that the two electrons are paired in a single orbital.

Quantum numbers for the orbitals of the first four shells are given in Table 5.4. The quantum number sets for the electrons in these orbitals can be obtained by indicating an m_s value (either $+\frac{1}{2}$ or $-\frac{1}{2}$) with the set of n, l, and m_s values that denote the orbital.

The maximum number of electrons that a shell can hold is given by $2n^2$. Each orbital can hold two electrons. The maximum number of electrons in a shell, therefore, is equal to twice the number of orbitals in the shell. The maximum number of electrons for a subshell can be calculated by multiplying the number of orbitals in the subshell by 2. The capacities of the shells and subshells from $n = 1$ to $n = 4$ are given in Table 5.5.

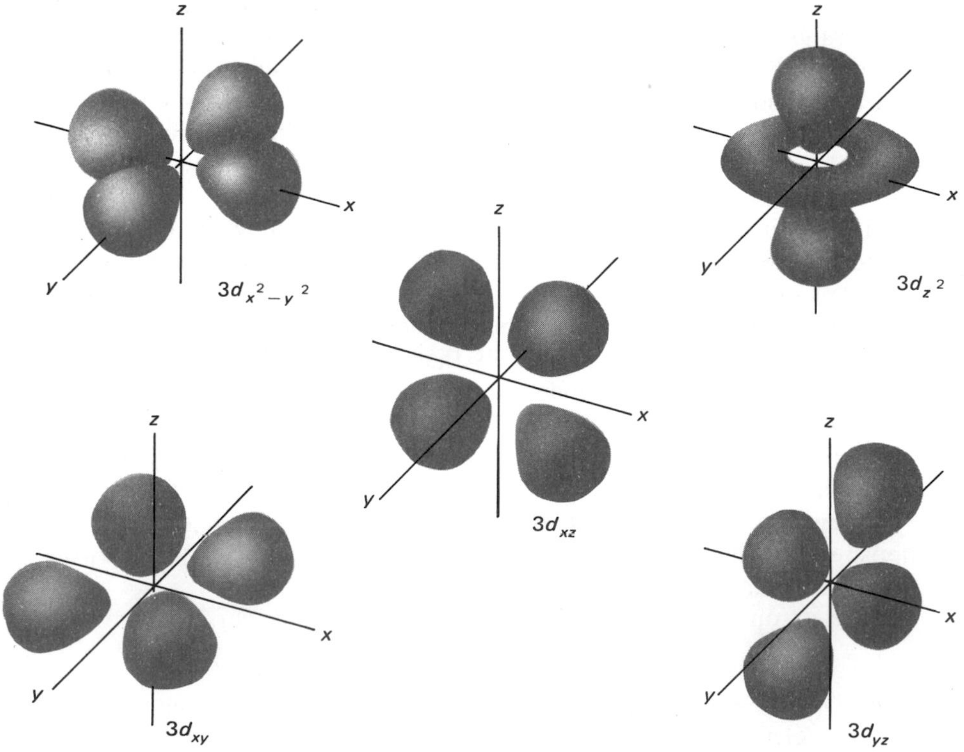

Figure 5.17 Boundary surface diagrams for the 3*d* orbitals.

Table 5.5 Maximum number of electrons for the subshells of the first four shells

SUBSHELL NOTATION	ORBITALS PER SUBSHELL	ELECTRONS PER SUBSHELL	ELECTRONS PER SHELL ($2n^2$)
1*s*	1	2	2
2*s*	1	2	8
2*p*	3	6	
3*s*	1	2	18
3*p*	3	6	
3*d*	5	10	
4*s*	1	2	32
4*p*	3	6	
4*d*	5	10	
4*f*	7	14	

5.12 ORBITAL FILLING AND HUND'S RULE

The way electrons are arranged in an atom is called the **electron configuration** of the atom. Several ways to indicate the electron configuration of the ground state of an atom are given in Table 5.6. For the first 18 elements, the electron configuration can be derived by assuming that the electrons occupy the shells in order of increasing value of *n*, and within a given shell in order of increasing value of *l*. The situation for elements with atomic numbers higher than 18 is slightly more complicated and is discussed in Section 5.13.

Each electron in an atom is described by a set of values for the four quantum numbers. In Table 5.6, a set is given for each electron of each atom listed. In the list, positive values of both m_s and m_l are arbitrarily used before negative values.

The single electron of the hydrogen atom is assigned the quantum set $(1, 0, 0, +\frac{1}{2})$. Since $n = 1$ and $l = 0$, the electron occupies a 1*s* orbital. In the orbital diagram shown in Table 5.6, the electron is indicated by an arrow (which denotes electron spin) in a blank for the 1*s* orbital. The electronic notation for the hydrogen atom is $1s^1$. The superscript indicates the number of electrons in the designated orbital—in this case, one electron in the 1*s* orbital.

There are two sets of quantum numbers listed for the helium atom since He has two electrons. The sets differ only in the value of m_s. The 1*s* orbital contains two electrons that have opposite spins, which is shown in the orbital diagram. The electronic notation is $1s^2$. Notice that the $n = 1$ shell of helium is filled.

The lithium and the beryllium atoms have electrons in the 2*s* orbital ($n = 2$, $l = 0$) as well as pairs of electrons in the 1*s* orbital. The boron atom has five electrons, and five sets of quantum numbers are listed for it. The first

Table 5.6 Electron configurations of the ground states of the first six elements

	QUANTUM NUMBERS OF ELECTRONS				ORBITAL DIAGRAM			ELECTRONIC NOTATION
	n	l	m_l	m_s	1s	2s	2p	
$_1$H	1	0	0	$+\frac{1}{2}$	↑	_	_ _ _	$1s^1$
$_2$He	1	0	0	$+\frac{1}{2}$	↑↓	_	_ _ _	$1s^2$
	1	0	0	$-\frac{1}{2}$				
$_3$Li	1	0	0	$+\frac{1}{2}$	↑↓	↑	_ _ _	$1s^2\ 2s^1$
	1	0	0	$-\frac{1}{2}$				
	2	0	0	$+\frac{1}{2}$				
$_4$Be	1	0	0	$+\frac{1}{2}$	↑↓	↑↓	_ _ _	$1s^2\ 2s^2$
	1	0	0	$-\frac{1}{2}$				
	2	0	0	$+\frac{1}{2}$				
	2	0	0	$-\frac{1}{2}$				
$_5$B	1	0	0	$+\frac{1}{2}$	↑↓	↑↓	↑ _ _	$1s^2\ 2s^2\ 2p^1$
	1	0	0	$-\frac{1}{2}$				
	2	0	0	$+\frac{1}{2}$				
	2	0	0	$-\frac{1}{2}$				
	2	1	+1	$+\frac{1}{2}$				
$_6$C	1	0	0	$+\frac{1}{2}$	↑↓	↑↓	↑ ↑ _	$1s^2\ 2s^2\ 2p^2$
	1	0	0	$-\frac{1}{2}$				
	2	0	0	$+\frac{1}{2}$				
	2	0	0	$-\frac{1}{2}$				
	2	1	+1	$+\frac{1}{2}$				
	2	1	0	$+\frac{1}{2}$				

two sets $(1, 0, 0, +\frac{1}{2})$ and $(1, 0, 0, -\frac{1}{2})$ indicate that the 1*s* orbital contains a pair of electrons. The next two sets $(2, 0, 0, +\frac{1}{2})$ and $(2, 0, 0, -\frac{1}{2})$ pertain to two electrons that occupy the 2*s* orbital. The last set $(2, 1, +1, +\frac{1}{2})$ shows that one electron is present in a 2*p* orbital ($n = 2$, $l = 1$).

There are three 2*p* orbitals of equal energy. We have arbitrarily assumed that the 2*p* orbitals are filled in the order of m_l value +1, 0, −1 (positive to negative). The m_l value given in the set, therefore, is +1. The orbital diagram shows an unpaired electron in one of the 2*p* orbitals, but no m_l distinction is made in a diagram of this type. The electronic notation is $1s^2\ 2s^2\ 2p^1$.

A question arises concerning the configuration of an atom of the sixth element, carbon. Does the last electron belong in the 2*p* orbital that already holds one electron, or does it belong in another 2*p* orbital? **Hund's rule of maximum multiplicity** provides an answer. Hund's rule states that the electrons are distributed among the orbitals in a subshell in a way that gives the maximum number of unpaired electrons with parallel spins. The term *parallel* *spins* means that all the unpaired electrons spin in the same direction, and all the m_s values of these electrons have the same sign.

In carbon, therefore, each of the $2p$ electrons occupies an orbital separately rather than pairing with the other in a single orbital. The m_l values in the last two sets of quantum numbers are $+1$ and 0, which show that different orbitals are employed. Both of the m_s values of the last two sets are $+\frac{1}{2}$, which shows that the spins of these two electrons are parallel. The two unpaired electrons are clearly shown in the orbital diagram. The distinction, however, is not apparent in the electronic notation, $1s^2\ 2s^2\ 2p^2$.

Notice that all the superscripts in this notation are even numbers. Such a situation does *not* mean that all the electrons are paired in orbitals. The unpaired electrons can be shown by using a separate designation for each orbital. The electronic configuration of carbon would be shown as $1s^2\ 2s^2\ 2p^1\ 2p^1$. This device is not necessary if we remember that there are three $2p$ orbitals and that Hund's rule states that the two electrons will occupy separate orbitals.

Figure 5.18 illustrates Hund's rule. The orbitals are represented by circles. The order of filling is shown by numbers entered in these circles. Since electrons are negatively charged and repel each other, they spread out and occupy the $2p$ orbitals singly before they begin to pair. The orbitals of a d subshell and of an f subshell are filled in the same way. Hund's rule has been confirmed by magnetic measurements.

The number of unpaired electrons in an atom, ion, or molecule can be determined by magnetic measurements. **Paramagnetic substances** are drawn into a magnetic field. Substances that contain unpaired electrons are paramagnetic. The magnetic moment depends upon the number of unpaired electrons present. Two effects contribute to the paramagnetism of an atom: the spin of the unpaired electrons and the orbital motion of these electrons. The effect of electron spin is the greater of the two, and in many cases, the effect of orbital motion is negligible.

Diamagnetic substances are weakly repelled by a magnetic field. A material is diamagnetic if all of its electrons are paired. Diamagnetism is a property

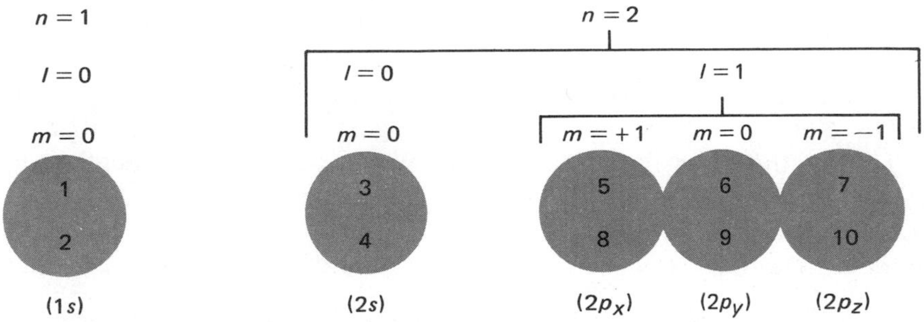

Figure 5.18 Order in which the orbitals of the $n = 1$ and $n = 2$ shells are filled.

Table 5.7 Electron configurations of the outer shells of the elements of the first three periods

I A							0
$_1H$ $1s^1$	II A	III A	IV A	V A	VI A	VII A	$_2He$ $1s^2$
$_3Li$ $2s^1$	$_4Be$ $2s^2$	$_5B$ $2s^2\ 2p^1$	$_6C$ $2s^2\ 2p^2$	$_7N$ $2s^2\ 2p^3$	$_8O$ $2s^2\ 2p^4$	$_9F$ $2s^2\ 2p^5$	$_{10}Ne$ $2s^2\ 2p^6$
$_{11}Na$ $3s^1$	$_{12}Mg$ $3s^2$	$_{13}Al$ $3s^2\ 3p^1$	$_{14}Si$ $3s^2\ 3p^2$	$_{15}P$ $3s^2\ 3p^3$	$_{16}S$ $3s^2\ 3p^4$	$_{17}Cl$ $3s^2\ 3p^5$	$_{18}Ar$ $3s^2\ 3p^6$

of all matter, but it is obscured by the stronger paramagnetic effect if unpaired electrons are present.*

In Table 5.7, the electronic notations for the *outer shells* of atoms of the elements of the first three periods are shown. In each atom that contains electrons in inner shells, these shells are complete. Notice the similarity between the configurations of elements of the same group. All of the elements of group I A, for example, have one electron in an *s* orbital in the outer shell. The similarity in electron configuration between elements of a given group accounts for their similarities in properties.

The outermost shells of these atoms are called **valence shells,** and electrons in them are called **valence electrons.** All the electrons in the valence shell, regardless of subshell, are counted as valence electrons. For elements that are members of A groups, the number of valence electrons is the same as the group number. The noble gases (group O) have eight electrons in their valence shell with the exception of helium, which has two.

5.13 DERIVATION OF ELECTRON CONFIGURATIONS

The data of Tables 5.6 and 5.7 indicate a way that the electron configurations of atoms can be derived. We start with the hydrogen atom, which has one electron in a $1s$ orbital. By adding one electron, we get the configuration of an atom of the next element, helium (which is $1s^2$). In this manner, we go from element to element until we derive the configuration of the atom that we desire. This method was first suggested by Wolfgang Pauli and is called the **aufbau method** (*aufbau* means "building up" in German).

The electron added in going from one element to the next is called the **differentiating electron.** It makes the configuration of an atom different from

* Ferromagnetic substances, such as iron, are strongly attracted into a magnetic field. Ferromagnetism is a form of paramagnetism that is shown by only a few solid substances.

that of the atom that precedes it. The differentiating electron is added in each step to the orbital of lowest energy available to it.

All the orbitals of a given subshell have equivalent energies. The energy of any one of the $3p$ orbitals, for example, is the same as the energy of either of the other two $3p$ orbitals. All five $3d$ orbitals are alike in terms of energy. Orbitals that belong to different subshells of the same shell, however, have different energies. For a given value of n, the energies increase in the order $s < p < d < f$. In the $n = 3$ shell, for example, the $3s$ orbital has the lowest energy, a $3p$ orbital has an intermediate energy, and a $3d$ orbital has the highest energy. At times, the energies of orbitals from different shells overlap. In some atoms, for example, the $4s$ orbital has a lower energy than a $3d$ orbital.

There is no standard order of orbitals, based on energy, that pertains to all the elements. In the *hypothetical* aufbau process, the character of the atom changes as protons and neutrons are added to the nucleus and more and more electrons are introduced. Fortunately, the orbital energy order varies from element to element in a slow and regular manner. As a result, the aufbau order shown in Figure 5.19 can be derived.

This order pertains *only* to the orbital position that the differentiating electron takes in the aufbau process. Electron configurations can be derived from the diagram shown in Figure 5.19 by successively filling the orbitals starting at the bottom of the chart and proceeding upward. Remember that

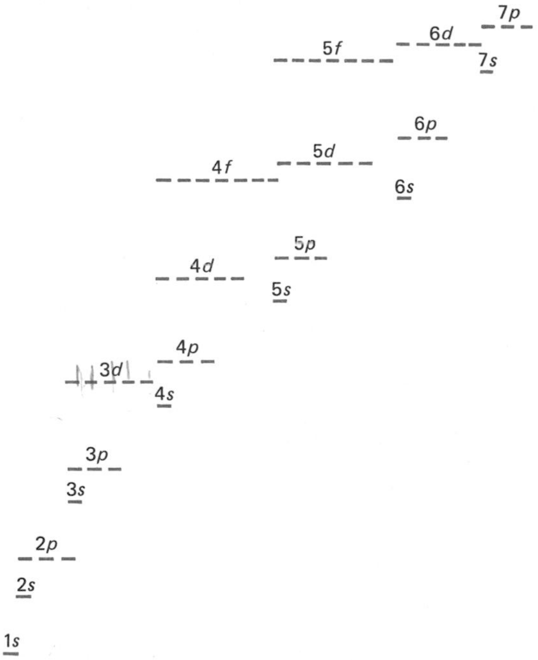

Figure 5.19 Aufbau order of atomic orbitals.

there are three orbitals in a *p* sublevel, five in a *d*, and seven in an *f*. A given sublevel is filled before electrons are added to the next sublevel.

The periodic table can be used to derive electron configurations. In Figure 5.20, the type of differentiating electron is related to the position of the element in the periodic chart. Notice that the table can be divided into an "*s* block," a "*p* block," a "*d* block," and an "*f* block."

The first period consists of hydrogen ($1s^1$) and helium ($1s^2$). The second period begins with lithium ($1s^2\ 2s^1$) and beryllium ($1s^2\ 2s^2$), for which electrons are added to the $2s$ orbital. For the six elements that complete this period—boron ($1s^2\ 2s^2\ 2p^1$) to the noble gas neon ($1s^2\ 2s^2\ 2p^6$)—electrons are gradually added to the three $2p$ orbitals.

The pattern of the second period is repeated in the third. The two "*s* block" elements are sodium ($1s^2\ 2s^2\ 2p^6\ 3s^1$) and magnesium ($1s^2\ 2s^2\ 2p^6\ 3s^2$). The six "*p* block" elements go from aluminum ($1s^2\ 2s^2\ 2p^6\ 3s^2\ 3p^1$) to the noble gas argon ($1s^2\ 2s^2\ 2p^6\ 3s^2\ 3p^6$).

In the discussion of the configurations of the remaining elements, only the outer orbitals will be indicated. The first overlap of orbital energies is observed with potassium ($Z = 19$), the first element of the fourth period. The configuration of potassium is . . . $3s^2\ 3p^6\ 4s^1$ despite the fact that $3d$ orbitals are vacant. In like manner, calcium ($Z = 20$) has the configuration . . . $3s^2\ 3p^6\ 4s^2$. Notice that potassium and calcium are "*s* block" elements.

With the next element, scandium ($Z = 21$), the $3d$ subshell comes into use (. . . $3s^2\ 3p^6\ 3d^1\ 4s^2$). In the series from scandium to zinc, the $3d$ subshell is gradually filled. The configuration of zinc ($Z = 30$) is . . . $3s^2\ 3p^6\ 3d^{10}\ 4s^2$. These "*d* block" elements are called **transition elements.** They are said to exhibit **inner building** since the last electron is added to the shell ($n = 3$) next

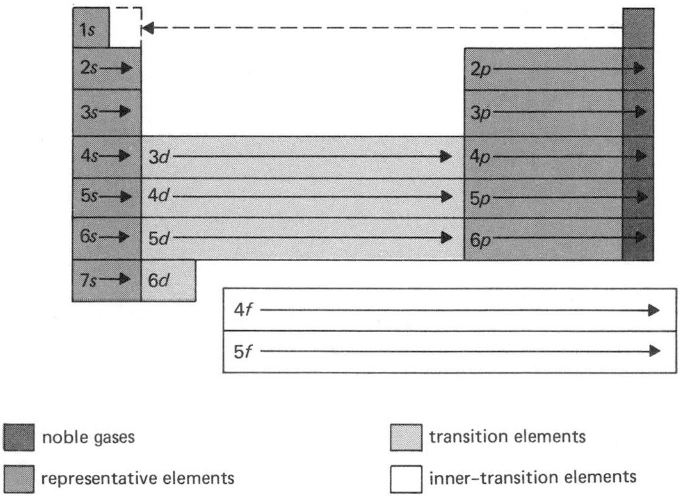

Figure 5.20 Type of differentiating electron related to position of the element in the periodic table.

to the outermost shell ($n = 4$). The elements from number 21 to 30, which belong to B families, are said to form the first transition series.

With element 31, gallium (. . . $3s^2\ 3p^6\ 3d^{10}\ 4s^2\ 4p^1$), the $4p$ subshell begins to be filled. The fourth period ends with krypton ($Z = 36$, . . . $3s^2\ 3p^6\ 3d^{10}\ 4s^2\ 4p^6$).

The fifth period starts with rubidium ($Z = 37$, . . . $4s^2\ 4p^6\ 5s^1$) and strontium ($Z = 38$, . . . $4s^2\ 4p^6\ 5s^2$) with electrons being added to the $5s$ subshell even though both $4d$ and $4f$ orbitals are vacant. Notice that these two elements are "s block" elements. A second transition series follows, in which electrons are added to the $4d$ subshell. It starts with yttrium ($Z = 39$, . . . $4s^2\ 4p^6\ 4d^1\ 5s^2$) and ends with cadmium ($Z = 48$, . . . $4s^2\ 4p^6\ 4d^{10}\ 5s^2$). The fifth period ends with the series from indium to xenon with electrons being added to the $5p$ subshell. Xenon ($Z = 54$) has the configuration . . . $4s^2\ 4p^6\ 4d^{10}\ 5s^2\ 5p^6$. The $4f$ subshell is still vacant at the end of this period.

The sixth period is far more complicated as far as orbital overlap is concerned. The first two elements, cesium ($Z = 55$, . . . $4d^{10}\ 5s^2\ 5p^6\ 6s^1$) and barium ($Z = 56$, . . . $4d^{10}\ 5s^2\ 5p^6\ 6s^2$) have electrons in the $6s$ subshell. Here, we come upon a complication of the aufbau procedure. The $4f$ and $5d$ subshells are close in energy. The next electron (for lanthanum, $Z = 57$) is added to the $5d$ subshell (thus lanthanum is a transition element), but the next electron (for cerium, $Z = 58$) is added to the $4f$ subshell and the electron added for lanthanum falls back into the $4f$ subshell. For La, the configuration is . . . $4d^{10}\ 5s^2\ 5p^6\ 5d^1\ 6s^2$. For Ce, the configuration is . . . $4d^{10}\ 4f^2\ 5s^2\ 5p^6\ 6s^2$. For the elements 58 to 70 (cerium to ytterbium), electrons are added to the $4f$ subshell.

These elements are called **inner-transition elements.** For elements of this type, electron addition occurs in the third subshell ($4f$) from the outermost subshell ($6s$). After the $4f$ subshell has been filled, the next electron is added to the $5d$ subshell. Thus, for lutetium ($Z = 71$) the configuration is . . . $4d^{10}\ 4f^{14}\ 5s^2\ 5p^6\ 5d^1\ 6s^2$. This third transition series is completed and the $5d$ subshell filled with element 80, mercury (. . . $4d^{10}\ 4f^{14}\ 5s^2\ 5p^6\ 5d^{10}\ 6s^2$). The period ends with the filling of the $6p$ sublevel in elements 81 to 86.

The seventh period is incomplete and includes many man-made elements. This period follows the pattern established by the sixth period: elements 87 and 88 have electrons added to the $7s$ sublevel, for element 89 an electron is added to the $6d$ sublevel, elements 90 to 103 constitute a second inner-transition series and exhibit an electron buildup of the $5f$ sublevel, and the transition elements 104, 105, and 106 (as yet unnamed) have electrons added to the $6d$ sublevel. The actual electron configurations of some atoms deviate slightly from the configurations predicted by the aufbau method (Section 5.14).

To determine the electronic configuration of any element, we start with hydrogen and on the basis of the periodic table account for every electron added until the desired element is reached. Thus, for tungsten ($Z = 74$), the first period gives $1s^2$; the second period, $2s^2\ 2p^6$; the third, $3s^2\ 3p^6$; the fourth,

$4s^2$ $3d^{10}$ $4p^6$; and the fifth, $5s^2$ $4d^{10}$ $5p^6$. The sixth period—in which tungsten is found—starts with $6s^2$ for cesium and barium, adds $4f^{14}$ for the inner-transition series, and concludes with $5d^4$, since tungsten is the fourth element in the $5d$ transition series. Rearranging these terms in sequence gives: $1s^2$ $2s^2$ $2p^6$ $3s^2$ $3p^6$ $3d^{10}$ $4s^2$ $4p^6$ $4d^{10}$ $4f^{14}$ $5s^2$ $5p^6$ $5d^4$ $6s^2$.

The aufbau order cannot be used to interpret processes that involve the loss of electrons (ionizations). The configuration of the iron atom, Fe, is $1s^2$ $2s^2$ $2p^6$ $3s^2$ $3p^6$ $3d^6$ $4s^2$, and that of the Fe^{2+} ion is $1s^2$ $2s^2$ $2p^6$ $3s^2$ $3p^6$ $3d^6$. Ionization, therefore, results in the loss of the $4s$ electrons even though $3d$ electrons are the last added by the aufbau method. The Fe atom has 26 protons in the nucleus and 26 electrons. The Fe^{2+} ion has 26 protons in the nucleus but only 24 electrons. The order of orbital energies is different in the atom and the ion. In general, the first electrons lost in an ionization are those with the highest value of n and l. Electron notations, therefore, should be written by increasing value of n and not by the hypothetical order of filling.

5.14 HALF-FILLED AND FILLED SUBSHELLS

In Table 5.8 the correct electronic configurations of the elements are listed. The configurations predicted by the aufbau procedure are confirmed by spectral and magnetic studies for most elements. There are a few, however, that exhibit slight variations from the standard pattern. In certain instances, it is possible to explain these variations on the basis of the stability of a filled or half-filled subshell.

The predicted configuration for the $3d$ and $4s$ subshells in the chromium atom is $3d^4$ $4s^2$, whereas the experimentally derived configuration is $3d^5$ $4s^1$. Presumably the stability gained by having one unpaired electron in each of the five $3d$ orbitals (a half-filled subshell) accounts for the fact that the $3d^5$ $4s^1$ configuration is the one observed. The existence of a half-filled subshell also accounts for the fact that the configuration for the $4d$ and $5s$ subshells of molybdenum is $4d^5$ $5s^1$ rather than the predicted $4d^4$ $5s^2$.

For copper, the predicted configuration for the last two subshells is $3d^9$ $4s^2$, whereas the accepted structure is $3d^{10}$ $4s^1$. The explanation for this deviation lies in the stability of the $3d^{10}$ $4s^1$ configuration that results from the completed $3d$ subshell. Silver and gold also have configurations with completely filled d subshells instead of the $(n-1)d^9$ ns^2 configurations predicted. In the case of palladium, two electrons are involved—the only case with a difference of more than one electron. The predicted configuration for the last two subshells of palladium is $4d^8$ $5s^2$; the observed configuration is $4d^{10}$ $5s^0$.

Other types of deviations are observed, particularly among elements with high atomic numbers. For our purposes these exceptions are not important. In general, the chemistry of the elements is satisfactorily explained on the basis of the predicted configurations.

Table 5.8 Electronic configurations of the elements

ELEMENT	Z	1s	2s	2p	3s	3p	3d	4s	4p	4d	4f	5s	5p	5d	5f	6s	6p	6d	7s
H	1	1																	
He	2	2																	
Li	3	2	1																
Be	4	2	2																
B	5	2	2	1															
C	6	2	2	2															
N	7	2	2	3															
O	8	2	2	4															
F	9	2	2	5															
Ne	10	2	2	6															
Na	11	2	2	6	1														
Mg	12	2	2	6	2														
Al	13	2	2	6	2	1													
Si	14	2	2	6	2	2													
P	15	2	2	6	2	3													
S	16	2	2	6	2	4													
Cl	17	2	2	6	2	5													
Ar	18	2	2	6	2	6													
K	19	2	2	6	2	6		1											
Ca	20	2	2	6	2	6		2											
Sc	21	2	2	6	2	6	1	2											
Ti	22	2	2	6	2	6	2	2											
V	23	2	2	6	2	6	3	2											
Cr	24	2	2	6	2	6	5	1											
Mn	25	2	2	6	2	6	5	2											
Fe	26	2	2	6	2	6	6	2											
Co	27	2	2	6	2	6	7	2											
Ni	28	2	2	6	2	6	8	2											
Cu	29	2	2	6	2	6	10	1											
Zn	30	2	2	6	2	6	10	2											
Ga	31	2	2	6	2	6	10	2	1										
Ge	32	2	2	6	2	6	10	2	2										
As	33	2	2	6	2	6	10	2	3										
Se	34	2	2	6	2	6	10	2	4										
Br	35	2	2	6	2	6	10	2	5										
Kr	36	2	2	6	2	6	10	2	6										
Rb	37	2	2	6	2	6	10	2	6			1							
Sr	38	2	2	6	2	6	10	2	6			2							
Y	39	2	2	6	2	6	10	2	6	1		2							
Zr	40	2	2	6	2	6	10	2	6	2		2							
Nb	41	2	2	6	2	6	10	2	6	4		1							
Mo	42	2	2	6	2	6	10	2	6	5		1							
Tc	43	2	2	6	2	6	10	2	6	6		1							
Ru	44	2	2	6	2	6	10	2	6	7		1							
Rh	45	2	2	6	2	6	10	2	6	8		1							
Pd	46	2	2	6	2	6	10	2	6	10									
Ag	47	2	2	6	2	6	10	2	6	10		1							

Table 5.8 Electronic configurations of the elements (Continued)

ELEMENT	*Z*	1*s*	2*s*	2*p*	3*s*	3*p*	3*d*	4*s*	4*p*	4*d*	4*f*	5*s*	5*p*	5*d*	5*f*	6*s*	6*p*	6*d*	7*s*
Cd	48	2	2	6	2	6	10	2	6	10		2							
In	49	2	2	6	2	6	10	2	6	10		2	1						
Sn	50	2	2	6	2	6	10	2	6	10		2	2						
Sb	51	2	2	6	2	6	10	2	6	10		2	3						
Te	52	2	2	6	2	6	10	2	6	10		2	4						
I	53	2	2	6	2	6	10	2	6	10		2	5						
Xe	54	2	2	6	2	6	10	2	6	10		2	6						
Cs	55	2	2	6	2	6	10	2	6	10		2	6			1			
Ba	56	2	2	6	2	6	10	2	6	10		2	6			2			
La	57	2	2	6	2	6	10	2	6	10		2	6	1		2			
Ce	58	2	2	6	2	6	10	2	6	10	2	2	6			2			
Pr	59	2	2	6	2	6	10	2	6	10	3	2	6			2			
Nd	60	2	2	6	2	6	10	2	6	10	4	2	6			2			
Pm	61	2	2	6	2	6	10	2	6	10	5	2	6			2			
Sm	62	2	2	6	2	6	10	2	6	10	6	2	6			2			
Eu	63	2	2	6	2	6	10	2	6	10	7	2	6			2			
Gd	64	2	2	6	2	6	10	2	6	10	7	2	6	1		2			
Tb	65	2	2	6	2	6	10	2	6	10	9	2	6			2			
Dy	66	2	2	6	2	6	10	2	6	10	10	2	6			2			
Ho	67	2	2	6	2	6	10	2	6	10	11	2	6			2			
Er	68	2	2	6	2	6	10	2	6	10	12	2	6			2			
Tm	69	2	2	6	2	6	10	2	6	10	13	2	6			2			
Yb	70	2	2	6	2	6	10	2	6	10	14	2	6			2			
Lu	71	2	2	6	2	6	10	2	6	10	14	2	6	1		2			
Hf	72	2	2	6	2	6	10	2	6	10	14	2	6	2		2			
Ta	73	2	2	6	2	6	10	2	6	10	14	2	6	3		2			
W	74	2	2	6	2	6	10	2	6	10	14	2	6	4		2			
Re	75	2	2	6	2	6	10	2	6	10	14	2	6	5		2			
Os	76	2	2	6	2	6	10	2	6	10	14	2	6	6		2			
Ir	77	2	2	6	2	6	10	2	6	10	14	2	6	7		2			
Pt	78	2	2	6	2	6	10	2	6	10	14	2	6	9		1			
Au	79	2	2	6	2	6	10	2	6	10	14	2	6	10		1			
Hg	80	2	2	6	2	6	10	2	6	10	14	2	6	10		2			
Tl	81	2	2	6	2	6	10	2	6	10	14	2	6	10		2	1		
Pb	82	2	2	6	2	6	10	2	6	10	14	2	6	10		2	2		
Bi	83	2	2	6	2	6	10	2	6	10	14	2	6	10		2	3		
Po	84	2	2	6	2	6	10	2	6	10	14	2	6	10		2	4		
At	85	2	2	6	2	6	10	2	6	10	14	2	6	10		2	5		
Rn	86	2	2	6	2	6	10	2	6	10	14	2	6	10		2	6		
Fr	87	2	2	6	2	6	10	2	6	10	14	2	6	10		2	6		1
Ra	88	2	2	6	2	6	10	2	6	10	14	2	6	10		2	6		2
Ac	89	2	2	6	2	6	10	2	6	10	14	2	6	10		2	6	1	2
Th	90	2	2	6	2	6	10	2	6	10	14	2	6	10		2	6	2	2
Pa	91	2	2	6	2	6	10	2	6	10	14	2	6	10	2	2	6	1	2
U	92	2	2	6	2	6	10	2	6	10	14	2	6	10	3	2	6	1	2
Np	93	2	2	6	2	6	10	2	6	10	14	2	6	10	4	2	6	1	2
Pu	94	2	2	6	2	6	10	2	6	10	14	2	6	10	6	2	6		2
Am	95	2	2	6	2	6	10	2	6	10	14	2	6	10	7	2	6		2

Table 5.8 Electronic configurations of the elements (Continued)

ELEMENT	Z	1s	2s	2p	3s	3p	3d	4s	4p	4d	4f	5s	5p	5d	5f	6s	6p	6d	7s
Cm	96	2	2	6	2	6	10	2	6	10	14	2	6	10	7	2	6	1	2
Bk	97	2	2	6	2	6	10	2	6	10	14	2	6	10	8	2	6	1	2
Cf	98	2	2	6	2	6	10	2	6	10	14	2	6	10	10	2	6		2
Es	99	2	2	6	2	6	10	2	6	10	14	2	6	10	11	2	6		2
Fm	100	2	2	6	2	6	10	2	6	10	14	2	6	10	12	2	6		2
Md	101	2	2	6	2	6	10	2	6	10	14	2	6	10	13	2	6		2
No	102	2	2	6	2	6	10	2	6	10	14	2	6	10	14	2	6		2
Lr	103	2	2	6	2	6	10	2	6	10	14	2	6	10	14	2	6	1	2

SCIENTISTS AND THE ATOM

Most of the scientific events described in this chapter took place within a relatively short period of time. During the first 30 years of the twentieth century, scientists developed an entirely new physics to describe the world of the atom. The quantum theory, the wave nature of matter, the use of probability, and the uncertainty principle were novel concepts—at odds with classical physics.

The scientists of this period were fascinating. To begin with, they all knew each other and met together at international conferences fairly frequently. They came from many countries—Austria, Britain, Denmark, France, Germany, Hungary, Italy, Russia, and the United States.

Many of these scientists were relatively young at the time they did their most important work. Einstein was 26 when he proposed the special theory of relativity. Bohr was 28 when he published his first paper on atomic structure. Heisenberg proposed the uncertainty principle when he was 25. Pauli was 25 when he stated the exclusion principle.

Some of these scientists did not perform laboratory experiments—they were theoretical physicists, not experimental physicists. Einstein is probably the most famous example, and Pauli is almost as famous. Pauli was not at home in the laboratory, which in time, got to be a joke among his friends. Whenever Pauli came near an experiment in progress, something would go wrong. This phenomenon became known as the "Pauli effect."

Once some laboratory equipment collapsed at the University of Göttingen. Pauli, who lived in Zurich, apparently was nowhere near. Later, however, it was learned that Pauli was traveling, and at the time of the mishap, his train was stopped in the Göttingen station. The Pauli effect worked at long distances!

PROBLEMS

5.1 Complete the following table.

Symbol	*Z*	*A*	*Protons*	*Neutrons*	*Electrons*
Sn	50	120			
Ag		109			
	53	127			53
Ar				22	18
W		184		110	
Cs^+				78	
	29			34	27

5.2 Complete the following table.

Symbol	*Z*	*A*	*Protons*	*Neutrons*	*Electrons*
Sr	38	88			
As		75			
	65	159			65
Rn				136	86
Mo		98		56	
Au^+				118	
	16			16	18

5.3 Complete the following table.

Symbol	*Z*	*A*	*Protons*	*Neutrons*	*Electrons*
Se	34	80			
Hg		202			
	48	114			48
Ce				82	
Bi		209		126	
Cr^{3+}				29	
	7			7	10

5.4 Describe the three types of rays emitted by radioactive substances that occur in nature.

5.5 Bohr calculated the radius of the hydrogen atom to be 5.29×10^{-9} cm. If the radius of a proton is 1.30×10^{-13} cm, what percentage of the total volume of the hydrogen atom is occupied by the nucleus? The volume of a sphere, V, is given by the formula $V = \frac{4}{3}\pi r^3$, where r is the radius of the sphere.

5.6 Assume that the radius of the hydrogen atom is 5.29×10^{-9} cm and that the radius of its nucleus is 1.30×10^{-13} cm. What would the diameter of the nucleus be (in cm) if the diameter of the atom were the same as the length of a football field, 100 yards? One inch is 2.54 cm.

5.7 The proton is thought to have a radius of 1.30×10^{-13} cm and a mass of 1.67×10^{-24} g. (a) What is the density of a proton, in g/cm^3? The volume of a sphere, V, is given by the formula $V = \frac{4}{3}\pi r^3$, where r is the radius of the sphere. (b) A basketball has a radius of 12.0 cm. What would the mass of a basketball be if it had the same density as a proton? Could you lift it?

5.8 The element boron consists of two isotopes: $^{10}_{5}B$ which has a mass of 10.01 u and $^{11}_{5}B$ which has a mass of 11.01 u. The atomic weight of boron is 10.81. What is the percent abundance of each of the two isotopes?

5.9 There are two naturally occurring isotopes of copper: ^{63}Cu, which has a mass of 62.930 u, and ^{65}Cu, which has a mass of 64.928 u. The atomic weight of copper is 63.546. What is the percent abundance of each of the isotopes?

5.10 There are two naturally occurring isotopes of rubidium: ^{85}Rb, which has a mass of 84.91 u, and ^{87}Rb, which has a mass of 86.92 u. The atomic weight of rubidium is 85.47. What is the percent abundance of each of the isotopes?

5.11 Naturally occurring bromine consists of two isotopes: ^{79}Br, which has a mass of 78.918 u, and ^{81}Br, which has a mass of 80.916 u. The atomic weight of bromine is 79.904. What is the percent abundance of each of the isotopes?

5.12 If element X consists of 78.7% of atoms with a mass of 24.0 u, 10.1% of atoms with a mass of 25.0 u, and 11.2% of atoms with a mass of 26.0 u, what is the atomic weight of X?

5.13 What is the atomic weight of element Y if Y consists of 57.25% of atoms with a mass of 120.90 u and 42.75% of atoms with a mass of 122.90 u?

5.14 If element X consists of 60.4% of atoms with a mass of 68.9 u each and 39.6% of atoms with a mass of 70.9 u each, what is the atomic weight of X?

5.15 If element X consists of 92.0% of atoms with a mass of 28.0 u each, 5.0% of atoms with a mass of 29.0 u each, and 3.0% of atoms with a mass of 30.0 u each, what is the atomic weight of X?

5.16 Which unit would be larger: an atomic mass unit based on the mass of a ^{19}F atom set at exactly 19 u or an atomic mass unit based on the current standard? Only one isotope of fluorine occurs in nature, ^{19}F.

5.17 The isotopes of neon were the first to be identified. What are the mass numbers and charges corresponding to the following e/m values for neon ions? 4.38×10^3 coul/g, 4.59×10^3 coul/g, 4.81×10^3 coul/g, 8.76×10^3 coul/g, 9.64×10^3 coul/g. The unit charge, e, is 1.60×10^{-19} coul, and 1 u is 1.66×10^{-24} g.

5.18 What are the frequency and energy per quantum of (a) red light with a wavelength of 700 nm and (b) blue light with a wavelength of 400 nm? The speed of light, c, is 3.00×10^8 m/s, and Planck's constant, h, is 6.63×10^{-34} J·s.

5.19 What is the frequency and energy per quantum of (a) an X ray with a wavelength of 1.50×10^{-10} m and (b) a radio wave with a wavelength of 5.00×10^2 m? The speed of light, c, is 3.00×10^8 m/s, and Planck's constant, h, is 6.63×10^{-34} J·s.

5.20 Moseley found that the frequency, ν, of a characteristic line of the X-ray spectrum of an element is related to the atomic number, Z, of the element by the formula $\sqrt{\nu} = a(Z - b)$, where a is approximately $5.0 \times 10^7/\sqrt{\text{sec}}$ and b is approximately 1.0. What is the atomic number of an element for which the corresponding line in the X-ray spectrum occurs at a wavelength of 5.64×10^{-11} m? What is the element?

5.21 What is the wavelength of the line in the X-ray spectrum of Fe that conforms to the formula given in Problem 5.20?

5.22 The Lyman series of lines in the hydrogen spectrum results from electron transitions to the $n = 1$ level from higher levels. What wavelengths correspond to transitions from $n = 2$ to $n = 1$ and from $n = \infty$ to $n = 1$? In what region of the electromagnetic spectrum do the Lyman lines occur?

5.23 The Paschen series of lines in the hydrogen spectrum results from electron transitions to the $n = 3$ level from higher levels. What wavelengths correspond to transitions from $n = 4$ to $n = 3$ and from $n = \infty$ to $n = 3$. In what region of the electromagnetic spectrum do the Paschen lines occur?

5.24 The spectral lines of hydrogen in the visible region (the Balmer series) represent electron transitions to the $n = 2$ level from higher levels. What is the electron transition that corresponds to the 434.0 nm spectral line?

5.25 Give the values for all four quantum numbers for each electron in the ground state of the oxygen atom. Use positive values of m_l and m_s first.

5.26 Each electron in an atom may be characterized by a set of four quantum numbers. For each of the following parts, tell how many sets of quantum numbers are possible such that each set contains all the values listed. (a) $n = 5, l = 4$; (b) $n = 4, l = 1$; (c) $n = 4, l = 0, m_l = 0$; (d) $n = 2, l = 0, m_l = +2$; (e) $n = 3, l = 2, m_l = +1$; (f) $n = 2, l = 2$; (g) $n = 5$.

5.27 Each electron in an atom may be characterized by a set of four quantum numbers. For each of the following parts, tell how many sets of quantum numbers are possible such that each set contains all the values listed (a) $n = 3, l = 2$; (b) $n = 3, l = 0, m_l = -1$; (c) $n = 4$; (d) $n = 4, l = 4$; (e) $n = 5, l = 1$; (f) $n = 5, l = 3$; (g) $n = 5, l = 3, m_l = 0$.

5.28 In the ground state of Kr: (a) how many electrons have $m_l = +1$ as one of their quantum numbers? (b) how many electrons have $l = 1$ as one of their quantum numbers?

5.29 In the ground state of Rb: (a) how many electrons have $m_l = 0$ as one of their quantum numbers? (b) how many electrons have $m_l = +2$ as one of their quantum numbers?

5.30 Give the values of l, m_l, and m_s for the last electron added according to the aufbau method to the electron configuration of an element in each of the following groups of the periodic table. (a) I A, (b) IV B, (c) IV A,

(d) VII A. Assume that positive values of m_l and m_s are used first.

5.31 Sketch boundary surface diagrams for 1s, 2p, and 3d orbitals.

5.32 Identify the atoms that have the following ground-state electronic configurations in their outer shell or shells: (a) $3s^2\ 3p^6\ 3d^8\ 4s^2$, (b) $4s^2\ 4p^4$, (c) $4s^2\ 4p^6\ 4d^2\ 5s^2$, (d) $5s^2\ 5p^3$, (e) $5s^2\ 5p^6\ 6s^1$, (f) $6s^2\ 6p^6$.

5.33 Identify the atoms that have the following ground-state electronic configurations in their outer shell or shells: (a) $4s^2\ 4p^5$, (b) $4s^2\ 4p^6\ 4d^1\ 5s^2$, (c) $4s^2\ 4p^6\ 4d^{10}\ 4f^3\ 5s^2\ 5p^6\ 6s^2$, (d) $5s^2\ 5p^6\ 6s^2$, (e) $5s^2\ 5p^6\ 5d^4\ 6s^2$, (f) $6s^2\ 6p^5$.

5.34 Identify the atoms that have the following ground-state electronic configurations in their outer shell or shells: (a) $5s^25p^2$, (b) $3s^23p^63d^54s^2$, (c) $4s^24p^64d^{10}5s^2$, (d) $4s^24p^65s^2$, (e) $5s^25p^5$.

5.35 Write the notations for the ground-state electronic configurations of the following atoms: (a) $_{30}$Zn, (b) $_{52}$Te, (c) $_{61}$Pm, (d) $_{72}$Hf, (e) $_{82}$Pb, (f) $_{88}$Ra.

5.36 State the number of unpaired electrons in each of the atoms listed in Problem 5.35.

5.37 Write the notations for the ground-state electronic configurations of the following atoms: (a) $_{37}$Rb, (b) $_{50}$Sn, (c) $_{54}$Xe, (d) $_{67}$Ho, (e) $_{76}$Os, (f) $_{80}$Hg.

5.38 State the number of unpaired electrons in each of the atoms listed in Problem 5.37.

5.39 Write spectroscopic notations for the ground-state electronic configurations of: (a) $_6$C, (b) $_{17}$Cl, (c), $_{20}$Ca, (d) $_{24}$Cr, (e) $_{27}$Co, (f) $_{29}$Cu, (g) $_{48}$Cd, (h) $_{55}$Cs, (i) $_{58}$Ce, (j) $_{96}$Cm, (k) $_{98}$Cf.

5.40 State the number of unpaired electrons in each of the atoms listed in Problem 5.39.

5.41 Write ground-state spectroscopic notations for the electronic configuration of (a) $_{13}$Al, (b) $_{18}$Ar, (c) $_{33}$As, (d) $_{51}$Sb, (e) $_{85}$At, (f) $_{89}$Ac, (g) $_{95}$Am.

5.42 State the number of unpaired electrons in each of the atoms listed in Problem 5.41.

5.43 Describe how the ground-state electronic configurations of the elements of a period compare to one another and how the ground-state electronic configurations of the elements of a group are related.

5.44 What number would you predict for the atomic number of the noble gas of the seventh period (as yet unknown)?

5.45 What is the interpretation of the meaning of the Bohr radius, a_0, in terms of the Bohr theory and in terms of wave mechanics?

5.46 Describe how each of the following contributed to the theory of atomic structure: Chadwick, Rutherford, J. J. Thomson, Millikan, Moseley, Bohr, and Schrödinger.

5.47 Compare and contrast: (a) mass number—atomic number, (b) wavelength—frequency, (c) period—group,

(d) ground state—excited state, (e) paramagnetism—diamagnetism, (f) Pauli's exclusion principle—Heisenberg's uncertainty principle, (g) continuous spectrum—line spectrum.

5.48 Define the following terms: (a) binding energy, (b) atomic weight, (c) Hund's rule of maximum multiplicity, (d) half-filled subshell, (e) photon, (f) orbital, (g) e/m.

Chapter 6

Properties of Atoms and the Ionic Bond

Science evolves. Currently, some scientists think that the atomic concepts developed in the first three decades of the twentieth century have been almost "mined out." Just as these concepts modified the old physics, perhaps a new set will emerge to revolutionize today's physics. To many of the scientists who took part in the changes at the beginning of this century, the ideas that were emerging then were almost too bizarre to be believed.

At a meeting of scientists in the 1950's, Pauli presented a new theory that he and Heisenberg had developed. The theory later proved to be of little value and was discarded. At the meeting, many in the audience were critical of the proposed theory. Bohr was in the audience, and his comments reflected his feelings toward the revolutionary changes in science that he had seen in his lifetime. "We are all agreed that your theory is crazy," he said. "The question which divides us is whether it is crazy enough to have a chance of being correct."

Modern theories of chemical bonding are based on the "crazy" theories of atomic structure that we have discussed. Chemical bonding is the topic of the next two chapters.

6.1 TYPES OF ELEMENTS

The classification of the elements into metals and nonmetals is based on certain properties. **Metals** are good conductors of heat and electricity, have a characteristic luster, are malleable (capable of being pounded into shapes), and are ductile (capable of being drawn into wire). **Nonmetals,** on the other hand, are very poor conductors of heat and electricity, are not lustrous, and are brittle in the solid state.

About 80% of the known elements are metals. The stepped, diagonal line that appears in the periodic table marks the approximate division between the metals and the nonmetals. The nonmetals are found to the right of this line. The division, however, is not sharp since the elements that appear close to this line (sometimes called **metalloids**) have properties that are intermediate between metallic and nonmetallic properties.

The elements may also be classified according to their electron configurations.

1. *The noble gases.* In the periodic table the noble gases are found at the end of each period in group 0. They are colorless monatomic gases, which are chemically unreactive, and diamagnetic. With the exception of helium (which has the configuration $1s^2$), all the noble gases have outer configurations of $ns^2\,np^6$, a very stable arrangement.

2. *The representative elements.* These elements are found in the A families of the periodic table and include metals and nonmetals. They exhibit a wide range of chemical behavior and physical characteristics. Some of the elements are diamagnetic and some are paramagnetic. The compounds of these elements, however, are generally diamagnetic and colorless. All of their electronic shells are either complete or stable (e.g., $ns^2\,np^6$) except the outer shell to which the last electron may be considered as having been added. This outer shell is termed the valence shell: electrons in it are valence electrons. The number of valence electrons for each atom is the same as the group number. The chemistry of these elements depends upon these valence electrons.

3. *The transition elements.* These elements are found in the B families of the periodic table. They are characterized by inner building—the differentiating electron added by the aufbau procedure is an inner *d* electron. Electrons from the two outermost shells are used in chemical reactions. All of these elements are metals. Most of them are paramagnetic and form highly colored, paramagnetic compounds.

4. *The inner-transition elements.* These elements are found at the bottom of the periodic table, but they belong to the sixth and seventh periods after the elements of group III B. The sixth-period series of 14 elements that fol-

lows lanthanum is called the **lanthanide series.** The seventh-period series that follows actinium is known as the **actinide series.** The differentiating electron in each atom is an *f* electron. It is added to the shell that is two shells in from the outermost shell. The outer three shells, therefore, may be involved in the chemistry of these elements. All inner-transition elements are metals. They are paramagnetic and their compounds are paramagnetic and highly colored.

6.2 ATOMIC SIZE

How an atom reacts depends upon many factors. Nuclear charge and electron configuration are the most important. The effective size of the atom is also of importance. Determination of this size, however, is a problem. The wave theory predicts that beyond a region of high density, the electron cloud of an atom thins out gradually and ends only at infinity. We cannot isolate and measure a single atom.

It is possible, however, to measure in several ways the distance between the nuclei of two atoms that are bonded together. **Atomic radii** are derived from these bond distances.* The Cl—Cl bond distance in the Cl_2 molecule, for example, is 198 pm.† One-half of this value, 99 pm, is assumed to be the atomic radius of chlorine. In turn, the atomic radius of Cl (99 pm) can be subtracted from the C—Cl bond distance (176 pm) to derive the atomic radius of C (77 pm). The effective size of an atom may vary slightly from bond to bond when the atom is bonded to other types of atoms. Such variations, however, are usually less than a few picometers. Data derived in this way, therefore, can be used in making comparisons.

Atomic radius is plotted against atomic number in Figure 6.1. Values for the noble gases are not available. Two trends should be noted.

1. The atomic radii of the representative elements decrease across a period from left to right. Notice the portion of the curve that pertains to the elements of the second period (from Li to F). As we move from one atom to the next, an electron is added to the *same shell* (the $n = 2$ shell) and a proton is added to the nucleus. The increase in nuclear charge causes an increase in the extent to which each electron of the $n = 2$ shell is attracted to the nucleus. As a result, the shell is drawn in toward the nucleus, and the atomic radius decreases.
2. Within a group of the periodic table, an increase in atomic radius is generally observed from top to bottom. The group I A elements (Li,

* The bond distance of *single* covalent bonds are employed. See Section 7.1.

† Atomic dimensions have been recorded in angstrom units in the past (1 Å = 10^{-10} m). In the International System, these dimensions are recorded in nanometers (1 nm = 10^{-9} m) or picometers (1 pm = 10^{-12} m). The Cl—Cl bond distance is

$$1.98\ \text{Å} = 0.198\ \text{nm} = 198\ \text{pm}$$

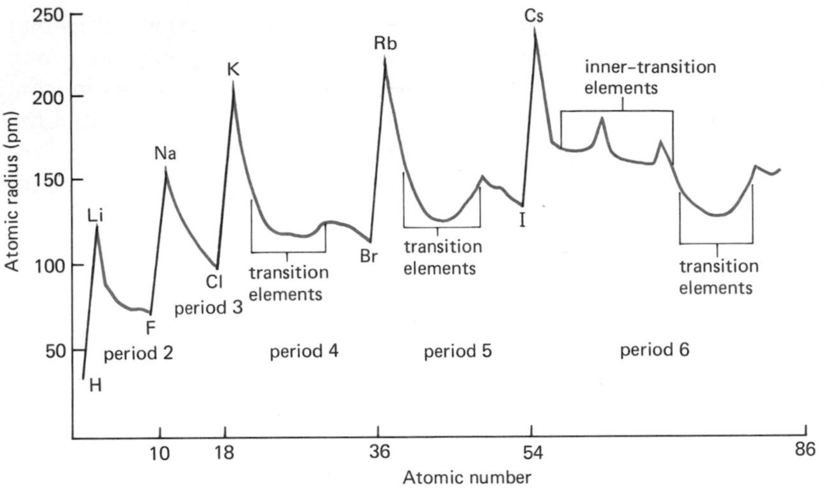

Figure 6.1 Atomic radii of the elements in picometers (1 pm = 10^{-12} m).

Na, K, Rb, and Cs) and the group VII A elements (F, Cl, Br, and I) are labeled in Figure 6.1. The increase in size is evident. The larger atoms of each group employ more electron shells than the smaller atoms. On the other hand, the larger atoms also have higher nuclear charges, which is a factor that tends to decrease the size.

The positive charge on the nucleus, however, is **shielded** from the outer electrons by electrons that occupy shells that are between the outer shell and the nucleus. The number of shielding electrons increases from atom to atom in a group at the same rate that the nuclear charge increases. As a result, an outer electron of each atom experiences about the same attraction for the nucleus. The size of the atom, therefore, is determined by the number of electron shells that are occupied.

The transition elements and inner-transition elements show some variations from this general pattern. The outer electrons of atoms of a given series occupy the same electron shell. The differentiating electrons of these atoms are added to inner shells where they are very effective in shielding the nuclear charge from the outer electrons. For a transition element series, this shielding at first slows down the rate of decrease in atomic radius. Then, when the inner *d* subshell is nearing completion at the end of the series, the shielding results in an increase in atomic radius. For the inner-transition series, atomic radius decreases slowly but is almost constant. (Europium, which has a half-filled *f* subshell is an exception.)

The general trends in atomic radii are summarized in Figure 6.2. As a rule, atoms of metals are larger than atoms of nonmetals. The atomic radii of most metals are larger than 120 pm. The atomic radii of most nonmetals are smaller than 120 pm.

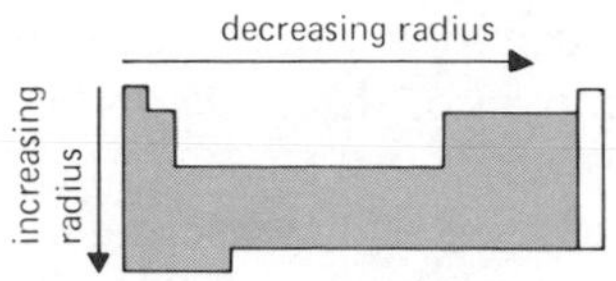

Figure 6.2 General trends in atomic radii in relation to the periodic classification.

6.3 IONIZATION ENERGY

The energy required to remove the most loosely held electron from an isolated atom is called the **first ionization energy.**

$$A(g) \rightarrow A^+(g) + e^-$$

The symbol A(g) stands for a gaseous atom of any element. Energy is *used* to pull the electron away from the atom where it is held by the attraction that the nucleus exerts. This process, therefore, is endothermic, and all ionization energies have positive signs.

Ionization energies can be derived from the line spectra of atoms. Values are given in electron volts (eV) for individual electrons or kilojoules per mole (kJ/mol) for a mole of electrons.* Ionization energy is plotted against atomic number in Figure 6.3.

1. In general, ionization energy increases across a period from left to right. Notice the portions of the curve that pertain to the second period elements (from Li to Ne), the third period elements (from Na to Ar), and

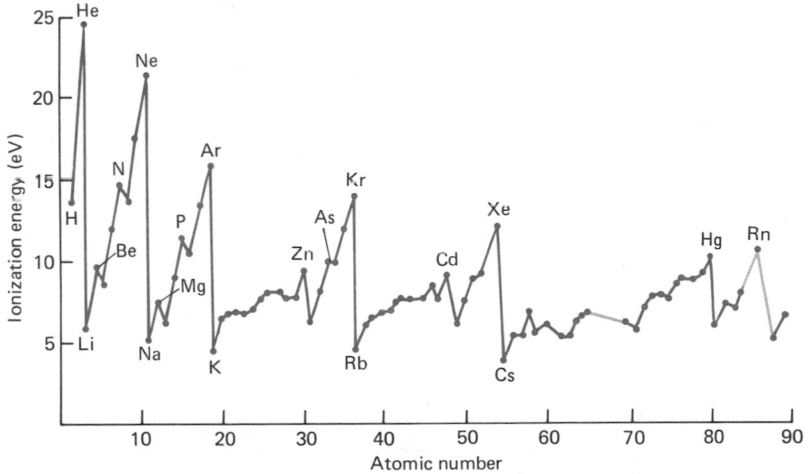

Figure 6.3 First ionization energies of the elements versus atomic number.

* One electron volt is the kinetic energy acquired by an electron in passing through a potential difference of 1 V in a vacuum.

$$1 \text{ eV} = 1.6022 \times 10^{-19}\text{J} = 96.487 \text{ kJ/mol}$$

so on. The ionization energy increases because the atoms become smaller and the nuclear charge increases. Removal of the electron becomes more and more difficult.

2. In general, ionization energy decreases within a group of representative elements from top to bottom. The elements of group I A (Li, Na, K, Rb, and Cs) and the elements of group O (He, Ne, Ar, Kr, Xe, and Rn) are labeled in Figure 6.3. As we go from atom to atom down a group, the nuclear charge increases, but the effect is largely canceled by the increase in the number of shielding electrons in the inner shells. The atoms become larger, and the electron that is removed comes from a higher and higher level. Removal of the electron becomes easier and ionization energy decreases.

The ionization energies of the transition elements do not increase across a period as rapidly as those of the representative elements. The ionization energies of the inner-transition elements remain almost constant. In these series, the differentiating electrons are being added to inner shells. The resulting increase in shielding accounts for the effects noted.

Atoms of metals tend to lose electrons and become positive ions in chemical reactions. Atoms of nonmetals do not usually behave in this way. Metals, therefore, have comparatively low ionization energies and nonmetals have comparatively high ones. The ionization energies of most metals are below 10 eV and of most nonmetals are above 10 eV.

Several features of the curve in Figure 6.3 relate to certain electron configurations. Consider the points on the curve that pertain to

1. The noble gases (He, Ne, Ar, Kr, Xe, and Rn), each of which has an $ns^2\ np^6$ configuration in the outer shell (except for He which has the configuration $1s^2$).
2. The elements Be, Mg, Zn, Cd, and Hg, each of which has a *filled s subshell* in the outermost shell (ns^2).
3. The elements N, P, and As, each of which has a *half-filled p subshell* in the outermost shell ($ns^2\ np^3$).

The ionization energy of each of these elements (particularly the noble gases) is higher than the ionization energy of the element that follows it in the periodic table. These three types of electron configuration, therefore, are comparatively stable, and it is relatively difficult to remove an electron from them. In each case, the differentiating electron added by the aufbau method to form the next element is more easily removed.

We have so far discussed only *first* ionization energies. The *second* ionization energy of an element refers to the removal of *one* electron from a 1+ ion of the element.

$$A^+(g) \rightarrow A^{2+}(g) + e^-$$

Table 6.1 Ionization energies of the third period metals

		IONIZATION ENERGIES (kJ/mol)			
METAL	GROUP	FIRST	SECOND	THIRD	FOURTH
Na	I A	+496	+4,563	+6,913	+9,541
Mg	II A	+738	+1,450	+7,731	+10,545
Al	III A	+577	+1,816	+2,744	+11,575

The *third* ionization energy pertains to the removal of *one* electron from a 2+ ion. Removing an electron, which has a negative charge, from an ion that has a positive charge becomes more and more difficult as the charge on the ion increases. Consequently, the ionization energies increase in the order: first $<$ second $<$ third, and so on. Since ionization energies above the third are extremely large for any element, ions with charges higher than 3+ seldom exist under ordinary conditions.

Ionization energies in kilojoules per mole for the first three elements of the third period are listed in Table 6.1. For each element, the ionization energies increase from the first to the fourth, as expected. Notice, however, that in each case there is a decided jump in the required energy after all the valence electrons have been removed. These points are marked in the table; the number of valence electrons equals the group number. After the valence electrons have been removed, it is necessary to break into the very stable $2s^2\ 2p^6$ noble-gas arrangement of the shell beneath the valence shell to remove the next electron.

The trends in first ionization energy are summarized in Figure 6.4. Notice that the most reactive metals are found in the lower left corner of the periodic table. This reactivity in terms of electron loss decreases as we move upward or to the right from this corner of the chart.

6.4 ELECTRON AFFINITY

The energy change associated with the process in which an electron is added to an isolated atom is called an **electron affinity.**

$$e^- + A(g) \rightarrow A^-(g)$$

Energy is usually *evolved* in these processes. Therefore, most first electron affinities have negative signs.*

The direct determination of electron affinity is difficult. Electron affinity values, therefore, are available for only a limited number of elements, and many of these values are not highly accurate. Some electron affinities are listed

* Electron affinity is frequently defined as energy released. In sources that use this definition, electron affinities that pertain to exothermic processes are given positive signs. This sign convention is the reverse of the one we employ. We shall consistently use a negative sign to indicate that the system loses energy and a positive sign to indicate that the system absorbs energy.

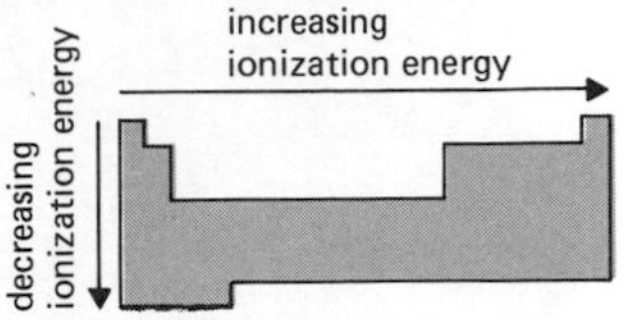

Figure 6.4 General trends in ionization energy in relation to the periodic classification.

in Table 6.2. Notice that some of these values have positive signs indicating that work must be done to force the atom under consideration to accept an additional electron.

A small atom should have a greater tendency to add an electron than a large atom. The added electron is, on the average, closer to the positively charged nucleus in a small atom. This trend is roughly followed from left to right in any period. In the second period, however, exceptions may be noted for beryllium (filled 2*s* subshell), nitrogen (half-filled 2*p* subshell), and neon (all subshells filled). Similar exceptions may be noted for corresponding elements of the third period. Elements with relatively stable electronic configurations do not accept additional electrons readily. Since the electronic configuration of each of the group VII A elements is one electron short of a noble gas configuration, each element of group VII A has the greatest tendency towards electron gain of any element in its period.

The values for the halogens illustrate what is believed to be the typical group trend. Electron attracting ability, except for that of fluorine, increases from bottom to top within the group in agreement with the decrease in atomic size. The effect of small size, however, may be partly canceled by the repulsion of electrons already present in the atom. The exception noted for fluorine,

Table 6.2 Electron affinities (kJ/mol)

H −72							He +54
Li −57	Be +66	B −15	C −121	N +31	O −142 (+702)*	F −333	Ne +99
Na −21	Mg +67	Al −26	Si −135	P −60	S −200 (+332)*	Cl −348	
						Br −324	
						I −295	

* Values in parentheses pertain to the *total* energy effect for the addition of *two* electrons.

as well as for other second period elements, may be explained by these repulsions. The negative charge is more concentrated in a very small shell than it is when the same number of electrons is placed in a larger shell.

Some second electron affinities have been determined. These values refer to processes in which an electron is added to a negative ion; for example,

$$e^- + O^-(g) \rightarrow O^{2-}(g)$$

Since a negative ion and an electron repel each other, energy is required, not released by this process. All second electron affinities have positive signs.

The production of a multicharged negative ion from a neutral atom, therefore, is endothermic. In the formation of the O^{2-} ion from the O atom, for example, less energy is *released* by the first electron affinity

$$e^- + O(g) \rightarrow O^-(g) \qquad \Delta H = -142 \text{ kJ}$$

than is *required* by the second electron affinity

$$e^- + O^-(g) \rightarrow O^{2-}(g) \qquad \Delta H = +844 \text{ kJ}$$

The overall process, therefore, is endothermic.

$$2e^- + O(g) \rightarrow O^{2-}(g) \qquad \Delta H = +702 \text{ kJ}$$

6.5 THE IONIC BOND

When a metal reacts with a nonmetal, electrons are transferred from the atoms of the metal to the atoms of the nonmetal, and an **ionic** (or **electrovalent**) **compound** is produced. The atoms that lose electrons become positive ions, called **cations.** Atoms that gain electrons become negative ions, called **anions.** These ions attract one another to form a crystal.

Compounds of the A family elements are often represented by using the symbols of the elements together with dots to indicate valence electrons. Valence electrons are the only electrons that are involved in the chemical reactions of the representative elements.

Consider the reaction of a sodium atom and a chlorine atom. Sodium is in group I A. Its atoms have one valence electron. Chlorine is a member of group VII A. Atoms of chlorine have seven valence electrons. The sodium atom loses one electron; the chlorine atom gains an electron.

$$\text{Na}\cdot + \cdot\ddot{\underset{..}{\text{Cl}}}\text{:} \rightarrow \text{Na}^+ + \text{:}\ddot{\underset{..}{\text{Cl}}}\text{:}^-$$

The number of sodium ions produced is the same as the number of chloride ions produced. The formula, NaCl, gives the simplest ratio of ions in the crystal.

The complete electronic configurations of the atoms and ions of this reaction are

$$\text{Na } (1s^2\ 2s^2\ 2p^6\ 3s^1) \rightarrow \text{Na}^+\ (1s^2\ 2s^2\ 2p^6) + e^-$$
$$e^- + \text{Cl } (1s^2\ 2s^2\ 2p^6\ 3s^2\ 3p^5) \rightarrow \text{Cl}^-\ (1s^2\ 2s^2\ 2p^6\ 3s^2\ 3p^6)$$

The sodium ion has an electronic configuration identical to that of neon, and the chloride ion has the same configuration as argon. The ions may be said to be **isoelectronic** (the same in electronic configuration) with neon and argon, respectively.

In ionic reactions, most A family elements lose or gain electrons in such a way as to produce ions that are isoelectronic with a noble gas. Most of these **noble gas ions** have eight electrons in the outermost shell (called an **s^2p^6 configuration**). A few, however, have the $1s^2$ configuration of helium (Li^+, Be^{2+}, and H^-), which is called an **s^2 configuration.**

An atom of oxygen (which is a group VI A nonmetal) has six valence electrons. It gains two electrons to attain the s^2p^6 configuration of neon.

$$2e^- + O(1s^2\ 2s^2\ 2p^4) \rightarrow O^{2-}(1s^2\ 2s^2\ 2p^6)$$

In a reaction between sodium and oxygen, two atoms of sodium are required for every one atom of oxygen since the number of electrons lost must equal the number of electrons gained.

$$2Na\cdot + \ddot{\underset{\cdot\cdot}{O}}: \rightarrow 2Na^+ + :\ddot{\underset{\cdot\cdot}{O}}:^{2-}$$

The simplest ratio of ions present in the product, sodium oxide, is indicated by the formula of the compound, Na_2O.

The formula of an ionic compound can be derived from the formulas of the ions it contains. The *total* positive charge of the cations must equal the *total* negative charge of the anions. The s^2p^6 ion derived from calcium (a group II A metal) has a charge of 2+.

$$Ca(1s^2\ 2s^2\ 2p^6\ 3s^2\ 3p^6\ 4s^2) \rightarrow Ca^{2+}(1s^2\ 2s^2\ 2p^6\ 3s^2\ 3p^6) + 2e^-$$

The compound formed from Ca^{2+} and Cl^- ions, calcium chloride, has the formula $CaCl_2$. Two Cl^- ions are required for every Ca^{2+} ion. Calcium oxide, which is composed of Ca^{2+} and O^{2-} ions, has the formula CaO. The formula gives the *simplest* ratio of ions.

An atom of Al (a group III A metal) attains a noble-gas configuration by the loss of three electrons. The oxide of aluminum consists of Al^{3+} and O^{2-} ions. In order to balance the charge, two Al^{3+} ions (total charge, 6+) and three O^{2-} ions (total charge, 6−) must be indicated in the formula for the compound, Al_2O_3.

6.6 LATTICE ENERGY

The energy change associated with the formation of a crystal from cations and anions is called a **lattice energy** or a **crystal energy.** Since energy is evolved in these processes, lattice energies have negative signs. The energy required to separate the ions of a crystal is the lattice energy with the sign changed.

The lattice energy of a crystal can be calculated by taking into account the attractions between the ions that have charges with opposite signs and the repulsions between ions that have charges with the same sign. The arrangement of a crystal is such that the attractions outweigh the repulsions (Figure 6.5). The net attraction holds the crystal together and accounts for the lattice energy.

The lattice energy can also be calculated by using a **Born–Haber cycle.** The calculation of the lattice energy of the sodium chloride crystal will serve as an example of this method. The preparation of NaCl(s) from Na(s) and Cl_2(g) is imagined to occur by a series of steps.

1. Sodium metal is sublimed into gaseous sodium atoms; 108 kJ of energy (the heat of sublimation) is *absorbed* per mole of Na(s).

$$\text{Na(s)} \rightarrow \text{Na(g)} \qquad \Delta H_{\text{subl.}} = +108 \text{ kJ}$$

2. Gaseous sodium atoms lose electrons and become gaseous sodium ions. The energy *required* is the first ionization energy of sodium (Table 6.1).

$$\text{Na(g)} \rightarrow \text{Na}^+\text{(g)} + e^- \qquad \Delta H_{\text{i.e.}} = +496 \text{ kJ}$$

3. *One-half mole* of gaseous Cl_2 molecules is dissociated into *one mole* of gaseous Cl atoms. The bond energy of Cl_2 is 243 kJ/mol (Table 3.2, Section 3.7). One-half of this amount is the energy *absorbed* in this step.

$$\tfrac{1}{2}\text{Cl}_2\text{(g)} \rightarrow \text{Cl(g)} \qquad \tfrac{1}{2}\Delta H_{\text{b.e.}} = \tfrac{1}{2}(+243 \text{ kJ}) = +122 \text{ kJ}$$

4. Electrons are added to the gaseous Cl atoms to produce gaseous Cl^- ions. The energy *liberated* per mole of Cl(g) is the electron affinity of chlorine.

$$e^- + \text{Cl(g)} \rightarrow \text{Cl}^-\text{(g)} \qquad \Delta H_{\text{e.a.}} = -348 \text{ kJ}$$

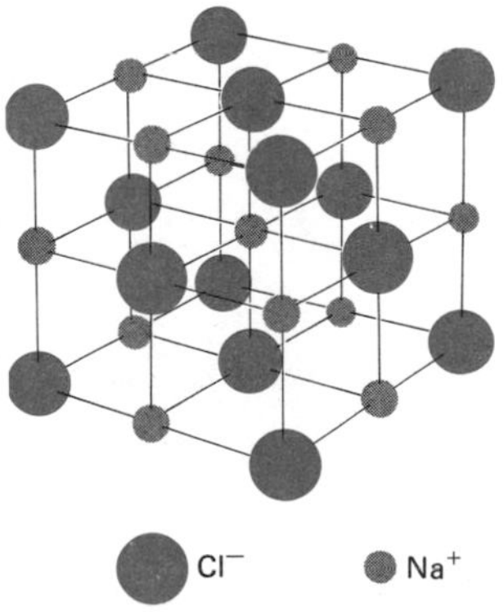

Figure 6.5 Sodium chloride crystal lattice.

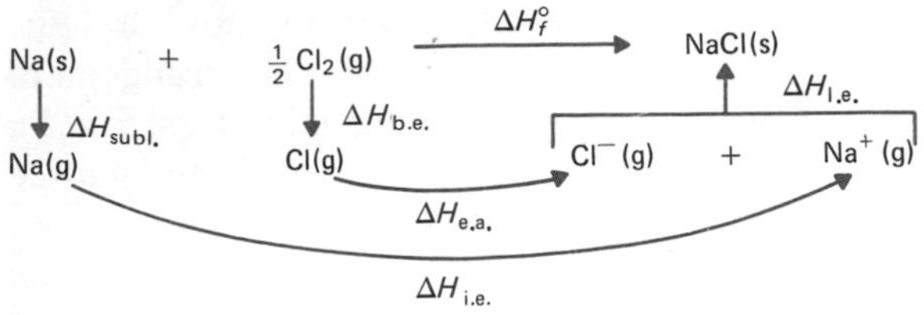

Figure 6.6 Born–Haber cycle for NaCl(s).

5. The gaseous ions form one mole of crystalline NaCl. The enthalpy change of this step is defined as the lattice energy.

$$Na^+(g) + Cl^-(g) \rightarrow NaCl(s) \qquad \Delta H_{l.e.} = ?$$

According to the law of Hess, the total ΔH for all the steps should equal the enthalpy of formation of NaCl(s), −411 kJ/mol.

$$Na(s) + \tfrac{1}{2}Cl_2(g) \rightarrow NaCl(s) \qquad \Delta H_f = -411 \text{ kJ}$$

Therefore,

$$\Delta H_f = \Delta H_{subl.} + \Delta H_{i.e.} + \tfrac{1}{2}\Delta H_{b.e.} + \Delta H_{e.a.} + \Delta H_{l.e.}$$
$$-411 \text{ kJ} = +108 \text{ kJ} + 496 \text{ kJ} + 122 \text{ kJ} - 348 \text{ kJ} + \Delta H_{l.e.}$$
$$-411 \text{ kJ} = +378 \text{ kJ} + \Delta H_{l.e.}$$
$$\Delta H_{l.e.} = -789 \text{ kJ}$$

The Born–Haber cycle for NaCl(s) is summarized in Figure 6.6.

The driving force of the ionic reaction is the attraction of the ions for one another. It is this attraction that results in the liberation of the lattice energy. The first three steps of the NaCl Born–Haber cycle require the addition of energy. Far more energy is required than the electron affinity of chlorine (step 4) can supply. The lattice energy, however, more than makes up the difference so that the process as a whole is exothermic.

Ions with charges that are higher than 1+ and 1− attract one another more strongly than ions with 1+ and 1− charges. The lattice energies listed in Table 6.3 reflect this fact.

Table 6.3 Lattice energies

COMPOUND	COMPONENT IONS	LATTICE ENERGY (kJ/mol)
NaCl	Na^+, Cl^-	−789
Na_2O	$2Na^+$, O^{2-}	−2570
$MgCl_2$	Mg^{2+}, $2Cl^-$	−2530
MgO	Mg^{2+}, O^{2-}	−3890

6.7 TYPES OF IONS

The types of ions that atoms form depend upon the energy considerations that we have discussed.

1. Why doesn't Na lose two electrons and become "Na^{2+}"? The energy required for this process is the sum of the first and second ionization energies of Na (Table 6.1).

   $$+496 \text{ kJ/mol} + 4563 \text{ kJ/mol} = +5059 \text{ kJ/mol}$$

   The lattice energy of an imaginary "$NaCl_2$" would be far too small to compensate for the energy required by this ionization. The lattice energy of "$NaCl_2$" would probably be approximately the same as that of $MgCl_2$, −2530 kJ/mol (Table 6.3). The removal of an electron beyond the noble-gas configuration requires too much energy.
2. Since Na cannot lose two electrons in ion formation, how is Mg able to lose two electrons to become Mg^{2+}? The Mg atom attains a noble-gas configuration through this loss. The sum of the first and second ionization energies of Mg (Table 6.1) is much less than the sum for Na.

   $$+738 \text{ kJ/mol} + 1450 \text{ kJ/mol} = +2188 \text{ kJ/mol}$$

   The lattice energy of $MgCl_2$, −2530 kJ/mol (Table 6.3) is more than enough to supply the energy required for this ionization.
3. Since less energy is required to remove one electron than is required to remove two, why doesn't Mg form a "Mg^+" ion? In this case, the energy required to form "Mg^+" would be only +738 kJ/mol (the first ionization energy of Mg, Table 6.1). On the other hand, the lattice energy of the imaginary "MgCl" would be only about −789 kJ/mol (similar to that of NaCl). The larger lattice energy of $MgCl_2$ (brought about by the higher charge on Mg^{2+}) favors the formation of $MgCl_2$.

We can see, therefore, why so many metals form cations with s^2p^6 (noble-gas) configurations. More than three electrons are never lost or gained in ion formation. The energy required to bring about a gain or loss of more than three electrons is not available.

All monatomic anions are noble-gas ions. Atoms of nonmetals add electrons until the noble-gas configuration is reached. Energy considerations favor the attainment of this limit. The noble-gas ions and their relation to the periodic table are shown in Figure 6.7.

Certain metals undergo ionic reactions even though they cannot possibly produce s^2p^6 cations. Zinc, for example, would have to lose 12 electrons to get a noble-gas configuration. In its reactions, zinc forms the Zn^{2+} ion by the loss of two electrons.

$$Zn(\ldots 3s^2\ 3p^6\ 3d^{10}\ 4s^2) \rightarrow Zn^{2+}(\ldots 3s^2\ 3p^6\ 3d^{10}) + 2e^-$$

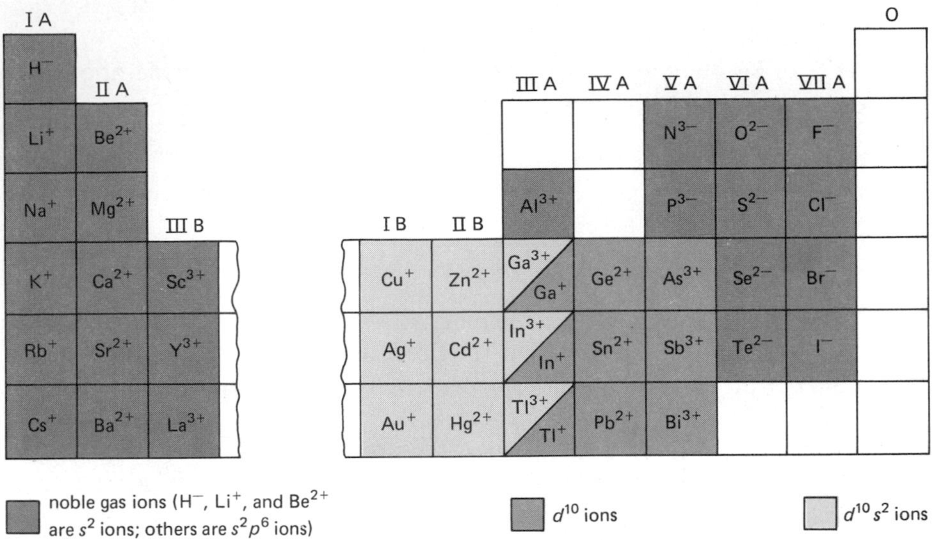

Figure 6.7 Types of ions and their relations to the periodic table.

The ionization energies of Zn and the lattice energies of Zn^{2+} compounds favor the formation of this ion. The electron configuration of the Zn^{2+} ion is a stable one. All of the subshells in the outer shell of the Zn^{2+} ion are filled. Other atoms form ions with $ns^2\ np^6\ nd^{10}$ configurations like that of Zn^{2+}. They are called **d^{10} ions** and are listed in Figure 6.7.

Tin forms a cation, Sn^{2+}, that will serve as an example of another type of ion.

$$Sn(\ldots 4s^2\ 4p^6\ 4d^{10}\ 5s^2\ 5p^2) \rightarrow Sn^{2+}(\ldots 4s^2\ 4p^6\ 4d^{10}\ 5s^2) + 2e^-$$

This configuration is also a stable one. Ions with similar configurations are called **$d^{10}s^2$ ions.** Nine of them are listed in Figure 6.7. The larger elements of group III A can form both $d^{10}s^2$ ions and d^{10} ions.

In the reactions of the transition metals, inner d electrons may be used as well as outer s electrons. Most transition elements cannot form ions with any of the regular configurations (s^2, s^2p^6, d^{10}, or $d^{10}s^2$). A list of some of these is given in Table 6.4. The configuration listed is that of the outer d subshell of the ion.

Table 6.4 Some first transition series cations

CONFIGURATION	EXAMPLE	CONFIGURATION	EXAMPLE
d^1	Ti^{3+}	d^6	Fe^{2+}, Co^{3+}
d^2	V^{3+}	d^7	Co^{2+}
d^3	Cr^{3+}, V^{2+}	d^8	Ni^{2+}
d^4	Cr^{2+}, Mn^{3+}	d^9	Cu^{2+}
d^5	Mn^{2+}, Fe^{3+}		

Table 6.5 Common polyatomic anions

FORMULA	NAME	FORMULA	NAME
$C_2H_3O_2^-$	acetate	ClO^-	hypochlorite
AsO_4^{3-}	arsenate	OH^-	hydroxide
AsO_3^{3-}	arsenite	NO_3^-	nitrate
CO_3^{2-}	carbonate	NO_2^-	nitrite
ClO_3^-	chlorate	ClO_4^-	perchlorate
ClO_2^-	chlorite	MnO_4^-	permanganate
CrO_4^{2-}	chromate	PO_4^{3-}	phosphate
CN^-	cyanide	SO_4^{2-}	sulfate
$Cr_2O_7^{2-}$	dichromate	SO_3^{2-}	sulfite

Many transition metals form more than one type of cation. Examples are Cu^+ and Cu^{2+}, Cr^{2+} and Cr^{3+}, and Fe^{2+} and Fe^{3+}. More energy is required to produce the Fe^{3+} ion than is required to produce the Fe^{2+} ion. The lattice energies of Fe^{3+} compounds, however, are larger than those of Fe^{2+} compounds. These factors balance to the point that it is possible to prepare compounds of both ions.

A large number of polyatomic anions exist. The names and formulas of some common ones are given in Table 6.5. The atoms of these ions are held together by covalent bonds (see Chapter 7). There are not many simple polyatomic cations, but the ammonium ion, NH_4^+, deserves mention.

6.8 IONIC RADIUS

The distance between the centers of two adjacent ions in a crystal can be determined by X-ray diffraction (Section 9.13). For most crystals, this distance is the sum of the radius of a cation and the radius of an anion. Dividing such a distance to obtain the two radii is a problem.

One solution to the problem is to study a crystal composed of very small cations and large anions, such as lithium iodide (Figure 6.8a). In the LiI

Figure 6.8 Determination of ionic radii. (See text for further discussion.)

crystal, I^- ions are assumed to touch each other. The distance between two I^- ions (d in Figure 6.8a) is divided in half to get the radius of the I^- ion.

$$\text{radius } I^- = 432 \text{ pm}/2 = 216 \text{ pm}$$

In most crystals, the anions do not touch each other. The distance d in Figure 6.8(b) could not be used in a similar way.

Once the radius of the I^- ion has been found, other ionic radii can be calculated. If the distance between the center of a K^+ ion and an I^- ion is determined (d' in Figure 16.8b), the radius of the K^+ ion can be found by subtracting the I^- radius from the K^+ to I^- distance.

$$\begin{aligned} d' &= \text{radius } K^+ + \text{radius } I^- \\ 349 \text{ pm} &= \text{radius } K^+ + 216 \text{ pm} \\ \text{radius } K^+ &= 133 \text{ pm} \end{aligned}$$

A positive ion is always smaller than the atom from which it is derived (Figure 6.9). The radius of K is 203 pm, and the radius of K^+ is 133 pm. In the formation of the K^+ ion, the loss of an electron represents the loss of the entire $n = 4$ shell of the K atom. Furthermore, the protons outnumber the electrons in the positive ion. The electrons of the ion are drawn in closer to the nucleus. For similar reasons, a 2+ ion is larger than a 3+ ion. For example,

$$\text{radius Fe} = 117 \text{ pm} \qquad \text{radius } Fe^{2+} = 75 \text{ pm} \qquad \text{radius } Fe^{3+} = 60 \text{ pm}$$

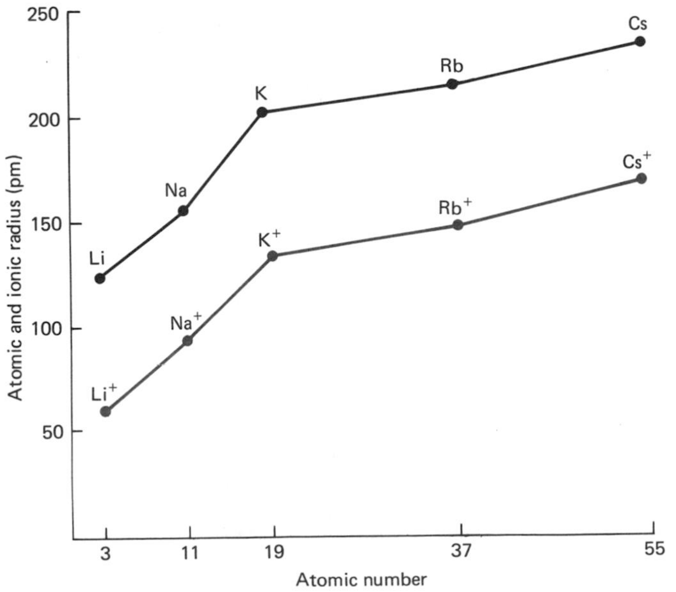

Figure 6.9 Atomic and ionic radii of the group IA elements.

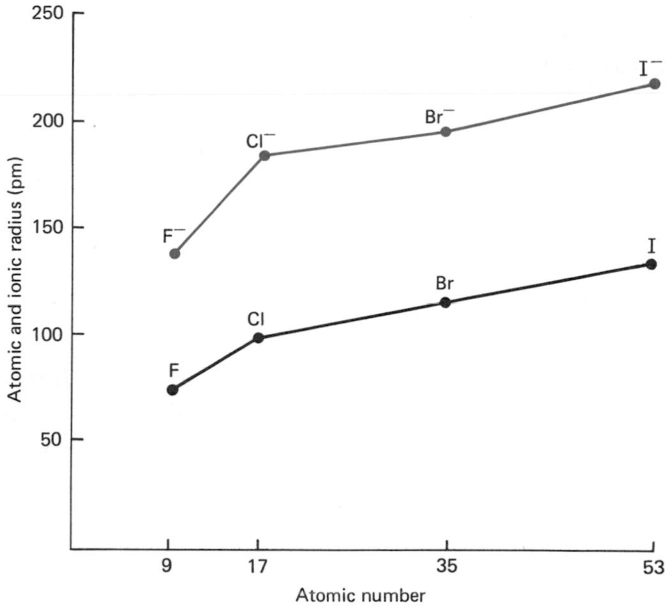

Figure 6.10 Atomic and ionic radii of the group VIIA elements.

A negative ion is always larger than the atom from which it is derived (Figure 6.10). The radius of the Cl atom is 99 pm, and the radius of the Cl^- ion is 181 pm. In the formation of the Cl^- ion, the addition of an electron causes an increase in the extent to which the valence electrons repel one another, and the valence shell expands. The chloride ion has 18 electrons and 17 protons.

PROBLEMS

6.1 Define, describe the characteristics of, and give examples of: (a) representative elements, (b) transition elements, (c) noble gases, (d) inner-transition elements.

6.2 Which member of each of the following pairs would you predict to be larger? (a) P or Cl, (b) P or Ge, (c) Mg or Ca, (d) Na or Mg, (e) K or Mg.

6.3 Which member of each of the following pairs would you predict to be larger? (a) Sn or Pb, (b) Sn or As, (c) As or Se, (d) Al or Cl, (e) Cl or Br.

6.4 Which member of each of the following pairs has the higher first ionization energy? (a) F or Ne, (b) Se or Cl, (c) S or Cl, (d) Cl or Br, (e) K or Ca.

6.5 Which member of each of the following pairs has the higher first ionization energy? (a) Sr or Ba, (b) Cs or Ba, (c) Cl or Ar, (d) In or Sn, (e) Xe or Kr.

6.6 Explain why the first ionization energy of O is lower than that of N.

6.7 Discuss the factors that influence ionization energy.

6.8 Why is the second ionization energy of an element always higher than the first ionization energy?

6.9 Use the Bohr equation $\nu = 3.29 \times 10^{15}/\text{s}\left(\frac{1}{n_i^2} - \frac{1}{n_o^2}\right)$ to evaluate the ionization energy of hydrogen in electron volts. Planck's constant, h, is 6.63×10^{-35} J·s, and 1 eV is 1.60×10^{-19} J.

6.10 Why are the electron affinities of Be, N, and Ne out of line in comparison to those of the other members of the second period?

6.11 Of all the elements of the third period (Na through Ar): (a) which has the largest atomic radius, (b) which has the highest first ionization energy, (c) which is the most reactive metal, (d) which is the most reactive nonmetal, (e) which is the least reactive, (f) how many are metals?

6.12 Which of the group II A elements, Be, Mg, Ca, Sr, and Ba, (a) has the largest atomic radius, (b) has the highest first ionization energy, (c) is the most reactive element?

6.13 Use the following data to calculate the lattice energy of LiF. The enthalpy of formation of LiF is −583.7 kJ/mol. The enthalpy of sublimation of Li is +155.2 kJ/mol, and the first ionization energy of Li is +520.1 kJ/mol. The bond energy of the F_2 bond is +158.2 kJ/mol F_2, and the first electron affinity of the fluorine atom is −333.0 kJ/mol F atoms.

6.14 Use the following data to calculate the lattice energy of CaO. The enthalpy of formation of CaO is −635.5 kJ/mol. The enthalpy of sublimation of Ca is +192.5 kJ/mol. The first ionization energy of Ca is +589.5 kJ/mol, and the second ionization energy of Ca is +1145.2 kJ/mol. The bond energy of the O_2 bond is +495.0 kJ/mol O_2. The first electron affinity of the oxygen atom is −141.8 kJ/mol O atoms, and the second electron affinity of the oxygen atom is +844.2 kJ/mol O atoms.

6.15 Use the following data to calculate the first electron affinity of iodine. The enthalpy of formation of KI is −327.0 kJ/mol, and the lattice energy of KI is −643.9 kJ/mol. The enthalpy of sublimation of K is +90.0 kJ/mol, and the first ionization energy of K is +415.0 kJ/mol. For the reaction: $I_2(s) \rightarrow 2I(g), \Delta H = +213.0$ kJ.

6.16 Use the following data to calculate the first electron affinity of chlorine. The enthalpy of formation of $MgCl_2$ is −642 kJ/mol, and the lattice energy of $MgCl_2$ is −2530 kJ/mol. The enthalpy of sublimation of Mg is +152 kJ/mol. The first ionization energy of Mg is +738 kJ/mol, and the second ionization energy of Mg is +1450 kJ/mol. The bond energy of the Cl_2 bond is +244 kJ/mol Cl_2.

6.17 Write notations for the ground-state electronic configurations of the following ions: (a) Ca^{2+}, (b) Cr^{2+}, (c) Co^{2+}, (d) Pd^{2+}, (e) Ag^+, (f) I^-.

6.18 (a) State the number of unpaired electrons in each of the ions listed in Problem 6.17. (b) Which of these ions would you predict to be paramagnetic and which diamagnetic?

6.19 Write notations for the ground-state electronic configurations of the following ions: (a) S^{2-}, (b) Cu^{+}, (c) Cu^{2+}, (d) Fe^{3+}, (e) F^{-}, (f) Hg^{2+}, (g) Pb^{2+}, (h) Cr^{3+}.

6.20 (a) State the number of unpaired electrons in each of the ions listed in Problem 6.19. (b) Which of these ions would you predict to be paramagnetic and which diamagnetic?

6.21 For each of the following, give the formulas of two ions that are isoelectronic with the atom or ion listed. (a) Ne, (b) Cd^{2+}, (c) Br^{-}, (d) Sc^{3+}, (e) Hg.

6.22 For each of the following, give the formulas of two ions that are isoelectronic with the atom or ion listed. (a) Zn^{2+}, (b) Au^{+}, (c) Cs^{+}, (d) Cd, (e) Kr.

6.23 Give examples of s^2 ions, s^2p^6 ions, d^{10} ions, and $d^{10}s^2$ ions.

6.24 Identify the s^2p^6, d^{10}, and $d^{10}s^2$ ions that are included in the following list: Al^{3+}, Ga^{3+}, Sc^{3+}, N^{3-}, Pb^{2+}, K^{+}, Ba^{2+}, Cu^{+}, Cd^{2+}, Tl^{+}, Bi^{3+}.

6.25 Using arguments based on the enthalpy changes involved in ion formation, explain why Na forms Na^{+}, not Na^{2+}, but Cu forms both Cu^{+} and Cu^{2+}.

6.26 Which member of each of the following pairs would you predict to be larger? (a) Cs^{+} or Ba^{2+}, (b) S or S^{2-}, (c) S^{2-} or Cl^{-}, (d) Cr^{2+} or Cr^{3+}, (e) Ag or Ag^{+}, (f) Cu^{2+} or Ag^{+}.

6.27 Which member of each of the following pairs would you predict to be larger? (a) Br or Br^{-}, (b) Cs or Cs^{+}, (c) O^{2-} or F^{-}, (d) Au^{+} or Au^{3+}, (e) In^{3+} or Tl^{+}, (f) In^{+} or Sn^{2+}.

6.28 Compare and contrast metals and nonmetals. Discuss physical properties, atomic size, ionization energies, and chemical reactivity.

6.29 Give formulas for the oxides, chlorides, phosphates, sulfates, and dichromates of sodium, magnesium, and aluminum.

Chapter 7

The Covalent Bond

When atoms of nonmetals interact, molecules are formed which are held together by covalent bonds. Since these atoms are similar in their attraction for electrons (identical when two atoms of the same element are considered), electron transfer does not occur. Instead, electrons are shared. A **covalent bond** consists of a pair of electrons (with opposite spins) that is shared by two atoms.

7.1 THE LEWIS THEORY

The simplest covalent bond is found in the H_2 molecule. An individual hydrogen atom has a single electron in a spherical 1*s* orbital. When two hydrogen atoms form a covalent bond, the atomic orbitals overlap. The electron clouds reinforce each other in the region between the nuclei, and there is an increased probability of finding an electron in this region. According to the Pauli ex-

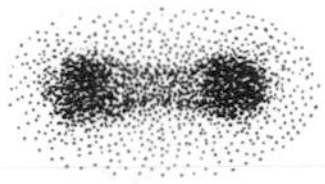

Figure 7.1 Representation of the electron distribution in a hydrogen molecule.

clusion principle, the two electrons of the bond must have opposite spins. The strength of the covalent bond comes from the attraction of the positively charged nuclei for the negative cloud of the bond (Figure 7.1).

The hydrogen molecule can be represented by the symbol H:H or H—H. Although the electrons belong to the molecule as a whole, each hydrogen atom can be considered to have the noble-gas configuration of helium (two electrons in the $n = 1$ level). This concept is based on the assumption that both shared electrons contribute to the stable configuration of each hydrogen atom.

Each of the diatomic molecules of the group VII A elements contains a covalent bond. An atom of group VII A has seven valence electrons. The F_2 molecule is assumed to have the structure

:F̤̈· + ·F̤̈: → :F̤̈:F̤̈:

Only the two electrons between the F atoms are shared and considered to form the covalent bond. A single covalent bond consists of two electrons with opposite spins. Each of the F atoms in the molecule may be considered to have a noble-gas configuration—eight electrons in the valence shell (an octet). This conclusion is reached by counting the shared electrons twice—once for each F atom.

More than one covalent bond may form between two atoms. A nitrogen atom (group V A) has five valence electrons.

:Ṇ̇· + ·Ṇ̇: → :N:::N:

In the molecule, N_2, six electrons are shared in three covalent bonds (usually called a triple bond). As a result of this formulation, each of the nitrogen atoms can be considered to have an octet of electrons.

The electron-dot formulas we have been using are called **valence-bond structures** or **Lewis structures,** named after Gilbert N. Lewis who proposed this theory of covalent bonding in 1916. The Lewis theory emphasizes the attainment of noble-gas configurations on the part of atoms in covalent molecules. Since the number of valence electrons is the same as the group number for the nonmetals, we might expect that VII A elements, such as Cl, would form one covalent bond to attain a stable octet; VI A elements, such as O and S, two covalent bonds; V A elements, such as N and P, three covalent bonds; and IV A elements, such as C, four covalent bonds. The expected numbers of covalent bonds are found in many molecules.

H· + ·C̤̈l: → H:C̤̈l:
hydrogen chloride

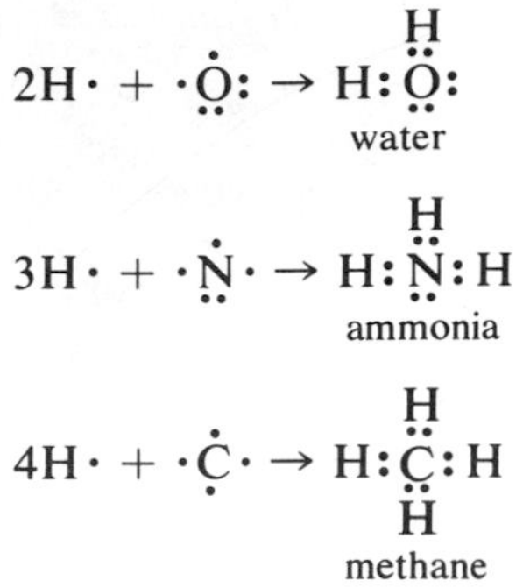

In these molecules, each hydrogen atom can be considered to have a complete $n = 1$ shell. The other atoms have characteristic noble-gas octets.

The covalent bonding of compounds can also be indicated by dashes. Each dash represents one bond, a pair of electrons.

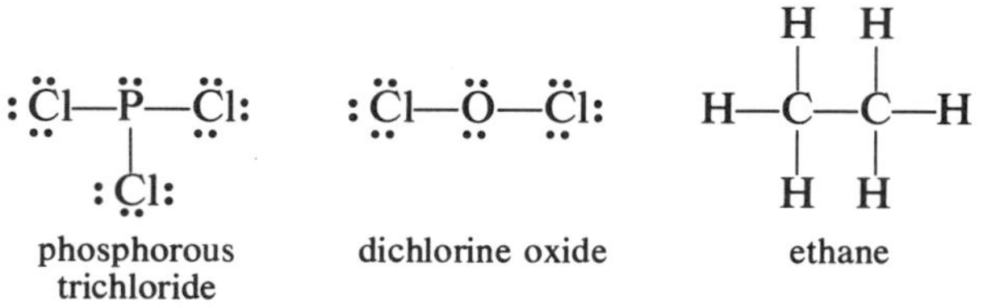

The following are examples of molecules that contain double and triple bonds.

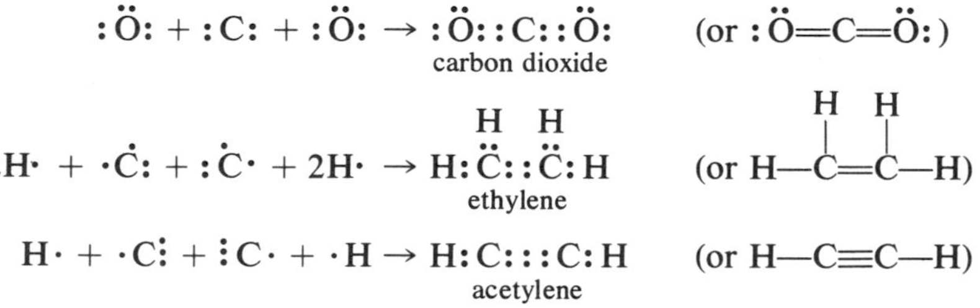

Notice that in each compound the number of covalent bonds on each atom agrees with the number predicted.

7.2 FORMAL CHARGE

In the formation of certain covalent bonds, *both* of the shared electrons are furnished by *one* of the bonded atoms. For example, in the reaction of ammonia with a proton (a hydrogen atom stripped of its electron), the unshared electron pair of the nitrogen atom of NH_3 is used to form a new covalent bond.

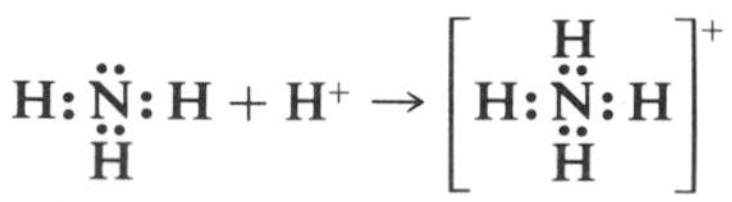

A bond formed in this way is frequently called a "coordinate covalent" bond, but it is probably unwise to do so. Labeling a specific bond as a "coordinate covalent" bond implies that it is different from other covalent bonds and has little justification. All electrons are alike no matter what their source. All the bonds in NH_4^+ are identical. It is impossible to distinguish between them.

Notice, however, that the number of covalent bonds on the N atom of NH_4^+ does not agree with the number predicted in the preceding section. Since a nitrogen atom has five valence electrons (group V A), it would be expected to satisfy the octet principle through the formation of three covalent bonds. This prediction is correct for NH_3; it is not correct for NH_4^+.

An answer to the question may be obtained by calculating the formal charges of the atoms in NH_4^+. A **formal charge** is calculated in the following way. For an A family atom, the number of valence electrons equals the group number. If we remove all of these electrons, the imaginary positive ion that we produce has a charge equal to

$$+(\textit{group no.})$$

An atom in a molecule, however, must have some valence electrons associated with it. We divide the electron pair of each bond equally between the bonded atoms. The division of each *bond,* therefore, gives the atoms a 1− charge. Each atom of the molecule also has 1− charges equal to the number of unshared *electrons* on that atom. For each atom in the structure, the total *negative* charge from these sources is

$$-[(\textit{no. bonds}) + (\textit{no. unshared } e^-)]$$

The formal charge of the bonded atom, therefore is

$$\textit{formal charge} = +(\textit{group no.}) - [(\textit{no. bonds}) + (\textit{no. unshared } e^-)]$$

Since the N atom in NH_4^+ has four bonds and no unshared electrons, its formal charge is

$$\textit{formal charge} = +5 - (4 + 0) = 1+$$

Each H atom in the NH_4^+ ion has a formal charge of zero.

$$\textit{formal charge} = +1 - (1 + 0) = 0$$

The formal charge of the N atom in NH_4^+ is indicated in the following way.

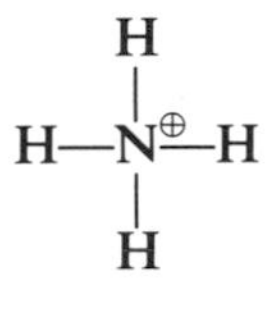

We can now explain the difference between the predicted and actual number of bonds on the N atom in NH_4^+. A hypothetical N^+ would have four valence electrons and be able to form four covalent bonds. An uncharged N atom, on the other hand, has five valence electrons and can form only three covalent bonds.

A formal charge is, as the name implies, only a formality. In the assignment of formal charges, the assumption is made that the electron pair of any covalent bond is shared equally by the bonded atoms. Such an assumption is usually not true, and formal charges must be interpreted carefully. The electron density on the N atom in NH_4^+ is less than that on the N atom in NH_3, but the actual charge is not a full positive charge because the bonding electrons are not equally shared.

As another example of the way in which formal charges are calculated, consider the $POCl_3$ molecule.

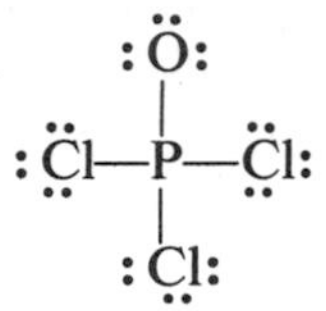

The formal charge of the O atom is

$$\textit{formal charge} = +(\textit{group no.}) - [(\textit{no. bonds}) + (\textit{no. unshared } e^-)]$$
$$= +6 - (1 + 6) = 1-$$

The formal charge of the P atom is

$$\textit{formal charge} = +5 - (4 + 0) = 1+$$

The formal charge of each Cl atom is

$$\textit{formal charge} = +7 - (1 + 6) = 0$$

The structure of the molecule is

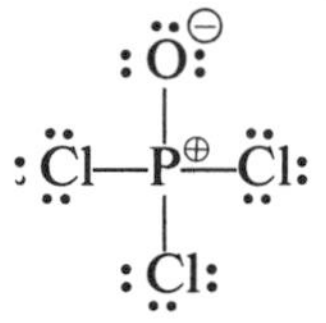

Notice that the sum of the formal charges in the $POCl_3$ structure is zero. The formal charges of any molecule add up to zero. The sum of the formal charges of the atoms in an ion equals the charge of the ion.

An atom in a Lewis structure that has the number of bonds expected on the basis of its group number has no formal charge. If possible, a Lewis structure should be drawn so that each atom has the number of bonds expected on the basis of its group number. Frequently, however, it is not possible.

Atoms that are bonded to each other in a structure should not have formal charges with the same sign. The repulsion between the charges would tend to break the bond between the atoms. A Lewis structure in which this **adjacent**

FORMAL CHARGES

1. The formal charge of an atom in a Lewis structure can be calculated by use of the formula:

$$formal\ charge = +(group\ no.) - [(no.\ bonds) + (no.\ unshared\ e^-)]$$

2. In a molecule, the sum of the formal charges is zero. In an ion, the formal charges add up to the charge of the ion.
3. An atom in a Lewis structure that has the number of bonds expected on the basis of its group number has no formal charge. If possible, a Lewis structure should be drawn so that all atoms have these numbers of bonds. Frequently, however, it is not possible.
4. Atoms that are bonded to each other in a structure should not have formal charges with the same sign. A Lewis structure in which this adjacent charge rule is violated is usually not an accurate description of the molecule or ion.

charge rule is violated is usually not an accurate representation of the molecule or ion (see Example 7.3).

The following examples illustrate how to draw Lewis structures.

Example 7.1 Diagram the Lewis structure of the SO_2 molecule. The molecule is angular, and the two O atoms are bonded to a central S atom.

Solution

1. Since both S and O are members of group VI A, each of the 3 atoms in the molecule supplies 6 valence electrons. The total number of valence electrons is

$$total\ no.\ e^- = 3(6) = 18$$

2. If each atom had its own octet, the number of electrons required would be

$$\begin{aligned} no.\ e^-\ for\ individual\ atoms &= 2(no.\ H\ atoms) + 8(no.\ other\ atoms) \\ &= 2(0) + 8(3) = 24 \end{aligned}$$

3. The number of electrons needed to complete the octet of each atom individually minus the total number of electrons is the number of electrons that must be shared (the number of bonding electrons).

$$\begin{aligned} no.\ bonding\ e^- &= (no.\ e^-\ for\ individual\ atoms) - (total\ no.\ e^-) \\ &= 24 - 18 = 6 \end{aligned}$$

4. The total number of bonds in the molecule is one-half the number of bonding electrons. It takes 2 electrons to make a bond.

$$\textit{no. bonds} = (\textit{no. bonding } e^-)/2$$
$$= 6/2 = 3$$

5. The atoms are arranged in the following way.

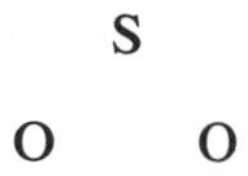

6. The 3 covalent bonds are indicated by dashes.

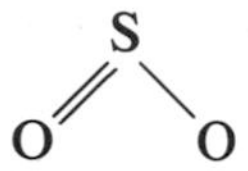

7. The number of unshared electrons is calculated from the total number of electrons minus the number of bonding *electrons* (from step 3).

$$\textit{no. unshared } e^- = (\textit{total no. } e^-) - (\textit{no. bonding } e^-)$$
$$= 18 - 6 = 12$$

These electrons are distributed in such a way as to give each atom an octet.

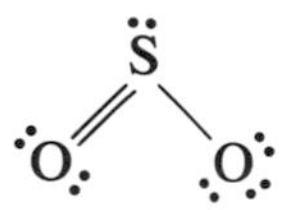

8. The formal charge of the S atom is

$$\textit{formal charge} = +(\textit{group no.}) - [(\textit{no. bonds}) + (\textit{no. unshared } e^-)]$$
$$= +6 - (3 + 2) = 1+$$

The formal charge of the left-hand O atom is

$$\textit{formal charge} = +6 - (2 + 4) = 0$$

The formal charge of the right-hand O atom is

$$\textit{formal charge} = +6 - (1 + 6) = 1-$$

The structure of the molecule is

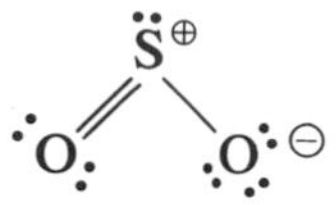

HOW TO DIAGRAM LEWIS STRUCTURES

1. Find the total number of valence electrons supplied by all the atoms in the structure. The number supplied by each A family element is the same as the group number of the element.
   (a) For a negative ion, increase the number by the charge of the ion.
   (b) For a positive ion, decrease the number by the charge of the ion.
2. Determine the number of electrons that would be required to give 2 electrons to each H atom individually and 8 electrons to each of the other atoms individually.

   $$no.\ e^-\ for\ individual\ atoms = 2(no.\ H\ atoms) + 8(no.\ other\ atoms)$$

3. The number obtained in step 2 minus the number obtained in step 1 is the number of electrons that must be shared in the final structure.

   $$no.\ bonding\ e^- = (no.\ e^-\ for\ individual\ atoms) - (total\ no.\ e^-)$$

4. One-half the number of bonding electrons (step 3) is the number of covalent bonds in the final structure.

   $$no.\ bonds = (no.\ bonding\ e^-)/2$$

5. Write the symbols for the atoms present in the structure, arranging them in the way that they are found in the structure.
6. Indicate covalent bonds by dashes written between the symbols. Indicate one bond between each pair of symbols, and then use any remaining from the total calculated in step 4 to make multiple bonds. Note that each H atom is limited to one bond.
7. The total number of electrons (step 1) minus the number of bonding *electrons* (step 3) is the number of unshared electrons.

   $$no.\ unshared\ e^- = (total\ no.\ e^-) - (no.\ bonding\ e^-)$$

   Complete the electron octet of each atom (other than the H atoms) by adding dots to represent unshared electrons.
8. Indicate the formal charges of the atoms where appropriate, and evaluate the structure.

Example 7.2 Diagram the Lewis structure of the carbonate ion, CO_3^{2-}. The C atom is the central atom to which the 3 O atoms are bonded.

Solution

1. The total number of valence electrons in the ion is

4	(from C)
18	(from the three O atoms)
2	(from the ionic charge)
24	

2. $\textit{no. } e^- \textit{ for individual atoms} = 2(\textit{no. H atoms}) + 8(\textit{no. other atoms})$
$= 2(0) + 8(4) = 32$

3. $\textit{no. bonding } e^- = (\textit{no. } e^- \textit{ for individual atoms}) - (\textit{total no. } e^-)$
$= 32 - 24 = 8$

4. $\textit{no. bonds} = (\textit{no. bonding } e^-)/2$
$= 8/2 = 4$

5. and **6.**

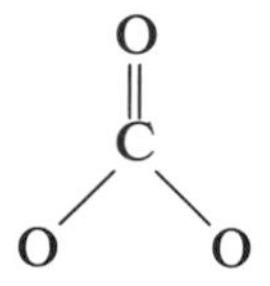

7. $\textit{no. unshared } e^- = (\textit{total no. } e^-) - (\textit{no. bonding } e^-)$
$= 24 - 8 = 16$

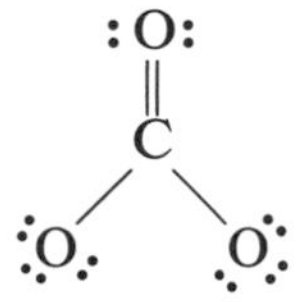

8. $\textit{formal charge} = +(\textit{group no.}) - [(\textit{no. bonds}) + (\textit{no. unshared } e^-)]$
For the C atom,

$$\textit{formal charge} = +4 - (4 + 0) = 0$$

For the top O atom

$$\textit{formal charge} = +6 - (2 + 4) = 0$$

For both the left-hand and right-hand bottom O atom,

$$\textit{formal charge} = +6 - (1 + 6) = 1-$$

The final structure is:

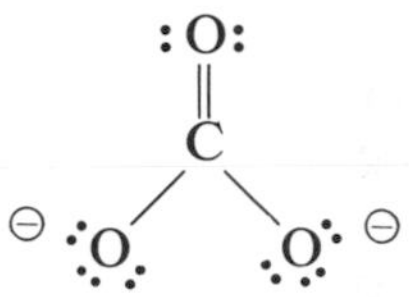

Notice that the formal charges add up to the charge of the ion.

Example 7.3 Diagram the Lewis structure of nitric acid, HNO_3. The N atom is the central atom to which the three O atoms are bonded. The H atom is bonded to one of the O atoms.

Solution

1. The total number of valence electrons in the molecule is

1	(from the H atom)
5	(from the N atom)
18	(from the three O atoms)
24	

2. $no.\ e^-\ for\ individual\ atoms = 2(no.\ H\ atoms) + 8(no.\ other\ atoms) = 2(1) + 8(4) = 34$

3. $no.\ bonding\ e^- = (no.\ e^-\ for\ individual\ atoms) - (total\ no.\ e^-) = 34 - 24 = 10$

4. $no.\ bonds = (no.\ bonding\ e^-)/2 = 10/2 = 5$

5. and **6.**

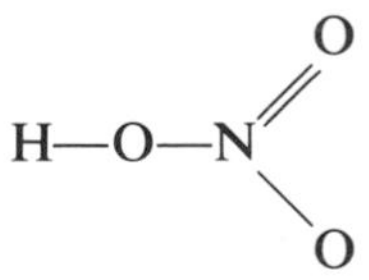

7. $no.\ unshared\ e^- = (total\ no.\ e^-) - (no.\ bonding\ e^-) = 24 - 10 = 14$

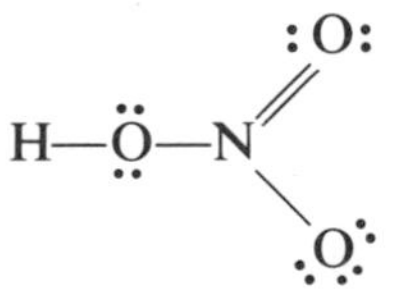

8. The final structure (including formal charges) is:

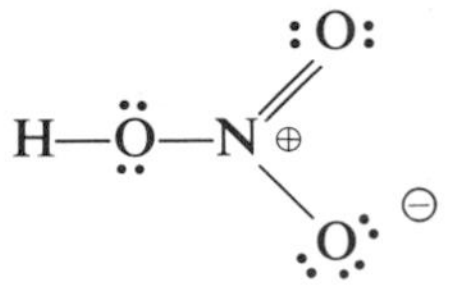

Another structure

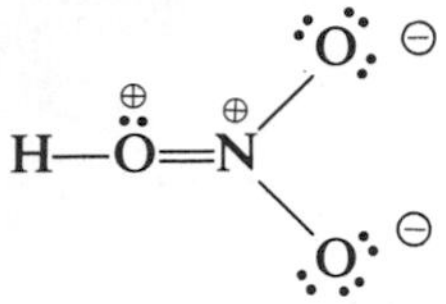

is discarded because adjacent atoms in the molecule carry formal charges with the same sign.

7.3 TRANSITION BETWEEN IONIC AND COVALENT BONDING

The bonding in most compounds is intermediate between purely ionic and purely covalent.

The best examples of ionic bonding are found in compounds formed by a metal with a very low ionization energy (for example, Cs) and a nonmetal that has a strong tendency toward electron gain (for example, F). In a compound such as CsF, the ions exist as separate units in the crystal.

A purely covalent bond is found only in molecules formed from two *identical* atoms, such as Cl_2. One Cl atom attracts electrons to the same extent as any other. The electron cloud of the bond is distributed symmetrically around the two nuclei. In other words, the bonding electrons are shared equally by the two identical atoms.

The bonding in most compounds is somewhere between these two extremes. The character of the bonding in a compound that contains a metal and a nonmetal can be interpreted in terms of the interactions between the ions. The positively charged ion is believed to attract and deform the electron cloud of the anion. The electron cloud of the negative ion is drawn toward the cation. In extreme cases, ion deformation can lead to compounds that are more covalent than ionic (Figure 7.2). The degree of covalent character that a compound has can be considered to correspond to the extent that the anion is distorted.

How easily an anion can be distorted depends upon its size and its charge. A *large anion,* in which the outer electrons are far away from the nucleus, is easily deformed. The iodide ion (I^-, ionic radius of 216 pm) is more easily distorted than the fluoride ion (F^-, ionic radius of 136 pm). If an anion has a *high negative charge,* so much the better. In a highly charged anion the electrons outnumber the protons to a greater extent than in an anion with a lower charge. The electron cloud of a highly charged anion is easily distorted. The S^{2-} ion is more easily distorted than the Cl^- ion.

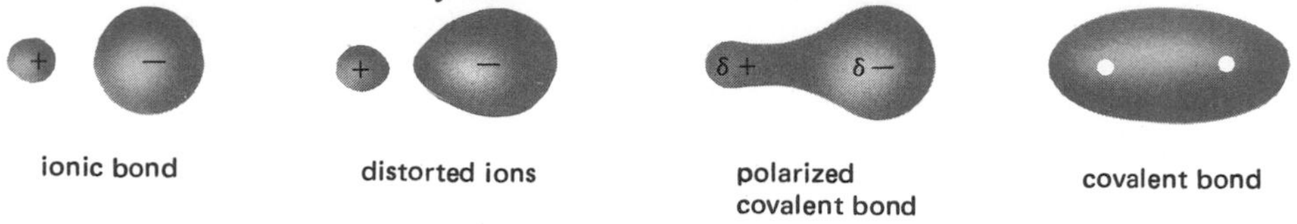

Figure 7.2 Transition between ionic and covalent bonding.

The ability of a cation to distort the electron cloud of a neighboring anion also depends upon size and charge. A *small cation* with a *high positive charge* is the most effective in bringing about anion distortion. A cation of this type has a high concentration of positive charge.

In any group of metals in the periodic table, the member that forms the smallest cation (for example, Li in group I A) has the greatest tendency toward the formation of compounds with a high degree of covalent character. All compounds of beryllium (Be^{2+} is the smallest cation of any group II A element) are significantly covalent. Boron (the smallest member of group III A) forms only covalent compounds. The hypothetical B^{3+} ion would have a comparatively high charge combined with a very small size, which would cause extensive anion distortion leading to covalent bonding.

The first four metals of the fourth period form chlorides in which each of the "cations" is isoelectronic with Ar. In this series, KCl, $CaCl_2$, $ScCl_3$, and $TiCl_4$, the covalent character increases with increasing charge and decreasing size of the "cation." KCl is strongly ionic, and $TiCl_4$, a liquid, is definitely covalent. Truly ionic compounds that contain cations with a charge of 3+ or higher are rare. Such compounds as $SnCl_4$, $PbCl_4$, $SbCl_5$, and BiF_5 are covalent.

Another approach to bonds of intermediate character considers the polarization of covalent bonds. A purely covalent bond results only when two *identical* atoms are bonded. Whenever two *different* atoms are joined by a covalent bond, the electron density of the bond is not symmetrically distributed around the two nuclei. The electrons of the bond are not shared equally. No matter how similar the atoms may be, there will be a difference in their ability to attract electrons.

Chlorine has a greater attraction for electrons than bromine has. In the BrCl molecule the electrons of the covalent bond are more strongly attracted by the chlorine atom than by the bromine atom. The electron cloud of the bond is denser in the vicinity of the chlorine atom. The chlorine end of the bond, therefore, has a partial negative charge. Since the molecule as a whole is electrically neutral, the bromine end is left with a partial positive charge of equal magnitude. Such a bond, with positive and negative poles, is called a **polar covalent bond.** The partial charges of the bond are indicated by the symbols δ^+ and δ^- to distinguish them from full ionic charges.

The greater the difference between the electron-attracting ability of two atoms joined by a covalent bond, the more polar the bond, and the larger the magnitude of the partial charges. If the unequal sharing of electrons were carried to an extreme, one of the bonded atoms would have all of the bonding electrons, and separate ions would result.

A polar covalent molecule turns itself in the electric field between the plates of a condenser in such a way that the negative end is toward the positive plate and the positive end is toward the negative plate (Figure 7.3). Polar molecules arranged in this way affect the amount of charge that a pair of elec-

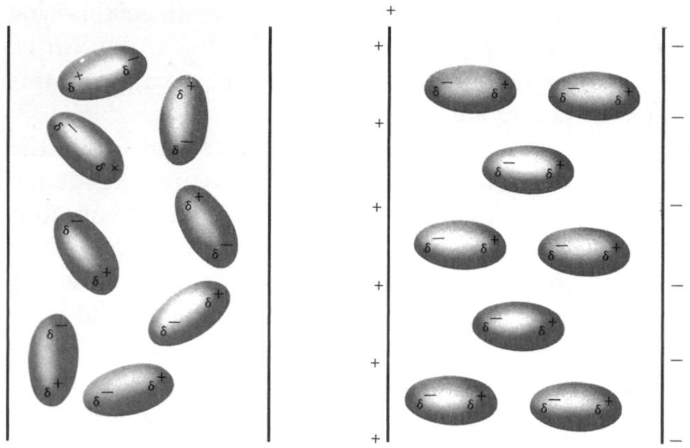

Figure 7.3 Effect of an electrostatic field on the orientation of polar molecules.

trically charged plates can hold. As a result, measurements can be made that allow the calculation of a value called the dipole moment.

If two equal charges, with opposite signs, are separated by a known distance, the **dipole moment** is

$$dipole\ moment = (charge)(distance)$$

The dipole moments of nonpolar molecules, such as H_2, Cl_2, and Br_2, are zero. The dipole moment of polar diatomic molecules increases as the polarity of the molecule increases.

Linus Pauling has used dipole moment to calculate the **partial ionic character** of a covalent bond. If HCl were completely ionic, the H^+ ion and the Cl^- ion would each have a unit charge (1.60×10^{-19} coul). The bond distance in the HCl molecule is 127 pm (which is 1.27×10^{-10} m). The dipole moment of the imaginary H^+Cl^- would be

$$\begin{aligned} dipole\ moment &= (charge)(distance) \\ &= (1.60 \times 10^{-19}\ \text{coul})(1.27 \times 10^{-10}\ \text{m}) \\ &= 2.03 \times 10^{-29}\ \text{coul·m} = 6.08\ \text{D} \end{aligned}$$

The debye unit, D, is 3.34×10^{-30} coul·m.

The experimentally determined dipole moment of HCl is 1.03 D. The correct dipole moment is

$$1.03\ \text{D}/6.08\ \text{D} = 0.169$$

times the value for the imaginary, ionic compound. The HCl bond appears to be about 17% ionic.

7.4 ELECTRONEGATIVITY

Electronegativity is a measure of the ability of an atom in a molecule to attract electrons. The polarity of the HCl molecule may be said to arise from the difference between the electronegativity of chlorine and that of hydrogen.

Since chlorine is more electronegative than hydrogen, the chlorine end of the molecule is the negative end of the dipole.

The concept of electronegativity is useful but inexact. There is no simple, direct way to measure electronegativities, and various methods of determining them have been proposed. Since this property depends not only upon the structure of the atom under consideration but also upon the number and nature of the atoms to which it is bonded, the electronegativity of an atom is not constant. The electronegativity of phosphorus is different in PCl_3 than in PF_5.

The electronegativity scale of Pauling is based on bond energies. The bond energies of the H—H and Br—Br bonds are

$$\text{H—H(g)} \rightarrow \text{2H(g)} \qquad \Delta H = +435 \text{ kJ}$$
$$\text{Br—Br(g)} \rightarrow \text{2Br(g)} \qquad \Delta H = +193 \text{ kJ}$$

Both these bonds are nonpolar. The average of the two values

$$(435 \text{ kJ/mol} + 193 \text{ kJ/mol})/2 = +314 \text{ kJ/mol}$$

may be considered to apply to a "nonpolar H—Br bond" in which the electrons are shared equally. The bond in HBr actually is polar. The measured bond energy is larger than the one we have calculated.

$$\text{H—Br(g)} \rightarrow \text{H(g)} + \text{Br(g)} \qquad \Delta H = +364 \text{ kJ}$$

The difference, Δ, between the measured and calculated values is

$$\Delta = 364 \text{ kJ/mol} - 314 \text{ kJ/mol} = +50 \text{ kJ/mol}$$

The HBr bond is 50 kJ/mol stronger than expected on the basis of equal electron sharing. The difference represents the amount of energy needed to separate the partial charges produced by unequal sharing.

The measured bond energy of a polar bond is always larger than the calculated value. The amount of the difference, Δ, depends upon the degree of polarity of the bond. Highly polar bonds are formed from atoms that have a large difference in electronegativity. A large value of Δ, therefore, indicates a large difference in electronegativity. Values of Δ were used to obtain the electronegativity values of Table 7.1.

In general, electronegativity increases from left to right across any period (with increasing number of valence electrons) and from bottom to top in any group (with decreasing size). The most highly electronegative elements are found in the upper-right corner of the periodic table (ignoring the noble gases). The least electronegative elements are found in the lower-left corner of the chart.

Metals are elements that have small attractions for valence electrons (low electronegativities). Nonmetals, except for the noble gases, have large attractions (high electronegativities). Electronegativities, therefore, can be used to rate the reactivity of metals and nonmetals. The positions of the ele-

Table 7.1 Electronegativities of the elements[a]

1 H 2.1																	2 He —
3 Li 1.0	4 Be 1.5											5 B 2.0	6 C 2.5	7 N 3.0	8 O 3.5	9 F 4.0	10 Ne —
11 Na 0.9	12 Mg 1.2											13 Al 1.5	14 Si 1.8	15 P 2.1	16 S 2.5	17 Cl 3.0	18 Ar —
19 K 0.8	20 Ca 1.0	21 Sc 1.3	22 Ti 1.5	23 V 1.6	24 Cr 1.6	25 Mn 1.5	26 Fe 1.8	27 Co 1.8	28 Ni 1.8	29 Cu 1.9	30 Zn 1.6	31 Ga 1.6	32 Ge 1.8	33 As 2.0	34 Se 2.4	35 Br 2.8	36 Kr —
37 Rb 0.8	38 Sr 1.0	39 Y 1.2	40 Zr 1.4	41 Nb 1.6	42 Mo 1.8	43 Tc 1.9	44 Ru 2.2	45 Rh 2.2	46 Pd 2.2	47 Ag 1.9	48 Cd 1.7	49 In 1.7	50 Sn 1.8	51 Sb 1.9	52 Te 2.1	53 I 2.5	54 Xe —
55 Cs 0.7	56 Ba 0.9	57–71 La-Lu 1.1–1.2	72 Hf 1.3	73 Ta 1.5	74 W 1.7	75 Re 1.9	76 Os 2.2	77 Ir 2.2	78 Pt 2.2	79 Au 2.4	80 Hg 1.9	81 Tl 1.8	82 Pb 1.8	83 Bi 1.9	84 Po 2.0	85 At 2.2	86 Rn —
87 Fr 0.7	88 Ra 0.9	89– Ac- 1.1–1.7															

[a] Based on Linus Pauling, *The Nature of the Chemical Bond,* © 1939, 1940, *Third Edition,* © 1960 by Cornell University Press. Used with the permission of Cornell University Press. British Commonwealth Rights by permission of Oxford University Press, Oxford, England.

ments in the periodic table are helpful in making predictions concerning chemical reactivity (Figure 7.4).

Electronegativities can be used to predict the type of bonding that a compound will have. When two elements of widely different electronegativity combine, an ionic compound results. The electronegativity difference between sodium and chlorine is 2.1, and NaCl is an ionic compound.

Covalent bonding occurs between nonmetals. The difference between the electronegativities of two nonmetals is not very large. In such cases the electronegativity differences give an indication of the degree of polarity of the covalent bonds. If the electronegativity difference is zero or very small, an essentially nonpolar bond with equal or almost equal sharing of electrons can be assumed. The larger the electronegativity difference, the more polar the covalent bond is. The bond is polarized in the direction of the atom with the larger electronegativity. By using electronegativities we can predict that HF is the most polar, and has the largest bond energy, of any hydrogen halide (see Table 7.2).

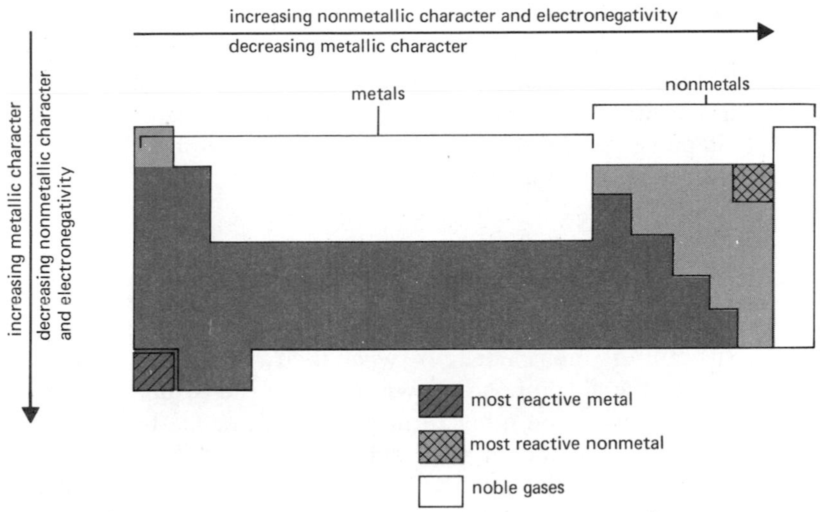

Figure 7.4 The relation between position in the periodic classification and metallic or nonmetallic reactivity.

The type of bonding that occurs between two metals (the metallic bond), in which electronegativity differences are also relatively small, will be considered in Section 19.1.

7.5 RESONANCE

In some cases the properties of a molecule or ion are not adequately represented by a single Lewis structure. For example, the structure for SO_2 that was derived in Example 7.1

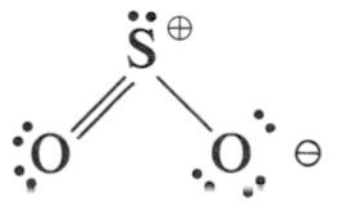

is unsatisfactory in two respects. The structure depicts the S atom to be bonded to one O atom by a double bond and to the other O atom by a single bond. Double bonds are shorter than single bonds, yet experimental evidence

Table 7.2 Some properties of the hydrogen halides

HYDROGEN HALIDE	DIPOLE MOMENT (D)	BOND ENERGY (kJ/mol)	ELECTRONEGATIVITY OF HALOGEN	ELECTRONEGATIVITY DIFFERENCE BETWEEN HYDROGEN AND HALOGEN
HF	1.91	565	F = 4.0	1.9
HCl	1.03	431	Cl = 3.0	0.9
HBr	0.78	364	Br = 2.8	0.7
HI	0.38	297	I = 2.5	0.4

shows that both S to O linkages are the same length. The structure also shows one of the O atoms to be more negative than the other. It is known, however, that both of the O atoms are the same in this respect.

In a case of this type, two or more valence-bond structures can be used in combination to depict the molecule. The molecule is said to be a **resonance hybrid** of the structures, which are called **resonance forms.** For SO_2,

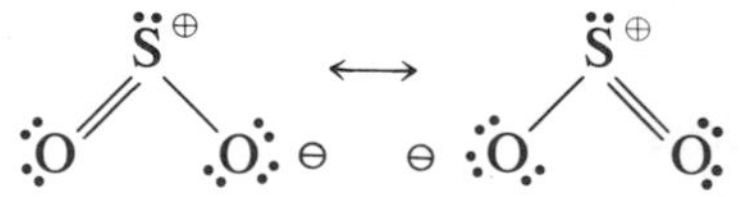

The actual structure does not correspond to either resonance form alone. Instead, the true structure is intermediate between the two forms. Each of the S to O bonds is neither the single bond shown in one of the resonance forms nor the double bond shown in the other form. It is intermediate between the two. Both bonds, therefore, are identical. Each O atom may be considered to have a formal charge of $\frac{1}{2}-$, since in one form it has a formal charge of 1− and in the other it has a formal charge of zero. The oxygens, therefore, are equally negative.

There is only one type of SO_2 molecule and only one structure. The electrons of the molecule do *not* flip back and forth so that the molecule is in one form at one moment and in the other form at the next. The molecule always has the same structure. The problem arises because the theory is limited—not because the SO_2 molecule is unusual.

In all the resonance forms of a molecule, the nuclei of the atoms *must be arranged in the same way*. The forms differ only in electron configuration. In the benzene molecule, C_6H_6, the six C atoms are arranged in a ring, and each C atom has an H atom bonded to it by a single bond. All the C to C bond distances are the same length, 139 pm. This bond distance is intermediate between the C—C single bond distance of 154 pm and the C=C double bond distance of 133 pm found in other molecules. The resonance forms of benzene are

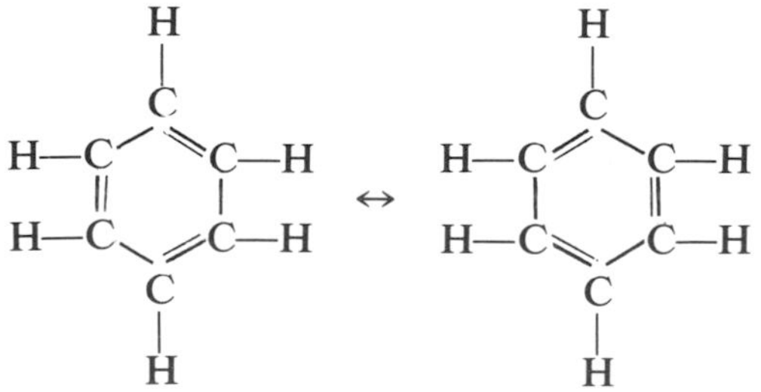

This representation shows that all the C to C bonds are the same length, between a single and a double bond distance.

The structures of ions may be represented by resonance forms. A Lewis

structure for the carbonate ion, CO_3^{2-}, was derived in Example 7.2. The ion is planar, all bonds are equivalent (between single and double bonds), and all oxygen atoms are equally negative. The formal charges add up to the charge on the ion.

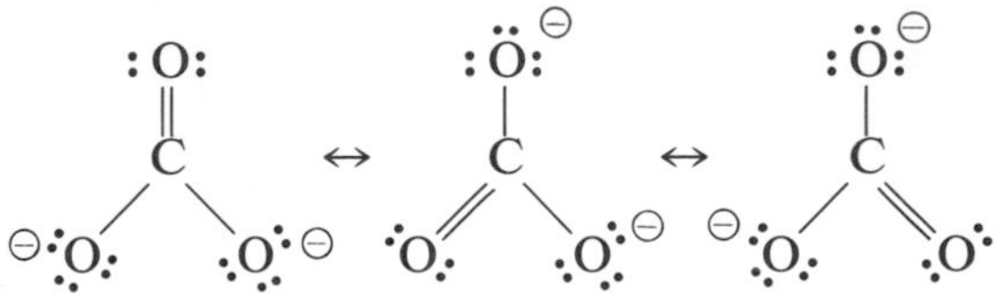

7.6 HYBRID ORBITALS

We have seen that certain ions exist that do not have noble-gas configurations and that some of these ions are relatively stable. Some molecules also exist in which atoms have configurations other than those that the octet principle would lead us to expect.

There are a few molecules (such as NO and NO_2) that have an odd number of valence electrons. It is impossible to divide an odd number of electrons so that each atom of the molecule has a configuration of eight electrons (an even number). There are not many stable odd electron molecules. Odd electron species are usually very reactive and consequently short-lived.

More common are the molecules that have an even number of valence electrons but contain atoms with valence shells of less than, or more than, eight electrons. In the BF_3 molecule, the B atom has six valence electrons

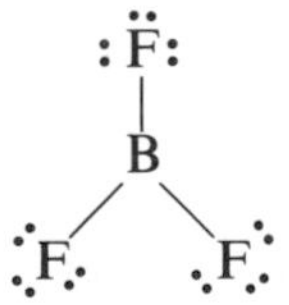

In the PCl_5 and SF_6 molecules the P and S atoms have ten and twelve valence electrons, respectively.

For the elements of the second period, only four bonding orbitals are available (2*s* and 2*p*). The number of covalent bonds on atoms of these elements is therefore limited to a maximum of four. The elements of the third and subsequent periods, however, have more orbitals available in their outer electron shells. Compounds are known in which these elements form four, five, six, or (infrequently) an even higher number of covalent bonds. In valence-bond structures representing compounds containing elements of the third and subsequent periods, therefore, the octet principle is frequently violated. Apparently, the criteria for covalent bond formation should be centered on the electron pair rather than on the attainment of an octet.

In the formation of a covalent bond, orbitals from two atoms overlap (Figure 7.5). The electron clouds reinforce each other in the bonding region between the two nuclei, and electronic charge is concentrated in this region.

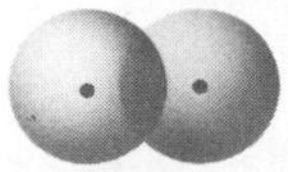

Figure 7.5 Overlap of the 1*s* orbitals of two hydrogen atoms.

The strength of the bond depends upon the extent of the overlap. The strongest bonding occurs when the orbitals overlap the most.

In the vapor state, beryllium chloride consists of *linear* $BeCl_2$ molecules.

$$:\ddot{\underset{..}{Cl}}:Be:\ddot{\underset{..}{Cl}}:$$

Both covalent bonds of the molecule are equivalent in strength and bond length. A *p* orbital of each Cl atom is used to form a bond with the central Be atom. But what type of Be orbitals are employed? The electronic configuration of Be is $1s^2 2s^2$, and it appears that in the ground state beryllium should not form covalent bonds at all. The excited state $1s^2\ 2s^1\ 2p^1$ provides two unpaired electrons for bonding, but it too is inadequate in explaining the bonding of the molecule.

If the Be atom used the 2*s* and 2*p* orbitals to form bonds, the two bonds would not be similar. The 2*s* and 2*p* orbitals would not overlap the orbitals of the two Cl atoms in the same way or to the same extent. The bond lengths would be different, and the bonds would have different bond strengths.

The structure of the molecule can be explained by assuming that the 2*s* orbital and a 2*p* orbital of Be are blended, or hybridized, into two **hybrid orbitals** that are used in the bonding (Figure 7.6). These hybrid orbitals (called ***sp* orbitals**) have one-half *s* character and one-half *p* character, are equivalent, and are directed 180° from one another (Figure 7.7). The *sp* orbitals extend farther along the *x* axis than either the *s* orbital or the *p* orbital. Greater orbital overlap results when an *sp* hybrid orbital is used for bonding than when a pure *s* or a pure *p* orbital is used. Calculations show that stronger bonds result from the use of hybrid orbitals.

The hypothetical formation of $BeCl_2$ may be summarized as

2*p* __ __ __	2*p* ↑ __ __	2*p* __ __
		sp ↑ ↑
2*s* ↑↓	2*s* ↑	
1*s* ↑↓	1*s* ↑↓	1*s* ↑↓
Be (ground state)	Be (excited state)	Be (hybrid state)

Figure 7.6 Formation of $BeCl_2$. Shapes of orbitals have been simplified for clarity.

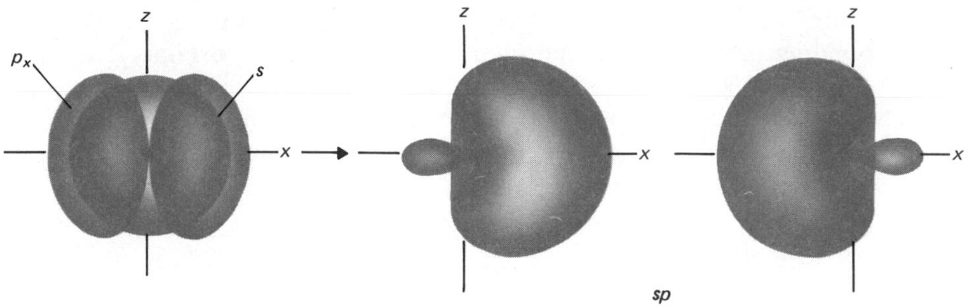

Figure 7.7 Hybridization of s and p orbitals to produce sp hybrid orbitals. Nuclei at origins; sp orbitals should be superimposed.

The concept of hybridization is a theoretical device that permits correlation of the ground-state electronic configuration of an atom with the observed structural properties of a molecule that it forms. The preceding analysis (and others that follow) should not be interpreted as a description of an actual physical process by which a molecule is formed.

Other types of hybrid orbitals are known (Figure 7.8). In the description

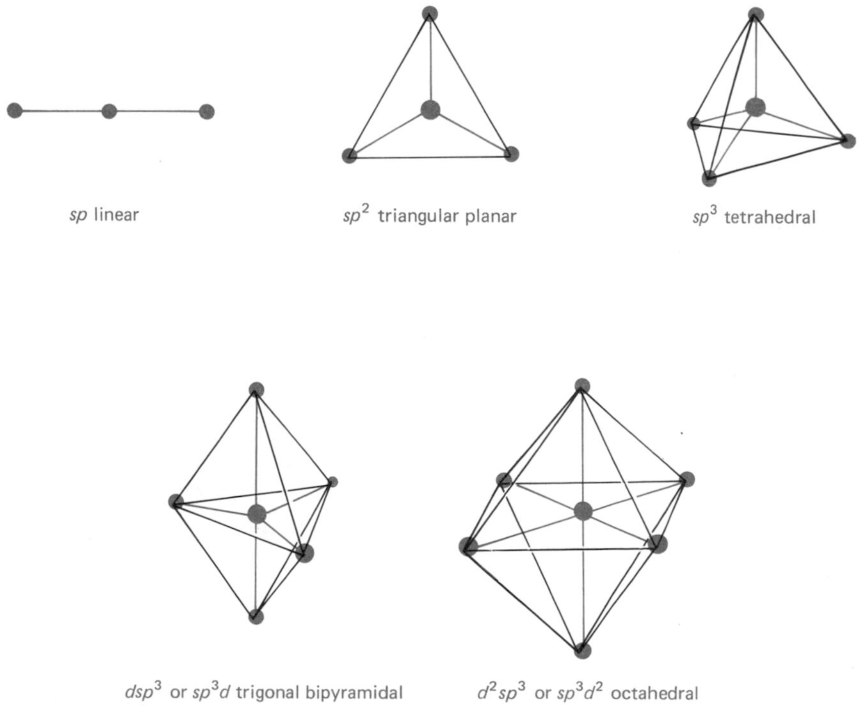

Figure 7.8 Directional characteristics of hybrid orbitals.

of molecules of the type AB_n, where n is two or more, attention is directed to the bonding orbitals of the central atom, A, since these orbitals determine the shape of the molecule.

The atoms of boron trifluoride lie in the same plane, the angle formed by any two bonds is 120°, and the three bonds are equivalent.

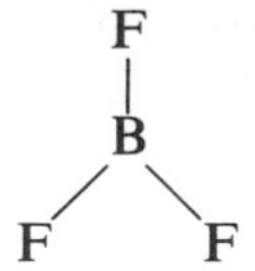

It is assumed that sp^2 **hybrid orbitals** of B are used in the formation of the molecule and that each orbital has $\frac{1}{3}s$ character and $\frac{2}{3}p$ character (Figure 7.9).

B (ground state)	B (excited state)	B (hybrid state)
2p ↑ _ _	2p ↑ ↑ _	2p _
		sp^2 ↑ ↑ ↑
2s ↑↓	2s ↑	
1s ↑↓	1s ↑↓	1s ↑↓

All molecules in which the central atom uses sp^2 hybrid orbitals for bonding are triangular (trigonal) planar, and all sp^2 hybrid orbitals have the orientation shown in Figure 7.9.

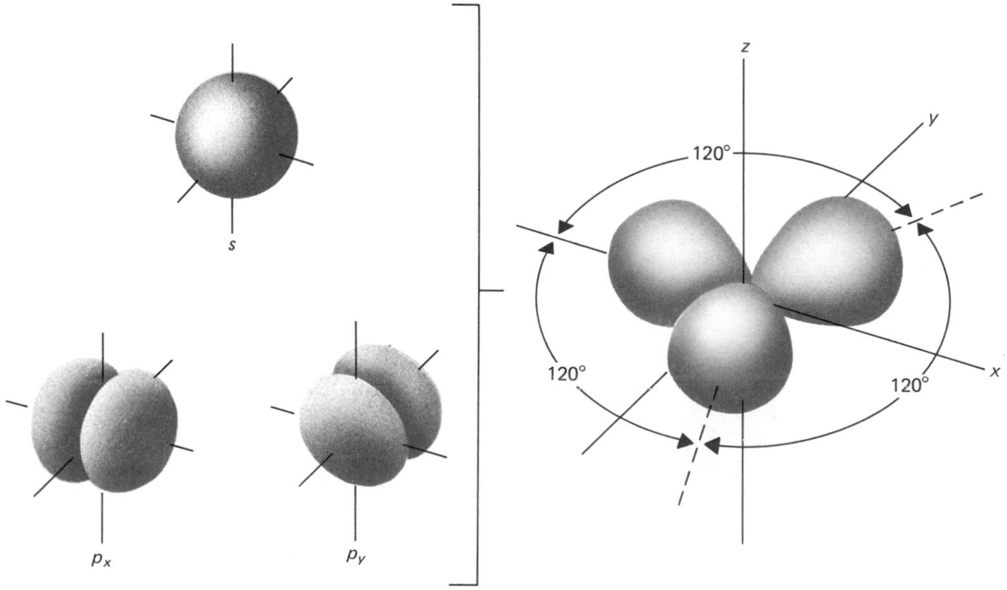

Figure 7.9 Formation of hybrid sp^2 orbitals. The hybrid orbitals are coplanar. Shapes of the hybrid orbitals have been simplified for clarity.

An important and relatively common type of hybridization is found in methane, CH_4. In this molecule an *s* and three *p* orbitals are mixed to produce four equivalent ***sp*³ hybrid orbitals,** each one with $\frac{1}{4}s$ character and $\frac{3}{4}p$ character.

$2p$ ↑ ↑ __	$2p$ ↑ ↑ ↑	
		sp^3 ↑ ↑ ↑ ↑
$2s$ ↑↓	$2s$ ↑	
$1s$ ↑↓	$1s$ ↑↓	$1s$ ↑↓
C (ground state)	C (excited state)	C (hybrid state)

The sp^3 hybrid orbitals are directed to the corners of a regular tetrahedron (Figure 7.10), and the CH_4 molecule has this geometry. In CH_4, all H—C—H bond angles are 109° 28′, the so-called tetrahedral angle. Chemical and physical evidence show that all the C—H bonds are equivalent in methane.

The nitrogen atom of NH_3 may be assumed to use sp^3 hybrid orbitals for bonding and to have one of these orbitals occupied by an unshared pair of electrons. The atoms of NH_3 form a trigonal pyramid, but the four N orbitals are arranged approximately in the form of a tetrahedron (see Figure 7.11, which appears in the next section).

$2p$ ↑ ↑ ↑	
	sp^3 ↑↓ ↑ ↑ ↑
$2s$ ↑↓	
$1s$ ↑↓	$1s$ ↑↓
N (ground state)	N (hybrid state)

The H—N—H bond angle of NH_3 (107.3°) is close to the tetrahedral angle (109.5°). The N atom could conceivably use its three $2p$ orbitals for bonding.

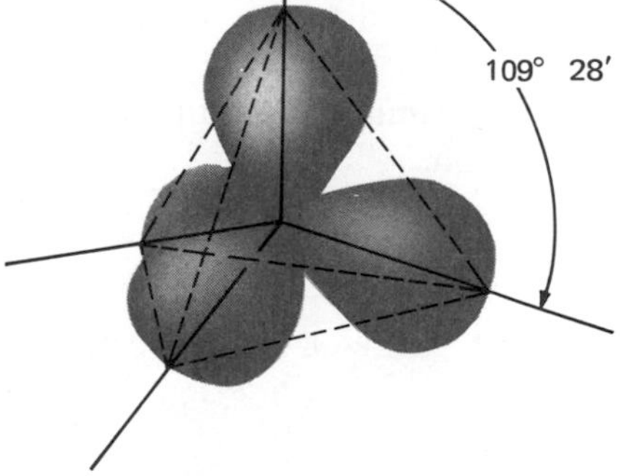

Figure 7.10 Orientation of the sp^2 hybrid orbitals.

In this instance, however, the bond angle would be 90°, since p orbitals are directed in space at right angles to one another along x, y, and z axes. The observed bond angle is closer to the tetrahedral angle than it is to 90°.

Similarly, the oxygen atom of the H_2O molecule may be assumed to use sp^3 hybrid orbitals in the bonding of the molecule and to have two of these hybrid orbitals occupied by unshared electron pairs.

O (ground state)	O (hybrid state)
$2p$ ↑↓ ↑ ↑	
	sp^3 ↑↓ ↑↓ ↑ ↑
$2s$ ↑↓	
$1s$ ↑↓	$1s$ ↑↓

The atoms of the H_2O molecule are arranged in angular form with a bond angle of 104.5°—an angle closer to the tetrahedral angle of 109.5° than to the 90° angle expected if pure p orbitals were used.

The orbitals employed in a given type of hybridization must be comparatively close in energy. The ones most commonly used in the bonding of molecules are the s, p, and d orbitals of the outer shell and, in some cases, the d orbitals of the shell next to the outer shell. With the exception of these inner d orbitals, electrons of inner orbitals are usually bound too tightly to be involved in covalent bond formation.

The sulfur atom of sulfur hexafluoride, SF_6, employs **d^2sp^3 hybrid orbitals** in the bonding of the molecule (Figure 7.8).

S (ground state)	S (excited state)	S (hybrid state)
$3d$ _ _ _ _ _	$3d$ ↑ ↑ _ _ _	$3d$ _ _ _
		d^2sp^3 ↑ ↑ ↑ ↑ ↑ ↑
$3p$ ↑↓ ↑ ↑	$3p$ ↑ ↑ ↑	
$3s$ ↑↓	$3s$ ↑	

The d^2sp^3 hybrid orbitals are equivalent and are directed to the corners of a regular octahedron.

Phosphorus pentafluoride, PF_5, is a trigonal bipyramidal molecule (Figure 7.8), and **dsp^3 hybrid orbitals** are used in its formation.

P (ground state)	P (excited state)	P (hybrid state)
$3d$ _ _ _ _ _	$3d$ ↑ _ _ _ _	$3d$ _ _ _ _
$3p$ ↑ ↑ ↑	$3p$ ↑ ↑ ↑	dsp^3 ↑ ↑ ↑ ↑ ↑
$3s$ ↑↓	$3s$ ↑	

7.7 ELECTRON PAIR REPULSION THEORY

The arrangement of atoms in molecules and ions may be predicted by a simple concept called the **electron pair repulsion theory.**

1. The theory applies to molecules and ions in which a *central atom* is bonded to two or more atoms.
2. Because the electron cloud of an electron pair is negatively charged, electron pairs repel one another. The electron pairs in the *valence shell of the central atom* are assumed to take positions as far apart as possible.
3. All the electron pairs of the *central atom* are considered—both the pairs that form covalent bonds and those that are unshared.
4. The unshared pairs help to determine the positions of the atoms in the molecule or ion. The shape of the molecule or ion, however, is described in terms of the positions of the nuclei and not in terms of the positions of the electron pairs.

Some examples will make these points clear. In the diagrams that follow, only the electrons of the valence shell of the central atom are shown.

1. *Two electron pairs.* Mercury has two electrons in its valence level ($6s^2$). These electrons are used to form two covalent bonds with two chlorine atoms in the $HgCl_2$ molecule. The molecule is **linear.**

Cl:Hg:Cl

The $HgCl_2$ molecule adopts a shape in which the two electron-pair bonds are as far apart as possible. Molecules of the type B:A:B are always linear. Beryllium, zinc, cadmium, and mercury form molecules of this type.

2. *Three electron pairs.* The boron trifluoride molecule is **triangular** and **planar.**

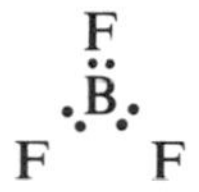

Each of the bond angles of the molecule is 120°. This arrangement provides the greatest possible separation between the three electron pairs.

Tin(II) chloride vapor, $SnCl_2$, consists of angular molecules.

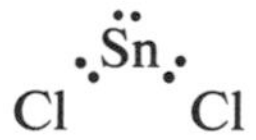

The tin atom has four electrons in its valence shell (Sn is a member of group IV A), and each chlorine atom supplies one electron to form a bond. These six electrons constitute three electron pairs (two bonding and one nonbonding). The pairs assume a triangular planar configuration because of electrostatic repulsion. The shape of a molecule is described in terms of the positions of its constituent atoms, not its electrons. $SnCl_2$ is therefore described as **angular.**

3. *Four electron pairs.* In the methane molecule, CH_4, the carbon atom has four bonding pairs of electrons.

$$\begin{array}{c} \mathrm{H} \\ \mathrm{H:\underset{\cdot\cdot}{\overset{\cdot\cdot}{C}}:H} \\ \mathrm{H} \end{array}$$

The bond pairs repel each other least when the bonds are directed toward the corners of a regular *tetrahedron* (Figure 7.11). In this configuration, all of the bonds are equidistant from one another, and all H—C—H bond angles are 109° 28′. The **tetrahedral** configuration is a common and important one. Many molecules and ions (for example, ClO_4^-, SO_4^{2-}, and PO_4^{3-}) are tetrahedral.

The structure of ammonia

$$\begin{array}{c} \mathrm{H:\underset{\cdot\cdot}{\overset{\cdot\cdot}{N}}:H} \\ \mathrm{H} \end{array}$$

can also be related to the tetrahedron (Figure 7.11). The N atom has three bonding electron pairs and one nonbonding electron pair. The four pairs assume a tetrahedral configuration which causes the atoms of the molecule to have a **trigonal pyramidal** arrangement. The unshared electron pair, which is under the influence of only one positive charge center, spreads out over a larger volume than a bonding pair, which is under the influence of two nuclei. As a result, the three bonds of the NH_3 molecule are forced slightly closer together than is normal for the tetrahedral arrangement. Thus, each of the H—N—H bond angles is 107° rather than the tetrahedral angle of 109.5°.

The water molecule has two bond pairs and two unshared pairs.

$$\begin{array}{c} \mathrm{:\underset{\cdot\cdot}{\overset{\cdot\cdot}{O}}:H} \\ \mathrm{H} \end{array}$$

The four electron pairs are arranged in an approximately tetrahedral manner (Figure 7.11) so that the atoms of the molecule have a **V-shaped (angular)** configuration. Since there are two nonbonding electron pairs in the molecule,

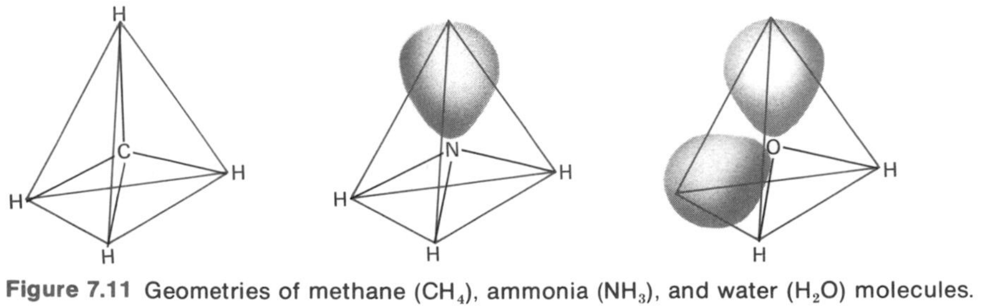

Figure 7.11 Geometries of methane (CH_4), ammonia (NH_3), and water (H_2O) molecules.

the bonds are forced together even more than those of the NH_3 molecule. Hence, the H—O—H bond angle of H_2O (105°) is less than the H—N—H bond angle of NH_3.

4. *Five electron pairs.* In the PCl_5 molecule, the five valence electrons of the P atom (phosphorus is a member of group V A) form five bond pairs.

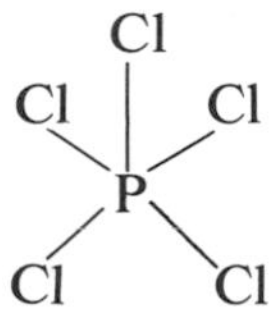

The shape that minimizes electron-pair repulsion is the **trigonal bipyramid** (Figure 7.12a). The five bonds in the structure, however, are not equivalent.

The positions that lie around the "equator" (numbers 2, 4, and 5 in the figure) are called **equatorial positions.** The positions at the "north pole" and "south pole" (numbers 1 and 3 in the figure) are called **axial positions.** Equatorial atoms lie in the same plane. Any bond angle formed by two equatorial atoms and the central atom is 120°. The axial atoms are on an axis that is at right angles to the equatorial plane. Any bond angle formed by an axial atom, the central atom, and an equatorial atom is 90°. In addition, the P—Cl axial bond length (219 pm) is slightly longer than the P—Cl equatorial bond length (204 pm).

When one or more nonbonding electron pairs are present in a molecule of this type, the nonbonding pairs occupy *equatorial* positions. In $TeCl_4$, four of the six valence electrons of Te (a group VI A element) form bond pairs; the remaining two constitute a single unshared pair.

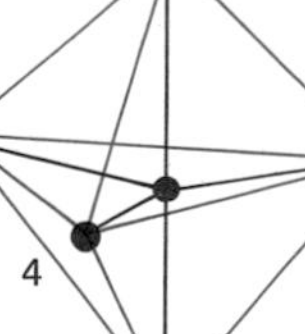

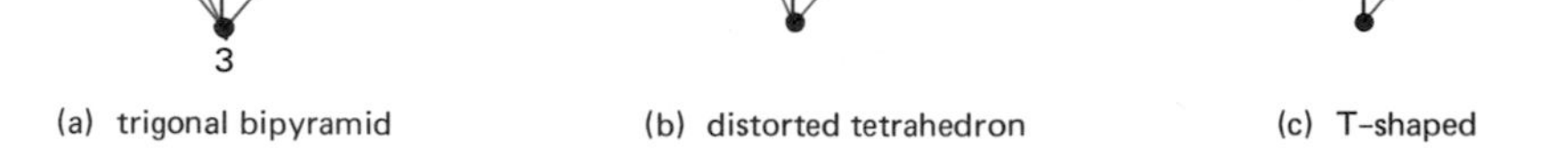

(a) trigonal bipyramid (b) distorted tetrahedron (c) T-shaped

Figure 7.12 Geometries of molecules in which the central atom has five electron pairs.

The nonbonding pair occupies an equatorial position, and the molecule is described as **distorted tetrahedral** (Figure 7.12b).

In ClF_3, three of the seven valence electrons of Cl form bonding pairs, and the remaining four make up two nonbonding pairs.

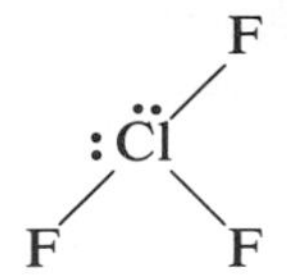

Placement of the unshared pairs in equatorial positions produces a **T-shaped** molecule (Figure 7.12c).

The Xe atom of XeF_2 has two bonding pairs and three nonbonding pairs since six of the eight valence electrons of Xe (group 0) remain unshared after two bonds have formed.

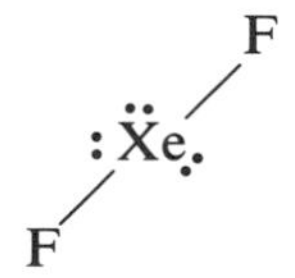

All three unshared electron pairs occupy equatorial positions. XeF_2 is a **linear** molecule.

5. *Six electron pairs.* In the SF_6 molecule, the S atom (a group VI A element) has six bonding pairs.

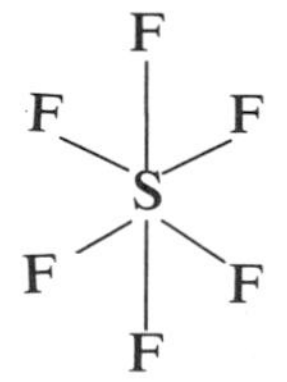

The form that minimizes electron repulsions is the regular **octahedron** (Figure 7.13a). All the positions are equivalent, and all the bond angles are 90°.

The I atom of IF_5 has five bonding pairs and one nonbonding pair.

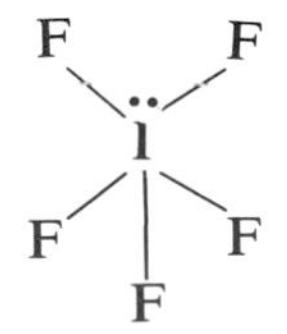

An I atom has seven valence electrons. In IF_5, five of these electrons are engaged in bonding five F atoms, and the other two constitute a nonbonding pair. The electron pairs are directed to the corners of an octahedron. The

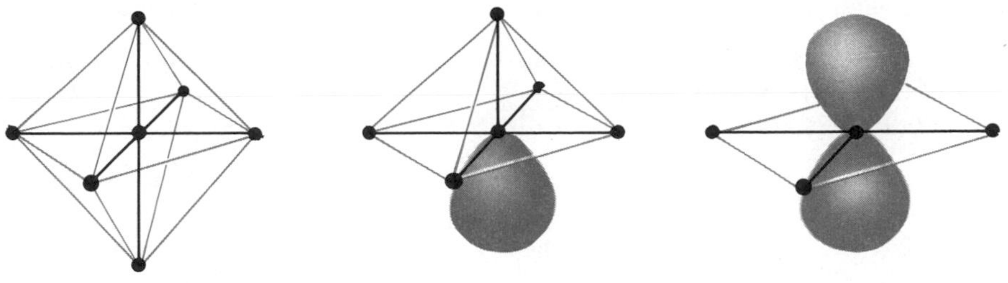

(a) octahedron (b) square pyramid (c) square planar

Figure 7.13 Geometries of molecules and ions in which the central atom has six electron pairs.

atoms of the molecule form a **square pyramid** (Figure 7.13b), which may be slightly distorted.

The Br atom of the BrF_4^- ion has four bonding pairs and two nonbonding pairs.

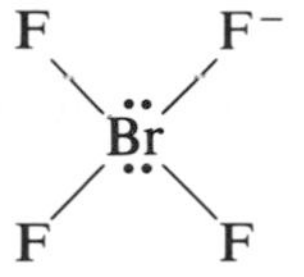

To account for the charge on the ion, we can consider that the central Br atom gains an electron to become a Br^- ion with eight valence electrons. Four of these electrons are used to form four bonding pairs, and the other four electrons constitute two nonbonding pairs. Since the total number of electron pairs is six, the electron pairs of this ion assume octahedral positions. The nonbonding pairs take positions opposite one another which minimizes electron repulsion (Figure 7.13c). The ion, therefore, has a **square planar** geometry. The relationships between molecular shape and valence-shell electron pairs are summarized in Table 7.3.

The concept of electron pair repulsions can be extended to molecules and ions that contain double bonds. The double bond is counted as a unit. A double bond, however, takes up more room than a single bond. For example, the molecule

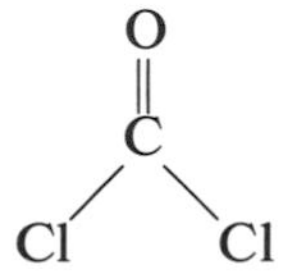

has a *triangular planar* structure (similar to BF_3), but the presence of the double bond forces the Cl—C—Cl bond angle to contract from 120° (the angle typical of the triangular planar arrangement) to 111°.

Table 7.3 Number of electron pairs in the valence shell of the central atom and molecular shape

NUMBER OF ELECTRON PAIRS				
TOTAL	BONDING	NONBONDING	SHAPE OF MOLECULE OR ION	EXAMPLES
2	2	0	linear	$HgCl_2$, $CuCl_2^-$
3	3	0	triangular planar	BF_3, $HgCl_3^-$
3	2	1	angular	$SnCl_2$, NO_2^-
4	4	0	tetrahedral	CH_4, BF_4^-
4	3	1	trigonal pyramidal	NH_3, PF_3
4	2	2	angular	H_2O, ICl_2^+
5	5	0	trigonal bipyramidal	PCl_5, $SnCl_5^-$
5	4	1	distorted tetrahedral	$TeCl_4$, IF_4^+
5	3	2	T shaped	ClF_3
5	2	3	linear	XeF_2, ICl_2^-
6	6	0	octahedral	SF_6, PF_6^-
6	5	1	square pyramidal	IF_5, SbF_5^{2-}
6	4	2	square planar	BrF_4^-, XeF_4

Predictions may also be made for resonance hybrids. The nitrite ion, NO_2^-,

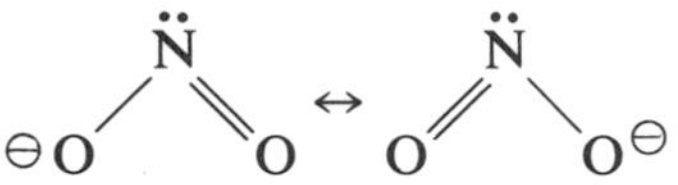

has an *angular* structure similar to $SnCl_2$.

Example 7.4 Use electron pair repulsion theory to predict the shapes of the following ions. All the bonds in these structures are single bonds. Assume that each halogen atom contributes one electron to the valence shell of the central atom for bond formation. (a) $TlCl_2^+$, (b) AsF_2^+, (c) IBr_2^-, (d) $SnCl_3^-$, (e) ClF_4^-.

Solution The number of valence electrons of the central atom (A), plus one electron for each substituent halogen atom (X), and an adjustment for the charge of the ion (chg.), gives the total number of electrons in the valence shell of the central atom. One-half of this number is the total number of electron pairs. Since each halogen atom is bonded by a single bond pair, the number of halogen atoms is also the number of bonding electron pairs. The number of nonbonding electron pairs is obtained by subtraction.

	Electrons	*Electron pairs*			*Shape*
	A + X + chg. = total	*total*	*bdg.*	*nonbdg.*	
(a) $TlCl_2^+$	$3 + 2 - 1 = 4$	2	2	0	linear
(b) AsF_2^+	$5 + 2 - 1 = 6$	3	2	1	angular
(c) IBr_2^-	$7 + 2 + 1 = 10$	5	2	3	linear

	Electrons	Electron pairs			Shape
	A + X + chg. = total	*total*	*bdg.*	*nonbdg.*	
(d) $SnCl_3^-$	4 + 3 + 1 = 8	4	3	1	trigonal pyramidal
(e) ClF_4^-	7 + 4 + 1 = 12	6	4	2	square planar

Example 7.5 Draw Lewis structures for the following and predict their shape. (a) CO_2, (b) SO_4^{2-}, (c) SO_3^{2-}, (d) NO_3^-.

Solution

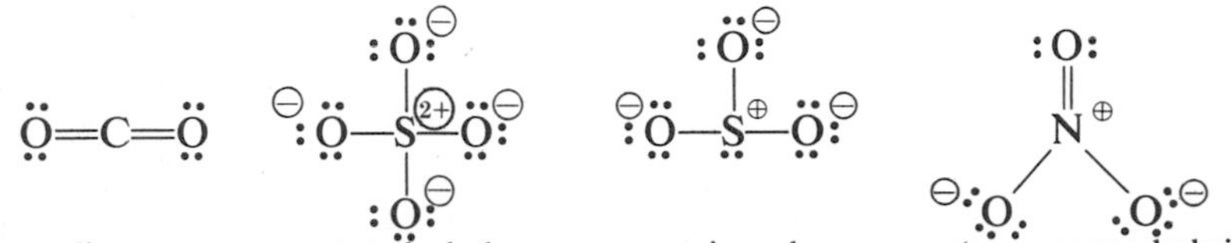

linear (a)

tetrahedral (b)

trigonal pyramidal (c)

(a resonance hybrid) triangular planar (d)

7.8 MOLECULAR ORBITALS

The theories of molecular structure that we have discussed so far have described the bonding in molecules in terms of *atomic* orbitals. The **method of molecular orbitals** is a different approach in which orbitals are associated with the molecule as a whole. The electronic structure of a molecule is derived by adding electrons to these molecular orbitals in an aufbau order. Corresponding to the practice of indicating atomic orbitals by the letters *s, p,* and so on, molecular orbitals are assigned the Greek letter designations σ (sigma), π (pi), and so on.

If two waves of the same wavelength (λ) and amplitude (a) are combined and are in phase, they reinforce each other (Figure 7.14a). The wavelength of the resultant wave remains the same, but the amplitude of the resultant wave is $a + a = 2a$. Two waves that are completely out of phase, however, cancel each other (Figure 7.14b); the "amplitude" of the resultant is $a + (-a) = 0$. The combination of waves can be *additive* or *subtractive*.

The molecular orbitals of the H_2 molecule can be imagined to result from the overlap of the 1*s* orbitals of two H atoms (Figure 7.5). If the overlap results in wave reinforcement (the additive combination), the electron density in the region between the nuclei is high. The attraction of the nuclei for this extra electronic charge holds the molecule together. The molecular orbital is called a **sigma bonding orbital** and is given the symbol $\boldsymbol{\sigma}$ (Figure 7.15).

In the combination, two atomic orbitals are used and therefore two molecular orbitals must be produced. The other molecular orbital is the result of the out-of-phase combination of waves (the subtractive combination). The electron density in the region between the nuclei is low. In this case, the nuclei repel each other since the low charge density between the nuclei does little to counteract this repulsion. This orbital is called a **sigma antibonding orbital**

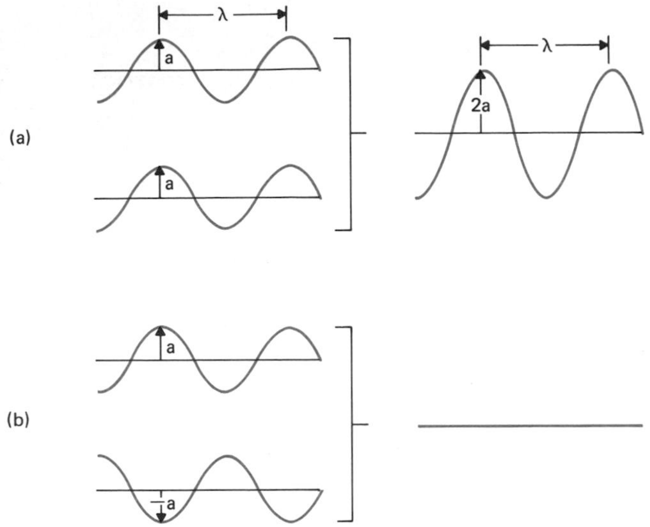

Figure 7.14 (a) Reinforcement of in-phase waves. (b) Cancellation of out-of-phase waves.

(given the symbol σ^*) since its net effect is disruptive (Figure 7.15). Sigma orbitals (both σ and σ^*) are symmetrical about a line joining the two nuclei. Rotation of the molecule about this axis causes no observable change in orbital shape.

An energy-level diagram for the formation of $\sigma 1s$ and $\sigma^* 1s$ molecular orbitals from the $1s$ atomic orbitals of two atoms is shown in Figure 7.16. The energy of the σ bonding orbital is lower than that of either atomic orbital from which it is derived, whereas the energy of the σ^* antibonding orbital is higher. When two atomic orbitals are combined, the resulting bonding molecular orbital represents a decrease in energy, and the antibonding molecular orbital represents an increase in energy.

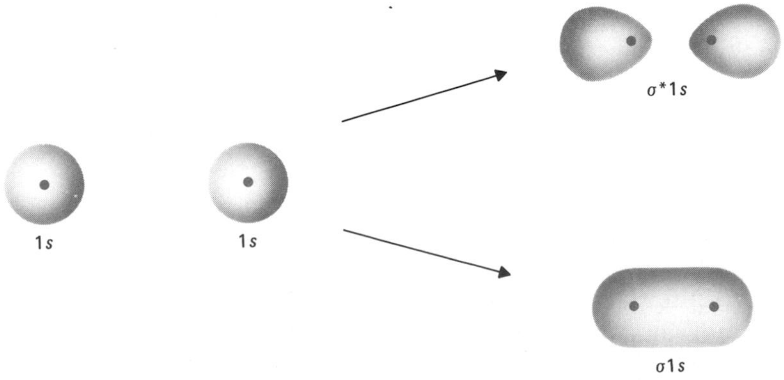

Figure 7.15 Formation of σ and σ^* molecular orbitals from $1s$ atomic orbitals.

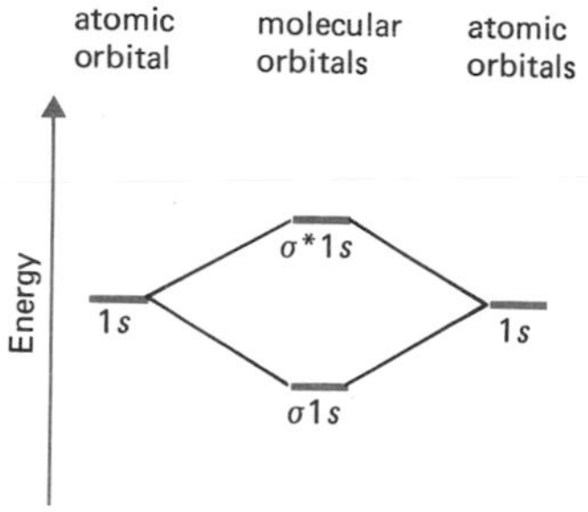

Figure 7.16 Energy-level diagram for the formation of σ and σ^* molecular orbitals from the 1s orbitals of two atoms.

Any orbital (atomic or molecular) can hold two electrons of opposed spin. In the hydrogen molecule, two electrons occupy the $\sigma 1s$ orbital. The $\sigma^* 1s$ orbital is unoccupied. One-half the difference between the number of bonding electrons and the number of antibonding electrons gives the number of bonds in the molecule (the **bond order**).

$$\text{bond order} = \tfrac{1}{2}[(\text{no. bonding } e^-) - (\text{no. antibonding } e^-)]$$

For H_2,

$$\text{bond order} = \tfrac{1}{2}(2 - 0) = 1$$

If an attempt is made to combine two helium atoms, a total of four electrons must be placed in the two molecular orbitals. Since the $\sigma 1s$ orbital is filled with two electrons, the other two must be placed in the higher $\sigma^* 1s$ orbital. The number of bonds in He_2, then, would be

$$\text{bond order} = \tfrac{1}{2}(2 - 2) = 0$$

He_2 does not exist.

The combination of two 2s orbitals produces σ and σ^* molecular orbitals similar to those formed from 1s orbitals. The molecular orbitals derived from 2p atomic orbitals, however, are slightly more complicated. The three 2p orbitals of an atom are directed along the x, y, and z coordinates. If we consider that a diatomic molecule is formed by the atoms approaching each other along the x axis, the p_x atomic orbitals approach each other head on and overlap to produce $\sigma 2p$ bonding and $\sigma^* 2p$ antibonding molecular orbitals (Figure 7.17). All sigma orbitals are symmetrical about the internuclear axis.

In the formation of a diatomic molecule (Figure 7.17), the p_z atomic orbitals approach each other side to side and produce a **pi bonding molecular orbital** (symbol, π) and a **pi antibonding molecular orbital** (symbol, π^*). Pi orbitals are not symmetrical about the internuclear axis. The electron density of a π orbital is concentrated in two regions that lie above and below the internuclear axis. The net effect is one that holds the molecule together. The π^* orbital has a low electron density in the region between the nuclei. Its effect is disruptive.

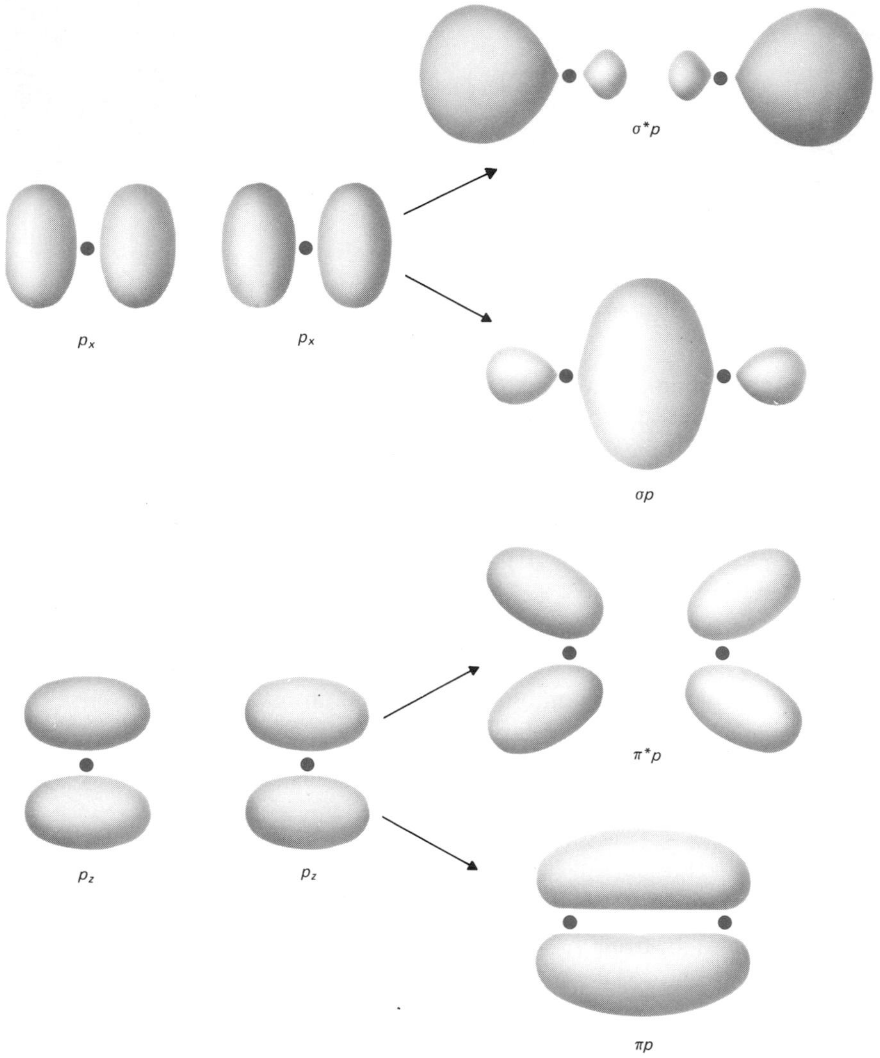

Figure 7.17 Formation of molecular orbitals from p atomic orbitals.

The p_y orbitals, which are not shown in Figure 7.17, also approach each other sideways. They produce another set of π and π^* orbitals, which lie at right angles to the first set. The two $\pi 2p$ orbitals have equal energies, and the two $\pi^* 2p$ orbitals have equal energies. Six molecular orbitals, therefore, arise from the two sets of $2p$ atomic orbitals—one $\sigma 2p$, one $\sigma^* 2p$, two $\pi 2p$, and two $\pi^* 2p$. These six together with the two derived from the $2s$ atomic orbitals make a total of eight molecular orbitals obtained from the $n = 2$ atomic orbitals of two atoms.

To illustrate the aufbau process for molecular orbitals, we will consider the diatomic molecules of the second period elements. The electrons in the inner shell of the atoms of these molecules (1*s* electrons) do not affect the bonding significantly. There are two aufbau orders for these molecules. The first order applies to the molecules from Li_2 to N_2.

$$\sigma 2s,\ \sigma^* 2s,\ (\pi 2p,\ \pi 2p),\ \sigma 2p,\ (\pi^* 2p,\ \pi^* 2p),\ \sigma^* 2p$$

The orbitals in parentheses have equal energies.

Lithium is a member of group I A, and each Li atom has one valence electron. The Li_2 molecule, therefore, has two electrons with opposed spins in the molecular orbital of lowest energy, the $\sigma 2s$ orbital. The bond order of Li_2 is $\frac{1}{2}(2-0)=1$.

If we attempt to make a Be_2 molecule, four electrons must be accommodated since each Be atom has two 2*s* electrons. The $\sigma 2s$ orbital is filled when two electrons are entered. The other two electrons are placed in the $\sigma^* 2s$ orbital. The net effect of two bonding electrons and two antibonding electrons is no bond. The bond order is $\frac{1}{2}(2-2)=0$. Be_2 does not exist.

Molecular orbital energy level diagrams for the molecules B_2, C_2, and N_2 are shown in Figure 7.18. The orbitals are arranged from bottom to top in the pertinent aufbau order. Each diagram is obtained by placing the correct

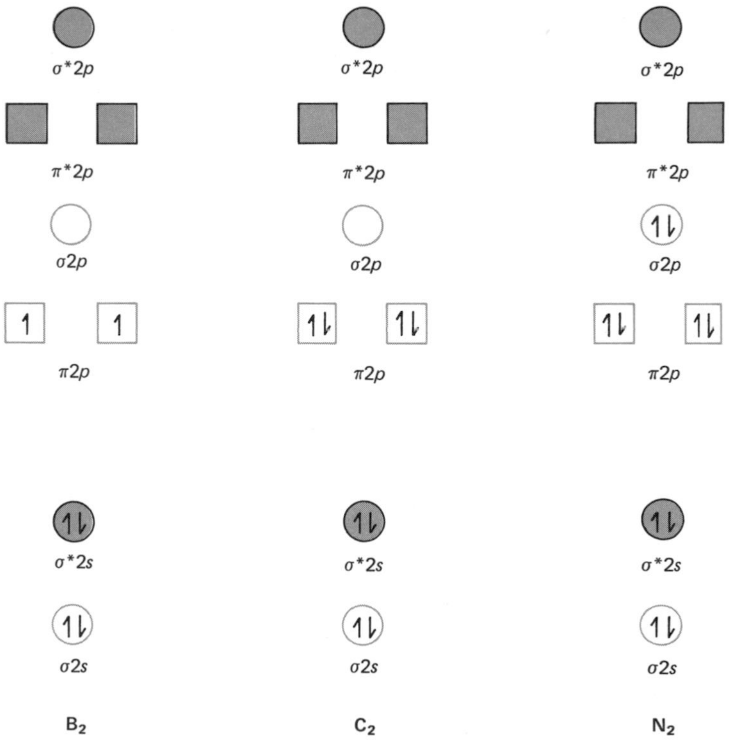

Figure 7.18 Molecular orbital energy-level diagrams for B_2, C_2, and N_2.

number of electrons one after the other into the lowest molecular orbital available. Hund's rule is followed in the filling. The two $\pi 2p$ orbitals have equal energies, and an electron enters each separately before electron pairing is begun.

In the case of B_2, the molecule has six valence electrons (three from each B atom since B is a member of group III A). The first two electrons enter the $\sigma 2s$ orbital, and the second two are placed in the $\sigma^* 2p$ orbital. The last two electrons enter separate $\pi 2p$ orbitals. As a result, the B_2 molecule has two unpaired electrons and is paramagnetic. The bond order of the B_2 molecule is $\frac{1}{2}(4-2)=1$.

The diagrams for C_2 and N_2 are obtained by adding eight and ten electrons, respectively. Notice that C_2 has a bond order of 2 and N_2 has a bond order of 3. Neither molecule contains unpaired electrons.

The aufbau order for the O_2 and F_2 molecules is different from the order that we have used so far (see Figure 7.19). The diagram for O_2 is obtained by entering twelve electrons (six from each O atom) into the molecular orbitals in the order shown. Notice that the last two electrons enter $\pi^* 2p$ orbitals separately. The O_2 molecule, therefore, has two unpaired electrons and is

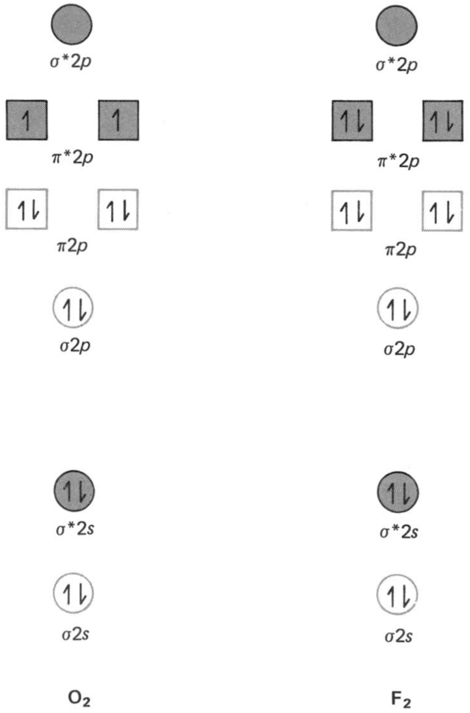

Figure 7.19 Molecular orbital energy-level diagrams for O_2 and F_2.

Table 7.4 Properties of diatomic molecules of the elements of the second period

MOLECULE	NUMBER OF ELECTRONS IN MOLECULAR ORBITALS						BOND ORDER	BOND LENGTH (pm)	BOND ENERGY (kJ/mol)	UNPAIRED ELECTRONS
	$\sigma 2s$	$\sigma^* 2s$	$\pi 2p$	$\sigma 2p$	$\pi^* 2p$	$\sigma^* 2p$				
[a]Li_2	2						1	267	106	0
[b]Be_2	2	2					0	—	—	0
[a]B_2	2	2	2				1	159	289	2
[a]C_2	2	2	4				2	131	627	0
N_2	2	2	4	2			3	110	941	0
	$\sigma 2s$	$\sigma^* 2s$	$\sigma 2p$	$\pi 2p$	$\pi^* 2p$	$\sigma^* 2p$				
O_2	2	2	2	4	2		2	121	494	2
F_2	2	2	2	4	4		1	142	155	0
[b]Ne_2	2	2	2	4	4	2	0	—	—	0

[a] Exists only in the vapor state at elevated temperatures.
[b] Does not exist.

paramagnetic. The bond order of O_2 is $\frac{1}{2}(8-4)=2$. The Lewis structure for O_2

$$:\ddot{O}::\ddot{O}:$$

is unsatisfactory. It shows the double bond of the O_2 molecule, but it does not show the two unpaired electrons.

A summary of the diatomic molecules of the second period elements is given in Table 7.4. As the number of bonds increases, the bond distance shortens and the bonding becomes stronger. The molecule bonded most strongly is N_2, which is held together by a triple bond (Figure 7.18). Molecules for which the method of molecular orbitals assigns a bond order of zero (Be_2 and Ne_2) do not exist.

For molecules such as CO and NO, the same types of molecular orbitals, although slightly distorted, are formed. Either aufbau order may be used in most cases with the same qualitative results. The actual order, however, is uncertain.

Since CO is isoelectronic with N_2 (each molecule has ten valence electrons), the molecular orbital energy-level diagram for CO is similar to that for N_2 (Figure 7.18). Carbon monoxide, therefore, has a bond order of 3. The dissociation energy of the CO molecule is about the same as that of the N_2 molecule.

We have said that it is impossible to diagram a Lewis structure for a molecule with an odd number of valence electrons. Nitrogen oxide, NO, is such a molecule. Since five valence electrons are contributed by the N atom and six valence electrons by the O atom, the total number of valence electrons is eleven. An energy-level diagram for the molecular orbitals of NO is given in Figure 7.20. Since there are eight bonding electrons and three antibonding electrons shown in the diagram, a bond order of $\frac{1}{2}(8-3)$, or $2\frac{1}{2}$, is indicated. Nitrogen oxide is paramagnetic. The NO molecule has one unpaired electron in a π^*2p orbital.

Figure 7.20 Molecular orbital energy-level diagram for NO.

7.9 MOLECULAR ORBITALS IN POLYATOMIC MOLECULES

Molecular orbitals can be derived for molecules that contain more than two atoms, such as H_2O and NH_3. In each case the number of molecular orbitals derived equals the number of atomic orbitals used, and the molecular orbitals encompass the whole molecule. For many purposes, however, it is convenient to think in terms of molecular orbitals that are localized between adjacent atoms. Consider the series

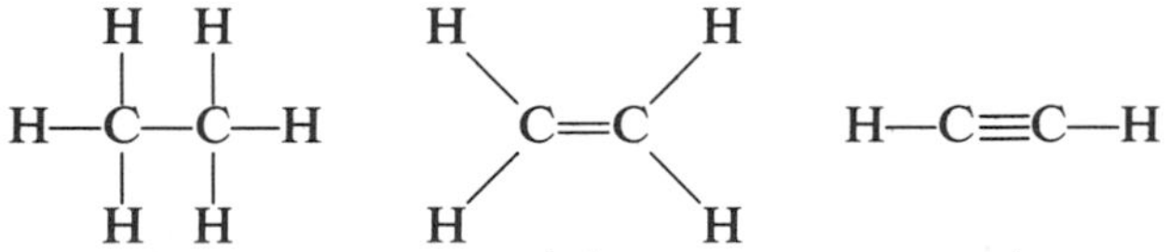

ethane ethylene acetylene

Each C atom in ethane utilizes sp^3 hybrid orbitals in the formation of σ bonds with the other C atom and the three H atoms (Figure 7.21). Thus, all bond angles are 109.5°, the tetrahedral angle. Since σ bonding orbitals are symmetrical about the internuclear axis, free rotation about each bond is possible. Rotation about the C—C bond causes a changing atomic configuration (Figure 7.21).

There can only be one σ bond between any two bonded atoms; any

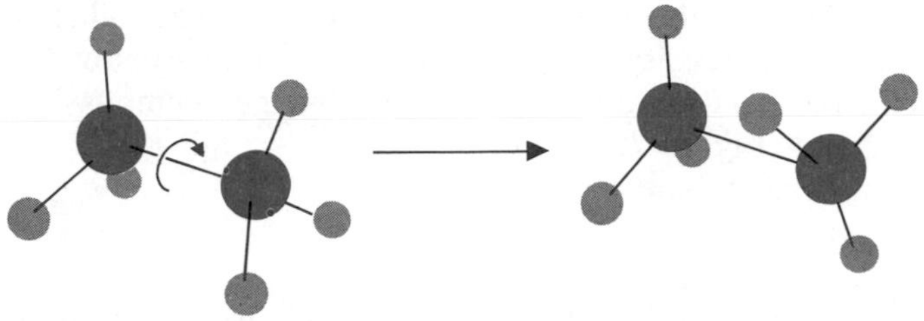

Figure 7.21 Rotation about the C—C bond in ethane.

additional bonds must be π bonds. The σ bonding skeleton of ethylene, therefore, is

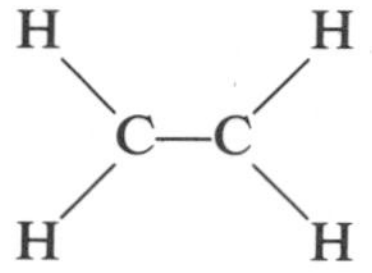

The C atoms utilize sp^2 hybrid orbitals to form this skeleton. This type of hybridization leads to a planar configuration of hybrid orbitals in which the orbitals form angles of 120° with each other. The ethylene molecule is planar with H—C—H bond angles of 118° and H—C—C bond angles of 121° (Figure 7.22).

The hybridization of the $2s$ and two of the $2p$ orbitals of carbon leaves one $2p$ orbital remaining.

C (ground state)	C (excited state)	C (hybrid state)
$2p$ ↿ ↿ __	$2p$ ↿ ↿ ↿	$2p$ ↿
		sp^2 ↿ ↿ ↿
$2s$ ⇅	$2s$ ↿	
$1s$ ⇅	$1s$ ⇅	$1s$ ⇅

Each C atom, therefore, has one electron in a $2p$ orbital that is not involved

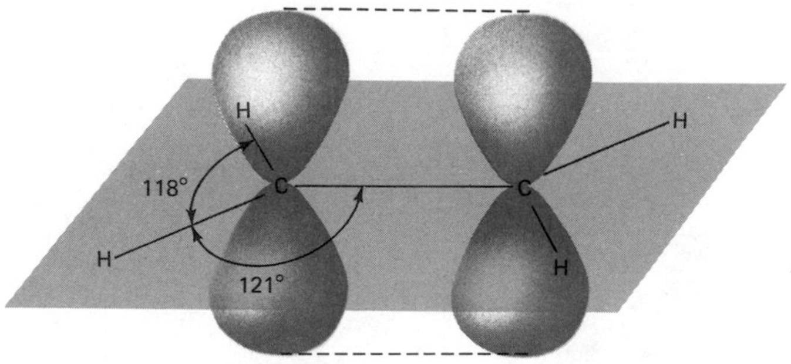

Figure 7.22 Geometric configuration of ethylene. Shapes of the p orbitals that overlap to form a π bond are simplified.

in σ bond formation. The $2p$ orbitals of the two C atoms overlap to produce a π bonding orbital. The electron density of the π bond is above and below the plane of the molecule. Free rotation about the C=C linkage is impossible without breaking this π bond (Figure 7.22).

The bonding skeleton of acetylene is

H—C—C—H

Each C atom forms two σ bonds, and *sp* (linear) hybrid orbitals of C are used to form these σ bonds.

$2p$ ↑ ↑ __	$2p$ ↑ ↑ ↑	$2p$ ↑ ↑
		sp ↑ ↑
$2s$ ↑↓	$2s$ ↑	
$1s$ ↑↓	$1s$ ↑↓	$1s$ ↑↓
C (ground state)	C (excited state)	C (hybrid state)

The $2p$ carbon orbitals that are not involved in the hybridization and σ bond formation overlap to form two π bonding molecular orbitals (Figure 7.23). The structure is much like that of N_2.

The multiple bonds of ethylene and acetylene are localized between two nuclei. In some molecules and ions **multicenter** (or **delocalized**) **bonding** exists in which some bonding electrons bond more than two atoms. The description of these species by the valence bond approach requires the use of resonance structures.

An example of delocalized bonding is found in the bonding in the carbonate ion (Figure 7.24). The ion is triangular planar, and each O—C—O bond angle is 120°. The C atom may be assumed to use sp^2 hybrid orbitals to form the σ bonding skeleton. One $2p$ orbital remains unhybridized. This unhybridized p orbital of C overlaps similar p orbitals of the O atoms to form a π bonding *system* that encompasses all the atoms of the ion.

The CO_3^{2-} ion has a total of twenty-four valence electrons. Six are used in the formation of the three σ bonds of the ion. Placement of two pairs of un-

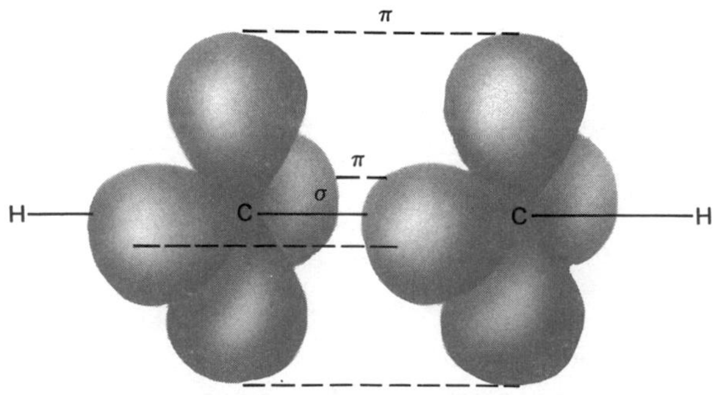

Figure 7.23 Formation of the π bonds of acetylene. Orbital shapes are simplified.

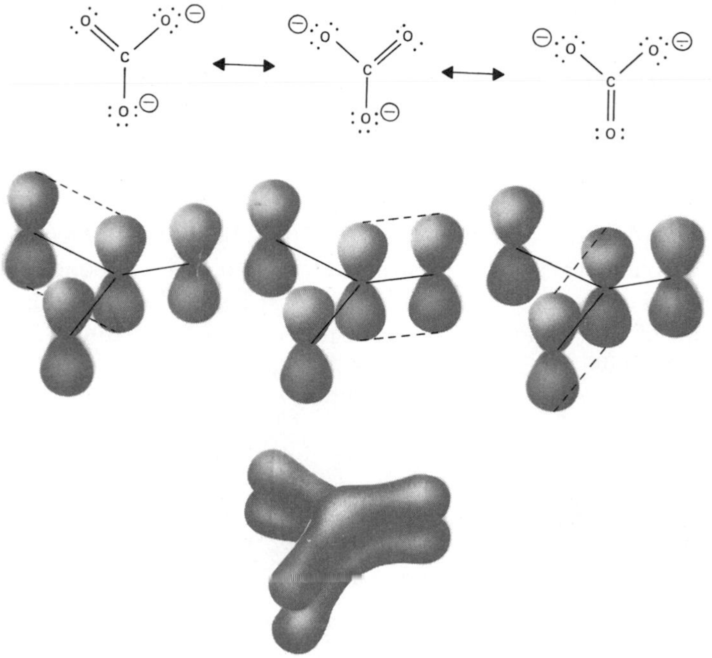

Figure 7.24 Multicenter π bonding system of the carbonate ion and the relationship to resonance structures.

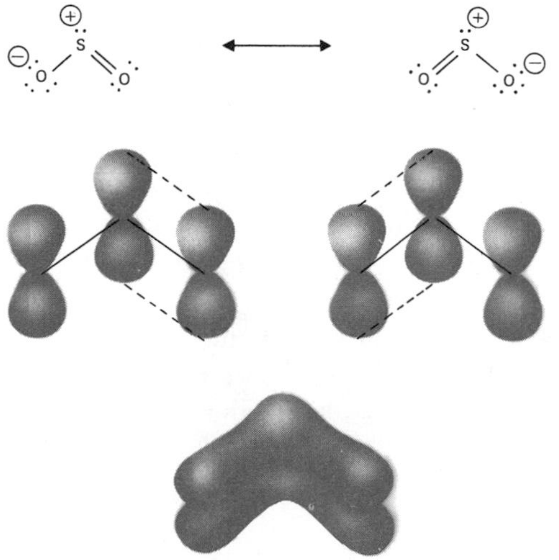

Figure 7.25 Relation of the π bonding system of sulfur dioxide to the resonance forms of the molecule. Orbital shapes are simplified.

shared electrons on each O atom accounts for twelve more. The remaining six electrons are placed in the molecular orbitals of the π bonding system. It must be understood that a molecular orbital, like an atomic orbital, can hold only two electrons. A π bonding system, therefore, involves two or more molecular orbitals. The structures of sulfur trioxide, SO_3, and the nitrate ion, NO_3^-, are similar to the structure of the carbonate ion, and the resonance forms of these species are also similar.

The S atom of the sulfur dioxide molecule, SO_2, may be assumed to employ sp^2 (triangular planar) hybrid orbitals in the formation of its σ bonding framework. An unshared electron pair occupies one of these hybrid orbitals. The π bonding system of this molecule is shown in Figure 7.25. Ozone, O_3, and the nitrite ion, NO_2^-, have similar structures.

THE DUALISTIC THEORY AND THE THEORY OF TYPES

The **dualistic theory** of Jöns Berzelius (1811) is one of the earliest theories of chemical bonding. Atoms were assumed to be held together in compounds by electrical attractions. Atoms of a negative element were thought to attract atoms of a positive element. In some ways, the dualistic theory is an early form of the ionic theory. There are, however, important differences between the two.

Organic chemistry is almost completely concerned with covalent molecules. At the time that the study of organic chemistry was begun (in the early decades of the nineteenth century) nothing was known about the structures of covalent molecules. The bonding in organic compounds was not well explained by the dualistic theory. Friedrich Wöhler was a chemist who made significant contributions to organic chemistry. In 1835, Wöhler wrote to Berzelius, his former teacher, "Organic chemistry just now is enough to drive one mad. It gives me the impression of a primeval tropical forest . . . into which one may well dread to enter."

In the years 1835–39, Jean Baptiste Dumas studied the reactions of some organic molecules with chlorine. In these reactions, one or more H atoms are replaced by Cl atoms. A simple example, although one that Dumas did not study, is

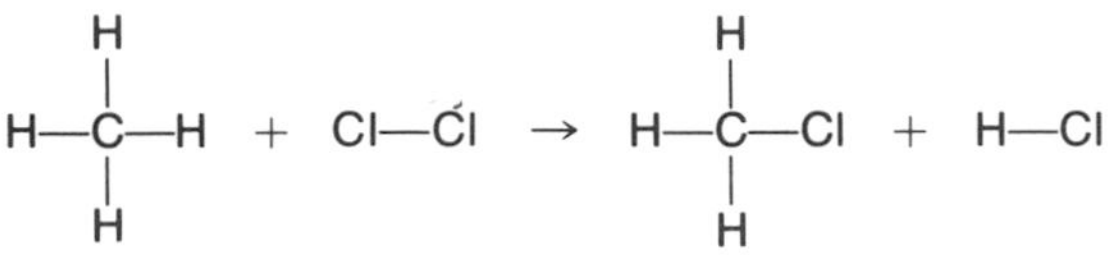

The properties of the molecules that contain Cl atoms (such as CH_3Cl) are similar to the properties of the molecules that contain only H atoms (such as CH_4). Although Dumas knew nothing about molecular structure, the form (or type) of the molecules appeared to him to be unchanged. Dumas proposed a **theory of substitution,** or a **theory of types,** to explain these reactions.

The supporters of the dualistic theory refused to accept Dumas' theory. How could chlorine (a negative element) replace hydrogen (a positive element) in a compound? A controversy followed in which Wöhler published a satirical letter that ridiculed Dumas' theory. Wöhler reported the "results" of a fabricated set of experiments in which the atoms of manganese(II) acetate were replaced one after another by chlorine atoms. Several fantastic compounds were obtained along the way—one was described as "a yellow salt with a violet tint," another as "silky yellow needles with a green cast."

In the end, all of the atoms of the original compound were replaced by chlorine atoms. Since the original salt, manganese(II) acetate, is a solid, and since the "type" had not changed, the final product (which was made up entirely of chlorine atoms) was reported to be a solid, not a gas! Wöhler signed the letter "S. C. H. Windler." The German word *Schwindler* means *swindler, fraud,* or *cheat.* He added a footnote that stated that fabrics of spun chlorine were on sale in London and "much sought after by hospitals and preferred above all others for night caps, drawers, etc."

Eventually, however, the theory of types won, and the dualistic theory was discarded. It might have been better if both theories had been retained, but scientists at that time did not recognize the difference between ionic and covalent compounds. Justus von Liebig, an early organic chemist, showed that he realized some distinction might exist when he wrote: "Up to a point . . . we follow the principles of inorganic chemistry, but beyond that point . . . we require new principles."

PROBLEMS

7.1 Draw Lewis structures for the following molecules (include formal charges): OCF_2, OSF_2, and OPF_3. In each case, each O and F atom is directly bonded to the central atom (C, S, and P, respectively).

7.2 Draw Lewis structures for the following molecules (include formal charges): H_2NOH, FNNF, NF_3, HCCH, H_2CCH_2, H_2NNH_2, HONO.

7.3 Draw Lewis structures for the following ions (include formal charges): SO_4^{2-}, SO_3^{2-}, ClO_2^-, CN^-, PO_4^{3-}, and O_2^{2-}.

7.4 The following species are isoelectronic: CO, NO^+, CN^-, and N_2. (a) Draw Lewis structures for each and include formal charges. (b) Each of the four species reacts to form complexes with metals or metal cations. In the formation of a complex, we can assume that an electron pair from one of the four species is used to form a covalent bond by occupying an empty orbital of the metal. For each heteronuclear species, tell which atom is bonded to the metal.

7.5 Draw Lewis structures for the following molecules (include formal charges): O_2NONO_2, PH_3, H_3COH, ClSCl, $H_2NC(O)NH_2$.

7.6 Draw Lewis structures for the following molecules (include formal charges): ClOCl, $OPCl_3$, H_2CO, H_2CCl_2, NCCN.

7.7 On the basis of anion deformation, predict which compound of each of the following pairs is the more covalent. (a) CuI or CuI_2, (b) Al_2O_3 or Al_2S_3, (c) $SnBr_2$ or $PbBr_2$, (d) $CdBr_2$ or CdI_2, (e) CrS or Cr_2S_3, (f) BeI_2 or MgI_2.

7.8 On the basis of anion deformation, predict which compound of each of the following pairs is the more covalent. (a) $SbCl_3$ or $BiCl_3$, (b) BeO or BeS, (c) Tl_2O or Tl_2O_3, (d) $CdBr_2$ or $HgBr_2$, (e) HgF_2 or HgI_2, (f) $CrCl_2$ or $CrCl_3$.

7.9 The bond distance in HF is 91.7 pm, and the dipole moment of HF is 1.91 D. Calculate the partial ionic character of HF. The unit charge, *e*, is 1.60×10^{-19} coul, and 1 D is 3.34×10^{-30} coul·m.

7.10 The bond distance in HI is 162 pm, and the dipole moment of HI is 0.38 D. Calculate the partial ionic character of HI. The unit charge, *e*, is 1.60×10^{-19} coul, and 1 D is 3.34×10^{-30} coul·m.

7.11 (a) Use electronegativities to estimate the degree of polarity of the bond in each of the molecules: ICl, BrCl, BrF, and ClF. (b) List the molecules in decreasing order according to the magnitude of their dipole moments.

7.12 For each of the following pairs, use electronegativities to determine which bond is the more polar. In each case tell the direction in which the bond is polarized. (a) O—H bond, O—F bond; (b) S—H bond, S—F bond; (c) O—H bond, S—H bond; (d) O—F bond, S—F bond.

7.13 On the basis of electronegativity, state whether you think the bonds formed between the following pairs of elements would be ionic or covalent. If covalent, estimate the degree of polarity of the bond. (a) Mg—O, (b) P—Cl, (c) P—H, (d) Hg—Cl, (e) C—Br, (f) Cs—Br, (g) N—Cl, (h) B—H, (i) Al—I, (j) N—H.

7.14 Arrange the following bonds in order of increasing ionic character. (a) C—H, (b) Si—H, (c) C—Cl, (d) Si—Cl, (e) C—O, (f) Si—O, (g) C—S, (h) Al—C, (i) Si—C.

7.15 Define and discuss the difference in meaning between the terms *electron affinity* and *electronegativity*.

7.16 Diagram the resonance forms for the linear cyanate ion, OCN^-.

7.17 In the O_2NF molecule, the N atom is the central atom to which the others are bonded. Diagram the resonance forms of O_2NF. What is the shape of the molecule?

7.18 The azide ion, N_3^-, is linear and symmetrical. Diagram the resonance forms of N_3^-.

7.19 Diagram the resonance forms of the formate ion, HCO_2^-, in which the C atom is the central atom and the two C to O bonds are of equal length.

7.20 In the oxalate ion, $C_2O_4^{2-}$, the two C atoms are bonded to each other, and each C atom has two O atoms bonded to it. The structure is symmetrical, and all C—O bond distances are equal. Diagram the resonance forms of $C_2O_4^{2-}$.

7.21 Diagram the resonance forms of nitramide, H_2NNO_2 (include formal charges). Is a resonance form with a double bond between the two N atoms a valid contributor to the resonance hybrid?

7.22 In the hyponitrite ion, $N_2O_2^{2-}$, the two N atoms are bonded to each other by a double bond and each N atom is bonded to an O atom by a single bond. Draw the Lewis structure (complete with formal charges) for $N_2O_2^{2-}$ and predict the shape of the ion by using electron-pair repulsion theory. A structure in which one of the O—N linkages is a double bond and the other bonds of the molecule (the N—N and other N—O) are single does *not* contribute to a resonance hybrid even though it conforms to the octet principle. Why not?

7.23 The structure of borazine, $B_3N_3H_6$, consists of a six-membered ring, composed of alternating B and N atoms, with a H atom bonded to each atom of the ring. Draw the resonance forms of borazine.

7.24 For each of the following molecules, indicate the type of hybrid orbitals employed by the central atom and state the geometric configuration of the molecule. (a) $CdBr_2$, (b) GeF_2, (c) SCl_2, (d) PCl_3, (e) GaI_3, (f) SnH_4, (g) $SbCl_5$, (h) SeF_6.

7.25 Interpret the configurations of the molecules given in Problem 7.24 on the basis of electron-pair repulsions.

7.26 The following species have been identified in aqueous solutions of mercury(II) chloride: Hg^{2+}, $HgCl^+$, $HgCl_2$, $HgCl_3^-$, and $HgCl_4^{2-}$. What type of hybrid orbitals are employed? What is the geometric configuration of $HgCl^+$, $HgCl_2$, $HgCl_3^-$, and $HgCl_4^{2-}$?

7.27 Crystalline PCl_5 consists of an ionic lattice formed from PCl_4^+ and PCl_6^- ions. In the vapor and liquid states, however, the compound exists as PCl_5 molecules. What types of hybrid orbitals does P employ in PCl_4^+, PCl_6^-, and PCl_5? What are the geometric configurations of these species?

7.28 For each of the following molecules, indicate the type of hybrid orbitals employed by the central atom and state the geometric configuration of the molecule. (a) Cl_2O, (b) AsH_3, (c) AsF_5, (d) CCl_4, (e) $SnCl_4$, (f) $SnCl_2$ (g) $HgBr_2$, (h) SF_6.

7.29 Interpret the configurations of the molecules given in Problem 7.28 on the basis of electron-pair repulsions.

7.30 The following molecules and ions of antimony are known: $SbCl_3$, $SbCl_4^-$, $SbCl_5^{2-}$, $SbCl_5$, and $SbCl_6^-$.

(a) On the basis of electron-pair repulsions, predict the shape of each of these species. (b) What type of hybrid orbitals does Sb employ in the bonding of each of these?

7.31 Use the concept of electron-pair repulsions to predict the geometric configuration of (a) SiF_5^-, (b) SbF_5^{2-}, (c) SeF_4, (d) NH_4^+, (e) NH_2^-, (f) $SnCl_6^{2-}$, (g) $TlBr_4^-$.

7.32 Use the concept of electron-pair repulsions to predict the geometric configuration of (a) $BeCl_2$, (b) $BeCl_4^{2-}$, (c) BeF_3^-, (d) SF_3^+, (e) TeF_5^-, (f) AsF_4^-, (g) SiF_6^{2-}.

7.33 Use the concept of electron-pair repulsions to predict the geometric configuration of (a) HgI_4^{2-}, (b) $SeBr_2$, (c) AsF_5, (d) BrF_3, (e) ClF_2^-, (f) BrF_5, (g) ICl_4^-.

7.34 Use the concept of electron-pair repulsions to predict the geometric configuration of (a) $CuCl_2^-$, (b) TeF_6, (c) BF_4^-, (d) XeF_4, (e) PF_3, (f) IF_3, (g) ClF_2^+.

7.35 Draw Lewis structures for the following ions. Predict the geometric shape on the basis of electron-pair repulsions. (a) ClO^-, (b) ClO_2^-, (c) ClO_3^-, (d) ClO_4^-.

7.36 In electron-pair repulsion theory, the effective volume of a bond pair is assumed to decrease with increasing electronegativity of the atom bonded to the central atom. In the light of this generalization, would you expect the Cl atoms of the trigonal bipyramidal PCl_2F_3 to occupy axial or equatorial positions?

7.37 In view of the opening statement of Problem 7.36, predict which of the PX_3 molecules (where X = F, Br, Cl, or I) would have the smallest X—P—X bond angles.

7.38 By means of molecular orbital energy-level diagrams, describe the bonding of the homonuclear diatomic molecules of the first ten elements.

7.39 Draw a molecular orbital energy-level diagram, and determine the bond order for each of the following. (a) H_2, (b) H_2^+, (c) HHe, (d) He_2, (e) He_2^+.

7.40 Oxygen forms compounds containing the dioxygenyl ion, O_2^+ (for example, O_2PtF_6); the superoxide ion, O_2^- (for example, KO_2); and the peroxide ion, O_2^{2-} (for example, Na_2O_2). (a) Draw molecular orbital energy-level diagrams for O_2, O_2^+, O_2^-, and O_2^{2-}. (b) State the bond order for each species. (c) Which of the four are paramagnetic?

7.41 The bond distances for N_2, N_2^+, O_2, and O_2^+ are: N_2, 109 pm; N_2^+, 112 pm; O_2, 121 pm; and O_2^+, 112 pm. Draw molecular orbital energy-level diagrams for these four species. Explain why the bond distances vary in the way described.

7.42 The anion of calcium carbide, CaC_2, should properly be called the acetylide ion, C_2^{2-}. Draw molecular orbital energy-level diagrams for C_2 and C_2^{2-} and state the bond order of each.

7.43 Draw molecular orbital energy-level diagrams for CO and NO. Use your diagrams to determine the bond order of CO, CO^+, CO^-, NO, NO^+, and NO^-. Which of these species are paramagnetic?

7.44 Discuss the structure of the sulfur trioxide molecule, SO_3, in terms of resonance and the concept of delocalized π bonding.

7.45 Discuss the structure of the ozone molecule, O_3, in terms of resonance and the concept of delocalized π bonding.

Chapter 8

Some Applications of Bonding Theory

In this chapter, we shall use the bonding theories discussed in Chapters 6 and 7 to study acids and bases as well as some important compounds of hydrogen, oxygen, and the group VII A elements. We will also discuss the systems used to name chemical compounds.

8.1 OXIDATION NUMBERS

In the assignment of formal charges to the atoms of a covalent molecule (Section 7.2), the bonding electrons are apportioned *equally* between the bonded atoms, and any bond polarity caused by unequal sharing of electrons is ignored. Formal charges are useful in interpreting the structure and some of the properties of covalent molecules (as we shall see in later sections), but the concept itself is merely a convention.

Another arbitrary but useful convention is the concept of oxidation numbers. **Oxidation numbers** are charges (fictitious charges in the case of covalent species) assigned to the atoms of a compound according to arbitrary rules that take into account bond polarity. The oxidation number of an atom in a binary, electrovalent compound is the same as the charge on the ion derived from that atom. In NaCl, for example, the oxidation number of sodium is 1+ and that of chlorine is 1−.

The oxidation numbers of the atoms in a covalent molecule can be derived by assigning the electrons of each bond to the more electronegative of the bonded atoms (rather than by dividing them equally as is done in deriving formal charges). For the molecule

$$\mathrm{H}\!:\!\underset{\cdot\cdot}{\overset{\cdot\cdot}{\mathrm{Cl}}}\!:$$

both electrons of the covalent bond are assigned to the chlorine atom since chlorine is more electronegative than hydrogen. The chlorine atom is said to have an oxidation number of 1− and the hydrogen atom 1+. In the case of a nonpolar bond between identical atoms, in which there is no electronegativity difference, the bonding electrons are divided equally between the bonded atoms in deriving oxidation numbers. Thus, the oxidation numbers of both chlorine atoms are zero in the molecule

$$:\!\underset{\cdot\cdot}{\overset{\cdot\cdot}{\mathrm{Cl}}}\!:\!\underset{\cdot\cdot}{\overset{\cdot\cdot}{\mathrm{Cl}}}\!:$$

The following rules, based on these ideas, can be used to assign oxidation numbers.

1. Any uncombined atom or any atom in a molecule of an element is assigned an oxidation number of zero.
2. The sum of the oxidation numbers of the atoms in a compound is zero since compounds are electrically neutral.
3. The oxidation number of a monatomic ion is the same as the charge on the ion. The sum of the oxidation numbers of the atoms that constitute a polyatomic ion equals the charge on the ion.
4. The oxidation number of fluorine, the most electronegative element, is 1− in all fluorine-containing compounds.
5. In most oxygen-containing compounds, the oxidation number of oxygen is 2−. There are, however, a few exceptions.
   a. In peroxides each oxygen has an oxidation number of 1−. The two oxygen atoms of the peroxide ion, O_2^{2-}, are equivalent. Each must be assigned an oxidation number of 1− so that the sum equals the charge on the ion.

b. In the superoxide ion, O_2^-, each oxygen has an oxidation number of $\frac{1}{2}-$.
c. In OF_2, the oxygen has an oxidation number of 2+ (see rule 4).

6. The oxidation number of hydrogen is 1+ in all its compounds except the metallic hydrides (for example, CaH_2 and NaH) in which hydrogen is in the 1− oxidation state.

Example 8.1 What is the oxidation number of the P atom in H_3PO_4?

Solution The oxidation numbers of the molecule must add up to zero. Therefore,

$$3(ox.\ no.\ H) + (ox.\ no.\ P) + 4(ox.\ no.\ O) = 0$$

Each H is assigned an oxidation number of 1+ (rule 6), and each O is assigned an oxidation number of 2− (rule 5).

$$\begin{aligned} 3(1+) + x + 4(2-) &= 0 \\ x &= 5+ \end{aligned}$$

Example 8.2 What is the oxidation number of Cr in the dichromate ion, $Cr_2O_7^{2-}$?

Solution The sum of the oxidation numbers must equal the charge of the ion, 2−. The oxidation number of O is 2− (rule 5).

$$\begin{aligned} 2(ox.\ no.\ Cr) + 7(ox.\ no.\ O) &= 2- \\ 2x + 7(2-) &= 2- \\ 2x &= 12+ \\ x &= 6+ \end{aligned}$$

Notice that the oxidation state of an element is reported on the basis of a single atom. It would be misleading to say that oxygen has an oxidation number of 2− in H_2O and 4− in SO_2. In both compounds, the O atom has an oxidation number of 2−.

Example 8.3 What is the oxidation number of Cl in calcium perchlorate, $Ca(ClO_4)_2$?

Solution

$$(ox.\ no.\ Ca) + 2(ox.\ no.\ Cl) + 8(ox.\ no.\ O) = 0$$

Since Ca is a member of group II A, its oxidation number is 2+, which is the charge on a Ca^{2+} ion. The oxidation number of O is 2−. Therefore,

$$\begin{aligned} (2+) + 2x + 8(2-) &= 0 \\ 2x &= 14+ \\ x &= 7+ \end{aligned}$$

There is another way to solve problems of this type. Since the charge on the Ca^{2+} ion is 2+, and since there are two perchlorate ions for every one Ca^{2+} ion in the compound, the charge on the perchlorate ion must be 1−. The ion has the formula ClO_4^-. Therefore,

$$\begin{aligned} (ox.\ no.\ Cl) + 4(ox.\ no.\ O) &= 1- \\ x + 4(2-) &= 1- \\ x &= 7+ \end{aligned}$$

These values should be interpreted with care. The ionic charges have physical significance; the oxidation numbers of O and Cl are merely conventions.

Fractional oxidation numbers are possible. The oxidation number of Fe in Fe_3O_4 is $2\frac{2}{3}+$. The sum of the oxidation numbers of all the atoms in the compound is zero. Since the oxidation numbers of the four oxygen atoms add up to 8−, the oxidation numbers of the three Fe atoms must add up to 8+. Each of the three Fe atoms, therefore, presumably has an oxidation number of $8+/3 = 2\frac{2}{3}+$.

The compound Fe_3O_4, however, may be considered to be made up of FeO and Fe_2O_3, in which Fe has oxidation numbers of 2+ and 3+, respectively. In Fe_3O_4, therefore, one Fe atom at 2+ and two Fe atoms at 3+ average out to an oxidation number of $2\frac{2}{3}+$. When an element has a fractional oxidation number in a particular compound, this often means that the compound contains two or more atoms of the element that are not equivalent.

Frequently an element exhibits a range of oxidation numbers in its compounds. Nitrogen, for example, exhibits oxidation numbers from 3− (as in NH_3) to 5+ (as in HNO_3).

1. The highest oxidation number of an A family element is the same as its group number. The number of valence electrons that an A family element has is equal to its group number. It is not logical to expect an atom to lose more valence electrons than it has. The highest possible positive charge—even a hypothetical one, therefore—is the same as the group number.
2. The lowest oxidation number of an A family element in a compound containing the element is the same as the charge of a monatomic ion of the element.

The highest oxidation number of sulfur (a member of group VI A) is 6+ (in H_2SO_4, for example). The lowest oxidation number of sulfur is 2− (as in Na_2S and H_2S). In its compounds, the highest oxidation number of sodium (found in group I A) is the same as the lowest oxidation number, 1+. The oxidation number of uncombined sodium is, of course, zero. There are, however, exceptions to these generalizations (fluorine and oxygen, for example).

8.2 NOMENCLATURE OF BINARY COMPOUNDS

A **binary compound** is one that is formed from only two elements. The **nomenclature** (system of naming) of binary compounds follows several simple rules.

1. Ionic binary compounds are named with the name of the metal first.
2. Covalent binary compounds are named with the name of the less electronegative element first.
3. For all binary compounds, the ending *-ide* is substituted for the usual ending of the element that appears last in the name. All compounds ending in *-ide* are not binaries, however. There are a few exceptions, such as the cyanides (NaCN, sodium cyanide, for example) and the hydroxides (NaOH, sodium hydroxide, for example).

If two elements form only one compound with each other, the preceding guidelines are sufficient to name that compound. Thus, the name of Na_2O is sodium oxide; $AlCl_3$, aluminum chloride; AgBr, silver bromide; ZnS, zinc sulfide; CaH_2, calcium hydride; and HF, hydrogen fluoride.

Some pairs of elements, however, form several compounds with each other (for example, $FeCl_2$ and $FeCl_3$). Each compound must have a name that identifies it. There are three ways to handle this problem.

1. In a system of nomenclature proposed by Albert Stock, the distinction between such compounds is made by indicating the oxidation state of the first element in Roman numerals immediately after the *English* name of that element. Thus, $FeCl_2$ is iron(II) chloride [Fe^{2+} is called the iron(II) ion]. Notice that the Roman numeral applies to the oxidation state of the first element of the name (the one in a positive oxidation state) and not to numbers that appear as subscripts in the formula.

Table 8.1 Names of some binary compounds

FORMULA	NAME	
FeO	iron(II) oxide	ferrous oxide
Fe_2O_3	iron(III) oxide	ferric oxide
Cu_2S	copper(I) sulfide	cuprous sulfide
CuS	copper(II) sulfide	cupric sulfide
SnF_2	tin(II) fluoride	stannous fluoride
SnF_4	tin(IV) fluoride	stannic fluoride
Hg_2Br_2[a]	mercury(I) bromide	mercurous bromide
$HgBr_2$	mercury(II) bromide	mercuric bromide

[a] The mercury(I) ion (or the mercurous ion) is unusual since it is diatomic (Hg_2^{2+}). The two atoms of mercury are covalently bonded. Since the charge on the ion is 2+, each atom has an oxidation number of 1+.

Table 8.2 Names of the nitrogen oxides

FORMULA		NAME
N_2O	dinitrogen oxide[a]	nitrogen(I) oxide
NO	nitrogen oxide[b]	nitrogen(II) oxide
N_2O_3	dinitrogen trioxide	nitrogen(III) oxide
NO_2	nitrogen dioxide	nitrogen(IV) oxide
N_2O_4	dinitrogen tetroxide	dimer of nitrogen(IV) oxide[c]
N_2O_5	dinitrogen pentoxide	nitrogen(V) oxide

[a] N_2O is often called nitrous oxide.
[b] NO is often called nitric oxide.
[c] A dimer is a molecule formed from two identical, simpler molecules (monomers). In this case an N_2O_4 molecule is formed from two NO_2 molecules.

2. Another method to distinguish between the compounds formed by an element that exhibits more than one positive oxidation state is to change the suffix of the name of the element. *Latin* names are employed when the symbol derives from the Latin. The ending *-ous* is used to indicate the lower oxidation state of a pair, and *-ic* is used for the higher oxidation state. Thus, $FeCl_2$ is ferrous chloride (and Fe^{2+} is the ferrous ion); $FeCl_3$ is ferric chloride (and Fe^{3+} is the ferric ion). Difficulties arise in the use of this method since some pairs of elements form more than two compounds. The first two systems of nomenclature are compared in Table 8.1.
3. A system of nomenclature that is particularly useful for series of compounds formed from two nonmetals employs Greek prefixes: *mono-* (1), *di-* (2), *tri-* (3), *tetra-* (4), *penta-* (5), *hexa-* (6), *hepta-* (7), *octa-* (8), and so forth. One of the prefixes is added to the name of each element to indicate the number of atoms of that element in the molecule that is being named. The prefix *mono-* is frequently omitted. The names of the oxides of nitrogen, according to this system and the Stock system, appear in Table 8.2.

Certain binary compounds have acquired nonsystematic names by which they are known exclusively. The list of such substances includes water (H_2O), ammonia (NH_3), hydrazine (N_2H_4), and phosphine (PH_3). Notice that the formulas of the last three compounds are customarily inverted. The symbol H should appear first in the formula since hydrogen is the element that occurs in a positive oxidation state.

8.3 ARRHENIUS ACIDS AND BASES

Several concepts of acids and bases are currently used (see Chapter 12). The oldest of these was first proposed by Svanté Arrhenius in 1884–87. The

Arrhenius acid–base system will be introduced in this section and used throughout this chapter.

An **acid** is defined as a substance that dissociates in water to produce H_3O^+ ions, which are called **hydronium ions.**

$$HCl(g) + H_2O \rightarrow H_3O^+(aq) + Cl^-(aq)$$

Pure HCl gas consists of covalent molecules. In an aqueous solution, ions are produced by the reaction of HCl molecules with H_2O molecules.

Every ion in an aqueous solution is closely associated with a number of water molecules. The symbol (aq) placed after the formula of an ion indicates that the ion exists as a **hydrated ion** in aqueous solution. This symbolism does not give the number of water molecules associated with each ion. The number in most cases is not known and in many cases is not a fixed value.

The H^+ ion is a special case. The H^+ ion (which is nothing more than a proton) does not exist independently in ordinary chemical systems. Its positive charge is not shielded by any electrons, and in comparison to other ions, it is extremely small. The H^+ ion, therefore, is strongly attracted by an electron pair of the water molecule.

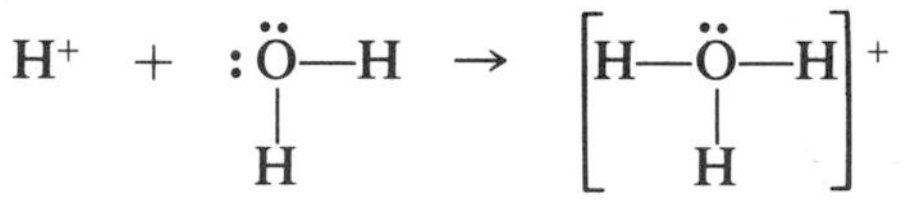

In this respect, the hydronium ion is like the ammonium ion.

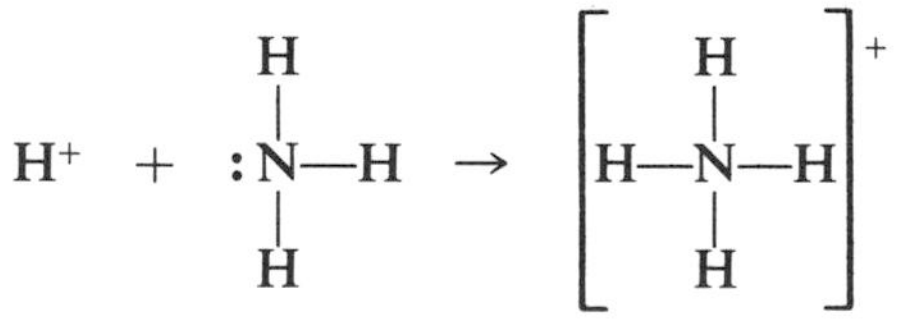

There is, however, evidence that the H_3O^+ ion has three additional water molecules associated with it in an ion that has the formula $H_9O_4^+$. Other evidence supports the idea that several types of hydrated ions exist simultaneously in water solution. Some chemists, therefore, prefer to represent the hydrated proton as $H^+(aq)$. The process in which HCl molecules dissolve in water would be indicated

$$HCl(g) \rightarrow H^+(aq) + Cl^-(aq)$$

In the Arrhenius system, a **base** is a substance that contains hydroxide ions, OH^-, or dissolves in water to produce hydrated hydroxide ions, $OH^-(aq)$.

$$NaOH(s) \rightarrow Na^+(aq) + OH^-(aq)$$
$$Ca(OH)_2(s) \rightarrow Ca^{2+}(aq) + 2OH^-(aq)$$

The only soluble metal hydroxides are those of the group I A elements and $Ba(OH)_2$, $Sr(OH)_2$, and $Ca(OH)_2$ of group II A. Insoluble hydroxides, however, react as bases with acids.

In a **neutralization,** an acid and a base react to produce a **salt** and water. A typical neutralization is

$$HCl(aq) + NaOH(aq) \rightarrow NaCl(aq) + H_2O$$

This type of equation, known as a **molecular equation,** is useful for some purposes but does not represent the reaction with complete accuracy. Both HCl and NaOH are ionic in water solution. The **ionic equation** for the reaction is

$$H^+(aq) + Cl^-(aq) + Na^+(aq) + OH^-(aq) \rightarrow Na^+(aq) + Cl^-(aq) + H_2O$$

We notice, however, that $Na^+(aq)$ and $Cl^-(aq)$ ions appear on both sides of this equation. These ions, therefore, do not enter into any chemical change. Ions of this type are often called **spectator ions.** If we eliminate these ions from the equation, we get the **net ionic equation**

$$H^+(aq) + OH^-(aq) \rightarrow H_2O$$

which can also be written

$$H_3O^+(aq) + OH^-(aq) \rightarrow 2H_2O$$

The Arrhenius system is summed up in this equation.

If 1 mol of HCl is dissolved in 1 liter of H_2O, 1 mol of $H_3O^+(aq)$ and 1 mol of $Cl^-(aq)$ are produced.

$$HCl(g) + H_2O \rightarrow H_3O^+(aq) + Cl^-(aq)$$

The formula of acetic acid, $HC_2H_3O_2$, distinguishes between two types of hydrogen atoms. The first one is acidic, the other three are not. If 1 mol of $HC_2H_3O_2$ is dissolved in 1 liter of H_2O, 0.004 mol of $H_3O^+(aq)$ and 0.004 mol of $C_2H_3O_2^-(aq)$ are produced.

$$HC_2H_3O_2(l) + H_2O \rightleftharpoons H_3O^+(aq) + C_2H_3O_2^-(aq)$$

The reversible arrow ($\rightleftharpoons$) is used in this equation to show that reactions occur in both directions. A balance is achieved with 0.4% of the acetic acid dissociated into ions. Acids that are 100% dissociated in water solution are called **strong acids.** Those that are less than 100% dissociated are called **weak acids.**

Weak acids are always represented in equations for neutralization reactions in molecular form. The *ionic* equation for the neutralization of acetic acid is

$$HC_2H_3O_2(aq) + OH^-(aq) \rightarrow C_2H_3O_2^-(aq) + H_2O$$

Even though acetic acid is not completely ionic in water solution, it is virtually completely neutralized in a neutralization reaction.

All the soluble metal hydroxides are **strong bases.** After all, these substances are 100% ionic when pure. There are a few **weak bases.** The most common one is an aqueous solution of ammonia, NH_3.

$$NH_3(g) + H_2O \rightleftharpoons NH_4^+(aq) + OH^-(aq)$$

In this reaction, the ammonia molecule accepts a proton from the water molecule to form the ammonium ion and the hydroxide ion. The reaction, however, is not complete.

Acids that can lose only one proton per molecule (such as HCl, $HC_2H_3O_2$, and HNO_3) are called **monoprotic acids.** Some acids can lose more than one proton per molecule; they are called **polyprotic acids.** A molecule of sulfuric acid, for example, can lose two protons.

$$H_2SO_4(l) + H_2O \rightarrow H_3O^+(aq) + HSO_4^-(aq)$$
$$HSO_4^-(aq) + H_2O \rightleftharpoons H_3O^+(aq) + SO_4^{2-}(aq)$$

Only one proton is neutralized if 1 mol of H_2SO_4 is reacted with 1 mol of NaOH.

$$H_2SO_4(aq) + NaOH(aq) \rightarrow NaHSO_4(aq) + H_2O$$

The salt obtained, $NaHSO_4$, is called an **acid salt** because it contains an acid hydrogen. If 1 mol of H_2SO_4 is reacted with 2 mol of NaOH, both acid hydrogens are neutralized. The **normal salt,** Na_2SO_4, is obtained.

$$H_2SO_4(aq) + 2NaOH(aq) \rightarrow Na_2SO_4(aq) + 2H_2O$$

The acid salt can be reacted with NaOH to produce the normal salt.

$$NaHSO_4(aq) + NaOH(aq) \rightarrow Na_2SO_4(aq) + H_2O$$

The products of the neutralization of a polyprotic acid, therefore, depend upon the quantities of acid and base employed.

Phosphoric acid, H_3PO_4, has three acid hydrogens, and three salts can be obtained from it.

$$NaH_2PO_4 \qquad Na_2HPO_4 \qquad Na_3PO_4$$

8.4 NOMENCLATURE OF ACIDS, BASES, AND SALTS

Some common acids and bases are listed in Table 8.3. Rules for naming these compounds and the salts derived from them follow.

1. Aqueous solutions of *binary compounds that function as acids* are named by modifying the root of the name of the element that is combined with hydrogen. The prefix *hydro-* and the suffix *-ic* are used followed by the word *acid.*

   HCl forms hydrochloric acid
   H_2S forms hydrosulfuric acid

2. *Metal hydroxides* are named in the manner described in Section 8.2.

   $Mg(OH)_2$ is magnesium hydroxide
   $Fe(OH)_2$ is iron(II) hydroxide or ferrous hydroxide

Table 8.3 Some common acids and bases (weak acids and bases are indicated by *)

	ACIDS		
	Binary Compounds		
Monoprotic Acids		Polyprotic Acids	
HF*	hydrofluoric acid	H_2S*	hydrosulfuric acid
HCl	hydrochloric acid		
HBr	hydrobromic acid		
HI	hydroiodic acid		
	Ternary Compounds		
Monoprotic Acids		Polyprotic Acids	
HNO_3	nitric acid	$H_2SO_4^a$	sulfuric acid
HNO_2^*	nitrous acid	$H_2SO_3^*$	sulfurous acid
$HClO_4$	perchloric acid	$H_3PO_4^*$	phosphoric acid
$HClO_3$	chloric acid	$H_2CO_3^*$	carbonic acid
$HClO_2^*$	chlorous acid		
$HOCl$*	hypochlorous acid	$H_3BO_3^*$	boric acid
$HC_2H_3O_2^*$	acetic acid		
	BASES		
$LiOH$	lithium hydroxide	$Ca(OH)_2$	calcium hydroxide
$NaOH$	sodium hydroxide	$Sr(OH)_2$	strontium hydroxide
KOH	potassium hydroxide	$Ba(OH)_2$	barium hydroxide
$RbOH$	rubidium hydroxide	NH_3^*	ammonia
$CsOH$	cesium hydroxide		

[a] The second ionization of H_2SO_4 is incomplete.

3. *Salts of binary acids* are themselves binary compounds, and their names have the customary *-ide* ending. They are named according to the rules given in Section 8.2.
4. *Ternary acids* are composed of three elements. When oxygen is the third element, the compound is called an *oxyacid.*
   a. If an element forms only one oxyacid, the acid is named by changing the ending of the name of the element to *-ic* and adding the word *acid.*

      H_3BO_3 is boric acid

   b. If there are two common oxyacids of the same element, the ending *-ous* is used in naming the oxyacid of the element in its lower oxidation state; the ending *-ic* is used to denote the higher oxidation state. See Table 8.3.

      HNO_2 is nitrous acid
      HNO_3 is nitric acid

c. There are a few series of oxyacids for which two names are not enough. See the names of the oxyacids of chlorine in Table 8.3. The prefix *hypo-* is added to the name of an *-ous* acid to indicate an oxidation state of the central element lower than that of the *-ous* acid.

$HClO_2$ is chlorous acid (Cl has an oxidation number of 3+)
HOCl is hypochlorous acid (Cl has an oxidation number of 1+)

The prefix *per-* is added to the name of an *-ic* acid to indicate an oxidation state of the central atom that is higher than that of the *-ic* acid.

$HClO_3$ is chloric acid (Cl has an oxidation number of 5+)
$HClO_4$ is perchloric acid (Cl has an oxidation number of 7+)

5. The names of the *anions of normal salts* are derived from the names of the acids from which the salts are obtained. The *-ic* ending is changed to *-ate*. The *-ous* ending is changed to *-ite*. Prefixes, if any, are retained.

SO_4^{2-} (from sulfuric acid) is the sulfate ion
OCl^- (from hypochlorous acid) is the hypochlorite ion

The name of the salt itself is obtained by combining the name of the cation with the name of the anion.

$NaNO_2$ is sodium nitrite
$Fe(ClO_4)_3$ is iron(III) perchlorate or ferric perchlorate

6. In naming the *anion of an acid salt,* the number of acid hydrogens retained by the anion must be indicated. The prefix *mono-* usually is omitted.

$H_2PO_4^-$ is the dihydrogen phosphate ion
HPO_4^{2-} is the hydrogen phosphate ion
PO_4^{3-} (the anion of the normal salt) is the phosphate ion

The prefix *bi-* may be used in place of the word *hydrogen* in the name of the anion of an acid salt derived from a diprotic acid.

HCO_3^- is the hydrogen carbonate ion or the bicarbonate ion
HSO_3^- is the hydrogen sulfite ion or the bisulfite ion

8.5 STOICHIOMETRY OF NEUTRALIZATION REACTIONS

The method discussed in Section 2.7 can be used to solve stoichiometric problems involving neutralization reactions if the amounts of reactants are given in units of mass. Neutralizations, however, are usually run in water solution and the quantities of reactants are usually given in terms of solution concentrations. The amount of a substance dissolved in a given volume of solution is called the **concentration** of the solution.

The **molarity,** *M,* of a solution is the number of moles of a substance (called a solute) dissolved in one liter of solution. Notice that the definition is based on one liter of *solution* and not on one liter of solvent. Because the definition is made in this way, it is relatively simple to determine the number of moles of solute in a measured volume of a solution on the basis of the molar concentration of the solution. One liter (which is 1000 ml) of a 3 *M* solution contains 3 mol of solute, 500 ml contains 1.5 mol, 250 ml contains 0.75 mol, and 100 ml contains 0.3 mol.

Example 8.4 How many grams of NaOH are required to prepare 250 ml of a 0.300 *M* solution of NaOH?

Solution The solution contains 0.300 mol NaOH per liter, or

$$0.300 \text{ mol NaOH} = 1000 \text{ ml NaOH sol'n.}$$

We find the number of moles of NaOH required to prepare 250 ml of solution.

$$? \text{ mol NaOH} = 250 \text{ ml NaOH sol'n.} \left(\frac{0.300 \text{ mol NaOH}}{1000 \text{ ml NaOH sol'n.}}\right)$$

$$= 0.0750 \text{ mol NaOH}$$

The molecular weight of NaOH is 40.0. Therefore,

$$40.0 \text{ g NaOH} = 1 \text{ mol NaOH}$$

The number of grams of NaOH needed is

$$? \text{ g NaOH} = 0.0750 \text{ mol NaOH} \left(\frac{40.0 \text{ g NaOH}}{1 \text{ mol NaOH}}\right) = 3.00 \text{ g NaOH}$$

Example 8.5 How many milliliters of 0.750 *M* NaOH solution are required to completely neutralize 50.0 ml of 0.150 *M* H_2SO_4 solution?

Solution We find the number of moles of H_2SO_4 in the sample of acid.

$$? \text{ mol } H_2SO_4 = 50.0 \text{ ml } H_2SO_4 \text{ sol'n.} \left(\frac{0.150 \text{ mol } H_2SO_4}{1000 \text{ ml } H_2SO_4 \text{ sol'n.}}\right)$$

$$= 0.00750 \text{ mol } H_2SO_4$$

The equation for the complete neutralization

$$H_2SO_4(aq) + 2NaOH(aq) \rightarrow Na_2SO_4(aq) + 2H_2O$$

shows that

$$2 \text{ mol NaOH} = 1 \text{ mol } H_2SO_4$$

$$? \text{ mol NaOH} = 0.00750 \text{ mol } H_2SO_4 \left(\frac{2 \text{ mol NaOH}}{1 \text{ mol } H_2SO_4}\right)$$

$$= 0.0150 \text{ mol NaOH}$$

Finally, we find the volume of 0.750 *M* NaOH solution that contains 0.0150 mol of NaOH.

$$? \text{ ml NaOH sol'n.} = 0.0150 \text{ mol NaOH} \left(\frac{1000 \text{ ml NaOH sol'n.}}{0.750 \text{ mol NaOH}}\right)$$

$$= 20.0 \text{ ml NaOH sol'n.}$$

Additional examples of problems that pertain to the preparations of solutions are found in Section 10.6.

8.6 EQUIVALENT WEIGHTS AND NORMAL SOLUTIONS

There is another way to solve stoichiometric problems involving neutralization reactions. This method employs quantities known as equivalent weights. The equivalent weight of a compound depends upon the type of reaction being considered. For neutralization reactions, equivalent weights are based on the fact that one H^+(aq) ion reacts with one OH^-(aq) ion.

$$H^+(aq) + OH^-(aq) \rightarrow H_2O$$

One **equivalent weight of an acid** is the amount of acid that supplies one mole of H^+(aq) ions. One **equivalent weight of a base** is the amount of base that supplies one mole of OH^-(aq) ions or reacts with one mole of H^+(aq) ions. Since 1 mol of H^+(aq) reacts with 1 mol of OH^-(aq), one equivalent weight of an acid will react with exactly one equivalent weight of a base.

One molecular weight (mole) of HCl supplies one mole of H^+(aq). The molecular weight of HCl, therefore, is also the equivalent weight of HCl. One mole of H_2SO_4 can supply two moles of H^+(aq) and, based on complete neutralization, is two equivalents. The equivalent weight of H_2SO_4, therefore, can be found by dividing the molecular weight by two. In general,

$$\textit{equivalent weight} = \frac{\textit{molecular weight}}{a}$$

where a is the number of moles of H^+(aq) supplied by one mole of an acid or the number of moles of H^+(aq) neutralized by one mole of base.

Polyprotic acids can undergo reactions in which not all of the acid hydrogens are neutralized. In these cases, the value of a is adjusted to equal the number of moles of H^+(aq) supplied in the reaction considered. In the case of H_2SO_4, $a = 2$ for reactions that produce SO_4^{2-}, and $a = 1$ for reactions that produce HSO_4^-.

Equivalent weights are used as the basis of a way to express the concentration of solutions. The **normality,** *N*, of a solution is the number of equivalent weights of solute dissolved in one liter of solution. The normality of a solution and its molarity are related. One liter of a 1 *M* solution of H_2SO_4 contains 1 mol of H_2SO_4. One mole of H_2SO_4, however, is two equivalent weights (for complete neutralization). A 1 *M* solution of H_2SO_4, therefore, is 2 *N*.
In general,

$$N = aM$$

The number of equivalents in a sample of a solution is obtained by multiplying the volume of the sample (in liters) by the normality of the solution (equivalents in one liter). Since a 4 *N* solution contains 4 equivalents per liter, 0.5 liter (which is 500 ml) contains 0.5(4) or 2 equivalents. The number of equivalents of A in a sample of a solution of A, n_A, is equal to the volume of the sample expressed in liters, V_A, times the normality of the solution, N_A.

$$n_A = V_A N_A \qquad (V_A \text{ in liters})$$

By design, one equivalent of A reacts with exactly one equivalent of B; *n* equivalents of A, therefore, react with the same number of equivalents of B.

$$n_A = n_B$$

Since $n_A = V_A N_A$ and $n_B = V_B N_B$,

$$V_A N_A = V_B N_B$$

Since a volume term appears on both sides of this equation, any volume unit can be used to express V_A and V_B provided that both volumes are expressed in the same unit.

Example 8.6 How many milliliters of 0.130 *N* HCl solution are required to completely neutralize 35.0 ml of 0.100 *N* $Ba(OH)_2$ solution?

Solution

$$V_A N_A = V_B N_B$$
$$x(0.130\ N) = (35.0\ \text{ml})(0.100\ N)$$
$$x = 26.9\ \text{ml}$$

(a) What is the normality of a solution of H_2SO_4 if 50.00 ml of the solution requires 37.52 ml of 0.1492 *N* NaOH solution for complete neutralization? (b) What is the molarity of the solution?

Solution (a)

$$V_A N_A = V_B N_B$$
$$(50.00\ \text{ml})x = (37.52\ \text{ml})(0.1492\ N)$$
$$x = 0.1120N$$

(b) Since

$$1 \text{ mol } H_2SO_4 = 2 \text{ equiv. } H_2SO_4$$

$$? \text{ mol } H_2SO_4 = 0.1120 \text{ equiv. } H_2SO_4 \left(\frac{1 \text{ mol } H_2SO_4}{2 \text{ equiv. } H_2SO_4}\right)$$

$$= 0.05600 \text{ mol } H_2SO_4$$

The solution is 0.05600 *M*.

Example 8.8 (a) Oxalic acid is a poisonous substance that occurs in many plants (including the leaves of the rhubarb plant) as the calcium salt. A 2.25 g sample of oxalic acid is completely neutralized by 30.3 ml of 1.65 *N* NaOH. What is the equivalent weight of oxalic acid? (b) Oxalic acid is diprotic. What is its molecular weight?

Solution (a) First we find the number of equivalents in the volume of NaOH solution employed. From the normality of the NaOH solution,

$$1.65 \text{ equiv. NaOH} = 1000 \text{ ml NaOH sol'n.}$$

$$? \text{ equiv. NaOH} = 30.3 \text{ ml NaOH sol'n.} \left(\frac{1.65 \text{ equiv. NaOH}}{1000 \text{ ml NaOH sol'n.}}\right)$$

$$= 0.0500 \text{ equiv. NaOH}$$

The sample of oxalic acid contains the same number of equivalents.

$$2.25 \text{ g oxalic acid} = 0.0500 \text{ equiv. oxalic acid}$$

$$? \text{ g oxalic acid} = 1 \text{ equiv. oxalic acid} \left(\frac{2.25 \text{ g oxalic acid}}{0.0500 \text{ equiv. oxalic acid}}\right)$$

$$= 45.0 \text{ g oxalic acid}$$

(b) Since the acid is diprotic, $a = 2$.

$$\textit{equivalent weight} = \frac{\textit{molecular weight}}{a}$$

$$45.0 = \frac{\textit{molecular weight}}{2}$$

$$\textit{molecular weight} = 90.0$$

Under some circumstances, another equation is useful. The number of equivalents in a sample is equal to the weight of the sample (g) divided by the equivalent weight (eq) of the substance. Thus,

$$n_B = \frac{g_B}{eq_B}$$

We can use this expression together with $n_A = V_A N_A$ to derive

$$V_A N_A = \frac{g_B}{eq_B} \text{ (volume in liters)}$$

If the volume is expressed in milliliters,

$$V_A N_A = \frac{1000 g_B}{eq_B} \text{ (volume in milliliters)}$$

This expression is convenient for calculations for reactions between a weighed substance (B) and another (A) in solution. If sample B is impure, the actual weight of the reactive part of the sample is equal to the fractional purity (p_B) times the total weight of the sample (g_B); thus,

$$V_A N_A = \frac{1000 p_B g_B}{eq_B}$$

Example 8.9 A 25.0 g sample of vinegar, which contains acetic acid ($HC_2H_3O_2$), requires 37.5 ml of 0.460 N NaOH for neutralization. What percentage of the vinegar is acetic acid?

Solution Since $HC_2H_3O_2$ is monoprotic, the equivalent weight is the same as the molecular weight, 60.0.

$$V_A N_A = \frac{1000 p_B g_B}{eq_B}$$

$$(37.5)(0.460) = \frac{1000 p_B (25.0)}{60.0}$$

$$p_B = 0.0414$$

The vinegar is 4.14% $HC_2H_3O_2$.

8.7 THE STRENGTH OF OXYACIDS AND STRUCTURE

The key to the acidity of compounds with the structure

$$\overset{\text{a}}{\text{H—O}}\overset{\text{b}}{\text{—Z}}$$

lies with the electronegativity of atom Z. If Z is an atom of a metal with a low electronegativity, the electron pair that is marked *b* will belong completely to the O atom, which has a high electronegativity. The compound will be an ionic hydroxide—a base. Sodium hydroxide (HO^-Na^+, usually written Na^+OH^-) falls into this category.

If Z is an atom of a nonmetal with a high electronegativity, the situation is different. The bond marked *b* will be a strong covalent bond, not an ionic bond. Instead of adding to the electron density around the O atom, Z will tend to reduce the electron density, even though oxygen is itself highly electronegative. The effect will be felt in bond *a*. The O atom will draw the electron density of this H—O bond away from the H atom, which will allow the proton to dissociate and make the compound acidic. Hypochlorous acid, HOCl, is an acid of this type.

The higher the electronegativity of Z, the more the electrons of the H—O bond are drawn away from the H atom and the more readily the proton is lost. In the series

$$\text{HOI} < \text{HOBr} < \text{HOCl}$$

the electronegativity of Z increases (I < Br < Cl), and the acid strength increases in the same order.

In some molecules, additional O atoms are bonded to Z. For example,

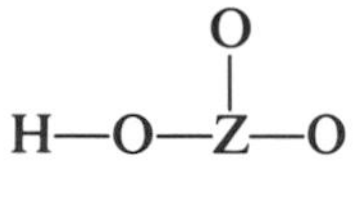

These O atoms draw electrons away from the Z atom and make it more positive. The Z atom, therefore, becomes more effective in withdrawing electron density away from the O atom that is bonded to H. In turn, the electrons of the H—O bond are drawn more strongly away from the H atom. The net effect makes it easier for the proton to dissociate and increases the acidity of the compound.

The more O atoms bonded to Z, the stronger the acid is. This effect is illustrated by the following series of acids, which are arranged by increasing order of acid strength.

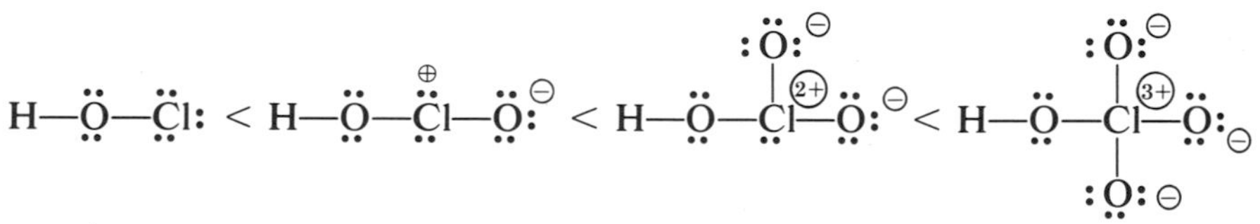

Notice that the formal charge on the central atom increases in this series. As the formal charge on the Cl increases, the electron density of the H—O bond shifts away from the H atom. As a result, the acidity increases.

We can rate the acid strength of compounds of this type by counting the number of O atoms bonded to Z but *not bonded to H atoms.*

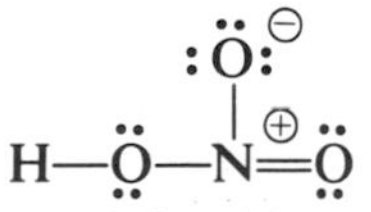

nitric acid
(a resonance hybrid)

is a stronger acid than

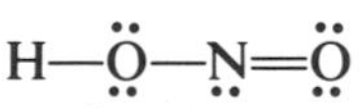

nitrous acid

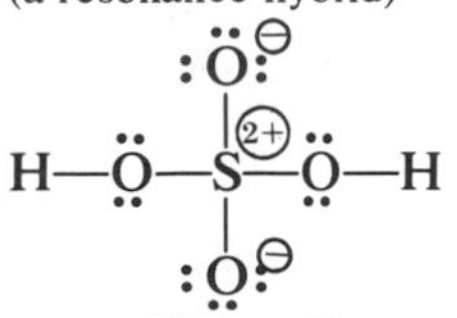

sulfuric acid

is a stronger acid than

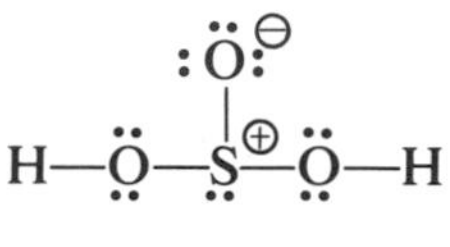

sulfurous acid

In general, the strengths of acids that have the general formula

$(HO)_mZO_n$

can be related to the value of n.

1. If $n = 0$, the acid is *very weak.* HOCl, $(HO)_3B$
2. If $n = 1$, the acid is *weak.* HOClO, HONO, $(HO)_2SO$, $(HO)_3PO$
3. If $n = 2$, the acid is *strong.* $HOClO_2$, $HONO_2$, $(HO)_2SO_2$
4. If $n = 3$, the acid is *very strong.* $HOClO_3$, $HOIO_3$

The first proton of each acid that belongs to the last two categories ($n = 2$ and $n = 3$) is virtually completely dissociated in water. The distinction between the two groups applies to dissociations in solvents other than water.

The effect of electron-withdrawing groups is also seen in organic acids. None of the H atoms of ethanol

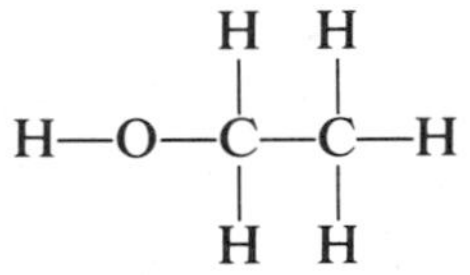

dissociates as a proton in water solution. The introduction of an O atom into the molecule

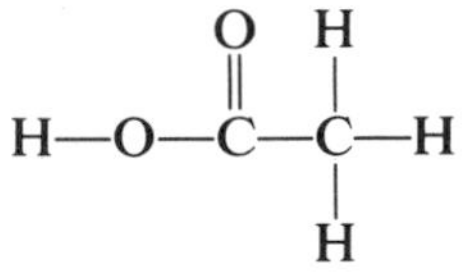

produces the compound called acetic acid (which we have previously written $HC_2H_3O_2$). Acetic acid is a weak monoprotic acid; only the H of the H—O group is acidic. There are a large number of organic acids that contain the

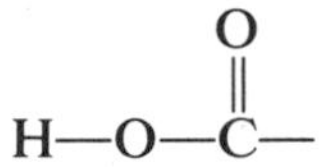

group.

If one or more of the H atoms that are bonded to C are replaced by highly electronegative atoms (such as Cl atoms), the acidity is increased. Trichloroacetic acid

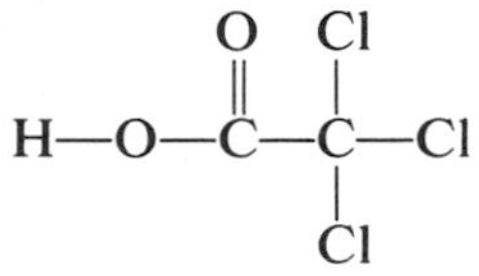

is a much stronger acid than acetic acid.

8.8 HYDRIDES

Hydrides of the representative elements are listed in Table 8.4.

Hydrogen reacts with the metals of group I A and the heavier metals of group II A (Ba, Sr, and Ca) to form **salt-like hydrides.** The hydride ion (H^-) of these compounds is isoelectronic with helium and achieves this stable configuration by the addition of an electron from the metal. Since the electron affinity of hydrogen is low (−72 kJ/mol), hydrogen reacts in this manner only with the most reactive metals.

$$2Na(s) + H_2(g) \rightarrow 2NaH(s)$$
$$Ca(s) + H_2(g) \rightarrow CaH_2(s)$$

The saltlike hydrides react with water to form hydrogen.

$$H^- + H_2O \rightarrow H_2(g) + OH^-(aq)$$

The other hydrides listed in Table 8.4 are covalent hydrides. These covalent compounds are formed by hydrogen with the metals Be, Mg, Al, Ge, and Sn as well as with the nonmetals. Some, but not all, of the covalent compounds can be prepared by the direct reaction of the elements.

Hydrogen reacts with the halogens to form colorless covalent gases.

$$H_2(g) + Cl_2(g) \rightarrow 2HCl(g)$$

Reactions with fluorine or chlorine will take place at room temperature or below, but the reactions with the less reactive bromine or iodine require higher temperatures (400°C to 600°C).

The reaction of hydrogen and oxygen is highly exothermic and is the basis

Table 8.4 Simple hydrides of the representative elements

I A							0
H_2	II A	III A	IV A	V A	VI A	VII A	
Li^+H^-	BeH_2	B_2H_6	CH_4	NH_3	H_2O	HF	
Na^+H^-	MgH_2	AlH_3	SiH_4	PH_3	H_2S	HCl	
K^+H^-	$Ca^{2+}2H^-$		GeH_4	AsH_3	H_2Se	HBr	
Rb^+H^-	$Sr^{2+}2H^-$		SnH_4	SbH_3	H_2Te	HI	
Cs^+H^-	$Ba^{2+}2H^-$						

of the high temperature (approximately 2800°C) produced by the oxyhydrogen torch.

$$2H_2(g) + O_2(g) \rightarrow 2H_2O(l)$$

The reaction of hydrogen with sulfur is more difficult and requires high temperatures.

$$H_2(g) + S(g) \rightarrow H_2S(g)$$

Hydrogen reacts with nitrogen at high pressures (300 to 1000 atm), at high temperatures (400°C to 600°C), and in the presence of a catalyst to produce ammonia (Haber process).

$$3H_2(g) + N_2(g) \rightleftharpoons 2NH_3(g)$$

Hydrogen does not react readily with carbon, but at high temperatures, hydrogen can be made to react catalytically with finely divided carbon (Bergius process). Hydrocarbons are produced in these reactions.

$$2H_2(g) + C(g) \rightarrow CH_4(g)$$

The hydrides of the group VI A and group VII A elements are Arrhenius acids. The acid strengths of the hydrides of the elements of a *period* increase from left to right across the period in the same order that the electronegativities of the elements increase. We would expect a highly electronegative element to withdraw electrons from the hydrogen and facilitate its release as a proton.

Consider the hydrides of nitrogen, oxygen, and fluorine of the second period. The electronegativity of these elements increases from nitrogen to fluorine. A water solution of ammonia is basic.

$$NH_3(g) + H_2O \rightleftharpoons NH_4^+(aq) + OH^-(aq)$$

Water dissociates to a very small extent and forms extremely low concentrations of both $H_3O^+(aq)$ and $OH^-(aq)$.

$$H_2O + H_2O \rightleftharpoons H_3O^+(aq) + OH^-(aq)$$

A water solution of hydrogen fluoride is acidic.

$$HF(g) + H_2O \rightleftharpoons H_3O^+(aq) + F^-(aq)$$

The acid strength of the hydrides of the third-period elements also increases in this order: PH_3 does not react with water, H_2S is a weak acid, and HCl is a strong acid.

$$PH_3 < H_2S < HCl$$

The electronegativities of the central atoms increase in the same order:

$$P < S < Cl$$

The acidity of the hydrides of the elements of a *group* increases with increasing size of the central atom.

$H_2O < H_2S < H_2Se < H_2Te$
$HF < HCl < HBr < HI$*

This order is the reverse of that expected on the basis of electronegativity. The first hydride of each series is the weakest acid and is formed by the element with the highest electronegativity.

Two factors that influence acid strength are the electronegativity of the central atom and the size of the central atom. When these factors work against each other, the effect of atomic size outweighs the electronegativity effect. A proton is more easily removed from a hydride in which the central atom is large than from one in which the central atom is small.

When the hydrogen compounds of the elements of a period are compared, the small differences in the sizes of the central atoms are unimportant. In each of the oxyacids, the acidic hydrogen is bonded to an O atom, and the variation in the size of this atom from compound to compound is very small. The size of the central atom, however, becomes very important when the hydrides of the elements of a group are compared. The atomic radii of the members of a group increase markedly from the lightest to the heaviest member. Fluorine has an atomic radius of 71 pm, and iodine has an atomic radius of 133 pm. HF is a weak acid and HI is a strong acid.

Consider the hydrides of carbon, sulfur, and iodine. The electronegativities of C, S, and I, which belong to different groups, are about the same (2.5). The atomic radius of C is 77 pm, of S is 103 pm, and of I is 133 pm. There is a marked increase in the acidity of the hydride with the increase in the size of the central atom. Methane, CH_4, does not dissociate in water, H_2S is a weak acid, and HI is a strong acid.

The covalent hydrides provide examples of the use of bonding theories. The structure of the B_2H_6 molecule is unusual. With this exception, the structures of the covalent hydrides listed in Table 8.4 can be explained by and predicted by the theories presented in Chapter 7. Many of these hydrides have been discussed in that chapter.

Boron might be expected to form BH_3 since the boron atom has three valence electrons. The BH_3 molecule, however, has never been isolated. The simplest hydride of boron is diborane, B_2H_6.

Diborane is sometimes called an **electron-deficient molecule** because there are not enough valence electrons to bond all the atoms of the molecule together by conventional covalent bonds. The bonding may be represented as

* In water solution, HCl, HBr, and HI are virtually completely dissociated.

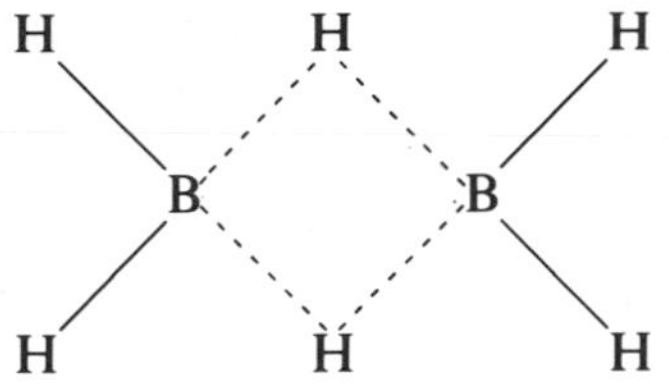

Each B atom has two H atoms (called *terminal* H atoms) bonded to it by regular electron-pair bonds, and the two B atoms of these BH_2 fragments are joined by two **hydrogen bridges** (indicated B---H---B). The hydrogen bridges are **three-center bonds.** Each three-center bond consists of a pair of electrons in a molecular orbital that extends over the nuclei of a B atom, the bridge H atom, and the other B atom.

The B atoms may be considered to use sp^3 hybrid orbitals in the formation of the molecule. Two of the four sp^3 orbitals of each B atom are used to form ordinary covalent bonds with the terminal H atoms. The remaining two sp^3 orbitals of each B atom are used to form three-center bonding orbitals. The molecular orbital for each three-center bond is formed by the overlap of one sp^3 orbital from each B atom and the $1s$ orbital of the bridge H atom (Figure 8.1).

The molecule has twelve valence electrons (six from the two B atoms and six from the six H atoms). Four pairs are used in forming the bonds to the four terminal H atoms. The other two pairs are used in the formation of the three-center bonds. The molecule is diamagnetic, which indicates that all the electrons are paired.

Boron forms many other hydrides (for example, B_4H_{10}, B_5H_9, and $B_{10}H_{14}$), each of which contains at least one three-center bond. Carbon, however, forms more hydrides (called **hydrocarbons**) than any other element—more hydrides, in fact, than all of the other elements put together. The ability of atoms of the same element to bond to each other to form chains is called **catenation,** from the Latin word *catena*, meaning *a chain*. Carbon has a pronounced ability to catenate—more so than any other element.

There appears to be no limit to carbon's ability to catenate. Compounds

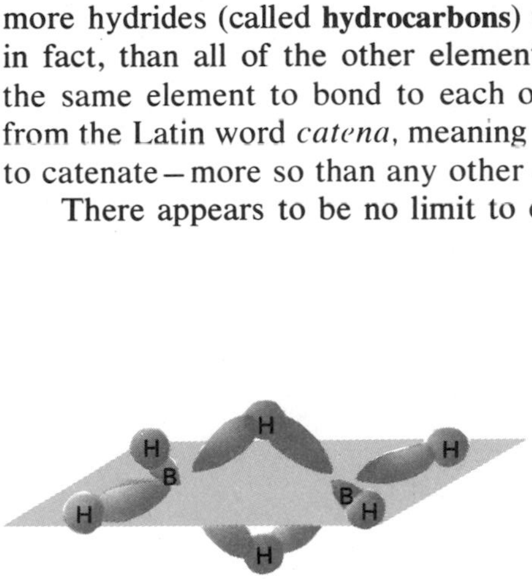

Figure 8.1 Formation of the bonding orbitals of diborane. Shapes of orbitals are simplified.

are known with chains that contain thousands of carbon atoms. The chains may or may not be branched.

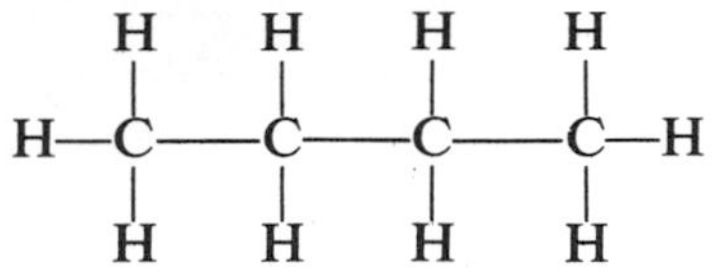

butane

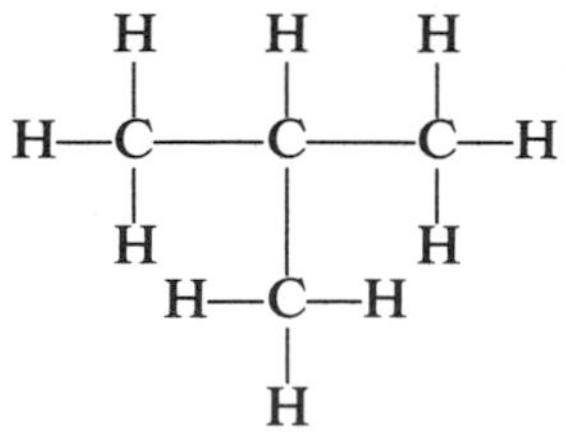

methylpropane

Many ring compounds are known.

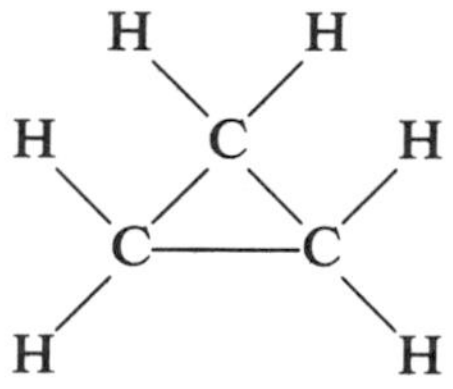

cyclopropane

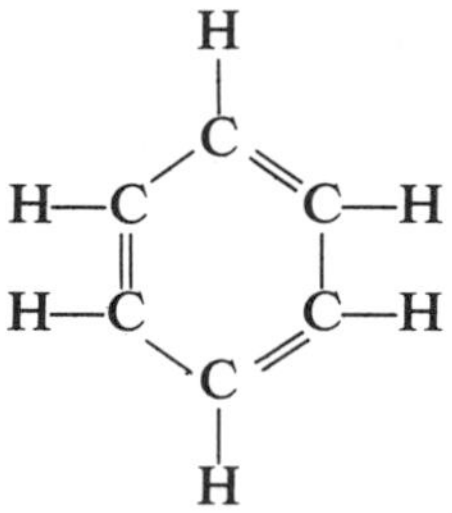

benzene
(a resonance hybrid)

Many of the hydrocarbons contain one or more multiple bonds.

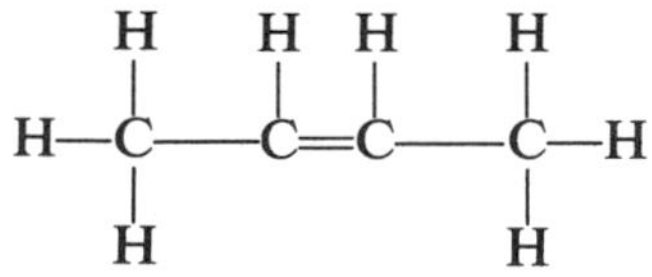

2-butene

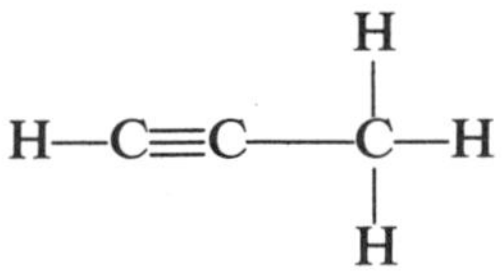

propyne

The hydrocarbons form an important group of organic compounds. Approximately one million organic compounds are known.

Catenated hydrides are formed by some other elements near to carbon in the periodic table (see Table 8.5). Hydrogen peroxide and hydrazine are examples.

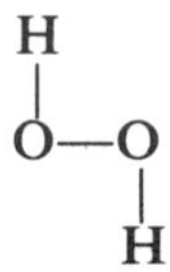

hydrogen peroxide

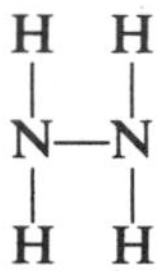

hydrazine

Table 8.5 Catenated hydrides of the representative elements. (Compounds with the highest number of catenated atoms are shown.)

I A	II A	III A	IV A	V A	VI A	VII A	0
			unlimited	N_2H_4	H_2O_2		
			Si_6H_{14}	P_3H_5	H_2S_6		
			Ge_9H_{20}				
			Sn_2H_6				

The atoms of no other element catenate to the extent that carbon atoms do.

The bond energy of the C—C single bond is 347 kJ/mol, which is higher than the bond energies of other single bonds formed between two atoms of the same element. Compare this value to the bond energy of the Si—Si bond (180 kJ/mol), the S—S bond (264 kJ/mol), and the Ge—Ge bond (188 kJ/mol). Furthermore, the bond energy of the C—C bond is not too different from the bond energies of the bonds that carbon forms with other elements. With most other elements, the bond that an atom forms with another atom of the same element is weaker than the bonds that it forms with atoms of other elements.

8.9 OXIDES

Some of the oxides of the representative elements are listed in Table 8.6. In this section, we will discuss a few of the properties of these compounds.

Oxygen forms three different anions: superoxide, peroxide, and oxide ions. Molecular orbital diagrams for oxygen, the superoxide ion, and the peroxide ion are given in Figure 8.2. The **superoxide ion,** O_2^-, can be formed by the addition of one electron to the π^*2p orbital of the O_2 molecule, which reduces the number of unpaired electrons to 1 and the bond order to $1\frac{1}{2}$. The **peroxide ion,** O_2^{2-}, contains two more electrons (in the π^*2p orbitals) than the O_2 molecule. The bond order, therefore, is reduced to 1, and the ion is diamagnetic. The **oxide ion,** O^{2-}, is isoelectronic with the neon atom and is diamagnetic.

All metals except a few of the less-reactive metals (for example, Ag and Au) react with oxygen. Oxides of all metals are known, but some must be made indirectly. The most reactive metals of group I A (and those with largest

Table 8.6 Some oxides of the representative elements

I A	II A	III A	IV A	V A	VI A	VII A	0
H_2O H_2O_2							
Li_2O	BeO	B_2O_3	CO_2 CO	N_2O_5 NO_2 NO	O_2 O_3	OF_2 O_2F_2	
Na_2O Na_2O_2	MgO	Al_2O_3	SiO_2	P_4O_{10} P_4O_6	SO_3 SO_2	Cl_2O_7 ClO_2 Cl_2O	
K_2O KO_2	CaO	Ge_2O_3	GeO_2 GeO	As_2O_5* As_4O_6	SeO_3 SeO_2	BrO_2 Br_2O	
Rb_2O RbO_2	SrO	In_2O_3	SnO_2 SnO	Sb_2O_5* Sb_4O_6	TeO_3	I_2O_7 I_2O_5	XeO_4 XeO_3
Cs_2O CsO_2	BaO BaO_2	Tl_2O_3 Tl_2O	PbO_2 PbO	Bi_2O_3*			

* Empirical formula, actual structure is unknown.

atomic radii)—Cs, Rb, and K—react with oxygen to produce superoxides. For example,

$$Cs(s) + O_2(g) \rightarrow CsO_2(s)$$

Sodium peroxide is produced by the reaction of sodium with oxygen.

$$2Na(s) + O_2(g) \rightarrow Na_2O_2(s)$$

Lithium metal forms a normal oxide with O_2, rather than a peroxide or a superoxide, because the small Li^+ ion cannot form a stable lattice with the larger O_2^{2-} or O_2^- ions.

$$4Li(s) + O_2(g) \rightarrow 2Li_2O(s)$$

The normal oxides of sodium, potassium, rubidium, and cesium can be obtained by heating oxygen with an excess of the metal.

With the exception of barium (which reacts with oxygen to yield barium peroxide), the remaining metals generally produce normal oxides in their reactions with oxygen.

$$2Mg(s) + O_2(g) \rightarrow 2MgO(s)$$
$$4Al(s) + 3O_2(g) \rightarrow 2Al_2O_3(s)$$

Except for the noble gases and the group VII elements, all nonmetals react with oxygen. Oxides of the halogens and the heavier members of the noble gas family have been prepared by indirect means. In general, both types of

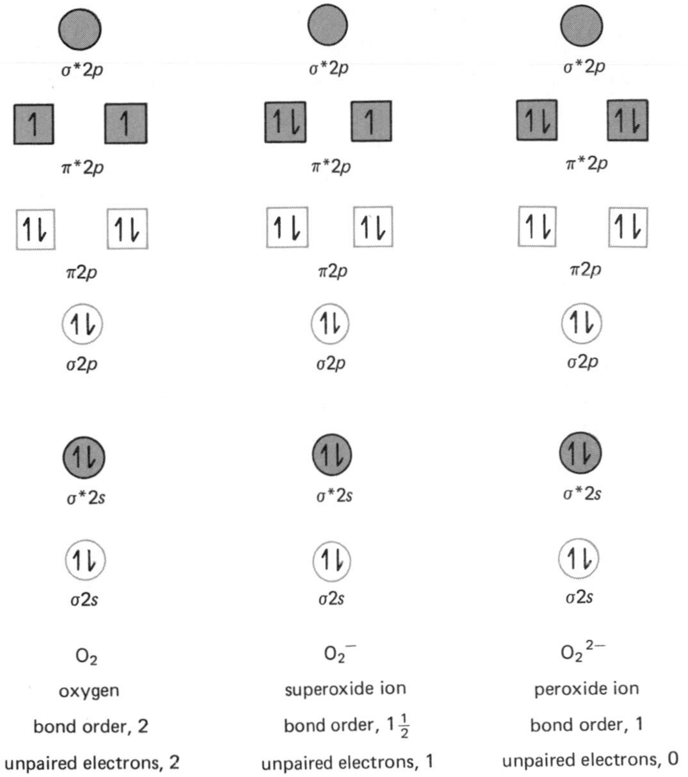

Figure 8.2 Molecular orbital energy-level diagrams for oxygen, the superoxide ion, and the peroxide ion.

these compounds are relatively unstable; they decompose fairly easily to form the elements.

The reaction of oxygen with hydrogen produces water. The product of the reaction of carbon with oxygen depends upon the proportion of carbon to oxygen employed.

$$2C(s) + O_2(g) \rightarrow 2CO(g)$$
$$C(s) + O_2(g) \rightarrow CO_2(g)$$

In like manner, the product of the reaction of phosphorus and oxygen depends upon whether phosphorus is reacted in a limited oxygen supply (P_4O_6) or in excess oxygen (P_4O_{10}). Sulfur reacts to produce SO_2.

$$S(s) + O_2(g) \rightarrow SO_2(g)$$

The reaction of nitrogen with oxygen requires high temperatures. The following reaction occurs in a high-energy electric arc.

$$N_2(g) + O_2(g) \rightarrow 2NO(g)$$

Additional important oxides of sulfur (for example, SO_3) and nitrogen (for example, NO_2 and N_2O_5) are prepared by means other than the direct combination of the elements. In addition to water, hydrogen and oxygen form hydrogen peroxide, H_2O_2, in which the oxygen atoms catenate.

The oxides of metals are called **basic oxides.** The oxides of the group I A metals and those of Ca, Sr, and Ba dissolve in water to produce hydroxides. All these oxides are ionic. When one of them dissolves in water, it is the oxide ion that reacts with the water.

$$O^{2-}(aq) + H_2O \rightarrow 2OH^-(aq)$$

The oxides (as well as the hydroxides) of other metals are insoluble in water.

Nevertheless, metal oxides and hydroxides are chemically related. When heated, most hydroxides are converted into oxides.

$$Mg(OH)_2(s) \rightarrow MgO(s) + H_2O(g)$$

Metal oxides, as well as hydroxides, can be neutralized by acids.

$$MgO(s) + 2H^+(aq) \rightarrow Mg^{2+}(aq) + H_2O$$
$$Mg(OH)_2(s) + 2H^+(aq) \rightarrow Mg^{2+}(aq) + 2H_2O$$

Most of the oxides of the nonmetals are **acidic oxides.** Many of them react with water to produce oxyacids.

$$Cl_2O + H_2O \rightarrow 2HOCl$$
$$Cl_2O_7 + H_2O \rightarrow 2HClO_4$$
$$N_2O_5 + H_2O \rightarrow 2HNO_3$$
$$P_4O_{10} + 6H_2O \rightarrow 4H_3PO_4$$
$$SO_3 + H_2O \rightarrow H_2SO_4$$
$$SO_2 + H_2O \rightleftharpoons H_2SO_3$$
$$CO_2 + H_2O \rightleftharpoons H_2CO_3$$

In the last two cases, both the oxides [$SO_2(g)$ and $CO_2(g)$] and the acids [$H_2SO_3(aq)$ and $H_2CO_3(aq)$] exist in the solutions. For some nonmetal oxides, there are no corresponding acids.

Oxides of nonmetals will neutralize alkalies. The same products are obtained from the reaction of an acidic oxide as are obtained from the reaction of the corresponding acid.

$$H_2SO_3(aq) + 2OH^-(aq) \rightarrow SO_3^{2-}(aq) + 2H_2O$$
$$SO_2(g) + 2OH^-(aq) \rightarrow SO_3^{2-}(aq) + H_2O$$

Mortar consists of lime, $Ca(OH)_2$, sand, SiO_2, and water. The initial hardening of mortar occurs when the mortar dries out. Over a long period of time, however, the mortar sets by absorbing $CO_2(g)$ from the air and forming insoluble $CaCO_3$.

$$Ca(OH)_2(s) + CO_2(g) \rightarrow CaCO_3(s) + H_2O$$

Some oxides have both acidic and basic properties (for example, BeO and Al_2O_3). They are called **amphoteric oxides** and are formed principally by the elements in the center of the periodic table, near to the borderline between the metals and the nonmetals.

$$Al_2O_3(s) + 6H^+(aq) \rightarrow 2Al^{3+}(aq) + 3H_2O$$
$$Al_2O_3(s) + 2OH^-(aq) + 3H_2O \rightarrow \underset{\text{aluminate ion}}{2Al(OH)_4^-(aq)}$$

The properties of the oxides of the third period elements reflect the types of bonding found in these compounds.

Na_2O, MgO, Al_2O_3, SiO_2, P_4O_{10}, SO_3, Cl_2O_7

Oxygen is highly electronegative (3.5 on the Pauling scale). The electronegativities of the third-period elements increase as we go from left to right across the period.

With Na and Mg (the two elements of the period that have the lowest electronegativities) oxygen forms ionic, crystalline oxides. The bonding in Al_2O_3 is ionic with a significant degree of covalent character. Crystals of SiO_2 consist of Si and O atoms joined by polar covalent bonds. These atoms do not form separate molecules. They are held together in a network and the whole crystal may be thought of as a giant molecule. The remaining three oxides, P_4O_{10}, SO_3, and Cl_2O_7 are molecular and covalent. The degree of polarity of the covalent bonds decreases from P_4O_{10} to Cl_2O_7.

The acid–base properties of these oxides reflect the bonding found in them. Sodium oxide is strongly basic; it dissolves in water to form NaOH. Magnesium oxide, MgO, is insoluble in water but has properties of a base. Aluminum oxide, Al_2O_3, is amphoteric. Silicon dioxide, SiO_2, is weakly acidic. It reacts with metal oxides to form complex silicates, but an acid corresponding to SiO_2 is not known. The remaining oxides form acids with water: P_4O_{10} forms a weak acid (H_3PO_4), SO_3 forms a strong acid (H_2SO_4), and Cl_2O_7 forms a very strong acid ($HClO_4$).

8.10 HALIDES

The elements of group VII A are called **halogens** (from Greek, meaning *salt former*). Each halogen is the most reactive nonmetal of its period. Fluorine is the most reactive of all the nonmetals. As a result, the halogens form a large number of compounds. Halides are known for every element except He, Ne, and Ar: Most halogen compounds can be made by the direct reaction of the elements. Some of the more important fluorides and chlorides are listed in Table 8.7.

All of the halides of the I A and II A elements (except those of Be) are ionic. The electronegativities of the halogens decrease in the order: F > Cl > Br > I. Fluorides, therefore, are the most ionic halides that a given metal forms,

Table 8.7 Fluorides and chlorides of the representative elements*

I A	II A	III A	IV A	V A	VI A	VII A	0
HX							
LiX	BeX_2	BX_3	CX_4	NX_3	OX_2	ClF	
NaX	MgX_2	AlX_3	SiX_4	PX_5 PX_3	SF_6 SF_4 S_2Cl_2	ClF ClF_3	
KX	CaX_2	GaX_3	GeX_4 GeX_2	AsX_5 AsX_3	SeF_4 $SeCl_2$	BrX BrF_5	KrF_2
RbX	SrX_2	InX_3	SnX_4 SnX_2	SbX_5 SbX_3	TeX_4	ICl IF_5	XeF_6 XeF_4 XeF_2
CsX	BaX_2	TlX_3 TlX	PbX_4 PbX_2	BiF_5 BiX_3			

* X = F or Cl.

and iodides are the least ionic. The structures of most of the covalent halides can be predicted by use of the electron pair repulsion theory, and many of them have served as examples in Chapter 7.

Only the simple halides of carbon are listed in Table 8.7. Halogen atoms, however, can replace hydrogen atoms in hydrocarbons. Consequently, a large number of catenated compounds of carbon are known that contain one or more halogen atoms. Some compounds contain more than one type of halogen atom. Freon has two chlorine atoms and two fluorine atoms bonded to a central carbon atom.

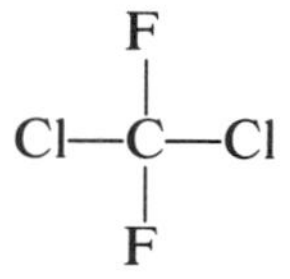

This compound, which is a gas, is used as a refrigerant and as a propellant in aerosol cans.

The fluorides of the noble gases Xe and Kr are interesting compounds. Before 1962, no true compounds of the noble gases were known, and they were called the "inert gases." Today, approximately twenty-five compounds of Xe and Kr are known.

The first reaction of a noble gas was reported by Neil Bartlett. Oxygen reacts with PtF_6 to form an unusual ionic compound, $[O_2^+][PtF_6^-]$. Since the first ionization energy of *molecular* oxygen (1180 kJ/mol)

$$O_2 \rightarrow O_2^+ + e^-$$

is close to the first ionization energy of xenon (1170 kJ/mol), Bartlett reasoned that Xe should react with PtF_6. Experiment showed this prediction to be true. A red, crystalline solid, originally reported as $[Xe^+][PtF_6^-]$ and now thought to be $[XeF]^+[Pt_2F_{11}]^-$ is obtained from the reaction.

The fluorides of Xe and Kr listed in Table 8.7 have been prepared by reacting the noble gas with fluorine under various reaction conditions. The xenon oxide XeO_3 is prepared by reacting XeF_6 with water.

$$XeF_6(s) + 3H_2O \rightarrow XeO_3(aq) + 6HF(aq)$$

NOTES ON NOMENCLATURE

In 1787, Lavoisier, together with Louis Guyton de Morveau, Antoine de Fourcroy, and Claude Berthollet, published *Methods of Chemical Nomenclature.* They proposed that **common names** (such as *powder of algaroth, salt of alembroth, pompholix,* and *colcothar*) be replaced by **systematic names**—names derived by use of a system of nomenclature. A systematic name tells something about the composition and structure of the compound. A common name tells nothing. The system for naming inorganic compounds that they proposed in 1787 is, with some modifications, in use today.

In the early days of organic chemistry, chemists frequently did not know the exact compositions of compounds that they had discovered, and they certainly had no idea of their structures. Common names were used of necessity.

The barbiturates are drugs that are used in sleeping pills. They are derivatives of barbituric acid, a compound that was discovered and named by Adolf von Baeyer in 1863. For a long time, it was believed that the name *barbituric acid* had been derived from a Latin word that means *bearded,* the same Latin word from which the term *barber* is derived. This explanation of the origin of the name seems reasonable since small, hairlike, white crystals of barbituric acid can be prepared. The only trouble with the explanation is that it is wrong. Baeyer recorded in his diary why he named the compound *barbituric acid:* at the time of his discovery, he was in love with a girl named Barbara!

Lavoisier gave *oxygen* its name, which he derived from Greek words that mean *acid former.* He assumed that the element is to be found in all acids. The oxyacids do, of course, contain oxygen, but many acids, such as HF and HCl, do not, so Lavoisier was only partly right. Lavoisier also named *hydrogen* (from Greek, meaning *water former*).

Most elements, however, were named by their discoverers. *Germanium, francium,* and *polonium* were named after Germany, France, and Poland. *Strontium* was named after a town in Scotland. The name of *lutetium* was derived from *Lutetia,* the ancient name for the city of Paris.

Glenn Seaborg in his book *Man-Made Transuranium Elements* tells a story about the naming of elements 97 and 98. These elements were discovered in 1949–50 at the University of California at Berkeley. Element 97 was named *berkelium,* and element 98 was named *californium.*

In a column in the *New Yorker* magazine, the discoverers were told that they lacked foresight. Undoubtedly they would soon discover elements 99 and 100, and the sequence of elements from 97 to 100 could have been named: *universitium* (97), *ofium* (98), *californium* (99), *berkelium* (100). The discoverers replied that they had not named elements 97 and 98 *universitium* and *ofium* because they were afraid that a New Yorker might discover 99 and 100 before they did and name them *newium* and *yorkium.* The staff of the *New Yorker* answered: "We are already at work in our office laboratories on 'newium' and 'yorkium.' So far we just have the names."

PROBLEMS

8.1 State the oxidation number of the atom other than O in each of the following ions: (a) CrO_4^{2-}, (b) $Cr_2O_7^{2-}$, (c) AsO_4^{3-}, (d) MnO_4^-, (e) SO_3^{2-}, (f) ClO_4^-, (g) NO_3^-.

8.2 State the oxidation number of: (a) Cl in Cl_2O_4, (b) Sn in K_2SnO_3, (c) N in $H_2N_2O_2$, (d) B in CaB_2O_4, (e) V in VO^{2+}, (f) Br in BrF_6^-, (g) U in $Li_2U_2O_7$, (h) S in SO_3F_2.

8.3 State the oxidation number of: (a) P in OPF_3, (b) Xe in $CsXeF_7$, (c) B in $LiBF_4$, (d) Te in H_6TeO_6, (e) P in P_4O_8, (f) Sb in $Sb(OH)_2^+$, (g) Bi in $(Bi_6O_6)^{6+}$, (h) Ge in $(Ge_3O_9)^{6-}$.

8.4 State the oxidation number of: (a) Sb in Sb_4O_6, (b) Ti in $K_2Ti_2O_5$, (c) I in H_5IO_6, (d) S in $S_2O_5Cl_2$, (e) Fe in $BaFeO_4$, (f) N in N_2H_4, (g) N in O_2NF, (h) Ge in Mg_2GeO_4.

8.5 State the oxidation number of: (a) Pb in $PbCl_6^{2-}$, (b) Sn in $Sn_2F_5^-$, (c) Re in ReO_4^-, (d) Xe in $HXeO_4^-$, (e) Bi in BiO^+, (f) N in $(NH_3OH)^+$, (g) Mo in $(Mo_6Cl_8)^{4+}$, (h) W in $(H_2W_{12}O_{40})^{6-}$.

8.6 State the oxidation numbers of: (a) P in H_3PO_3, (b) Ge in GeO_4^{4-}, (c) As in $Mg_2As_2O_7$, (d) C in C_2H_4O, (e) S in $Na_2S_2O_3$, (f) Sb in SbO_4^{3-}, (g) B in $Na_2B_4O_7$, (h) P in $Na_4P_2O_7$, (i) Mo in $H_2Mo_3O_{10}$, (j) Sn in $SnCl_6^{2-}$.

8.7 Give formulas for (a) iron(III) sulfate, (b) chromium(II) sulfite, (c) cupric sulfide, (d) silver(I) dihydrogen phosphate, (e) tellurium hexafluoride, (f) mercury(I) acetate, (g) disulfur dichloride, (h) ammonium nitrate.

8.8 Give formulas for (a) cobalt(III) sulfide, (b) bromine pentafluoride, (c) ferrous carbonate, (d) gold(I) sulfate,

(e) calcium bicarbonate, (f) tetraphosphorus trisulfide, (g) cuprous cyanide, (h) lanthanum(III) phosphate.

8.9 Give formulas for: (a) iron(II) oxide, (b) potassium oxide, (c) magnesium nitride, (d) magnesium nitrite, (e) pentasulfur dinitride, (f) thallium(I) sulfide, (g) calcium permanganate, (h) ferric sulfite, (i) lead(II) chromate, (j) silver(I) dichromate.

8.10 Name the following: (a) N_2F_4, (b) $(NH_4)_2Cr_2O_7$, (c) $Co(CN)_2$, (d) Cl_2O_6, (e) $KMnO_4$, (f) $FeSO_3$, (g) $Sr(HSO_4)_2$, (h) TlOH.

8.11 Name the following: (a) InN, (b) $PbCrO_4$, (c) SF_4, (d) $Cr(NO_3)_3$, (e) $NaNO_2$, (f) $Ni_3(AsO_3)_2$, (g) $Ba(HPO_4)$, (h) HNO_2.

8.12 Name the following: (a) $CuCrO_4$, (b) $Fe(ClO_4)_3$, (c) BaO, (d) Au_2O_3, (e) P_4S_3, (f) Cl_2O_7, (g) ZnS, (h) HOCl, (i) $Ca_3(AsO_3)_2$, (j) $CrCl_2$.

8.13 Give the formulas for the sulfides, fluorides, arsenates, permanganates, and sulfates of potassium, calcium, and iron(III).

8.14 (a) Give names for the oxyacids of bromine: HOBr, $HBrO_2$, $HBrO_3$, and $HBrO_4$. (b) What is the name of the potassium salt of each of these acids?

8.15 How many grams of H_2SO_4 does 500 ml of a 0.600 *M* H_2SO_4 solution contain?

8.16 A 200 ml solution of $Ba(OH)_2$ contains 6.16 g of $Ba(OH)_2$. What is the molarity of the solution?

8.17 How many grams of KOH should be used to prepare 750 ml of a 6.00 *M* solution?

8.18 How many milliliters of 2.50 *M* KOH solution will completely neutralize 10.0 ml of 1.00 *M* H_3PO_4?

8.19 How many milliliters of 0.250 *M* H_2SO_4 solution will completely neutralize 45.0 ml of 0.400 *M* KOH?

8.20 What fraction of a mole is one equivalent weight of each of the following substances? (a) H_3PO_4 for a reaction in which Na_3PO_4 is prepared, (b) H_3PO_3 for a reaction in which Na_2HPO_3 is prepared, (c) H_5IO_6 for a reaction in which $Na_3H_2IO_6$ is prepared, (d) H_2SO_4 for a reaction in which $NaHSO_4$ is prepared.

8.21 How many milliliters of 0.150 *N* NaOH would be required to neutralize 35.0 ml of 0.13 *N* H_2SO_4?

8.22 A 15.0 ml sample of an acid requires 37.3 ml of 0.303 *N* NaOH for neutralization. What is the normality of the acid?

8.23 How many ml of 0.300 *N* H_2SO_4 would be required to neutralize 150 ml of 0.250 *N* NaOH?

8.24 A 25.0 ml sample of a solution of an acid requires 36.2 ml of 0.132 *N* NaOH for neutralization. What is the normality of the acid?

8.25 A 0.300 g sample of a pure solid acid requires 27.3 ml of 0.150 *N* NaOH for neutralization. What is the equivalent weight of the acid?

8.26 An antacid tablet containing magnesium hydroxide requires 30.5

ml of 0.450 *N* HCl for neutralization. How many grams of $Mg(OH)_2$ did the tablet contain?

8.27 Citric acid, which may be obtained from lemon juice, has the molecular formula $C_6H_8O_7$. A 0.250 g sample of pure citric acid requires 37.2 ml of 0.105 *N* NaOH for complete neutralization. (a) What is the equivalent weight of citric acid? (b) How many acidic hydrogens per molecule does citric acid have?

8.28 Aluminum hydroxide is used as an antacid. What volume of 0.800 *N* HCl solution will 0.400 g of $Al(OH)_3$ neutralize? Assume the formation of the normal salt.

8.29 The equivalent weight of a diprotic acid is 52.0 for the production of the normal salt and 104 for the production of the acid salt. If 2.81 g of the acid and 45.0 ml of 0.600 *N* NaOH solution are reacted, has the normal or the acid salt been prepared?

8.30 A 0.250 g sample of impure oxalic acid, $H_2C_2O_4$, requires 30.1 ml of 0.120 *N* NaOH for complete neutralization. What is the percentage of $H_2C_2O_4$ in the material?

8.31 Potassium acid phthalate, $KHC_8H_4O_4$, functions as a monoprotic acid. If a 1.46 g sample of impure $KHC_8H_4O_4$ requires 34.3 ml of 0.145 *N* NaOH solution for neutralization, what percentage of $KHC_8H_4O_4$ is in the material?

8.32 Sodium carbonate reacts with acids,

$$Na_2CO_3 + 2HCl \rightarrow 2NaCl + H_2O + CO_2$$

Since 1 mol of Na_2CO_3 neutralizes 2 mol of $H^+(aq)$, the equivalent weight of Na_2CO_3 is one-half the molecular weight. If a 0.283 g sample of impure Na_2CO_3 reacts with 32.1 ml of 0.123 *N* HCl, what is the weight percent of Na_2CO_3 in the material?

8.33 A 0.200 g sample of pure Na_2CO_3 requires 41.2 ml of a solution of HCl (see Problem 8.32). What is the normality of the acid?

8.34 How many grams of pure Na_2CO_3 should be taken so that the sample will neutralize 40.0 ml of 0.100 *N* HCl solution? See Problem 8.32.

8.35 Which compound of each of the following pairs is the stronger acid?
(a) $(HO)_3PO$ or $(HO)_3AsO$
(b) $(HO)_3As$ or $(HO)_3AsO$
(c) $(HO)_2SO_2$ or $(HO)_2SO$
(d) $(HO)_3B$ or $(HO)_2CO$
(e) H_2Te or HI

8.36 Perchloric acid, $HClO_4$, is a stronger acid than chloric acid, $HClO_3$. On the other hand, periodic acid, H_5IO_6, is a weaker acid than iodic acid, HIO_3. (a) Draw Lewis structures for the four compounds, and assign formal charges to the atoms of the molecules. (b) Explain the differences in acid strength on the basis of your structures of part (a).

8.37 Draw Lewis structures (include formal charges) for selenic acid, H_2SeO_4, and telluric acid, H_6TeO_6. (b) Which is the stronger acid?

8.38 Draw Lewis structures for H_3PO_4, H_3PO_3, and H_3PO_2. In H_3PO_4, there are three HO- groups and one O atom attached to a central

P atom. In H_3PO_3, two HO- groups, one H atom, and one O atom are bonded to a central P atom. In H_3PO_2, one HO- group, two H atoms, and one O atom are bonded to a central P atom. Compare the acid strengths of the compounds. Notice the formal charge of the P atom in each case.

8.39 Consider a series of compounds formed from acetic acid,

$$H{-}O{-}\overset{\overset{\displaystyle O}{\|}}{C}{-}CH_3,$$

in which the three H atoms bonded to C are successively replaced by Cl atoms. How does the acidity vary in the series: acetic acid, chloroacetic acid, dichloroacetic acid, and trichloroacetic acid?

8.40 Compare the acidity of AsH_3, H_2Se, and HBr. How does acidity vary among the hydrides of the elements of a period?

8.41 Compare the acidity of H_2S, H_2Se, and H_2Te. How does acidity vary among the hydrides of elements of a group?

8.42 Write chemical equations for the reactions of hydrogen with: (a) Na, (b) Ca, (c) Cl_2, (d) O_2, (e) N_2, (f) C.

8.43 What is a three-center bond?

8.44 In a molecule of B_4H_{10}, the four boron atoms form a puckered square and are joined by B—H—B three-center bonds. Two of the boron atoms, which are diagonally across the square from one another, are joined by an electron-pair bond, and the remaining hydrogens are bonded to boron atoms by electron-pair bonds. Diagram the structure of the molecule.

8.45 What is catenation? Why does carbon exhibit this property more than any other element?

8.46 Write chemical equations for the reactions of oxygen with: (a) K, (b) Na, (c) Li, (d) Al, (e) C, (f) S, (g) N_2.

8.47 What are the anhydrides of (a) KOH, (b) $Ca(OH)_2$, (c) $Al(OH)_3$, (d) H_2CO_3, (e) H_3PO_3, (f) $HClO_4$, (g) H_3BO_3, (h) HNO_2, (i) HOCl.

8.48 Write a chemical equation for the reaction of each of the following substances with water: (a) CaH_2, (b) CaO, (c) SO_3, (d) P_4O_{10}, (e) N_2O_5.

8.49 Complete and balance the following equations. Assume complete neutralization for the acid–alkali reactions.

(a) $Sc_2O_3(s) + H^+(aq) \rightarrow$
(b) $Ca(OH)_2(s) + CO_2(g) \rightarrow$
(c) $CaO(s) + SiO_2(s) \xrightarrow{\text{heat}}$
(d) $SO_2(g) + OH^-(aq) \rightarrow$
(e) $Ag_2O(s) + HCl(g) \rightarrow$

8.50 Discuss the bonding in the compounds: Na_2O, MgO, Al_2O_3, SiO_2, P_4O_{10}, SO_3, and Cl_2O_7. Describe the acid–base properties of these compounds.

8.51 Use the electron-pair repulsion theory to determine the shapes of XeF_4 and XeF_2.

8.52 What explanation can you give for the fact that the only binary compounds of the noble gases known are fluorides and oxides of Xe, Kr, and Rn?

Chapter 9

Liquids and Solids

In gases, the molecules are far apart and are moving rapidly. The attractions between the molecules play a very minor role in the behavior of gases because the molecules have high kinetic energies.

If a gas is cooled, the kinetic energy of the molecules decreases. When the molecules have slowed down enough, the intermolecular attractive forces cause them to condense into a liquid. The molecules are held together by these forces, but they are still free to move about throughout the liquid.

Additional cooling causes the kinetic energies of the molecules to decrease still further. Ultimately, the molecules take positions in a crystal lattice and a solid results. The molecules are held in the lattice by the intermolecular forces of attraction. The motion of the molecules in crystals is restricted to vibration about the fixed lattice positions.

9.1 INTERMOLECULAR FORCES OF ATTRACTION

Atoms are held together in molecules by covalent bonds, but what forces attract molecules to each other in the liquid and solid state? There are several types of attractive forces that hold molecules together which, taken as a group, are called intermolecular attractive forces. Two types are discussed in this section, and a third is presented in the next section.

Dipole–dipole forces occur between polar molecules. Molecules of this type have dipoles and line up in an electric field (see Section 7.3). Dipole–dipole forces are caused by the attractions of the positive and negative poles of the molecules for one another. In a crystal of a polar molecular substance, the molecules are lined up in a way that reflects the dipole–dipole forces (Figure 9.1).

Electronegativity differences between atoms can be used to predict the degree of polarity of a diatomic molecule as well as the positions of the positive and negative poles. A prediction concerning the polarity of a molecule that contains more than two atoms, however, must be based upon a knowledge of the geometry of the molecule *and the arrangement of unshared electron pairs*.

Consider the three molecules (CH_4, NH_3, and H_2O) represented in Figure 9.2. The dipole moment of a molecule is the result of the individual bond dipoles and the unshared electron pairs of the molecule. In each of the molecules under consideration, the central atom is more electronegative than the H atoms bonded to it. The negative end of each *bond* dipole, therefore, points toward the central atom. In CH_4, the tetrahedral arrangement of the four polar C—H bonds produces a molecule that is not polar; CH_4 has no dipole moment. The center of positive charge of the *molecule* (derived by considering all four bonds) falls in the center of the C atom and coincides with the center of negative charge for the molecule.

On the other hand, the trigonal pyramidal NH_3 molecule is polar (its dipole moment is 1.49 D). The three polar bonds and the unshared electron pair are arranged so that the molecule has a dipole with the negative end directed toward the top of the trigonal pyramid and the positive end toward the bottom. Similarly, the angular H_2O molecule is polar (its dipole moment is 1.85 D). The polar bonds and the unshared electron pairs contribute to a dipole with the negative end directed toward the O atom and the positive end directed toward a point halfway between the two H atoms.

The influence that an unshared pair of electrons has on the dipole moment of a molecule is seen in the case of NF_3. The NF_3 molecule has a structure similar to that of NH_3 (Figure 9.2), but the direction of the polarity of the

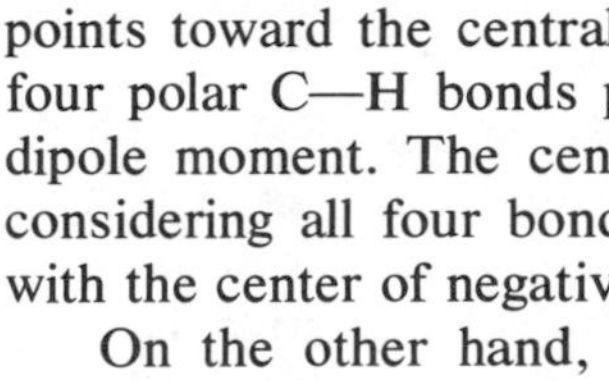

Figure 9.1 Orientation of polar molecules in a crystal.

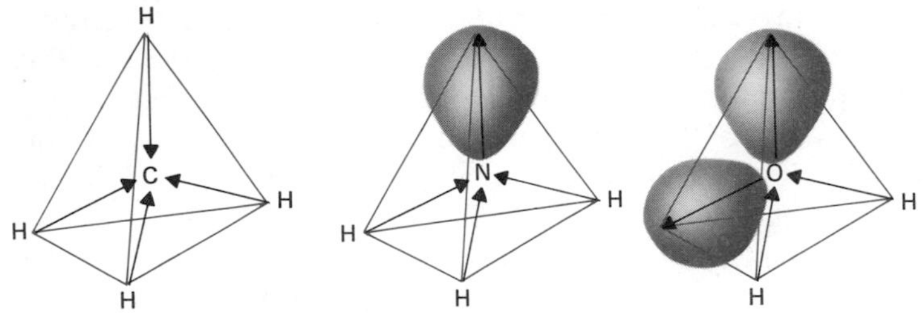

Figure 9.2 Analysis of the polarities of the methane (CH_4), ammonia (NH_3), and water (H_2O) molecules. Arrows point toward the negative end of the individual dipoles that compose the dipole moment of the molecule.

bonds is the reverse of that in NH_3 since F is more electronegative than N. Nitrogen trifluoride has a dipole moment of 0.24 D, a surprisingly low value in view of the highly polar nature of the N—F bonds. The N—F bond dipoles combine to give the molecule a dipole with the negative end in the direction of the base of the pyramid, but the contribution of the unshared electron pair works in the opposite direction and reduces the total polarity of the molecule.

What intermolecular forces attract *nonpolar* molecules to each other in the liquid and solid state? Such molecules do not have permanent dipoles, but nevertheless, they can be liquefied. Some type of intermolecular force, therefore, must exist in addition to the dipole–dipole force.

The existence of **London forces (dispersion forces)** is postulated.* These forces are thought to arise from the motion of electrons. At one instant of time, the electron cloud of a molecule may be distorted so that a dipole is produced in which one part of the molecule is slightly more negative than the rest. At the next instant, the positions of the negative and positive poles of the dipole will be different because the electrons have moved. Over a period of time (a very short period of time—electrons move rapidly), the effects of these **instantaneous dipoles** cancel so that a nonpolar molecule has no permanent dipole moment.

The instantaneous, fluctuating dipole of a molecule, however, induces matching dipoles in neighboring molecules (lined up in the same way that permanent dipoles are aligned). The motion of the electrons of neighboring molecules is synchronized (see Figure 9.3). The force of attraction between these instantaneous dipoles constitutes the London force. The strongest London forces occur between large, complex molecules, which have large electron clouds that are easily distorted, or polarized.

Since all molecules contain electrons, London forces also exist between

* Johannes van der Waals postulated the existence of intermolecular attractive forces between gas molecules in 1873 (Section 4.13). The explanation of the origin of the type of intermolecular force discussed here was proposed by Fritz London in 1930. Although there is a lack of uniformity in the use of terms, current usage appears to favor calling these specific forces *London forces* and intermolecular forces in general *van der Waals forces*.

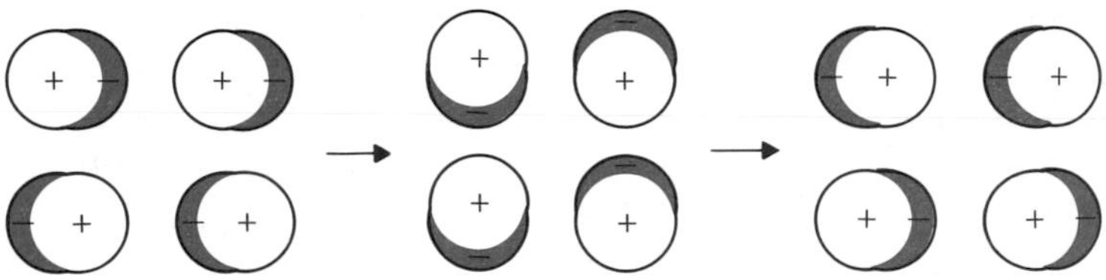

Figure 9.3 Instantaneous dipoles.

polar molecules; in the case of nonpolar molecular substances, London forces are the *only* intermolecular forces that exist. The values listed in Table 9.1 show that London forces are the principal intermolecular forces for most molecular substances. The hydrogen bond, a special type of dipole–dipole interaction that is discussed in the next section, is responsible for the magnitude of the dipole–dipole energy listed for H_2O, NH_3, and (to a lesser extent) HCl.

The dipole moments of the molecules listed in Table 9.1 increase in the order given, and the dipole–dipole energies increase in the same order. The London energies, however, depend upon the sizes of the molecules. The largest molecule listed is HI, and it has the strongest London forces. HCl is a much more polar molecule than HI; the electronegativity of Cl is 3.5, and the electronegativity of I is 2.5. The dipole–dipole energy of HCl is higher than that of HI. The London energy of HI, however, is so much higher than the London energy of HCl, that the *total* effect causes the molecules of HI to be more strongly attracted to one another than the molecules of HCl are. The boiling point of HI is −35°C, which is higher than the boiling point of HCl (−85°C).

9.2 THE HYDROGEN BOND

The intermolecular attractions of certain hydrogen-containing compounds are unusually strong. These strong attractions occur in compounds in which hydrogen is covalently bonded to highly electronegative elements of small atomic size. In these compounds the electronegative element exerts such a

Table 9.1 Intermolecular attractive energies in some simple molecular crystals

		ATTRACTIVE ENERGIES (kJ/mol)	
MOLECULE	DIPOLE MOMENT (D)	DIPOLE–DIPOLE	LONDON
Ar	0	0	8.49
CO	0.12	0.0004	8.74
HI	0.38	0.025	27.9
HBr	0.78	0.69	21.9
HCl	1.03	3.31	16.8
NH_3	1.49	13.3	14.7
H_2O	1.85	36.4	9.0

strong attraction on the bonding electrons that the hydrogen is left with a significant δ^+ charge. In fact, the hydrogen is almost an exposed proton since this element has no screening electrons.

The hydrogen of one molecule and a pair of unshared electrons on the electronegative atom of another molecule are attracted to one another and form what is called a **hydrogen bond.** Because of its small size, each hydrogen atom is capable of forming only one hydrogen bond. The association of HF, H_2O, and NH_3 by hydrogen bonds (indicated by dotted lines) can be roughly diagrammed as follows:

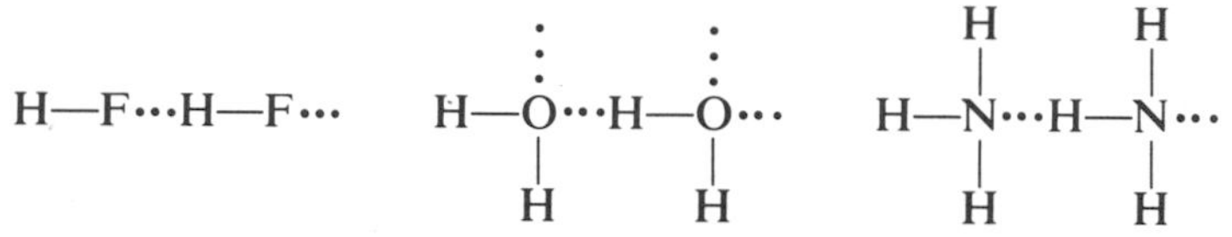

Compounds in which hydrogen bonding occurs have unusual properties. In Figure 9.4, the normal boiling points of the hydrogen compounds of the elements of groups IV A, V A, VI A, and VII A are plotted. The series CH_4, SiH_4, GeH_4, and SnH_4 illustrates the expected trend in boiling point for nonpolar compounds in which the only intermolecular forces are London forces; the boiling point increases as the molecular size and number of electrons increase. The hydrogen compounds of the IV A elements are nonpolar molecules.

In each compound of the other three series of Figure 9.4, however, London forces are aided by dipole–dipole forces in holding the molecules together. Nevertheless, the boiling point of the first member of each series (HF, H_2O, and NH_3) is unusually high in comparison to those of the other members of the series. In each of these three compounds, hydrogen bonding increases the difficulty of separating the molecules of a liquid. Strong hydrogen bonding is not found in any of the other compounds for which a boiling point is plotted in Figure 9.4.

Hydrogen bonding occurs not only between the identical molecules of some pure compounds but also between the different molecules that make up certain solutions. There are two requirements for strong hydrogen bonding. First, the molecule that supplies the proton for the formation of the hydrogen bond must be highly polar so that the hydrogen atom will have a relatively high positive charge. The strength of the hydrogen bonds N—H···N < O—H···O < F—H···F parallels the increasing electronegativity of the atom bonded to hydrogen, N < O < F. The high positive charge on the hydrogen atom attracts the electron pair from another molecule strongly, and the small size of the hydrogen atom permits the second molecule to approach closely.

The second requirement for the formation of a strong hydrogen bond is that the atom that supplies the electron pair for the hydrogen bond must be relatively small. Really effective hydrogen bonds are formed only by fluorine,

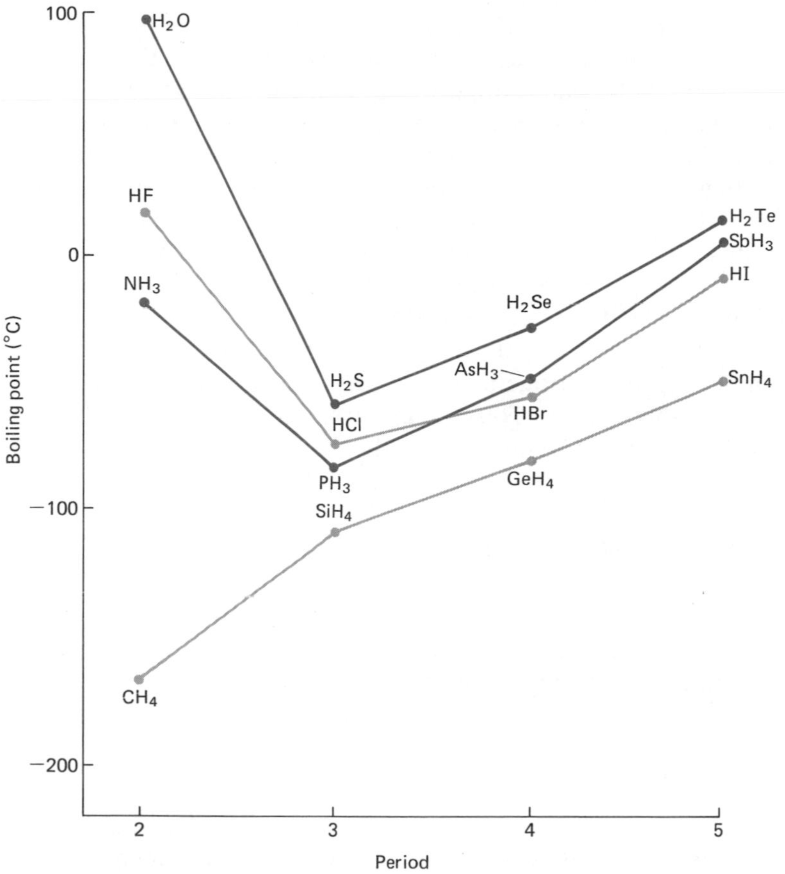

Figure 9.4 Boiling points of the hydrogen compounds of the elements of groups IVA, VA, VIA, and VIIA.

oxygen, and nitrogen compounds. Chlorine compounds form weak hydrogen bonds, as shown by the slight displacement of the boiling point of HCl (Figure 9.4). Chlorine has approximately the same electronegativity as nitrogen. A chlorine atom, however, is larger than a nitrogen atom, and the electron cloud of the chlorine atom is, therefore, more diffuse than that of the nitrogen atom.

An examination of Figure 9.4 will show that hydrogen bonding has a greater effect on the boiling point of water than on the boiling point of hydrogen fluoride. This effect is observed even though the O—H···O bond is only about two-thirds as strong as the F—H···F bond. There are, *on the average,* twice as many hydrogen bonds *per molecule* in H_2O as there are *per molecule* in HF. The oxygen atom of each water molecule has two hydrogen atoms and two unshared electron pairs. The fluorine atom of the hydrogen fluoride molecule has three free electron pairs that can bond with hydrogen atoms but only one hydrogen atom with which it can form a hydrogen bond.

Other properties of water are affected to an unusual degree by hydrogen bonding. The tetrahedral arrangement of the hydrogen atoms and the unshared electron pairs of oxygen in water (Figure 9.2) cause the hydrogen bonds of the ice crystal to be arranged in this manner and leads to the open structure of the ice crystal. Ice, therefore, has a comparatively low density, and ice floats on water. In water at the freezing point, the molecules are arranged more closely together. The liquid (water), therefore, has a higher density than the solid (ice)—an unusual situation. It should be noted that H_2O molecules are associated by hydrogen bonds in the liquid state but not to the same extent, nor in the rigid manner, as they are associated in ice.

Hydrogen bonding also accounts for the high solubilities of some compounds containing oxygen, nitrogen, and fluorine in certain hydrogen-containing solvents, notably water. Thus, ammonia and methanol dissolve in water through the formation of hydrogen bonds.

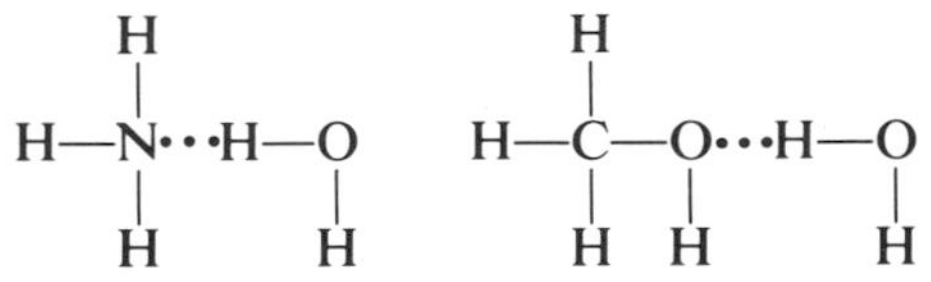

In addition, certain oxygen-containing anions (for example, sulfate, SO_4^{2-}) dissolve in water through hydrogen-bond formation.

9.3 THE LIQUID STATE

In gases, the molecules move rapidly in a completely random manner. In solids, the molecules are held together in the orderly arrangements typical of crystals. The liquid state is intermediate between the gaseous state and the solid state.

In liquids, the molecules move more slowly than they do in gases. The intermolecular attractive forces are able, therefore, to hold them together into a definite volume. The molecular motion, however, is too rapid for the attractive forces to fix the molecules into definite positions in a crystal lattice. A liquid, consequently, retains its volume but not its shape. Liquids flow and assume the shapes of their containers.

A change in pressure has almost no effect on the volume of a liquid since there is little free space between the molecules. An increase in temperature, however, increases the volume of most liquids slightly and, consequently, decreases the liquid density. When the temperature of a liquid is increased, the average kinetic energy of the molecules increases, and this increased molecular motion works against the attractive forces. The expansion, however, is much less than that observed for gases, in which the effect of attractive forces is negligible.

Two liquids that are soluble in one another will diffuse into each other

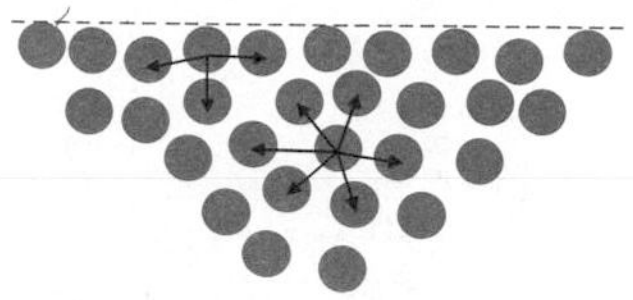

Figure 9.5 Schematic diagram indicating the unbalanced intermolecular forces on the surface molecules of a liquid as compared to the balanced intermolecular forces on the interior molecules.

when placed together. If one liquid is carefully poured on top of another, more dense liquid, the boundary between the two liquids will be sharp and easily seen. This boundary will gradually become less distinct and, in time, will disappear as the molecules of the two liquids diffuse into each other.

The diffusion of liquids is a much slower process than the diffusion of gases. Since the molecules of liquids are relatively close together, a molecule undergoes a tremendous number of collisions in a given time period. The average distance it travels between collisions, the mean free path, is much shorter for a molecule of a liquid than it is for a molecule of a gas. Gases, therefore, diffuse much more rapidly than liquids.

Surface tension is another property of liquids caused by the intermolecular attractive forces. A molecule in the center of a liquid is attracted equally in all directions by surrounding molecules. Molecules on the surface of a liquid, however, are attracted only toward the interior of the liquid (Figure 9.5). The surface molecules, therefore, are pulled inward, and the surface area of a liquid tends to be minimized. This behavior accounts for the spherical shape of liquid drops. Surface tension is a measure of this inward force on the surface of a liquid, the force which must be overcome to expand the surface area. The surface tension of a liquid decreases with increasing temperature since the increased molecular motion tends to decrease the effect of the intermolecular attractive forces.

9.4 EVAPORATION

The kinetic energies of the molecules of a liquid follow a Maxwell-Boltzmann distribution similar to the distribution of kinetic energy among gas molecules (Figure 9.6). The kinetic energy of a given molecule of a liquid is continually changing as the molecule collides with other molecules. At any given instant, however, some of the molecules of the total collection have relatively high energies and some have relatively low energies. The molecules with kinetic energies high enough to overcome the attractive forces of surrounding molecules can escape from the liquid and enter the gas phase if they are close to the surface and are moving in the right direction. They use part of their energy to work against the attractive forces when they escape.

In time, the loss of a number of high-energy molecules causes the average kinetic energy of the molecules remaining in the liquid to decrease, and the

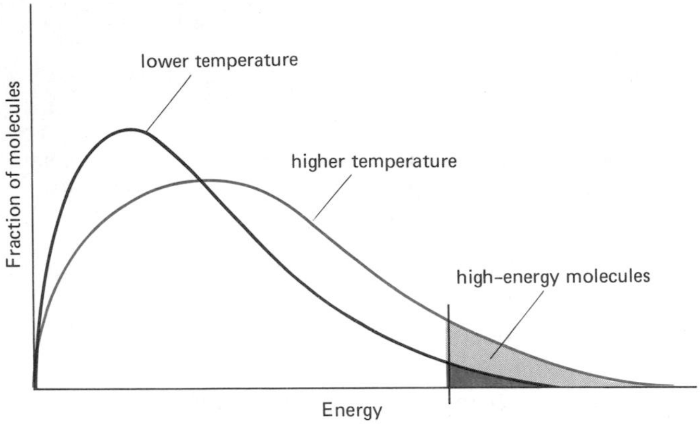

Figure 9.6 Distribution of kinetic energy among molecules of a liquid.

temperature of the liquid falls. When liquids evaporate from an open container, heat flows into the liquid from the surroundings to maintain the temperature of the liquid. In this way the supply of high-energy molecules is replenished, and the process continues until all of the liquid has evaporated. The total quantity of heat required to vaporize a mole of liquid at a given temperature is called the **molar heat of vaporization** of that liquid. At 25°C, for example,

$$H_2O(l) \rightarrow H_2O(g) \qquad \Delta H_v = +43.8 \text{ kJ}$$

The transfer of heat from the surroundings explains why swimmers emerging from the water become chilled as the water evaporates from their skin. Likewise, the regulation of body temperature is, in part, accomplished by the evaporation of perspiration from the skin. Various cooling devices have made use of this principle. A water cooler of the Middle East consists of a jar of unglazed pottery filled with water. The water saturates the clay of the pottery and evaporates from the outer surface of the jar, thus cooling the water remaining in the jar.

The rate of evaporation increases as the temperature of a liquid is raised. When the temperature is increased, the average kinetic energy of the molecules increases, and the number of molecules with energies high enough for them to escape into the vapor phase increases (Figure 9.6).

9.5 VAPOR PRESSURE

When a liquid in a closed container evaporates, the vapor molecules cannot escape from the vicinity of the liquid. In the course of their random motion, some of the vapor molecules return to the liquid. We can represent the process for water by using a double arrow.

$$H_2O(l) \rightleftharpoons H_2O(g)$$

The rate of return of the vapor molecules to the liquid depends upon the concentration of the molecules in the vapor. The more molecules that there are in a given volume of vapor, the greater the chance that some of them will strike the liquid and be recaptured.

At the start, the rate of return of molecules from the vapor to the liquid is low since there are few molecules in the vapor. The continued vaporization, however, causes the concentration of the molecules in the vapor to increase. The rate of condensation, therefore, also increases. Eventually the system reaches a point in which the rate of condensation equals the rate of vaporization.

This condition, in which the rates of two opposite tendencies are equal, is called a state of **equilibrium.** At equilibrium, the concentration of molecules in the vapor state is constant because molecules leave the vapor through condensation at the same rate that molecules add to the vapor through vaporization. Similarly, the quantity of liquid is a constant because molecules are returning to the liquid at the same rate that they are leaving it.

It is important to note that a condition of equilibrium does not imply that nothing is going on. In any system, the numbers of molecules present in the liquid and in the vapor are constant because the two opposing changes are taking place at the same rates and *not* because vaporization and condensation have stopped.

Since the concentration of the molecules in the vapor is a constant at equilibrium, the pressure that the vapor exerts is a constant too. The pressure of vapor in equilibrium with a liquid at a given temperature is called the **vapor pressure** of the liquid. The vapor pressure of a given liquid is determined by the temperature and increases when the temperature is increased. Trace the reasoning through the following statements.

1. An increase in temperature of a liquid causes more molecules to have the energy needed to overcome the attractive forces and escape from the liquid. The *temperature*, therefore, determines the *rate of vaporization*.
2. At equilibrium, the rate of vaporization and the *rate of condensation* are equal.
3. The rate of condensation depends upon the *concentration of vapor molecules*.
4. The concentration of vapor molecules determines the *vapor pressure*.
5. The vapor pressure, therefore, is fixed by the *temperature*.

Figure 9.7 shows the temperature–vapor pressure curves for diethyl ether, ethyl alcohol, and water. The curves show the increase in vapor pressure that accompanies an increase in temperature. The curve for each substance could be extended to the critical temperature of that substance. At the critical temperature, the vapor pressure equals the critical pressure, and the curves

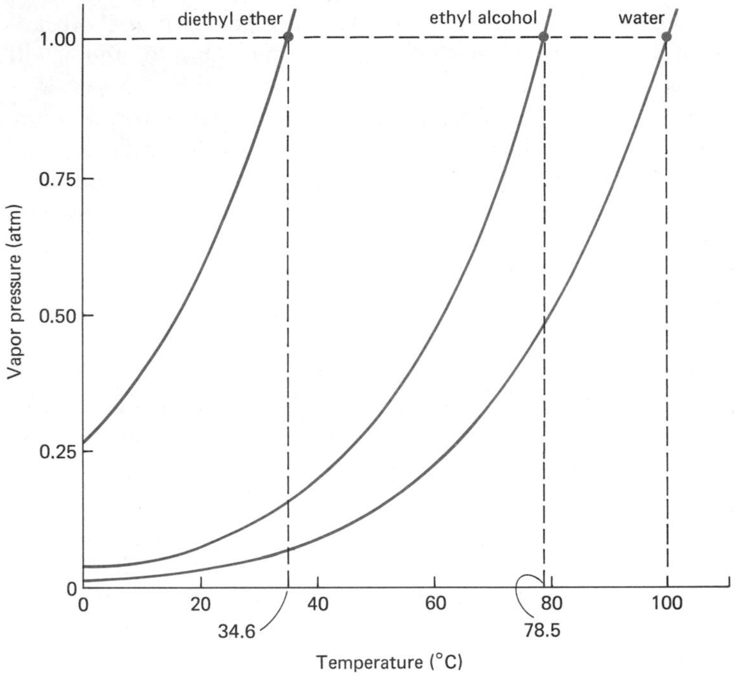

Figure 9.7 Vapor pressure curves for diethyl ether, ethyl alcohol, and water.

end at this point. Above the critical temperature only one phase can exist—the gas phase.

The magnitude of the vapor pressure of a liquid gives an indication of the strength of the intermolecular attractive forces of that liquid. Liquids that have strong attractive forces have low vapor pressures. At 20°C, the vapor pressure of water is 0.023 atm, of ethyl alcohol is 0.058 atm, and of diethyl ether is 0.582 atm. The forces of attraction are strongest in water and weakest in diethyl ether. A list of the vapor pressures of water at various temperatures is given in Table 4.4.

9.6 BOILING POINT

The temperature at which the vapor pressure of a liquid equals the external pressure is called the **boiling point** of the liquid. At this temperature, vapor produced in the interior of a liquid results in the bubble formation and turbulence that is characteristic of boiling. Bubble formation is impossible at temperatures below the boiling point. The atmospheric pressure on the surface of the liquid prevents the formation of bubbles with internal pressures that are less than the pressure of the atmosphere.

The temperature of a boiling liquid remains constant until all the liquid has been vaporized. In an open container the maximum vapor pressure that can be attained by any liquid is the atmospheric pressure. This vapor pressure corresponds to the boiling point. Heat must be added to a boiling liquid to maintain the temperature because in the boiling process, the high-energy molecules are lost by the liquid. The higher the rate at which heat is added to a boiling liquid, the faster it boils. The temperature of the liquid, however, does not rise.

The boiling point of a liquid changes with changes in external pressure. Water, for example, will boil at 98.6°C at a pressure of 0.950 atm and at 101.4°C at a pressure of 1.05 atm. Only at a pressure of 1.00 atm will water boil at 100°C. The **normal boiling point** of a liquid is defined as the temperature at which the vapor pressure of the liquid equals 1 atm. Boiling points given in reference books are understood to be normal boiling points.

The normal boiling points of diethyl ether (34.6°C), ethyl alcohol (78.5°C), and water are indicated on the vapor pressure curves of Figure 9.7. The boiling point of a liquid can be read from its vapor pressure curve by finding the temperature at which the vapor pressure of the liquid equals the prevailing pressure.

The changes in atmospheric pressure at any one geographic location cause a maximum variation of about 2°C in the boiling point of water. The variations from place to place, however, can be greater than this. The average barometric pressure at sea level is 1 atm. At higher elevations, average barometric pressures are less. At an elevation of 1524 m above sea level, for example, the average barometric pressure is 0.836 atm; at this pressure, water boils at 95.1°C. Water boils at 90.1°C at 0.695 atm, which is the average atmospheric pressure at 3048 m above sea level.

If a liquid has a high normal boiling point or decomposes when heated, it can be made to boil at lower temperatures by reducing the pressure. This procedure is followed in vacuum distillation. Water can be made to boil at 10°C, which is considerably below room temperature, by adjusting the pressure to 0.0121 atm (see Table 4.4). Unwanted water is removed from many food products by boiling it away under reduced pressure. In these procedures, the product is not subjected to temperatures that bring about decomposition or discoloration.

9.7 HEAT OF VAPORIZATION

The quantity of energy that must be supplied to vaporize a mole of a liquid at a specified temperature is called the **molar heat of vaporization,** ΔH_v. Heats of vaporization are usually recorded at the normal boiling point in kilojoules per mole (kJ/mol) (see Table 9.2).

The heat of vaporization includes both the energy required to overcome the intermolecular attractive forces and the energy needed to expand the vapor. The volume of a gas is considerably larger than the volume of the liquid from

Table 9.2 Heats of vaporization of liquids at their normal boiling points

LIQUID	t_b NORMAL BOILING POINT (°C)	ΔH_v HEAT OF VAPORIZATION (kJ/mol)	$\Delta S_v = \Delta H_v/T_b$ ENTROPY OF VAPORIZATION (J/K mol)
water	100.0	40.7	109
benzene	80.1	30.8	87
ethyl alcohol	78.5	38.6	110
carbon tetrachloride	76.7	30.0	86
chloroform	61.3	29.4	88
carbon disulfide	46.3	26.8	84
diethyl ether	34.6	26.0	84

which it is derived. A volume of about 1700 ml of steam, for example, is produced by the vaporization of 1 ml of water at 100°C. Energy must be supplied to do the work of pushing back the atmosphere to make room for the vapor.

When a mole of vapor is condensed into a liquid, energy is released, not absorbed. The heat effect is called the **molar heat of condensation.** It has a negative sign, but it is numerically equal to the molar heat of vaporization at the same temperature.

The magnitude of the molar heat of vaporization gives an indication of the strength of the intermolecular attractive forces. A high heat of vaporization indicates that these forces are strong. In 1884, Frederick Trouton discovered that for many liquids the molar heat of vaporization (at the normal boiling point) divided by the normal boiling point (on the Kelvin scale) is a constant, +88 J/K mol. **Trouton's rule** may be expressed in the following way.

$$\frac{\Delta H_v}{T_b} = +88 \text{ J/K mol}$$

The values given in Table 9.2 show that the rule is only approximate.

The value ($\Delta H_v/T_b$) of Trouton's rule is called the **molar entropy of vaporization** and is given the symbol ΔS_v. The **entropy,** S, of a system is a measure of the system's randomness, or disorder. A positive change in entropy, ΔS, indicates an increase in disorder. A negative value of ΔS indicates that the system has become more ordered.

The change from liquid to vapor involves an increase in molecular randomness. The extent of the increase is about the same for all nonpolar liquids. The values of ΔS_v, therefore, are about the same, +88 J/K mol. In water and ethyl alcohol, the intermolecular forces of attraction are unusually strong because of hydrogen bonding. The ΔS_v values for these liquids are comparatively large because they correspond to changes for liquids that are more highly ordered than usual (see Table 9.2).

Example 9.1 The heat of vaporization of N_2O at the normal boiling point is +376 J/g. Use Trouton's rule to find the normal boiling point of N_2O.

Solution The molecular weight of N_2O is 44.0. The molar heat of vaporization, therefore, is

$$? \text{ J} = 1 \text{ mol } N_2O \left(\frac{44.0 \text{ g } N_2O}{1 \text{ mol } N_2O}\right)\left(\frac{+376 \text{ J}}{1 \text{ g } N_2O}\right) = +16{,}500 \text{ J}$$

We use this value to find T_b.

$$\frac{\Delta H_v}{T_b} = +88 \text{ J/K mol}$$

$$T_b = \frac{\Delta H_v}{+88 \text{ J/K mol}}$$

$$T_b = \frac{+16{,}500 \text{ J/mol}}{+88 \text{ J/K mol}} = 188 \text{ K}$$

The answer can be converted to the Celsius scale.

$$\begin{aligned} t &= T - 273 \\ &= 188 - 273 = -85°\text{C} \end{aligned}$$

The handbook lists the normal boiling point of N_2O as −88.5°C.

Example 9.2 A compound has a normal boiling point of 88°C, a heat of vaporization of +322 J/g, and an empirical formula of CH_2. Use Trouton's rule to find the molecular formula of the compound.

Solution The normal boiling point on the Kelvin scale is

$$\begin{aligned} T &= t + 273 \\ &= 88 + 273 = 361 \text{ K} \end{aligned}$$

We find the approximate molar heat of vaporization by using Trouton's rule.

$$\frac{\Delta H_v}{T_b} = +88 \text{ J/K mol}$$

$$\begin{aligned} \Delta H_v &= (+88 \text{ J/K mol})(T_b) \\ &= (+88 \text{ J/K mol})(361 \text{ K}) = 31{,}800 \text{ J/mol} \end{aligned}$$

To get the approximate molecular weight of the unknown compound, X, we use this value and the fact that the heat of vaporization for 1 g of X is +322 J.

$$1 \text{ g X} = +322 \text{ J}$$

$$? \text{ g X} = 1 \text{ mol X}\left(\frac{+31{,}800 \text{ J}}{1 \text{ mol X}}\right)\left(\frac{1 \text{ g X}}{+322 \text{ J}}\right) = 98.8 \text{ g X}$$

The formula weight of CH_2 is 14.0. Consequently, 1 mol of X contains 98.8/14.0 = 7.06 formula weights. Trouton's rule is not exact. There must be a whole number of CH_2 formula weights in one molecular weight. We conclude that this number must be 7 and that the molecular formula of X is C_7H_{14}.

9.8 THE FREEZING POINT

When a liquid is cooled, the molecules move more and more slowly. Eventually a temperature is reached at which some of the molecules have kinetic energies that are low enough to allow the intermolecular attractions to hold them in a crystal lattice. The substance then starts to freeze. Gradually the low-energy molecules assume positions in the crystal lattice. The molecules remaining in the liquid have a higher temperature because of the loss of these low-energy molecules. Heat must be removed from the liquid to maintain the temperature.

The **normal freezing point** of a liquid is the temperature at which solid and liquid are in equilibrium under a total pressure of 1 atm. At the freezing point the temperature of the solid–liquid system remains constant until all of the liquid is frozen. The quantity of heat that must be removed to freeze a mole of a substance at the freezing point is called the **molar heat of crystallization.** This quantity represents the difference between the enthalpies of the liquid and the solid.

At times the molecules of a liquid, as they are cooled, continue the random motion characteristic of the liquid state at temperatures below the freezing point. Such liquids are referred to as **undercooled** or **supercooled liquids.** These systems can usually be caused to revert to the freezing temperature and the stable solid–liquid equilibrium by scratching the interior walls of the container with a stirring rod or by adding a seed crystal around which crystallization can occur. The crystallization process supplies heat, and the temperature is brought back to the freezing point until normal crystallization is complete.

Some supercooled liquids can exist for long periods, or even permanently, in this state. When these liquids are cooled, molecules solidify in a random arrangement typical of the liquid state rather than in an orderly geometric pattern of a crystal. Substances of this type have complex molecular forms for which crystallization is difficult. They are frequently called **amorphous solids, vitreous materials,** or **glasses;** examples include glass, tar, and certain plastics. Amorphous solids have no definite freezing or melting points. These transitions take place over a temperature range. They break into fragments that have curved, shell-like surfaces. Crystalline materials break into fragments that resemble the parent crystals.

When a crystalline substance is heated, the temperature at which solid–liquid equilibrium is attained under air at 1 atm pressure is called the **melting point.** It is, of course, the same temperature as the freezing point of the sub-

Table 9.3 Heats of fusion of solids at the melting point

SOLID	t_f MELTING POINT (°C)	ΔH_f HEAT OF FUSION (kJ/mol)	$\Delta S_f = \Delta H_f/T_f$ ENTROPY OF FUSION (J/K mol)
water	0.0	6.02	22.0
benzene	5.5	9.83	35.3
ethyl alcohol	−117.2	4.60	29.7
carbon tetrachloride	−22.9	2.51	10.0
chloroform	−63.5	9.20	43.9
carbon disulfide	−112.1	4.39	27.3

stance. The quantity of heat that must be *added* to melt a mole of the material at the melting point is called the **molar heat of fusion** and is numerically equal to the heat of crystallization but opposite in sign.

The molar heats of fusion and **molar entropies of fusion** of several substances are listed in Table 9.3. The transition from solid to liquid represents an increase in disorder (and consequently in entropy) in the same way that the transition from liquid to gas does. Entropies of fusion, however, are generally of a lower order than entropies of vaporization because molecular disorder is greater in the gaseous state than in the liquid or solid states. The variation in ΔS_f values from substance to substance is indicative of the differences in the lattice energies of the crystalline solids; ΔS_f values are not nearly so constant as ΔS_v values.

9.9 VAPOR PRESSURE OF A SOLID

Molecules in a crystal vibrate about their lattice positions. A distribution of kinetic energies exist among these molecules similar to the distribution for liquids and gases but on a lower level. Energy is transmitted from molecule to molecule within a crystal; the energy of any one molecule, therefore, is not constant. High-energy molecules on the surface of the crystal can overcome the attractive forces of the crystal and escape into the vapor phase. If the crystal is in a closed container, an equilibrium is eventually reached in which the rate of the molecules leaving the solid equals the rate at which the vapor molecules return to the crystal. The vapor pressure of a solid at a given temperature is a measure of the number of molecules in a given volume of the vapor at equilibrium.

Every solid has a vapor pressure, although some pressures are very low. The magnitude of the vapor pressure is inversely proportional to the strength of the attractive forces. Ionic crystals, therefore, have very low vapor pressures.

Since the ability of molecules to overcome the intermolecular forces of attraction depends upon their kinetic energies, the vapor pressure of a solid increases as the temperature increases. The temperature–vapor pressure

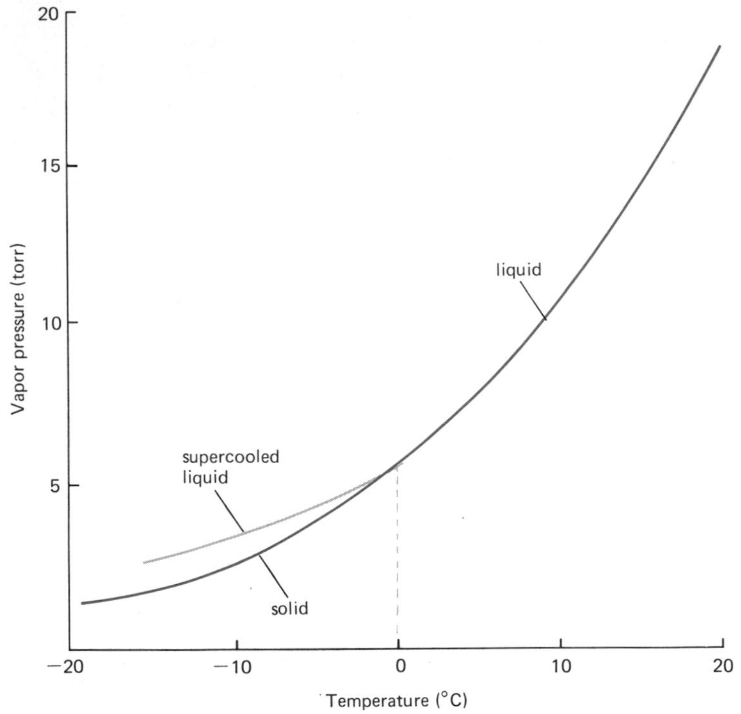

Figure 9.8 Vapor pressure curves for ice and water near the freezing point. Vapor pressures are partial pressures of H_2O under air at a total pressure of 1.00 atm.

curve for ice is illustrated in Figure 9.8. This curve intersects the vapor pressure curve for water at the freezing point. At the freezing point, the vapor pressures of solid and liquid are equal.

In the absence of air, the normal freezing point of water (1 atm total pressure) is 0.0025°C. *In air,* however, and under a total pressure of 1 atm, the freezing point of water is 0.0000°C, which is the commonly reported value. The difference in freezing point is caused by the presence of dissolved air in the water (Section 10.8). The vapor pressures plotted in Figure 9.8 are the partial pressures of H_2O in air with the total pressure equal to 1 atm. Freezing points are usually determined in air; however, in any event, any change in freezing point of a given substance caused by the presence of air is generally very small.

9.10 PHASE DIAGRAMS

The temperature–pressure **phase diagram** for water conveniently illustrates the conditions under which water can exist as solid, liquid, or vapor, as well as the conditions that bring about changes in the state of water. Figure 9.9 is a schematic representation of the water system. It is not drawn to scale, and

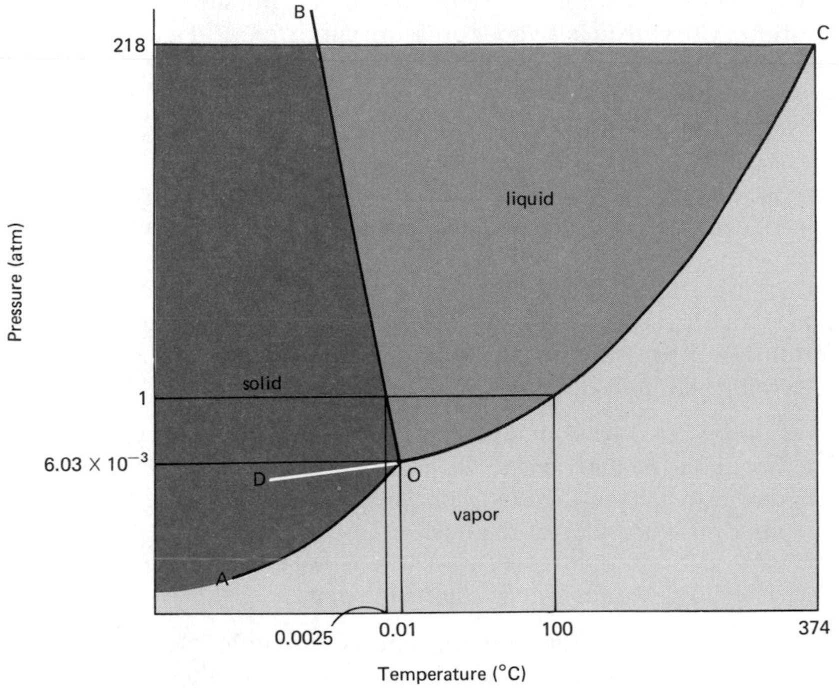

Figure 9.9 Phase diagram for water (not drawn to scale).

some of its features are exaggerated so that important details can be easily seen. Every substance has its own phase diagram, which describes only systems in equilibrium. These diagrams are derived from experimental observations.

The diagram of Figure 9.9 relates to what is called a **one-component system;** that is, it pertains to the behavior of water in the absence of any other substance. No part of the total pressure of any system described by the diagram is due to the pressure of a gas other than water vapor. The vapor pressure curves plotted in Figure 9.8 (measured in air under a constant total pressure of 1 atm) therefore deviate slightly, but only very slightly, from the vapor pressure curves of Figure 9.9 (for which the vapor pressure of water is the *total* pressure). The easiest way to interpret the phase diagram for water is to visualize the total pressure acting on a system in mechanical terms, for example, as a piston acting on the material of the system contained in a cylinder.

In Figure 9.9 curve OC is a vapor pressure curve for liquid and terminates at the critical point, C. Any point on this line describes a set of temperature and pressure conditions under which liquid and vapor can exist in equilibrium. The extension DO is the curve for supercooled liquid; systems between liquid and vapor described by points on this line are **metastable.** (The term *metastable* is applied to systems that are not in the most stable state possible at the temperature in question.) Curve AO is a vapor pressure curve for solid and represents a set of points that describe the possible temperature and pressure conditions for solid–vapor equilibria. The line BO, the melting point curve, represents conditions for equilibria between solid and liquid.

These three curves intersect at point O, a **triple point.** Solid, liquid, and vapor can exist together in equilibrium under the conditions represented by this point: 0.01°C (which is 273.16 K) and a pressure of 0.00603 atm (or 4.58 torr).

The phases (solid, liquid, and vapor) that exist in equilibrium under a set of temperature and pressure conditions can be read from the phase diagram. The temperature and pressure define a point on the diagram. The phases can be read from the position of the point. If the point falls

1. *in a region* labeled solid, liquid, or vapor, only *one phase* exists—the phase noted on the diagram.
2. *on a line, two phases* exist; the phases are those that are marked in the regions on both sides of the line.
3. *on a point,* all *three phases* exist. There is only one such point in the phase diagram given for water—the triple point.

The slope of the melting point (or freezing point) curve, BO, shows that the freezing point decreases as the pressure is increased. A slope of this type is observed for only a few substances such as gallium, bismuth, and water. It indicates an unusual situation in which the liquid expands upon freezing. At 0°C a mole of water occupies 18.00 cm^3, and a mole of ice occupies 19.63 cm^3. The system expands, therefore, when one mole of liquid water freezes into ice. An increase in pressure on the system would oppose this expansion and the freezing process. Hence, the freezing point of water is lowered when the total pressure is increased. In Figure 9.9 the slope of the line BO is exaggerated.

Phase changes brought about by temperature changes at constant pressure may be read from a phase diagram by interpreting a horizontal line drawn at the reference pressure (like the line drawn at 1.00 atm in Figure 9.9). The point where this line intersects curve BO indicates the normal melting point (or freezing point), and the point where the 1.00 atm line intersects curve CO represents the normal boiling point. Beyond this point, only vapor exists.

Phase changes brought about by pressure changes at constant temperature may be read from a vertical line drawn at the reference temperature. If the pressure is increased, for example, at 0.0025°C (Figure 9.9), the point where the vertical line crosses AO is the pressure where vapor changes to solid, and the point where the vertical line crosses BO represents the pressure where solid changes to liquid. Above this point, only liquid exists.

For materials that contract upon freezing (that is, the solid phase is more dense than the liquid phase), the freezing-point curve slants in the opposite direction, and the freezing point increases as the pressure is increased. This

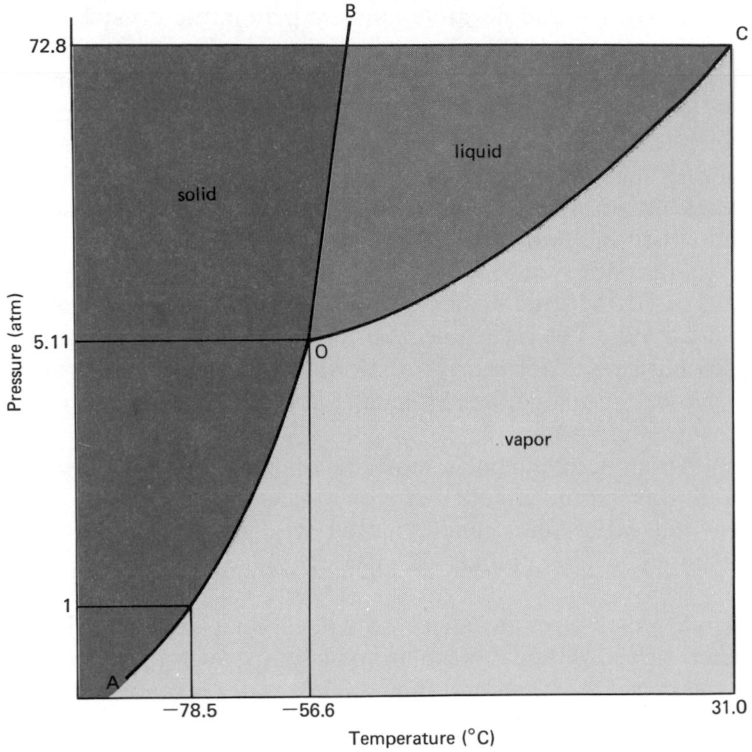

Figure 9.10 Phase diagram for carbon dioxide (not drawn to scale).

behavior is characteristic of most substances. The freezing-point curves of most phase diagrams slant to the right as is seen in the phase diagram for carbon dioxide in Figure 9.10.

The process in which a solid goes directly into a vapor without going through the liquid state is known as **sublimation;** this process is reversible. The phase diagram for carbon dioxide is typical for substances that sublime at ordinary pressures rather than melt and then boil. The triple point of the carbon dioxide system is −55.6°C at a pressure of 5.11 atm. Liquid carbon dioxide exists only at pressures greater than 5.11 atm. When solid carbon dioxide (dry ice) is heated at 1 atm pressure, it is converted directly into gas at −78.5°C. This relationship is shown in Figure 9.10. The **molar heat of sublimation** is the heat that must be added to a mole of solid to convert it directly into a gas.

9.11 PROPERTIES OF SOLIDS

Crystals are formed by atoms, ions, and molecules. We can classify crystals into four types according to the kind of particles that make up the crystal lattice and the forces that hold them together.

1. *Ionic crystals.* Positive and negative ions are held in the crystal lattice by electrostatic attraction. Because these forces are strong, ionic substances have high melting points. Ionic crystals are hard and brittle. Figure 9.11 shows what happens if an attempt is made to deform an ionic crystal. Because of the movement of one plane of ions over another, ions with the same charge are brought next to one another. The crystal breaks into fragments. Ionic compounds are good conductors of electricity when molten or in solution but not in the crystalline state where the ions are not free to move.
2. *Molecular crystals.* Molecules occupy the lattice points in crystals of covalent compounds. The intermolecular forces that hold the molecules in the crystal lattice are not nearly so strong as the electrostatic forces that hold ionic crystals together. Molecular crystals, therefore, are soft and have low melting points.

   London forces hold nonpolar molecules in the lattice. In crystals of polar molecules, dipole–dipole forces as well as London forces occur. Polar compounds, therefore, generally melt at slightly higher temperatures than nonpolar compounds of *comparable molecular size and shape.*

   In general, molecular substances do not conduct electricity in the solid or liquid states. A few molecular compounds, such as water, dissociate to a very slight extent and produce low concentrations of ions; these liquids are poor electrical conductors.
3. *Network crystals.* In these crystals, atoms occupy the lattice positions

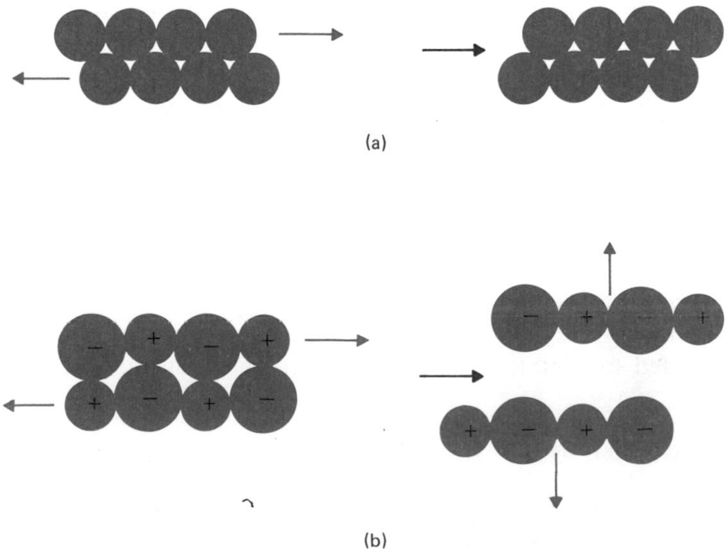

Figure 9.11 Effect of deformation on (a) a metallic crystal and (b) an ionic crystal.

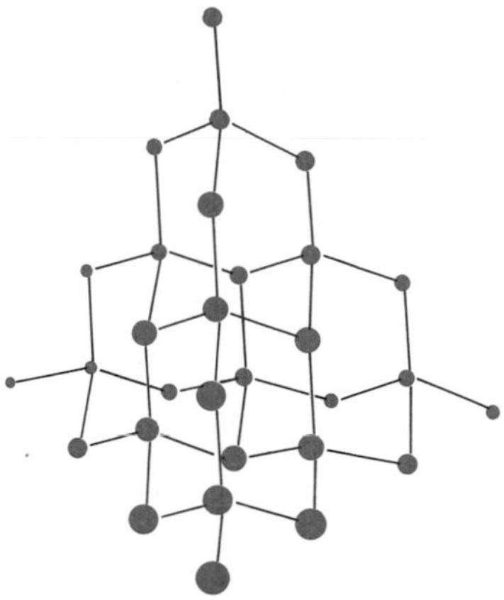

Figure 9.12 Arrangement of atoms in a diamond crystal.

and they are joined by a network of covalent bonds. The entire crystal can be looked at as one giant molecule. In diamond, an example of this type of crystal, carbon atoms are bonded by covalent bonds into a three-dimensional structure (see Figure 9.12). Materials of this type have high melting points and are extremely hard because of the large number of covalent bonds that would have to be broken to destroy the crystal structure. Network crystals do not conduct electricity.

4. *Metallic crystals*. The outer electrons of metal atoms are loosely held and move freely throughout a metallic crystal. The remainder of the metal atoms, positive ions, occupy fixed lattice positions in the crystal. The negative cloud of the freely moving electrons, sometimes called an electron gas or a sea of electrons, binds the crystal together. This binding force, called a **metallic bond,** is described more fully in Section 19.1.

   The metallic bond is strong. Most metals have high melting points, high densities, and crystal lattices in which the positive ions are packed together closely (called **close-packed** arrangements). Unlike ionic crystals, the positions of the positive ions can be altered without destroying the crystal because of the uniform cloud of negative charge provided by the freely moving electrons (see Figure 9.11). Most metallic crystals, therefore, are easily deformed, and most metals are malleable and ductile. The freely moving electrons are also responsible for the fact that most metals are good conductors of electricity.

The properties of the four types of crystals are summarized in Table 9.4.

9.12 CUBIC UNIT CELLS

The particles of a crystal are arranged in a repeating three-dimensional pattern called a **crystal** (or **space**) **lattice.** The smallest section of a crystal lattice

Table 9.4 Types of crystalline solids

CRYSTAL	PARTICLES	ATTRACTIVE FORCES	PROPERTIES	EXAMPLES
ionic	positive and negative ions	electrostatic attractions	high m.p. hard, brittle good electrical conductor in fused state	NaCl, BaO, KNO_3
molecular	polar molecules	London and dipole–dipole	low m.p. soft nonconductor or extremely poor conductor of electricity in liquid state	H_2O, NH_3, SO_2
	nonpolar molecules	London		H_2, Cl_2, CH_4
network	atoms	covalent bonds	very high m.p. very hard nonconductor of electricity	C (diamond) SiC, AlN, SiO_2
metallic	positive ions and mobile electrons	metallic bonds	fairly high m.p. hard or soft malleable and ductile good electrical conductor	Ag, Cu, Na, Fe, K

that can be used to describe the lattice is the **unit cell.** A crystal can be reproduced, in theory, by stacking its unit cells in three dimensions.

Crystal lattices may be grouped into six **crystal systems** (Figure 9.13). A crystal system may be described by the dimensions of a unit cell along its three axes (a, b, c) and the sizes of the three angles between the axes (α, β, γ). Crystals have the same symmetry as their constituent unit cells.

In illustrations of crystal lattices, points are customarily used to indicate the centers of the ions, atoms, or molecules. Six point lattices can be drawn by placing points at the corners of the unit cells shown in Figure 9.13. However, it is possible to have points at positions other than the corners. Thus, three cubic lattices are known (Figure 9.14): simple cubic, body-centered cubic, and face-centered cubic. Altogether, there are 14 possible lattices in which each point is surrounded in an identical manner by other points.

In counting the number of atoms per unit cell, one must keep in mind that atoms on corners or faces are shared with adjoining cells. Eight unit cells share each corner atom, and two unit cells share each face-centered atom (Figure 9.15).

1. The simple cubic unit cell contains the equivalent of only one atom (8 corners at $\frac{1}{8}$ atom each).

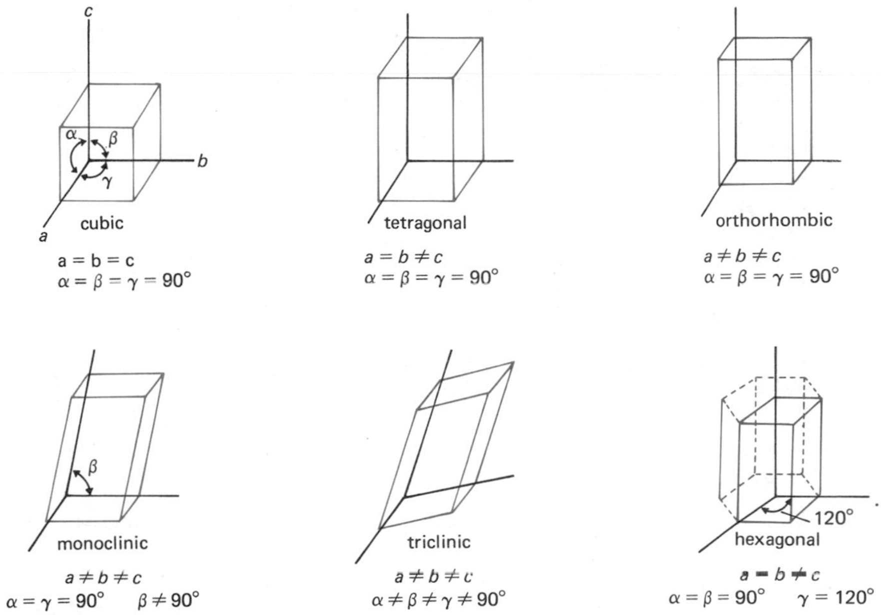

Figure 9.13 Crystal systems.

2. The body-centered unit cell contains two atoms (8 corners at $\frac{1}{8}$ atom each and one unshared atom in the center).
3. The face-centered unit cell contains the equivalent of four atoms (8 corners at $\frac{1}{8}$ atom each and 6 face-centered atoms at $\frac{1}{2}$ atom each).

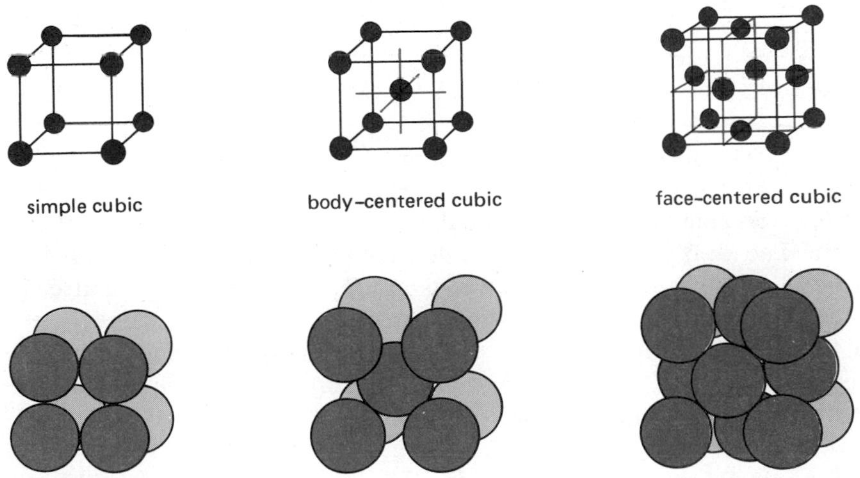

Figure 9.14 Cubic lattices.

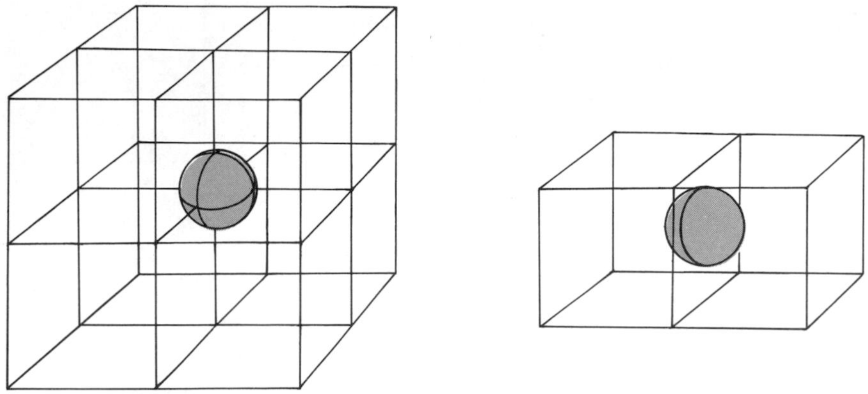

(a) (b)

Figure 9.15 In cubic crystals (a) a corner atom is shared by eight unit cells and (b) a face-centered atom is shared by two unit cells.

Example 9.3 Nickel crystallizes in a face-centered cubic crystal. The edge of a unit cell is 352 pm. The atomic weight of nickel is 58.7, and its density is 8.94 g/cm³. Calculate Avogadro's number from these data.

Solution Since 1 pm $= 10^{-10}$ cm,

$$352 \text{ pm} = 3.52 \times 10^{-8} \text{ cm}$$

The volume of one unit cell is $(3.52 \times 10^{-8} \text{ cm})^3$, or 4.36×10^{-23} cm³. Since the unit cell is face-centered, it contains 4 atoms. Therefore,

$$4 \text{ atoms} = 4.36 \times 10^{-23} \text{ cm}^3$$

We derive from the density,

$$1 \text{ cm}^3 = 8.94 \text{ g Ni}$$

The number of atoms in 58.7 g of Ni is Avogadro's number.

$$? \text{ atoms} = 58.7 \text{ g Ni} \left(\frac{1 \text{ cm}^3}{8.94 \text{ g Ni}}\right)\left(\frac{4 \text{ atoms}}{4.36 \times 10^{-23} \text{ cm}^3}\right) = 6.02 \times 10^{23} \text{ atoms}$$

Example 9.4 Sodium crystallizes in a cubic lattice, and the edge of a unit cell is 430 pm. The density of sodium is 0.963 g/cm³, and the atomic weight of sodium is 23.0. How many atoms of sodium are contained in one unit cell? What type of cubic unit cell does sodium form?

Solution The edge of the unit cell is 4.30×10^{-8} cm. The volume of the unit cell, therefore, is $(4.30 \times 10^{-8} \text{ cm})^3$, or 7.95×10^{-23} cm³. We must find the number of Na atoms in this volume.

We derive our conversion factors from the density of Na

$$0.963 \text{ g Na} = 1 \text{ cm}^3$$

and the fact that 1 mol of Na (which is 23.0 g of Na) contains Avogadro's number of Na atoms.

$$6.02 \times 10^{23} \text{ atoms Na} = 23.0 \text{ g Na}$$

The solution is

$$? \text{ atoms Na} = 7.95 \times 10^{-23} \text{ cm}^3 \left(\frac{0.963 \text{ g Na}}{1 \text{ cm}^3}\right)\left(\frac{6.02 \times 10^{23} \text{ atoms Na}}{23.0 \text{ g Na}}\right)$$
$$= 2.00 \text{ atoms Na}$$

Sodium crystallizes in a body-centered cell since the *body-centered cubic unit cell* contains 2 atoms.

9.13 X-RAY DIFFRACTION OF CRYSTALS

Much of what is known about the internal structure of crystals has been learned from X-ray diffraction experiments. When two X rays that have the same wavelength are in phase, they reinforce each other and produce a wave that is stronger than either of the original waves. Two waves that are completely out of phase cancel each other (see Figure 7.14).

Figure 9.16 illustrates the way that crystal spacings can be determined by use of X rays of a single wavelength, λ. The rays strike the parallel planes of the crystal at an angle θ. Some of the rays are reflected from the upper plane, some from the second plane, and some from the lower planes. A strong reflected beam will result only if all the reflected rays are in phase.

In Figure 9.16, the lower ray travels farther than the upper ray by an amount equal to EF + FG. The rays will be in phase at BG only if the difference is equal to a whole number of wavelengths.

$$\text{EF} + \text{FG} = n\lambda$$

where n is a simple integer.

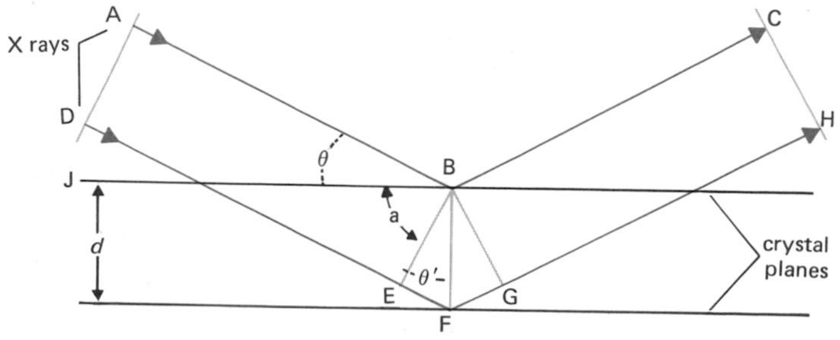

Figure 9.16 Derivation of the Bragg equation.

Since angle ABE is a right angle,

$$\theta + a = 90°$$

Angle JBF is also a right angle, and

$$\theta' + a = 90°$$

The angle θ' therefore is equal to θ. The sine of angle θ' is equal to EF/BF (the ratio of the side opposite the angle to the hypotenuse). Since line BF is equal to d,

$$\sin \theta = \frac{EF}{d}$$

or

$$EF = d \sin \theta$$

The expression

$$FG = d \sin \theta$$

can be derived in the same way. Therefore,

$$EF + FG = 2d \sin \theta$$

since EF + FG is equal to $n\lambda$,

$$n\lambda = 2d \sin \theta$$

This equation, derived by William Henry Bragg and his son William Lawrence Bragg in 1913, is called the **Bragg equation.**

With X rays of a definite wavelength, reflections at various angles will be observed for a given set of planes separated by a given distance, d. These reflections correspond to $n = 1, 2, 3$, and so on, and are spoken of as first order, second order, third order, and so on. With each successive order, the angle θ increases, and the intensity of the reflected beam weakens.

Figure 9.17 is a schematic representation of an X-ray spectrometer. An X-ray beam defined by a slit system impinges upon a crystal that is mounted on a turntable. A detector (photographic plate, ionization chamber, or Geiger counter) is positioned as shown in the figure. As the crystal is rotated, strong signals flash out as angles are passed that satisfy the Bragg equation. Any set of regularly positioned planes that contain atoms can give rise to reflections—not only those that form the faces of the unit cells. Thus, the value of d is not necessarily the edge of the unit cell, although the two are always mathematically related.

Example 9.5 The diffraction of a crystal of barium with X radiation of wavelength 229 pm gives a first-order reflection at 27° 8′. What is the distance between the diffracted planes?

Solution Substitution into the Bragg equation gives

$$n\lambda = 2d \sin \theta$$
$$1(229 \text{ pm}) = 2d(0.456)$$
$$d = 251 \text{ pm}$$

9.14 METALLIC AND IONIC CRYSTAL LATTICES

In the large majority of cases, metal crystals belong to one of three classifications: body-centered cubic (Figure 9.14), face-centered cubic (Figure 9.14), and hexagonal close-packed (Figure 9.18). The geometric arrangement of atoms in the face-centered cubic (also called cubic close-packed) and the hexagonal close-packed crystals is such that each atom has a coordination number of 12 (each is surrounded by 12 other atoms at equal distances). If the atoms are viewed as spheres, there is a minimum of empty space in these two types of crystals (about 26%), and both of the crystal lattices are called close-packed structures. The body-centered cubic arrangement is slightly more open than either of the close-packed arrangements (about 32% empty space). Each atom of a body-centered cubic crystal has a coordination number of 8.

The crystal structures of a number of metals are summarized in Table 9.5. The division between the three structures (hexagonal close packed, face-centered cubic, and body-centered cubic) is about even. The close packing of most metallic crystals helps to explain the relatively high densities of metals. The structures of a few metals (for example, manganese and mercury) do not

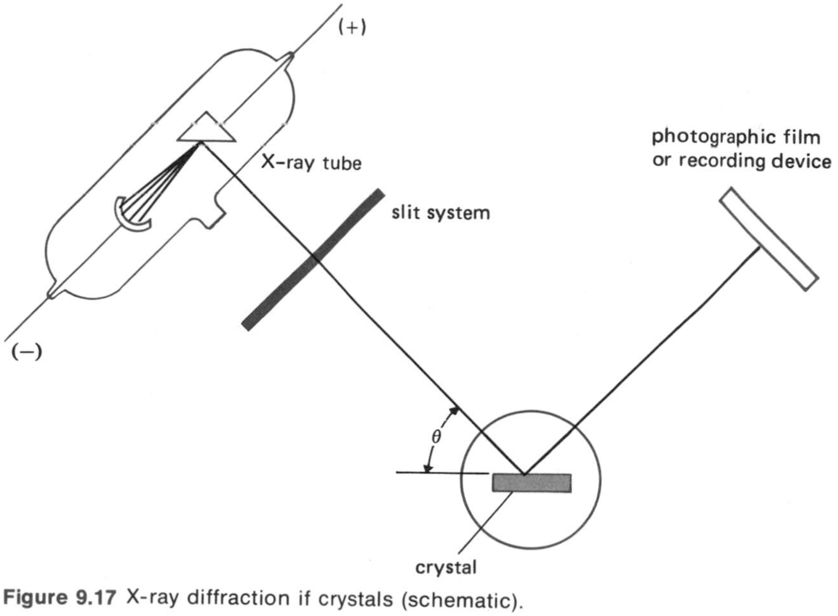

Figure 9.17 X-ray diffraction if crystals (schematic).

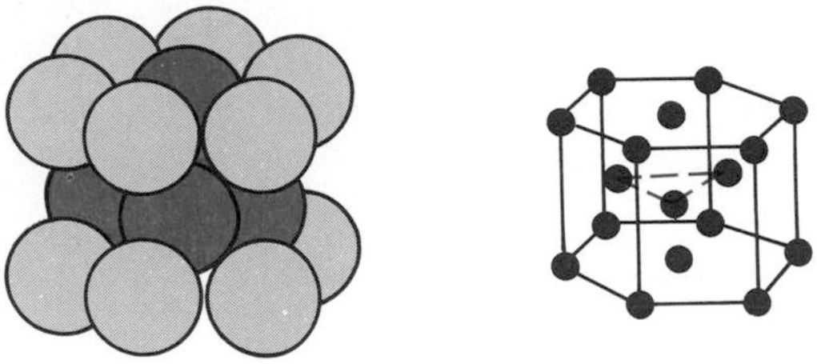

Figure 9.18 Hexagonal close-packed structure.

fall into any of the three categories, and the symbols for these metals do not appear in the table. Some metals exhibit **crystal allotropy;** that is, different crystal forms of the same metal are stable under different conditions. A modification of calcium, for example, exists in each of the three structures. The form of each metal that is indicated in Table 9.5 is the one that is stable under ordinary conditions. In addition to many metals, the noble gases also crystallize in the face-centered cubic lattice (except for He).

Ionic lattices are more complicated than metallic lattices. An arrangement must accommodate ions of opposite charge and different size in the proper stoichiometric ratio and in such a way that electrostatic attractions outweigh electrostatic repulsions.

We will use the lattices of compounds in which the cation to anion ratio is one to one as examples. The three most common crystal types for compounds of formula MX are shown in Figure 9.19. In each diagram there are

Table 9.5 Crystal structures of metals

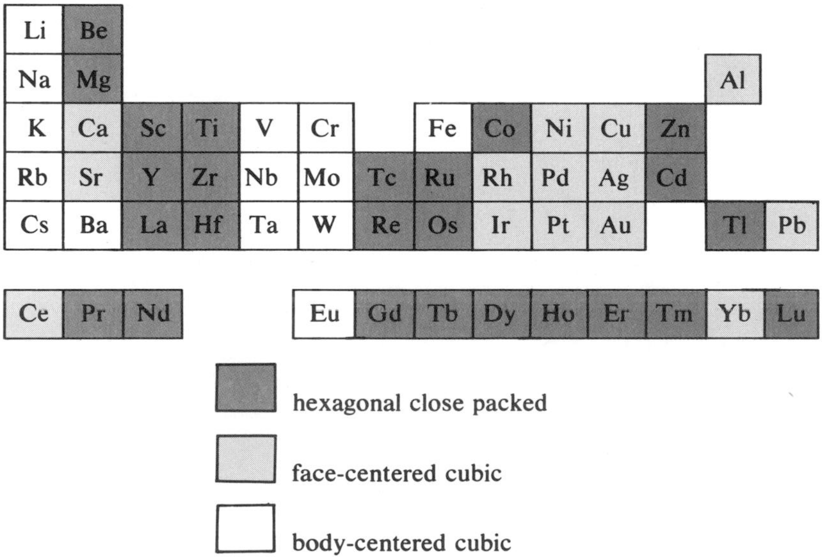

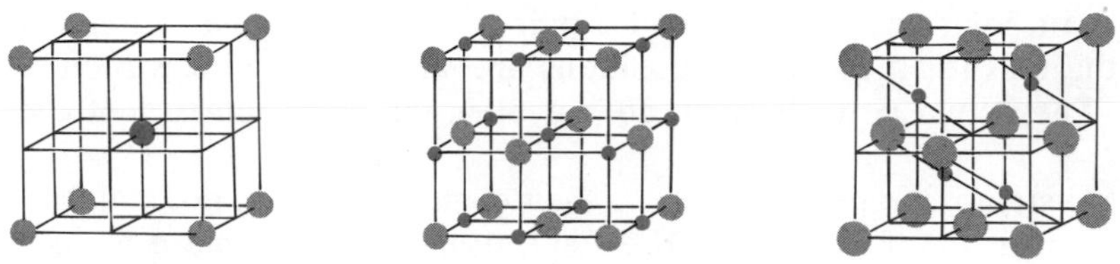

cesium chloride
CsCl

sodium chloride
NaCl

zinc blende
ZnS

Figure 9.19 Crystal structures of ionic compounds of the MX type. Colored circles represent cations.

as many cations shown belonging to a unit cell as there are anions (one each in CsCl, four each in both NaCl and ZnS, taking into account sharing by adjacent unit cells). Therefore, a one to one stoichiometric ratio is represented in each case.

The type of crystal that an ionic compound forms depends largely on the ratio of the size of cation to the size of the anion. If the cation is comparatively large, the compound crystallizes in the **cesium chloride** lattice in which each cation is surrounded by eight anions. A smaller cation leads to the formation of a **sodium chloride** lattice in which six anions surround each cation. The smallest cations form **zinc blende** crystals in which the coordination number of each cation is four. The anions of most ionic crystals are larger than the cations. Examples of compounds that form crystals of these three types are listed in Table 9.6.

9.15 CRYSTAL DEFECTS

Few crystals are perfect; many have some type of lattice defect. **Dislocations** are crystal imperfections that occur when planes of atoms are out of alignment. For example, one type of dislocation is caused by the insertion (perpendicular to a face of the crystal) of an extra plane of atoms part way through a crystal. Atoms within the part of the crystal containing the extra plane are compressed.

Point defects are caused by missing or misplaced ions. One type of defect consists of a cation that has been moved from its proper position (thus creating a vacancy) to a place between regular lattice sites (an **interstitial** position).

Table 9.6 Crystal structures of some ionic compounds

STRUCTURE	EXAMPLES
cesium chloride	CsCl, CsBr, CsI, TlCl, TlBr, TlI, NH_4Cl, NH_4Br
sodium chloride	halides of Li^+, Na^+, K^+, Rb^+ oxides and sulfides of Mg^{2+}, Ca^{2+}, Sr^{2+}, Ba^{2+}, Mn^{2+}, Ni^{2+} AgF, AgCl, AgBr, NH_4I
zinc blende	sulfides of Be^{2+}, Zn^{2+}, Cd^{2+}, Hg^{2+} CuCl, CuBr, CuI, AgI, ZnO

Another type of point defect consists of a pair of vacancies – one cation and one anion; the ions for these lattice positions are missing from the structure completely. These two point defects do not alter the stoichiometry of the crystal.

Certain crystals are imperfect because their compositions are not stoichiometric. Thus, samples of iron(II) oxide, FeO, usually contain more oxygen atoms than iron atoms, whereas zinc oxide, ZnO, usually has more zinc atoms than oxygen atoms. These departures from stoichiometry are usually quite small, on the order of 0.1%.

There are several causes for **nonstoichiometry,** and in each the electrical neutrality of the crystal is preserved. Extra metal *atoms* or nonmetal *atoms* may be included in interstitial positions between the *ions* of the crystal (such as extra Zn atoms in ZnO). In some crystals, metal *atoms* or nonmetal *atoms* assume regular lattice positions in place of *ions*. In these cases, ionic vacancies exist in the crystal lattice so that the whole lattice is electrically neutral. In FeO, for example, oxygen atoms assume positions normally occupied by oxide ions, and there are missing Fe^{2+} ions so that a 1 to 1 cation–anion ratio is maintained. In some crystals of this type (for example, KCl), electrons from the "extra" metal atoms occupy the holes created by the anion vacancies.

Nonstoichiometric substances are sometimes referred to as **berthollides,** named for Claude Louis Berthollet, who believed that the compositions of compounds varied continuously within limits. Joseph Proust, who upheld the law of definite proportions, engaged Berthollet in an argument of eight years' duration (1799–1807) concerning the composition of compounds.

The presence of **impurities** frequently accounts for crystal defects. A Mg^{2+} ion, for example, may occur in a lattice position in an NaCl crystal in place of an Na^+ ion. Electrical neutrality requires that another cation lattice position be vacant since the charge of the Mg^{2+} ion is twice that of the Na^+ ion.

THE CASE OF THE DISAPPEARING DIAMONDS

In 1768, Jean Darcet was using a new type of furnace to study the effect of high temperatures on minerals, when he made an amazing discovery. Two diamonds vanished completely upon intense heating! This effect had been observed before. Boyle, for example, had reported it in "An Essay about the Origin and Virtues of Gems" (1672). Darcet's results, however, created a mild sensation among French scientists, jewelers, and the public at large. Diamonds were thought to be indestructible.

Other scientists repeated the work, and several public demonstrations were held (with the lieutenant general of police present at one of them). Since the French Academy of Sciences encouraged scientists to work together on important questions, Pierre Macquer, Louis Cadet de Gassicourt, and Antoine Lavoisier decided in 1772 to investigate the matter jointly. A jeweler, who was anxious to show that diamonds do not vaporize in the absence of air, contributed three diamonds for the experiments. The job of reporting the results was given to Lavoisier, the youngest member of the team.

The work continued off and on for almost two years, and Lavoisier wrote several reports. The early work was inconclusive. The investigators could not decide whether the diamonds were volatile and vaporized, whether they burned in air, or whether the effect of the cold air on the heated diamonds caused them to burst apart into an invisible powder. Finally, in 1773, they showed that diamonds burn in air and that "fixed air" (CO_2) is the product of the combustion. The work of Smithson Tennant (in 1797) and Humphry Davy and Michael Faraday (in 1813) proved conclusively that diamonds and graphite are both forms of pure carbon.

Throughout the latter part of the nineteenth century, many scientists attempted to make diamonds from graphite. There were several claims of success—all of them false. The most famous one was made by Henri Moissan in 1893. Since the density of diamond (3.51 g/cm^3) is much higher than the density of graphite (2.22 g/cm^3), it is obvious that high pressures are necessary for the conversion. High temperatures are also necessary to make the conversion occur at a satisfactory rate.

Moissan dissolved carbon in molten iron, then chilled the mass quickly by plunging it into cold water. He believed that the contraction of the surface that occurred when the outer portion solidified would create a high pressure on the interior, liquid portion and cause diamonds to form. He claimed that very small, impure diamonds were left after he dissolved away the cooled iron in acid. We know now that the conditions that he (as well as others of this period) employed could not have produced diamonds.

Artificial diamonds were first produced in 1955 after the important features of the phase diagram for carbon had been worked out (some details are still uncertain). The conditions required for the conversion of graphite into diamond can be predicted from the phase diagram. The success also rested upon the development of materials able to withstand the severe conditions required as well as the development of equipment for producing high pressures.

The stable form of carbon under ordinary conditions of temperature and pressure is graphite, not diamond. Synthetic diamonds are produced from graphite at temperatures around 2000°C and pressures around 70,000 atm. A small amount of a transition metal (such as Fe, Cr, or Pt) is used as a catalyst. Small, synthetic diamonds of industrial quality (for polishing, grinding, and cutting) are commercially available. Gem quality stones can be made, but the process required to produce them is so expensive that they cost more than natural diamonds.

Four gems are classified as precious stones: diamonds, emeralds, rubies, and sapphires. The diamond was the last of these gems to be made artificially. The emerald ($Be_3Al_2Si_6O_{18}$) is made from beryllium oxide, aluminum oxide, and silicon dioxide. Both rubies and sapphires are forms of aluminum oxide. The ruby is Al_2O_3 with a trace of chromium(III) oxide as an impurity. The sapphire is Al_2O_3 that contains very small amounts of iron(II) oxide, iron(III) oxide, and titanium(IV) oxide. These artificial stones are chemically identical to their natural counterparts.

PROBLEMS

9.1 How could measurement of the dipole moment of the trigonal bipyramidal molecule PCl_2F_3 help to determine whether the Cl atoms occupy axial or equatorial positions?

9.2 The dipole moment of PF_3 is 1.03 D, whereas PF_5 has no dipole moment. Explain.

9.3 The dipole moment of NH_3 (1.49 D) is greater than that of NF_3 (0.24 D). On the other hand, the dipole moment of PH_3 (0.55 D) is less than that of PF_3 (1.03 D). Explain these results.

9.4 Explain why the dipole moment of SCO is 0.72 D, whereas the dipole moment of CO_2 is zero. Would CS_2 have a dipole moment?

9.5 In Tablc 7.1 the electronegativity of C is given as 2.5 and that of O is given as 3.5. On the other hand, in Table 9.1 the dipole moment of CO is listed as only 0.12 D (the dipole–dipole forces of CO are negligible). Draw the Lewis structure of CO and offer an explanation for the low dipole moment of CO.

9.6 Explain why the following melting points fall in the order given: F_2 (−233°C), Cl_2 (−103°C), Br_2 (−7°C), and I_2 (113.5°C).

9.7 Consider the following molecules, each of which is tetrahedral with the C atom as the central atom: CH_4, CH_3Cl, CH_2Cl_2, $CHCl_3$, CCl_4. In which of these compounds would dipole–dipole forces exist in the liquid state? In what order would you expect the boiling points of the compounds to fall?

9.8 Which molecules (not ions) given as examples in Table 7.3 would you expect to have a dipole moment of zero?

9.9 The effect of hydrogen bonding on the properties of the following hydrides falls in the order: H_2O > HF > NH_3. Explain this observation.

9.10 The compound KHF_2 can be prepared from the reaction of KF and HF in water solution. Explain the structure of the HF_2^- ion.

9.11 Diagram the structures of the following molecules and explain how the water solubility of each compound is enhanced by hydrogen bonding. (a) NH_3, (b) $H_2N—OH$, (c) H_3COH, (d) H_2CO.

9.12 Although there are exceptions, most acid salts are more soluble in water than the corresponding normal salts. Offer an explanation for this generalization.

9.13 Offer an explanation as to why chloroform, $HCCl_3$, and acetone, $CH_3—\overset{\overset{\displaystyle O}{\|}}{C}—CH_3$, mixtures have higher boiling points than either pure component.

9.14 (a) Calculate the molar entropy of vaporization at the boiling point for formic acid, HCO_2H. The normal boiling point of formic acid is 101°C and the heat of vaporization is 502 J/g. How does your answer compare to the Trouton's rule value? (b) The formic acid molecule consists of a central C atom to which a double-bonded O atom, an —O—H group, and a H atom are bonded. Formic acid vapor at temperatures near the boiling point consists of double molecules, or dimers, that are joined by hydrogen bonds. Diagram the formic acid dimer. How does the existence of these dimers in the vapor explain the result obtained in part (a)?

9.15 Briefly explain how and why each of the following gives an indication of the strength of intermolecular forces of attraction: (a) critical temperature, (b) surface tension, (c) vapor pressure, (d) heat of vaporization, (e) heat of fusion, (f) normal boiling point.

9.16 The normal boiling point of HBr is −67°C. Use Trouton's rule to estimate the molar heat of vaporization of HBr.

9.17 Calculate the molar entropy of vaporization, ΔS_v, for hydrazine, H_2NNH_2. The heat of vaporization of hydrazine is 41.8 kJ/mol at the normal boiling point, 113.5°C. How well does this value compare with the one predicted by Trouton's rule? Can you explain the difference?

9.18 A hydride of silicon has a normal boiling point of 53°C and a heat of vaporization at the boiling point of 308 J/g. Use Trouton's rule to estimate the molecular weight of the compound.

9.19 The heat of vaporization of tin(IV) chloride, $SnCl_4$, at the normal boiling point is 130 J/g. Use Trouton's rule to estimate the normal boiling point of $SnCl_4$.

9.20 A compound has a normal boiling point of 80°C, a heat of vaporization of 394 J/g, and an empirical formula of CH. Use Trouton's rule to find the molecular formula of the compound.

9.21 A compound has a normal boiling point of 81°C, a heat of vaporization of 358 J/g, and an empirical formula of CH_2. Use Trouton's rule to find the molecular formula of the compound.

9.22 A compound has a normal boiling point of 161°C, a heat of vaporization of 320 J/g, and an empirical formula of C_3H_4. Use Trouton's rule to find the molecular formula of the compound.

9.23 Explain why the entropy of vaporization of a compound is larger than the entropy of fusion of that compound. See Tables 9.2 and 9.3.

9.24 Use the following data to draw a rough phase diagram for hydrogen. Normal melting point, 14.01 K; normal boiling point, 20.38 K; triple point, 13.95 K, 7×10^{-2} atm; critical point, 33.3 K, 12.8 atm; vapor pressure of solid at 10 K, 1×10^{-3} atm.

9.25 Use the following data to draw a rough phase diagram for krypton. Normal boiling point, −152°C; normal melting point, −157°C; triple point, −169°C, 0.175 atm; critical point, −63°C, 54.2 atm; vapor pressure of solid at −199°C, 1.3×10^{-3} atm. Which has the higher density at a pressure of 1 atm: solid Kr or liquid Kr?

9.26 Figure 9.9 is the phase diagram for water. Describe the phase changes that occur, and the approximate pressures at which they occur, when the pressure on a H_2O system is gradually increased (a) at a constant temperature of −1°C, (b) at a constant temperature of 50°C, (c) at a constant temperature of −50°C.

9.27 Refer to Figure 9.9 and describe the phase changes that occur, and the approximate temperatures at which they occur, when water is heated from −10°C to 110°C (a) under a pressure of 1×10^{-3} atm, (b) under a pressure of 0.5 atm, (c) under a pressure of 1.1 atm.

9.28 Refer to Figure 9.10 and describe the phase changes that occur and the approximate pressures at which they occur when the pressure on a CO_2 system is gradually increased (a) at a constant temperature of −60°C, (b) at a constant temperature of 0°C.

9.29 Refer to Figure 9.10 and describe the phase changes that occur, and the approximate temperatures at which they occur, when carbon dioxide is heated (a) at a constant pressure of 2.0 atm, (b) at a constant pressure of 6.0 atm.

9.30 Briefly explain how and why the slope of the melting-point curve of a phase diagram for a substance depends upon the relative densities of the solid and liquid forms of the substance.

9.31 Sublimation is sometimes used to purify solids. The impure material is heated, and the pure crystalline product condenses on a cold surface. Is it possible to purify ice by sublimation? What conditions would have to be employed?

9.32 What type of forces must be overcome in order to melt crystals of the following? (a) Si, (b) Ba, (c) F_2, (d) BaF_2, (e) BF_3, (f) PF_3.

9.33 What type of forces must be overcome in order to melt crystals of the following? (a) O_2, (b) Br_2, (c) Br_2O, (d) Ba, (e) $BaBr_2$, (f) BaO.

9.34 Which substance of each of the following pairs would you expect to have the higher melting point? Why? (a) ClF or BrF, (b) BrCl or Cl_2, (c) CsBr or BrCl, (d) Cs or Br_2, (e) C (diamond) or Cl_2.

9.35 Which substance of each of the following pairs would you expect to have the higher melting point? Why? (a) Sr or Cl_2, (b) $SrCl_2$ or $SiCl_4$, (c) $SiCl_4$ or $SiBr_4$, (d) $SiCl_4$ or SCl_4, (e) SiC (carborundum) or $SiCl_4$.

9.36 Which substance of each of the following pairs would you expect to have the higher melting point? Why? (a) Li or H_2, (b) LiH or H_2, (c) Li or LiH, (d) H_2 or Cl_2, (e) H_2 or HCl.

9.37 Xenon crystallizes in a face-centered cubic lattice, and the edge of the unit cell is 620 pm. What is the density of crystalline xenon?

9.38 Vanadium crystallizes in a body-centered cubic lattice, and the edge of the unit cell is 305 pm. What is the density of vanadium?

9.39 Silver crystallizes in a cubic system, and the edge of the unit cell is 408 pm. The density of silver is 10.6 g/cm^3. How many atoms of Ag are contained in one unit cell? What type of cubic unit cell does Ag form?

9.40 Tantalum crystallizes in a cubic system, and the edge of the unit cell is 330 pm. The density of tantalum is 16.6 g/cm^3. How many atoms of Ta are contained in one unit cell? What type of cubic unit cell does Ta form?

9.41 The density of platinum is 21.5 g/cm^3, and Pt crystallizes in a face-centered cubic unit cell with an edge equal to 392 pm. Calculate the atomic weight of Pt using these data.

9.42 An element crystallizes in a body-centered cubic lattice, and the edge of the unit cell is 305 pm. The density of the crystalline solid is 5.96 g/cm^3. What is the atomic weight of the element?

9.43 Calcium crystallizes in a face-centered cubic lattice. The density of Ca is 1.55 g/cm^3. What is the length of an edge of the unit cell?

9.44 Copper crystallizes in a face-centered cubic lattice. The density of Cu is 8.93 g/cm^3. What is the length of the edge of a unit cell?

9.45 Lithium crystallizes in a body-centered cubic lattice, and the length of the edge of a unit cell is 350 pm. The atomic weight of Li is 6.94, and the density of Li is 0.534 g/cm^3. Use these data to calculate Avogadro's number.

9.46 Sodium crystallizes in a body-centered cubic unit cell with an edge equal to 430 pm. Calculate the dimensions of a cube that would contain 1 mol of Na (23.0 g).

9.47 Indium crystallizes in a face-centered tetragonal lattice (see Figure 9.13). The base of the unit cell is 458 pm on a side, and the height of a unit cell is 494 pm. What is the density of In?

9.48 Xenon difluoride, XeF_2, *molecules* crystallize in a body-centered tetragonal lattice (see Figure 9.13). The base of the unit cell is a square 432 pm on a side, and the height of the unit cell is 699 pm. In counting the number of *molecules* per unit cell, assume that *molecules* occupy lattice positions. What is the density of XeF_2?

9.49 In the diffraction of a crystal using X rays with a wavelength of 71.0 pm, a first-order reflection was obtained at an angle of 12°. What is the distance between the diffracted planes? The sine of 12° is 0.208.

9.50 In the diffraction of a gold crystal using X rays with a wavelength of 154 pm, a first-order reflection was obtained at an angle of 22°10′. What is the distance between the diffracted planes? The sine of 22°10′ is 0.377.

9.51 In the X-ray diffraction of a set of crystal planes for which *d* is 180 pm, a first-order reflection was obtained at an angle of 22°. What is the wavelength of the X rays that were employed? The sine of 22° is 0.375.

9.52 In the diffraction of a crystal using X rays with a wavelength of 154 pm, a first-order reflection is found at an angle of 11°. What is the wavelength of X rays that show the same reflection at an angle of 13°55′? The sine of 11° is 0.191; the sine of 13°55′ is 0.241.

9.53 In the diffraction of a crystal using X rays with a wavelength of 154 pm, a first-order reflection is found at an angle of 16°. What is the wavelength of X rays that show the same reflection at an angle of 20°20′? The sine of 16° is 0.276; the sine of 20°20′ is 0.347.

9.54 At what angle would a first-order reflection be observed in the X-ray diffraction of a set of crystal planes for which *d* is 204 pm if the X rays used have a wavelength of 154 pm? At what angle would a second-order reflection be observed?

9.55 At what angle would a first-order reflection be observed in the X-ray diffraction of a set of crystal planes for which *d* is 303 pm if the X rays used have a wavelength of 71.0 pm? At what angle would a second-order reflection be observed?

9.56 Aluminum crystallizes in a face-centered cubic unit cell with the length of an edge equal to 405 pm. Assume that the atoms are identical hard spheres and that a face-centered atom touches each of the four corner atoms of its face. (a) Calculate the length of the diagonal of a face. (b) By means of your answer to part (a), calculate the radius of the hard-sphere atoms that make up the crystal.

9.57 Sodium crystallizes in a body-centered cubic unit cell with the length of an edge equal to 430 pm. Assume that the atoms are identical hard spheres and that the body-centered atom touches all eight corner atoms. (a) Calculate the length of the body diagonal of the cube. (b) By means of your answer to part (a), calculate the radius of the hard-sphere atoms that make up the crystal.

9.58 How many ions of each type are shown in the unit cell of the cesium chloride crystal depicted in Figure 9.19? The density of TlCl, which crystallizes in a cesium chloride lattice, is 7.00 g/cm^3. What is the length of the edge of the unit cell shown in the figure? What is the shortest distance between a Tl^+ ion and a Cl^- ion?

9.59 How many ions of each type are shown in the unit cell of zinc sulfide depicted in Figure 9.19? The density of BeS, which crystallizes in a zinc blende lattice, is 2.36 g/cm^3. What is the length of the edge of the unit cell shown in the figure? What is the shortest distance between a Be^{2+} ion and a S^{2-} ion?

9.60 (a) How many Na^+ ions and Cl^- ions belong to the cell lattice of the NaCl crystal shown in Figure 9.19? (b) Calcium sulfide, CaS, crystallizes in the sodium chloride lattice. The shortest distance between a Ca^{2+} ion and a S^{2-} ion is 283 pm. What is the length of the edge of the cell similar to that shown in Figure 9.19? What is the density, in g/cm^3, of CaS?

9.61 List and describe the types of lattice defects found in crystals.

Chapter 10

Solutions

Solutions are homogeneous mixtures. They are usually classified according to their physical state; gaseous, liquid, and solid solutions can be prepared. Dalton's law of partial pressures describes the behavior of gaseous solutions, of which air is the most common example. Certain alloys are solid solutions; coinage silver is copper dissolved in silver, and brass is a solid solution of zinc in copper. Not all alloys are solid solutions, however. Some are heterogeneous mixtures, and some are intermetallic compounds. Liquid solutions are the most common and are probably the most important to the chemist.

An understanding of the laws governing the behavior of gaseous solutions (as well as pure gases) is mainly the product of scientific investigations of the seventeenth and eighteenth centuries. The three theories upon which our understanding of liquid solutions has largely been built were all proposed within a brief period of time. They are Raoult's law of ideal solutions (1878–87), van't Hoff's theory of dilute solutions (1886), and Arrhenius' theory of ionization (1887). In the late nineteenth century significant work was also done on the nature of solid solutions. The phase rule of J. Willard Gibbs (developed in 1873–78) is central to these investigations.

10.1 NATURE OF SOLUTIONS

The component of a solution that is present in greatest quantity is usually called the **solvent,** and all other components are called **solutes.** This terminology is loose and arbitrary. It is sometimes convenient to designate a component as the solvent even though it is present in only small amounts. At other times, the assignment of the terms solute and solvent has little significance (for example, in describing gaseous solutions).

Certain pairs of substances will dissolve in each other in all proportions. Complete miscibility is characteristic of the components of all gaseous solutions and some pairs of components of liquid and solid solutions. For most materials, however, there is a limit on the amount of the substance that will dissolve in a given solvent. The **solubility** of a substance in a particular solvent at a specified temperature is the maximum amount of the solute that will dissolve in a definite amount of the solvent and produce a stable system.

For a given solution, the amount of solute dissolved in an amount of solvent given is the **concentration** of the solute. Solutions containing a relatively low concentration of solute are called **dilute solutions;** those of relatively high concentrations are called **concentrated solutions.**

If an excess of solute (more than will normally dissolve) is added to a quantity of a liquid solvent, an equilibrium is established between the pure solute and the dissolved solute.

$$\text{solute}_{\text{pure}} \rightleftharpoons \text{solute}_{\text{dissolved}}$$

The pure solute may be a solid, liquid, or gas. At equilibrium in such a system, the rate at which the pure solute dissolves equals the rate at which the dissolved solute comes out of solution. The concentration of the dissolved solute, therefore, is a constant. A solution of this type is called a **saturated solution,** and its concentration is the solubility of the solute in question.

That such dynamic equilibria exist has been shown experimentally. If small crystals of a solid solute are placed in contact with a saturated solution of the solute, the crystals are observed to change in size and shape. Throughout this experiment, however, the concentration of the saturated solution does not change, nor does the quantity of excess solute decrease or increase.

An **unsaturated solution** has a lower concentration of solute than a saturated solution. On the other hand, it is sometimes possible to prepare a **supersaturated solution,** one in which the concentration of solute is higher than that of a saturated solution. A supersaturated solution, however, is metastable, and if a very small amount of pure solute is added to it, the solute that is in excess of that needed to saturate the solution will precipitate.

Reactions that are run in solution are usually rapid since the reactants are in a small state of subdivision (solutes exist in solution as molecules, atoms, or ions), the attractions between the particles (molecules, atoms, or ions) of the pure solute have been at least partially overcome, and the particles are free to move about through the solution.

10.2 THE SOLUTION PROCESS

London forces are the only intermolecular forces between nonpolar covalent molecules. On the other hand, the intermolecular attractions between polar covalent molecules are due to dipole–dipole forces as well as to London forces. In substances in which there is hydrogen bonding the intermolecular forces are unusually strong.

Nonpolar substances and polar substances are generally insoluble in one another. Carbon tetrachloride (a nonpolar substance) is insoluble in water (a polar substance). The attraction of one water molecule for another water molecule is much greater than an attraction between a carbon tetrachloride molecule and a water molecule. Hence, carbon tetrachloride molecules are "squeezed out," and these two substances form a two-liquid-layer system.

Iodine, a nonpolar material, is soluble in carbon tetrachloride. The attractions between I_2 molecules in solid iodine are approximately of the same type and magnitude as those between CCl_4 molecules in pure carbon tetrachloride. Hence, significant iodine–carbon tetrachloride attractions are possible, and iodine molecules can mix with carbon tetrachloride molecules. The resulting solution is a random molecular mixture.

Methyl alcohol, CH_3OH, like water, consists of polar molecules that are highly associated. In both pure liquids the molecules are attracted to one another through hydrogen bonding.

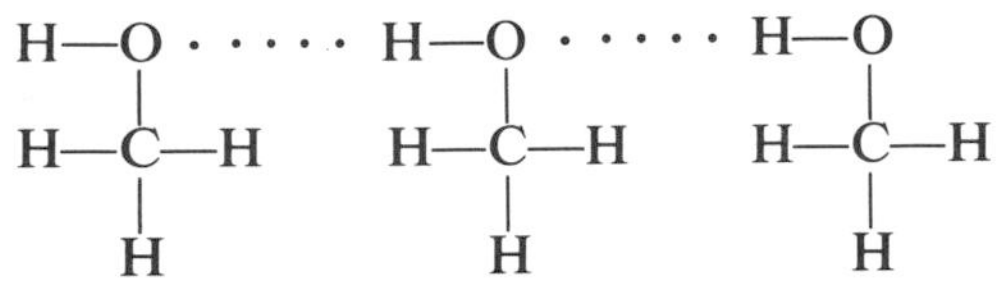

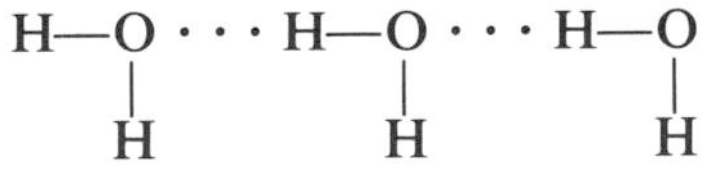

Methyl alcohol and water are miscible in all proportions. In solutions of methyl alcohol in water, CH_3OH and H_2O molecules are associated through hydrogen bonding.

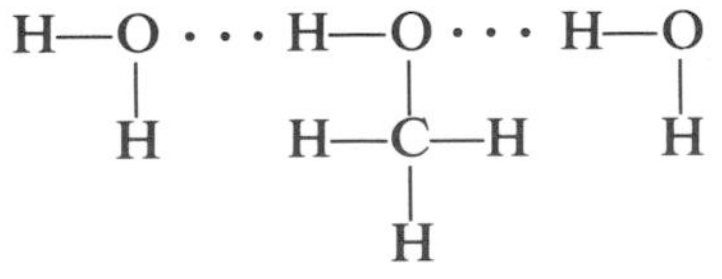

Methyl alcohol does not dissolve in nonpolar solvents. The strong intermolecular attractions of pure methyl alcohol are not overcome unless the solvent molecules can form attractions of equal, or almost equal, strength with the methyl alcohol molecules.

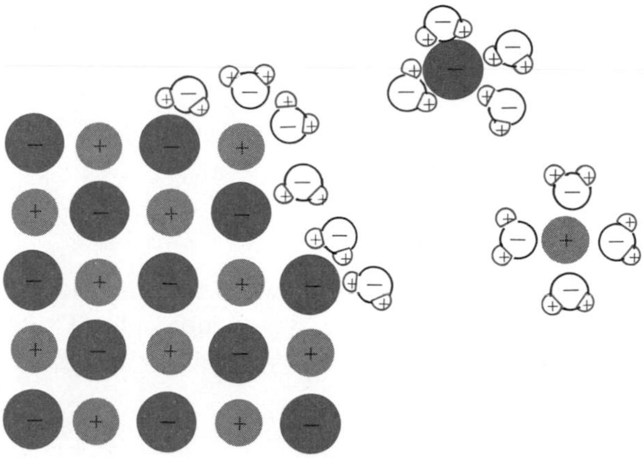

Figure 10.1 Solution of an ionic crystal in water.

In general, polar materials dissolve only in polar solvents, and nonpolar substances are soluble in nonpolar solvents. This is the first rule of solubility: "like dissolves like." Network crystals (the diamond, for example), in which the atoms that make up the crystal are held together by covalent bonds, are insoluble in all liquids. This crystalline structure is far too stable to be broken down by a solution process. Any potential solute–solvent attractions cannot approach the strength of the covalent bonding of the crystal.

Polar liquids (water, in particular) can function as solvents for many ionic compounds. The ions of the solute are electrostatically attracted by the polar solvent molecules—negative ions by the positive poles of the solvent molecules, positive ions by the negative poles of the solvent molecules. These ion–dipole attractions can be relatively strong.

Figure 10.1 diagrams the solution of an ionic crystal in water. The ions in the center of the crystal are attracted equally in all directions by the oppositely charged ions of the crystal. The electrostatic attractions on the ions of the surface of the crystal, however, are unbalanced. Water molecules are attracted to these surface ions, the positive ends of the water molecules to the anions and the negative ends of the water molecules to the cations. The ion–dipole attractions formed in this way allow the ions to escape from the crystal and drift into the liquid phase. The dissolved ions are **hydrated** and move through the solution surrounded by a sheath of water molecules. All ions are hydrated in water solution.

10.3 HYDRATED IONS

Negative ions are hydrated in water solution by means of attractions between the ion and the hydrogen atoms of the water molecule. In some cases (the sulfate ion, for example) these attractions may be one or more hydrogen bonds.

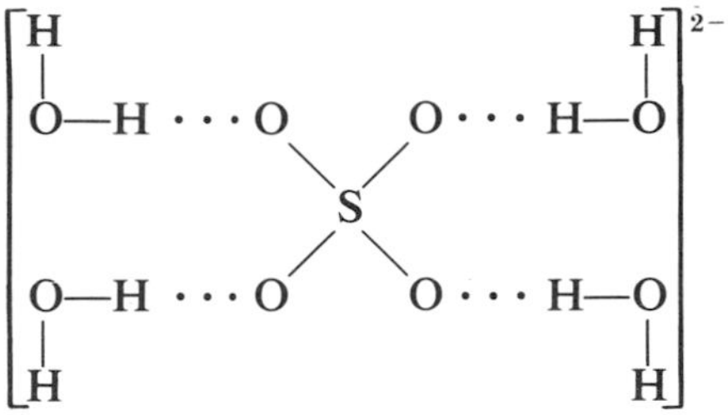

Positive ions are hydrated by means of attractions between the ion and the unshared electron pairs of the oxygen atom of the water molecule. These attractions are strong. In some cases, each cation is hydrated by a definite number of H_2O molecules.

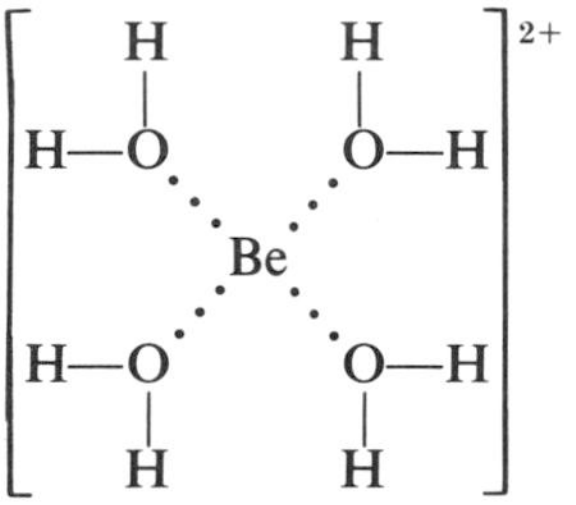

Additional water molecules form hydrogen bonds with those molecules that are bonded to the cation or anion. These outer layers of water molecules, however, are more loosely held.

What factors lead to the formation of strong interactions between the ion and water molecules?

1. Ions with *high charges* strongly attract the H or O atoms of the H_2O molecule.
2. *Small ions* are more effective than large ones because the charge is more highly concentrated in small ions.

A number of *covalent* compounds of metals produce hydrated *ions* in water solution. The compounds of beryllium, for example, are covalent when pure. The same factor that is largely responsible for the covalent character of the compounds of beryllium (a high ratio of ionic charge to ion size) also leads to the formation of very stable hydrated ions.

$$BeCl_2(s) + 4H_2O \rightarrow Be(H_2O)_4^{2+}(aq) + 2Cl^-(aq)$$

The formation of a bond always liberates energy, and the breaking of a bond always requires energy. The energy *released* by a hypothetical process in which hydrated ions are formed from gaseous ions is called the **heat of hydration** of the ions. For example,

$$K^+(g) + Cl^-(g) \rightarrow K^+(aq) + Cl^-(aq) \qquad \Delta H = -684.1 \text{ kJ}$$

The size of the heat of hydration depends upon the concentration of the final solution. If no distinction is made (as in the example previously given), it is assumed that the heat effect pertains to a process in which the ions are hydrated to the greatest possible extent. This high degree of hydration would occur only if the solution were *very dilute*. The corresponding ΔH values are called heats of hydration at **infinite dilution.**

The value of a heat of hydration gives an indication of the strength of the attractions between the ions and the water molecules that hydrate them. A large negative value (which indicates a large quantity of energy evolved) shows that the ions are strongly hydrated.

10.4 HEAT OF SOLUTION

The enthalpy change associated with the process in which a solute dissolves in a solvent is called the **heat of solution.** The value of a heat of solution (given in kJ/mol of solute) depends upon the concentration of the final solution in the same way that a heat of hydration does. Unless otherwise noted, a heat of solution is considered to apply to the preparation of a solution that is infinitely dilute. The heat of solution is virtually constant for all the dilute solutions of a given solute–solvent pair.

The enthalpy change observed when a solution is prepared is the net result of the energy *required* to break apart certain chemical bonds or attractions (solute–solute and solvent–solvent) and the energy *released* by the formation of new ones (solute–solvent). The heat of solution for the preparation of a solution of KCl in water, for example, can be considered to be the sum of two heat effects:

1. the *lattice energy* of KCl, which is the energy required to break apart the KCl crystal lattice and form gaseous ions.

   $$KCl(s) \rightarrow K^+(g) + Cl^-(g) \qquad \Delta H = +701.2 \text{ kJ}$$

2. the *heat of hydration* of KCl, which is the energy released when the gaseous ions are hydrated.

   $$K^+(g) + Cl^-(g) \rightarrow K^+(aq) + Cl^-(aq) \qquad \Delta H = -684.1 \text{ kJ}$$

   This enthalpy change is actually the sum of two enthalpy changes: the energy required to break the hydrogen bonds between some of the water molecules and the energy released when these water molecules hydrate the ions. It is, however, difficult to investigate these two effects separately.

In this example, the overall process is endothermic. The heat of solution is positive because more energy is required in step 1 than is released in step 2.

$$KCl(s) \rightarrow K^+(aq) + Cl^-(aq) \qquad \Delta H = +17.1 \text{ kJ}$$

Some heats of solution are negative because more heat is liberated by the hydration of the ions of the solute than is required to break apart the crystal lattice.

$$\begin{array}{rlr} AgF(s) \rightarrow Ag^+(g) + F^-(g) & & \Delta H = +910.9 \text{ kJ} \\ Ag^+(g) + F^-(g) \rightarrow Ag^+(aq) + F^-(aq) & & \Delta H = -931.4 \text{ kJ} \\ \hline AgF(s) \rightarrow Ag^+(aq) + F^-(aq) & & \Delta H = \ \ -20.5 \text{ kJ} \end{array}$$

The factor that leads to a high lattice energy (high charge to size ratio for the ions) also produces high heats of hydration. Consequently, the values for the steps are usually numerically close and the heat of solution itself is a much lower value than either of the values that go into it. A relatively small error in the lattice energy or the heat of hydration can, therefore, lead to a proportionately large error in the heat of solution.

If a solvent other than water is employed in the preparation of a solution, the same type of analysis can be made. The values for the second step are called **heats of solvation.**

Similar considerations apply to the solution of nonionic materials. The lattice energies of molecular crystals are not so large as those of ionic crystals since the forces holding molecular crystals together are not so strong as those holding ionic crystals together. Solvation energies for these nonionic materials, however, are also low. For molecular substances that dissolve in nonpolar solvents without ionization and without appreciable solute–solvent interaction, the heat of solution is endothermic and has about the same magnitude as the heat of fusion of the solute.

Gases generally dissolve in liquids with the evolution of heat. Since no energy is required to separate the molecules of a gas, the predominant heat effect of such a solution process is the solvation of the gas molecules; therefore, the process is exothermic. Calculation of the heat of solution of hydrogen chloride gas in water, however, requires that the heat evolved by the formation of hydrated ions be reduced by the energy required for the ionization of the HCl molecules.

10.5 EFFECT OF TEMPERATURE AND PRESSURE ON SOLUBILITY

The effect of a temperature change on the solubility of a substance depends upon whether heat is absorbed or evolved *when a saturated solution is prepared.* Suppose that a small quantity of solute dissolves in a nearly saturated solution with the absorption of heat. We can represent the equilibrium between excess solid solute and the saturated solution as

$$\text{energy} + \text{solute} + H_2O \rightleftharpoons \text{saturated solution}$$

The effect of a temperature change on this system can be predicted by means of a principle proposed by Henri Le Chatelier in 1884. **Le Chatelier's principle** states that a system in equilibrium reacts to a stress in a way that counteracts the stress and establishes a new equilibrium state.

Suppose that we have a beaker containing a saturated solution of the type previously described together with some excess solid solute in equilibrium with it. What will be the effect if we raise the temperature? According to Le

Chatelier's principle, the system will react in a way that lowers the temperature. It can do that if it shifts to the right—the direction in which heat is absorbed (note the equation given previously). The shift means that more solute will dissolve. We conclude that increasing the temperature increases the solubility of this particular solute.

What will happen if we lower the temperature? Le Chatelier's principle predicts that the system will react in a way that raises the temperature—a way that liberates energy. The reaction will shift to the left; solute will precipitate out of solution. We conclude that lowering the temperature decreases the solubility of this particular solute. Our two conclusions are actually the same.

The solubilities of substances that *absorb heat* when they dissolve in nearly saturated solutions *increase* with *increasing temperature.* Most ionic solutes behave in this way. The heat of solution of many ionic compounds for solutions that are *infinitely dilute* are *exothermic.* The same compounds, however, usually dissolve in *nearly saturated solutions* with the *absorption* of energy. When a solute dissolves in a nearly saturated solution, the heat of hydration is less than when it dissolves in a very dilute solution. An ion has a lower degree of hydration (is hydrated by fewer water molecules) in a concentrated solution than in a dilute solution. For this reason, heats of solution for nearly saturated solutions are more positive than heats of solution for dilute solutions.

When a substance dissolves in a nearly saturated solution with the evolution of heat,

$$\text{solute} + H_2O \rightleftharpoons \text{saturated solution} + \text{energy}$$

Le Chatelier's principle indicates that the solubility of the solute increases when the temperature is lowered, and the solubility decreases when the temperature is raised. A few ionic compounds (such as Li_2CO_3 and Na_2SO_4) behave in this fashion. In addition, the solubility of all gases decreases as the temperature is raised. Warming a soft drink causes bubbles of carbon dioxide gas to come out of solution.

How much the solubility changes when the temperature changes depends upon the magnitude of the heat of solution. The solubilities of substances with small heats of solution do not change much with changes in temperature.

Changes in pressure ordinarily have little effect upon the solubility of solid and liquid solutes. However, increasing or decreasing the pressure on a solution that contains a dissolved gas has a definite effect. William Henry in 1803 discovered that the amount of a gas that dissolves in a given quantity of a liquid at constant temperature is directly proportional to the partial pressure of the gas above the solution. **Henry's law** is valid only for dilute solutions and relatively low pressures. Gases that are extremely soluble generally react chemically with the solvent (for example, hydrogen chloride gas in water reacts to produce hydrochloric acid); these solutions do not follow Henry's law.

10.6 CONCENTRATIONS OF SOLUTIONS

The concentration of a solute in a solution can be expressed in several different ways. We have discussed some of these in previous sections, and we will review the common ways to express solution concentrations here.

1. The **percentage by mass** of a solute in a solution is 100 times the mass of the solute divided by the *total mass* of the solution. A 10% aqueous solution of sodium chloride contains 10 g of NaCl and 90 g of H_2O. Volume percentages are seldom used. Figures recorded as percentages should be understood to be based on mass unless a specific notation to the contrary is made.

2. The **mole fraction,** X, of a component of a solution is the ratio of the number of moles of that component to the *total number* of moles of all the substances present in the solution (Section 4.10).

$$X_A = \frac{n_A}{n_A + n_B + n_C + \cdots}$$

where X_A is the mole fraction of A and $n_A, n_B, n_C, \cdots$, are the number of moles of A, B, C, $\cdots$. The sum of the mole fractions of all the components present in the solution must equal one.

Example 10.1 A gaseous solution contains 2.00 g of He and 4.00 g of O_2. What are the mole fractions of He and O_2 in the solution?

Solution We first find the number of moles of each component present in the solution.

$$? \text{ mol He} = 2.00 \text{ g He}\left(\frac{1 \text{ mol He}}{4.00 \text{ g He}}\right) = 0.500 \text{ mol He}$$

$$? \text{ mol } O_2 = 4.00 \text{ g } O_2\left(\frac{1 \text{ mol } O_2}{32.0 \text{ g } O_2}\right) = 0.125 \text{ mol } O_2$$

We use these values to find the mole fractions.

$$X_{He} = \frac{n_{He}}{n_{He} + n_{O_2}}$$

$$= \frac{0.500 \text{ mol}}{0.500 \text{ mol} + 0.125 \text{ mol}} = \frac{0.500 \text{ mol}}{0.625 \text{ mol}} = 0.800$$

$$X_{O_2} = \frac{n_{O_2}}{n_{He} + n_{O_2}}$$

$$= \frac{0.125 \text{ mol}}{0.625 \text{ mol}} = 0.200$$

Notice that the sum of the mole fractions is equal to one.

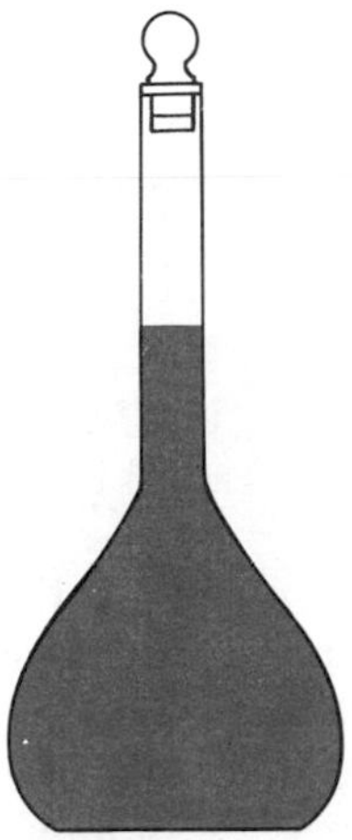

Figure 10.2 Volumetric flask.

3. The **molarity,** M, of a solution is the number of moles of solute per liter of *solution* (Section 8.5). Thus, a 6 M solution of sulfuric acid is prepared by taking 6 mol of H_2SO_4 ($6 \times 98.08 = 588.48$ g) and adding the acid, with careful mixing, to sufficient water to make exactly 1 liter of solution.

Notice that the definition is based on the *total volume of the solution.* When a liquid solution is prepared, the volume of the solution rarely equals the sum of the volumes of the pure components. Usually the final volume of the solution is larger or smaller than the total of the volumes of the materials used to prepare it. It is not practical, therefore, to attempt to predict the amount of solvent that should be employed to prepare a given solution. Molar solutions (as well as others that are based on total volume) are generally prepared by the use of the volumetric flasks (Figure 10.2). In the preparation of a solution, the correct amount of solute is placed in the flask, and water is then added, with careful and constant mixing, until the solution fills the flask to the calibration mark on the neck.

Example 10.2 What is the molarity of concentrated HCl if the solution contains 37.0% HCl by mass and if the density of the solution is 1.18 g/ml? The molecular weight of HCl is 36.5.

Solution In order to find the molarity of the solution, we must determine the number of moles of HCl in 1 liter of solution. The mass of 1 liter of the solution is found by means of the density.

$$? \text{ g sol'n} = 1000 \text{ ml sol'n} \left(\frac{1.18 \text{ g sol'n}}{1 \text{ ml sol'n}}\right) = 1180 \text{ g sol'n}$$

The mass of pure HCl in this quantity of solution is obtained by use of the percent composition.

$$? \text{ g HCl} = 1180 \text{ g sol'n} \left(\frac{37.0 \text{ g HCl}}{100 \text{ g sol'n}}\right) = 437 \text{ g HCl}$$

The molecular weight of HCl is 36.5, and, therefore,

$$? \text{ mol HCl} = 437 \text{ g HCl} \left(\frac{1 \text{ mol HCl}}{36.5 \text{ g HCl}}\right) = 12.0 \text{ mol HCl}$$

Since 1 liter of the solution contains 12.0 mol HCl, the solution is 12.0 *M*.

Often solutions must be prepared by diluting concentrated reagents. The molarities of common concentrated reagents are listed in Table 10.1. These values can be used to determine the proportions of reagent and water required to prepare a solution of a desired concentration.

The number of moles of solute in a sample of a solution may be obtained by multiplying the volume of the solution (V_1, in liters) by the molarity of the solution (M_1, the number of moles of solute in 1 liter).

$$\text{number of moles of solute} = V_1M_1$$

When the solution is diluted to a new volume, V_2, *it still contains the same number of moles of solute*. The concentration has decreased to M_2, but the product V_2M_2 equals the same number of moles. Therefore,

$$V_1M_1 = V_2M_2$$

Since a volume term is found on both sides of this equation, any volume unit can be used to express V_1 and V_2 provided the same unit is used for each. Note that this equation is used for dilution problems only.

Example 10.3 What volume of concentrated HCl should be used to prepare 500 ml of a 3.00 *M* HCl solution?

Solution From Table 10.1 (as well as from Example 10.2) we note that concentrated HCl is 12.0 *M*.

$$V_1M_1 = V_2M_2$$
$$V_1(12.0\ M) = (500 \text{ ml})(3.00 \text{ M})$$
$$V_1 = 125 \text{ ml}$$

4. The **normality** of a solution, *N*, is the number of equivalent weights of solute per liter of *solution* (Section 8.6). Since the normality of a solution is based upon its *total volume*, the volumetric flasks are employed in the preparation of normal solutions in the same way that they are used in the preparation of molar solutions.

The equivalent weight of a substance depends upon the reaction that

Table 10.1 Composition of some common concentrated reagents

REAGENT	FORMULA	MOLECULAR WEIGHT	PERCENT BY MASS	DENSITY (g/ml)	MOLARITY
acetic acid	$HC_2H_3O_2$	60.05	100	1.05	17.5
hydrochloric acid	HCl	36.46	37	1.18	12.0
nitric acid	HNO_3	63.01	70	1.42	15.8
phosphoric acid	H_3PO_4	98.00	85	1.70	14.7
sulfuric acid	H_2SO_4	98.07	96	1.84	18.0
ammonia	NH_3	17.03	28	0.90	14.8

the substance undergoes. The equivalent weight of an acid or base is defined in terms of the number of moles of H^+(aq) supplied or neutralized by the acid or base (a). The normality of a solution (N), therefore, is related to its molarity (M) by the equation

$$N = aM$$

If, for example, one mole of an acid supplies two moles of H^+(aq), $a = 2$. One gram molecular weight of the solute is equal to two gram equivalent weights, and a 1 M solution is 2 N.

Example 10.4 What is the normality of a 0.50 M solution of H_3PO_4 in terms of complete neutralization?

Solution For the complete neutralization of H_3PO_4, $a = 3$. Therefore,

$$\begin{aligned} N &= aM \\ &= 3(0.50) \\ &= 1.50\ N \end{aligned}$$

Example 10.5 How many milliliters of concentrated H_2SO_4 should be used to prepare 1500 ml of 6.00 N H_2SO_4, calculated on the basis of complete neutralization of the acid?

Solution The molarity of a 6.00 N H_2SO_4 solution is found by noting that $a = 2$.

$$\begin{aligned} N &= aM \\ 6.00 &= 2M \\ M &= 3.00\ M \end{aligned}$$

We can use this value to find the volume of concentrated H_2SO_4 required. Concentrated H_2SO_4 is 18.0 M (see Table 10.1).

$$\begin{aligned} V_1M_1 &= V_2M_2 \\ V_1(18.0\ M) &= (1500\ \text{ml})(3.00\ M) \\ V_1 &= 250\ \text{ml} \end{aligned}$$

The normality of a solution prepared in this way will be approximately correct. The concentrations of solutions used in analyses, however, must be accurately known. Such solutions are usually prepared in the manner described and then standardized. In the standardization, a sample of the solution in question is reacted with a solution of known concentration or a solid of known purity. The correct normality is obtained by calculation (see Example 8.7).

5. The **molality** of a solution, *m*, is defined as the number of moles of solute dissolved in 1 kg of *solvent*. A 1 *m* solution of urea, $CO(NH_2)_2$, is made by dissolving 1 mol of urea (60.06 g) in 1000 g of water. Notice that the definition is *not* based on the total volume of solution. The final volume of a molal solution is of no importance. One molal solutions of different solutes, each containing 1000 g of water, will have different volumes. All these solutions, however, will have the same mole fractions of solute and solvent (see Example 10.6).

Example 10.6 What are the mole fractions of solute and solvent in a 1.00 *m* aqueous solution?

Solution The molecular weight of H_2O is 18.0. We find the number of moles of water in 1000 g of H_2O.

$$? \text{ mol } H_2O = 1000 \text{ g } H_2O \left(\frac{1 \text{ mol } H_2O}{18.0 \text{ g } H_2O}\right) = 55.5 \text{ mol } H_2O$$

A 1.00 *m* aqueous solution contains:

$$n_{\text{solute}} = 1.0 \text{ mol}$$
$$n_{H^2O} = 55.5 \text{ mol}$$
$$n_{\text{total}} = 56.5 \text{ mol}$$

The mole fractions are:

$$X_{\text{solute}} = \frac{n_{\text{solute}}}{n_{\text{total}}} = \frac{1.00 \text{ mol}}{56.5 \text{ mol}} = 0.018$$

$$X_{H_2O} = \frac{n_{H_2O}}{n_{\text{total}}} = \frac{55.5 \text{ mol}}{56.5 \text{ mol}} = 0.982$$

These mole fractions pertain to all 1.00 *m* aqueous solutions.

Molal solutions that employ solvents other than water are often prepared. In each case, all the 1 *m* solutions that employ the same solvent have the same mole fractions of solute and solvent. The mole fraction of solute in any 1 *m* carbon tetrachloride solution is 0.133; the mole fraction of solvent is 0.867. These numbers differ from those in Example 10.6 because the molecular weight of carbon tetrachloride is different from that of water. A kilogram of CCl_4, therefore, is a different number of moles than a kilogram of H_2O.

10.7 VAPOR PRESSURE OF SOLUTIONS

The vapor pressure of any solution (P_{total}) is the sum of the partial pressures of the components of the solution ($p_A, p_B, \ldots$). For a solution of two components, A and B,

$$P_{\text{total}} = p_A + p_B$$

The partial pressure of the vapor of A, p_A, that is in equilibrium with an **ideal solution** is equal to the mole fraction of A present in the solution, X_A, times the vapor pressure of pure A at the same temperature, P_A°.

$$p_A = X_A P_A^\circ$$

This relationship was proposed by François Raoult in 1886 and is called **Raoult's law.**

In a solution prepared from 4 mol of A and 1 mol of B, the total number of moles present is 5 and the mole fraction of A in the solution is $\frac{4}{5}$. If the solution is *ideal,* the partial pressure of A is $\frac{4}{5}$ of the vapor pressure of pure A at the same temperature. The partial pressure of B is similarly given by the equation

$$p_B = X_B P_B^\circ$$

Since the mole fraction of B in the solution is $\frac{1}{5}$, the partial pressure of B is $\frac{1}{5}$ of the vapor pressure of pure B at the same temperature. The vapor pressure of the solution is equal to the sum of these two partial pressures.

$$P_{\text{total}} = X_A P_A^\circ + X_B P_B^\circ$$

The vapor pressure of an *ideal solution,* therefore, can be derived from the vapor pressures of the pure components by taking into account the proportion of the components (by moles) present in the solution.

Actually, a solution containing A and B will be ideal only if the intermolecular attractive forces between A and A, B and B, and A and B are all alike. In such a solution, the A molecules on the surface have the same tendency to escape to the vapor that they have in pure A. The partial pressure of A above the solution is less than the vapor pressure of pure A, however, because of the smaller proportion of A molecules present on the surface (which is given by the mole fraction of A). The situation for B is similar.

Few solutions are ideal. Two types of deviations from Raoult's law are observed.

1. *Positive deviations.* The partial pressures of A and B and the total vapor pressure are *higher* than predicted. This type of deviation is observed when the attractive forces between A and B molecules are *weaker* than those between two A molecules or two B molecules. In such a situation, A molecules find it easier to escape from the liquid and the partial pressure of A is higher than predicted. The behavior of B molecules is similar.
2. *Negative deviations.* The partial pressures of A and B and the total vapor pressure are *lower* than predicted. The A–B attractions are *stronger* than A–A or B–B attractions. The A molecules find it more difficult to leave the liquid, and the partial pressure of A is lower than predicted. The behavior of B molecules is similar.

In most *dilute* solutions, the behavior of the *solvent* approximately follows Raoult's law. If A is the solvent and the solution is dilute, most A molecules are surrounded by other A molecules. The strength of the forces of attraction between A and B molecules does not greatly influence the tendency of the A molecules to escape.

Consider a dilute solution prepared from a solute (which we will designate as B) that is *nonvolatile* ($P^\circ_B = 0$, for all practical purposes) and that does not dissociate in solution. The vapor pressure of the solution is caused by solvent molecules alone (A molecules). Such solutions usually follow Raoult's law.

$$P_{total} = X_A P^\circ_A$$

Since $X_A + X_B = 1$, $X_A = 1 - X_B$. Therefore,

$$P_{total} = (1 - X_B)P^\circ_A$$

or

$$P_{total} = P^\circ_A - X_B P^\circ_A$$

which means that the vapor pressure of pure A, P°_A, is lowered by an amount equal to $X_B P^\circ_A$.

The vapor pressure of a solution prepared from 1 mol of a nonvolatile, nondissociating solute and 99 mol of solvent is 99% of the vapor pressure of the pure solvent at the same temperature. The escape of the solvent molecules to the vapor is reduced because only 99% of the surface molecules in the solution are solvent molecules. The other 1% of the surface molecules are not volatile. The vapor pressure of the solvent is reduced by an amount that is proportional to the mole fraction of nonvolatile solute present.

Example 10.7 Heptane (C_7H_{16}) and octane (C_8H_{18}) form ideal solutions. What is the vapor pressure at 40°C of a solution that contains 3.00 mol of heptane and 5.00 mol of octane? At 40°C, the vapor pressure of heptane is 0.121 atm and the vapor pressure of octane is 0.041 atm.

Solution The total number of moles is 8.00. Therefore,

$$X_{\text{heptane}} = \frac{3.00 \text{ mol}}{8.00 \text{ mol}} = 0.375$$

$$X_{\text{octane}} = \frac{5.00 \text{ mol}}{8.00 \text{ mol}} = 0.625$$

The vapor pressure is

$$\begin{aligned} P_{\text{total}} &= X_{\text{heptane}} P^{\circ}_{\text{heptane}} + X_{\text{octane}} P^{\circ}_{\text{octane}} \\ &= 0.375(0.121 \text{ atm}) + 0.625(0.041 \text{ atm}) \\ &= 0.045 \text{ atm} + 0.026 \text{ atm} \\ &= 0.071 \text{ atm} \end{aligned}$$

Example 10.8 Assuming ideality, calculate the vapor pressure of a 1.00 m solution of a nonvolatile, nondissociating solute in water at 50°C. The vapor pressure of water at 50°C is 0.122 atm.

Solution The mole fraction of water in a 1.00 m solution is 0.982 (from Example 10.6). The vapor pressure of a 1.00 m solution of this type at 50°C is

$$\begin{aligned} P_{\text{total}} &= X_{H_2O} P^{\circ}_{H_2O} \\ &= (0.982)(0.122 \text{ atm}) \\ &= 0.120 \text{ atm} \end{aligned}$$

Example 10.9 A solution is prepared from 10.6 g of a nonvolatile, nondissociating solute and 90.0 g of water. The vapor pressure of the solution at 60°C is 0.1867 atm. Use Raoult's law to find the approximate molecular weight of the solute. The vapor pressure of water at 60°C is 0.1966 atm.

Solution We use Raoult's law to find the mole fraction of water in the solution.

$$\begin{aligned} P_{\text{total}} &= X_{H_2O} P^{\circ}_{H_2O} \\ 0.1867 \text{ atm} &= X_{H_2O}(0.1966 \text{ atm}) \\ X_{H_2O} &= 0.9496 \end{aligned}$$

The mole fraction of solute in the solution, therefore, is

$$\begin{aligned} X_{\text{solute}} &= 1 - X_{H_2O} \\ &= 1 - 0.9496 \\ &= 0.0504 \end{aligned}$$

Next, we find the number of moles of water used in the preparation of the solution.

$$? \text{ mol } H_2O = 90.0 \text{ g } H_2O \left(\frac{1 \text{ mol } H_2O}{18.0 \text{ g } H_2O} \right) = 5.00 \text{ mol } H_2O$$

The mole fraction of solute is equal to the number of moles of solute present divided by the total number of moles.

$$X_{\text{solute}} = \frac{n_{\text{solute}}}{n_{H_2O} + n_{\text{solute}}}$$

$$0.0504 = \frac{n_{\text{solute}}}{5.00 + n_{\text{solute}}}$$

$$0.252 + 0.0504n_{\text{solute}} = n_{\text{solute}}$$

$$n_{\text{solute}} = 0.265 \text{ mol}$$

The 10.6 g of solute is 0.265 mol.

$$? \text{ g solute} = 1.00 \text{ mol solute}\left(\frac{10.6 \text{ g solute}}{0.265 \text{ mole solute}}\right) = 40.0 \text{ g solute}$$

Notice that by using the mole fraction of solute rather than the mole fraction of water in the second to last step, we obtain a higher number of significant figures in the answer.

10.8 BOILING POINT AND FREEZING POINT OF SOLUTIONS

The lowering of the vapor pressure in solutions of nonvolatile solutes affects the boiling points and freezing points of these solutions.

The boiling point of a liquid is defined as the temperature at which the vapor pressure of the liquid is equal to the prevailing atmospheric pressure. Boiling points measured under 1 atmosphere pressure are called normal boiling points (Section 9.6.) Since the addition of a nonvolatile solute decreases the vapor pressure of a liquid, a solution will not boil at the normal boiling point of the solvent. It is necessary to increase the temperature above this point in order to attain a vapor pressure over the solution of 1 atmosphere. The boiling point of a solution containing a nonvolatile molecular solute, therefore, is higher than that of the pure solvent. The elevation is proportional to the concentration of solute in the solution.

This effect is illustrated by the vapor pressure curves plotted in Figure 10.3. The extent to which the vapor pressure curve of the solution lies below the vapor pressure curve of the solvent is proportional to the mole fraction of solute in the solution. The elevation of the boiling point, ΔT_b, reflects this displacement of the vapor pressure curve. For a given solvent the elevation is a constant for all solutions of the same concentration.

Concentrations are customarily expressed in molalities, rather than mole fractions, for problems involving boiling-point elevations. The boiling point of a 1 *m* aqueous solution, for example, is 0.512°C higher than the boiling point of water. Table 10.2 lists molal boiling-point elevation constants for several solvents.

The boiling point of a 0.5 *m* solution would be expected to be elevated by an amount equal to $\frac{1}{2}$ the molal constant. Thus, the boiling-point elevation, ΔT_b, of a solution can be calculated by multiplying the molal boiling-point elevation constant of the solvent, K_b, by the molality of the solution, *m*.

$$\Delta T_b = mK_b$$

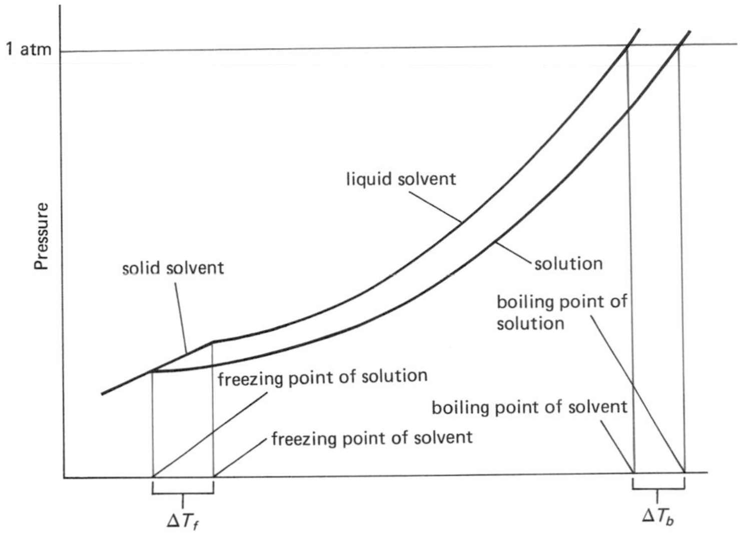

Figure 10.3 Vapor pressure curves of a pure solvent and a solution of a nonvolatile solute (not drawn to scale).

In reality, this relationship is only approximate. A more exact statement would require that the concentration be expressed in mole fraction of solute, not in molality. However, molalities of **dilute** solutions are proportional (at least with sufficient accuracy) to mole fractions of solute. And since Raoult's law only describes the behavior of most real solutions satisfactorily if they are dilute, the use of molalities in these calculations is justified.

At the freezing point the vapor pressure of solid and liquid are equal (Section 9.8). In Figure 10.3 the vapor pressure curves of the liquid solvent and the solid solvent intersect at the freezing point of the solvent. At this

Table 10.2 Molal boiling-point elevation and freezing-point depression constants

SOLVENT	BOILING POINT (°C)	K_b (°C/m)	FREEZING POINT (°C)	K_f (°C/m)
acetic acid	118.1	3.07	16.6	−3.90
benzene	80.1	2.53	5.5	−5.12
camphor	—	—	179.	−39.7
carbon tetrachloride	76.8	5.02	−22.8	−29.8
chloroform	61.2	3.63	−63.5	−4.68
ethyl alcohol	78.4	1.22	−114.6	−1.99
naphthalene	—	—	80.2	−6.80
water	100.0	0.512	0.0	−1.86

temperature, however, the vapor pressure of the solution is lower than the equilibrium vapor pressure of the pure solvent. The vapor pressure curve of the solution intersects the vapor pressure curve of the solid solvent at a lower temperature. The freezing point of the solution, therefore, is lower than that of the pure solvent. As in the case of boiling-point elevations, freezing-point depressions depend upon the concentration of the solution and the solvent employed. Molal freezing-point depression constants for some solvents are listed in Table 10.2. The freezing-point depression, ΔT_f, of a solution can be calculated from the molality of the solution and the constant for the solvent K_f.

$$\Delta T_f = mK_f$$

This statement assumes that the solute does not form a solid solution with the solvent. If this is not the case, the relationship is not valid.

Example 10.10 What are the boiling point and freezing point of a solution prepared by dissolving 2.40 g of biphenyl ($C_{12}H_{10}$) in 75.0 g of benzene? The molecular weight of biphenyl is 154.

Solution The molality of the solution is the number of moles of biphenyl dissolved in 1000 g of benzene.

$$\begin{aligned} ?\ \text{mol } C_{12}H_{10} &= 1000\ \text{g benzene}\left(\frac{2.40\ \text{g } C_{12}H_{10}}{75.0\ \text{g benzene}}\right)\left(\frac{1\ \text{mol } C_{12}H_{10}}{154\ \text{g } C_{12}H_{10}}\right) \\ &= 0.208\ \text{mol } C_{12}H_{10} \end{aligned}$$

The molal boiling-point elevation constant for benzene solutions is 2.53°C/m (Table 10.2).

$$\begin{aligned} \Delta T_b &= mK_b \\ &= (0.208\ m)(2.53^\circ\text{C}/m) \\ &= 0.526^\circ\text{C} \end{aligned}$$

The normal boiling point of benzene is 80.1°C (Table 10.2). The boiling point of the solution, therefore, is

$$80.1^\circ\text{C} + 0.5^\circ\text{C} = 80.6^\circ\text{C}$$

The molal freezing-point depression constant for benzene solutions is −5.12°C/m (Table 10.2).

$$\begin{aligned} \Delta T_f &= mK_f \\ &= (0.208\ m)(-5.12^\circ\text{C}/m) \\ &= -1.06^\circ\text{C} \end{aligned}$$

The normal freezing point of benzene is 5.5°C (Table 10.2). The freezing point of the solution, therefore, is

$$5.5^\circ\text{C} - 1.1^\circ\text{C} = 4.4^\circ\text{C}$$

Example 10.11 A solution prepared by dissolving 0.300 g of an unknown non-volatile solute in 30.0 g of carbon tetrachloride has a boiling point that is

0.392°C higher than that of pure CCl_4. What is the molecular weight of the solute?

Solution For CCl_4 solutions, K_b is 5.02°C (Table 10.2). We find the molality of the solution from the boiling-point elevation.

$$\Delta T_b = mK_b$$
$$0.392°\text{C} = m(5.02°\text{C}/m)$$
$$m = 0.0781\ m$$

Next, we find the number of grams of solute dissolved in 1000 g of CCl_4.

$$?\ \text{g solute} = 1000\ \text{g CCl}_4\left(\frac{0.300\ \text{g solute}}{30.0\ \text{g CCl}_4}\right) = 10.0\ \text{g solute}$$

Since the solution is 0.0781 m, 10.0 g of solute is 0.0781 mol of solute.

$$?\ \text{g solute} = 1\ \text{mol solute}\left(\frac{10.0\ \text{g solute}}{0.0781\ \text{mol solute}}\right) = 128\ \text{g solute}$$

10.9 OSMOSIS

The properties of solutions that depend principally upon the concentration of dissolved particles, rather than upon the nature of these particles, are called **colligative properties.** For solutions of nonvolatile solutes these properties include: vapor-pressure lowering, freezing-point depression, boiling-point elevation, and osmotic pressure. The last of these, osmotic pressure, is the topic of this section.

A membrane, such as cellophane or parchment, that permits some molecules or ions, but not all, to pass through it is called a **semipermeable membrane.** Figure 10.4 shows a membrane that is permeable to water but not to

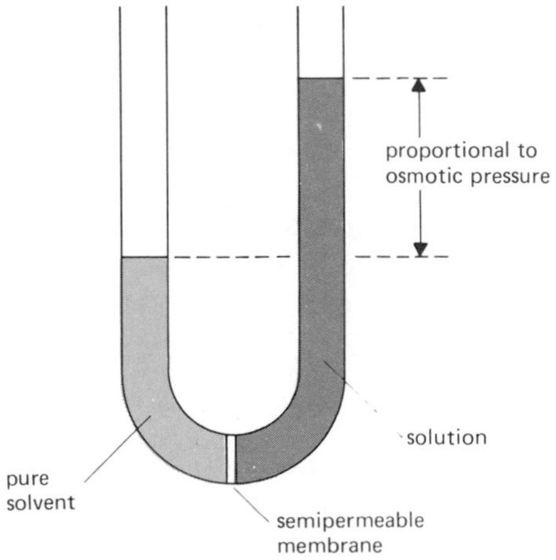

Figure 10.4 Osmosis.

sucrose (cane sugar) placed between pure water and a sugar solution. Water molecules, but not sugar molecules, can go through the membrane in either direction. However, there are more water molecules per unit volume on the left side (the side containing pure water) than on the right. The rate of passage through the membrane from left to right, therefore, exceeds the rate in the opposite direction.

As a result, the number of water molecules on the right increases, the sugar solution becomes more dilute, and the height of the solution in the right arm of the U tube increases. This process is called **osmosis.** The difference in height between the levels of the liquids in the two arms of the U tube is a measure of the **osmotic pressure.**

The increased hydrostatic pressure on the right tends to force water molecules through the membrane from right to left, so that ultimately the rate of passage to the left equals the rate of passage to the right. The final condition, therefore, is an equilibrium state with equal rates of passage of water molecules through the membrane in both directions. If a pressure that is higher than the equilibrium, osmotic pressure is applied to the solution in the right arm, water is forced in a direction contrary to that normally observed. This process, called **reverse osmosis,** is used to secure pure water from salt water.

Similarities exist between the behavior of water molecules in osmosis and the behavior of gas molecules in diffusion. In both processes molecules diffuse from regions of high concentrations to regions of low concentration. In 1887 Jacobus van't Hoff discovered the relation

$$\pi V = nRT$$

where π is the osmotic pressure (in atm), n is the number of moles of solute dissolved in volume V (in liters), T is the absolute temperature, and R is the gas constant (0.08206 liter atm/K mol). The similarity between this equation and the equation of state for an ideal gas is striking. The equation may be written in the form

$$\pi = \left(\frac{n}{V}\right)RT$$

$$\pi = MRT$$

where M is the molarity of the solution. Osmosis plays an important role in plant and animal physiological processes; the passage of substances through the semipermeable walls of a living cell, the action of the kidneys, and the rising of sap in trees are prime examples.

Example 10.12 Find the osmotic pressure of blood at body temperature (37°C) if the blood behaves as if it were a 0.296 M solution of a nonionizing solute.

Solution

$$\begin{aligned}\pi &= MRT\\ &= (0.296\ \text{mol/liter})(0.0821\ \text{liter atm/K mol})(310\ \text{K})\\ &= 7.53\ \text{atm}\end{aligned}$$

Example 10.13 One liter of an aqueous solution contains 30.0 g of a protein. The osmotic pressure of the solution is 0.0167 atm at 25°C. What is the approximate molecular weight of the protein?

Solution We use the van't Hoff equation to find the number of moles of protein present in the solution.

$$\pi = \left(\frac{n}{V}\right)RT$$

$$0.0167\ \text{atm} = \left(\frac{n}{1\ \text{liter}}\right)(0.0821\ \text{liter atm/K mol})(298\ \text{K})$$

$$n = 6.83 \times 10^{-4}\ \text{mol}$$

Since there are 30.0 g of protein in the solution,

$$\begin{aligned}?\ \text{g protein} &= 1\ \text{mol protein}\left(\frac{30.0\ \text{g protein}}{6.83 \times 10^{-4}\ \text{mol protein}}\right)\\ &= 4.39 \times 10^{4}\ \text{g protein}\end{aligned}$$

The molecular weight of the protein is approximately 43,900.

Since proteins have high molecular weights, the molal and molar concentrations of saturated solutions of proteins are very low. For the solution described in this example, the following effects can be calculated.

vapor pressure lowering	0.000000385 atm
boiling-point elevation	0.000350°C
freezing-point depression	−0.00127°C
osmotic pressure	0.0167 atm

Clearly, the first three are too small for accurate measurement. The osmotic pressure of this solution, however, would make the difference in the height of the two columns shown in Figure 10.4 approximately 17 cm, an amount which is easily measured.

10.10 SOLUTIONS OF ELECTROLYTES

If an aqueous solution contains ions, it will conduct electricity. Pure water itself is slightly ionized and is a poor conductor.

$$2H_2O \rightleftharpoons H_3O^+(aq) + OH^-(aq)$$

The solute of an aqueous solution that is a better electrical conductor than pure water alone is called an **electrolyte.** An electrolyte is wholly or partly ionized in water solution. Covalent solutes that are exclusively molecular in solution and therefore do nothing to enhance the conductivity of the solvent are called **nonelectrolytes;** sucrose (cane sugar) is an example.

Electrolytes can be further divided into two groups.

1. **Strong electrolytes** are virtually completely ionic in water solution. The strong acids and strong bases that were discussed in Chapter 8 are examples of strong electrolytes. Most salts are ionic compounds; these substances, if soluble, are strong electrolytes in water solution. A number of covalent compounds that contain metals and nonmetals (aluminum chloride, for example) are ionized almost completely in water and are classed as strong electrolytes.
2. **Weak electrolytes** are polar covalent substances that are incompletely dissociated in water solution. The conductivity of a 1 *m* solution of a weak electrolyte is lower than the conductivity of a 1 *m* solution of a strong electrolyte. The weak acids and bases discussed in Chapter 8 are examples of weak electrolytes. A small number of covalent compounds of metals dissolve in water but do not ionize completely. Solutions of mercury(II) chloride, for example, contain $HgCl_2$ molecules in addition to $HgCl^+$, Hg^{2+}, and Cl^- ions and other more complicated species.

The boiling-point elevations and freezing-point depressions of dilute solutions of electrolytes are different from those of solutions of nonelectrolytes with the same concentrations. Since 1 mol of NaCl contains 2 mol of ions (1 mol of Na^+ and 1 mol of Cl^-) and since colligative properties depend upon the number of dissolved particles and not their nature, we would expect the freezing-point depression of a 1 *m* solution of NaCl to be twice that of a 1 *m* solution of a nonelectrolyte. We would also expect the freezing-point depres-

Table 10.3 Observed freezing-point depressions for some aqueous solutions compared to calculated depressions[a]

	CONCENTRATION OF SOLUTION		
SOLUTE	0.001 *m*	0.01 *m*	0.1 *m*
nonelectrolyte	0.00186°C	0.0186°C	0.186°C
sucrose	0.00186	0.0186	0.188
2 ions/formula	0.00372	0.0372	0.372
NaCl	0.00366	0.0360	0.348
3 ions/formula	0.00558	0.0558	0.558
K_2SO_4	0.00528	0.0501	0.432
4 ions/formula	0.00744	0.0744	0.744
$K_3[Fe(CN)_6]$	0.00710	0.0626	0.530

[a] Calculated on the assumptions that K_f for water is 1.86°C/molal over the entire range of concentrations, the salts are 100% ionic in solution, and the ions act independently of one another in their effect on the freezing point of the solution.

sion of a solution of K_2SO_4 (which contains 3 mol of ions per mole of K_2SO_4) to be three times that of a solution that has the same concentration and that contains a solute that does not dissociate into ions.

The observed freezing-point depressions listed in Table 10.3 agree approximately with these predictions. The observed values agree best with the calculated values when the solutions are most dilute. Data such as these, together with data from electrical conductivity experiments led Svanté Arrhenius to propose his "chemical theory of electrolytes" in 1887. The boiling-point elevations of solutions of electrolytes are proportionately higher than the boiling-point elevations of solutions of nonelectrolytes with the same concentrations.

10.11 INTERIONIC ATTRACTIONS IN SOLUTIONS

The **van't Hoff factor,** *i,* is defined as the ratio of the observed freezing-point depression for a solution (ΔT_f) to the calculated depression (mK_f).

$$i = \frac{\Delta T_f}{mK_f}$$

Since a similar relation is used for boiling-point elevations,

$$i = \frac{\Delta T_b}{mK_b}$$

we will use the symbols ΔT and K, without the subscripts, in the discussion that follows.

The preceding expressions can be rearranged to

$$\Delta T = imK$$

The expressions that we used in Section 10.8,

$$\Delta T_b = mK_b \qquad \text{and} \qquad \Delta T_f = mK_f$$

were derived on the assumption that the solute *does not dissociate* ($i = 1$).

When dissociation of the solute occurs, it is necessary to correct the molality of the solution for freezing-point and boiling-point calculations. Thus, if AB is a strong electrolyte and therefore dissociates completely in solution, in a 0.001 m solution of AB there would be 0.001 mol of A^+ and 0.001 mol of B^- contained in 1000 g of water. On the assumption that each ion acts independently in its effect on the boiling point and freezing point, the effective concentration of this hypothetical solution would be 0.002 m. Hence, in this simplified example, $i = 2$, $m = 0.001$, and

$$\Delta T = 2(0.001\ m)K$$

Inspection of the i values recorded in Table 10.4 (which were derived from freezing-point determinations) reveals that they do not exactly equal the number of ions per formula unit for each of the strong electrolytes listed. Thus,

Table 10.4 Van't Hoff factor, *i*, for various strong electrolytes in solution[a]

	CONCENTRATION OF SOLUTION		
ELECTROLYTE	0.001*m*	0.01*m*	0.1*m*
NaCl	1.97	1.94	1.87
$MgSO_4$	1.82	1.53	1.21
K_2SO_4	2.84	2.69	2.32
$K_3[Fe(CN)_6]$	3.82	3.36	2.85

[a] From freezing-point determinations.

for 0.001 *m* solutions, the *i* value of NaCl is 1.97 (not 2); that of K_2SO_4, 2.84 (not 3); and that of $K_3[Fe(CN)_6]$, 3.82 (not 4). Furthermore, the *i* value changes with concentration of the solution and approaches the value expected for complete dissociation as the solution becomes more and more dilute.

In 1923 Peter Debye and Erich Hückel proposed a quantitative theory to explain the behavior of dilute solutions of electrolytes.* According to the **Debye–Hückel theory,** interionic attractions occur in solutions of electrolytes so that the ions are not completely independent of one another as in the case for uncharged molecules in solution. Thus, the electrical forces that operate between the oppositely charged ions reduce the effectiveness of these ions.

The *i* values in Table 10.4 show that the interionic attractions in solutions of $MgSO_4$ produce a stronger effect than those in solutions of NaCl of corresponding concentration even though both solutes contain two moles of ions per mole of compound. For 0.001*m* NaCl, $i = 1.97$, whereas for 0.001*m* $MgSO_4$, $i = 1.82$. Both the ions of magnesium sulfate are doubly charged (Mg^{2+}, SO_4^{2-}), whereas the ions of sodium chloride are singly charged (Na^+, Cl^-); hence, the interionic attractions are stronger in solutions of magnesium sulfate.

The effect of dilution upon *i* values can also be explained on the basis of these interionic attractions. As a solution is diluted, the ions spread farther and farther apart, their influence upon one another diminishes accordingly, and the *i* factor approaches its limiting value. Hence, in *very* dilute solutions the Debye–Hückel effect is negligible, and the ions essentially behave independently.

Interionic attractions, therefore, make a solution behave as though its ionic concentrations were less than they actually are. The **activity** of an ion, *a*,

* It should be noted that the Debye-Hückel theory accounts only for the behavior of dilute solutions of electrolytes. Concentrated solutions are imperfectly understood at present. Among the factors that have been postulated to explain the anomalous behavior of concentrated solutions are the lack of sufficient water to hydrate the ions and the presence of ion pairs and ion triplets (ions closely associated with one another).

Table 10.5 Characteristics of solutes

SOLUTE	FORM OF SOLUTE IN 1m SOLUTION	VAN'T HOFF FACTOR FOR ΔT_f; DILUTE SOLUTIONS OF MOLALITY, m $\Delta T_f = iK_f m$	EXAMPLES
nonelectrolytes	molecules	$i = 1$	$C_{12}H_{22}O_{11}$ (sucrose) CON_2H_4 (urea) $C_3H_5(OH)_3$ (glycerol)
strong electrolytes	ions	$i \approx n^a$	NaCl KOH
weak electrolytes	molecules and ions	$1 < i < n^a$	$HC_2H_3O_2$ NH_3 $HgCl_2$

[a] n = number of moles of ions per mole of solute.

is defined as the "effective concentration" of that ion; it is related to the actual concentration of the ion, c, by an **activity coefficient,** γ.

$$a = \gamma c$$

The activity coefficient is always less than one and approaches unity as the concentration of the solution decreases. The calculation of activity coefficients from experimental data is complicated. It is important to recognize that the activity of a given ion depends upon the concentrations of all ions present in the solution whether they are all derived from the same compound or not, since all ions can enter into interionic attractions.

Some characteristics of solutes are summarized in Table 10.5.

THE STORY OF WATER

Water is one of the four terrestial elements of the ancient Greeks (earth, air, fire, and water). To Johann van Helmont (1579–1644), water had special significance. In his famous tree experiment, van Helmont planted a 5 pound willow sapling in a tub that contained 200 pounds of soil. For five years he watered the tree with distilled water and took precautions so that dust would not add to the weight of the soil in the tub. At the end of that time, the tree weighed 169 pounds, but the weight of the soil had not changed. Since van Helmont believed that earth (the new growth of the willow tree) had been made from water alone, he concluded that water was an element but that earth was not.

Many chemists after the time of van Helmont thought that they too had converted water into earth. The change was believed to occur because an earthy residue is left when distilled water is evaporated to dryness in a vessel made of soft glass. In 1770, Lavoisier showed that a glass vessel loses weight when it is used in an evaporation of this type. He concluded, correctly, that the earthy residue is actually material dissolved away from the glass by the water and is not formed from water.

The work of several chemists in the years 1781 to 1784 established the fact that water is a compound, not an element. Joseph Priestley noted that when "inflammable air" (H_2) is exploded in "common air," moisture is produced. Henry Cavendish studied the preparation of water from "inflammable air" (H_2) and "dephlogisticated air" (O_2). Cavendish showed that the two gases combine in the ratio of 2 volumes to 1 volume and that the weight of the water produced equals the sum of the weights of the gases used. He thought, however, that water was an element that existed in both of the gases.

James Watt (of steam engine fame) was the first to propose that water is a compound. He believed it to be composed of the elements phlogiston (H_2) and "dephlogisticated air" (O_2). The modern interpretation of the composition of water was published in 1784 by Lavoisier, who gave hydrogen and oxygen their names. The name *hydrogen,* in fact, means *water former.*

Fifty years after Lavoisier's publication, the question of who discovered the composition of water became an issue. Claims were made for Cavendish, Watt, and Lavoisier, and each man had his supporters. The debate lasted for over twenty years and has become known as the *water controversy.* Few people today would ascribe the discovery to any one of the three men alone.

In recent years, water has again served as a center of controversy. In the 1960's and early 1970's, many articles appeared about a new form of water, which was called *polywater.* It was reported to boil far above 100°C, freeze far below 0°C, and have a lower vapor pressure and a higher density than ordinary water. Polywater was made by condensing ordinary water vapor in capillary tubes made of silicon dioxide or glass. Investigation of the question was hampered by the extremely small size of the samples that could be prepared and by the inability to duplicate experimental results. It appears today that polywater was probably water contaminated by silicates and other impurities. In any event, publications about the matter no longer appear.

PROBLEMS

10.1 Define the following: (a) heat of hydration, (b) heat of solution, (c) mole fraction, (d) ideal solution, (e) activity coefficient, (f) DeBye-Hückle effect, (g) semipermeable membrane.

10.2 What factors cause a solute to have a high heat of hydration?

10.3 Which ion of each of the following pairs would be expected to be the more strongly hydrated? (a) Fe^{2+} or Fe^{3+}, (b) Li^+ or Na^+, (c) F^- or Cl^-, (d) Sn^{2+} or Pb^{2+}, (e) Mg^{2+} or Al^{3+}, (f) Mg^{2+} or Ba^{2+}.

10.4 The heat of solution of NaCl at 298 K for the preparation of very dilute aqueous solutions is +3.9 kJ/mol. The lattice energy of NaCl is −789 kJ/mol. What is the heat of hydration of the ions of this compound? Describe the processes to which this value pertains.

10.5 The heat of solution of RbF at 298 K for the preparation of very dilute aqueous solutions is −24 kJ/mol. The lattice energy of RbF is −768 kJ/mol. What is the heat of hydration of the ions of this compound? Describe the processes to which this value pertains.

10.6 The lattice energy of $SrCl_2$ is −2114 kJ/mol. The heat of hydration of $SrCl_2$ (infinite dilution) at 298 K is −2161 kJ/mol. What is the heat of solution of $SrCl_2$ for the preparation of very dilute solutions?

10.7 The lattice energy of KF is −804 kJ/mol. The heat of hydration of KF (infinite dilution) at 298 K is −819 kJ/mol. What is the heat of solution of KF for the preparation of very dilute solutions?

10.8 Which compound of each of the following pairs is the more soluble in water? (a) CH_3OH or CH_3CH_3, (b) CCl_4 or NaCl, (c) CH_3F or CH_3Cl.

10.9 (a) The molecules in pure ethanol, CH_3CH_2OH, are extensively associated through hydrogen bonding. Those, however, in pure acetone, $CH_3{-}\overset{\overset{\displaystyle O}{\|}}{C}{-}CH_3$, and in pure chloroform, $CHCl_3$, are not. Explain why this is so. (b) On the basis of your answer to part (a), explain why the vapor pressures of chloroform and acetone solutions exhibit negative deviations from Raoult's law, whereas the vapor pressures of chloroform and ethanol solutions exhibit positive deviations.

10.10 How does the sign of a heat of solution determine the effect of a temperature change on the solubility of a solute?

10.11 Henry's law may be stated as $p = KX$, where p is the partial pressure of a gas over a saturated solution, X is the mole fraction of the dissolved gas in the solution, and K is a constant. For aqueous solutions of $CO_2(g)$ at 10°C, K is 1.04×10^3 atm. How many moles of CO_2 will dissolve in 900 g of H_2O if the partial pressure of CO_2 over the solution is 1.00 atm? How many grams?

10.12 Henry's law may be stated as $p = KX$, where p is the partial pressure of a gas over a saturated solution, X is the mole fraction of the dissolved gas in the solution, and K is a constant. For aqueous solutions of $N_2O(g)$ at 20°C, K is 2.04×10^3 atm. How many moles of N_2O will dissolve in 100 g of H_2O if the partial pressure over the solution is 1.50 atm? How many grams?

10.13 (a) Air is approximately 80.0% N_2 and 20.0% O_2 by volume. What are the partial pressures of N_2 and O_2 in air when the total pressure is 1.00 atm? (b) Henry's law may be stated as $p = KX$, where p is the partial pressure of the gas over a saturated solution, X is the mole fraction of the dissolved gas, and K is a constant. At 0°C the values of K for N_2 and O_2 are 5.38×10^4 atm and 2.51×10^4 atm, respectively. What are the mole fractions of N_2 and O_2 in a saturated aqueous solution at 0°C under air at a pressure of 1.00 atm? (c) What is the freezing point of the solution?

10.14 The volume of an aqueous solution changes as the temperature changes. Thus, a concentration of a given solution determined at one temperature may not be correct for that same solution at another temperature. Which methods of expressing concentration yield results that are temperature-independent, and which yield results that are temperature-dependent?

10.15 A given aqueous solution has a density of x g/ml and is y% solute by mass. Derive a mathematical equation relating the molarity of the solution, M, to the molality of the solution, m.

10.16 What are the mole fractions of ethyl alcohol, C_2H_5OH, and water in a solution that contains 11.5 g C_2H_5OH and 27.0 g H_2O?

10.17 What are the mole fractions of carbon tetrachloride, CCl_4, and benzene, C_6H_6, in a solution that contains 53.9 g CCl_4 and 46.8 g C_6H_6?

10.18 (a) What is the molarity of concentrated hydrobromic acid if the solution is 48.0% HBr by mass and has a density of 1.50 g/ml? (b) What is the molality of the solution?

10.19 (a) What is the molarity of concentrated hydrofluoric acid if the solution is 50.0% HF by mass and has a density of 1.17 g/ml? (b) What is the molality of the solution?

10.20 What volume of concentrated acetic acid should be used to prepare 250 ml of 6.00 M $HC_2H_3O_2$ solution? See Table 10.1.

10.21 What volume of concentrated nitric acid should be used to prepare 1.00 liter of 0.150 M HNO_3? See Table 10.1.

10.22 A 100 ml sample of concentrated acetic acid is diluted until the final volume is 500 ml. The concentrated acid is 100% $HC_2H_3O_2$ and has a density of 1.05 g/ml. What is the molarity of the final solution?

10.23 A 25.0 ml sample of concentrated HI solution is diluted to a final volume of 500 ml. The concentrated HI solution is 47.0% HI and has a density of 1.50 g/ml. What is the molarity of the final solution?

10.24 Concentrated H_2SO_4 is 96.0% H_2SO_4 and has a density of 1.84 g/ml. How many grams of concentrated H_2SO_4 should be used to prepare 500 ml of 0.300 *M* H_2SO_4? How many milliliters of concentrated H_2SO_4 should be used to prepare this solution?

10.25 Concentrated H_3PO_4 is 85.0% H_3PO_4 and has a density of 1.70 g/ml. How many grams of concentrated H_3PO_4 should be used to prepare 250 ml of 0.400 *M* H_3PO_4? How many milliliters of concentrated H_3PO_4 should be used to prepare this solution?

10.26 How many grams of $CaCl_2$ should be used to prepare 750 ml of 0.150 *M* $CaCl_2$ solution if the solid $CaCl_2$ is 96.0% pure?

10.27 A 10.0% solution of $AgNO_3$ has a density of 1.09 g/ml. What is (a) the molarity and (b) the molality of this solution?

10.28 A quantity of Na is added to water and 140 ml of $H_2(g)$, measured dry and at STP, and 250 ml of a solution of NaOH are produced. Write the equation for the reaction. What is the molarity of the NaOH solution?

10.29 How many grams of solid NaOH should be used to prepare 500 ml of a 0.100 *M* solution of NaOH if the solid is 95.0% NaOH by mass?

10.30 What are the concentrations of the reagents listed in Table 10.1 in terms of normality?

10.31 What is the normality of 6.00 *M* H_3PO_4, 6.00 *M* HCl, and 6.00 *M* H_2SO_4? Assume complete neutralization of the acids.

10.32 What is the molarity of 6.00 *N* H_3PO_4, 6.00 *N* HCl, and 6.00 *N* H_2SO_4? Assume complete neutralization of the acids.

10.33 Use the data of Table 10.1 to determine the number of milliliters of each reagent listed required to prepare 500 ml of 1.50 *N* solution.

10.34 Benzene, C_6H_6, and toluene, C_7H_8, form ideal solutions. At 60°C the vapor pressure of pure benzene is 0.507 atm and the vapor pressure of pure toluene is 0.184 atm. What is the vapor pressure at 60°C of a solution containing 6.5 g of benzene and 23.0 g of toluene?

10.35 Use the data given in Problem 10.34 to calculate the mole fraction of benzene in a solution that has a vapor pressure of 0.350 atm at 60°C.

10.36 Benzene, C_6H_6, and octane, C_8H_{18}, form ideal solutions. At 60°C, the vapor pressure of pure benzene is 0.507 atm, and the vapor pressure of octane is 0.103 atm. What is the vapor pressure at 60°C of a solution that contains 9.75 g of benzene and 57.0 g of octane?

10.37 Use the data of Problem 10.35 to calculate the mole fraction of benzene in a solution that has a vapor pressure of 0.461 atm at 60°C.

10.38 Liquids A and B form ideal solutions. The vapor pressure of pure B is 0.650 atm at the boiling point of a solution prepared from

0.200 mol of B and 0.600 mol of A. What is the vapor pressure of pure A at this temperature?

10.39 A solution containing 1 mol of chloroform and 4 mol of acetone has a vapor pressure of 0.400 atm at 35°C. At this temperature the vapor pressure of pure chloroform is 0.359 atm and that of acetone is 0.453 atm. What would the vapor pressure of the solution be if chloroform and acetone formed ideal solutions? Does the vapor pressure of the solution show a positive or negative deviation from that predicted by Raoult's law?

10.40 A solution containing 1 mol of carbon disulfide and 4 mol of acetone has a vapor pressure of 0.750 atm at 35°C. At this temperature the vapor pressure of pure carbon disulfide is 0.674 atm and that of acetone is 0.453 atm. What would the vapor pressure of the solution be if it were an ideal solution? Does the vapor pressure of the solution show a positive or negative deviation from that predicted by Raoult's law?

10.41 A solution prepared from 30.0 g of a nonvolatile, nondissociating solute in 10.0 mol of H_2O has a vapor pressure of 0.117 atm at 50°C. What is the approximate molecular weight of the solute? The vapor pressure of water at 50°C is 0.122 atm.

10.42 A solution prepared from 20.0 g of a nonvolatile, nondissociating solute in 4.00 mol of CCl_4 has a vapor pressure of 0.663 atm at 65°C. What is the approximate molecular weight of the solute? The vapor pressure of carbon tetrachloride at 65°C is 0.699 atm.

10.43 Which of the following measurements will give the best value for the molecular weight of a substance, such as a protein, that has a high molecular weight: vapor–pressure lowering, freezing-point depression, boiling-point elevation, or osmotic pressure? Why?

10.44 The antifreeze commonly used in car radiators is ethylene glycol, $C_2H_4(OH)_2$. How many grams of ethylene glycol must be added to 1000 g of water to produce a solution that will freeze at −15°C?

10.45 What is the freezing point of a solution containing 1.00 g of naphthalene, $C_{10}H_8$, in 250 g of camphor?

10.46 What weight of sucrose, $C_{12}H_{22}O_{11}$, should be added to 250 g of water to produce a solution that freezes at −2.50°C?

10.47 A solution containing 22.0 g of ascorbic acid (vitamin C) in 100 g of water freezes at −2.33°C. What is the molecular weight of ascorbic acid?

10.48 A solution containing 5.23 g of a nonvolatile, nonelectrolyte in 168 g of water freezes at −0.510°C. What is the molecular weight of the solute?

10.49 (a) What is the boiling point of a solution prepared from 9.66 g of glycerol $C_3H_5(OH)_3$, and 300 g of CCl_4? (b) What is the freezing point of this solution?

10.50 Lauryl alcohol is obtained from coconut oil and is used to make detergents. A solution of 5.00 g of lauryl alcohol in 100.0 g of benzene boils at 80.78°C. The normal boiling point of benzene is 80.10°C and K_b for benzene is +2.53°C/*m*. What is the molecular weight of lauryl alcohol?

10.51 A solution containing 9.66 g of a nonvolatile solute in 300 g of CCl_4 boils at 78.6°C. What is the molecular weight of the solute?

10.52 What is the osmotic pressure of an aqueous solution containing 1.00 g of urea, CON_2H_4, in 100 ml of solution at 20°C?

10.53 An aqueous solution containing 1.00 g/liter of the ethoxylate of lauryl alcohol, which is a detergent, has an osmotic pressure of 0.0448 atm at 20°C. What is the molecular weight of the compound?

10.54 An aqueous solution that contains 10.0 g of a sugar in 100 ml has an osmotic pressure of 13.6 atm at 25°C. What is the molecular weight of the sugar?

10.55 A solution containing 4.50 g of a protein in 100 ml has an osmotic pressure of 0.0157 atm at 25°C. What is the molecular weight of the protein?

10.56 One of the problems of producing fresh water from salt water by reverse osmosis is developing semipermeable membranes strong enough to withstand the high pressure required. Sea water contains dissolved salts (3.50% by mass). Assume that the solute is entirely NaCl (over 90% is), that the density of sea water is 1.03 g/ml, and that the *i* factor for a NaCl solution of this concentration is 1.83. For solutions of electrolytes, $\pi = iMRT$. What is the osmotic pressure of sea water at 20°C? What pressure would have to be applied to effect reverse osmosis?

10.57 A solution prepared by adding a given weight of a solute to 20.0 g of benzene freezes at a temperature 0.384°C below the freezing point of pure benzene. The freezing point of a solution prepared from the same weight of the solute and 20.00 g of water freezes at −0.4185°C. Assume that the solute is undissociated in benzene solution but is completely dissociated into ions in water solution. How many ions result from the dissociation of one molecule in water solution?

10.58 A solution containing 3.81 g of $MgCl_2$ in 400 g of water freezes at −0.497°C. What is the van't Hoff factor, *i*, for the freezing point of this solution?

10.59 A solution is prepared from 3.00 g of NaOH and 75.0 g of water. If the van't Hoff factor, *i*, for this solution is 1.83, at what temperature will the solution freeze?

10.60 What is the freezing point of a 0.125 *m* aqueous solution of a weak acid, HX, if the acid is 4.00% ionized? Ignore interionic attractions.

10.61 A 0.0200 *m* solution of a weak acid, HY, in water freezes at −0.0385°C. Ignore interionic attractions and calculate the percentage ionization of the weak acid.

Chapter 11

Reactions in Aqueous Solution

In Section 8.1, the assignment of oxidation numbers to atoms, combined and uncombined, was discussed. An atom may change its oxidation number in the course of a reaction. Reactions in which oxidation-number changes occur are called oxidation–reduction reactions (Section 11.1).

Reactions that occur in aqueous solution can be divided into two principal types: those that are oxidation–reduction reactions and those that are not. Examples of these two types are discussed in this chapter.

11.1 OXIDATION–REDUCTION REACTIONS

The term oxidation was originally applied to reactions in which substances combined with oxygen, and reduction was defined as the removal of oxygen from an oxygen-containing compound. The meanings of the terms have gradually been broadened. Today oxidation and reduction are defined on the basis of change in oxidation number (Section 8.1).

Oxidation is the process in which an atom undergoes an algebraic increase in oxidation number, and **reduction** is the process in which an atom undergoes an algebraic decrease in oxidation number. On this basis, oxidation–reduction is involved in the reaction

$$\overset{0}{S} + \overset{0}{O}_2 \rightarrow \overset{4+}{S}\ \overset{2-}{O}_2$$

The oxidation number of each type of atom is written above its symbol. Since the oxidation number of the S atom *increases* from 0 to 4+, sulfur is said to be *oxidized.* The oxidation number of the O atom *decreases* from 0 to 2−, and oxygen is said to be *reduced.* Oxidation–reduction is not involved in the reaction

$$\overset{4+}{S}\ \overset{2-}{O}_2 + \overset{1+}{H}_2\overset{2-}{O} \rightarrow \overset{1+}{H}_2\ \overset{4+}{S}\ \overset{2-}{O}_3$$

since no atom undergoes a change in oxidation number.

It is apparent, from the way in which oxidation numbers are assigned, that *neither oxidation nor reduction can occur by itself.* Furthermore, *the total increase in oxidation number must equal the total decrease in oxidation number.* In the reaction of sulfur and oxygen, the sulfur undergoes an increase of 4. Each oxygen atom undergoes a decrease of 2. Since two O atoms appear in the equation, the total decrease is 4.

Since one substance cannot be reduced unless another is simultaneously oxidized, the substance that is reduced is responsible for the oxidation. This substance is called, therefore, the **oxidizing agent** or **oxidant.** Because of the interdependence of the two processes, the opposite is also true. The material that is itself oxidized is called the **reducing agent** or **reductant.** Therefore,

$$\underset{\substack{\text{oxidized} \\ \text{reducing agent}}}{\overset{0}{S}} + \underset{\substack{\text{reduced} \\ \text{oxidizing agent}}}{\overset{0}{O}_2} \rightarrow \overset{4+}{S}\ \overset{2-}{O}_2$$

Equations for oxidation–reduction reactions are usually more difficult to balance than those for reactions that do not entail oxidation and reduction. Two methods are commonly used to balance oxidation–reduction equations: the oxidation-number method and the ion–electron method. Both of these methods will be discussed. For clarity, the physical state of the reactants and products will not be indicated in the examples that follow. In addition, the symbol H^+, instead of H_3O^+ or H^+ (aq), will be used.

There are three steps in the **oxidation-number method** of balancing oxidation–reduction equations. We will use the equation for the reaction of nitric acid and hydrogen sulfide to illustrate this method. The unbalanced expression for the reaction is

$$HNO_3 + H_2S \rightarrow NO + S + H_2O$$

1. The oxidation numbers of the atoms in the equation are determined in order to identify those undergoing oxidation or reduction. Thus,

$$H\overset{5+}{N}O_3 + H_2\overset{2-}{S} \rightarrow \overset{2+}{N}O + \overset{0}{S} + H_2O$$

Nitrogen is reduced (from 5+ to 2+, a decrease of 3), and sulfur is oxidized (from 2− to 0, an increase of 2).

2. Coefficients are added so that the total decrease and the total increase in oxidation number will be equal. We have a decrease of 3 and an increase of 2 indicated in the unbalanced expression. The lowest common multiple of 3 and 2 is 6. We therefore indicate $2HNO_3$ and $2NO$ (for a total decrease of 6) and $3H_2S$ and $3S$ (for a total increase of 6).

$$2HNO_3 + 3H_2S \rightarrow 2NO + 3S + H_2O$$

3. Balancing is completed by inspection. This method takes care of only those substances that are directly involved in oxidation-number change. In this example, the method does not assign a coefficient to H_2O. We note, however, that there are now 8 H atoms on the left of the equation. We can indicate the same number of H atoms on the right by showing $4H_2O$.

$$2HNO_3 + 3H_2S \rightarrow 2NO + 3S + 4H_2O$$

The final, balanced equation should be checked to ensure that there are as many atoms of each element on the right as there are on the left.

The oxidation-number method can be used to balance net ion equations, in which only those ions and molecules that take part in the reaction are shown. Consider the reaction between $KClO_3$ and I_2.

$$H_2O + I_2 + ClO_3^- \rightarrow IO_3^- + Cl^- + H^+$$

The K^+ ion does not take part in the reaction and is not shown in the equation. The steps in balancing the equation follow.

1. $$H_2O + \overset{0}{I_2} + \overset{5+}{Cl}O_3^- \rightarrow \overset{5+}{I}O_3^- + \overset{1-}{Cl^-} + H^+$$

2. *Each* iodine atom undergoes an increase of 5 (from 0 to 5+), but there are *two* iodine atoms in I_2. The increase in oxidation number is therefore 10. Chlorine undergoes a decrease of 6 (from 5+ to 1−). The lowest common multiple of 6 and 10 is 30. Therefore, $3I_2$ molecules must be indicated (a total increase of 30) and $5ClO_3^-$ ions are needed (a total decrease of 30). The coefficients of the products, IO_3^- and Cl^-, follow from this assignment.

$$H_2O + 3I_2 + 5ClO_3^- \rightarrow 6IO_3^- + 5Cl^- + H^+$$

3. If H_2O is ignored, there are now 15 oxygen atoms on the left and 18 oxygen atoms on the right. To make up 3 oxygen atoms on the left, we

must indicate $3H_2O$ molecules. It then follows that the coefficient of H^+ must be 6 to balance the hydrogens of the H_2O molecules.

$$3H_2O + 3I_2 + 5ClO_3^- \rightarrow 6IO_3^- + 5Cl^- + 6H^+$$

An ionic equation must indicate charge balance as well as mass balance. Since the algebraic sum of the charges on the left (5−) equals that on the right (5−), the equation is balanced.

Reactions in which electrons are transferred are clearly examples of oxidation–reduction reactions. In the reaction of sodium and chlorine, a sodium atom loses its valence electron to a chlorine atom.

$$2\overset{0}{Na} + \overset{0}{Cl_2} \rightarrow 2\overset{1+}{Na^+} + 2\overset{1-}{Cl^-}$$

For simple ions the oxidation number is the same as the charge on the ion. It follows, then, that electron loss represents a type of oxidation, and electron gain represents a type of reduction. This equation can be divided into two **partial equations** that represent **half reactions.**

Oxidation: $2Na \rightarrow 2Na^+ + 2e^-$
Reduction: $2e^- + Cl_2 \rightarrow 2Cl^-$

The **ion–electron method** of balancing oxidation–reduction equations employs partial equations. One partial equation is used for the oxidation (in which electrons are lost), and another partial equation is used for the reduction (in which electrons are gained). The final equation is obtained by combining the partial equations in such a way that the number of electrons lost equals the number of electrons gained.

Two slightly different procedures are employed to balance equations by the ion–electron method. One is used for reactions that take place in acid solution; the other, for reactions that occur in alkaline solution. We will give examples of both procedures.

An example of a reaction that occurs in *acid solution* is

$$Cr_2O_7^{2-} + Cl^- \rightarrow Cr^{3+} + Cl_2$$

In this unbalanced expression, H_2O and H^+ are not shown. The proper numbers of H_2O molecules and H^+ ions, as well as their positions in the final equation (whether on the left or the right), are determined in the course of the balancing.

1. Two skeleton partial equations for the half reactions are written. The central elements of the partial equations (Cr and Cl) are balanced.

$$Cr_2O_7^{2-} \rightarrow 2Cr^{3+}$$
$$2Cl^- \rightarrow Cl_2$$

2. The H and O atoms are now balanced. Since the reaction occurs in

acid solution, H_2O and H^+ can be added where needed. For each O atom that is needed, one H_2O is added to the side that is deficient. The hydrogen is then brought into balance by the addition of H^+.

Seven O atoms are needed on the right side of the first partial equation, and, therefore, $7H_2O$ are added to this side. The H atoms of the first partial equation are then balanced by the addition of $14H^+$ to the left side. The second partial equation is already in material balance.

$$14H^+ + Cr_2O_7^{2-} \rightarrow 2Cr^{3+} + 7H_2O$$
$$2Cl^- \rightarrow Cl_2$$

3. The next step is to balance the partial equations electrically. In the first partial equation, the net charge is 12+ on the left side of the equation (14+ and 2−) and 6+ on the right. Six electrons must be added to the left. In this way, the net charge on both sides of the equation will be 6+. The second partial equation is balanced electrically by the addition of 2 electrons to the right.

$$6e^- + 14H^+ + Cr_2O_7^{2-} \rightarrow 2Cr^{3+} + 7H_2O$$
$$2Cl^- \rightarrow Cl_2 + 2e^-$$

4. The number of electrons gained must equal the number of electrons lost. The second partial equation, therefore, is multiplied through by 3.

$$6e^- + 14H^+ + Cr_2O_7^{2-} \rightarrow 2Cr^{3+} + 7H_2O$$
$$6Cl^- \rightarrow 3Cl_2 + 6e^-$$

5. Addition of the two partial equations gives the final equation. In the addition, the electrons cancel.

$$14H^+ + Cr_2O_7^{2-} + 6Cl^- \rightarrow 2Cr^{3+} + 3Cl_2 + 7H_2O$$

As a second example, consider the reaction

$$MnO_4^- + As_4O_6 \rightarrow Mn^{2+} + H_3AsO_4$$

which also occurs in *acid solution.* The same steps are followed.

1. The equation is divided into two skeleton partial equations. The As atoms in the second partial equation are balanced.

$$MnO_4^- \rightarrow Mn^{2+}$$
$$As_4O_6 \rightarrow 4H_3AsO_4$$

2. The first partial equation can be brought into material balance by the addition of $4H_2O$ to the right side and $8H^+$ to the left side. In the second partial equation, $10H_2O$ must be added to the left side to make up the needed 10 oxygens. If we stopped at this point, there would be 20 hydrogen atoms on the left and 12 on the right. Therefore, $8H^+$ must be added to the right.

$$8H^+ + MnO_4^- \rightarrow Mn^{2+} + 4H_2O$$
$$10H_2O + As_4O_6 \rightarrow 4H_3AsO_4 + 8H^+$$

3. To balance the net charges, electrons are added.

$$5e^- + 8H^+ + MnO_4^- \rightarrow Mn^{2+} + 4H_2O$$
$$10H_2O + As_4O_6 \rightarrow 4H_3AsO_4 + 8H^+ + 8e^-$$

4. The first partial equation must be multiplied through by 8 and the second by 5 so that the same number of electrons are lost in the oxidation partial equation as are gained in the reduction partial equation.

$$40e^- + 64H^+ + 8MnO_4^- \rightarrow 8Mn^{2+} + 32H_2O$$
$$50H_2O + 5As_4O_6 \rightarrow 20H_3AsO_4 + 40H^+ + 40e^-$$

5. When these two partial equations are added, water molecules and hydrogen ions must be canceled as well as electrons. It is poor form to leave an equation with $64H^+$ on the left and $40H^+$ on the right.

$$24H^+ + 18H_2O + 5As_4O_6 + 8MnO_4^- \rightarrow 20H_3AsO_4 + 8Mn^{2+}$$

Equations for reactions that take place in *alkaline solution* are balanced in a manner slightly different from those that occur in acidic solution. All the steps are the same except the second one; H^+ cannot be used to balance equations for reactions that occur in alkaline solution. As an example, consider the reaction

$$MnO_4^- + N_2H_4 \rightarrow MnO_2 + N_2$$

that takes place in alkaline solution.

1. The equation is divided into two partial equations.

$$MnO_4^- \rightarrow MnO_2$$
$$N_2H_4 \rightarrow N_2$$

2. For reactions occurring in alkaline solution, OH^- and H_2O are used to balance oxygen and hydrogen. For each oxygen that is needed, $2OH^-$ ions are added to the side of the partial equation that is deficient, and one H_2O molecule is added to the opposite side. For each hydrogen that is needed, one H_2O molecule is added to the side that is deficient, and one OH^- ion is added to the opposite side.

   In the first partial equation, the right side is deficient by 2 oxygen atoms. We add, therefore, $4OH^-$ to the right side and $2H_2O$ to the left.

$$2H_2O + MnO_4^- \rightarrow MnO_2 + 4OH^-$$

   In order to bring the second partial equation into material balance, we must add four hydrogen atoms to the right side. For *each* hydrogen atom needed, we add one H_2O to the side deficient in hydrogen and one OH^- to the opposite side. In the present case we add $4H_2O$ to the right side and $4OH^-$ to the left to make up the four hydrogen atoms needed on the right.

$$4OH^- + N_2H_4 \rightarrow N_2 + 4H_2O$$

3. Electrons are added to effect charge balances.

$$3e^- + 2H_2O + MnO_4^- \rightarrow MnO_2 + 4OH^-$$
$$4OH^- + N_2H_4 \rightarrow N_2 + 4H_2O + 4e^-$$

4. The lowest common multiple of 3 and 4 is 12. Therefore, the first partial equation is multiplied through by 4 and the second by 3 so that the number of electrons gained equals the number lost.

$$12e^- + 8H_2O + 4MnO_4^- \rightarrow 4MnO_2 + 16OH^-$$
$$12OH^- + 3N_2H_4 \rightarrow 3N_2 + 12H_2O + 12e^-$$

THE ION–ELECTRON METHOD FOR BALANCING OXIDATION–REDUCTION EQUATIONS

1. Divide the equation into two skeleton partial equations.
2. Balance the atoms that change their oxidation numbers in each partial equation.
3. Balance the O and H atoms in each partial equation.
   a. For reaction in *acid solution:*
      i. For each O atom that is needed, add one H_2O to the side of the partial equation that is deficient in oxygen.
      ii. Add H^+ where needed to bring the hydrogen into balance.
   b. For reactions in *alkaline solution:*
      i. For each O atom that is needed, add *two* OH^- ions to the side of the partial equation that is deficient in O, and add *one* H_2O to the opposite side.
      ii. For each H atom that is needed, add *one* H_2O to the side of partial equation that is deficient in H, and add *one* OH^- ion to the opposite side.
4. To each partial equation, add electrons in such a way that the net charge on the left side of the equation equals the net charge on the right side.
5. If necessary, multiply one or both partial equations by numbers that will make the number of electrons lost in one partial equation equal the number of electrons gained in the other partial equation.
6. Add the partial equations. In the addition, cancel terms common to both sides of the final equation.

5. Addition of these partial equations, with cancellation of OH^- ions and H_2O molecules as well as electrons, gives the final equation.

$$4MnO_4^- + 3N_2H_4 \rightarrow 4MnO_2 + 3N_2 + 4H_2O + 4OH^-$$

As a final example, consider the following skeleton equation for a reaction in *alkaline solution.*

$$Br_2 \rightarrow BrO_3^- + Br^-$$

In this reaction the same substance, Br_2, is both oxidized and reduced. Such reactions are called **disproportionations** or **auto-oxidation–reduction reactions.**

1. $$Br_2 \rightarrow 2BrO_3^-$$
$$Br_2 \rightarrow 2Br^-$$

2. $$12OH^- + Br_2 \rightarrow 2BrO_3^- + 6H_2O$$
$$Br_2 \rightarrow 2Br^-$$

3. $$12OH^- + Br_2 \rightarrow 2BrO_3^- + 6H_2O + 10e^-$$
$$2e^- + Br_2 \rightarrow 2Br^-$$

4. $$12OH^- + Br_2 \rightarrow 2BrO_3^- + 6H_2O + 10e^-$$
$$10e^- + 5Br_2 \rightarrow 10Br^-$$

5. $$12OH^- + 6Br_2 \rightarrow 2BrO_3^- + 10Br^- + 6H_2O$$

When the ion-electron method is applied to a disproportionation reaction, the coefficients of the resulting equation usually are divisible by some common number since one reactant was used in both partial equations. The coefficients of this equation are all divisible by 2 and should be reduced to the lowest possible terms.

$$6OH^- + 3Br_2 \rightarrow BrO_3^- + 5Br^- + 3H_2O$$

11.2 EQUIVALENT WEIGHTS OF OXIDANTS AND REDUCTANTS

In an oxidation–reduction reaction, the total increase in oxidation number of one element must equal the total decrease in oxidation number of another. Therefore, the system of equivalent weights for oxidizing agents and reducing agents is based on oxidation number changes.

The equivalent weight of a substance that functions as an oxidant or reductant is defined as

$$\textit{equiv. wt.} = \frac{\textit{molecular wt.}}{\textit{total change in oxidation no.}}$$

In the reaction

$$\overset{0}{I_2} + 10H\overset{5+}{N}O_3 \rightarrow 2H\overset{5+}{I}O_3 + 10\overset{4+}{N}O_2 + 4H_2O$$

each iodine atom undergoes an increase in oxidation number of 5. The total increase per molecule of I_2, therefore, is 10. Since the molecular weight of I_2 is 253.81, the equivalent weight of I_2 is

$$\frac{253.81}{10} = 25.381$$

The N atom of HNO_3 undergoes a decrease of 1. The equivalent weight of HNO_3 (which has a molecular weight of 63.01) is

$$\frac{63.01}{1} = 63.01$$

In this reaction, therefore, 25.381 g of I_2 (1 equiv. wt.) will react with 63.01 g of HNO_3 (1 equiv. wt.).

The stoichiometric ratio of reductant (I_2) to oxidant (HNO_3) derived from equivalent weights is the same as that derived from the balanced chemical equation. Thus, the balanced chemical equation for the reaction of I_2 and HNO_3 shows 1 mol of I_2, which is 10 equivalents, reacting with 10 mol of HNO_3, which is also 10 equivalents.

Notice that the equivalent weight of a compound may vary from reaction to reaction. If HNO_3 is used in a reaction in which NO is produced, the change in oxidation number of N is 3 (from 5+ to 2+). The equivalent weight of HNO_3 for this reaction is

$$\frac{63.01}{3} = 21.00$$

It is also possible to use half equations to derive equivalent weights. The increase or decrease in oxidation number is the same as the number of electrons indicated as being lost or gained in the partial equations. When potassium permanganate serves as an oxidizing agent in acid solution, the partial equation for the reduction of permanganate ion is

$$5e^- + 8H^+ + \overset{7+}{Mn}O_4^- \rightarrow Mn^{2+} + H_2O$$

The equivalent weight of $KMnO_4$ for this reaction is the molecular weight (158.04) divided by the change in the oxidation number of Mn (5), which is 31.61.

The partial equation for the use of $KMnO_4$ as an oxidizing agent in alkaline solution is

$$3e^- + 2H_2O + \overset{7+}{Mn}O_4^- \rightarrow \overset{4+}{Mn}O_2 + 4OH^-$$

The equivalent weight of $KMnO_4$ for this reaction is 158.04/3, or 52.68.

Example 11.1 What is the equivalent weight of As_4O_6 for a reaction in which H_3AsO_4 is produced?

Solution When As_4O_6 is transformed into H_3AsO_4,

$$\overset{3+}{As_4O_6} \rightarrow H_3\overset{5+}{As}O_4$$

each As atom undergoes an increase in oxidation number of 2. The *total* change for the four As atoms is 8. Since the molecular weight of As_4O_6 is 395.6, the equivalent weight is

$$\frac{395.6}{8} = 49.45$$

The partial equation for the half reaction can also be used to find the total change in oxidation number (it is equal to the number of electrons lost).

$$10H_2O + As_4O_6 \rightarrow 4H_3AsO_4 + 8H^+ + 8e^-$$

Example 11.2 What quantity of $KMnO_4$ is needed to oxidize 15.0 g of As_4O_6 in acidic solution? The products of the reaction are H_3AsO_4 and Mn^{2+}.
Solution The equivalent weight of $KMnO_4$ (to Mn^{2+}) is 31.6 (see text). The equivalent weight of As_4O_6 (to H_3AsO_4) is 49.5 (see Example 11.1). Therefore,

$$? \text{ g } KMnO_4 = 15.0 \text{ g } As_4O_6 \left(\frac{31.6 \text{ g } KMnO_4}{49.5 \text{ g } As_4O_6}\right) = 9.58 \text{ g } KMnO_4$$

11.3 SOLUTIONS OF OXIDIZING AND REDUCING AGENTS

The concentration of a solution of an oxidant or reductant can be expressed in terms of molarity or normality. As in the case of solutions of acids and bases, the normality of a solution is the number of gram equivalent weights of solute per liter of solution (Section 8.6).

Example 11.3 In acidic solution, potassium dichromate acts as an oxidizing agent. The half reaction for the reduction of the $Cr_2O_7^{2-}$ ion is

$$6e^- + 14H^+ + Cr_2O_7^{2-} \rightarrow 2Cr^{3+} + 7H_2O$$

What quantity of $K_2Cr_2O_7$ should be used to prepare 500.0 ml of a 0.1000 *N* solution?
Solution The total change in oxidation number is 6 (given by the number of electrons gained in the partial equation). Since the molecular weight of $K_2Cr_2O_7$ is 294.2, the equivalent weight of $K_2Cr_2O_7$ is

$$\frac{294.2}{6} = 49.03$$

Therefore,

$$? \text{ g } K_2Cr_2O_7 = 500.0 \text{ ml soln.} \left(\frac{0.1000 \text{ eqiv. wt.}}{1000 \text{ ml soln.}}\right)\left(\frac{49.03 \text{ g } K_2Cr_2O_7}{1 \text{ eqiv. wt.}}\right)$$

$$= 2.452 \text{ g } K_2Cr_2O_7$$

For the reaction of $K_2Cr_2O_7$ described in Example 11.3, the total change in oxidation number is 6. One molecular weight of $K_2Cr_2O_7$, therefore, contains 6 equivalent weights. It follows that a 1 *M* solution of $K_2Cr_2O_7$ is 6 *N*. In general, the normality of a solution of an oxidant or reductant is related to the molarity of the solution by the equation

$$N = aM$$

where a is the total change in oxidation number that the solute undergoes.

Analyses are sometimes run using solutions of oxidizing agents and reducing agents. The equations derived in Section 8.6 are useful for calculating the results of these analyses.

$$V_A N_A = V_B N_B$$

where V_A and V_B are the volumes of the solutions employed and N_A and N_B are the normalities of the solutions.

$$V_A N_A = \frac{1000 p_B g_B}{eq_B} \qquad \text{(volume in milliliters)}$$

where V_A and N_A are the volume and normality of a solution, p_B is the purity of a sample, g_B is the mass of the sample, and eq_B is the equivalent weight of the substance present in the sample.

Example 11.4 A 0.4308 g sample of iron ore is dissolved in acid and the iron converted into the iron(II) state. This solution is titrated against permanganate. The titration employs 32.31 ml of 0.1248 *N* $KMnO_4$ which oxidizes Fe^{2+} to Fe^{3+}. What is the percentage of iron in the ore?

Solution Since the oxidation number of iron increases by 1 unit, the equivalent weight of iron is the same as the atomic weight, 55.85.

$$V_A N_A = \frac{1000\ p_B g_B}{eq_B}$$

$$(32.31\ \text{ml})(0.1248\ \text{equiv./liter}) = \frac{(1000\ \text{ml/liter})\ x\ (0.4308\ g)}{55.85\ g/\text{equiv.}}$$

$$x = 0.5228$$

The ore is 52.28% iron.

11.4 METATHESIS REACTIONS

A **metathesis reaction** has the general form

$$AB + CD \rightarrow AD + CB$$

In the reaction, the cations and anions of the substances involved exchange partners. This type of reaction is a common one, particularly in aqueous solution. *Metathesis reactions are not oxidation–reduction reactions.* No atom involved in them changes oxidation number.

A specific example of a metathesis reaction is

$$AgNO_3(aq) + NaCl(aq) \rightarrow AgCl(s) + NaNO_3(aq)$$

Since we are talking about reactions in aqueous solution, the symbol (aq) will be omitted from future examples. The symbols (g) for gaseous and (s) for solid, however, will be used.

The preceding equation is written in *molecular form* (Section 8.3). These substances, however, are ionic. A solution of a soluble ionic substance in water contains hydrated ions. The *ionic form* of the equation is

$$Ag^+ + NO_3^- + Na^+ + Cl^- \rightarrow AgCl(s) + Na^+ + NO_3^-$$

Why did the reaction occur? Collisions of the Ag^+ and Cl^- ions result in the formation of AgCl, which is insoluble and falls out of solution. An insoluble substance formed in this way is called a **precipitate,** and the process is called a **precipitation.** Sodium nitrate ($NaNO_3$), the other product of the reaction, is soluble in water and remains in solution in the form of hydrated ions.

If we eliminate the spectator ions (ions that appear on both sides of the equation and hence do not enter into the reaction), we get the *net ionic form* of the equation (Section 8.3).

$$Ag^+ + Cl^- \rightarrow AgCl(s)$$

This is the most general form of the equation. It tells us that a solution of any soluble Ag^+ salt and a solution of any soluble Cl^- salt will produce insoluble AgCl when mixed.

Suppose that we mix solutions of NaCl and NH_4NO_3. The ionic equation for the "reaction" is

$$Na^+ + Cl^- + NH_4^+ + NO_3^- \rightarrow Na^+ + NO_3^- + NH_4^+ + Cl^-$$

All of the compounds involved in the "reaction" are soluble in water. If we eliminate spectator ions from the equation, nothing is left. There is, therefore, no reaction, which is usually indicated in the following way.

$$Na^+ + Cl^- + NH_4^+ + NO_3^- \rightarrow \text{N. R.}$$

There are other reasons for a metathesis reaction to occur in addition to the formation of an insoluble solid. Consider the reaction that occurs when a hydrochloric acid solution and a solution of sodium sulfide are mixed.

molecular equation:	$2HCl + Na_2S \rightarrow H_2S(g) + 2NaCl$
ionic equation:	$2H^+ + 2Cl^- + 2Na^+ + S^{2-} \rightarrow H_2S(g) + 2Na^+ + 2Cl^-$
net ionic equation:	$2H^+ + S^{2-} \rightarrow H_2S(g)$

In this reaction, collisions of the H^+ and S^{2-} ions have formed a gas, H_2S, which is only slightly soluble and which escapes from the solution.

In a third type of metathesis reaction, a weak electrolyte is produced

(Section 10.10). Soluble weak electrolytes do not dissociate completely into ions in aqueous solution. They exist principally in the form of molecules. A neutralization reaction is a metathesis reaction of this type.

molecular equation:	$HCl + NaOH \rightarrow H_2O + NaCl$
ionic equation:	$H^+ + Cl^- + Na^+ + OH^- \rightarrow H_2O + Na^+ + Cl^-$
net ionic equation:	$H^+ + OH^- \rightarrow H_2O$

The reaction occurs because the H^+ and OH^- ions form H_2O molecules, and water is a weak electrolyte.

Metathesis reactions, then, occur when a precipitate, an insoluble gas, or a weak electrolyte is formed. In ionic equations for these reactions,

1. a *soluble salt* is indicated by the formulas of the ions that make up the compound.
2. an *insoluble* or *slightly soluble compound* is indicated by the molecular formula of the compound followed by the symbol (s).
3. an *insoluble* or *slightly soluble gas* is indicated by the molecular formula of the gas followed by the symbol (g).
4. a *weak electrolyte* is indicated by the molecular formula of the compound. Weak electrolytes are partly dissociated into ions in aqueous solution, but they exist principally in molecular form.

In order to write equations for metathesis reactions, we need to identify these types of compounds. The following rules are designed for this purpose.

1. **Solubility rules.** The classification of ionic substances according to their solubility in water is difficult. Nothing is completely "insoluble" in water. The degree of solubility varies greatly from one "soluble" substance to another. Nevertheless, a solubility classification scheme is useful even though it must be regarded as approximate. The rules given in Table 11.1 apply to compounds of the following cations.
   - **(a) 1+ cations.** Li^+, Na^+, K^+, Rb^+, Cs^+, NH_4^+, Ag^+
   - **(b) 2+ cations.** Mg^{2+}, Ca^{2+}, Sr^{2+}, Ba^{2+}, Mn^{2+}, Fe^{2+}, Co^{2+}, Ni^{2+}, Cu^{2+}, Zn^{2+}, Cd^{2+}, Hg^{2+}, Hg_2^{2+}, Sn^{2+}, Pb^{2+}
   - **(c) 3+ cations.** Fe^{3+}, Al^{3+}, Cr^{3+}

   Compounds that dissolve to the extent of at least 10 g/liter are listed as *soluble*. Those that fail to dissolve to the extent of 1 g/liter are classed as *insoluble*. Those compounds with solubilities that are intermediate between these limits are listed as *slightly soluble* (and marked with a star in the table). These standards are common but arbitrary. Most acids are water soluble.

2. **Gases.** The common insoluble gases formed in metathesis reactions are listed in Table 11.2. Reactions in which three of the gases (CO_2, SO_2, and NH_3) are produced may be viewed as involving the initial formation of a substance that then breaks down to give the gas and H_2O. The reaction of Na_2SO_3 and HCl, for example, produces H_2SO_3.

Table 11.1 Solubilities of some ionic compounds in water[a]

MAINLY WATER-SOLUBLE	
NO_3^-	All nitrates are soluble.
$C_2H_3O_2^-$	All acetates are soluble.
ClO_3^-	All chlorates are soluble.
Cl^-	All chlorides are soluble except AgCl, Hg_2Cl_2, and $PbCl_2$*.
Br^-	All bromides are soluble except AgBr, Hg_2Br_2, $PbBr_2$*, and $HgBr_2$*.
I^-	All iodides are soluble except AgI, Hg_2I_2, PbI_2, and HgI_2.
SO_4^{2-}	All sulfates are soluble except $CaSO_4$*, $SrSO_4$, $BaSO_4$, $PbSO_4$, Hg_2SO_4, and Ag_2SO_4*.
MAINLY WATER-INSOLUBLE	
S^{2-}	All sulfides are insoluble except those of the I A and II A elements and $(NH_4)_2S$.
CO_3^{2-}	All carbonates are insoluble except those of the I A elements and $(NH_4)_2CO_3$.
SO_3^{2-}	All sulfites are insoluble except those of the I A elements and $(NH_4)_2SO_3$.
PO_4^{3-}	All phosphates are insoluble except those of the I A elements and $(NH_4)_3PO_4$.
OH^-	All hydroxides are insoluble except those of the I A elements, $Ba(OH)_2$, $Sr(OH)_2$*, and $Ca(OH)_2$*.

[a] The following cations are considered: those of the I A and II A families, NH_4^+, Ag^+, Al^{3+}, Cd^{2+}, Co^{2+}, Cr^{3+}, Cu^{2+}, Fe^{2+}, Fe^{3+}, Hg_2^{2+}, Hg^{2+}, Mn^{2+}, Ni^{2+}, Pb^{2+}, Sn^{2+}, and Zn^{2+}.

Soluble compounds dissolve to the extent of at least 10 g/liter. Slightly soluble compounds (marked with an *) dissolve to the extent of from 1 g/liter to 10 g/liter.

$$2Na^+ + SO_3^{2-} + 2H^+ + 2Cl^- \rightarrow H_2SO_3 + 2Na^+ + 2Cl^-$$

The H_2SO_3 is unstable and decomposes to give SO_2 and H_2O.

$$H_2SO_3 \rightarrow H_2O + SO_2(g)$$

The ionic equation for the complete reaction, therefore, is

$$2Na^+ + SO_3^{2-} + 2H^+ + 2Cl^- \rightarrow H_2O + SO_2(g) + 2Na^+ + 2Cl^-$$

Table 11.2 Rules for the formation of some common gases by metathesis reactions

GAS	
H_2S	Any sulfide (salt of S^{2-}) and any acid form $H_2S(g)$ and a salt.
CO_2	Any carbonate (salt of CO_3^{2-}) and any acid form $CO_2(g)$, H_2O, and a salt.
SO_2	Any sulfite (salt of SO_3^{2-}) and any acid form $SO_2(g)$, H_2O, and a salt.
NH_3	Any ammonium salt (salt of NH_4^+) and any soluble strong hydroxide form $NH_3(g)$, H_2O, and a salt.

A typical reaction of a carbonate and an acid is

$$2Na^+ + CO_3^{2-} + 2H^+ + 2Cl^- \rightarrow H_2O + CO_2(g) + 2Na^+ + 2Cl^-$$

Ammonium salts and soluble strong bases react in the following way.

$$NH_4^+ + Cl^- + Na^+ + OH^- \rightarrow H_2O + NH_3(g) + Na^+ + Cl^-$$

3. **Weak electrolytes.** Electrolytes are discussed in Section 10.10. A list of weak and strong acids and bases appears in Table 8.3 found in Section 8.4. A simplified list of rules follows.
   (a) **Acids.** Most common acids are weak electrolytes except $HClO_4$, $HClO_3$, HCl, HBr, HI, HNO_3, and H_2SO_4 (first ionization only).
   (b) **Bases.** The hydroxides of the I A elements and of Ca^{2+}, Sr^{2+}, and Ba^{2+} are soluble and strong. Most others are insoluble and weak.
   (c) **Salts.** Most common salts are strong electrolytes.

Example 11.5 Write balanced chemical equations for the reactions that occur when aqueous solutions of the following pairs of compounds are mixed. Show all substances (both reactants and products) in proper form.

(a) $FeCl_3$ and $(NH_4)_3PO_4$
(b) Na_2SO_4 and $CuCl_2$
(c) $ZnSO_4$ and $Ba(OH)_2$
(d) $Fe(C_2H_3O_2)_2$ and HBr
(e) $CaCO_3$ and HNO_3

Solution

(a) $Fe^{3+} + 3Cl^- + 3NH_4^+ + PO_4^{3-} \rightarrow FePO_4(s) + 3NH_4^+ + 3Cl^-$
(b) $2Na^+ + SO_4^{2-} + Cu^{2+} + 2Cl^- \rightarrow$ N. R.
(c) $Zn^{2+} + SO_4^{2-} + Ba^{2+} + 2OH^- \rightarrow Zn(OH)_2(s) + BaSO_4(s)$
(d) $Fe^{2+} + 2C_2H_3O_2^- + 2H^+ + 2Br^- \rightarrow 2HC_2H_3O_2 + Fe^{2+} + 2Br^-$
(e) $CaCO_3(s) + 2H^+ + 2NO_3^- \rightarrow H_2O + CO_2(g) + Ca^{2+} + 2NO_3^-$

Metathesis reactions are reversible to some extent. Aqueous equilibrium systems of this type are discussed in Chapter 15.

Metathesis reactions also occur in the absence of water. Examples are,

$$CaF_2(s) + H_2SO_4(l) \rightarrow CaSO_4(s) + 2HF(g)$$
$$2NaNO_3(s) + H_2SO_4(l) \rightarrow Na_2SO_4(s) + 2HNO_3(g)$$

The reactants in these reactions are heated, and the acids (HF and HNO_3, which are soluble in water) are driven off as gases.

CLEOPATRA'S PEARLS

In his book *Natural History* (written in 77), the Roman author Pliny the Elder says that Cleopatra would drop a pearl into a goblet of wine before she drank the wine. Cleopatra's cocktail may seem strange, but it is based on sound chemical principles.

Wine is produced by the fermentation of grape juice or other fruit juices. In the fermentation, glucose, a sugar found in fruit juice, is converted into ethyl alcohol by the action of an enzyme of yeast.

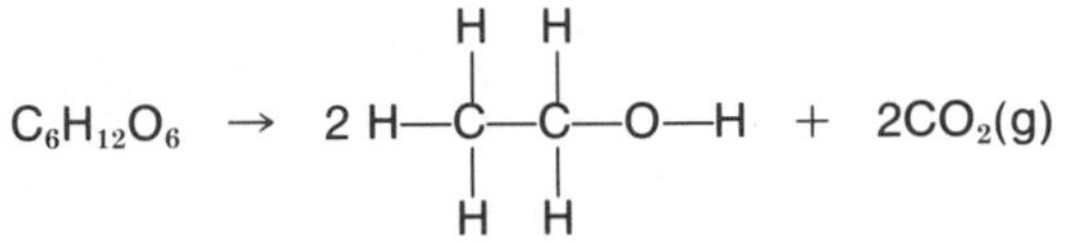

glucose ethyl alcohol

Wine can be changed into vinegar by a second fermentation process. In the presence of oxygen from air and a bacterial enzyme (from a microorganism that is sometimes called *mother of vinegar*), ethyl alcohol is converted into acetic acid.

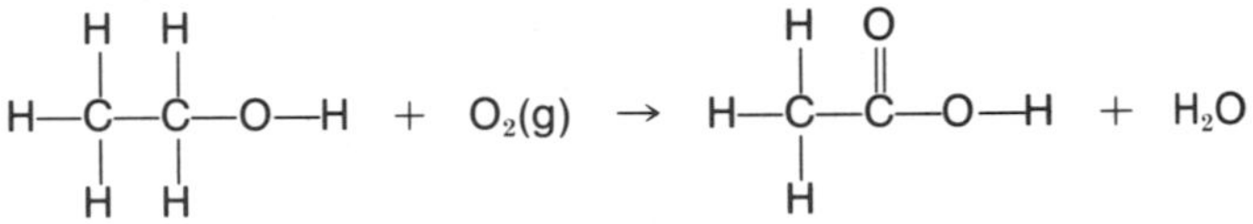

ethyl alcohol acetic acid

A careful wine maker will take pains to exclude air from a vessel in which grape juice is fermenting. In this way, the fermentation that produces acetic acid is prevented. Even a small amount of acetic acid can make wine taste sour and vinegary.

Natural pearls are composed principally of calcium carbonate ($CaCO_3$). When Cleopatra added a pearl to wine that contained acetic acid, the $CaCO_3$ neutralized the acid. If we use the formula $HC_2H_3O_2$ for acetic acid,

$$CaCO_2(s) + 2HC_2H_3O_2 \rightarrow Ca^{2+} + 2C_2H_3O_2^- + CO_2(g) + H_2O$$

The treated wine probably did taste better.

The reaction, however, dissolved away a part of the pearl. Instead of a pearl, Cleopatra could have used a chunk of limestone or a piece of marble, both of which are forms of calcium carbonate. Adding limestone, however, wouldn't have been as regal a gesture.

PROBLEMS

11.1 For each of the following reactions, identify the substance oxidized, the substance reduced, the oxidizing agent, and the reducing agent.
(a) $Zn + Cl_2 \rightarrow ZnCl_2$
(b) $2ReCl_5 + SbCl_3 \rightarrow 2ReCl_4 + SbCl_5$
(c) $Mg + CuCl_2 \rightarrow MgCl_2 + Cu$
(d) $2NO + O_2 \rightarrow 2NO_2$
(e) $WO_3 + 3H_2 \rightarrow W + 3H_2O$

11.2 For each of the following reactions, identify the substance oxidized, the substance reduced, the oxidizing agent, and the reducing agent.
(a) $Cl_2 + 2NaBr \rightarrow 2NaCl + Br_2$
(b) $Zn + 2HCl \rightarrow ZnCl_2 + H_2$
(c) $Fe_2O_3 + Al \rightarrow Al_2O_3 + 2Fe$
(d) $OF_2 + H_2O \rightarrow O_2 + 2HF$
(e) $2HgO \rightarrow 2Hg + O_2$

11.3 Balance the following by the change-in-oxidation-number method.
(a) $H_2O + MnO_4^- + ClO_2^- \rightarrow MnO_2 + ClO_4^- + OH^-$
(b) $H^+ + Cr_2O_7^{-2} + H_2S \rightarrow Cr^{3+} + S + H_2O$
(c) $H_2O + P_4 + HOCl \rightarrow H_3PO_4 + Cl^- + H^+$
(d) $Cu + H^+ + NO_3^- \rightarrow Cu^{2+} + NO + H_2O$
(e) $PbO_2 + HI \rightarrow PbI_2 + I_2 + H_2O$

11.4 Balance the following by the change-in-oxidation-number method.
(a) $Sb + H^+ + NO_3^- \rightarrow Sb_4O_6 + NO + H_2O$
(b) $NaI + H_2SO_4 \rightarrow H_2S + I_2 + Na_2SO_4 + H_2O$
(c) $IO_3^- + H_2O + SO_2 \rightarrow I_2 + SO_4^{2-} + H^+$
(d) $NF_3 + AlCl_3 \rightarrow N_2 + Cl_2 + AlF_3$
(e) $As_4O_6 + Cl_2 + H_2O \rightarrow H_3AsO_4 + HCl$

11.5 Balance the following by the change-in-oxidation-number method.
(a) $Fe^{2+} + H^+ + ClO_3^- \rightarrow Fe^{3+} + Cl^- + H_2O$
(b) $Pt + H^+ + NO_3^- + Cl^- \rightarrow PtCl_6^{2-} + NO + H_2O$
(c) $Cu + H^+ + SO_4^{2-} \rightarrow Cu^{2+} + SO_2 + H_2O$
(d) $Pb + PbO_2 + H^+ + SO_4^{2-} \rightarrow PbSO_4 + H_2O$
(e) $MnO_2 + HI \rightarrow MnI_2 + I_2 + H_2O$

11.6 Complete and balance the following equations by the ion–electron method. All reactions occur in acid solution.
(a) $AsH_3 + Ag^+ \rightarrow As_4O_6 + Ag$
(b) $Mn^{2+} + BiO_3^- \rightarrow MnO_4^- + Bi^{3+}$
(c) $NO + NO_3^- \rightarrow N_2O_4$
(d) $MnO_4^- + HCN + I^- \rightarrow Mn^{2+} + ICN$
(e) $Zn + H_2MoO_4 \rightarrow Zn^{2+} + Mo^{3+}$

11.7 Complete and balance the following equations by the ion–electron method. All reactions occur in acid solution.
(a) $S_2O_3^{2-} + IO_3^- + Cl^- \rightarrow SO_4^{2-} + ICl_2^-$
(b) $Se + BrO_3^- \rightarrow H_2SeO_3 + Br^-$
(c) $H_3AsO_3 + MnO_4^- \rightarrow H_3AsO_4 + Mn^{2+}$
(d) $H_5IO_6 + I^- \rightarrow I_2$
(e) $Pb_3O_4 \rightarrow Pb^{2+} + PbO_2$

11.8 Complete and balance the following equations by the ion–electron method. All reactions occur in acid solution.
(a) $ClO_3^- + I^- \rightarrow Cl^- + I_2$
(b) $Zn + NO_3^- \rightarrow Zn^{2+} + NH_4^+$
(c) $H_3AsO_3 + BrO_3^- \rightarrow H_3AsO_4 + Br^-$
(d) $H_2SeO_3 + H_2S \rightarrow Se + HSO_4^-$

(e) $ReO_2 + Cl_2 \rightarrow HReO_4 + Cl^-$

11.9 Complete and balance the following equations by the ion–electron method. All reactions occur in acid solution.

(a) $Fe^{2+} + Cr_2O_7^{2-} \rightarrow Fe^{3+} + Cr^{3+}$
(b) $HNO_2 + MnO_4^- \rightarrow NO_3^- + Mn^{2+}$
(c) $As_2S_3 + ClO_3^- \rightarrow$
$H_3AsO_4 + S + Cl^-$
(d) $IO_3^- + N_2H_4 \rightarrow I^- + N_2$
(e) $Cu + NO_3^- \rightarrow Cu^{2+} + NO$

11.10 Complete and balance the following equations by the ion–electron method. All reactions occur in acid solution.

(a) $P_4 + HOCl \rightarrow H_3PO_4 + Cl^-$
(b) $XeO_3 + I^- \rightarrow Xe + I_3^-$
(c) $UO^{2+} + Cr_2O_7^{2-} \rightarrow$
$UO_2^{2+} + Cr^{3+}$
(d) $H_2C_2O_4 + BrO_3^- \rightarrow CO_2 + Br^-$
(e) $Te + NO_3^- \rightarrow TeO_2 + NO$

11.11 Complete and balance the following equations by the ion–electron method. All reactions occur in acid solution.

(a) $Hg_5(IO_6)_2 + I^- \rightarrow HgI_4^{2-} + I_2$
(b) $MnO_4^- + Mn^{2+} + H_2P_2O_7^{2-} \rightarrow$
$Mn(H_2P_2O_7)_3^{3-}$
(c) $CS(NH_2)_2 + BrO_3^- \rightarrow$
$CO(NH_2)_2 + SO_4^{2} + Br$
(d) $Co(NO_2)_6^{3-} + MnO_4^- \rightarrow$
$Co^{2+} + NO_3^- + Mn^{2+}$
(e) $CNS^- + IO_3^- + Cl^- \rightarrow$
$CN^- + SO_4^{2-} + ICl_2^-$
(f) $CrI_3 + Cl_2 \rightarrow$
$CrO_4^{2-} + IO_3^- + Cl^-$

11.12 Complete and balance the following equations by the ion–electron method. All reactions occur in alkaline solution.

(a) $S^{2-} + I_2 \rightarrow SO_4^{2-} + I^-$
(b) $CN^- + MnO_4^- \rightarrow$
$CNO^- + MnO_2$
(c) $Au + CN^- + O_2 \rightarrow$
$Au(CN)_2^- + OH^-$
(d) $Si + OH^- \rightarrow SiO_3^{2-} + H_2$
(e) $Cr(OH)_3 + BrO^- \rightarrow$
$CrO_4^{2-} + Br^-$

11.13 Complete and balance the following equations by the ion–electron method. All reactions occur in alkaline solution.

(a) $Al + H_2O \rightarrow Al(OH)_4^- + H_2$
(b) $S_2O_3^{2-} + OCl^- \rightarrow SO_4^{2-} + Cl^-$
(c) $I_2 + Cl_2 \rightarrow H_3IO_6^{2-} + Cl^-$
(d) $Bi(OH)_3 + Sn(OH)_4^{2-} \rightarrow$
$Bi + Sn(OH)_6^{2-}$
(e) $NiO_2 + Fe \rightarrow$
$Ni(OH)_2 + Fe(OH)_3$

11.14 Complete and balance the following equations by the ion–electron method. All reactions occur in alkaline solution.

(a) $HClO_2 \rightarrow ClO_2 + Cl^-$
(b) $MnO_4^- + I^- \rightarrow MnO_4^{2-} + IO_4^-$
(c) $P_4 \rightarrow HPO_3^{2-} + PH_3$
(d) $SbH_3 + H_2O \rightarrow Sb(OH)_4^- + H_2$
(e) $CO(NH_2)_2 + OBr^- \rightarrow$
$CO_2 + N_2 + Br^-$

11.15 Complete and balance the following equations by the ion–electron method. All reactions occur in alkaline solution.

(a) $Mn(OH)_2 + O_2 \rightarrow Mn(OH)_3$
(b) $Cl_2 \rightarrow ClO_3^- + Cl^-$
(c) $HXeO_4^- \rightarrow XeO_6^{4-} + Xe + O_2$
(d) $As + OH^- \rightarrow AsO_3^{3-} + H_2$
(e) $S_2O_4^{2-} + O_2 \rightarrow SO_3^{2-} + OH^-$

11.16 Complete and balance the following equations by the ion–electron method. All reactions occur in alkaline solution.

(a) $Al + NO_3^- \rightarrow Al(OH)_4^- + NH_3$
(b) $Ni^{2+} + Br_2 \rightarrow NiO(OH) + Br^-$
(c) $S \rightarrow SO_3^{2-} + S^{2-}$

(d) $S_2O_3^{2-} + I_2 \rightarrow SO_4^{2-} + I^-$
(e) $S^{2-} + HO_2^- \rightarrow SO_4^{2-} + OH^-$

11.17 Because the oxygen of H_2O_2 can be either oxidized (to O_2) or reduced (to H_2O), hydrogen peroxide can function as a reducing agent or as an oxidizing agent. Write and balance equations (by the ion–electron method) to show the following reactions of H_2O_2. (a) the oxidization of PbS to $PbSO_4$ in acid solution, (b) the oxidation of $Cr(OH)_3$ to CrO_4^{2-} in alkaline solution, (c) the reduction of MnO_4^- to Mn^{2+} in acid solution, (d) the reduction of Ag_2O to Ag in alkaline solution.

11.18 What fraction of a mole is one equivalent weight of each of the following substances? (a) As_4O_6 in a reaction in which H_3AsO_4 is produced, (b) N_2H_4 in a reaction in which N_2 is produced, (c) $H_2S_2O_8$ in a reaction in which SO_4^{2-} is produced, (d) H_2SO_4 in a reaction in which H_2S is produced.

11.19 What fraction of a mole is one equivalent weight of each of the following substances? (a) KIO_3 in a reaction in which I_2 is produced, (b) KIO_3 in a reaction in which ICl_2^- is produced, ~~(c)~~ HIO_3 in a reaction in which H_5IO_6 is produced, (d) $Ca(C_2O_4)$ in a reaction in which CO_2 is produced.

11.20 The reaction of I_2O_5 and CO yields I_2 and CO_2. (a) What are the equivalent weights of I_2O_5 and CO for this reaction? (b) How many grams of CO are present in a gas sample that requires 1.00 g of I_2O_5 for complete reaction?

11.21 (a) What is the equivalent weight of $KBrO_3$ (in a transformation producing Br^-), of Se (in a transformation producing H_2SeO_3), and of Se (in a transformation producing H_2SeO_4)? (b) In a reaction between $KBrO_3$ and Se, 1.02 g of $KBrO_3$ reacts with 0.725 g of Se. What is the formula of the selenium-containing product of the reaction?

11.22 (a) What is the equivalent weight of H_2O_2 for a reaction in which H_2O is produced? (b) What is the equivalent weight of N_2H_4 for a reaction in which N_2 is produced? (c) How many grams of H_2O_2 are needed to react with 4.00 g of N_2H_4?

11.23 (a) What is the equivalent weight of $K_2Cr_2O_7$ for a reaction in which Cr^{3+} is produced? (b) $K_2Cr_2O_7$ oxidizes Fe^{2+} to Fe^{3+}. How many grams of Fe^{2+} will react with 1.000 g of $K_2Cr_2O_7$?

11.24 Ethylene glycol, $H_4C_2(OH)_2$, which is used as an antifreeze, is oxidized to formaldehyde, H_2CO, by periodic acid, HIO_4. The periodic acid is reduced to the IO_3^- ion. (a) Write the partial equations for the half reactions and determine the equivalent weights of $H_4C_2(OH)_2$ and HIO_4. (b) How many grams of ethylene glycol were present in a sample that required 7.68 g of HIO_4 for complete reaction?

11.25 The reaction of sodium tellurate, Na_2TeO_4, with hydrogen iodide, HI, yields Te and I_2. (a) What are the equivalent weights of Na_2TeO_4 and HI for this reaction? (b) What quantity of HI is needed to react with 5.00 g of Na_2TeO_4?

11.26 The reaction of AsH_3 and $AgNO_3$ yields As_4O_6 and Ag. (a) What are the equivalent weights of AsH_3 and $AgNO_3$ for this reaction? (b) How many grams of AsH_3 are needed to react with 5.00 g $AgNO_3$?

11.27 A 4.00 g sample of an ore of Bi_2S_3 was reacted with acid and a quantity of H_2S released equivalent to the amount of Bi_2S_3 present in the ore sample. The H_2S was reacted with I_2 yielding S and I^-. The reaction required 3.04 g of I_2. (a) What are the equivalent weights of H_2S and I_2 for the second reaction? (b) What quantity of H_2S is equivalent to the 3.04 g of I_2 required? (c) What percent of the ore is Bi_2S_3?

11.28 (a) Complete and balance the partial equation

$$[CO(NO_2)_6]^{3-} \rightarrow Co^{2+} + NO_3^-$$

(b) What is the equivalent weight of $K_3[Co(NO_2)_6]$ with regard to this transformation? (c) This change can be brought about by reaction with MnO_4^- (which goes to Mn^{2+}). What quantity of $K_3[Co(NO_2)_6]$ will react with 1.00 mole of $KMnO_4$?

11.29 (a) What is the normality of a 0.03 *M* $K_2Cr_2O_7$ solution for a reaction in which Cr^{3+} is produced? (b) What is the molarity of a 0.1 *N* $Ca(OCl)_2$ solution (calculated on the basis of a reaction in which Cl^- is produced)? (c) What is the normality of a 0.1 *M* solution of $Na_2S_2O_3$ for a reaction in which $S_4O_6^{2-}$ is produced? (d) What is the molarity of a 0.4 *N* solution of $KBrO_3$ (calculated on the basis of a reaction in which Br_2 is produced)? (e) What is the normality of a 0.8 *M* solution of H_2O_2 for a reaction in which O_2 is produced?

11.30 A 5.00 g sample of hemoglobin is treated in such a way as to produce small water-soluble molecules and ions by destroying the hemoglobin molecule. The iron in the aqueous solution that results from this procedure is reduced to Fe^{2+} and titrated against standard $KMnO_4$ solution. In the titration, Fe^{2+} is oxidized to Fe^{3+} and MnO_4^- is reduced to Mn^{2+}. The sample requires 30.5 ml of 0.0100 *N* $KMnO_4$. (a) What is the percent iron in hemoglobin? (b) The molecular weight of hemoglobin is 65,600. How many iron atoms are there in one hemoglobin molecule?

11.31 Iron wire containing 99.8% Fe is used to standardize a solution of $K_2Cr_2O_7$. A 0.185 g sample of iron wire is dissolved in acid and converted to the Fe^{2+} ion. This solution requires 29.3 ml of the dichromate solution for complete reaction. In the reaction, Fe^{2+} is oxidized to Fe^{3+} and $Cr_2O_7^{2-}$ is reduced to Cr^{3+}. (a) What is the normality of the $K_2Cr_2O_7$ solution? (b) What is the molarity of the $K_2Cr_2O_7$ solution?

11.32 Hydrazine (N_2H_4) reacts with the BrO_3^- ion in acid solution to produce $N_2(g)$ and Br^-. A 0.132 g sample of impure hydrazine requires 38.3 ml of 0.103 *N* $KBrO_3$ for complete reaction. What percent of the sample is hydrazine?

11.33 A 0.502 g sample of impure BaO_2 was dissolved in acid to produce an amount of H_2O_2 equivalent to the amount of BaO_2 present in the sample. The resulting solution was titrated against standard $KMnO_4$. In the titration the H_2O_2 was oxidized to O_2 and the MnO_4^- was reduced to Mn^{2+}. The titration required 33.3 ml of 0.130 *N* $KMnO_4$ solution. What percent of the material is BaO_2?

11.34 Write balanced chemical equations for the aqueous metathesis reactions that occur between the following pairs of compounds: (a) zinc sulfide and hydrochloric acid, (b) sodium carbonate and strontium acetate, (c) tin(II) chloride and ammonium sulfate, (d) magnesium nitrate and barium hydroxide, (e) sodium phosphate and hydrobromic acid.

11.35 Write balanced chemical equations for the aqueous metathesis reactions that occur between the following pairs of compounds: (a) silver nitrate and cadmium iodide, (b) zinc sulfite and hydrochloric acid, (c) aluminum chloride and lithium chlorate, (d) strontium hydroxide and nickel sulfate, (e) magnesium hydroxide and nitric acid.

11.36 Write balanced chemical equations for the aqueous metathesis reactions that occur between the following pairs of compounds: (a) lead(II) nitrate and hydrogen sulfide, (b) barium sulfide and zinc sulfate, (c) sodium phosphate and ammonium acetate, (d) mercury(I) carbonate and hydrochloric acid, (e) iron(III) hydroxide and phosphoric acid.

11.37 Write balanced chemical equations for the aqueous metathesis reactions that occur between the following pairs of compounds: (a) ammonium sulfate and calcium hydroxide, (b) manganese(II) chloride and cobalt(II) sulfate, (c) cadmium chlorate and potassium sulfide, (d) iron(III) carbonate and nitric acid, (e) lead(II) nitrate and magnesium sulfate.

Chapter 12

Acids and Bases

Throughout the history of chemistry various acid–base concepts have been proposed and used. In this chapter four concepts in current use are reviewed. Each of the definitions can be applied with advantage in appropriate circumstances. In a given situation chemists use the concept that best suits their purpose.

The earliest criteria for the characterization of acids and bases were the experimentally observed properties of aqueous solutions. An acid was defined as a substance that in water solution tastes sour, turns litmus red, neutralizes bases, and so on. A substance was a base if its aqueous solution tastes bitter, turns litmus blue, neutralizes acids, and so on. Concurrent with the development of generalizations concerning the structure of matter, scientists searched for a correlation between acidic and basic properties and the structure of compounds that exhibit these properties.

12.1 THE ARRHENIUS CONCEPT

Acids, bases, and salts were classified as electrolytes by Faraday in 1834. Justus von Liebig proposed (1838) that acids are compounds containing hydrogen that can be replaced by metals. The role of water as an ionizing solvent was emphasized by Svanté Arrhenius in his theory of electrolytic dissociation (1884), which led to the definition of acids and bases in terms of the ions of water. Arrhenius' views were vigorously championed and developed by Wilhelm Ostwald.

At the time of Arrhenius the distinction between ionic and covalent compounds was not clear, and the nature of $H^+(aq)$ was not understood. Consequently, the **water system** (or **Arrhenius concept**) has been modified throughout its history. In present-day terms an **acid** is a substance that ionizes in water solution to produce $H^+(aq)$, which is called the hydronium ion. A **base** is a substance that dissolves in water to produce hydroxide ion, $OH^-(aq)$. The strength of an acid is defined in terms of the concentration of $H^+(aq)$ that is present in a water solution of given concentration of acid. The strength of a base depends upon the relative concentration of $OH^-(aq)$ in an aqueous solution of base.

Neutralization, then, may be represented by the equation

$$H^+(aq) + OH^-(aq) \rightarrow H_2O \qquad \Delta H = -57.3 \text{ kJ}$$

As indicated in the preceding equation, the enthalpy for the neutralization of a dilute solution of any strong acid by a dilute solution of any strong base is a constant value. If a weak acid or a weak base is involved in the neutralization, or if a slightly soluble salt is formed from the cation derived from the base and the anion derived from the acid, the enthalpy of the reaction is different. In such cases the total enthalpy includes heat effects due to the ionization of the weak acid or base or due to the precipitation of a salt, as well as the enthalpy of neutralization.

The properties that all Arrhenius acids have in common are those of $H^+(aq)$. Chemical properties include such reactions as those of acids with reactive metals and with the bicarbonate ion.

$$2H^+(aq) + Zn(s) \rightarrow H_2(g) + Zn^{2+}(aq)$$
$$H^+(aq) + HCO_3^-(aq) \rightarrow H_2O + CO_2(g)$$

The properties that are characteristic of Arrhenius bases are due to the $OH^-(aq)$ ion. Among the reactions of strong bases are

$$2OH^-(aq) + H_2O + Si(s) \rightarrow SiO_3^{2-}(aq) + 2H_2(g)$$
$$OH^-(aq) + NH_4^+(aq) \rightarrow H_2O + NH_3(g)$$

The oxides of many nonmetals react with water to form acids and are called **acidic oxides** or **acid anhydrides.**

$$N_2O_5(s) + H_2O \rightarrow 2H^+(aq) + 2NO_3^-(aq)$$

Many oxides of metals dissolve in water to form hydroxides. Such compounds are called **basic oxides.**

$$Na_2O(s) + H_2O \rightarrow 2Na^+(aq) + 2OH^-(aq)$$

Acidic oxides and basic oxides react to produce salts in the absence of water. It must be noted, however, that not all acids and bases may be derived from oxides (for example, HCl and NH_3).

The Arrhenius concept is severely limited by its emphasis on water and reactions in aqueous solution. Later definitions are more general, serve to correlate more reactions, and are applicable to reactions in nonaqueous media.

12.2 THE SOLVENT SYSTEM CONCEPT

The principles of the Arrhenius-water concept can be used to devise acid–base schemes for many solvents. In a **solvent system** an **acid** is a substance that gives the cation characteristic of the solvent, and a **base** is a substance that gives the anion characteristic of the solvent. Thus, the reaction of an acid and a base, a **neutralization,** yields the solvent as one of its products. Many solvent systems of acids and bases have been developed (Table 12.1). The water concept is but a single example of a solvent system.

The ammonia system has been investigated more extensively than any other with the exception of the water system. The properties of liquid ammonia (boiling point, −33.4°C) are strikingly similar to those of water. Liquid ammonia is associated through hydrogen bonding, and the NH_3 molecule is polar. Hence, liquid ammonia is an excellent solvent for ionic and polar compounds, and it functions as an ionizing solvent for electrolytes. Many compounds form ammoniates, which are analogous to hydrates (for example, $BaBr_2 \cdot 8NH_3$ and $CaCl_2 \cdot 6NH_3$), and ions are solvated in liquid ammonia solu-

Table 12.1 Some solvent systems

SOLVENT	ACID ION	BASE ION	TYPICAL ACID	TYPICAL BASE
H_2O	H_3O^+ ($H^+ \cdot H_2O$)	OH^-	HCl	NaOH
NH_3	NH_4^+ ($H^+ \cdot NH_3$)	NH_2^-	NH_4Cl	$NaNH_2$
NH_2OH	NH_3OH^+ ($H^+ \cdot NH_2OH$)	$NHOH^-$	$NH_2OH \cdot HCl$ (NH_3OH^+, Cl^-)	K(NHOH)
$HC_2H_3O_2$	$H_2C_2H_3O_2^+$ ($H^+ \cdot HC_2H_3O_2$)	$C_2H_3O_2^-$	HCl	$NaC_2H_3O_2$
SO_2	SO^{2+}	SO_3^{2-}	$SOCl_2$	Cs_2SO_3
N_2O_4	NO^+	NO_3^-	NOCl	$AgNO_3$
$COCl_2$	$COCl^+$	Cl^-	$(COCl)AlCl_4$	$CaCl_2$

tions [for example, $Ag(NH_3)_2^+$ and $Cr(NH_3)_6^{3+}$]. Whereas solutions of electrolytes in ammonia are good conductors of electricity, pure liquid ammonia, like water, has a relatively low conductance.

The autoionization of ammonia

$$2NH_3 \rightleftharpoons NH_4^+ + NH_2^-$$

which occurs only to a low degree, is responsible for the electrical conductivity of pure solvent, just as the autoionization of water

$$2H_2O \rightleftharpoons H_3O^+ + OH^-$$

is responsible for the electrical properties of this compound.

Any compound that produces ammonium ion, NH_4^+, in liquid ammonia solution is an acid, and any compound that yields amide ion, NH_2^-, is a base. Thus, the neutralization reaction is the reverse of the autoionization reaction.

$$NH_4^+ + NH_2^- \rightarrow 2NH_3$$

Indicators may be used to follow an acid-base reaction in liquid ammonia. For example, phenolphthalein is red in a liquid ammonia solution of potassium amide, KNH_2, and becomes colorless after a stoichiometrically equivalent amount of ammonium chloride has been added.

In addition to the neutralization reaction, the ammonium ion in liquid ammonia undergoes other reactions analogous to the reactions of the hydronium ion in water. For example, metals such as sodium react with the ammonium ion to liberate hydrogen.

$$2Na(s) + 2NH_4^+ \rightarrow 2Na^+ + H_2(g) + 2NH_3(l)$$

The reactions of the amide ion are analogous to those of the hydroxide ion.

$$Zn(OH)_2(s) + 2OH^- \rightarrow Zn(OH)_4^{2-}$$
$$Zn(NH_2)_2(s) + 2NH_2^- \rightarrow Zn(NH_2)_4^{2-}$$
$$Hg^{2+} + 2OH^- \rightarrow HgO(s) + H_2O$$
$$3Hg^{2+} + 6NH_2^- \rightarrow Hg_3N_2(s) + 4NH_3$$

Many properties and reactions of compounds belonging to the ammonia system have been predicted and correlated by comparison to the better known chemistry of the compounds of the water system. In fact, the study of various solvent systems has been responsible for greatly increasing our knowledge of the reactions that occur in solvents other than water. The solvent system approach, however, it compartmentalized and does not offer extensive correlation or generalization.

12.3 THE BRØNSTED-LOWRY CONCEPT

In 1923 Johannes Brønsted and Thomas Lowry independently proposed a broader concept of acids and bases. According to the Brønsted–Lowry definitions, an **acid** is a substance that can donate protons, and a **base** is a substance that can accept protons. In these terms, the reaction of an acid with a base constitutes a transfer of a proton from the acid to the base; this is the only characteristic reaction of these substances treated by the Brønsted–Lowry concept. We shall see, however, that many reactions are included in this classification.

The dissolution of ammonia in water may be represented by the equation

$$H_2O + NH_3 \rightleftharpoons NH_4^+ + OH^-$$

In the reaction as written, H_2O is serving as an acid and is releasing a proton to the base, NH_3. Although the solution resulting from the addition of NH_3 to H_2O is alkaline, conductivity and *i* factor measurements show that the ionization is incomplete. Furthermore, if a solution of an ammonium salt is made strongly alkaline, ammonia gas is released. The reaction, therefore, is reversible, and the system exists in equilibrium.

According to the Brønsted–Lowry concept, acids and bases may be molecules or ions. In the reverse reaction, NH_4^+ is functioning as an acid and OH^- is functioning as a base since the equation, read from right to left, shows NH_4^+ releasing a proton to OH^-. It follows, then, that in this Brønsted acid–base reaction, two acids (H_2O and NH_4^+) and two bases (OH^- and NH_3) are involved. The reaction is actually a competition between the two bases for a proton.

The base NH_3 gains a proton and thereby forms the acid NH_4^+. The acid NH_4^+ upon the loss of a proton forms the base NH_3. Such an acid–base pair, which is related through the loss or gain of a proton, is called a **conjugate pair;** NH_4^+ is the conjugate acid of the base NH_3, and NH_3 is the conjugate base of the acid NH_4^+. In like manner, the acid H_2O and the base OH^- constitute a second conjugate pair in the preceding reaction. We may indicate conjugate relationships by the use of subscripts in the following manner:

$$\underset{\text{Acid}_1}{H_2O} + \underset{\text{Base}_2}{NH_3} \rightleftharpoons \underset{\text{Acid}_2}{NH_4^+} + \underset{\text{base}_1}{OH^-}$$

There are many molecules and ions that can function as acids in certain reactions and as bases in other reactions. Such species are called **amphiprotic.** For example, in the reaction with ammonia, water acts as an acid (conjugate base, OH^-). In the reaction with acetic acid, water is a base (conjugate acid, H_3O^+).

$$\underset{\text{Acid}_1}{HC_2H_3O_2} + \underset{\text{Base}_2}{H_2O} \rightleftharpoons \underset{\text{Acid}_2}{H_3O^+} + \underset{\text{Base}_1}{C_2H_3O_2^-}$$

In like manner, NH_3 acts as a base (conjugate acid, NH_4^+) in its reaction with water. In the reaction of the hydride ion, H^-, with liquid ammonia, NH_3 acts as an acid (conjugate base, NH_2^-).

Table 12.2 Some amphiprotic substances

AMPHIPROTIC SUBSTANCE	TYPICAL REACTION ACID$_1$	BASE$_2$	ACID$_2$	BASE$_1$
H_2O	H_2O	$+ H^-$	$\rightleftharpoons H_2$	$+ OH^-$
	HCN	$+ H_2O$	$\rightleftharpoons H_3O^+$	$+ CN^-$
NH_3	H_2O	$+ NH_2^-$	$\rightleftharpoons NH_3$	$+ OH^-$
	HCl	$+ NH_3$	$\rightleftharpoons NH_4^+$	$+ Cl^-$
H_2SO_4	H_2SO_4	$+ H_2O$	$\rightleftharpoons H_3O^+$	$+ HSO_4^-$
	$HClO_4$	$+ H_2SO_4$	$\rightleftharpoons H_3SO_4^+$	$+ ClO_4^-$
HSO_4^-	H_2SO_4	$+ H_2O$	$\rightleftharpoons H_3O^+$	$+ HSO_4^-$
	HSO_4^-	$+ H_2O$	$\rightleftharpoons H_3O^+$	$+ SO_4^{2-}$

$$\underset{\text{Acid}_1}{NH_3} + \underset{\text{Base}_2}{H^-} \rightarrow \underset{\text{Acid}_2}{H_2} + \underset{\text{Base}_1}{NH_2^-}$$

Several amphiprotic substances are listed in Table 12.2.

The neutralization reactions of the Arrhenius system and certain solvent systems may therefore be interpreted in terms of the Brønsted definitions. Such neutralizations are merely acid–base reactions between the conjugate acid and the conjugate base of an amphiprotic solvent. Thus,

$$\underset{\text{Acid}_1}{H_3O^+} + \underset{\text{Base}_2}{OH^-} \rightleftharpoons \underset{\text{Acid}_2}{H_2O} + \underset{\text{Base}_1}{H_2O}$$

$$NH_4^+ + NH_2^- \rightleftharpoons NH_3 + NH_3$$

12.4 STRENGTHS OF BRØNSTED ACIDS AND BASES

In Brønsted terms the strength of an acid is determined by its tendency to donate protons, and the strength of a base is dependent upon its tendency to receive protons. The reaction

$$\underset{\text{Acid}_1}{HCl} + \underset{\text{Base}_2}{H_2O} \rightleftharpoons \underset{\text{Acid}_2}{H_3O^+} + \underset{\text{Base}_1}{Cl^-}$$

proceeds virtually to completion (from left to right). We must conclude, therefore, that HCl is a stronger acid than H_3O^+, since it has the stronger tendency to lose protons and the equilibrium is displaced far to the right. In addition, it is apparent that H_2O is a stronger base than Cl^-, since, in the competition for protons, water molecules succeed in holding practically all of them. The strong acid, HCl, has a weak conjugate base, Cl^-.

A strong acid, which has a great tendency to lose protons, is necessarily conjugate to a weak base, which has a small tendency to gain and hold protons. Hence, the stronger the acid, the weaker its conjugate base. In like manner, a strong base attracts protons strongly and is necessarily conjugate to a weak acid, one that does not readily lose protons. The stronger the base, the weaker its conjugate acid.

Acetic acid in 1.0 *M* solution is 0.42% ionized at 25°C. The equilibrium

$$\underset{\text{Acid}_1}{HC_2H_3O_2} + \underset{\text{Base}_2}{H_2O} \rightleftharpoons \underset{\text{Acid}_2}{H_3O^+} + \underset{\text{Base}_1}{C_2H_3O_2^-}$$

is displaced to the left. This equation may be said to represent a competition between bases (acetate ions and water molecules) for protons. The position of the equilibrium shows that the $C_2H_3O_2^-$ ion is a stronger base than H_2O. At equilibrium more protons form $HC_2H_3O_2$ molecules than form H_3O^+ ions. We may also conclude that H_3O^+ is a stronger acid than $HC_2H_3O_2$. At equilibrium more H_3O^+ ions than $HC_2H_3O_2$ molecules have lost protons. In the preceding example we note again that the stronger acid, H_3O^+, is conjugate to the weaker base, H_2O, and the stronger base, $C_2H_3O_2^-$, is conjugate to the weaker acid, $HC_2H_3O_2$.

One further conclusion should be stated. In a given reaction, the position of equilibrium favors the formation of the weaker acid and the weaker base. Thus, in the reaction between HCl and H_2O, the equilibrium concentrations of H_3O^+ and Cl^- (the *weaker* acid and base, respectively) are *high,* whereas in the solution of acetic acid, the equilibrium concentrations of H_3O^+ and $C_2H_3O_2^-$ (the *stronger* acid and base, respectively) are *low.*

Notice that the Brønsted concept is an extension of the Arrhenius concept. According to the Arrhenius scheme, strong acids, such as HCl, are strong electrolytes—they are virtually 100% ionized in water solution and produce solutions with high concentrations of H_3O^+. According to the Arrhenius concept, weak acids, such as acetic acid, are weak electrolytes—they are incompletely ionized in water solution and produce solutions with relatively low concentrations of H_3O^+. Although the Brønsted system classifies many more substances as acids, compounds that qualify as acids according to the Arrhenius definition are also acids in the Brønsted concept; that is, they are proton donors.

In the Brønsted system, acids are classified according to their ability to donate protons to the specific base under consideration. If water is used as a reference base, the acid strengths of the Arrhenius-water concept may be explained in terms of the Brønsted concept. Thus, strong Arrhenius acids are those compounds that are stronger acids than H_3O^+, and weak Arrhenius acids are those compounds that are weaker acids than H_3O^+.

It is apparent that acid strengths are influenced by the solvent water. Acids that are stronger than H_3O^+ are essentially completely ionized in water solution.

$$HClO_4 + H_2O \rightleftharpoons H_3O^+ + ClO_4^-$$
$$HCl + H_2O \rightleftharpoons H_3O^+ + Cl^-$$
$$HNO_3 + H_2O \rightleftharpoons H_3O^+ + NO_3^-$$

Aqueous solutions of $HClO_4$, HCl, and HNO_3 of the same concentration appear to be of the same acid strength. The acid properties of the solutions are due to the H_3O^+ ion, which the compounds produce to an equivalent extent in their reactions with water. Water is said to have a **leveling effect** on acids stronger than H_3O^+. The strongest acid that can exist in water solution is the conjugate acid of water, H_3O^+. Acids that are weaker than H_3O^+ are not leveled by water. Thus, $HC_2H_3O_2$, H_3PO_4, HNO_2, H_2S, and other weak acids show a wide variation in their degree of ionization—the extent to which they form H_3O^+ in their reactions with water.

The leveling effect is also observed for solvents other than water. The strongest acid in liquid ammonia solutions is the conjugate acid of ammonia, NH_4^+. Acetic acid in liquid ammonia solution is essentially completely ionized, since $HC_2H_3O_2$ is a stronger acid than NH_4^+.

$$HC_2H_3O_2 + NH_3 \rightleftharpoons NH_4^+ + C_2H_3O_2^-$$

Nitric acid in methanol (CH_3OH) solution is incompletely dissociated, since HNO_3 is a weaker acid than the conjugate acid of methanol, $CH_3OH_2^+$.

$$HNO_3 + CH_3OH \rightleftharpoons CH_3OH_2^+ + NO_3^-$$

Solvents that can function as acids exert a leveling effect on bases. The strongest base capable of existing in water solution is the conjugate base of water, OH^-. Many substances, such as NH_2^- and H^-, are stronger bases than OH^-. However, in water solution these strongly basic substances accept protons from water to form OH^- ions. These reactions are essentially complete. The apparent basicity of strongly basic materials in water solution is reduced to the level of the OH^- ion.

$$H_2O + NH_2^- \rightleftharpoons NH_3 + OH^-$$
$$H_2O + H^- \rightleftharpoons H_2 + OH^-$$

Substances, such as ammonia, that are less basic than OH^- are not leveled by water and show varying degrees of ionization in aqueous solution.

Other solvents, in addition to water, level bases. Since H^- and NH_2^- appear equally strong in water solution, we must examine their reactions in other solvents to compare their relative basicities. It would be surprising if both ions had exactly the same basic character. In liquid ammonia the hydride ion reacts rapidly and essentially completely to form hydrogen and the amide ion.

$$NH_3 + H^- \rightleftharpoons H_2 + NH_2^-$$

We conclude that the hydride ion is a stronger base than the amide ion. Liquid ammonia reduces H^- to the level of the conjugate base of ammonia, NH_2^-—the strongest base possible in liquid ammonia solution.

The Brønsted definition enlarges the Arrhenius concept even more for bases than it does for acids. In essence, the Arrhenius concept treats only one Brønsted base—the hydroxide ion. According to the Arrhenius concept, strong bases are completely ionic in water solution and form solutions with high concentrations of OH^- ion. In fact, a compound such as NaOH is ionic when pure. Molecules of NaOH do not exist. Solutions of weak bases, according to the

Arrhenius concept, are prepared from compounds that produce low concentrations of OH^- ion in water (such as NH_3).

The Brønsted system classifies many more substances as bases, and in this system it is the OH^- ion of NaOH that is the base, not the compound itself. From the Arrhenius view, solutions resulting from the addition of such compounds as NaH or Na_2O to water owe their basic character to the fact that they are merely solutions of NaOH after the reaction of H^- or O^{2-} with water.

$$H_2O + H^- \rightarrow H_2 + OH^-$$
$$H_2O + O^{2-} \rightarrow OH^- + OH^-$$

12.5 HYDROLYSIS

The reaction of an ion with water in which either H_3O^+ or OH^- is produced is called **hydrolysis.** Such reactions are Brønsted acid–base reactions. Anions that function as bases in water solution (B^- in the equation that follows) are hydrolyzed.

$$\underset{\text{Acid}_1}{H_2O} + \underset{\text{Base}_2}{B^-} \rightleftharpoons \underset{\text{Acid}_2}{HB} + \underset{\text{Base}_1}{OH^-}$$

In the reaction water acts as an acid, and a proton is transferred to the anion. The extent of hydrolysis of a given anion depends upon the base strength of the anion. An indication of the degree of hydrolysis is given by the concentration of OH^- relative to the concentration of anion present in the solution. Anions that are extensively hydrolyzed are strong bases with weak conjugate acids.

Hydrolysis is complete in the reaction

$$H_2O + H^- \rightarrow H_2 + OH^-$$

The anion undergoing hydrolysis, H^-, is a stronger base than OH^-, and the conjugate acid of the anion, H_2, is a much weaker acid than H_2O. Complete hydrolysis is characteristic of such anions. This equation, in fact, illustrates the leveling effect of water on bases stronger than OH^-.

At the other extreme, some anions, such as Cl^-, do not hydrolyze at all.

$$H_2O + Cl^- \leftarrow HCl + OH^-$$

The Cl^- ion is a very weak base, and its conjugate acid, HCl, is not only stronger than H_2O but is also stronger than H_3O^+. In fact, HCl is leveled in water solution to H_3O^+. Therefore, HCl molecules do not exist in measurable quantities in aqueous solutions of moderate concentrations. The chloride ion is simply too weak a base to accept a proton from water to a significant extent. Anions that have conjugate acids stronger than H_3O^+ do not hydrolyze.

The behavior of many anions in water solution is intermediate between the two extremes represented by H^- and Cl^-. For example,

$$H_2O + C_2H_3O_2^- \rightleftharpoons HC_2H_3O_2 + OH^-$$

The $C_2H_3O_2^-$ ion is a weaker base than OH^- but not so weak as Cl^-. Even though $HC_2H_3O_2$, the conjugate acid of $C_2H_3O_2^-$, is a stronger acid than water, it is a weaker acid than HCl or H_3O^+. Acetic acid is not leveled by water. Molecules of the weak acid $HC_2H_3O_2$ can exist in aqueous solution. The hydrolysis represented by the equation occurs only to a slight extent. The equilibrium, as written, is displaced to the left because the substances written on the left are the weaker acid and the weaker base of the reaction. Nevertheless, in a solution of sodium acetate there is a measurable concentration of OH^- from the hydrolysis of the acetate ion—enough to cause red litmus to turn blue. Anions, such as $C_2H_3O_2^-$, CN^-, NO_2^-, and S^{2-}, that have conjugate acids stronger than water but weaker than H_3O^+ are measurably, but not completely, hydrolyzed in water solution.

There are some hydrogen-containing cations, of which the ammonium ion is the principal example, that hydrolyze to produce H_3O^+.

$$NH_4^+ + H_2O \rightleftharpoons H_3O^+ + NH_3$$

Ammonia is a stronger base than water, and therefore the position of equilibrium is to the left. Water, however, is a sufficiently strong base to gain some of the protons in the competition. Although the hydrolysis is far from complete, a solution of NH_4Cl gives an acid reaction toward litmus because of the hydrolysis of the NH_4^+ ion.

There are a few relatively unimportant cations that hydrolyze completely. For example, the phosphonium ion, PH_4^+, is a stronger acid than the hydronium ion. In water solution the phosphonium ion is leveled.

$$PH_4^+ + H_2O \rightarrow H_3O^+ + PH_3$$

Metal cations are hydrated in water solution, and some of these hydrated cations hydrolyze. For example, solutions of aluminum chloride, zinc chloride, and iron(III) chloride show acidic characteristics because of the hydrolysis of the cations. The hydrolysis of the zinc ion [which we shall indicate as $Zn(H_2O)_4^{2+}$] may be represented as

$$Zn(H_2O)_4^{2+} + H_2O \rightleftharpoons H_3O^+ + Zn(H_2O)_3(OH)^+$$

In the hydrolysis, $Zn(H_2O)_4^{2+}$ functions as an acid and donates a proton to H_2O, which acts as a base. A water molecule coordinated to the Zn^{2+} ion has an enhanced acidic character because the O—H bonds of

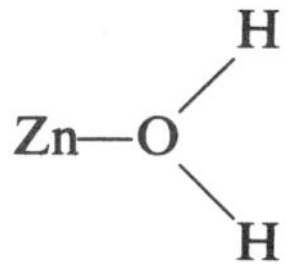

are weakened by the displacement of electrons toward the positively charged central ion. Thus, the coordinated water molecules release protons and act as Brønsted acids.

In general, small ions with high positive charges form hydrated species that act as Brønsted acids in hydrolysis reactions. The larger cations of the group I A and group II A metals have comparatively low charges for their relatively large sizes. The hydrated species of these ions have no significant acid character and consequently do not hydrolyze appreciably. The hydrolysis of ions is discussed in more detail in Section 15.12.

The amphoteric nature of some hydroxides [for example, $Zn(OH)_2$ and $Al(OH)_3$] may be explained by means of the Brønsted theory. These substances, which are normally insoluble in water, dissolve in both acidic solution and alkaline solution. In acid a typical reaction is

$$H_3O^+ + Zn(OH)_2(H_2O)_2(s) \rightleftharpoons Zn(OH)(H_2O)_3^+ + H_2O$$

A proton is transferred from H_3O^+ to the hydrated hydroxide, which acts as a base. A coordinated OH of the zinc complex accepts the proton and is transformed into a coordinated H_2O. The other coordinated OH groups can also accept protons, and the reaction continues until $Zn(H_2O)_4^{2+}$ is formed.

In alkaline solution the reaction is

$$Zn(OH)_2(H_2O)_2(s) + OH^- \rightleftharpoons H_2O + Zn(OH)_3(H_2O)^-$$

The hydrated zinc hydroxide acts as an acid, and a coordinated H_2O of the zinc compound loses a proton to the OH^- ion. The loss of a proton transforms this H_2O into an OH group, and the $Zn(OH)_3(H_2O)^-$ ion results. Whereas all hydroxides react with H_3O^+, only those hydroxides that are classed as amphoteric react with OH^-. Amphoterism is discussed further in Section 15.11.

12.6 THE LEWIS CONCEPT

In reality, the Brønsted concept enlarges the definition of a base much more than it does that of an acid. In the Brønsted system a base is a molecule or ion that has an unshared electron pair with which it can attract and hold a proton, and an acid is a substance that can supply a proton to a base. If a molecule or ion can share an electron pair with a proton, it can do the same thing with other substances as well.

Gilbert N. Lewis proposed a broader concept of acids and bases that liberated acid–base phenomena from the proton. Although Lewis first proposed his system in 1923, he did little to develop it until 1938. Lewis defined a **base** as a substance that has an unshared electron pair with which it can form a covalent bond with an atom, molecule, or ion. An **acid** is a substance that can form a covalent bond by accepting an electron pair from a base. The emphasis has been shifted by the Lewis concept from the proton to the electron pair and covalent-bond formation.

An example of an acid–base reaction that is not treated as such by any other acid–base concept is

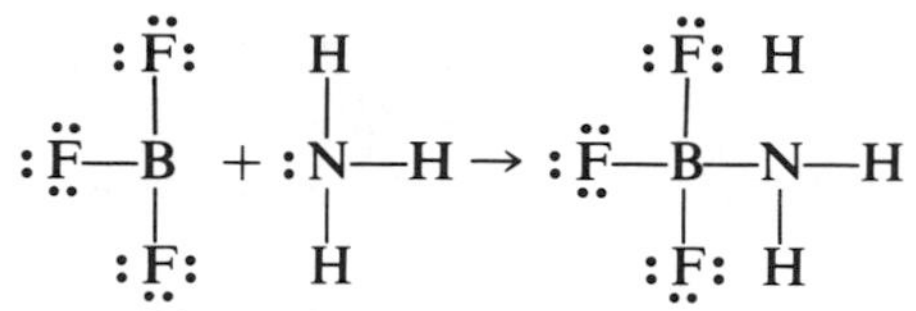

acid base

Many Lewis acids and bases of this type can be titrated against one another by the use of suitable indicators in the same way that Arrhenius acids and bases can be titrated.

Substances that are bases in the Brønsted system are also bases according to the Lewis concept. However, the Lewis definition of an acid considerably expands the number of substances that are classified as acids. A Lewis acid must have an empty orbital capable of receiving the electron pair of the base. The proton is but a single example of a Lewis acid.

Lewis acids include molecules or atoms that have incomplete octets.

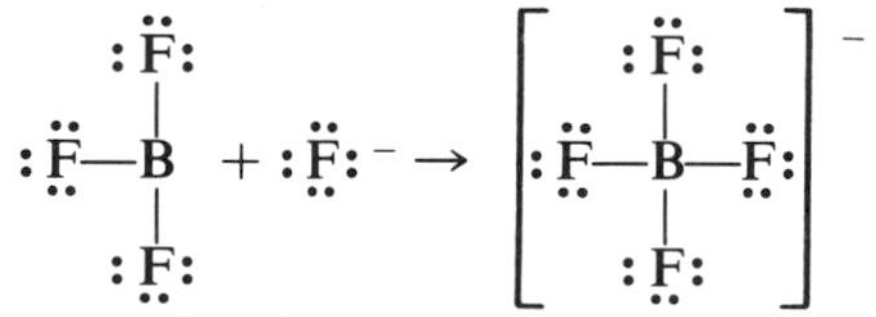

acid base

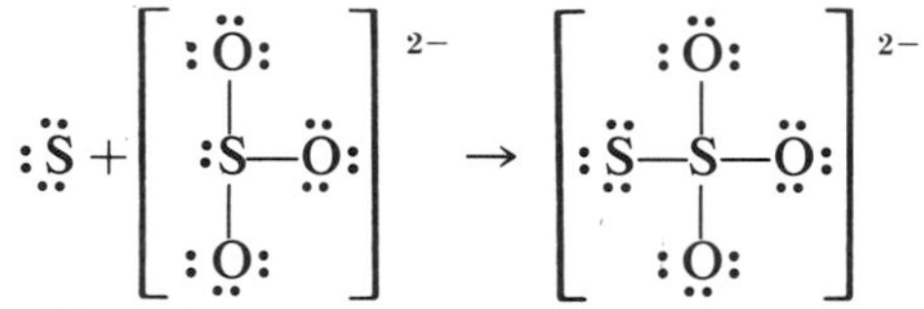

acid base

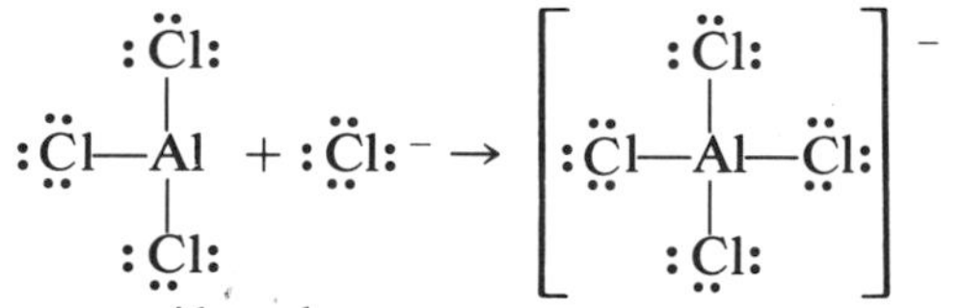

acid base

Aluminum chloride, although it reacts as $AlCl_3$, is actually a dimer—Al_2Cl_6. The formation of the dimer from the monomer may be regarded as a Lewis acid–base reaction in itself, since a chlorine atom in each $AlCl_3$ unit supplies an electron pair to the aluminum atom of the other $AlCl_3$ unit to complete the octet of the aluminum atom. These bonds are indicated by ··· in the diagram

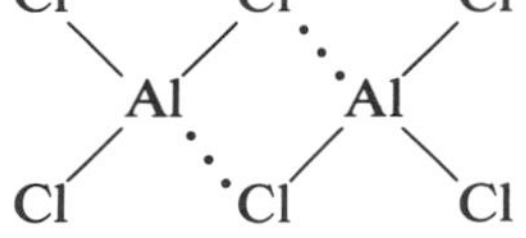

Many simple cations can function as Lewis acids. For example,

$$Cu^{2+} + 4{:}NH_3 \rightarrow Cu({:}NH_3)_4^{2+}$$
$$Fe^{3+} + 6{:}C{\equiv}N{:}^- \rightarrow Fe({:}C{\equiv}N{:})_6^{3-}$$

Some metal atoms can function as acids in the formation of compounds such as the carbonyls, which are produced by the reaction of the metal with carbon monoxide.

$$Ni + 4{:}C{\equiv}O{:} \rightarrow Ni({:}C{\equiv}O{:})_4$$

Compounds that have central atoms capable of expanding their valence shells are Lewis acids in reactions in which this expansion occurs. Examples are

$$SnCl_4 + 2Cl^- \rightarrow SnCl_6^{2-}$$
$$SiF_4 + 2F^- \rightarrow SiF_6^{2-}$$
$$PF_5 + F^- \rightarrow PF_6^-$$

In each of the first two reactions, the valence shell of the central atom (Sn and Si) is expanded from eight to twelve electrons, and in the third reaction the valence shell of P goes from ten to twelve electrons.

In addition, some compounds have an acidic site because of one or more multiple bonds in the molecule. Examples are

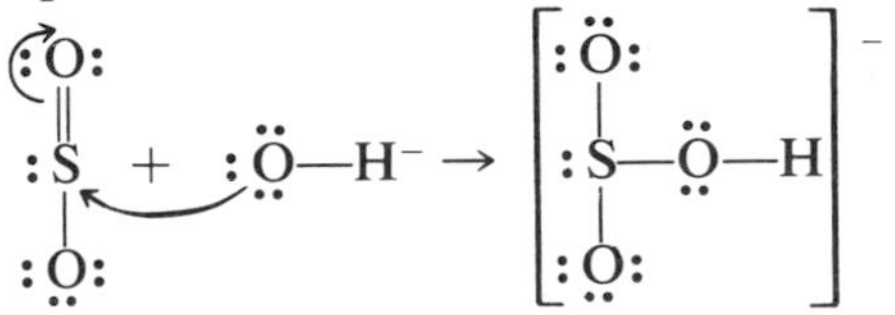

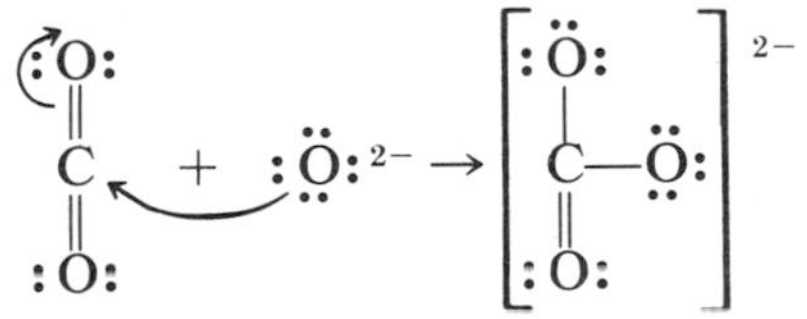

The reactions of silica, SiO_2, with metal oxides are analogous to the reaction of carbon dioxide with the oxide ion, although both silica and the silicate products (compounds of SiO_3^{2-}) are polymeric. This reaction is important in high-temperature metallurgical processes in which a basic oxide is added to an ore to remove silica in the form of silicates (slag). Many of the processes used in the manufacture of glass, cement, and ceramics involve the reaction of the base O^{2-} (from metal oxides, carbonates, and so forth) with acid oxides (such as SiO_2, Al_2O_3, and B_2O_3).

Arrhenius and Brønsted acid–base reactions may be interpreted in Lewis terms by focusing attention on the proton, as a Lewis acid

$$H^+(aq) + OH^-(aq) \rightarrow H_2O$$

in which case the Brønsted acid, H_3O^+, is termed a **secondary** Lewis acid since it serves to provide the **primary** Lewis acid, the proton. Probably a better interpretation is that which classifies Brønsted acid–base reactions as Lewis base displacements. The Brønsted acid is interpreted as a complex in which the Lewis acid (proton) is already combined with a base. The reaction is viewed as a displacement of this base by another, stronger base.

$$H_3O^+ + OH^- \rightarrow H_2O + H_2O$$

In this reaction the base OH^- displaces the weaker base H_2O from its combination with the acid, the proton.

All Brønsted acid–base reactions are Lewis base displacements. In the reaction

$$HCl + H_2O \rightarrow H_3O^+ + Cl^-$$

the base H_2O displaces the weaker base, Cl^-. A Lewis base supplies an electron pair to a nucleus and is therefore called **nucleophilic** (from Greek, meaning *nucleus loving*). Base displacements are nucleophilic displacements.

Nucleophilic displacements may be identified among reactions that are not Brønsted acid–base reactions. The formation of $Cu(NH_3)_4^{2+}$ has been previously used as an illustration of a Lewis acid–base reaction. Since the reaction occurs in water, the formation of this complex is more accurately interpreted as the displacement of the base H_2O from the complex $Cu(H_2O)_4^{2+}$ by the stronger base NH_3.

$$Cu(H_2O)_4^{2+} + 4NH_3 \rightarrow Cu(NH_3)_4^{2+} + 4H_2O$$

Lewis acids accept an electron pair in a reaction with a base. They are **electrophilic** (from Greek, meaning *electron loving*). Acid displacements, or electrophilic displacements, are not so common as base displacements, but this type of reaction is known. For example, if $COCl_2$ is viewed as a combination of $COCl^+$ (an acid) with Cl^- (a base), the reaction

$$COCl_2 + AlCl_3 \rightarrow COCl^+ + AlCl_4^-$$

is an electrophilic displacement in which the stronger acid $AlCl_3$ displaces the weaker acid $COCl^+$ from its complex with the base Cl^- (Table 12.1). The reaction

$$SeOCl_2 + BCl_3 \rightarrow SeOCl^+ + BCl_4^-$$

may be similarly interpreted.

PROBLEMS

12.1 Give one-sentence descriptions of an *acid,* a *base,* and a *neutralization reaction* based on (a) the Arrhenius concept, (b) the solvent system concept, (c) the Brønsted–Lowry concept, and (d) the Lewis concept.

12.2 Write equations analogous to the following but employing compounds of the water system instead of compounds of the ammonia system.

(a) $KNH_2 + NH_4NO_3 \rightarrow KNO_3 + 2NH_3$

(b) $Ca_3N_2 + 4NH_3 \rightarrow 3Ca(NH_2)_2$

(c) $Cl_2 + 2NH_3 \rightarrow NH_4Cl + H_2NCl$

(d) $2Na + 2NH_3 \rightarrow 2NaNH_2 + H_2$

(e) $Zn(NH_2)_2 + 2NaNH_2 \rightarrow Na_2[Zn(NH_2)_4]$

12.3 What is the conjugate base of (a) H_3AsO_3, (b) $H_2AsO_4^-$, (c) HNO_2, (d) HS^-, (e) $Cr(H_2O)_6^{3+}$?

12.4 What is the conjugate base of (a) NH_3, (b) $HC_2O_4^-$, (c) HOBr, (d) H_3PO_4, (e) $Zn(OH)(H_2O)_5^+$?

12.5 What is the conjugate acid of (a) $H_2AsO_4^-$, (b) PH_3, (c) $C_2H_3O_2^-$, (d) S^{2} , (e) PO_4^{3} ?

12.6 What is the conjugate acid of (a) NH_3, (b) $HC_2O_4^-$, (c) OCl^-, (d) NH_2^-, (e) $Zn(OH)(H_2O)_5^+$?

12.7 Identify all the Brønsted acids and bases in the following equations.

(a) $HF + HF \rightleftharpoons H_2F^+ + F^-$ (in liquid HF)

(b) $HNO_3 + HF \rightleftharpoons H_2NO_3^+ + F^-$ (in liquid HF)

(c) $HF + CN^- \rightleftharpoons HCN + F^-$ (in liquid HF)

(d) $H_2PO_4^- + CO_3^{2-} \rightleftharpoons HPO_4^{2-} + HCO_3^-$ (in water)

(e) $N_2H_4 + HSO_4^- \rightleftharpoons N_2H_5^+ + SO_4^{2-}$ (in water)

(f) $HC_2O_4^- + HS^- \rightleftharpoons C_2O_4^{2-} + H_2S$ (in water)

12.8 Write chemical equations to illustrate the behavior of the following as Brønsted acids: (a) HOCl, (b) H_2O, (c) HCO_3^-, (d) $Fe(H_2O)_6^{3+}$.

12.9 Write chemical equations to illustrate the behavior of the following as Brønsted bases: (a) H_2NOH, (b) N^{3-}, (c) H^-, (d) HSO_3^-.

12.10 (a) What is an amphiprotic substance? (b) Give four examples of such substances. Select both molecules and ions as your examples. (c) Write chemical equations to illustrate the characteristic behavior of the substances that you listed in part (b).

12.11 What is the leveling effect of a solvent? Use chemical equations in your answer, and describe the leveling of both acids and bases.

12.12 What is the difference in meaning between the terms amphiprotic and amphoteric?

12.13 Each of the following reactions is displaced to the right. (a) Arrange all the Brønsted acids that appear in these equations according to decreasing acid strength. (b) Make a similar list for the Brønsted bases.

(a) $H_3O^+ + H_2PO_4^- \rightleftharpoons H_3PO_4 + H_2O$

(b) $HCN + OH^- \rightleftharpoons H_2O + CN^-$

(c) $H_3PO_4 + CN^- \rightleftharpoons HCN + H_2PO_4^-$

(d) $H_2O + NH_2^- \rightleftharpoons NH_3 + OH^-$

12.14 On the basis of your lists from Problem 12.13, would you expect an appreciable reaction (over 50%) between the species listed in each of the following?
(a) $H_3O^+ + CN^- \rightarrow$
(b) $NH_3 + CN^- \rightarrow$
(c) $HCN + H_2PO_4^- \rightarrow$
(d) $H_3PO_4 + NH_2^- \rightarrow$

12.15 Each of the following reactions is displaced to the right. (a) Arrange all the Brønsted acids that appear in these equations according to decreasing acid strength. (b) Make a similar list for the Brønsted bases.
(a) $HCO_3^- + OH^- \rightleftharpoons H_2O + CO_3^{2-}$
(b) $HC_2H_3O_2 + HS^- \rightleftharpoons H_2S + C_2H_3O_2^-$
(c) $H_2S + CO_3^{2-} \rightleftharpoons HCO_3^- + HS^-$
(d) $HSO_4^- + C_2H_3O_2^- \rightleftharpoons HC_2H_3O_2 + SO_4^{2-}$

12.16 On the basis of your lists from Problem 12.15, would you expect an appreciable reaction (over 50%) between the species listed in each of the following?
(a) $HCO_3^- + C_2H_3O_2^- \rightarrow$
(b) $HSO_4^- + HS^- \rightarrow$
(c) $HC_2H_3O_2 + CO_3^{2-} \rightarrow$
(d) $H_2S + C_2H_3O_2^- \rightarrow$

12.17 Briefly discuss, using chemical equations, how the compound H_2O is classified according to the (a) Arrhenius concept, (b) the Brønsted–Lowry concept, and (c) the Lewis concept.

12.18 Do acid–base reactions ever involve oxidation and reduction? Consider all of the acid–base theories presented in this chapter.

12.19 Write equations for the hydrolysis of (a) $C_2H_3O_2^-$, (b) CN^-, (c) NO_2^-, (d) NH_4^+, (e) PH_4^+, (f) $Cu(H_2O)_4^{2+}$.

12.20 Ammonium chloride (NH_4Cl) reacts with sodium amide ($NaNH_2$) in liquid ammonia to produce sodium chloride and ammonia. Interpret this reaction in terms of the solvent system, Brønsted, and Lewis theories of acids and bases. State clearly what acid(s) and base(s) are involved in each case.

12.21 Interpret the following reactions in terms of the Lewis theory:
(a) $S{=}C{=}S + SH^- \rightarrow S_2CSH^-$
(b) $Fe(CO)_5 + H^+ \rightarrow FeH(CO)_5^+$
(c) $Ag^+ + 2NH_3 \rightarrow Ag(NH_3)_2^+$
(d) $Mn(CO)_5^- + H^+ \rightarrow HMn(CO)_5$
(e) $HF + F^- \rightarrow HF_2^-$

12.22 Interpret the following reactions in terms of the Lewis theory.
(a) $BeF_2 + 2F^- \rightarrow BeF_4^{2-}$
(b) $H_2O + BF_3 \rightarrow H_2OBF_3$
(c) $S + S^{2-} \rightarrow S_2^{2-}$
(d) $H^- + H_2C{=}O \rightarrow H_3CO^-$
(e) $AuCN + CN^- \rightarrow Au(CN)_2^-$

12.23 Interpret the following as Lewis displacement reactions. For each, state the type of displacement, what is displaced, and the agent for the displacement.
(a) $NH_3 + H^- \rightarrow H_2 + NH_2^-$
(b) $HONO_2 + H_2SO_4 \rightarrow NO_2^+ + H_2O + HSO_4^-$
(c) $[Co(NH_3)_5(H_2O)]^{3+} + Cl^- \rightarrow [Co(NH_3)_5Cl]^{2+} + H_2O$
(d) $Ge + GeS_2 \rightarrow 2GeS$
(e) $O^{2-} + H_2O \rightarrow 2OH^-$
(f) $Br_2 + FeBr_3 \rightarrow Br^+ + FeBr_4^-$
(g) $CH_3Cl + AlCl_3 \rightarrow CH_3^+ + AlCl_4^-$
(h) $CH_3I + OH^- \rightarrow CH_3OH + I^-$

12.24 Sulfur dioxide is a significant contributor to atmospheric pollution. The sulfurous acid produced by the reaction of SO_2 and atmospheric moisture adversely effects the eyes, skin, and respiratory systems of humans and causes the corrosion of metals and the deterioration of other materials. The reaction of SO_2 and water can be roughly indicated as

$$H_2O + SO_2 \rightarrow H_2OSO_2 \rightarrow (HO)_2SO$$

(a) Draw Lewis structures (complete with formal charges) for the compounds given. (b) Interpret the two steps of the sequence in terms of the Lewis acid–base theory. (c) Why does the proton migration of the second step occur?

Chapter 13

Chemical Kinetics

Chemical kinetics is the study of the speeds, or rates, of chemical reactions. A small number of factors control how fast a reaction will occur. Investigation of these factors provides clues to the ways in which reactants are transformed into products in chemical reactions. The detailed description of the way a reaction occurs, based on the behavior of atoms, molecules, and ions is called a **reaction mechanism.** Most chemical changes take place by mechanisms that consist of several steps. We can never be sure that a proposed mechanism represents reality—mechanisms are only educated guesses based on kinetic studies.

Reactions are frequently classified as homogeneous or heterogeneous. A **homogeneous reaction** occurs in a single phase—solid, liquid, or gas. For example,

$$H_3O^+(aq) + OH^-(aq) \rightarrow 2H_2O$$
$$2NO(g) + O_2(g) \rightarrow 2NO_2(g)$$

A **heterogeneous reaction** takes place between reactants in different phases. Examples are

$$Zn(s) + 2H^+(aq) \rightarrow Zn^{2+}(aq) + H_2(g)$$
$$2Mg(s) + O_2(g) \rightarrow 2MgO(s)$$

13.1 REACTION RATES

Consider the hypothetical reaction

$$A_2(g) + B_2(g) \rightarrow 2AB(g)$$

Throughout the time that the reaction is occurring, A_2 and B_2 are being gradually used up. The concentrations of these two substances, which are usually expressed in moles per liter, are decreasing. Since AB is being produced at the same time, the concentration of AB is increasing.

The rate of the reaction is a measure of how fast these changes are taking place. Reaction rates are usually expressed in terms of the change in the concentration of either a reactant or a product that occurs in a given time interval. The rate of the reaction between A_2 and B_2 could be expressed, for example, in terms of the increase in the concentration of the product, AB, that occurs in one second. The units of the value would be

$$\frac{\text{mol/liter}}{1\ \text{s}} = \text{mol/liter·s}$$

The rate of the reaction could also be expressed in terms of the decrease in the concentration of A_2 or B_2 that takes place in one second. The rate based on the concentration of AB would not be numerically the same as the rate based on the concentration of A_2 or B_2, however. Suppose that at a given instant the concentration of A_2 is decreasing by 0.02 mol/liter in one second. The rate of decrease in the concentration of A_2 would be expressed as 0.02 mol/liter·s.*

The chemical equation shows two moles of AB produced for every one mole of A_2 used. In the same time interval that the concentration of A_2 decreases by 0.02 mol/liter, the concentration of AB must increase by 0.04 mol/liter. The rate of increase in the concentration of AB, therefore, is 0.04 mol/liter·s. These two values—the rate of disappearance of A_2 (0.02 mol/liter·s) and the rate of appearance of AB (0.04 mol/liter·s)—describe the rate of the

* The rate of *change* in the concentration of A_2 would be −0.02 mol/liter·s since the concentration of A_2 is becoming smaller. Reaction rates, however, are expressed as positive values. We can say that the concentration of A_2 *changes* by −0.02 mol/liter or that the concentration of A_2 *decreases* by 0.02 mol/liter.

rate of *decrease* = −(rate of *change*)

same reaction at the same instant. The rate of a reaction may be described in terms of the rate of appearance of a product or the rate of disappearance of a reactant, but *the basis of the rate measurement must be specified.*

The rate of a reaction usually changes as the reaction proceeds. In Figure 13.1, the concentrations of A_2, B_2, and AB are plotted against time. The symbol for the concentration of a substance consists of the formula of the substance enclosed in brackets. The symbol [AB], for example, stands for the concentration of AB in moles per liter.

Figure 13.1 shows that the concentration of the product, AB, starts at zero and rises rapidly at the beginning of the reaction. During this period, the concentration of the reactant A_2 drops rapidly. The curves show, however, that the concentrations change more and more slowly as the reaction goes on. The velocity of most chemical reactions depends upon the concentrations of the reactants. As these substances are used up, the reaction slows down (see Section 13.6).

The rate of increase in the concentration of AB at a given time can be obtained from the slope of a tangent to the [AB] curve at the point that corresponds to the time of interest. The slope is the increase in y (the concentration of AB) per unit of x (the time), which is the reaction rate.

Obtaining the data for a concentration curve is often a very difficult thing to do. The concentrations must be determined at definite times throughout the course of the reaction without disturbing the reaction. The best methods for such determinations are based on the continuous measurement of a property that changes as the reaction occurs. Changes in pressure, color (the appearance or disappearance of a colored substance), acidity, and volume have been used.

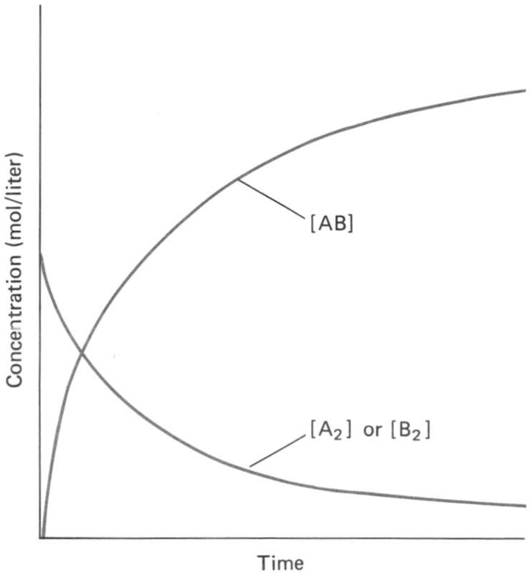

Figure 13.1 Curves showing changes in concentrations of materials with time for the reaction $A_2 + B_2 \rightarrow 2AB$.

13.2 MOLECULAR COLLISIONS AND REACTION RATE

Metathesis reactions between ions in water solution are usually extremely rapid. They are over by the time the solutions are mixed.

$$H_3O^+(aq) + OH^-(aq) \rightarrow 2H_2O$$
$$Ag^+(aq) + Cl^-(aq) \rightarrow AgCl(s)$$

In comparison, molecular reactions in which covalent bonds are broken and made usually occur more slowly. Some reactions of this type are extremely slow. An example is the reaction

$$2SO_2(g) + O_2(g) \rightarrow 2SO_3(g)$$

at temperatures below about 900°C. A mixture of SO_2 and O_2 can be kept for a long period of time at room temperature without detectable change. The rate of a reaction, therefore, depends upon the nature of the reacting substances.

Other factors also influence the rates of chemical reactions—factors that we can control. The most important are

1. the temperature
2. the presence of substances known as catalysts
3. the surface area of solids in heterogeneous reactions
4. the concentrations of reactants

The discussions of these factors in the following sections are based in part on the **collision theory,** which describes the progress of a reaction in terms of collisions between molecules.

Assume that the reaction

$$A_2(g) + B_2(g) \rightarrow 2AB(g)$$

takes place through collisions between A_2 and B_2 molecules as shown in Figure 13.2. Two molecules strike one another. The old A-A and B-B bonds break while simultaneously two new A-B bonds form, and two AB molecules leave the scene of the collision.

Not every collision between A_2 and B_2 molecules results in a reaction. Calculations show that in a situation such as this, the number of collisions that produce reactions, called **effective collisions,** is a very small fraction of the total number of collisions that occur between A_2 and B_2 molecules.

There are many reasons why a collision may not be effective. First, the molecules may be improperly aligned (see Figure 13.3). Second, the collision may be so gentle that the molecules rebound unchanged. The electron cloud of a molecule is, of course, negatively charged. When two slow-moving molecules approach one another closely, they rebound because of the repulsion due to the charges of their electron clouds. Faster-moving molecules, however, are not deterred by this repulsion; the impact of the collision causes the reaction to occur. For an effective collision, the sum of the energies of the colliding molecules must equal, or exceed, some minimum value.

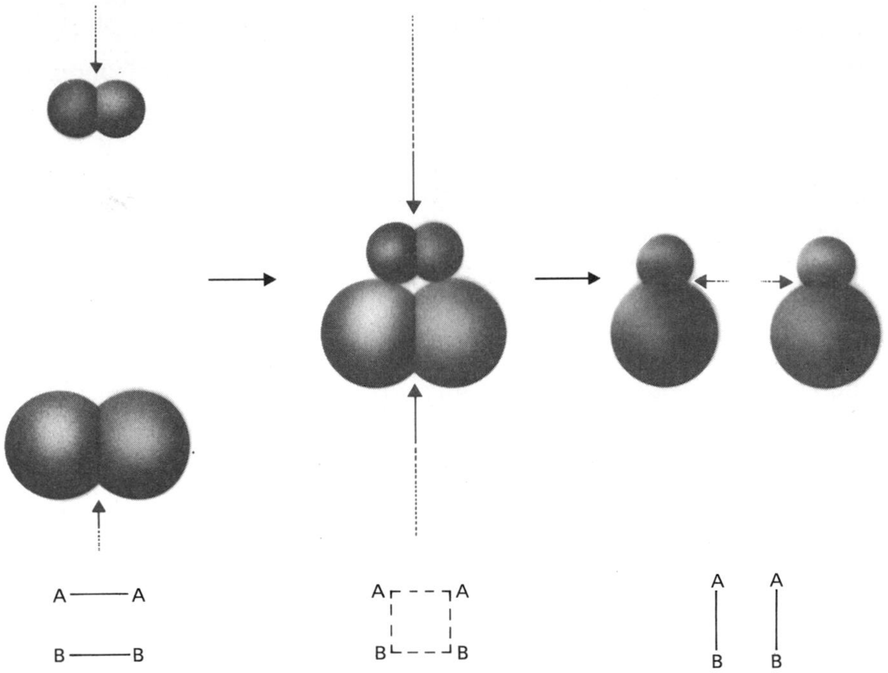

Figure 13.2 Collision between an A_2 molecule and a B_2 molecule resulting in a reaction.

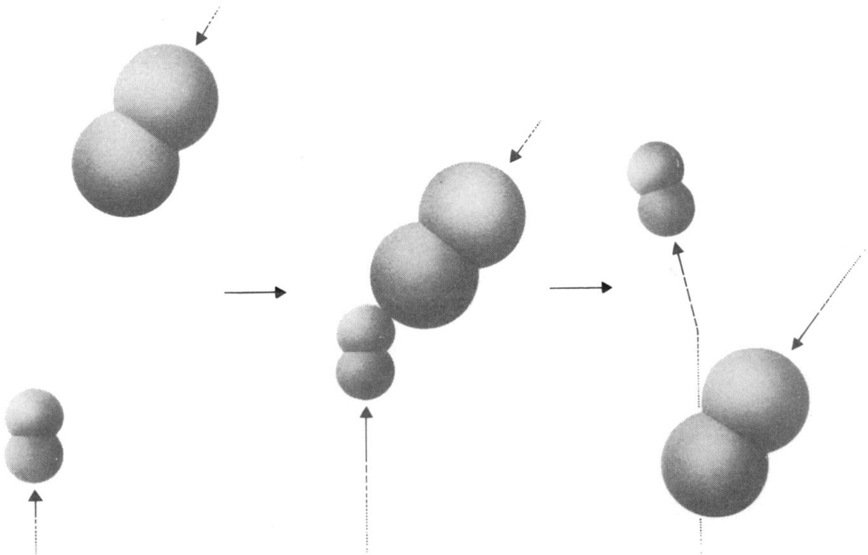

Figure 13.3 Collision between an A_2 molecule and a B_2 molecule producing no reaction.

The rate of the reaction can be increased, therefore, by increasing one or both of the following:

1. the total number of collisions per unit time
2. the proportion of the total number of collisions that are effective.

13.3 TEMPERATURE AND REACTION RATE

The rates of almost all chemical reactions increase when the temperature is raised. The effect is observed for endothermic as well as exothermic reactions. An increase of 10°C in the temperature, at temperatures near room temperature, is often found to increase the reaction rate from 200% to 400%. The more rapid molecular motion that results from an increase in temperature brings about a larger number of molecular collisions per unit time. This factor alone, however, cannot account for the increase in speed. Raising the temperature from 25°C to 35°C causes an increase in the total number of molecular collisions of only about 2%. Obviously, increasing the temperature must increase the fraction of molecular collisions that are effective, and this factor must be the more important of the two.

By an examination of Figure 13.4 we can understand why proportionately more molecular collisions result in reactions at a higher temperature than at a lower temperature. Two molecular energy distribution curves are shown—one

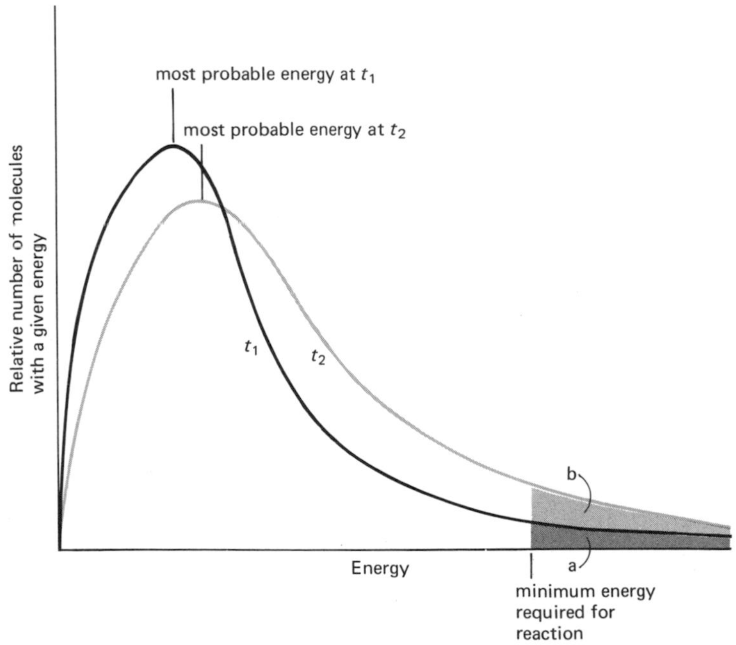

Figure 13.4 Molecular energy distributions at temperatures t_1 and t_2.

for a temperature, t_1, and another for a higher temperature, t_2. The minimum energy required for reaction is indicated on Figure 13.4. The number of molecules at t_1 with energies equal to or greater than this minimum energy is proportional to the area, a, under the curve for t_1.

The curve for temperature t_2 is shifted only slightly in the direction of higher energy. At t_2, however, the number of molecules possessing sufficient energy to react successfully upon collision is greatly increased and is proportional to the area $a + b$. An increase in temperature, therefore, produces an increase in reaction rate principally because the proportion of effective collisions is increased. The increase in the total number of collisions per unit time is only a minor factor. The influence of temperature on reaction rates is analyzed mathematically in Section 13.7.

In a successful collision between reacting molecules, the molecules form an unstable intermediate called an **activated complex.** Let us again consider the reaction between A_2 and B_2. In a gentle collision the A_2 and B_2 molecules are repelled by the charges of their electron clouds and never get close enough for A—B bonds to form. A collision between high-energy molecules, however, results in the formation of a short-lived activated complex for the reaction A_2B_2. The atoms of the complex assume the square configuration diagrammed in Figure 13.2. The A_2B_2 complex may split to form two AB molecules or may decompose to re-form A_2 and B_2 molecules.

$$A_2 + B_2 \rightleftharpoons \begin{array}{ccc} A & \cdots & A \\ \vdots & & \vdots \\ B & \cdots & B \end{array} \rightarrow 2AB$$

The bonding arrangement of the activated complex is not so stable as that of either the reactants or the products. In the activated complex, the A—A and B—B bonds are weakened and partly broken and the A—B bonds are only partly formed. The activated complex is a state of relatively high potential energy.

A potential energy diagram for the reaction between A_2 and B_2 is shown in Figure 13.5. The figure illustrates how the potential energy of the substances involved in the reaction changes in the course of the reaction. Distances along the reaction coordinate indicate how far the formation of products from reactants has progressed.

The difference between the potential energy of the reactants, A_2 and B_2, and the potential energy of the activated complex, A_2B_2, is called the **energy of activation** and is given the symbol E_a. The energy of activation is the minimum energy required for an effective collision. In any collision between an A_2 and a B_2 molecule, the total energy of the molecules remains constant, but kinetic energy (energy of motion) and potential energy may be converted from one form into the other. In a successful collision, part of the kinetic energy of fast-moving A_2 and B_2 molecules is used to provide the energy of activation and produce the high-energy activated complex.

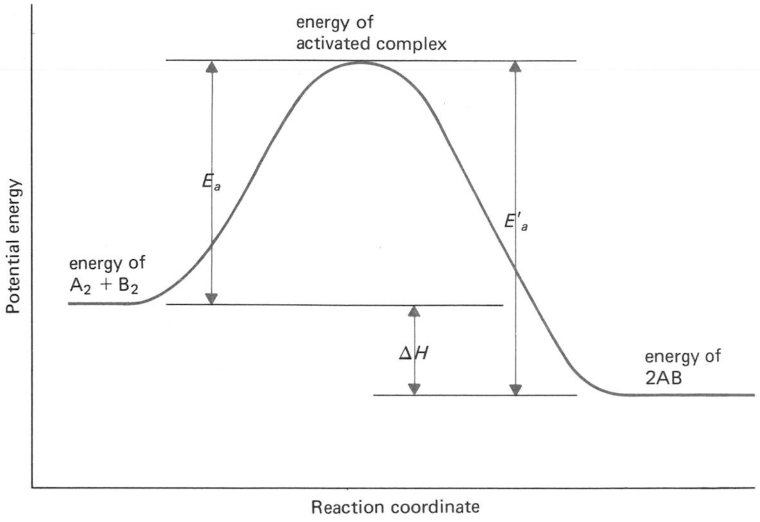

Figure 13.5 Potential energy diagram for the hypothetical reaction $A_2 + B_2 \rightleftharpoons A_2B_2 \rightleftharpoons 2AB$.

The activated complex can split in two ways. If it re-forms the reactants, A_2 and B_2, the energy of activation, E_a, is released in the form of kinetic energy to the A_2 and B_2 molecules. In this case, there is no net reaction. If the complex splits into the product, two AB molecules, the energy indicated as E'_a on Figure 13.5 is released as kinetic energy to the AB molecules. The difference between the energy absorbed, E_a, and the energy evolved, E'_a, is the enthalpy change, ΔH, for the reaction.

$$\Delta H = E_a - E'_a$$

Since E'_a is larger than E_a, ΔH is negative and the reaction is exothermic.

The energy of activation is a potential energy barrier between the reactants and products. Even though the energy of the reactant molecules is higher than the energy of the product molecules, the system must climb a potential energy hill before it can coast down to a state of lower energy. When A_2 and B_2 molecules that have relatively low kinetic energies approach each other, they do not have enough energy between them to produce the activated complex. The repulsion between their electron clouds prevents them from approaching each other closely enough to form the complex. In this case the molecules have only enough energy to get part way up the hill. Then, repelling each other, they coast back down the hill and fly apart unchanged.

Suppose that the reaction diagrammed in Figure 13.5 were reversible—reactions of this type usually are. The reverse reaction can be interpreted by reading the figure from right to left. The energy of activation for the reverse reaction is E'_a, and the energy released by the formation of the products (in this case A_2 and B_2) from the activated complex is E_a. The enthalpy change for the reverse reaction is

$$\Delta H = E'_a - E_a$$

The enthalpy change is positive since E'_a is larger than E_a. The reverse reaction is endothermic.

13.4 CATALYSTS

A **catalyst** is a substance that increases the rate of a chemical reaction without being used up in the reaction. A catalyst may be recovered unchanged at the end of the reaction. Oxygen can be prepared by heating potassium chlorate ($KClO_3$) by itself. Or, a small amount of manganese dioxide (MnO_2) can be used as a catalyst for this reaction. When MnO_2 is present, the reaction is much more rapid and the decomposition of $KClO_3$ takes place at a satisfactory rate at a lower temperature.

$$2KClO_3(s) \xrightarrow{MnO_2} 2KCl(l) + 3O_2(g)$$

The catalyst is written over the arrow in the chemical equation since a catalyst does not affect the overall stoichiometry of the reaction. The MnO_2 may be recovered unchanged at the conclusion of the reaction.

The mere *presence* of a catalyst (presumably acting as a cheering section) does not cause the effect on the reaction rate. A catalyzed reaction takes place by a pathway, or mechanism, that is different from the one that the uncatalyzed reaction follows. Suppose, for example, that an uncatalyzed reaction occurs by collisions between X and Y molecules.

$$X + Y \rightarrow XY$$

The catalyzed reaction might follow a two-step mechanism consisting of

1. $X + C \rightarrow XC$
2. $XC + Y \rightarrow XY + C$

where C is the catalyst. Notice that the catalyst is used in the first step and regenerated in the second. It is, therefore, used over and over again. Consequently, only a small amount of a catalyst is needed to do the job.

The mechanism of a reaction is never known with absolute certainty. Mechanisms are proposals based on data derived from kinetic studies (see Section 13.8). Many different types of mechanisms—for both catalyzed and uncatalyzed reactions—have been proposed.

A catalyst, therefore, works by opening a new path by which the reaction can take place. The catalyzed path has a lower energy of activation than the uncatalyzed path does (Figure 13.6), which accounts for the more rapid reaction rate. When a catalyst is used, proportionately more molecules have the energy required for reaction (Figure 13.7).

Two additional points can be derived from Figure 13.6.

1. The enthalpy change, ΔH, for the catalyzed reaction is the same as the ΔH for the uncatalyzed reaction.

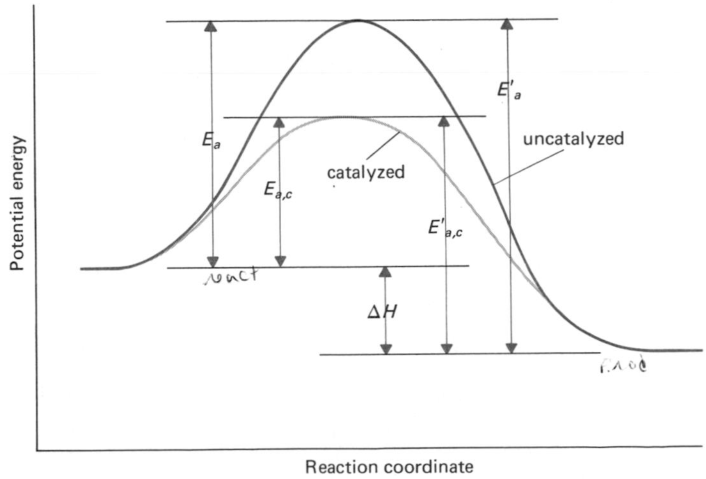

Figure 13.6 Potential energy diagrams for catalyzed and uncatalyzed reactions.

2. For reversible reactions, the catalyst has the same effect on the reverse reaction that it has on the forward reaction. The energy of activation for the reverse reaction, E'_a, is lowered by the catalyst to the same extent that the energy of activation for the forward reaction, E_a, is lowered.

A **homogeneous catalyst** is present in the same phase as the reactants. An example of homogeneous catalysis in the gas phase is the effect of chlorine gas on the decomposition of dinitrogen oxide gas. Dinitrogen oxide, N_2O, is relatively unreactive at room temperature, but at temperatures near 600°C it decomposes according to the equation

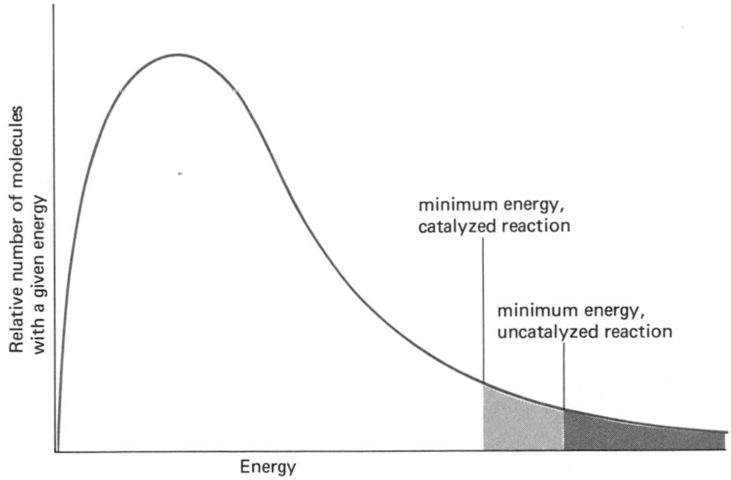

Figure 13.7 Molecular energy distribution showing the effect of a catalyst on the number of molecules possessing sufficient energy to react.

$$2N_2O(g) \rightarrow 2N_2(g) + O_2(g)$$

The uncatalyzed reaction is thought to occur by means of a complex mechanism that includes the following steps.

1. Through collisions between N_2O molecules, some N_2O molecules gain enough energy to split apart.

$$N_2O(g) \rightarrow N_2(g) + O(g)$$

2. The oxygen atoms are very reactive. They readily react with N_2O molecules.

$$O(g) + N_2O(g) \rightarrow N_2(g) + O_2(g)$$

The final products of the reaction are N_2 and O_2. The O atom is a **reaction intermediate** and not a final product. The energy of activation for the uncatalyzed reaction is about 240 kJ/mol.

The reaction is catalyzed by a trace of chlorine gas. The path that has been proposed for the catalyzed reaction consists of the following steps.

1. At the temperature of the decomposition, and particularly in the presence of light, some chlorine molecules dissociate into chlorine atoms.

$$Cl_2(g) \rightarrow 2Cl(g)$$

2. The chlorine atoms readily react with N_2O molecules.

$$N_2O(g) + Cl(g) \rightarrow N_2(g) + ClO(g)$$

3. The decomposition of the unstable ClO molecules follows.

$$2ClO(g) \rightarrow Cl_2(g) + O_2(g)$$

Notice that the catalyst (Cl_2) is returned to its original state in the last step. The final products of the catalyzed reaction ($2N_2$ and O_2) are the same as those of the uncatalyzed reaction. Cl and ClO are not products because they are used in steps that follow the ones in which they are produced. The energy of activation for the reaction catalyzed by chlorine is about 140 kJ/mol, which is considerably lower than E_a for the uncatalyzed reaction (240 kJ/mol).

Homogeneous catalysis also occurs in aqueous solution. Many reactions are catalyzed by acids and bases. The decomposition of hydrogen peroxide, H_2O_2, is catalyzed by a trace of iodide ion, I^-.

$$2H_2O_2(aq) \rightarrow 2H_2O + O_2(g)$$

The mechanism proposed for the catalyzed reaction is

1. $H_2O_2(aq) + I^-(aq) \rightarrow H_2O + IO^-(aq)$
2. $H_2O_2(aq) + IO^-(aq) \rightarrow H_2O + O_2(g) + I^-(aq)$

In **heterogeneous catalysis** the reactants and catalyst are present in different phases. Reactant molecules are *adsorbed* on the surface of the catalyst in these processes, and the reaction takes place on that surface. **Adsorption** is a process in which molecules adhere to the surface of a solid. Charcoal, for example, is used in gas masks as an adsorbent for noxious gases. In ordinary **physical adsorption,** the molecules are held to the surface by London forces.

Heterogeneous catalysis, however, usually takes place through **chemical adsorption** (or **chemisorption**) in which the adsorbed molecules are held to the surface by bonds that are similar in strength to those in chemical compounds. When these bonds form, the chemisorbed molecules undergo changes in the arrangement of their electrons. Some bonds of the molecules may be stretched and weakened, and in some cases, even broken. Hydrogen molecules, for example, are believed to be adsorbed as hydrogen atoms on the surface of platinum, palladium, nickel, and other metals. The chemisorbed layer of molecules or atoms, therefore, functions as an activated complex in a surface catalyzed reaction.

The decomposition of N_2O is catalyzed by gold. A proposed mechanism for the gold-catalyzed decomposition is diagrammed in Figure 13.8. The steps are:

1. Molecules of $N_2O(g)$ are chemisorbed on the surface of the gold.

   $N_2O(g) \rightarrow N_2O(\text{on Au})$

2. The bond between the O atom and the adjacent N atom of a N_2O molecule is weakened when the O atom bonds to the gold. This N—O bond breaks and an N_2 molecule leaves.

   $N_2O(\text{on Au}) \rightarrow N_2(g) + O(\text{on Au})$

3. Two O atoms on the surface of the gold combine to form an O_2 molecule which enters the gas phase.

   $O(\text{on Au}) + O(\text{on Au}) \rightarrow O_2(g)$

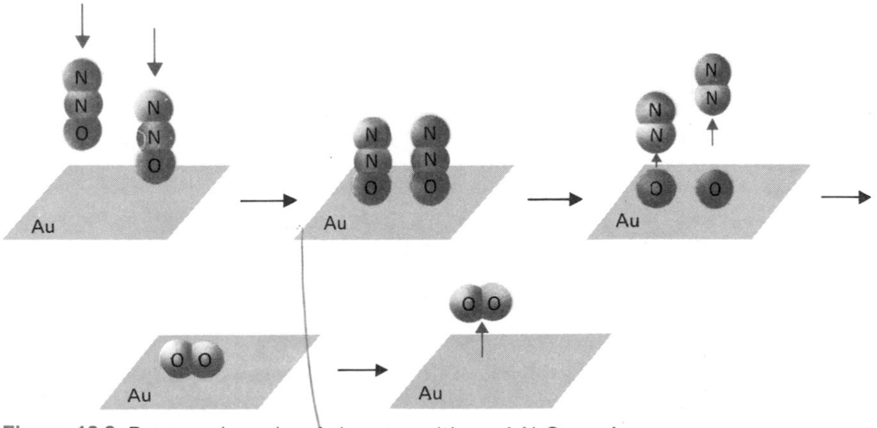

Figure 13.8 Proposed mode of decomposition of N_2O on Au.

The energy of activation for the gold-catalyzed decomposition is about 120 kJ/mol, which is lower than E_a for either the uncatalyzed decomposition (240 kJ/mol) or the chlorine-catalyzed decomposition (140 kJ/mol).

The electronic structure and arrangement of atoms on the surface of a catalyst determine its activity. Lattice defects and irregularities are thought to be active sites for catalysis. The surface of some catalysts can be changed by the addition of substances called **promoters,** which enhance the catalytic activity. In the synthesis of ammonia

$$N_2(g) + 3H_2(g) \xrightarrow{Fe} 2NH_3(g)$$

an iron catalyst is made more effective when traces of potassium or vanadium are added to it.

Catalytic **poisons** are substances that inhibit the activity of catalysts. For example, small amounts of arsenic destroy the power of platinum to catalyze the preparation of sulfur trioxide from sulfur dioxide.

$$2SO_2(g) + O_2(g) \xrightarrow{Pt} 2SO_3(g)$$

Presumably, platinum arsenide forms on the surface of the platinum and destroys its catalytic activity.

Catalysts are generally highly specific in their activity. In some cases a given substance will catalyze the synthesis of one set of products from certain reagents, whereas another substance will catalyze the synthesis of completely different products from the same reactants. In these cases, both reactions are possible and the products obtained are those that are produced most rapidly. Carbon monoxide and hydrogen can be made to yield a wide variety of products depending upon the catalyst employed and the conditions of the reaction.

If a cobalt or nickel catalyst is used, CO and H_2 produce mixtures of hydrocarbons. One hydrocarbon produced, for example, is methane, CH_4.

$$CO(g) + 3H_2(g) \xrightarrow{Ni} CH_4(g) + H_2O(g)$$

On the other hand, methanol is the product of the reaction of CO and H_2 when a mixture of zinc and chromic oxides is employed as a catalyst.

$$CO(g) + 2H_2(g) \xrightarrow{ZnO/Cr_2O_3} CH_3OH(g)$$

The catalytic converter installed on car mufflers is a recent application of surface catalysis. Carbon monoxide and hydrocarbons from unburned fuel are present in automobile engine exhaust and are serious air pollutants. In the converter, the exhaust gases and additional air are passed over a catalyst that consists of metal oxides. The CO and hydrocarbons are converted into

CO_2 and H_2O, which are relatively harmless and are released to the atmosphere. Since the catalyst is poisoned by lead, unleaded gasoline must be used in automobiles equipped with catalytic converters.

Many industrial processes depend upon catalytic procedures, but the natural catalysts known as **enzymes** are even more important to man. These extremely complicated substances catalyze life processes such as digestion, respiration, and cell synthesis. The large number of complex chemical reactions that occur in the body, and are necessary for life, can occur at the relatively low temperature of the body because of the action of enzymes. Thousands of enzymes are known, and each serves a specific function. Research into the structure and action of enzymes may lead to a better understanding of the causes of disease and the mechanism of growth.

13.5 SURFACE AREA IN HETEROGENEOUS REACTIONS

The rate of a heterogeneous reaction can be increased by *increasing the surface area* on which the reaction takes place. The surface area of a given mass of a solid is increased when the solid is powdered. The reaction of zinc metal with an acid, for example,

$$Zn(s) + 2H^+(aq) \rightarrow Zn^{2+}(aq) + H_2(g)$$

is much more rapid when zinc powder is used than when large pieces of zinc are used. The increase in surface area results in an increase in the total number of collisions per unit time and, therefore, an increase in the reaction rate. A finely divided combustible material may burn with such rapidity that an explosion results.

The rate of a heterogeneous process can also be increased by *agitation of the reaction mixture,* since agitation keeps the reactive surface continuously exposed. Sugar dissolves more quickly in coffee, for example, if the mixture is stirred.

Surface catalysts are usually prepared in a way that will produce a large surface area for a given amount of solid. Solids used as catalysts are usually porous and extremely finely divided.

13.6 CONCENTRATION AND REACTION RATES

Reaction rates usually depend upon the concentrations of the reacting substances. When the concentration of a reactant is increased, more molecules are crowded into a given volume, and the total number of molecular collisions per unit time is increased. An increase in reaction rate follows. If all other conditions are held constant, there is no change in the proportion of collisions that are effective.

For each chemical reaction there is a mathematical expression, called a

rate equation or a **rate law,** that relates the concentrations of the reactants to the reaction rate. For the reaction

$$2N_2O_5(g) \rightarrow 4NO_2(g) + O_2(g)$$

the rate equation is

$$\text{rate} = k[N_2O_5]$$

The expression tells us that the rate is directly proportional to the concentration of N_2O_5. If the concentration is doubled, the rate is doubled. If the concentration is tripled, the rate is tripled. The proportionality constant, k, is called the **rate constant.** *The form of the rate equation and the value of k must be determined by experiment.* The numerical value of k depends upon the temperature and the terms in which the rate is expressed.

The rate of the reaction

$$NO_2(g) + 2HCl(g) \rightarrow NO(g) + H_2O(g) + Cl_2(g)$$

is proportional to the concentration of NO times the concentration of HCl.

$$\text{rate} = k[NO_2][HCl]$$

Doubling the concentration of NO would double the reaction rate. Doubling the concentration of HCl would also double the reaction rate. If the concentrations of both reactants were doubled at the same time, the reaction rate would increase fourfold.

For the reaction

$$2NO(g) + 2H_2(g) \rightarrow N_2(g) + 2H_2O(g)$$

the rate equation is

$$\text{rate} = k[NO]^2[H_2]$$

The rate of the reaction is directly proportional to the *square* of the concentration of NO times the concentration of H_2. When the concentration of NO is increased by a factor of *two,* the rate increases by a factor of *four* (since 2^2 is 4). When the concentration of H_2 is increased by a factor of *two,* the rate increases by the factor of *two.* If the concentrations of both NO and H_2 are doubled, the rate would increase eightfold (since $2^2 \times 2 = 8$).

The **order** of a reaction is given by the *sum of the exponents* of the concentration terms in the rate equation. The decomposition of N_2O_5 is said to be first order since the exponent of $[N_2O_5]$ is one in the rate equation.

$$\text{rate} = k[N_2O_5]$$

The reaction between NO_2 and HCl is said to be first order in NO_2, first order in HCl, and second order overall.

$$\text{rate} = k[NO_2][HCl]$$

The reaction between NO and H_2 is second order in NO, first order in H_2, and third order overall.

$$\text{rate} = k[\text{NO}]^2[\text{H}_2]$$

The rate equation for a reaction, and consequently the order of a reaction, *must* be determined experimentally. They cannot be derived from the chemical equation for the reaction. The order of a reaction need not be a whole number. Reactions of fractional order, as well as zero order, are known. The mechanism of the reaction determines the form of the rate equation (see Section 13.8).

Example 13.1 The data given in the following table pertain to the reaction

$$2\text{NO(g)} + \text{O}_2\text{(g)} \rightarrow 2\text{NO}_2\text{(g)}$$

run at 25°C. Determine the form of the rate equation and the value of the rate constant, k.

	Initial concentration		Initial rate
Experiment	NO (mol/liter)	O_2 (mol/liter)	appearance of NO_2 (mol/liter·s)
A	0.01	0.01	0.007
B	0.01	0.02	0.014
C	0.01	0.03	0.021
D	0.02	0.03	0.084
E	0.03	0.03	0.189

Solution The rate equation is in the form

$$\text{rate of appearance of NO}_2 = k[\text{NO}]^x[\text{O}_2]^y$$

Data from the table are used to find the values of the exponents x and y.

In the first three experiments (A, B, and C) the concentration of NO is held constant and the concentration of O_2 is changed. Any change in the rate observed in this series of experiments, therefore, is caused by the change in the concentration of O_2. The concentration of O_2 in experiment B is double that in experiment A, and the rate observed in experiment B is twice the rate in experiment A. Comparison of the data from experiment C with the data from experiment A shows that when the concentration of O_2 is increased three times, the rate increases three times. The value of y, therefore, is 1. The rate is directly proportional to the first power of $[O_2]$.

In the last three experiments (C, D, and E), the concentration of O_2 is held constant (at 0.03 mol/liter), and the concentration of NO is changed. The increase in rate that is observed in this series of experiments is caused by the increase in the concentration of NO. The concentration of NO in experiment

D is *two* times the concentration of NO in experiment C. The rate observed in experiment D, however, is *four* times the rate observed in experiment C. It appears that the square of [NO] must appear in the rate equation since 2^2 is 4.

We can check this conclusion by comparing the data from experiment E with the data from experiment C. The concentration of NO increases by a factor of 3,

$$\frac{0.03 \text{ mol/liter}}{0.01 \text{ mol/liter}} = 3$$

The rate increases by a factor of 9

$$\frac{0.189 \text{ mol/liter·s}}{0.021 \text{ mol/liter·s}} = 9$$

Since 3^2 is 9, the exponent x must be 2; the term for the concentration of NO is squared. The rate equation is

$$\text{rate of appearance of } NO_2 = k[NO]^2[O_2]$$

The reaction, therefore, is second order in NO, first order in O_2, and third order overall.

The value of k can be obtained by using the data from any of the experiments. The same value should be obtained in each case. The data from experiment A are used in the following way.

$$\text{rate of appearance of } NO_2 = k[NO]^2[O_2]$$

$$0.007 \text{ mol/liter·s} = k(0.01 \text{ mol/liter})^2(0.01 \text{ mol/liter})$$
$$7 \times 10^{-3} \text{ mol/liter·s} = k(1 \times 10^{-4} \text{ mol}^2\text{/liter}^2)(1 \times 10^{-2} \text{ mol/liter})$$

$$k = \frac{7 \times 10^{-3} \text{ mol/liter·s}}{1 \times 10^{-6} \text{ mol}^3\text{/liter}^3}$$

$$k = 7 \times 10^3 \text{ liter}^2\text{/mol}^2\text{·s}$$

13.7 RATE EQUATIONS AND TEMPERATURE

The rate constant, k, varies with temperature in a manner described by the equation

$$\log k = \log A - \frac{E_a}{2.303RT}$$

where A is a constant that is characteristic of the reaction being studied, E_a is the energy of activation of the reaction (in J/mol), R is the molar gas constant (8.3143 J/mol·K), and T is the absolute temperature. (The use of logarithms is discussed in Appendix D.) The relationship, called the **Arrhenius**

equation, was proposed in 1889 by Svanté Arrhenius. The equation is only approximate, but in most cases the approximation is a good one.

The values of E_a and A for a reaction can be found from the rate constants for two different temperatures. If the rate constant at T_1 is k_1 and at T_2 is k_2, then

$$\log k_2 = \log A - \frac{E_a}{2.303RT_2}$$

and

$$\log k_1 = \log A - \frac{E_a}{2.303RT_1}$$

Subtraction of the second equation from the first gives

$$\log k_2 - \log k_1 = -\frac{E_a}{2.303RT_2} + \frac{E_a}{2.303RT_1}$$

Since $\log x - \log y$ is $\log (x/y)$,

$$\log \left(\frac{k_2}{k_1}\right) = \frac{E_a}{2.303R}\left(\frac{1}{T_1} - \frac{1}{T_2}\right)$$

or

$$\log \left(\frac{k_2}{k_1}\right) = \frac{E_a}{2.303R}\left(\frac{T_2 - T_1}{T_1T_2}\right)$$

If this equation is solved for the energy of activation, E_a, the following relationship is obtained.

$$E_a = 2.303R\left(\frac{T_1T_2}{T_2 - T_1}\right)\log\left(\frac{k_2}{k_1}\right)$$

The uses of the last two equations are illustrated in the following examples.

Example 13.2 For the reaction

$$2NOCl(g) \rightarrow 2NO(g) + Cl_2(g)$$

the rate equation is

$$\text{rate of production of } Cl_2 = k[NOCl]^2$$

The rate constant, k, is 3.0×10^{-8} liter/mol·s at 300 K and 7.0×10^{-4} liter/mol·s at 400 K. What is the energy of activation, E_a, for the reaction?

Solution Let

$T_1 = 300$ K
$T_2 = 400$ K

$k_1 = 3.0 \times 10^{-8}$ liter/mol·s
$k_2 = 7.0 \times 10^{-4}$ liter/mol·s

R is 8.31 J/mol·K, therefore,

$$E_a = 2.30R \left(\frac{T_1T_2}{T_2 - T_1}\right) \log \left(\frac{k_2}{k_1}\right)$$

$$= 2.30(8.31 \text{ J/mol·K}) \left(\frac{(300 \text{ K})(400 \text{ K})}{400 \text{ K} - 300 \text{ K}}\right) \log \left(\frac{7.0 \times 10^{-4} \text{ liter/mol·s}}{3.0 \times 10^{-8} \text{ liter/mol·s}}\right)$$

$$= (19.1 \text{ J/mol·K})(1200 \text{ K}) \log (2.33 \times 10^4)$$

$$= (22{,}900 \text{ J/mol})(4.37)$$

$$= 100{,}000 \text{ J/mol} = 100 \text{ kJ/mol}$$

Example 13.3 Given the data found in Example 13.2, find the value of k at 500 K.

Solution Let

$T_1 = 400$ K
$T_2 = 500$ K
$k_1 = 7.0 \times 10^{-4}$ liter/mol·s
$k_2 =$ unknown
$E_a = 1.00 \times 10^5$ J/mol

We use the equation

$$\log \left(\frac{k_2}{k_1}\right) = \frac{E_a}{2.30R} \left(\frac{T_2 - T_1}{T_1T_2}\right)$$

$$= \frac{1.00 \times 10^5 \text{ J/mol}}{2.30(8.31 \text{ J/mol·K})} \left(\frac{500 \text{ K} - 400 \text{ K}}{(400 \text{ K})(500 \text{ K})}\right)$$

$$= (5.23 \times 10^3 \text{ K})(5.00 \times 10^{-4}/\text{K})$$

$$= 2.62$$

Therefore,

$$\frac{k_2}{k_1} = \text{antilog } 2.62 = 4.2 \times 10^2$$

or

$k_2 = 4.2 \times 10^2(k_1)$
$k_2 = 4.2 \times 10^2(7.0 \times 10^{-4}$ liter/mol·s)
$= 2.9 \times 10^{-1}$ liter/mol·s $= 0.29$ liter/mol·s

These examples show that a temperature increase causes the rate constant to increase decidedly. Any change in the rate constant of a reaction is, of course, reflected in the rate of the reaction. For the reaction considered in the examples, a 100° increase in the temperature causes the following effects

300 K to 400 K	rate increases 23,300 times
400 K to 500 K	rate increases 414 times

The marked effect of a temperature increase is obvious. Notice, however, that the reaction rate is affected more at low temperatures than at high temperatures.

The energies of activation of many reactions range from 60 kJ/mol to 250 kJ/mol, values which are on the same scale as bond energies. For a 10° rise in temperature, from 300 K to 310 K, the rate of reaction varies with the energy of activation in the following way.

$E_a = 60$ kJ/mol	rate increases about 2 times
$E_a = 250$ kJ/mol	rate increases about 25 times

13.8 REACTION MECHANISMS AND RATE EQUATIONS

The chemical equation for a reaction gives the stoichiometric relationship between the initial reactants and the final products. Very often, however, a reaction occurs by way of a mechanism that consists of several steps. A product of one step may be a reactant in the next step. Reaction intermediates are substances that are produced and used in the course of the reaction and are therefore neither initial reactants nor final products of the reaction.

The formation of nitrosyl fluoride (ONF), for example,

$$2NO(g) + F_2(g) \rightarrow 2ONF(g)$$

is believed to follow a two-step mechanism

1. $NO(g) + F_2(g) \rightarrow ONF(g) + F(g)$
2. $NO(g) + F(g) \rightarrow ONF(g)$

Notice that the steps of the mechanism add up to the chemical equation for thc overall reaction. The F atoms, which are reaction intermediates, cancel in the addition.

Each step of a mechanism proceeds at its own rate—some are rapid and some are slow. The rate of a given step depends upon the concentrations of the substances that react in that step; they may be initial reactants or reaction intermediates. The rate of the overall reaction, therefore, depends upon how the rates of the steps fit together.

The rate equation for a reaction must be obtained experimentally; it cannot be derived from the chemical equation for the overall change. The rate equation for a *step* in a mechanism, however, can be derived from the chemical equation for the step. Suppose that the following equation represents one step of a reaction mechanism

$$A_2(g) + B_2(g) \rightarrow 2AB(g)$$

The reaction occurs through collisions between A_2 and B_2 molecules. If the concentration of A_2 is doubled, the number of A_2 molecules in the system is doubled. The number of A_2–B_2 collisions per second doubles, therefore, and a doubling of the reaction rate follows. The rate is proportional to the first power of $[A_2]$.

If the concentration of B_2 is doubled and the concentration of A_2 is held constant, the same effect on the rate occurs. The rate equation for this step, therefore, is

$$\text{rate} = k[A_2][B_2]$$

Let us assume that the equation

$$2DE(g) \rightarrow D_2(g) + E_2(g)$$

describes a step of a reaction mechanism. The step occurs by means of collisions between two DE molecules. Suppose that there are n molecules of DE in a container. The number of collisions per second for a *single* DE molecule is proportional to the number of other DE molecules present, $n - 1$. The *total* number of collisions per second for all n molecules may incorrectly be expected to be proportional to n times $(n - 1)$. It is, however, proportional to $\frac{1}{2}n(n - 1)$. The factor $\frac{1}{2}$ is included so that a given collision is not counted twice—once for a collision in which molecule 1 hits molecule 2 and again for a collision in which molecule 2 hits molecule 1.

Since n is a very large number $(n - 1)$ is equal to n for all practical purposes. We can say that the total number of collisions per second is proportional to $\frac{1}{2}n^2$. Since the rate of the reaction is proportional to the total number of collisions per second,

$$\text{rate} \propto \tfrac{1}{2}n^2$$

The number of molecules in the container, n, is proportional to the concentration of DE, [DE], and n^2 is proportional to $[DE]^2$. The constant $\frac{1}{2}$ can be incorporated into the proportionality constant, k. Thus,

$$\text{rate} = k[DE]^2$$

In general, if the chemical equation for a *single step* of a mechanism is

$$wW + xX \rightarrow yY + zZ$$

the rate equation is

$$\text{rate} = k[W]^w[X]^x$$

This equation is a statement of the **law of mass action.** The form in which it is given here was proposed by Cato Guldberg and Peter Waage in 1864 and used by them in their studies of equilibrium systems (see Section 14.1). The law itself, in more general terms, was first proposed by Carl F. Wenzel in 1777 and developed by many chemists afterward.

It is important to note, however, that the mechanism of a reaction is inferred from the rate equation for the overall reaction and not the other way around. The overall rate equation is determined experimentally. The mechanism is proposed on the basis of this rate equation and any other available evidence—such as the detection of reaction intermediates.

The following rate equation for the formation of nitrosyl fluoride was obtained experimentally.

$$2NO(g) + F_2(g) \rightarrow 2ONF(g) \qquad \text{rate} = k[NO][F_2]$$

The suggested mechanism and the rate equations for the steps (derived by use of law of mass action) are

1. $NO(g) + F_2(g) \rightarrow ONF(g) + F(g) \qquad \text{rate}_1 = k[NO][F_2]$
2. $NO(g) + F(g) \rightarrow ONF(g) \qquad \text{rate}_2 = k[NO][F]$

The first step is assumed to be much slower than the second step. Step 1 produces F atoms slowly. As soon as they are produced, they are rapidly used by step 2. Step 1, therefore, is the bottleneck of the reaction; the overall reaction cannot be faster than it is. Since it controls the overall rate, it is called the **rate-determining step.** For this reason, the rate equation for step 1 is the rate equation for the overall change.

A few reactions are believed to occur by a single step. For example,

$$NO_2(g) + CO(g) \rightarrow NO(g) + CO_2(g) \qquad \text{rate} = k[NO_2][CO]$$

The exponents of the rate equation *for a reaction of this type* correspond to the coefficients of the chemical equation. We should not assume, however, that every reaction in which the rate equation and the chemical equation are related in this way occurs by a single step.

The rate equation for the reaction of NO and O_2 is

$$2NO(g) + O_2(g) \rightarrow 2NO_2(g) \qquad \text{rate} = k[NO]^2[O_2]$$

The exponents of the rate equation are the same as the coefficients of the chemical equation. We might assume—incorrectly—that the reaction occurs by a single step. This step, however, will be a three-body collision involving two NO molecules and one O_2 molecule. Such collisions, in which three bodies must come together simultaneously, are rare.

The mechanism proposed for this reaction does not involve a three-body collision.

1. $2NO(g) \xrightarrow{k_1} N_2O_2(g) \qquad \text{rate}_1 = k_1[NO]^2$

2. $N_2O_2(g) \xrightarrow{k_2} 2NO(g) \qquad \text{rate}_2 = k_2[N_2O_2]$

3. $N_2O_2(g) + O_2(g) \xrightarrow{k_3} 2NO_2(g) \qquad \text{rate}_3 = k_3[N_2O_2][O_2]$

In the first step, the reaction intermediate N_2O_2 is formed from NO. This complex can decompose back into NO (step 2) or react with O_2 to form the product (step 3). The third step is the rate-determining step.

We assume that the concentration of the reaction intermediate N_2O_2 reaches a **steady state** after the reaction has been going on for a while. A steady state occurs when the intermediate is used as fast as it is produced so that its concentration is constant. The intermediate is produced in step 1

$$\text{rate of appearance of } N_2O_2 = k_1[NO]^2$$

It is used in steps 2 and 3.

$$\text{rate of disappearance of } N_2O_2 = k_2[N_2O_2] + k_3[N_2O_2][O_2]$$

Since the third step is *much* slower than the second, k_3 is *much* smaller than k_2. We can, therefore, neglect the term $k_3[N_2O_2][O_2]$ since it is much smaller than the term $k_2[N_2O_2]$.

When the steady state is reached,

$$\text{rate of appearance of } N_2O_2 = \text{rate of disappearance of } N_2O_2$$
$$k_1[NO]^2 = k_2[N_2O_2]$$

Therefore,

$$[N_2O_2] = \frac{k_1[NO]^2}{k_2}$$

Since the third step is the rate-determining step, the overall rate depends on it.

$$\text{rate} = \text{rate}_3 = k_3[N_2O_2][O_2]$$

If we substitute the expression for the steady-state concentration of N_2O_2 into this equation, we get

$$\text{rate} = k_3\left(\frac{k_1[NO]^2}{k_2}\right)[O_2]$$

$$\text{rate} = \frac{k_3k_1}{k_2}[NO]^2[O_2]$$

The constants can be combined into a single constant to give

$$\text{rate} = k[NO]^2[O_2]$$

which is the experimentally determined rate equation.

We can never be absolutely certain that a proposed mechanism is correct. Some chemists believe that the mechanism of the reaction between NO and O_2 is

1. $NO(g) + O_2(g) \rightarrow NO_3(g)$
2. $NO_3(g) \rightarrow NO(g) + O_2(g)$
3. $NO_3(g) + NO(g) \rightarrow 2NO_2(g)$

The reaction intermediate in this scheme is NO_3. If it is assumed that NO_3 attains a steady state, this mechanism also gives the correct rate equation for the overall change.

The reaction between H_2 gas and Br_2 vapor at temperatures around 200°C provides an example of an important type of reaction mechanism called a **chain mechanism.**

$$H_2(g) + Br_2(g) \rightarrow 2HBr(g)$$

The mechanism for the reaction can be described in four parts.

1. **Chain initiation.** Some Br_2 molecules dissociate into atoms.

   $$Br_2(g) \rightarrow 2Br(g)$$

2. **Chain propagation.** The Br atoms are reactive intermediates called **chain carriers.** A Br atom reacts with a H_2 atom.

   $$Br(g) + H_2(g) \rightarrow HBr(g) + H(g)$$

   The reaction produces a molecule of product, HBr, and a second chain carrier, an H atom. The H atom reacts with a Br_2 molecule

   $$H(g) + Br_2(g) \rightarrow HBr(g) + Br(g)$$

   producing another molecule of HBr and the original chain carrier, a Br atom. The Br atom reacts with a H_2 molecule, and the cycle starts again. These two steps are repeated over and over again.
3. **Chain inhibition.** If an H atom collides with an HBr molecule, the reaction that results is said to inhibit the overall reaction.

   $$H(g) + HBr(g) \rightarrow H_2(g) + Br(g)$$

   Since a molecule of product is used (HBr) and a molecule of reactant is produced (H_2), this step slows down the overall reaction. It does not, however, break the chain or stop the reaction since a chain carrier is also produced.
4. **Chain termination.** Two chains are terminated when two Br atoms come together to form a Br_2 molecule.

   $$2Br(g) \rightarrow Br_2(g)$$

The reaction of $H_2(g)$ and $Cl_2(g)$ is believed to follow a similar mechanism. A mixture of these two gases at room temperature can be kept for a long period of time *in the dark* without reacting. If the mixture is exposed to light, a violently rapid reaction occurs. Light is believed to start the chain reaction by dissociating some Cl_2 molecules into atoms. The reaction of H_2 and Br_2 is also light sensitive, but the reaction at room temperature is slower.

Many reactions are thought to occur by a chain mechanism. Chain reactions are usually very rapid; some of them are explosive.

THE EXACT TRUTH

Prior to 1968, perbromates (compounds of the BrO_4^- ion) could not be prepared although perchlorates and periodates were well known. Many papers were published to explain theoretically why perbromates did not exist and could not be made. Today, several methods for the preparation of these compounds are known. The problem probably arose because the energy of activation for the preparation of the BrO_4^- ion from BrO_3^- is thought to be unusually high.

Before 1963, the group 0 elements were called *inert gases* rather than *noble gases* (Section 8.10). Up until that time, all attempts to prepare ordinary chemical compounds of these elements had failed. Chemists stopped trying because they became convinced that the gases were inert and could not be made to undergo ordinary chemical reactions. In 1963, the first reaction of a noble-gas element was reported.

We sometimes get the idea that chemistry has arrived at such a state of perfection that all the answers are either at hand or will be as soon as they can be worked out. During the nineteenth century, many physicists thought that most of the theoretical work of physics had been accomplished and that all that remained to be done was to fill in the details. This attitude was severely shaken by the end of the century when atomic studies showed that classical physics could not be used to explain the world of the atom.

During the nineteenth century, many chemists thought that inorganic chemistry was approaching the end of its task. The attitude of the age is probably best typified by the patent official of the late nineteenth century who resigned his post because he thought that everything of significance had already been invented.

Science, however, grows and changes. The final word on any phenomenon is never written. As Bertrand Russell said: "When a man tells you that he knows the exact truth about anything, you are safe in inferring that he is an in-exact man."

PROBLEMS

13.1 List four factors that influence the rate of a chemical reaction and explain on the atomic-molecular level how each exerts its influence.

13.2 Define the following: (a) activated complex, (b) energy of activation, (c) reaction order, (d) catalyst, (e) chemisorption, (f) rate-determining step, (g) reaction intermediate, (h) zero-order reaction, (i) steady state.

13.3 The rate equation for the reaction

$$2NO(g) + 2H_2(g) \rightarrow N_2(g) + 2H_2O(g)$$

is second order in NO(g) and first order in $H_2(g)$. (a) Write an equation for the rate of appearance of $N_2(g)$. (b) If concentrations are expressed in mol/liter, what units would the rate constant, k, have? (c) Write an equation for the rate of disappearance of NO(g). Would k in this equation have the same numerical value as k in the equation of part (a)?

13.4 Why are some collisions between molecules of reactants not effective?

13.5 Draw a potential energy diagram for the hypothetical reaction $A(g) + B(g) \rightleftharpoons 2C(g)$ for which E_a is 37 kcal and E'_a is 25 kcal. Indicate the magnitude of ΔH on the diagram. Is the reaction exothermic or endothermic? Draw a curve for a catalyzed reaction on the diagram.

13.6 What is the difference between homogeneous and heterogeneous catalysis? Describe the way in which each type of catalyst functions.

13.7 The rate equation for the reaction $A \rightarrow B + C$ is expressed in the form

$$\text{rate of disappearance of A} = k[A]^x$$

The value of the rate constant, k, is 0.100 (units unspecified) and [A] = 0.050 mol/liter. What are the units of k and the rate of the reaction (in mol/liter·s) if the reaction is: (a) zero order in A, (b) first order in A, (c) second order in A?

13.8 The rate equation for the reaction $A \rightarrow B + C$ is expressed in terms of the concentration of A only. The rate of disappearance of A is 0.0080 mol/liter·s when [A] = 0.20 mol/liter. Calculate the value of k if the reaction is: (a) zero order in A, (b) first order in A, (c) second order in A.

13.9 For a reaction in which A and B form C, the following data were obtained from three experiments.

[A] (mol/liter)	[B] (mol/liter)	*Rate of formation of C* (mol/liter·s)
0.30	0.15	7.0×10^{-4}
0.60	0.30	2.8×10^{-3}
0.30	0.30	1.4×10^{-3}

(a) What is the rate equation for the reaction? (b) What is the numerical value of the rate constant, k?

13.10 For a reaction in which A and B form C, the following data were obtained from three experiments.

[A] (mol/liter)	[B] (mol/liter)	*Rate of formation of C* (mol/liter·s)
0.03	0.03	0.3×10^{-4}
0.06	0.06	1.2×10^{-4}
0.06	0.09	2.7×10^{-4}

(a) What is the rate equation for the reaction? (b) What is the numerical value of the rate constant, k?

13.11 For the reaction

$$2NO(g) + Cl_2(g) \rightarrow 2NOCl(g)$$

the rate equation is second order in NO(g), first order in $Cl_2(g)$, and third order overall. Compare the initial rate of reaction of a mixture of 0.02 mol NO(g) and 0.02 mol $Cl_2(g)$ in a

1 liter container with (a) the rate of the reaction when half the NO(g) has been consumed, (b) the rate of the reaction when half the $Cl_2(g)$ has been consumed, (c) the rate of the reaction when two-thirds of NO(g) has been consumed, (d) the initial rate of a mixture of 0.04 mol NO(g) and 0.02 mol $Cl_2(g)$ in a 1 liter container, (e) the initial rate of a mixture of 0.02 mol NO(g) and 0.02 mol $Cl_2(g)$ in a 0.5 liter container.

13.12 The reaction: $HI(g) + CH_3I(g) \rightarrow CH_4(g) + I_2(g)$ is first order in each of the reactants and second order overall. The rate constant is 1.7×10^{-5} liter/mol·s at 430 K and 9.6×10^{-5} liter/mol·s at 450 K. What is the energy of activation for the reaction?

13.13 The reaction: $C_2H_5Cl(g) \rightarrow C_2H_4(g) + HCl(g)$ is first order in C_2H_5Cl. The rate constant is 3.5×10^{-8}/s at 600 K and 1.6×10^{-6}/s at 650 K. What is the energy of activation for the reaction?

13.14 The reaction: $C_2H_4(g) + H_2(g) \rightarrow C_2H_6(g)$ is first order in C_2H_4, first order in H_2, and second order overall. The energy of activation for the reaction is 181 kJ/mol and k is 1.3×10^{-3} liter/mol·s at 700 K. What is the value of k for the reaction conducted at 730 K?

13.15 The reaction: $NO(g) + N_2O(g) \rightarrow NO_2(g) + N_2(g)$ is first order in each of the reactants and second order overall. The energy of activation of the reaction is 209 kJ/mol, and k is 0.77 liter/mol·s at 950 K. What is the value of k for a reaction conducted at 1000 K?

13.16 The reaction: $C_2H_5Br(g) \rightarrow C_2H_4(g) + HBr(g)$ is first order in C_2H_5Br. The rate constant is 2.0×10^{-5}/s at 650 K, and the energy of activation is 226 kJ/mol. At what temperature is the rate constant 6.0×10^{-5}/s?

13.17 What is the energy of activation of a reaction that increases tenfold in rate when the temperature is increased from 300 K to 310 K?

13.18 At 400 K a certain reaction is 50.0% complete in 1.50 min. At 430 K the same reaction is 50.0% complete in 0.50 min. Calculate the activation energy for the reaction.

13.19 The reaction

$$CH_4(g) + Cl_2(g) \rightarrow CH_3Cl(g) + HCl(g)$$

proceeds by a chain mechanism. The chain propagators are Cl atoms and CH_3 radicals, and it is believed that free H atoms are not involved. Write a series of equations showing the mechanism and identify the chain-initiating, chain-sustaining, and chain-terminating steps.

13.20 The rate equation for the reaction:

$$2O_3 \rightarrow 3O_2$$

has been determined experimentally as

$$\text{rate of disappearance of } O_3 = k\frac{[O_3]^2}{[O_2]}$$

The following mechanism has been postulated for the decomposition of ozone.

$$O_3 \xrightarrow{k_1} O_2 + O$$

$$O_2 + O \xrightarrow{k_2} O_3$$

$$O + O_3 \xrightarrow{k_3} 2O_2$$

The third step of the mechanism is the rate-determining step. Assume that the concentration of O reaches a steady state and find an expression for [O]. Remember that k_3 is much smaller than either k_1 or k_2. Show how the suggested mechanism leads to the observed rate expression.

Chapter 14

Chemical Equilibrium

Under suitable conditions, nitrogen and hydrogen react to form ammonia.

$$N_2(g) + 3H_2(g) \rightarrow 2NH_3(g)$$

On the other hand, ammonia decomposes at high temperatures to yield nitrogen and hydrogen.

$$2NH_3(g) \rightarrow N_2(g) + 3H_2(g)$$

The reaction is reversible, and the equation for the reaction can be written

$$N_2(g) + 3H_2(g) \rightleftharpoons 2NH_3(g)$$

The double arrow ($\rightleftharpoons$) indicates that the equation can be read in either direction.

All reversible processes tend toward a state of equilibrium (Section 9.5). In this chapter, we will study equilibrium systems that involve reversible chemical reactions.

14.1 REVERSIBLE REACTIONS AND CHEMICAL EQUILIBRIUM

Consider a hypothetical reversible reaction

$$A_2(g) + B_2(g) \rightleftharpoons 2AB(g)$$

This equation may be *read either forward or backward.* If A_2 and B_2 are mixed, they will react to produce AB. A sample of pure AB, on the other hand, will decompose to form A_2 and B_2.

Suppose that we place a mixture of A_2 and B_2 in a container. They will react to produce AB, and their concentrations will gradually decrease as the forward reaction occurs (see Figure 14.1). Since the concentrations of A_2 and B_2 decrease, the rate of the forward reaction decreases.

At the start of the experiment, the backward reaction cannot occur since there is no AB present. As soon as the forward reaction produces some AB, however, the backward reaction begins. The backward reaction starts slowly (since the concentration of AB is low) and gradually picks up speed.

As time passes, the rate of the forward reaction decreases and the rate of the backward reaction increases until the two rates are equal. At this point, **chemical equilibrium** is established. An equilibrium condition is one in which the rates of two opposing tendencies are equal.

At equilibrium, the concentrations of all the substances are constant.

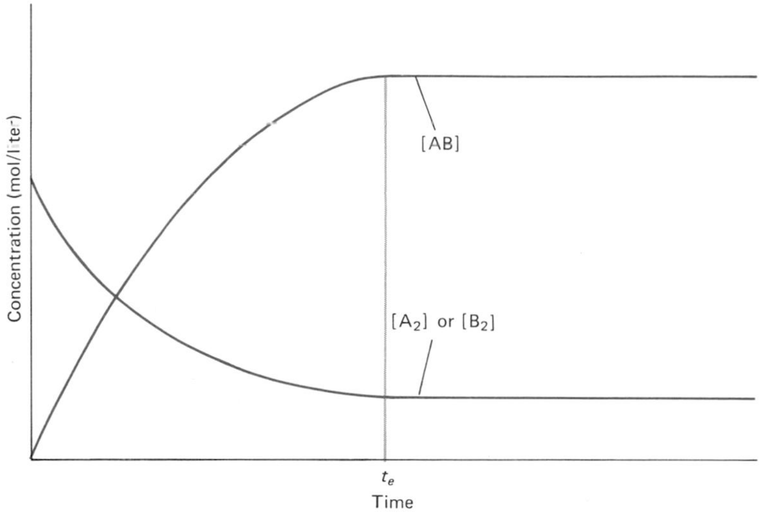

Figure 14.1 Curves showing changes in concentrations of materials with time for the reaction $A_2 + B_2 \rightleftharpoons 2AB$. Equilibrium is attained at time t_e.

The concentration of AB is constant since AB is produced by the forward reaction at the same rate that it is used by the backward reaction. In like manner, A_2 and B_2 are being made (by the backward reaction) at the same rate that they are being used (by the forward reaction). It is important to note that the concentrations are constant because the rates of the opposing reactions are equal and *not* because all activity has stopped. Data for the experiment are plotted in Figure 14.1. Equilibrium is attained at time t_e.

If we assume that the backward and forward reactions occur by simple one-step mechanisms, the rate of the forward reaction is

$$\text{rate}_f = k_f[A_2][B_2]$$

and the rate of the backward reaction is

$$\text{rate}_b = k_b[AB]^2$$

At equilibrium, these two rates are equal, and therefore

$$\text{rate}_f = \text{rate}_b$$
$$k_f[A_2][B_2] = k_b[AB]^2$$

This equation can be rearranged to

$$\frac{k_f}{k_b} = \frac{[AB]^2}{[A_2][B_2]}$$

The rate constant of the forward reaction, k_f, divided by the rate constant of the backward reaction, k_b, is equal to a third constant, which is called the **equilibrium constant,** K.

$$\frac{k_f}{k_b} = K$$

Therefore,

$$K = \frac{[AB]^2}{[A_2][B_2]}$$

The numerical value of K varies with temperature. There are an unlimited number of possible equilibrium systems for this reaction. The concentrations, however, of A_2, B_2, and AB for *any* system in equilibrium at a given temperature will, when expressed in the preceding manner, equal the same value of K.

In general, for any reversible reaction

$$wW + xX \rightleftharpoons yY + zZ$$

at equilibrium,

$$K = \frac{[Y]^y[Z]^z}{[W]^w[X]^x}$$

By convention, the concentration terms for the materials on the *right* of the chemical equation are written in the *numerator* of the expression for the equilibrium constant.

If the equation is written in the reverse form,

$$yY + zZ \rightleftharpoons wW + xX$$

the expression for the equilibrium constant (which we will indicate as K') is

$$K' = \frac{[W]^w[X]^x}{[Y]^y[Z]^z}$$

Notice that K' is the reciprocal of K.

$$K' = \frac{1}{K}$$

The equations that we have derived were first derived in 1864 by Cato Guldberg and Peter Waage in much the same way that we have derived them. These equations are statements of the **law of chemical equilibrium.**

In our derivation, we assumed that the forward and reverse reactions occurred by mechanisms each consisting of a single step. Does the law of chemical equilibrium hold for reactions that occur by mechanisms of more than one step? The answer to the question is that it does. It is, however, not immediately apparent why it does.

Consider the reaction

$$2NO_2Cl(g) \rightleftharpoons 2NO_2(g) + Cl_2(g)$$

for which the expression for the equilibrium constant is

$$K = \frac{[NO_2]^2[Cl_2]}{[NO_2Cl]^2}$$

The reaction is believed to occur by means of a mechanism consisting of two steps.

1. $$NO_2Cl \underset{k_1'}{\overset{k_1}{\rightleftharpoons}} NO_2 + Cl$$

2. $$NO_2Cl + Cl \underset{k_2'}{\overset{k_2}{\rightleftharpoons}} NO_2 + Cl_2$$

Symbols for the rate constants appear above and below the arrows in these equations. The symbol k is used for the rate constant of the forward reaction, and k' is used for the backward reaction. The subscripts of these symbols designate the steps.

Since the overall reaction is reversible, each step of the mechanism must also be reversible. When equilibrium is established for the overall reaction, each step of the mechanism must be in equilibrium. Therefore,

$$K_1 = \frac{k_1}{k'_1} = \frac{[NO_2][Cl]}{[NO_2Cl]}$$

and

$$K_2 = \frac{k_2}{k'_2} = \frac{[NO_2][Cl_2]}{[NO_2Cl][Cl]}$$

The product of these expressions is

$$K_1K_2 = \frac{k_1k_2}{k'_1k'_2} = \frac{[NO_2][Cl]}{[NO_2Cl]}\frac{[NO_2][Cl_2]}{[NO_2Cl][Cl]} = \frac{[NO_2]^2[Cl_2]}{[NO_2Cl]^2}$$

which is the same as the expression for the equilibrium constant that we derived directly from the equation for the overall change. In this case, the equilibrium constant for the overall change is the product of the equilibrium constants of each of the steps.

$$K = K_1K_2$$

14.2 EQUILIBRIUM CONSTANTS

For the reaction

$$H_2(g) + I_2(g) \rightleftharpoons 2HI(g)$$

the equilibrium constant at 425°C is

$$K = \frac{[HI]^2}{[H_2][I_2]} = 54.8$$

The numerical value of K must be determined experimentally. If the concentrations of the materials (in mol/liter) present in any equilibrium mixture at 425°C are substituted into the expression for the equilibrium constant, the result will equal 54.8. If any value other than 54.8 is obtained, the mixture is not in equilibrium.

The equilibrium condition may be approached from either direction. That is, an equilibrium mixture can be obtained by mixing hydrogen and iodine, by allowing pure hydrogen iodide to dissociate, or by mixing all three materials.

The magnitude of the value of the equilibrium constant gives an indication of the position of equilibrium. Recall that the concentration terms of substances on the right of the equation are written in the numerator of the expression for the equilibrium constant. For the reaction

$$CO(g) + Cl_2(g) \rightleftharpoons COCl_2(g)$$

at 100°C,

$$K = \frac{[COCl_2]}{[CO][Cl_2]} = 4.57 \times 10^9 \text{ liter/mol}$$

From this relatively large value of *K*, we conclude that equilibrium concentrations of CO and Cl_2 are small and that the synthesis of $COCl_2$ is virtually complete. In other words, the reaction to the right is fairly complete at equilibrium.

For the reaction

$$N_2(g) + O_2(g) \rightleftharpoons 2NO(g)$$

at 2000°C,

$$K = \frac{[NO]^2}{[N_2][O_2]} = 4.08 \times 10^{-4}$$

We conclude from this small value of *K* that NO is largely dissociated into N_2 and O_2 at equilibrium. The reaction to the left is fairly complete.

Equilibria between substances in two or more phases are called **heterogeneous equilibria.** The concentration of a pure solid or a pure liquid is constant when the temperature and pressure are constant. For any heterogeneous equilibrium, therefore, the values of the concentrations of solids or liquids involved are included in the value of *K*, and concentration terms for these substances do not appear in the expression for the equilibrium constant.

For example, for the reaction

$$CaCO_3(s) \rightleftharpoons CaO(s) + CO_2(g)$$

the values for the concentrations of CaO and $CaCO_3$ are included in the value of *K*, and the expression for the equilibrium constant is

$$K = [CO_2]$$

Hence, at any fixed temperature the equilibrium concentration of CO_2 over a mixture of the solids is a definite value. The equilibrium constant for the reaction

$$3Fe(s) + 4H_2O(g) \rightleftharpoons Fe_3O_4(s) + 4H_2(g)$$

is expressed in the following terms

$$K = \frac{[H_2]^4}{[H_2O]^4}$$

Some facts dealing with expressions for equilibrium constants may be summarized as follows.

1. The concentration terms for substances that appear on the *right side* of the chemical equation are written in the *numerator* of the expression for *K*. The concentration terms for substances that appear on the *left side* are written in the *denominator.*
2. No concentration terms are included for pure solids or pure liquids. The value of *K* includes these terms.

3. The value of K for a given equilibrium is constant if the equilibrium temperature does not change. At a different temperature, the value of K is different.
4. The magnitude of K for a given equilibrium indicates the position of equilibrium. A *large* value of K indicates that the reaction to the *right* is fairly complete. A *small* value of K indicates that the reaction to the *left* is fairly complete. A value of K that is neither very large nor very small indicates an intermediate situation.*

The following examples illustrate how some simple problems involving equilibrium constants may be solved.

Example 14.1 For the reaction

$$N_2O_4(g) \rightleftharpoons 2NO_2(g)$$

the concentrations of the substances present in an equilibrium mixture at 25°C are

$$[N_2O_4] = 4.27 \times 10^{-2} \text{ mol/liter}$$
$$[NO_2] = 1.41 \times 10^{-2} \text{ mol/liter}$$

What is the value of K for this temperature?

Solution

$$K = \frac{[NO_2]^2}{[N_2O_4]}$$

$$= \frac{(1.41 \times 10^{-2} \text{ mol/liter})^2}{(4.27 \times 10^{-2} \text{ mol/liter})}$$

$$= 4.66 \times 10^{-3} \text{ mol/liter}$$

Example 14.2 At 500 K, 1.00 mol of NOCl(g) is introduced into a 1 liter container. At equilibrium, the NOCl(g) is 9.0% dissociated.

$$2NOCl(g) \rightleftharpoons 2NO(g) + Cl_2(g)$$

Calculate the value of K for the equilibrium at 500 K.

Solution By considering exactly 1 liter, the problem is simplified since the number of moles of a gas is the same as the concentration of the gas (in mol/liter).

Since the NOCl is 9.0% dissociated,

$$\text{number of moles dissociated} = 0.090(1.00 \text{ mol}) = 0.090 \text{ mol NOCl}$$

* To be strictly accurate, equations for equilibrium constants should be written in terms of activities rather than concentrations or pressures (Section 17.7). At low concentrations and pressures up to a few atmospheres, however, concentrations may be used with reasonable accuracy.

We must subtract this quantity from the number of moles of NOCl present initially. The concentration of NOCl at equilibrium, therefore, is

$$[NOCl] = 1.00 \text{ mol/liter} - 0.090 \text{ mol/liter} = 0.91 \text{ mol/liter}$$

The NOCl that dissociates produces NO and Cl_2. We can derive the amounts of these substances produced from the coefficients of the chemical equation.

$$\underset{}{2NOCl} \rightleftharpoons \underset{0.090 \text{ mol}}{2NO} + \underset{0.045 \text{ mol}}{Cl_2}$$

Since 2 mol of NOCl produces 2 mol of NO, 0.090 mol of NOCl will produce 0.090 mol of NO. The equation shows that 2 mol of NOCl produces only 1 mol of Cl_2, therefore, 0.090 mol of NOCl will produce 0.045 mol of Cl_2. The equilibrium concentrations are

$$[NOCl] = 0.91 \text{ mol/liter}$$
$$[NO] = 0.090 \text{ mol/liter}$$
$$[Cl_2] = 0.045 \text{ mol/liter}$$

Therefore,

$$K = \frac{[NO]^2[Cl_2]}{[NOCl]^2}$$

$$= \frac{(0.090 \text{ mol/liter})^2(0.045 \text{ mol/liter})}{(0.91 \text{ mol/liter})^2}$$

$$= 4.4 \times 10^{-4} \text{ mol/liter}$$

Example 14.3 K for the HI equilibrium at 425°C is 54.8.

$$H_2(g) + I_2(g) \rightleftharpoons 2HI(g)$$

A quantity of HI(g) is placed in a 1.00 liter container and allowed to come to equilibrium at 425°C. What are the concentrations of $H_2(g)$ and $I_2(g)$ in equilibrium with 0.50 mol/liter of HI(g)?

Solution The concentrations of $H_2(g)$ and $I_2(g)$ must be equal since they are produced in equal amounts by the decomposition of HI(g). Therefore, let

$$[H_2] = [I_2] = x$$

We are told that the equilibrium concentration of HI is

$$[HI] = 0.50 \text{ mol/liter}$$

We substitute these values into the equation for the equilibrium constant and solve for x.

$$K = \frac{[HI]^2}{[H_2][I_2]} = 54.8$$

$$\frac{(0.50 \text{ mol/liter})^2}{x^2} = 54.8$$

$$54.8\,x^2 = 0.25 \text{ mol}^2/\text{liter}^2$$
$$x^2 = 0.00456 \text{ mol}^2/\text{liter}^2$$
$$x = 0.068 \text{ mol/liter}$$

The equilibrium concentrations are

$$[HI] = 0.50 \text{ mol/liter}$$
$$[H_2] = [I_2] = 0.068 \text{ mol/liter}$$

Example 14.4 For the reaction

$$H_2(g) + CO_2(g) \rightleftharpoons H_2O(g) + CO(g)$$

K is 0.771 at 750°C. If 0.0100 mol of H_2 and 0.0100 mol of CO_2 are mixed in a 1 liter container at 750°C, what are the concentrations of all substances present at equilibrium?

Solution If x mol of H_2 reacts with x mol of CO_2 out of the total amounts supplied, x mol of H_2O and x mol of CO will be produced. Since the container has a volume of 1 liter, at equilibrium the concentrations are (in mol/liter):

$$\underset{(0.0100-x)}{H_2(g)} + \underset{(0.0100-x)}{CO_2(g)} \rightleftharpoons \underset{x}{H_2O(g)} + \underset{x}{CO(g)}$$

$$K = \frac{[H_2O][CO]}{[H_2][CO_2]} = 0.771$$

$$= \frac{x^2}{(0.0100-x)(0.0100-x)} = 0.771$$

If we find the square root of both sides of this equation, we get

$$\frac{x}{(0.0100-x)} = 0.878$$

$$x = 0.00468$$

Therefore, at equilibrium

$$[H_2] = [CO_2] = (0.0100 - x)$$
$$= 0.0053 \text{ mol/liter}$$
$$= 5.3 \times 10^{-3} \text{ mol/liter}$$

$$[H_2O] = [CO] = x = 0.00468 \text{ mol/liter}$$
$$= 4.68 \times 10^{-3} \text{ mol/liter}$$

14.3 EQUILIBRIUM CONSTANTS EXPRESSED IN PRESSURES

The partial pressure of a gas is a measure of its concentration. Equilibrium constants for reactions involving gases, therefore, may be written in terms of the partial pressures of the reacting gases. An equilibrium constant of this type is given the designation K_p.

For the calcium carbonate equilibrium

$$CaCO_3(s) \rightleftharpoons CaO(s) + CO_2(g)$$

the equilibrium constant in terms of partial pressures is

$$K_p = p_{CO_2}$$

For the equilibrium

$$N_2(g) + 3H_2(g) \rightleftharpoons 2NH_3(g)$$

the K_p is

$$K_p = \frac{(p_{NH_3})^2}{(p_{N_2})(p_{H_2})^3}$$

There is a simple relation between the K_p for a reaction and the equilibrium constant derived from concentrations. Consider the reaction

$$wW + xX \rightleftharpoons yY + zZ$$

If *all* these materials are *gases,*

$$K_p = \frac{(p_Y)^y(p_Z)^z}{(p_W)^w(p_X)^x}$$

Assume that each of the gases follows the ideal gas law

$$PV = nRT$$

Then the partial pressure of any gas, p, is

$$p = \frac{n}{V}RT$$

The concentration of a gas in mol/liter is equal to n/V. Therefore, for gas W

$$p_W = [W]RT$$

$$(p_W)^w = [W]^w(RT)^w$$

If we substitute expressions such as this for the partial pressure terms in the expression for K_p, we get

$$K_p = \frac{[Y]^y(RT)^y[Z]^z(RT)^z}{[W]^w(RT)^w[X]^x(RT)^x}$$

$$= \frac{[Y]^y[Z]^z}{[W]^w[X]^x}(RT)^{+y+z-w-x}$$

The fractional term in the last equation is equal to K.

$$K_p = K(RT)^{+y+z-w-x}$$

If we read the chemical equation for the reaction

$$wW + xX \rightleftharpoons yY + zZ$$

in molar quantities,

$y + z =$ number of moles of gases on the right
$w + x =$ number of moles of gases on the left

We let Δn equal the change in the number of moles of gases when the equation is read from left to right.

$$\Delta n = (y + z) - (w + x) = +y + z - w - x$$

Therefore,

$$K_p = K(RT)^{\Delta n}$$

Partial pressures are expressed in atmospheres, concentrations are expressed in moles per liter, R is 0.08206 liter atm/K mol, T is the absolute temperature in K.

For the reaction

$$PCl_5(g) \rightleftharpoons PCl_3(g) + Cl_2(g)$$

Δn is +1. Therefore,

$$K_p = K(RT)^{+1}$$

For the reaction

$$CO(g) + Cl_2(g) \rightleftharpoons COCl_2(g)$$

$\Delta n = -1$. Therefore,

$$K_p = K(RT)^{-1} \qquad \text{or} \qquad K_p = \frac{K}{(RT)}$$

For the reaction

$$H_2(g) + I_2(g) \rightleftharpoons 2HI(g)$$

$\Delta n = 0$. Therefore

$$K_p = K(RT)^0 \qquad \text{or} \qquad K_p = K$$

Example 14.5 At 1100 K, the equilibrium constant for the reaction

$$2SO_3(g) \rightleftharpoons 2SO_2(g) + O_2(g)$$

is 0.0271 mol/liter. What is K_p at this temperature?

Solution Two moles of $SO_3(g)$ produce a total of 3 mol of gases. Therefore,

$$\Delta n = +1$$
$$K_p = K(RT)^{+1}$$
$$= (0.0271 \text{ mol/liter})[(0.0821 \text{ liter atm/K mol})(1100 \text{ K})]$$
$$= 2.45 \text{ atm}$$

Example 14.6 What is K for the reaction

$$N_2(g) + 3H_2(g) \rightleftharpoons 2NH_3(g)$$

at 500°C if K_p is $1.50 \times 10^{-5}/\text{atm}^2$ at this temperature?

Solution There are 4 mol of gases indicated on the left of the chemical equation and 2 mol of gases indicated on the right. Therefore,

$$\Delta n = -2$$

The temperature is 500°C, which is

$$T = 773 \text{ K}$$

Therefore,

$$K_p = K(RT)^{\Delta n}$$

$$(1.50 \times 10^{-5}/\text{atm}^2) = K[(0.0821 \text{ liter atm/K mol})(773 \text{ K})]^{-2}$$

$$(1.50 \times 10^{-5}/\text{atm}^2) = \frac{K}{(63.5 \text{ liter atm/mol})^2}$$

$$K = (1.50 \times 10^{-5}/\text{atm}^2)(4.03 \times 10^3 \text{ liter}^2 \text{ atm}^2/\text{mol}^2)$$

$$= 6.04 \times 10^{-2} \text{ liter}^2/\text{mol}^2$$

Example 14.7 For the reaction

$$C(s) + CO_2(g) \rightleftharpoons 2CO(g)$$

K_p is 167.5 atm at 1000°C. What is the partial pressure of CO(g) in an equilibrium system in which the partial pressure of CO_2(g) is 0.100 atm?

Solution

$$K_p = \frac{p_{CO}^2}{p_{CO_2}} = 167.5 \text{ atm}$$

$$\frac{p_{CO}^2}{0.100 \text{ atm}} = 167.5 \text{ atm}$$

$$p_{CO}^2 = 16.8 \text{ atm}^2$$

$$p_{CO} = 4.10 \text{ atm}$$

Example 14.8 K_p for the equilibrium

$$FeO(s) + CO(g) \rightleftharpoons Fe(s) + CO_2(g)$$

at 1000°C is 0.403. If CO(g), at a pressure of 1.000 atm, and excess FeO(s) are placed in a container at 1000°C, what are the pressures of CO(g) and CO_2(g) when equilibrium is attained?

Solution Let x equal the partial pressure of CO_2 when equilibrium is attained.

Then,

$$\underset{(1.000 - x)\text{ atm}}{FeO(s) + CO(g)} \rightleftharpoons \underset{x\text{ atm}}{Fe(s) + CO_2(g)}$$

$$K_p = \frac{p_{CO_2}}{p_{CO}} = 0.403$$

$$\frac{x\text{ atm}}{(1.000 - x)\text{ atm}} = 0.403$$

$$x = p_{CO_2} = 0.287\text{ atm}$$

$$1.000 - x = p_{CO} = 0.713\text{ atm}$$

14.4 LE CHATELIER'S PRINCIPLE

What happens to an equilibrium system if an experimental condition (such as temperature or pressure) is changed? The effects of such changes were summarized in 1884 by Henri Le Chatelier. Le Chatelier's principle states that a system in equilibrium reacts to a stress in a way that counteracts the stress and establishes a new equilibrium state. This important generalization is very simple to apply.

1. Concentration changes If the concentration of a substance is *increased,* the equilibrium will shift in a way that will *decrease* the concentration of the substance that was added. Suppose that we have a system in equilibrium

$$H_2(g) + I_2(g) \rightleftharpoons 2HI(g)$$

and we *increase* the concentration of H_2. The equilibrium is upset, and the system will react in a way that will *decrease* the concentration of H_2. It can do that by using up some of the H_2 (and some I_2 as well) to form more HI. When equilibrium is established again, the concentration of HI will be higher than it was initially. The position of equilibrium is said to have shifted to the right.

If the concentration of HI is increased, the position of equilibrium will shift to the left. In this way some HI will be used up. When equilibrium is established again, the concentrations of H_2 and I_2 will be higher than they were initially.

Removal of one of the substances from an equilibrium system will also cause the position of equilibrium to shift. If, for example, HI could be removed, the position of equilibrium would shift to the right. Additional HI would be produced and the concentrations of H_2 and I_2 would decrease.

By the continuous removal of a product it is possible to drive some reversible reactions "to completion." Complete conversion of $CaCO_3(s)$ into $CaO(s)$

$$CaCO_3(s) \rightleftharpoons CaO(s) + CO_2(g)$$

can be accomplished by removing the CO_2 gas as fast as it is produced.

2. *Pressure changes* Le Chatelier's principle may also be used to make qualitative predictions of the effect of pressure changes on systems in equilibrium. Consider the effect of a pressure increase on an equilibrium mixture of SO_2, O_2, and SO_3.

$$2SO_2(g) + O_2(g) \rightleftharpoons 2SO_3(g)$$

In the forward reaction two gas molecules ($2SO_3$) are produced by the disappearance of three gas molecules ($2SO_2 + O_2$). Two gas molecules do not exert as high a pressure as three gas molecules. When the pressure on an equilibrium mixture is increased (or the volume of the system decreased), the position of equilibrium shifts to the right. In this way the system counteracts the change. Alternatively, decreasing the pressure (or increasing the volume) causes the position of this equilibrium to shift to the left.

For reactions in which $\Delta n = 0$, pressure changes have no effect on the position of equilibria. Equilibria involving the systems

$$H_2(g) + I_2(g) \rightleftharpoons 2HI(g)$$
$$N_2(g) + O_2(g) \rightleftharpoons 2NO(g)$$
$$H_2(g) + CO_2(g) \rightleftharpoons H_2O(g) + CO(g)$$

are not influenced by changing the pressure, since there is no difference in the total volume in either the forward or backward direction for any one of these reactions.

For a system that involves only liquids and solids, the effect of pressure on the position of equilibrium is slight and may usually be ignored for ordinary changes in pressure. Large pressure changes, however, can significantly alter such equilibria, and, at times, even slight changes in such equilibria are of interest. For example, the position of equilibrium

$$H_2O(s) \rightleftharpoons H_2O(l)$$

is forced to the right by an increase in pressure because a given quantity of water occupies a smaller volume in the liquid state than in the solid state (its density is higher in the liquid state).

Pressure changes affect equilibria involving gases to a much greater degree. For example, a high pressure would favor the production of a high yield of ammonia from the equilibrium

$$N_2(g) + 3H_2(g) \rightleftharpoons 2NH_3(g)$$

Hence, Le Chatelier's principle is of practical importance as an aid in determining favorable reaction conditions for the production of a desired substance.

For heterogeneous equilibria the effect of pressure is predicted by counting the number of *gas* molecules on each side of the equation. For example, the position of equilibrium

$$3Fe(s) + 4H_2O(g) \rightleftharpoons Fe_3O_4(s) + 4H_2(g)$$

is virtually unaffected by pressure because there are four gas molecules on each side of the equation.

3. Temperature changes In order to predict the effect of a temperature change on a system in equilibrium, the nature of the heat effect that accompanies the reaction must be known. At 25°C, the thermochemical equation for the synthesis of ammonia is

$$N_2(g) + 3H_2(g) \rightleftharpoons 2NH_3(g) \qquad \Delta H = -92.4 \text{ kJ}$$

Since ΔH is negative, the reaction to the right evolves heat. We can write the equation to indicate heat as a product.

$$N_2(g) + 3H_2(g) \rightleftharpoons 2NH_3(g) + 92.4 \text{ kJ}$$

The forward reaction is *exothermic,* and the backward reaction is *endothermic.* In other words, the forward reaction produces heat, and the backward reaction uses heat. If heat is added (the temperature of the system is raised), the position of equilibrium will shift to the left—the direction in which heat is absorbed. If the mixture is cooled, the position of equilibrium will shift to the right—the direction in which heat is evolved. The highest yields of NH_3 will be obtained at the lowest temperatures. Unfortunately, if the temperature is too low, the reaction will be extremely slow. High pressures and temperatures around 500°C are employed in the commercial process.

Consider the reaction

$$CO_2(g) + H_2(g) \rightleftharpoons CO(g) + H_2O(g) \qquad \Delta H = +41.1 \text{ kJ}$$

Since ΔH is positive, the forward reaction is endothermic. We write the equation

$$41.1 \text{ kJ} + CO_2(g) + H_2(g) \rightleftharpoons CO(g) + H_2O(g)$$

Increasing the temperature *always* favors the endothermic change, and decreasing the temperature *always* favors the exothermic change. In this case, the reaction is forced to the right by an increase in temperature. A decrease in temperature causes the position of equilibrium to shift to the left.

The numerical value of the equilibrium constant changes when the temperature is changed. The reaction of CO_2 and H_2 is shifted to the right by an increase in temperature. The concentrations of the substances on the right are increased by this shift. The terms for the concentrations of these substances appear in the numerator of the expression for K. As the temperature is increased, therefore, the value of K increases. For this reaction, at 700°C, $K = 0.63$, and at 1000°C, $K = 1.66$.

4. Addition of a catalyst The presence of a catalyst has no effect on the position of a chemical equilibrium since a catalyst affects the rates of the forward and backward reactions equally (see Section 13.4). A catalyst will, however, cause a system to attain equilibrium more rapidly than it otherwise would.

POISONED BRANDY AND SODIUM CARBONATE

Claude Louis Berthollet (1748–1822) was a prominent and productive French chemist, one of the first to accept Lavoisier's new system of chemistry, and a friend of Napoleon. During the reign of terror he was asked to analyze some brandy which had been sent to the French troops. The liquor was thought to have been poisoned, and the wine merchants were to be executed for having sold it. Berthollet reported that the brandy was safe to drink.

Few students are as confident of their laboratory results as Berthollet was of his. When summoned by Robespierre, who insisted that the brandy was poisonous, Berthollet drank a glass. The act infuriated the French leader, but the wine merchants were released.

During this period of French history, the government began to use the services of scientists extensively. In 1798, Berthollet accompanied Napoleon as a scientific adviser on the expedition to Egypt. Berthollet noted that deposits of sodium carbonate (Na_2CO_3) occur along the banks of Egyptian salt lakes. He concluded that the carbonate had been produced by the reaction of limestone ($CaCO_3$), present in the sand of the banks, with sodium chloride (NaCl) from the lake water.

$$2NaCl + CaCO_3 \rightarrow Na_2CO_3 + CaCl_2$$

This reaction, however, was known to proceed in the opposite direction! When solutions of Na_2CO_3 and $CaCl_2$ (both of which are soluble in water) are mixed, $CaCO_3$ (which is insoluble in water) precipitates.

Berthollet claimed that the reaction is reversible. He thought that the high concentration of NaCl in the evaporating water along the banks of the lakes caused the reaction that produced Na_2CO_3 to occur. He was, therefore, the first to state that reactions could be reversed by concentration changes. Later, he found many examples of reversible systems that supported his claims.

Eventually, however, Berthollet carried the idea of reversibility too far. He stated that the compositions of compounds are not constant but vary with the amounts of substances used to prepare them. He, therefore, refused to accept the law of definite proportions, which had been proposed and was being developed at this time by Louis Proust. Berthollet and Proust engaged in an argument over the matter that lasted for many years (see Section 9.15).

PROBLEMS

14.1 For each of the following reactions, write an expression for the equilibrium constant, K.

(a) $2CO_2(g) \rightleftharpoons 2CO(g) + O_2(g)$

(b) $2Pb_3O_4(s) \rightleftharpoons 6PbO(s) + O_2(g)$

(c) $2Ag_2O(s) \rightleftharpoons 4Ag(s) + O_2(g)$

(d) $CH_4(g) + 2H_2S(g) \rightleftharpoons CS_2(g) + 4H_2(g)$

(e) $C(s) + CO_2(g) \rightleftharpoons 2CO(g)$

14.2 Indicate in which direction each of the equilibria given in Problem 14.1 will shift with an increase in pressure.

14.3 For the equilibrium

$$CO_2(g) + H_2(g) \rightleftharpoons CO(g) + H_2O(g)$$

K is 0.08 at 400°C and 0.41 at 600°C. Is the reaction as written exothermic or endothermic?

14.4 For the equilibrium

$$2SO_2(g) + O_2(g) \rightleftharpoons 2SO_3(g)$$

K_p is 1.0×10^5/atm at 700 K and 1.1×10^3/atm at 800 K. Is the reaction as written exothermic or endothermic?

14.5 The reaction $A(g) + B(g) \rightleftharpoons C(g)$ is exothermic as written. (a) What effect would an increase in temperature have on the numerical value of K? (b) How would the numerical values of K and K_p vary with an increase in total pressure?

14.6 The reaction $A(g) + B(g) \rightleftharpoons C(g)$ is exothermic as written. Assume that an equilibrium system is established. How would the equilibrium concentration of C(g) change with (a) an increase in temperature, (b) an increase in pressure, (c) the addition of A(g), (d) the addition of a catalyst, (e) the removal of B(g)? How would the numerical value of K change with (f) an increase in temperature, (g) an increase in pressure, (h) the addition of a catalyst, (i) the addition of A(g)?

14.7 What is the value of the equilibrium constant, K, for the following system at 395°C?

$$H_2(g) + I_2(g) \rightleftharpoons 2HI(g)$$

The equilibrium concentrations for such a system are: $[H_2] = 0.0064$ mol/liter, $[I_2] = 0.0016$ mol/liter, $[HI] = 0.0250$ mol/liter.

14.8 Solid NH_4HS was introduced into an evacuated container at 24°C. When equilibrium was established

$$NH_4HS(s) \rightleftharpoons NH_3(g) + H_2S(g)$$

the total pressure (of NH_3 and H_2S taken together) was 0.614 atm. What is K_p for the equilibrium at 24°C?

14.9 At 250°C, 0.110 mol of $PCl_5(g)$ was introduced into a one-liter container. Equilibrium was established.

$$PCl_5(g) \rightleftharpoons PCl_3(g) + Cl_2(g)$$

At equilibrium, the concentration of $PCl_3(g)$ was 0.050 mol/liter. (a) What were the equilibrium concentrations of $Cl_2(g)$ and $PCl_5(g)$? (b) What is the value of K at 250°C?

14.10 A mixture of 0.0080 mol of $SO_2(g)$ and 0.0056 mol of $O_2(g)$ is placed in a one-liter container at 1000 K. When equilibrium is established, 0.0040 mol of $SO_3(g)$ is present.

$$2SO_2(g) + O_2(g) \rightleftharpoons 2SO_3(g)$$

(a) What are the equilibrium concentrations of $SO_2(g)$ and $O_2(g)$? (b) What is the value of K for the equilibrium at 1000 K?

14.11 A mixture consisting of 1.000 mol of CO(g) and 1.000 mol of $H_2O(g)$ is placed in a 10.00 liter container at 800 K. At equilibrium, 0.665 mol of $CO_2(g)$ and 0.665 mol of $H_2(g)$ are present.

$$CO(g) + H_2O(g) \rightleftharpoons CO_2(g) + H_2(g)$$

(a) What are the equilibrium concentrations of all four gases? (b) What is the value of K at 800 K? (c) What is the value of K_p at 800 K?

14.12 If 0.025 mol of $COCl_2(g)$ is placed in a one-liter container at 400°C, 16.0% of the $COCl_2$ is dissociated when equilibrium is established.

$$COCl_2(g) \rightleftharpoons CO(g) + Cl_2(g)$$

Calculate the value of K for the equilibrium at 400°C.

14.13 At 585 K and a total pressure of 1.00 atm, NOCl(g) is 56.4% dissociated.

$$2NOCl(g) \rightleftharpoons 2NO(g) + Cl_2(g)$$

Assume that 1.00 mol of NOCl(g) was present before dissociation. (a) How many moles of NOCl(g), NO(g), and $Cl_2(g)$ are present at equilibrium? (b) What is the total number of moles of gas present at equilibrium? (c) What are the equilibrium partial pressures of the three gases? (d) What is the numerical value of K_p at 585 K?

14.14 For the equilibrium

$$CO(g) + 2H_2(g) \rightleftharpoons CH_3OH(g)$$

at 225°C, K is 10.2 liter²/mol². What concentration of $CH_3OH(g)$ is in equilibrium with CO(g) at a concentration of 0.020 mol/liter and $H_2(g)$ at a concentration of 0.020 mol/liter?

14.15 For the equilibrium

$$2NO(g) \rightleftharpoons N_2(g) + O_2(g)$$

at 1800 K, K is 8.36×10^3. What concentration of NO(g) is in equilibrium with $N_2(g)$ at a concentration of 0.0500 mol/liter and $O_2(g)$ at a concentration of 0.0500 mol/liter?

14.16 For the equilibrium

$$Br_2(g) + Cl_2(g) \rightleftharpoons 2BrCl(g)$$

at 400 K, K is 7.0. If 0.060 mol of $Br_2(g)$ and 0.060 mol of $Cl_2(g)$ are introduced into a one-liter container at 400 K, what is the concentration of BrCl(g) when equilibrium is established?

14.17 At 425°C, K is 54.8 for the equilibrium

$$H_2(g) + I_2(g) \rightleftharpoons 2HI(g)$$

If 1.000 mol of $H_2(g)$, 1.000 mol of $I_2(g)$, and 1.000 mol of HI(g) are placed in a one-liter container at 425°C, what are the concentrations of all gases present at equilibrium?

14.18 At 425°C, K is 1.82×10^{-2} for the equilibrium

$$2HI(g) \rightleftharpoons H_2(g) + I_2(g)$$

Assume that an equilibrium is established at 425°C by adding only HI(g) to the reaction flask. (a) What are the concentrations of $H_2(g)$ and $I_2(g)$ in equilibrium with 0.0100 mol/liter of HI(g)? (b) What was the initial concentration of HI(g) before equilibrium was established? (c) What percent of the HI(g) added is dissociated at equilibrium?

14.19 For the equilibrium

$$2IBr(g) \rightleftharpoons I_2(g) + Br_2(g)$$

K is 8.5×10^{-3} at 150°C. If 0.0300 mol of IBr(g) is introduced into a one-liter container, what is the concentration of this substance after equilibrium is established?

14.20 For the equilibrium

$$3Fe(s) + 4H_2O(g) \rightleftharpoons Fe_3O_4(s) + 4H_2(g)$$

at 900°C, K is 5.1. If 0.050 mol of $H_2O(g)$ and excess solid Fe are placed in a one-liter container, what is the concentration of $H_2(g)$ when equilibrium is established at 900°C?

14.21 At 100°C, K is 4.57×10^9 liter/mol for the equilibrium

$$CO(g) + Cl_2(g) \rightleftharpoons COCl_2(g)$$

What is (a) K and (b) K_p for the equilibrium

$$COCl_2(g) \rightleftharpoons CO(g) + Cl_2(g)$$

at this temperature?

14.22 For the equilibrium

$$4HCl(g) + O_2(g) \rightleftharpoons 2Cl_2(g) + 2H_2O(g)$$

K is 889 liter/mol at 480°C. (a) What is the value of K_p at this temperature? (b) Calculate the values of K and K_p at 480°C for the equilibrium

$$2Cl_2(g) + 2H_2O(g) \rightleftharpoons 4HCl(g) + O_2(g)$$

14.23 For the equilibrium

$$CH_3OH(g) \rightleftharpoons CO(g) + 2H_2(g)$$

at 275°C, K_p is 1.14×10^3 atm^2. (a) What is the value of K? (b) What are K and K_p for the following equilibrium at 275°C?

$$CO(g) + 2H_2(g) \rightleftharpoons CH_3OH(g)$$

14.24 At 800 K, K_p is 0.220 atm for the equilibrium

$$CaCO_3(s) \rightleftharpoons CaO(s) + CO_2(g)$$

What is the concentration of $CO_2(g)$, in mol/liter, that is, in equilibrium with solid $CaCO_3$ and CaO at this temperature?

14.25 For the equilibrium

$$N_2(g) + O_2(g) \rightleftharpoons 2NO(g)$$

K is 2.5×10^{-3} at 2400 K. The partial pressure of $N_2(g)$ is 0.50 atm, and the partial pressure of $O_2(g)$ is 0.50 atm in a mixture of these two gases. What is the partial pressure of NO(g) when equilibrium is established at 2400 K?

14.26 For the equilibrium

$$N_2(g) + 3H_2(g) \rightleftharpoons 2NH_3(g)$$

at 350°C, K_p is 7.73×10^{-4}/atm^2. If the partial pressure of $N_2(g)$ is 9.4 atm and the partial pressure of $H_2(g)$ is 28.0 atm in an equilibrium mixture at 350°C, what is the partial pressure of $NH_3(g)$? What is the total pressure? What is the mole fraction of $NH_3(g)$ present?

14.27 For the equilibrium

$$PCl_5(g) \rightleftharpoons PCl_3(g) + Cl_2(g)$$

at a given temperature, K_p is 2.25 atm. A quantity of $PCl_5(g)$ is introduced into an evacuated flask at the reference temperature. When equilibrium is established, the partial pressure of $PCl_5(g)$ is 0.25 atm. What are the equilibrium partial pressures of $PCl_3(g)$ and $Cl_2(g)$? What was the initial pressure of $PCl_5(g)$ before any of it had dissociated into $PCl_3(g)$ and $Cl_2(g)$? What mole percent of $PCl_5(g)$ has dissociated in this system at equilibrium?

14.28 At a given temperature and a total pressure of 1.00 atm, the partial pressures of an equilibrium mixture

$$N_2O_4(g) \rightleftharpoons 2NO_2(g)$$

are $p_{N_2O_4} = 0.50$ atm and $p_{NO_2} = 0.50$ atm. (a) What is K_p at this temperature? (b) If the total pressure is increased to 2.00 atm and the temperature is constant, what are the partial pressures of the components of an equilibrium mixture? Note that the quadratic formula must be used to solve this problem.

14.29 What effect does the addition of a catalyst have on a system in equilibrium?

14.30 State the direction in which each of the following equilibrium systems would be shifted upon the application of the stress listed beside the equation.

(a) $2SO_2(g) + O_2(g) \rightleftharpoons 2SO_3(g)$ (exothermic)	decrease temperature
(b) $C(s) + CO_2(g) \rightleftharpoons 2CO(g)$ (endothermic)	increase temperature
(c) $N_2O_4(g) \rightleftharpoons 2NO_2(g)$	increase total pressure
(d) $CO(g) + H_2O(g) \rightleftharpoons CO_2(g) + H_2(g)$	decrease total pressure
(e) $2NOBr(g) \rightleftharpoons 2NO(g) + Br_2(g)$	decrease total pressure
(f) $3Fe(s) + 4H_2O(g) \rightleftharpoons Fe_3O_4(s) + 4H_2(g)$	add Fe(s)
(g) $2SO_2(g) + O_2(g) \rightleftharpoons 2SO_3(g)$	add a catalyst
(h) $CaCO_3(s) \rightleftharpoons CaO(s) + CO_2(g)$	remove $CO_2(g)$
(i) $N_2(g) + 3H_2(g) \rightleftharpoons 2NH_3(g)$	increase the concentration of $H_2(g)$

Chapter 15

Ionic Equilibria

The principles of chemical equilibrium can be applied to equilibrium systems that involve molecules and ions in water solution. In pure water, H_3O^+ and OH^- ions exist in equilibrium with water molecules from which they are derived. Other molecular substances (weak electrolytes) are partially ionized in water and exist in equilibrium with their ions. Another type of ionic equilibrium consists of the heterogeneous systems in which a slightly soluble solid is in equilibrium with its ions in water solution. An understanding of equilibria of these types is important to the study of analytical chemistry.

15.1 WEAK ELECTROLYTES

Strong electrolytes are completely ionized in water solution. A 0.01 *M* solution of $CaCl_2$, for example, contains Ca^{2+} ions (at a concentration of 0.01 *M*),

Cl^- ions (at a concentration of 0.02 *M*), and no $CaCl_2$ molecules at all. Weak electrolytes, however, are *incompletely* ionized in water solution. Dissolved molecules exist in equilibrium with their ions in such solutions. The following chemical equation represents the dissociation of acetic acid in water.

$$HC_2H_3O_2 + H_2O \rightleftharpoons H_3O^+ + C_2H_3O_2^-$$

The equilibrium constant for this reaction as written is

$$\frac{[H_3O^+][C_2H_3O_2^-]}{[HC_2H_3O_2][H_2O]} = K'$$

In dilute solutions, the concentration of water may be considered to be a constant. The number of moles of water used in forming H_3O^+ (which is about 0.001 mol in 1 liter of a 0.1 *M* solution of acetic acid) is very small in comparison to the large number of moles of water present (which is about 55.5 mol in 1 liter of the solution). If we combine $[H_2O]$ with K', we get

$$\frac{[H_3O^+][C_2H_3O_2^-]}{[HC_2H_3O_2]} = K'[H_2O] = K$$

We shall follow the practice of representing the concentration of hydronium ion, $[H_3O^+]$, by the symbol $[H^+]$. The simplified expression for the equilibrium constant *K*, therefore, corresponds to the simplified chemical equation.

$$HC_2H_3O_2 \rightleftharpoons H^+ + C_2H_3O_2^-$$

$$\frac{[H^+][C_2H_3O_2^-]}{[HC_2H_3O_2]} = K$$

By convention, the *ions* are written on the *right* of the chemical equation for a reversible ionization. Consequently, the terms for the concentrations of the *ions* appear in the *numerator* of the expression for *K*.

Example 15.1 At 25°C, a 0.1000 *M* solution of acetic acid ($HC_2H_3O_2$) is 1.34% ionized. What is the ionization constant, *K*, for acetic acid?

Solution Let us assume that we have 1 liter of solution. Since the acid is 1.34% ionized, the number of moles of $HC_2H_3O_2$ in *ionic form* is

$$(0.0134)(0.1000 \text{ mol } HC_2H_3O_2) = 0.00134 \text{ mol } HC_2H_3O_2$$

We subtract this number from the total number of moles of $HC_2H_3O_2$ to get the number of moles of acid in *molecular form*.

$$(0.1000 \text{ mol } HC_2H_3O_2) - (0.00134 \text{ mol } HC_2H_3O_2) = 0.0987 \text{ mol } HC_2H_3O_2$$

According to the chemical equation for the ionization, 1 mol of H^+ and 1 mol of $C_2H_3O_2^-$ are produced for every 1 mol of $HC_2H_3O_2$ that ionizes. Therefore, the concentrations at equilibrium are

$$\underset{0.0987\,M}{HC_2H_3O_2} \rightleftharpoons \underset{0.00134\,M}{H^+} + \underset{0.00134\,M}{C_2H_3O_2^-}$$

These concentrations may be used to find the numerical value of the equilibrium constant.

$$K = \frac{[H^+][C_2H_3O_2^-]}{[HC_2H_3O_2]}$$

$$= \frac{(0.00134)(0.00134)}{(0.0987)}$$

$$= 1.82 \times 10^{-5}$$

In future problem work, we shall express equilibrium constants to two significant figures. Higher accuracy is usually not warranted.*

Example 15.2 What are the concentrations of all species present in 1.00*M* acetic acid at 25°C?

Solution If we let x equal the number of moles of $HC_2H_3O_2$ in ionic form in 1 liter of solution, the equilibrium concentrations are

$$\underset{(1.00 - x)\,M}{HC_2H_3O_2} \rightleftharpoons \underset{xM}{H^+} + \underset{x\,M}{C_2H_3O_2^-}$$

We substitute these values into the expression for K

$$K = \frac{[H^+][C_2H_3O_2^-]}{[HC_2H_3O_2]}$$

$$1.8 \times 10^{-5} = \frac{x^2}{(1.00 - x)}$$

The result can be rearranged to

$$x^2 + (1.8 \times 10^{-5})x - 1.8 \times 10^{-5} = 0$$

An equation in the form

$$ax^2 + bx + c = 0$$

is called a **quadratic equation** (see Appendix C). Solutions to a quadratic equation can be found by substitution in the **quadratic formula.**

$$x = \frac{-b \pm \sqrt{b^2 - 4ac}}{2a}$$

* Equilibrium constants should be expressed in terms of activities instead of concentrations. The concentrations of ions present in dilute solutions of weak electrolytes are so small, however, that interionic attractions are negligible. Under these conditions molar concentrations may be used rather than activities. Reasonably accurate results are obtained from this practice.

In the present problem, $a = 1$, $b = 1.8 \times 10^{-5}$, and $c = -1.8 \times 10^{-5}$. If we substitute these values into the quadratic formula, we get

$$x = 4.2 \times 10^{-3} M^*$$

Therefore,

$$[H^+] = [C_2H_3O_2^-] = x = 4.2 \times 10^{-3}\ M$$
$$[HC_2H_3O_2] = (1.00 - x)\ M$$
$$= (1.00 - 0.0042)\ M$$
$$= 0.9958\ M$$

Notice that the value of K that we employed has two significant figures. Our answer, therefore, cannot be given to more than two significant figures. The concentration of $HC_2H_3O_2$ must be rounded off.

$$[HC_2H_3O_2] = 1.0\ M$$

This value is the same as the initial concentration of $HC_2H_3O_2$.

The use of the quadratic formula in problem solving may be avoided by making an approximation that is frequently employed in calculations involving aqueous equilibria. The subtraction of a very small number from a large number does not significantly change the value of the large number and may be neglected.

In the preceding example such a small amount of acetic acid is ionized (x) that the quantity $(1.00 - x)$, which is the concentration of acetic acid molecules, is for all practical purposes equal to 1.00 (as previously noted). By using 1.00 instead of $(1.00 - x)$ for the concentration of $HC_2H_3O_2$, we find

$$1.8 \times 10^{-5} = \frac{[H^+][C_2H_3O_2^-]}{[HC_2H_3O_2]}$$

$$= \frac{x^2}{1.00}$$

$$x = 4.2 \times 10^{-3} M$$

which is the same as the result obtained through the use of the quadratic formula.

The subtraction of a *small* number from another *small* number may *not*

* There are always two solutions to a quadratic equation. One of them must be discarded because it represents a physical impossibility. The two solutions obtained in the present case are

$$x = 4.2 \times 10^{-3}\ M \quad \text{and} \quad x = -4.2 \times 10^{-3}\ M$$

The second solution is impossible. The concentration of $HC_2H_3O_2$ is $(1.00 - x)$. Substitution of the second value (which is a negative number) for x gives a final $HC_2H_3O_2$ concentration that would require more $HC_2H_3O_2$ than was supplied.

Table 15.1 Ion concentrations and percent ionization of solutions of acetic acid at 25°C

CONCENTRATION OF SOLUTION (*M*)	$[H^+]$ OR $[C_2H_3O_2^-]$ (*M*)	PERCENT IONIZATION
1.00	0.00426	0.426
0.100	0.00134	1.34
0.0100	0.000418	4.18
0.00100	0.000126	12.6

be neglected. Consequently, the quadratic formula *must* be used to solve some problems. If the simplified method is used, the answers should be checked to make sure that the use of the simplified procedure was justified.

The percent ionization of a weak electrolyte is found by dividing the number of moles of solute in the ionic form by the total number of moles of solute and multiplying the fraction obtained by 100. The percent ionization of the acetic acid solution described in Example 15.2 is

$$\frac{4.2 \times 10^{-3}\ M}{1.00\ M} \times 100 = 0.42\%$$

The ion concentration and percent ionization of solutions of acetic acid of various concentrations are listed in Table 15.1. The percent ionization increases with dilution, which is consistent with LeChatelier's principle. Addition of water to an equilibrium system of a weak electrolyte

$$HC_2H_3O_2 + H_2O \rightleftharpoons H_3O^+ + C_2H_3O_2^-$$

shifts the equilibrium to the right. Proportionately more of the solute is found in the ionic form.

The values of ionization constants change with temperature. At 25°C, K for acetic acid is 1.8×10^{-5}; at 100°C, K is 1.1×10^{-5}. Ionization constants are usually reported for equilibrium systems at 25°C.

An equilibrium system of a weak electrolyte can be prepared from compounds that supply the ions of the weak electrolyte. The following example is an illustration of this procedure.

Example 15.3 What are the concentrations of all species present in a solution made by diluting 0.10 mol of HCl and 0.50 mol of $NaC_2H_3O_2$ to 1.0 liter?

Solution Hydrochloric acid and sodium acetate are strong electrolytes. We may assume, therefore, that before equilibrium is attained, the concentration of H^+ is 0.10 *M*, and the concentration of $C_2H_3O_2^-$ is 0.50 *M*. If we let x equal the number of moles per liter of acetic acid at equilibrium, the equilibrium concentrations are

$$\underset{(x)\,M}{HC_2H_3O_2} \rightleftharpoons \underset{(0.10-x)\,M}{H^+} + \underset{(0.50-x)\,M}{C_2H_3O_2^-}$$

If we solve the problem using these quantities, the quadratic formula must be employed since x is not negligible in comparison to 0.10 or 0.50.

A simpler way to solve the problem is to assume that the reaction goes as far to the left as is possible and that then a portion of the acetic acid that has been formed dissociates into ions. In the backward reaction, 1 mol of H^+ will react with 1 mol of $C_2H_3O_2^-$ to form 1 mol of $HC_2H_3O_2$. Therefore, all the H^+ will be used, 0.40 mol of $C_2H_3O_2^-$ will be left of the original 0.50 mol supplied, and 0.10 mol of $HC_2H_3O_2$ will be formed. We then assume that y mol of $HC_2H_3O_2$ dissociates into ions.

	$HC_2H_3O_2$	$\rightleftharpoons$	H^+	+	$C_2H_3O_2^-$
after mixing:	–		$0.10\ M$		$0.50\ M$
reaction to left:	$0.10\ M$		–		$0.40\ M$
equilibrium:	$(0.10 - y)\ M$		$(y)\ M$		$(0.40 + y)\ M$

Now, y is indeed negligible in comparison to both 0.10 and 0.40, and we may simplify the problem by neglecting y in the terms for the concentrations of acetic acid and acetate ion. Thus,

$$1.8 \times 10^{-5} = \frac{[H^+][C_2H_3O_2^-]}{[HC_2H_3O_2]}$$

$$= \frac{y(0.40)}{(0.10)}$$

$$y = 4.5 \times 10^{-6}\ M$$

We can see from the value obtained for y that the approximations applied to $[C_2H_3O_2^-]$ and $[HC_2H_3O_2]$ are justified. To two significant figures, the equilibrium concentrations are

$$[H^+] = 4.5 \times 10^{-6} M$$
$$[C_2H_3O_2^-] = 0.40 M$$
$$[HC_2H_3O_2] = 0.10 M$$

The hydroxides of most metals are either strong electrolytes or slightly soluble compounds. There are some water-soluble compounds, however, that produce alkaline solutions in which equilibria exist between molecules and ions. The most important such compound is ammonia, NH_3, and the chemical equation for the reversible reaction is

$$H_2O + NH_3 \rightleftharpoons NH_4^+ + OH^-$$

The expression for the equilibrium constant for this reaction

$$K' = \frac{[NH_4^+][OH^-]}{[NH_3][H_2O]}$$

Table 15.2 Ionization constants at 25°C

	WEAK ACIDS	
acetic	$HC_2H_3O_2 \rightleftharpoons H^+ + C_2H_3O_2^-$	1.8×10^{-5}
benzoic	$HC_7H_5O_2 \rightleftharpoons H^+ + C_7H_5O_2^-$	6.0×10^{-5}
chlorous	$HClO_2 \rightleftharpoons H^+ + ClO_2^-$	1.1×10^{-2}
cyanic	$HOCN \rightleftharpoons H^+ + OCN^-$	1.2×10^{-4}
formic	$HCHO_2 \rightleftharpoons H^+ + CHO_2^-$	1.8×10^{-4}
hydrazoic	$HN_3 \rightleftharpoons H^+ + N_3^-$	1.9×10^{-5}
hydrocyanic	$HCN \rightleftharpoons H^+ + CN^-$	4.0×10^{-10}
hydrofluoric	$HF \rightleftharpoons H^+ + F^-$	6.7×10^{-4}
hypobromous	$HOBr \rightleftharpoons H^+ + BrO^-$	2.1×10^{-9}
hypochlorous	$HOCl \rightleftharpoons H^+ + ClO^-$	3.2×10^{-8}
nitrous	$HNO_2 \rightleftharpoons H^+ + NO_2^-$	4.5×10^{-4}
	WEAK BASES	
ammonia	$NH_3 + H_2O \rightleftharpoons NH_4^+ + OH^-$	1.8×10^{-5}
aniline	$C_6H_5NH_2 + H_2O \rightleftharpoons C_6H_5NH_3^+ + OH^-$	4.6×10^{-10}
dimethylamine	$(CH_3)_2NH + H_2O \rightleftharpoons (CH_3)_2NH_2^+ + OH^-$	7.4×10^{-4}
hydrazine	$N_2H_4 + H_2O \rightleftharpoons N_2H_5^+ + OH^-$	9.8×10^{-7}
methylamine	$CH_3NH_2 + H_2O \rightleftharpoons CH_3NH_3^+ + OH^-$	5.0×10^{-4}
pyridine	$C_5H_5N + H_2O \rightleftharpoons C_5H_5NH^+ + OH^-$	1.5×10^{-9}
trimethylamine	$(CH_3)_3N + H_2O \rightleftharpoons (CH_3)_3NH^+ + OH^-$	7.4×10^{-5}

simplifies to the following if the concentration of water is assumed to be a constant,

$$K = K'[H_2O] = \frac{[NH_4^+][OH^-]}{[NH_3]}$$

Formerly it was assumed that ammonia gas dissolves in water to produce ammonium hydroxide molecules, NH_4OH, and that these molecules dissociate into NH_4^+ and OH^- ions. Currently, however, the existence of such ammonium hydroxide molecules is questioned. The ammonia equilibrium, therefore, is probably best represented in the manner employed in the preceding equation.

Table 15.2 lists the ionization constants of some weak acids and some weak bases. Since the terms for the ion concentrations appear in the numerator of the ionization-constant expression, the magnitude of the value of K gives a qualitative idea of the strength of the electrolyte. Thus, HOCN ($K = 1.2 \times 10^{-4}$) is a stronger acid than HCN ($K = 4.0 \times 10^{-10}$).

15.2 THE IONIZATION OF WATER

Pure water is itself a very weak electrolyte and ionizes according to the equation

$$H_2O + H_2O \rightleftharpoons H_3O^+ + OH^-$$

In simplified form this equation is

$$H_2O \rightleftharpoons H^+ + OH^-$$

The expression for the ionization constant derived from the simplified equation is

$$K = \frac{[H^+][OH^-]}{[H_2O]}$$

In dilute solutions the concentration of water is virtually a constant, and we may combine $[H_2O]$ with the constant K. Thus,

$$K[H_2O] = [H^+][OH^-]$$

This constant, $K[H_2O]$, is called the ion product of water, or the **water constant,** and is given the symbol K_W. At 25°C,

$$K_w = 1.0 \times 10^{-14} = [H^+][OH^-]$$

In pure water,

$$[H^+] = [OH^-] = x$$
$$[H^+][OH^-] = 1.0 \times 10^{-14}$$
$$x^2 = 1.0 \times 10^{-14}$$
$$x = 1.0 \times 10^{-7}\ M$$

The concentrations of both of the ions of water are equal to $1.0 \times 10^{-7}\ M$ in pure water or in any neutral solution at 25°C. In 1 liter, only 10^{-7} mol of water is in ionic form out of a total of approximately 55.5 mol.

Both hydronium and hydroxide ions exist in any aqueous solution. In an acid solution the concentration of hydronium ions is larger than $1.0 \times 10^{-7}\ M$ and larger than the hydroxide ion concentration. In an alkaline solution the hydroxide ion concentration is larger than $1.0 \times 10^{-7}\ M$ and larger than the hydronium ion concentration.

Example 15.4 What are $[H^+]$ and $[OH^-]$ in a 0.020 M solution of HCl?

Solution The quantity of hydronium ion obtained from the ionization of water is negligible compared to that derived from the hydrochloric acid. Since HCl is a strong electrolyte, $[H^+] = 0.020\ M$.

$$[H^+][OH^-] = 1.0 \times 10^{-14}$$

$$(2.0 \times 10^{-2})[OH^-] = 1.0 \times 10^{-14}$$

$$[OH^-] = \frac{1.0 \times 10^{-14}}{2.0 \times 10^{-2}}$$

$$[OH^-] = 5.0 \times 10^{-13}\ M$$

Notice that $[OH^-]$ is extremely small. In this solution there would be one hydroxide ion for every 40 billion hydronium ions.

Example 15.5 What are $[H^+]$ and $[OH^-]$ in a 0.0050 *M* solution of NaOH?

Solution Sodium hydroxide is a strong electrolyte, and therefore $[OH^-] = 5.0 \times 10^{-3}M$.

$$[H^+][OH^-] = 1.0 \times 10^{-14}$$

$$[H^+](5.0 \times 10^{-3}) = 1.0 \times 10^{-14}$$

$$[H^+] = \frac{1.0 \times 10^{-14}}{5.0 \times 10^{-3}}$$

$$[H^+] = 2.0 \times 10^{-12}\ M$$

15.3 pH

The hydronium ion concentration of a solution may be expressed in terms of the pH scale. The pH of a solution is defined as

$$p\text{H} = \log \frac{1}{[H^+]} = -\log\ [H^+]$$

The common logarithm of a number is the power to which 10 must be raised in order to get the number (see Appendix A). If

$$a = 10^n$$

then

$$\log a = n$$

Therefore, if

$$a = 10^{-3}$$
$$\log a = -3$$

The pH is the *negative* logarithm of the hydronium ion concentration. Therefore, if

$$[H^+] = 10^{-3} M$$
$$\log\ [H^+] = -3$$
$$p\text{H} = -\log\ [H^+] = 3$$

For a neutral solution, therefore,

$$[H^+] = 1.0 \times 10^{-7} M$$
$$p\text{H} = -\log(10^{-7}) = -(-7)$$
$$p\text{H} = 7$$

The pOH of a solution is defined in the same terms.

$$p\text{OH} = -\log\ [OH^-]$$

The water constant is

$$[H^+][OH^-] = 10^{-14}$$

We take the logarithm of each term

$$\log [H^+] + \log [OH^-] = \log (10^{-14})$$

and multiply through by -1

$$-\log [H^+] - \log [OH^-] = -\log(10^{-14})$$

or

$$pH + pOH = 14$$

For a 0.01*M* solution of NaOH, which is a strong base,

$$[OH^-] = 10^{-2}M$$
$$pOH = -\log [OH^-] = 2$$

Since the sum of the *p*H and *p*OH of a solution at 25°C equals 14,

$$pH = 12$$

Example 15.6 What is the *p*H of a solution that is 0.050*M* in H^+?

Solution

$$[H^+] = 5.0 \times 10^{-2}M$$
$$\log [H^+] = \log 5.0 + \log 10^{-2}$$

The logarithm of 5.0 can be found in the logarithm table found in Appendix D.

$$\log [H^+] = 0.70 - 2.00 = -1.30$$
$$pH = 1.30$$

Example 15.7 What is the *p*H of a solution for which $[OH^-] = 0.030M$?

Solution

$$[OH^-] = 3.0 \times 10^{-2}$$
$$\log[OH^-] = \log 3.0 + \log 10^{-2}$$
$$= 0.48 - 2.00 = -1.52$$
$$pOH = 1.52$$
$$pH = 14.00 - pOH$$
$$= 14.00 - 1.52 = 12.48$$

An alternative solution is

$$[H^+][OH^-] = 1.0 \times 10^{-14}$$
$$[H^+] = \frac{1.0 \times 10^{-14}}{3.0 \times 10^{-2}} = 3.3 \times 10^{-13}$$
$$\log[H^+] = \log 3.3 + \log 10^{-13}$$
$$= 0.52 - 13.00 = -12.48$$
$$pH = 12.48$$

Example 15.8 What is the $[H^+]$ of a solution with a pH of 10.60?

Solution Since

$$pH = -\log [H^+]$$
$$\log [H^+] = -pH$$

In the present example, $pH = 10.60$, and therefore

$$\log [H^+] = -10.60$$

This logarithm must be divided into two parts: a decimal portion (which is called the mantissa and which *must* be positive) and a negative whole number (which is called the characteristic).

$$\log [H^+] = 0.40 - 11.00$$
$$[H^+] = \text{antilog } 0.40 \times \text{antilog } (-11)$$

The antilogarithm of 0.40 is obtained by finding the number that corresponds to this logarithm in the logarithm table (it is 2.5). The antilogarithm of -11 is 10^{-11}. Therefore,

$$[H^+] = 2.5 \times 10^{-11} M$$

It should be kept in mind that pH relates to a power of 10. Hence, a solution of $pH = 1$ has a hydronium ion concentration 100 times that of a solution of $pH = 3$ (not three times). Furthermore, since the pH is related to a *negative* exponent, the lower the pH value, the larger the concentration of hydronium ion. At $pH = 7$, a solution is neutral. Solutions with pH's below 7 are acidic. Those with pH's above 7 are alkaline. These relationships are summarized in Table 15.3.

Table 15.3 The *p*H scale

*p*H	$[H^+]$	$[OH^-]$	
14	10^{-14}	10^{0}	↑
13	10^{-13}	10^{-1}	
12	10^{-12}	10^{-2}	
11	10^{-11}	10^{-3}	increasing
10	10^{-10}	10^{-4}	alkalinity
9	10^{-9}	10^{-5}	
8	10^{-8}	10^{-6}	
7	10^{-7}	10^{-7}	neutrality
6	10^{-6}	10^{-8}	
5	10^{-5}	10^{-9}	
4	10^{-4}	10^{-10}	
3	10^{-3}	10^{-11}	increasing
2	10^{-2}	10^{-12}	acidity
1	10^{-1}	10^{-13}	
0	10^{0}	10^{-14}	↓

The ionization constant of a weak acid or a weak base can be determined by measuring the pH of a solution of known concentration of the weak electrolyte.

Example 15.9 The pH of a 0.10 M solution of a weak acid HX is 3.30. What is the ionization constant of HX?

Solution

$$HX \rightleftharpoons H^+ + X^-$$
$$pH = 3.30$$
$$\log [H^+] = -3.30$$
$$[H^+] = \text{antilog } (-3.30)$$
$$= \text{antilog } 0.70 \times \text{antilog } (-4.00)$$
$$= 5.0 \times 10^{-4}\ M$$

Since the solution was prepared from HX alone,

$$[H^+] = [X^-] = 5.0 \times 10^{-4}\ M$$

The concentration of HX in the solution is for all practical purposes equal to $0.10M$. The small quantity of HX that dissociated need not be considered. The equilibrium concentrations, therefore, are

$$\underset{0.10\ M}{HX} \rightleftharpoons \underset{5.0 \times 10^{-4}\ M}{H^+} + \underset{5.0 \times 10^{-4}\ M}{X^-}$$

The value of the ionization constant is

$$K = \frac{[H^+][X^-]}{[HX]}$$

$$= \frac{(5.0 \times 10^{-4})^2}{1.0 \times 10^{-1}} = \frac{25 \times 10^{-8}}{1.0 \times 10^{-1}}$$

$$K = 2.5 \times 10^{-6}$$

15.4 INDICATORS

Indicators are organic compounds of complex structure that change color in solution as the pH changes. Methyl orange, for example, is red in solutions of pH below 3.1 and yellow in solutions of pH above 4.5. The color of this indicator is a varying mixture of yellow and red in the pH range between 3.1 and 4.5. Many indicators have been described and used. A few are listed in Table 15.4.

By the use of indicators the pH of a solution may be determined. If thymol blue is yellow in a test solution and methyl orange is red in another sample of the same solution, the pH of the solution is between 2.8 and 3.1. Reference to

Table 15.4 Some indicators

INDICATOR	ACID COLOR	*p*H RANGE OF COLOR CHANGE	ALKALINE COLOR
thymol blue	red	1.2–2.8	yellow
methyl orange	red	3.1–4.5	yellow
bromcresol green	yellow	3.8–5.5	blue
methyl red	red	4.2–6.3	yellow
litmus	red	5.0–8.0	blue
bromthymol blue	yellow	6.0–7.6	blue
thymol blue	yellow	8.0–9.6	blue
phenolphthalein	colorless	8.3–10.0	red
alizarin yellow	yellow	10.0–12.1	lavender

Table 15.4 shows that thymol blue is yellow only in solutions of *p*H greater than 2.8, and methyl orange is red only in solutions of *p*H less than 3.1. If enough indicators are employed, it is possible to determine *p*H values that are accurate to the first decimal place.

Indicators are weak acids or weak bases. Since they are intensely colored, only a few drops of a dilute solution of an indicator need be employed in any determination. Hence, the acidity of the solution in question is not significantly altered by the addition of the indicator.

If we let the symbol HIn stand for the litmus molecule (which is red) and the symbol In^- stand for the anion (which is blue) derived from the weak acid, the equation for the litmus equilibrium may be written

$$\underset{\text{red}}{\mathrm{HIn}} \rightleftharpoons \mathrm{H^+} + \underset{\text{blue}}{\mathrm{In^-}}$$

According to the principle of Le Chatelier, increasing the concentration of H^+ shifts the equilibrium to the left, and the red (or acid) color of HIn is observed. On the other hand, addition of OH^- decreases the concentration of H^+. The equilibrium shifts to the right, and the blue (or alkaline) color of In^- is observed.

The ionization constant for litmus is approximately equal to 10^{-7}.

$$10^{-7} = \frac{[\mathrm{H^+}][\mathrm{In^-}]}{[\mathrm{HIn}]}$$

We may rearrange this expression in the following manner

$$\frac{10^{-7}}{[\mathrm{H^+}]} = \frac{[\mathrm{In^-}]}{[\mathrm{HIn}]}$$

At a *p*H of 5 or below, the red color of litmus is observed. If we substitute $[H^+] = 10^{-5}$, which corresponds to $pH = 5$, into the preceding expression, we get

$$\frac{10^{-7}}{10^{-5}} = \frac{1}{100} = \frac{[\mathrm{In^-}]\ \leftarrow \text{blue}}{[\mathrm{HIn}]\ \leftarrow \text{red}}$$

Thus, the mixture appears red to the eye when the concentration of the red HIn is 100 times (or more) that of the blue In^-.

The blue color of litmus is observed in solutions of $pH = 8$ or higher. If $[H^+] = 10^{-8}$,

$$\frac{10^{-7}}{10^{-8}} = \frac{10}{1} = \frac{[In^-]}{[HIn]} \begin{matrix} \leftarrow \text{blue} \\ \leftarrow \text{red} \end{matrix}$$

When the concentration of the blue In^- ion is 10 times that of the red HIn, or approximately 91% of the indicator is in ionic form, the blue color of the In^- ions completely masks the red color of the HIn molecules, and the mixture appears blue.

Thus, the blue color predominates when the concentration of the blue species is 10 times that of the red, whereas the red color predominates only when the concentration of the red form is 100 times that of the blue. This difference is not surprising since the blue color of litmus is a much stronger color than the red. Only when $[H^+] = K = 10^{-7}$ will the factor $K/[H^+] = 1$ and $[In^-] = [HIn]$. Hence, at $pH = 7$, litmus exhibits its "neutral" color (purple).

The pH range over which a given indicator changes color depends, therefore, upon the ionization constant of the indicator. For indicators that are weak acids, the smaller the value of K, the higher is the pH range of the color change.

15.5 THE COMMON-ION EFFECT

In a 0.1 M solution of acetic acid, methyl orange assumes its acid color, red. If sodium acetate is added to this solution, the color changes to yellow, showing that the addition causes the acidity of the solution to decrease. This experimental observation is readily explained on the basis of Le Chatelier's principle. The equilibrium

$$HC_2H_3O_2 \rightleftharpoons H^+ + C_2H_3O_2^-$$

is shifted to the left by the addition of the acetate ion from the sodium acetate, and the concentration of hydronium ion correspondingly decreases. Since acetic acid and sodium acetate have the acetate ion in common, this phenomenon is called the **common-ion effect.**

The concentration of hydronium ion in a solution of acetic acid to which acetate ion has been added can be calculated by means of the ionization constant of acetic acid.

Example 15.10 What is the concentration of hydronium ion in a 0.10 M solution of acetic acid that has been made 0.15 M in sodium acetate ($NaC_2H_3O_2$)?

Solution Sodium acetate is a strong electrolyte. Therefore, the concentration of $C_2H_3O_2^-$ from this source is 0.15 *M*. The addition of this excess $C_2H_3O_2^-$ shifts the acetic acid equilibrium to the left.

$$HC_2H_3O_2 \rightleftharpoons H^+ + C_2H_3O_2^-$$

Consequently, the concentration of $HC_2H_3O_2$ may be taken as equal to the total $HC_2H_3O_2$ concentration (0.10 *M*) since so little is ionized. The concentration of $C_2H_3O_2^-$ may be taken as equal to the concentration of $C_2H_3O_2^-$ from the $NaC_2H_3O_2$ (0.15 *M*) since very little is supplied by the ionization of $HC_2H_3O_2$.

$$\underset{0.10M}{HC_2H_3O_2} \rightleftharpoons \underset{(x)M}{H^+} + \underset{0.15M}{C_2H_3O_2^-}$$

$$1.8 \times 10^{-5} = \frac{[H^+][C_2H_3O_2^-]}{[HC_2H_3O_2]}$$

$$= \frac{[H^+](0.15)}{(0.10)}$$

$$[H^+] = 1.2 \times 10^{-5} M$$

The concentration of hydronium ion in a 0.10 *M* solution of pure $HC_2H_3O_2$ is 1.3×10^{-3} *M* (see Table 15.1).

Example 15.11 What is the concentration of hydroxide ions in a solution made by dissolving 0.020 mol of ammonium chloride in 100 ml of 0.15 *M* ammonia? Assume that the addition of the solid does not cause the volume of the solution to change.

Solution Ammonium chloride is a strong electrolyte. The concentration of NH_4^+ from the NH_4Cl added is

$$? \text{ mol } NH_4^+ = 1000 \text{ ml soln.} \left(\frac{0.020 \text{ mol } NH_4^+}{100 \text{ ml soln.}}\right) = 0.20 \text{ mol } NH_4^+$$

Therefore, the concentration of NH_4^+ is 0.20 *M*.

$$\underset{0.15\,M}{NH_3} + H_2O \rightleftharpoons \underset{0.20\,M}{NH_4^+} + \underset{(x)\,M}{OH^-}$$

$$1.8 \times 10^{-5} = \frac{[NH_4^+][OH^-]}{[NH_3]}$$

$$= \frac{(0.20)[OH^-]}{(0.15)}$$

$$[OH^-] = 1.4 \times 10^{-5}\,M$$

15.6 BUFFERS

It is sometimes necessary that a solution with a definite *p*H be prepared and stored. The preservation of such a solution is even more difficult than its preparation. If the solution comes in contact with the air, it will absorb carbon dioxide (an acid anhydride) and become more acidic. If the solution is stored in a glass bottle, alkaline impurities leached from the glass may alter the pH. **Buffer solutions** are capable of maintaining their pH at some fairly constant value even when small amounts of acid or base are added.

A buffer may be prepared from a weak acid or a weak base and a salt of the weak electrolyte. For example, a buffer can be prepared from acetic acid and sodium acetate. If both these materials are present in the same concentration, for example, 1.00M,

$$1.8 \times 10^{-5} = \frac{[H^+][C_2H_3O_2^-]}{[HC_2H_3O_2]}$$

$$[H^+] = 1.8 \times 10^{-5} \left(\frac{[HC_2H_3O_2]}{[C_2H_3O_2^-]}\right) = 1.8 \times 10^{-5}\ M$$

Since $[HC_2H_3O_2] = [C_2H_3O_2^-]$,

$$\frac{[HC_2H_3O_2]}{[C_2H_3O_2^-]} = 1$$

and, therefore,

$$[H^+] = 1.8 \times 10^{-5}\ M$$

$$pH = -\log(1.8 \times 10^{-5}) = 4.74$$

The pK of a weak electrolyte may be defined in a manner analogous to pH or pOH,

$$pK = -\log K$$

A solution of a weak acid in which the concentration of the anion is the same as the concentration of the undissociated acid has a pH equal to the pK of the acid.

In a sample of the buffer previously described, the quantities of acetic acid molecules and acetate ion are much larger than the quantity of hydronium ion present—approximately 50,000 times larger.

$$\underset{1.0\,M}{HC_2H_3O_2} \rightleftharpoons \underset{1.8 \times 10^{-5}\,M}{H^+} + \underset{1.0\,M}{C_2H_3O_2^-}$$

If a small quantity of hydronium ion is added, the large reserve of acetate ion will quickly convert it to acetic acid. If a small amount of hydroxide ion is added to the buffer, it will neutralize hydronium ion, but the large quantity of acetic acid present will by dissociation replace any hydronium ion removed and maintain the pH at a fairly constant value.

Buffers cannot withstand the addition of large amounts of acids or alkalies. The addition of 0.01 mol per liter of H^+ or OH^- is about the maximum that any buffer can be expected to withstand. How successfully a buffer withstands such an addition is illustrated in the following example.

Example 15.12 The ionization constant of acetic acid to three significant figures is 1.82×10^{-5}. A buffer containing 1.00 M concentrations of acetic acid and sodium acetate has a pH of 4.742. (a) What is the pH of the solution after 0.01 mol of HCl has been added to 1 liter of the buffer? (b) What is the pH of the solution after the addition of 0.01 mol of NaOH?

Solution (a) The 0.01 mol of H^+ from HCl converts an equivalent amount of acetate ion into acetic acid. Thus, the concentrations are

	$HC_2H_3O_2 \rightleftharpoons$	H^+	$+\ C_2H_3O_2^-$
buffer:	1.00 M	$1.82 \times 10^{-5}\ M$	1.00 M
after H^+ addition:	1.01 M	?	0.99 M

$$\frac{[H^+][C_2H_3O_2^-]}{[HC_2H_3O_2]} = 1.82 \times 10^{-5}$$

$$\frac{[H^+](0.99)}{(1.01)} = 1.82 \times 10^{-5}$$

$$[H^+] = 1.82 \times 10^{-5}\,\frac{(1.01)}{(0.99)}$$

$$= 1.86 \times 10^{-5}\ M$$

$$pH = 4.731$$

Thus, the addition causes the pH to change 0.009 pH units. A similar addition to pure water would change the pH from 7.0 to 2.0—a change of 5.0 pH units.

(b) The addition of 0.0100 mol of OH^- would change the concentration of acetate ion to 1.01 M and the concentration of acetic acid to 0.99 M. Thus, the concentrations are

	$HC_2H_3O_2 \rightleftharpoons$	H^+	$+\ C_2H_3O_2^-$
buffer:	1.00 M	$1.81 \times 10^{-5}\ M$	1.00 M
after OH^- addition:	0.99 M	?	1.01 M

$$\frac{[H^+][C_2H_3O_2^-]}{[HC_2H_3O_2]} = 1.82 \times 10^{-5}$$

$$\frac{[H^+](1.01)}{(0.99)} = 1.82 \times 10^{-5}$$

$$[H^+] = 1.82 \times 10^{-5}\,\frac{(0.99)}{(1.01)}$$

$$= 1.78 \times 10^{-5}\ M$$

$$pH = 4.749$$

A similar addition to water would cause the pH to rise from 7.0 to 12.0.

Alkaline buffers may also be prepared. If the base and its derived ion are present in equal concentrations,

$$pOH = pK$$
$$pH = 14.0 - pK$$

A solution with $[NH_3] = 0.10\ M$ and $[NH_4^+] = 0.10\ M$ is an example of this type of buffer.

$$\underset{0.10\,M}{NH_3} + H_2O \rightleftharpoons \underset{0.10\,M}{NH_4^+} + \underset{?}{OH^-}$$

$$\frac{[NH_4^+][OH^-]}{[NH_3]} = 1.8 \times 10^{-5}$$

$$[OH^-] = 1.8 \times 10^{-5}\ M$$

$$pOH = -\log(1.8 \times 10^{-5}) = 4.74$$

$$pH = 9.26$$

Buffers may also be prepared in which the concentration of the common ion is not equal to the concentration of the weak acid or base. The pH (or pOH) of these buffers does not equal the pK of the acid (or base). In general, the ratio of the concentrations of ion to weak electrolyte should be between 1/10 to 10/1 for an effective buffer. This range is equivalent to a pH range of plus or minus 1 pH unit applied to the pK. An acetic acid–acetate buffer, for example, can be prepared that has any desired pH between 3.7 and 5.7 since the pK of acetic acid is 4.7.

Example 15.13 What concentrations should be used to prepare a cyanic acid–cyanate buffer with a pH of 3.50?

Solution

$$pH = 3.50$$

$$\log[H^+] = -3.50 = 0.50 - 4.00$$

$$[H^+] = 3.1 \times 10^{-4}\ M$$

$$HOCN \rightleftharpoons H^+ + OCN^-$$

$$1.2 \times 10^{-4} = \frac{[H^+][OCN^-]}{[HOCN]}$$

$$= \frac{3.1 \times 10^{-4}[OCN^-]}{[HOCN]}$$

$$\frac{[HOCN]}{[OCN^-]} = \frac{3.1 \times 10^{-4}}{1.2 \times 10^{-4}}$$

$$\frac{[HOCN]}{[OCN^-]} = 2.6$$

Any solution in which [HOCN]/[OCN^-] is 2.6 will have a pH of 3.5. For example, if [HOCN] = 0.26 *M*, [OCN^-] must be 0.10 *M*; or if [HOCN] = 0.52 *M*, [OCN^-] = 0.20 *M*.

The use of buffers is an important part of many industrial processes. Examples are electroplating and the manufacture of leather, photographic materials, and dyes. In bacteriological research, culture media are generally buffered to maintain the pH required for the growth of the bacteria being studied. Buffers are used extensively in analytical chemistry and are used to calibrate pH meters. Human blood is buffered to a pH of 7.4 by means of bicarbonate, phosphate, and complex protein systems.

15.7 POLYPROTIC ACIDS

Polyprotic acids are acids that contain more than one acid hydrogen per molecule. Examples include sulfuric acid (H_2SO_4), oxalic acid ($H_2C_2O_4$), phosphoric acid (H_3PO_4), and arsenic acid (H_3AsO_4). Polyprotic acids ionize in a stepwise manner, and there is an ionization constant for each step. Subscripts are added to the symbol K in order to specify the step to which the constant applies.

Phosphoric acid is triprotic and ionizes in three steps.

$$H_3PO_4 \rightleftharpoons H^+ + H_2PO_4^- \qquad \frac{[H^+][H_2PO_4^-]}{[H_3PO_4]} = K_1 = 7.5 \times 10^{-3}$$

$$H_2PO_4^- \rightleftharpoons H^+ + HPO_4^{2-} \qquad \frac{[H^+][HPO_4^{2-}]}{[H_2PO_4^-]} = K_2 = 6.2 \times 10^{-8}$$

$$HPO_4^{2-} \rightleftharpoons H^+ + PO_4^{3-} \qquad \frac{[H^+][PO_4^{3-}]}{[HPO_4^{2-}]} = K_3 = 1 \times 10^{-12}$$

Thus, in a solution of phosphoric acid three equilibria occur together with the water equilibrium.

The ionization of phosphoric acid is typical of all polyprotic acids in that the primary ionization is stronger than the secondary, and the secondary ionization is stronger than the tertiary. This trend in the value of the ionization constant is consistent with the nature of the particle that releases a proton in each step. One would predict that a proton would be released more readily by an uncharged molecule than by an uninegative ion and more readily by a uninegative ion than by a binegative ion.

No polyprotic acid is known for which all ionizations are complete. The primary ionization of sulfuric acid is essentially complete,

$$H_2SO_4 \rightarrow H^+ + HSO_4^-$$

but the secondary ionization is not,

$$HSO_4^- \rightleftharpoons H^+ + SO_4^{2-} \qquad \frac{[H^+][SO_4^{2-}]}{[HSO_4^-]} = K_2 = 1.3 \times 10^{-2}$$

Solutions of carbon dioxide are acidic. Carbon dioxide reacts with water to form carbonic acid, H_2CO_3. The reaction, however, is not complete—most of the carbon dioxide exists in solution as CO_2 molecules. Therefore, we shall indicate the primary ionizations as

$$CO_2 + H_2O \rightleftharpoons H^+ + HCO_3^- \qquad \frac{[H^+][HCO_3^-]}{[CO_2]} = K_1 = 4.2 \times 10^{-7}$$

where the symbol $[CO_2]$ is used to represent the total concentration of $CO_2(aq)$ and H_2CO_3. The second ionization step is

$$HCO_3^- \rightleftharpoons H^+ + CO_3^{2-} \qquad \frac{[H^+][CO_3^{2-}]}{[HCO_3^-]} = K_2 = 4.8 \times 10^{-11}$$

An analogous situation exists for solutions of sulfur dioxide in water. The acidity of aqueous SO_2 has been attributed to the ionization of sulfurous acid, H_2SO_3. However, H_2SO_3 has never been isolated in pure form. In solution it apparently exists in equilibrium with $SO_2(aq)$

$$SO_2(aq) + H_2O \rightleftharpoons H_2SO_3(aq)$$

We shall represent the primary ionization of sulfurous acid as

$$SO_2 + H_2O \rightleftharpoons H^+ + HSO_3^-$$

The ionization constants of some polyprotic acids are listed in Table 15.5.

Polyprotic acids form more than one salt. Depending upon the stoichiometric ratio of reactants, the reaction of NaOH and H_2SO_4 yields either the

Table 15.5 Ionization constants of some polyprotic acids

arsenic	H_3AsO_4	$\rightleftharpoons H^+ + H_2AsO_4^-$	$K_1 = 2.5 \times 10^{-4}$
	$H_2AsO_4^-$	$\rightleftharpoons H^+ + HAsO_4^{2-}$	$K_2 = 5.6 \times 10^{-8}$
	$HAsO_4^{2-}$	$\rightleftharpoons H^+ + AsO_4^{3-}$	$K_3 = 3 \times 10^{-13}$
carbonic	$CO_2 + H_2O$	$\rightleftharpoons H^+ + HCO_3^-$	$K_1 = 4.2 \times 10^{-7}$
	HCO_3^-	$\rightleftharpoons H^+ + CO_3^{2-}$	$K_2 = 4.8 \times 10^{-11}$
hydrosulfuric	H_2S	$\rightleftharpoons H^+ + HS^-$	$K_1 = 1.1 \times 10^{-7}$
	HS^-	$\rightleftharpoons H^+ + S^{2-}$	$K_2 = 1.0 \times 10^{-14}$
oxalic	$H_2C_2O_4$	$\rightleftharpoons H^+ + HC_2O_4^-$	$K_1 = 5.9 \times 10^{-2}$
	$HC_2O_4^-$	$\rightleftharpoons H^+ + C_2O_4^{2-}$	$K_2 = 6.4 \times 10^{-5}$
phosphoric	H_3PO_4	$\rightleftharpoons H^+ + H_2PO_4^-$	$K_1 = 7.5 \times 10^{-3}$
	$H_2PO_4^-$	$\rightleftharpoons H^+ + HPO_4^{2-}$	$K_2 = 6.2 \times 10^{-8}$
	HPO_4^{2-}	$\rightleftharpoons H^+ + PO_4^{3-}$	$K_3 = 1 \times 10^{-12}$
phosphorus (diprotic)	H_3PO_3	$\rightleftharpoons H^+ + H_2PO_3^-$	$K_1 = 1.6 \times 10^{-2}$
	$H_2PO_3^-$	$\rightleftharpoons H^+ + HPO_3^{2-}$	$K_2 = 7 \times 10^{-7}$
sulfuric	H_2SO_4	$\rightarrow H^+ + HSO_4^-$	strong
	HSO_4^-	$\rightleftharpoons H^+ + SO_4^{2-}$	$K_2 = 1.3 \times 10^{-2}$
sulfurous	$SO_2 + H_2O$	$\rightleftharpoons H^+ + HSO_3^-$	$K_1 = 1.3 \times 10^{-2}$
	HSO_3^-	$\rightleftharpoons H^+ + SO_3^{2-}$	$K_2 = 5.6 \times 10^{-8}$

normal salt, Na_2SO_4 (sodium sulfate) or the acid salt $NaHSO_4$ (sodium bisulfate or sodium hydrogen sulfate). Three salts may be derived from phosphoric acid: NaH_2PO_4 (sodium dihydrogen phosphate), Na_2HPO_4 (sodium hydrogen phosphate), and Na_3PO_4 (sodium phosphate).

Example 15.14 Calculate $[H^+]$, $[H_2PO_4^-]$, $[HPO_4^{2-}]$, $[PO_4^{3-}]$, and $[H_3PO_4]$ in a 0.10 M solution of phosphoric acid.

Solution The principal source of H^+ is the primary ionization. The H^+ produced by the other ionizations, as well as that from the ionization of water, is negligible in comparison. Furthermore, the concentration of $H_2PO_4^-$ derived from the primary ionization is not significantly diminished by the secondary ionization. Thus, we write

$$\underset{(0.10-x)\,M}{H_3PO_4} \rightleftharpoons \underset{(x)\,M}{H^+} + \underset{(x)\,M}{H_2PO_4^-}$$

The problem must be solved by means of the quadratic formula.

$$\frac{[H^+][H_2PO_4^-]}{[H_3PO_4]} = 7.5 \times 10^{-3}$$

$$\frac{x^2}{(0.10 - x)} = 7.5 \times 10^{-3}$$

$$x = [H^+] = [H_2PO_4^-] = 2.4 \times 10^{-2}\ M$$

$$(0.10 - x) = [H_3PO_4] = 7.6 \times 10^{-2}\ M$$

The $[H^+]$ and $[H_2PO_4^-]$ apply to the secondary ionization. Therefore,

$$\underset{2.4 \times 10^{-2}\,M}{H_2PO_4^-} \rightleftharpoons \underset{2.4 \times 10^{-2}\,M}{H^+} + \underset{?}{HPO_4^{2-}}$$

$$\frac{[H^+][HPO_4^{2-}]}{[H_2PO_4^-]} = 6.2 \times 10^{-8}$$

$$\frac{(2.4 \times 10^{-2})[HPO_4^{2-}]}{(2.4 \times 10^{-2})} = 6.2 \times 10^{-8}$$

$$[HPO_4^{2-}] = 6.2 \times 10^{-8}\ M$$

In any solution of H_3PO_4 that does not contain ions derived from another electrolyte, the concentration of the secondary ion is equal to K_2.

For the tertiary ionization,

$$\underset{6.2 \times 10^{-8}M}{HPO_4^{2-}} \rightleftharpoons \underset{2.4 \times 10^{-2}M}{H^+} + \underset{?}{PO_4^{3-}}$$

$$\frac{[H^+][PO_4^{3-}]}{[HPO_4^{2-}]} = 1 \times 10^{-12}$$

$$\frac{(2.4 \times 10^{-2})[PO_4^{3-}]}{(6.2 \times 10^{-8})} = 1 \times 10^{-12}$$

$$[PO_4^{3-}] = 3 \times 10^{-18}\ M$$

Example 15.15 What are $[H^+]$, $[HS^-]$, $[S^{2-}]$, and $[H_2S]$ in a $0.10\,M$ solution of H_2S?

Solution K_1 for H_2S is 1.1×10^{-7}. Therefore, the small amount of H_2S that ionizes is negligible in comparison to the original concentration of H_2S. In addition, the concentrations of H^+ and HS^- are not significantly altered by the secondary ionization ($K_2 = 1.0 \times 10^{-14}$). Thus,

$$\underset{0.10\,M}{H_2S} \rightleftharpoons \underset{(x)\,M}{H^+} + \underset{(x)\,M}{HS^-}$$

$$\frac{[H^+][HS^-]}{[H_2S]} = 1.1 \times 10^{-7}$$

$$\frac{x^2}{0.10} = 1.1 \times 10^{-7}$$

$$x = [H^+] = [HS^-] = 1.0 \times 10^{-4}\ M$$

These concentrations also apply to the secondary ionization.

$$\underset{1.0 \times 10^{-4}\,M}{HS^-} \rightleftharpoons \underset{1.0 \times 10^{-4}\,M}{H^+} + \underset{?}{S^{2-}}$$

$$\frac{[H^+][S^{2-}]}{[HS^-]} = 1.0 \times 10^{-14}$$

$$\frac{(1.0 \times 10^{-4})[S^{2-}]}{(1.0 \times 10^{-4})} = 1.0 \times 10^{-4}\ M$$

$$[S^{2-}] = 1.0 \times 10^{-14} M$$

The concentration of the secondary ion is equal to K_2 in any solution of H_2S that does not contain ions derived from another electrolyte.

The product of the expressions for the two ionizations of H_2S is

$$\left(\frac{[H^+][HS^-]}{[H_2S]}\right)\left(\frac{[H^+][S^{2-}]}{[HS^-]}\right) = K_1 K_2$$

$$\frac{[H^+]^2\,[S^{2-}]}{[H_2S]} = (1.1 \times 10^{-7})(1.0 \times 10^{-14}) = 1.1 \times 10^{-21}$$

This very convenient relationship can be misleading. Superficially it looks as though it applies to a process in which one sulfide ion is produced for every two hydronium ions. However, the ionization of H_2S does not proceed in this manner. In any solution of H_2S, the concentration of hydronium ion is much larger than the concentration of sulfide ion (Example 15.15). The majority of the H_2S molecules that ionize do so only to the HS^- stage, and S^{2-} ions result only from the small ionization of the secondary ion.

At 25°C a saturated solution of H_2S is 0.10 *M*. For a *saturated solution,* therefore,

$$\frac{[H^+]^2[S^{2-}]}{(0.10)} = 1.1 \times 10^{-21}$$

$$[H^+]^2\,[S^{2-}] = 1.1 \times 10^{-22}$$

This relation can be used to calculate the sulfide ion concentration of a solution of known *p*H that has been saturated with H_2S. We shall see later that acidic solutions of H_2S are frequently employed as laboratory reagents.

Example 15.16 What is the sulfide ion concentration of a dilute HCl solution that has been saturated with H_2S if the *p*H of the solution is 3.00?
Solution Since the *p*H = 3.00,

$$[H^+] = 1.0 \times 10^{-3} M$$

Therefore,

$$[H^+]^2\,[S^{2-}] = 1.1 \times 10^{-22}$$
$$(1.0 \times 10^{-3})^2\,[S^{2-}] = 1.1 \times 10^{-22}$$
$$[S^{2-}] = 1.1 \times 10^{-16} M$$

In a saturated solution of pure H_2S (Example 15.15), $[S^{2-}] = 1.0 \times 10^{-14} M$. In the H_2S solution described in the problem, the common ion, H^+, has repressed the ionization of H_2S. In addition, since the solution contains H^+ ions from a source other than H_2S, $[H^+]$ does not equal $[HS^-]$, and consequently $[S^{2-}]$ does not equal K_2.

15.8 THE SOLUBILITY PRODUCT

Most substances are soluble in water to at least some slight extent. If an "insoluble" or "slightly soluble" material is placed in water, an equilibrium is established when the rate of dissolution of ions from the solid equals the rate of precipitation of ions from the *saturated solution.* Thus, an equilibrium exists between solid silver chloride and a saturated solution of silver chloride.

$$AgCl(s) \rightleftharpoons Ag^+(aq) + Cl^-(aq)$$

The equilibrium constant is

$$K' = \frac{[Ag^+][Cl^-]}{[AgCl]}$$

Since the concentration of a pure solid is a constant, [AgCl] may be combined with K' to give

$$K_{SP} = K'[AgCl] = [Ag^+][Cl^-]$$

The constant K_{SP} is called a **solubility product.** The ionic concentrations of the expression are those for a saturated solution at the reference temperature. Since the solubility of a salt usually varies widely with temperature, the numerical value for K_{SP} for a salt changes with temperature. A table of solubility products at 25°C is given in Appendix F.

The numerical value of K_{SP} for a salt may be found from the molar solubility of the salt.

Example 15.17 At 25°C, 0.00188 g of AgCl dissolves in 1 liter of water. What is the K_{SP} of AgCl?

Solution The molar solubility of AgCl (molecular weight, 143) is

$$? \text{ mol AgCl} = 0.00188 \text{ g AgCl} \left(\frac{1 \text{ mol AgCl}}{143 \text{ g AgCl}}\right)$$

$$= 1.31 \times 10^{-5} \text{ mol AgCl}$$

For each mole of AgCl dissolving, 1 mol of Ag^+ and 1 mol of Cl^- are formed.

$$\underset{}{AgCl(s)} \rightleftharpoons \underset{1.31 \times 10^{-5}\,M}{Ag^+} + \underset{1.31 \times 10^{-5}\,M}{Cl^-}$$

$$\begin{aligned} K_{SP} &= [Ag^+][Cl^-] \\ &= (1.31 \times 10^{-5})^2 \\ &= 1.7 \times 10^{-10} \end{aligned}$$

For salts that have more than two ions per formula unit, the ion concentrations must be raised to the powers indicated by the coefficients of the balanced chemical equation.

$$Mg(OH)_2(s) \rightleftharpoons Mg^{2+} + 2OH^- \qquad K_{SP} = [Mg^{2+}][OH^-]^2$$

$$Bi_2S_3(s) \rightleftharpoons 2Bi^{3+} + 3S^{2-} \qquad K_{SP} = [Bi^{3+}]^2\,[S^{2-}]^3$$

$$Hg_2Cl_2(s) \rightleftharpoons Hg_2^{2+} + 2Cl^- \qquad K_{SP} = [Hg_2^{2+}][Cl^-]^2$$

For a salt of this type the calculation of the K_{SP} from the molar solubility requires that the chemical equation representing the dissociation process be carefully interpreted.

Example 15.18 At 25°C, 7.8×10^{-5} mol of silver chromate dissolves in 1 liter of water. What is the K_{SP} of Ag_2CrO_4?

Solution For each mole of Ag_2CrO_4 that dissolves, 2 mol of Ag^+ and 1 mol of CrO_4^{2-} are formed. Therefore,

$$\underset{}{Ag_2CrO_4(s)} \rightleftharpoons \underset{2(7.8 \times 10^{-5})M}{2Ag^+} + \underset{7.8 \times 10^{-5}M}{CrO_4^{2-}}$$

$$\begin{aligned} K_{SP} &= [Ag^+]^2\,[CrO_4^{2-}] \\ &= (1.56 \times 10^{-4})^2\,(7.8 \times 10^{-5}) \\ &= 1.9 \times 10^{-12} \end{aligned}$$

Example 15.19 The K_{SP} of CaF_2 is 3.9×10^{-11} at 25°C. What is the concentration of Ca^{2+} and of F^- in the saturated solution? How many grams of calcium fluoride will dissolve in 100 ml of water at 25°C?
Solution Let x equal the molar solubility of CaF_2.

$$CaF_2(s) \rightleftharpoons \underset{x}{Ca^{2+}} + \underset{2x}{2F^-}$$

$$\begin{aligned} K_{SP} = [Ca^{2+}][F^-]^2 &= 3.9 \times 10^{-11} \\ x(2x)^2 &= 3.9 \times 10^{-11} \\ 4x^3 &= 3.9 \times 10^{-11} \\ x &= 2.1 \times 10^{-4}\ M \end{aligned}$$

Therefore,

$$\begin{aligned} [Ca^{2+}] &= x = 2.1 \times 10^{-4}\ M \\ [F^-] &= 2x = 4.2 \times 10^{-4}\ M \end{aligned}$$

$$\begin{aligned} ?\ g\ CaF_2 &= 100\ ml\ H_2O \left(\frac{2.1 \times 10^{-4}\ mol\ CaF_2}{1000\ ml\ H_2O}\right)\left(\frac{78\ g\ CaF_2}{1\ mol\ CaF_2}\right) \\ &= 1.6 \times 10^{-3}\ g\ CaF_2 \end{aligned}$$

15.9 PRECIPITATION AND THE SOLUBILITY PRODUCT

The numerical value of the solubility product of a salt is a quantitative statement of the limit of solubility of the salt. For a specific solution of a salt, the product of the concentrations of the ions, each raised to the proper power, is called the **ion product.** Thus, for a saturated solution in equilibrium with excess solid, the ion product equals the K_{SP}. If the ion product of a solution is less than the K_{SP}, the solution is unsaturated; additional solid can dissolve in this solution. On the other hand, if the ion product is greater than the K_{SP}, the solution is momentarily supersaturated; precipitation will occur until the ion product equals the K_{SP}.

Example 15.20 Will a precipitate form if 10 ml of 0.010 M $AgNO_3$ and 10 ml of 0.00010M NaCl are mixed? Assume that the final volume of the solution is 20 ml. For AgCl, $K_{SP} = 1.7 \times 10^{-10}$.

Solution Diluting a solution to twice its original volume reduces the concentrations of ions in the solution to half their original value. Therefore, if there were no reaction, the ion concentrations would be

$$[Ag^+] = 5.0 \times 10^{-3}\ M$$
$$[Cl^-] = 5.0 \times 10^{-5}\ M$$

The ion product is

$$[Ag^+][Cl^-] = ?$$
$$(5.0 \times 10^{-3})(5.0 \times 10^{-5}) = 2.5 \times 10^{-7}$$

Therefore, the ion product is larger than the K_{SP} (1.7×10^{-10}), and precipitation of AgCl should occur.

Example 15.21 Will a precipitate of $Mg(OH)_2$ form in a 0.0010 M solution of $Mg(NO_3)_2$ if the pH of the solution is adjusted to 9.0? The K_{SP} of $Mg(OH)_2$ is 8.9×10^{-12}.

Solution If the pH = 9.0, the pOH = 5.0

$$[OH^-] = 1.0 \times 10^{-5}\ M$$

Since $[Mg^{2+}] = 1.0 \times 10^{-3}$, the ion product is

$$[Mg^{2+}][OH^-]^2 = ?$$
$$(1.0 \times 10^{-3})(1.0 \times 10^{-5})^2 = 1.0 \times 10^{-13}$$

Since the ion product is less than 8.9×10^{-12}, no precipitate will form.

The common-ion effect pertains to solubility equilibria. As an example, consider the system

$$BaSO_4(s) \rightleftharpoons Ba^{2+}(aq) + SO_4^{2-}(aq)$$

The addition of sulfate ion, from sodium sulfate, to a saturated solution of barium sulfate will cause the equilibrium to shift to the left. The concentration of Ba^{2+} will decrease, and $BaSO_4$ will precipitate. Since the product $[Ba^{2+}][SO_4^{2-}]$ is a constant, increasing $[SO_4^{2-}]$ will cause $[Ba^{2+}]$ to decrease.

The amount of barium ion in a solution may be determined by precipitating the Ba^{2+} as $BaSO_4$. The precipitate is then removed by filtration and is dried and weighed. The concentration of Ba^{2+} left in solution after the precipitation may be reduced to a very low value if excess sulfate ion is employed in the precipitation.

Example 15.22 At 25°C a saturated solution of $BaSO_4$ is $3.9 \times 10^{-5}\ M$. The K_{SP} of $BaSO_4$ is 1.5×10^{-9}. What is the solubility of $BaSO_4$ in 0.050 M Na_2SO_4?

Solution The sulfate ion derived from $BaSO_4$ may be ignored. Thus,

$$\underset{}{BaSO_4(s)} \rightleftharpoons \underset{?}{Ba^{2+}} + \underset{5.0 \times 10^{-2}}{SO_4^{2-}}$$

$$[Ba^{2+}][SO_4^{2-}] = 1.5 \times 10^{-9}$$
$$[Ba^{2+}](5.0 \times 10^{-2}) = 1.5 \times 10^{-9}$$
$$[Ba^{2+}] = 3.0 \times 10^{-8}\,M$$

The solubility of $BaSO_4$ has been reduced from $3.9 \times 10^{-5}\,M$ to $3.0 \times 10^{-8}\,M$ by the common-ion effect.

At times, the common-ion effect is used to *prevent* the formation of a precipitate. Consider the precipitation of magnesium hydroxide from a solution that contains Mg^{2+} ions.

$$Mg(OH)_2(s) \rightleftharpoons Mg^{2+}(aq) + 2OH^-(aq)$$

The precipitation can be prevented by holding the concentration of OH^- to a low value. If the hydroxide ion is supplied by the weak base ammonia

$$NH_3 + H_2O \rightleftharpoons NH_4^+ + OH^-$$

the concentration of OH^- can be controlled by the addition of NH_4^+. When the common ion NH_4^+ is added, the ammonia equilibrium shifts to the left, which reduces the concentration of OH^-. In this way, the concentration of OH^- can be held to a level that will not cause $Mg(OH)_2$ to precipitate.

Example 15.23 What concentration of NH_4^+, derived from NH_4Cl, is necessary to prevent the formation of an $Mg(OH)_2$ precipitate in a solution that is $0.050M$ in Mg^{2+} and $0.050M$ in NH_3? The K_{SP} of $Mg(OH)_2$ is 8.9×10^{-12}.

Solution We first calculate the maximum concentration of hydroxide ion that can be present in the solution without causing $Mg(OH)_2$ to precipitate.

$$[Mg^{2+}][OH^-]^2 = 8.9 \times 10^{-12}$$
$$(5.0 \times 10^{-2})[OH^-]^2 = 8.9 \times 10^{-12}$$
$$[OH^-]^2 = 1.8 \times 10^{-10}$$
$$[OH^-] = 1.3 \times 10^{-5}\,M$$

The concentration of OH^-, therefore, can be no higher than $1.3 \times 10^{-5}\,M$. From the expression for the ionization constant for NH_3, we can derive the concentration of NH_4^+ that will maintain the concentration of OH^- at this level.

$$\frac{[NH_4^+][OH^-]}{[NH_3]} = 1.8 \times 10^{-5}$$

$$\frac{[NH_4^+](1.3 \times 10^{-5})}{(5.0 \times 10^{-2})} = 1.8 \times 10^{-5}$$

$$[NH_4^+] = 6.9 \times 10^{-2} M$$

Thus, the *minimum* concentration of NH_4^+ that must be present is 0.069 M.

The concentration of sulfide ion in a solution of H_2S can be controlled by adjusting the pH (Example 15.16). When the hydronium ion concentration is increased, the sulfide ion concentration decreases. If the sulfide ion concentration is made low enough, a sulfide precipitate can be prevented from forming.

Example 15.24 What must be the hydronium ion concentration of a solution that is 0.050 M in Ni^{2+} to prevent the precipitation of NiS when the solution is saturated with H_2S? The K_{SP} of NiS is 3×10^{-21}.

Solution

$$[Ni^{2+}][S^{2-}] = 3 \times 10^{-21}$$
$$(0.050)[S^{2-}] = 3 \times 10^{-21}$$
$$[S^{2-}] = 6 \times 10^{-20}\ M$$

Therefore, the $[S^{2-}]$ must be less than 6×10^{-20} M if NiS is not to precipitate.

For a solution saturated with H_2S,

$$[H^+]^2[S^{2-}] = 1.1 \times 10^{-22}$$
$$[H^+]^2(6 \times 10^{-20}) = 1.1 \times 10^{-22}$$
$$[H^+] = 0.04\ M$$

The $[H^+]$ must be greater than 0.04 M to prevent the precipitation of NiS.

Suppose that a solution contains two cations both of which are capable of forming insoluble sulfides. In some cases, it is possible to adjust the sulfide ion concentration (by means of the pH) in such a way that one cation will precipitate as the sulfide and the other cation will not.

Example 15.25 A solution that is 0.30 M in H^+, 0.050 M in Pb^{2+}, and 0.050 M in Fe^{2+} is saturated with H_2S. Should PbS and/or FeS precipitate? The K_{SP} of PbS is 7×10^{-29}, and the K_{SP} of FeS is 4×10^{-19}.

Solution For any saturated solution of H_2S,

$$[H^+]^2[S^{2-}] = 1.1 \times 10^{-22}$$

Since this solution is 0.30 M in H^+,

$$(3.0 \times 10^{-1})^2[S^{2-}] = 1.1 \times 10^{-22}$$
$$[S^{2-}] = 1.2 \times 10^{-21}\ M$$

Both Pb^{2+} and Fe^{2+} are 2+ ions, and the form of the ion product is

$$[M^{2+}][S^{2-}]$$

where M^{2+} stands for either metal ion. Since both are present in concentrations of 0.050 M,

$$[M^{2+}][S^{2-}]$$
$$(5.0 \times 10^{-2})(1.2 \times 10^{-21}) = 6.0 \times 10^{-23}$$

This ion product is greater than the K_{SP} of PbS; therefore, PbS will precipitate. The ion product, however, is less than the K_{SP} of FeS. The solubility of FeS has not been exceeded. No FeS will form.

The preceding examples illustrate an important analytical technique. In the usual qualitative analysis scheme, certain cations are separated into groups on the basis of whether their sulfides precipitate in acidic solution. In a solution that has an H^+ concentration of $0.3M$, the sulfides of Hg^{2+}, Pb^{2+}, Cu^{2+}, Bi^{3+}, Cd^{2+}, and Sn^{2+} are insoluble, whereas the sulfides of Fe^{2+}, Co^{2+}, Ni^{2+}, Mn^{2+}, and Zn^{2+} are soluble.

In considering systems that involve sulfide precipitation, one must note that the H^+ concentration of a solution increases when a sulfide precipitates from the solution.

Example 15.26 A solution that is $0.050M$ in Cd^{2+} and $0.10M$ in H^+ is saturated with H_2S. What concentration of Cd^{2+} remains in solution after CdS has precipitated? The K_{SP} of CdS is 1.0×10^{-28}.

Solution For each Cd^{2+} ion precipitated, two H^+ ions are added to the solution.

$$Cd^{2+}(aq) + H_2S(aq) \rightleftharpoons CdS(s) + 2H^+(aq)$$

We shall assume that virtually all the Cd^{2+} precipitates as CdS. Hence, the precipitation introduces 0.10 mol of H^+ per liter of solution, and the final $[H^+]$ is $0.20M$. Therefore,

$$[H^+]^2[S^{2-}] = 1.1 \times 10^{-22}$$
$$(0.20)^2[S^{2-}] = 1.1 \times 10^{-22}$$
$$[S^{2-}] = 2.8 \times 10^{-21}$$

The concentration of Cd^{2+} after the CdS has precipitated may be derived from the K_{SP} of CdS.

$$[Cd^{2+}][S^{2-}] = 1.0 \times 10^{-28}$$
$$[Cd^{2+}](2.8 \times 10^{-21}) = 1.0 \times 10^{-28}$$
$$[Cd^{2+}] = 3.6 \times 10^{-8}\ M$$

From our answer we can see that our assumption was justified. The value of $[H^+]$ that we used was well within our limits of accuracy.

15.10 EQUILIBRIA INVOLVING COMPLEX IONS

Complex ions are discussed in Chapter 20. These ions, however, take part in some aqueous equilibria that properly constitute a part of this chapter. A **complex ion** is an aggregate consisting of a central metal cation (usually a transition-metal ion) surrounded by a number of ligands. The **ligands** of a complex may be anions, molecules, or a combination of the two.

A ligand must have an unshared pair of electrons with which it can bond to the central ion. Thus, the ammonia molecule, NH_3, functions as a ligand in the formation of complex ions—the ammonium ion, NH_4^+, does not.

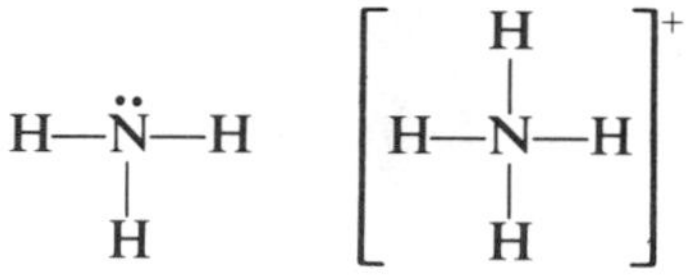

In addition to ammonia, the requirement is met by many molecules and anions.

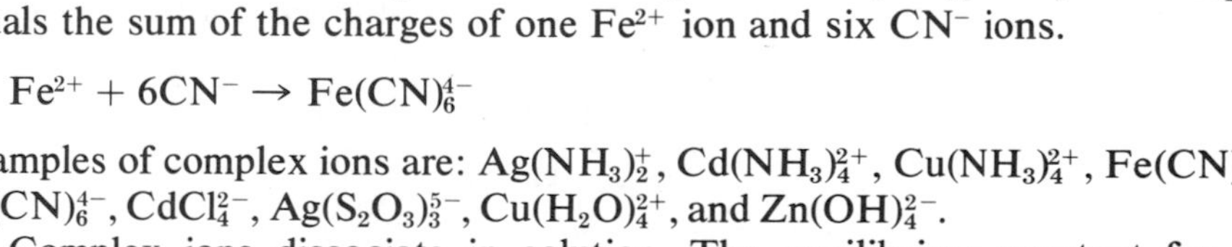

The charge of a complex ion can be found by adding the charges of all the particles of which it is composed. The charge of the $Fe(CN)_6^{4-}$ ion, for example, equals the sum of the charges of one Fe^{2+} ion and six CN^- ions.

$$Fe^{2+} + 6CN^- \rightarrow Fe(CN)_6^{4-}$$

Examples of complex ions are: $Ag(NH_3)_2^+$, $Cd(NH_3)_4^{2+}$, $Cu(NH_3)_4^{2+}$, $Fe(CN)_6^{3-}$, $Fe(CN)_6^{4-}$, $CdCl_4^{2-}$, $Ag(S_2O_3)_3^{5-}$, $Cu(H_2O)_4^{2+}$, and $Zn(OH)_4^{2-}$.

Complex ions dissociate in solution. The equilibrium constant for the complete dissociation is called the **instability constant** and is given the symbol K_{inst}.

$$Ag(NH_3)_2^+ \rightleftharpoons Ag^+ + 2NH_3 \qquad K_{inst} = \frac{[Ag^+][NH_3]^2}{[Ag(NH_3)_2^+]} = 6.0 \times 10^{-8}$$

Dissociations, however, take place by a series of steps in which the ligands are lost one at a time. The instability constant pertains to a chemical equation that is a summary of the steps.

At times, it is convenient to write the chemical equation in reverse form. The equilibrium constant that pertains to this equation is called a **formation constant,** which has the symbol, K_{form}.

$$Ag^+ + 2NH_3 \rightleftharpoons Ag(NH_3)_2^+ \qquad K_{form} = \frac{[Ag(NH_3)_2^+]}{[Ag^+][NH_3]^2}$$

A formation constant is, of course, the reciprocal of the corresponding instability constant.

$$K_{form} = \frac{1}{K_{inst}}$$

$$= \frac{1}{6.0 \times 10^{-8}} = 1.7 \times 10^7$$

Slightly soluble substances can often be dissolved through the formation of complex ions. The solubility of AgCl, for example, is increased if the solution contains NH_3. The equilibrium

$$AgCl(s) \rightleftharpoons Ag^+(aq) + Cl^-(aq)$$

is forced to the right by the removal of Ag^+ through the formation of $Ag(NH_3)_2^+$.

Example 15.27 What is the solubility of AgCl in $0.10M$ NH_3?

Solution We write the equation for the complete transformation and let x equal the molar solubility of AgCl.

$$\underset{}{AgCl(s)} + \underset{0.10-2x}{2NH_3} \rightleftharpoons \underset{x}{Ag(NH_3)_2^+} + \underset{x}{Cl^-}$$

Notice that since two molecules of NH_3 are required for every $Ag(NH_3)_2^+$ formed, the concentration of NH_3 is $(0.10 - 2x)$. An equilibrium constant corresponding to this equation may be derived by dividing the K_{SP} of AgCl by the K_{inst} of $Ag(NH_3)_2^+$.

$$K_{SP}\left(\frac{1}{K_{inst}}\right) = [Ag^+][Cl^-]\left(\frac{[Ag(NH_3)_2^+]}{[Ag^+][NH_3]^2}\right)$$

$$(1.7 \times 10^{-10})\left(\frac{1}{6.0 \times 10^{-8}}\right) = 2.8 \times 10^{-3}$$

Therefore,

$$\frac{[Ag(NH_3)_2^+][Cl^-]}{[NH_3]^2} = 2.8 \times 10^{-3}$$

$$\frac{x^2}{(0.10 - 2x)^2} = 2.8 \times 10^{-3}$$

$$\frac{x}{(0.10 - 2x)} = 5.3 \times 10^{-2}$$

$$x = 4.8 \times 10^{-3}\,M$$

The solubility of AgCl in pure water is $1.3 \times 10^{-5}\,M$.

A few metals form complex ions with S^{2-} (for example, HgS_2^{2-}, AsS_3^{3-}, SbS_2^{3-}, SnS_3^{2-}). The true structure of these anions, which are called thio complex ions, may involve the HS^- ion. The formation of a thio complex ion may be used to separate sulfide precipitates. For example, a mixture of CuS (which does not form a thio complex) and As_2S_3 may be separated by dissolving the As_2S_3 in an alkaline solution containing S^{2-} ion.

$$CuS(s) + S^{2-}(aq) \rightarrow \text{no reaction}$$
$$As_2S_3(s) + 3S^{2-}(aq) \rightleftharpoons 2AsS_3^{3-}(aq)$$

If a solution containing Al^{3+} and Zn^{2+} is treated with a buffer of NH_3 at a controlled alkaline *p*H, $Al(OH)_3$ will precipitate, but Zn^{2+} will stay in solution as $Zn(NH_3)_4^{2+}$. The precipitation of $Zn(OH)_2$ is prevented by the formation of the complex ion; Al^{3+} does not form an ammonia complex.

$$Al^{3+}(aq) + 3OH^-(aq) \rightleftharpoons Al(OH)_3(s)$$
$$Zn^{2+}(aq) + 4NH_3(aq) \rightleftharpoons Zn(NH_3)_4^{2+}(aq)$$

Complex ions are frequently highly colored; $Fe(SCN)^{2+}$ is deep red. Thus, the production of a deep red color when SCN^- is added to a solution serves as a test for the Fe^{3+} ion.

15.11 AMPHOTERISM

The hydroxides of certain metals, called **amphoteric hydroxides,** can function as both acids and bases. These water-insoluble compounds dissolve in solutions of high *p*H or low *p*H. For example, zinc hydroxide dissolves in excess hydrochloric acid to produce solutions of zinc chloride, $ZnCl_2$.

$$Zn(OH)_2(s) + 2H^+(aq) \rightarrow Zn^{2+}(aq) + 2H_2O$$

Zinc hydroxide will also dissolve in sodium hydroxide solutions. The reaction yields sodium zincate, $Na_2Zn(OH)_4$.

$$Zn(OH)_2(s) + 2OH^-(aq) \rightarrow Zn(OH)_4^{2-}(aq)$$

Other examples of amphoteric hydroxides are: $Al(OH)_3$, $Sn(OH)_2$, $Cr(OH)_3$, $Be(OH)_2$, $Sb(OH)_3$ (or Sb_2O_3), and $As(OH)_3$ (or As_2O_3). Many other compounds including $Cu(OH)_2$ and AgOH (or Ag_2O) exhibit this property to a lesser degree. The oxides that correspond to amphoteric hydroxides react in the same way.

The following series of equations can be used to describe the reactions that occur when the OH^- concentration of a $Zn^{2+}(aq)$ solution is gradually increased.

$$Zn^{2+}(aq) + OH^-(aq) \rightleftharpoons Zn(OH)^+(aq)$$
$$Zn(OH)^+(aq) + OH^-(aq) \rightleftharpoons Zn(OH)_2(s)$$
$$Zn(OH)_2(s) + OH^-(aq) \rightleftharpoons Zn(OH)_3^-(aq)$$
$$Zn(OH)_3^-(aq) + OH^-(aq) \rightleftharpoons Zn(OH)_4^{2-}(aq)$$

The precipitation of $Zn(OH)_2$ is observed when the proper *p*H is reached, and upon further increase in *p*H, the precipitate dissolves. The formulas of the amphoterate anions used here are based upon the fact that $NaZn(OH)_3$ and $NaZn(OH)_4$ have been isolated from such solutions.

The process can be reversed. Acidification of solutions of the zincate ion results in the stepwise removal of OH^-. For example,

$$Zn(OH)_4^{2-}(aq) + H^+(aq) \rightleftharpoons Zn(OH)_3^+(aq) + H_2O$$

In general, amphoteric systems are very complicated and much remains to be learned about them. The equilibrium constants are of doubtful validity. Indeed, there is evidence that not all the ions and equilibria in systems of this type have been identified. Probably the reactions of most systems are much more complicated than we have depicted them.

Advantage is taken of the amphoteric nature of some hydroxides in analytical chemistry and in some commercial processes. For example, Mg^{2+} and Zn^{2+} may be separated from a solution containing the two ions by making the solution alkaline.

$$Mg^{2+} + 2OH^- \rightleftharpoons Mg(OH)_2(s)$$
$$Zn^{2+} + 4OH^- \rightleftharpoons Zn(OH)_4^{2-}$$

The insoluble magnesium hydroxide may be removed by filtration from the solution containing the zincate ion.

In the production of aluminum metal from bauxite (impure hydrated Al_2O_3), the ore is purified prior to its reduction to aluminum metal. This purification is accomplished by dissolving the aluminum oxide in a solution of sodium hydroxide and removing the insoluble impurities by filtration.

$$Al_2O_3(s) + 2OH^- + 3H_2O \rightarrow 2Al(OH)_4^-$$

When the filtered solution is acidified, pure aluminum hydroxide precipitates and is recovered.

15.12 HYDROLYSIS

Anions derived from weak acids react with water to increase the concentration of OH^- ions in the solution. The acetate ion, for example, reacts in the following way.

$$C_2H_3O_2^- + H_2O \rightleftharpoons HC_2H_3O_2 + OH^-$$

This reaction of the acetate ion is similar to that of any other weak base, such as NH_3, with water.

$$NH_3 + H_2O \rightleftharpoons NH_4^+ + OH^-$$

The fact that the acetate ion has a charge and the ammonia molecule does not is unimportant. The reaction of an ion with water, however, is usually called a **hydrolysis reaction.**

Problems involving the hydrolysis of an ion such as the acetate ion are solved in the same way that other base-equilibrium problems are solved. We need, however, an equilibrium constant with which to work. The equilibrium constant for the acetate ion hydrolysis (given the symbol K_H) is obtained by multiplying K_w by the reciprocal of the ionization constant of acetic acid.

$$C_2H_3O_2^- + H_2O \rightleftharpoons HC_2H_3O_2 + OH^-$$

$$K_w\left(\frac{1}{K_{HC_2H_3O_2}}\right) = K_H$$

$$[H^+][OH^-]\left(\frac{[HC_2H_3O_2]}{[H^+][C_2H_3O_2^-]}\right) = \frac{[HC_2H_3O_2][OH^-]}{[C_2H_3O_2^-]}$$

$$(1.0 \times 10^{-14})\left(\frac{1}{1.8 \times 10^{-5}}\right) = 5.6 \times 10^{-10}$$

As we will see, cations derived from weak bases also hydrolyze. Anions derived from *strong* acids (such as Cl^- from HCl) and cations derived from *strong* bases (such as Na^+ from NaOH), however, *do not hydrolyze.* A hydrolysis equilibrium results only when the ion can form a *molecule* of a weak electrolyte in the reaction with water. Strong acids and bases do not exist as molecules in water solution.

Example 15.28 What is the pH of a $0.10M$ solution of $NaC_2H_3O_2$?

Solution The salt $NaC_2H_3O_2$ completely dissociates into Na^+ and $C_2H_3O_2^-$ ions in water solution. The Na^+ ion does not hydrolyze. Let x equal the equilibrium concentration of $HC_2H_3O_2$ that results from the hydrolysis of $C_2H_3O_2^-$ ions,

$$\underset{0.10M}{C_2H_3O_2^-} + H_2O \rightleftharpoons \underset{(x)\,M}{HC_2H_3O_2} + \underset{(x)\,M}{OH^-}$$

Since the value of K_H is very small (5.6×10^{-10}), x is small and $[C_2H_3O_2^-]$ may be assumed to be equal to 0.10 M rather than $(0.10 - x)\,M$.

$$\frac{[HC_2H_3O_2][OH^-]}{[C_2H_3O_2^-]} = K_H$$

$$\frac{x^2}{0.10} = 5.6 \times 10^{-10}$$

$$x^2 = 5.6 \times 10^{-11}$$

$$x = [HC_2H_3O_2] = [OH^-] = 7.5 \times 10^{-6}\ M$$

$$pOH = -\log(7.5 \times 10^{-6}) = 5.12$$

$$pH = 14.00 - 5.12 = 8.88$$

The hydrolysis of a cation derived from a weak base increases the concentration of hydronium ion.

$$NH_4^+ + H_2O \rightleftharpoons NH_3 + H_3O^+$$

This reaction of NH_4^+ is similar to the ionization of any other weak acid such as $HC_2H_3O_2$. The equilibrium constant is obtained by multiplying K_w by the reciprocal of the ionization constant of NH_3.

$$K_w\left(\frac{1}{K_{NH_3}}\right) = K_H$$

$$[H_3O^+][OH^-]\left(\frac{[NH_3]}{[NH_4^+][OH^-]}\right) = \frac{[NH_3][H_3O^+]}{[NH_4^+]}$$

$$(1.0 \times 10^{-14})\left(\frac{1}{1.8 \times 10^{-5}}\right) = 5.6 \times 10^{-10}$$

Example 15.29 What is the pH of a 0.30M solution of NH_4Cl?

Solution In water solution, NH_4Cl completely dissociates into NH_4^+ and Cl^- ions. The Cl^- ion since it is derived from a strong acid, does not hydrolyze. Let x equal the equilibrium concentration of NH_3 from the hydrolysis of NH_4^+.

$$\underset{}{H_2O} + \underset{0.30M}{NH_4^+} \rightleftharpoons \underset{(x)M}{NH_3} + \underset{(x)M}{H_3O^+}$$

$$\frac{[NH_3][H_3O^+]}{[NH_4^+]} = 5.6 \times 10^{-10}$$

$$\frac{x^2}{0.30} = 5.6 \times 10^{-10}$$

$$x^2 = 1.7 \times 10^{-10}$$

$$x = [NH_3] = [H_3O^+] = 1.3 \times 10^{-5} M$$

$$pH = -\log (1.3 \times 10^{-5}) = 4.89$$

The cations of the IA metals, as well as Ca^{2+}, Sr^{2+}, and Ba^{2+}, do not hydrolyze since they are derived from strong bases. Most other metal cations, however, do hydrolyze. In the hydrolysis of a metal cation a coordinated water molecule of the hydrated cation donates a proton to a free water molecule.

$$Fe(H_2O)_6^{3+}(aq) + H_2O \rightleftharpoons Fe(OH)(H_2O)_5^{2+}(aq) + H_3O^+(aq)$$

Such equations are sometimes written without indicating the coordinated water.

$$Fe^{3+}(aq) + H_2O \rightleftharpoons Fe(OH)^{2+}(aq) + H^+(aq)$$

Additional steps in the hydrolysis of the iron(III) ion produce $Fe(OH)_2^+$, $[Fe_2(OH)_2(H_2O)_8]^{4+}$, polynuclear ions of a higher order, and ultimately a precipitate of hydrous Fe_2O_3.

The weaker the electrolyte from which an ion is derived, the more extensive is its hydrolysis. A $0.1M$ solution of sodium acetate has a pH of 8.9, and the pH of a $0.1M$ solution of sodium cyanide is 11.2. In each case it is the hydrolysis of the anion that causes the solution to be alkaline. The sodium ion does not undergo hydrolysis.

$$C_2H_3O_2^- + H_2O \rightleftharpoons HC_2H_3O_2 + OH^-$$
$$CN^- + H_2O \rightleftharpoons HCN + OH^-$$

Since HCN (ionization constant, 4.0×10^{-10}) is a weaker electrolyte than $HC_2H_3O_2$ (ionization constant, 1.8×10^{-5}), HCN does a better job of tying up protons than $HC_2H_3O_2$ does. Therefore, the hydrolysis of CN^- is more complete than that of $C_2H_3O_2^-$, and the concentration of OH^- is higher in the NaCN solution than in the $NaC_2H_3O_2$ solution. Note, however, that in both of these systems the position of equilibrium is such that the reverse reaction, which may be regarded as a neutralization, proceeds to a greater extent than the forward reaction since H_2O is a weaker electrolyte than either HCN or $HC_2H_3O_2$.

The pH of a solution of a normal salt can be predicted on the basis of the strengths of the acid and base from which the salt is derived.

1. *Salt of a strong base and a strong acid.* Examples are NaCl, KNO_3, and $Ba(ClO_3)_2$. Neither cation nor anion hydrolyzes. The solution has a pH of 7.
2. *Salt of a strong base and a weak acid.* Examples are: KNO_2, $Ca(C_2H_3O_2)_2$, and NaCN. The anion hydrolyzes to produce OH^- ions. The solution has a pH that is higher than 7.
3. *Salt of a weak base and a strong acid.* Examples are: NH_4NO_3, $FeBr_2$,

and $AlCl_3$. The cation hydrolyzes to produce H_3O^+ ions. The pH of the solution is below 7.

4. *Salt of a weak base and a weak acid.* Examples are: $NH_4C_2H_3O_2$, NH_4CN, and $Cu(NO_2)_2$. Both cation and anion hydrolyze. The pH of the solution depends upon the extent to which each ion hydrolyzes. The pH of a solution of $NH_4C_2H_3O_2$ is 7 since NH_3 ($K = 1.8 \times 10^{-5}$) and $HC_2H_3O_2$ ($K = 1.8 \times 10^{-5}$) are equally weak. The pH of a solution of NH_4CN, on the other hand, is above 7 because HCN ($K = 4.0 \times 10^{-10}$) is a weaker acid than NH_3 ($K = 1.8 \times 10^{-5}$) is a base. As a consequence, the CN^- hydrolyzes to a greater extent (producing OH^-) than the NH_4^+ does (producing H_3O^+).

15.13 ACID-BASE TITRATIONS

We are now in a position to study acid–alkali titrations in some detail. Let us consider the titration of a 50.0 ml sample of 0.100N HCl with a solution of 0.100N NaOH. Since HCl and NaOH are both strong electrolytes, the only equilibrium to be considered is the water equilibrium. Furthermore, for either solution the molarity is equal to the normality.

The concentration of H^+ in the original 50.0 ml sample of acid in the titration flask is 0.100 M (or 10^{-1} M). The pH, therefore, is 1.00.

After 10.0 ml of 0.100N NaOH has been added from the buret, the total volume of the solution is 60.0 ml, and the equivalent of 40.0 ml of 0.100 N HCl remains unneutralized.

$$VN = VN$$
$$40.0(0.100) = 60.0x$$
$$x = [H^+] = 0.0667N$$
$$pH = 1.18$$

When 50.0 ml of 0.100N NaOH has been added, the equivalence point of the titration is reached. All the acid is neutralized, and the pH is 7.00.

As NaOH solution is added beyond the equivalent point, the solution in the titration flask becomes increasingly alkaline. For example, when 60.0 ml of 0.100N NaOH has been added (total volume, 110 ml), the solution contains the equivalent of 10.0 ml of 0.100N NaOH.

$$VN = VN$$
$$10.0(0.100) = 110.0x$$
$$x = [OH^-] = 0.00909N$$
$$pOH = 2.04$$
$$pH = 11.96$$

The values in Table 15.6 were obtained from calculations such as these. The data of Table 15.6 are plotted in Figure 15.1. Notice that the curve rises sharply in the section around the equivalence point. The first 49.9 ml of NaOH

Table 15.6 Titration of 50.0 ml of 0.100*N* HCl with 0.100*N* NaOH

VOLUME OF 0.100*N* NaOH ADDED (ml)	*p*H
0.0	1.00
10.0	1.18
20.0	1.37
30.0	1.60
40.0	1.96
49.0	3.00
49.9	4.00
50.0	7.00
50.1	10.00
51.0	11.00
60.0	11.96
70.0	12.22
80.0	12.36
90.0	12.46
100.0	12.52

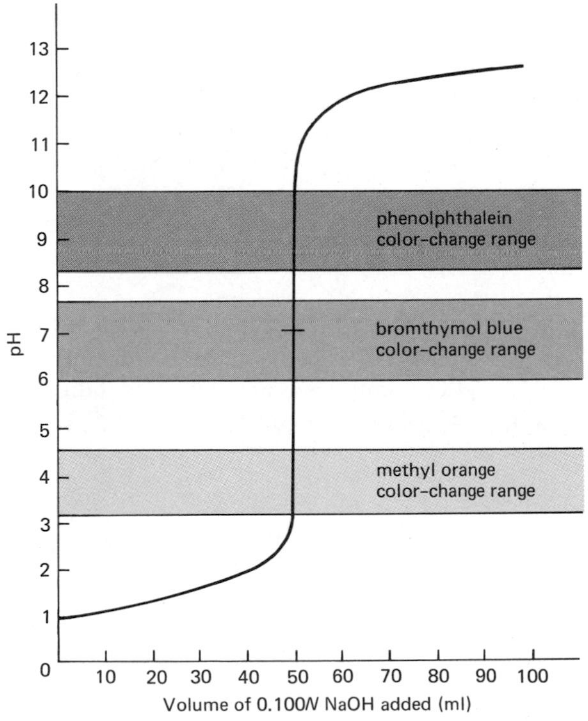

Figure 15.1 Titration of 50.0 ml of 0.100 *N* HCl with 0.100 *N* NaOH.

solution added cause the pH to change by three units but the next 0.2 ml added causes a change of *six* units in the pH.

If an indicator is to be used in this titration to locate the equivalence point, any indicator may be employed that changes color in the pH range of the straight portion of the curve. Around this point, the addition of one drop of NaOH solution causes a sharp increase in the pH. Any one of the three indicators in Figure 15.1 would be satisfactory.

Let us now consider the titration of 50.0 ml of 0.100N acetic acid – a weak acid – with 0.100N sodium hydroxide. The concentration of H^+ in the original 50.0 ml sample of acid may be calculated by using the ionization constant of $HC_2H_3O_2$. From Table 15.1 we see that $[H^+] = 1.34 \times 10^{-3}\ M$. The pH of the solution is therefore 2.87.

The solution resulting from the addition of 10.0 ml of 0.100 N NaOH to the 50.0 ml sample of 0.100N $HC_2H_3O_2$ is, in effect, a buffer. It contains a mixture of $C_2H_3O_2^-$ ions (produced by the neutralization) and unneutralized $HC_2H_3O_2$ molecules. The expression for the equilibrium constant for the ionization of acetic acid

$$K = \frac{[H^+][C_2H_3O_2^-]}{[HC_2H_3O_2]}$$

can be solved for $[H^+]$

$$[H^+] = K\left(\frac{[HC_2H_3O_2]}{[C_2H_3O_2^-]}\right)$$

The ratio $([HC_2H_3O_2]/[C_2H_3O_2^-])$ that appears in this equation is easily calculated. After 10.0 ml of NaOH has been added, 40/50 of the acid in the original 50.0 ml sample remains unneutralized and 10/50 has been converted into sodium acetate. The acetic acid–acetate ratio, therefore, is 40/50 to 10/50, or 4 to 1.

$$[H^+] = 1.8 \times 10^{-5}\left(\frac{4}{1}\right) = 7.2 \times 10^{-5}\ M$$

$$pH = 4.14$$

At the equivalence point all of the acid has been neutralized, and the solution (100 ml) is effectively 0.0500 M in sodium acetate. The calculation of the pH of this solution must take into account the hydrolysis of the $C_2H_3O_2^-$ ion (Example 15.28).

$$\underset{0.0500\ M}{C_2H_3O_2^-} + H_2O \rightleftharpoons \underset{(x)\ M}{HC_2H_3O_2} + \underset{(x)\ M}{OH^-}$$

$$\frac{[HC_2H_3O_2][OH^-]}{[C_2H_3O_2^-]} = 5.6 \times 10^{-10}$$

$$\frac{x^2}{5.00 \times 10^{-2}} = 5.6 \times 10^{-10}$$

$$x = [OH^-] = 5.3 \times 10^{-6}$$

$$pOH = 5.28 \qquad pH = 8.72$$

Notice that the equivalence point in this titration does not occur at the "neutral" pH, 7.

Table 15.7 Titration of 50.0 ml of 0.100*N* $HC_2H_3O_2$ with 0.100*N* NaOH

VOLUME OF 0.100*N* NaOH ADDED (ml)	*p*H
0.0	2.87
10.0	4.14
20.0	4.56
30.0	4.92
40.0	5.34
49.0	6.44
49.9	7.45
50.0	8.72
50.1	10.00
51.0	11.00
60.0	11.96
70.0	12.22
80.0	12.36
90.0	12.46
100.0	12.52

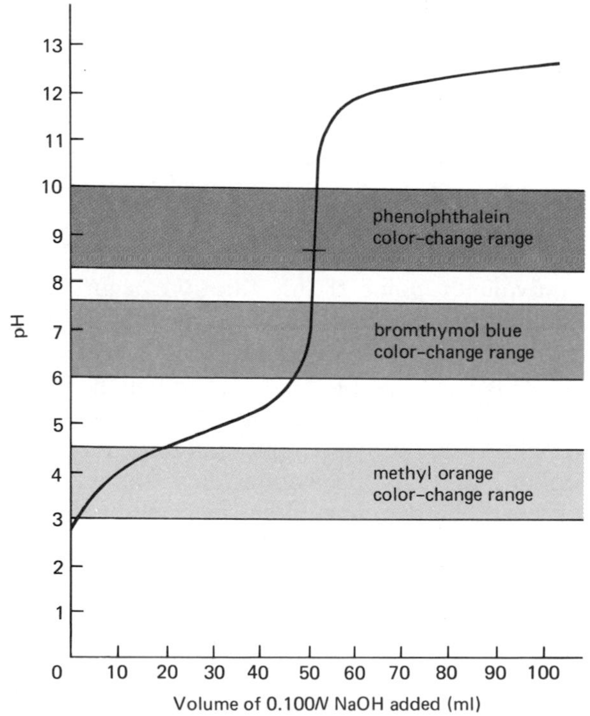

Figure 15.2 Titration of 50.0 ml of 0.100 *N* $HC_2H_3O_2$ with 0.100 *N* NaOH.

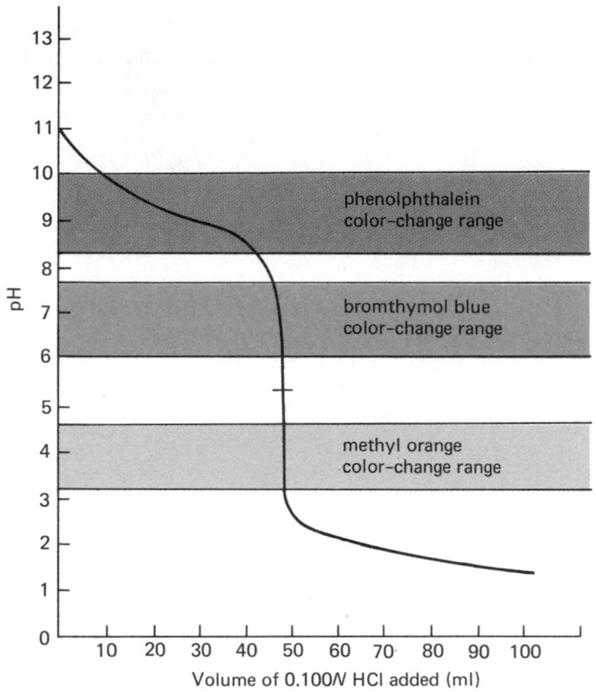

Figure 15.3 Titration of 50.0 ml of 0.100 *N* NH_3 with 0.100 *N* HCl.

After the equivalence point the addition of NaOH causes the solution to become increasingly alkaline. The added OH^- shifts the hydrolysis equilibrium to the left. The effect of the hydrolysis on the *p*H is negligible. Thus, the calculations from this point on are identical to those for the HCl–NaOH titration.

Data for a $HC_2H_3O_2$–NaOH titration are summarized in Table 15.7 and plotted in Figure 15.2. The equivalence point of this titration occurs at a higher *p*H than that of the preceding titration, and all *p*H values on the acid side of the equivalence point are higher. Therefore, the rapidly ascending portion of the curve around the equivalence point is reduced in length. From the curve of Figure 15.2 we can see that methyl orange is not a suitable indicator for this titration. Neither is bromthymol blue suitable. Its color change would start after 47.34 ml of NaOH has been added and continue until 49.97 ml has been added, which would hardly constitute a sharp end point for a titration. Phenolphthalein, however, would be a satisfactory indicator to employ.

Other titration curves may be drawn following the general line of approach outlined here. Figure 15.3 represents the titration curve for the titration of 50.0 ml of 0.100 *N* NH_3 with 0.100 *N* HCl. In this instance methyl orange could be used as the indicator. Figure 15.4 is the titration curve for the titration of 50.0 ml of 0.100 *N* $HC_2H_3O_2$ with 0.100 *N* NH_3. Both solutes are weak

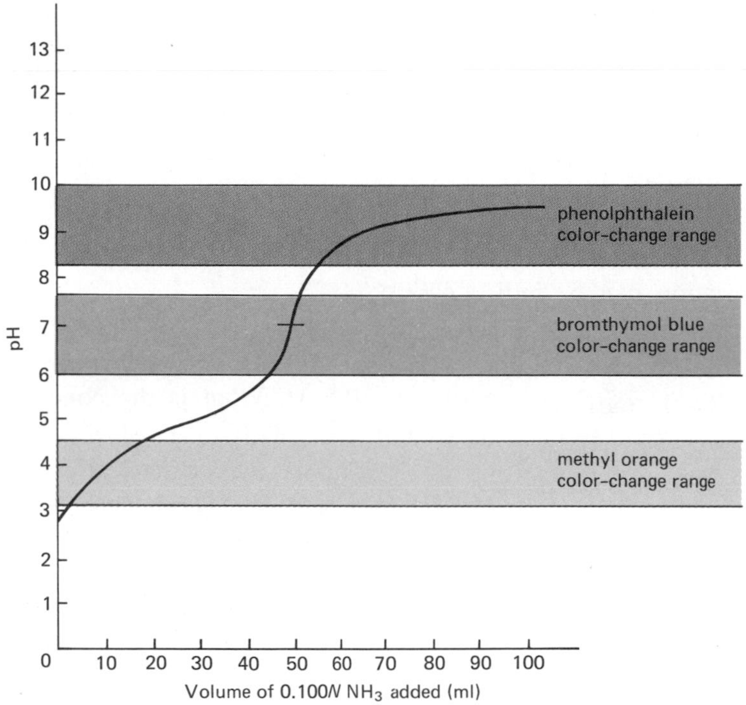

Figure 15.4 Titration of 50.0 ml of 0.100 *N* $HC_2H_3O_3$ with 0.100 *N* NH_6.

electrolytes. No indicator can be found that would function satisfactorily for this titration, and such titrations—between weak electrolytes—are not usually run.

PROBLEMS

15.1 Lactic acid, $HC_3H_5O_3$, is 2.4% ionized in 0.25*M* solution. What is the ionization constant of this acid?

15.2 A 0.200*M* solution of dichloroacetic acid, $H(O_2CCHCl_2)$, a weak monobasic acid, is 33% ionized. What is the ionization constant of this acid?

15.3 A solution is prepared by dissolving 0.300 mol of cyanoacetic acid, $H(O_2CCH_2CN)$, in sufficient water to make exactly one liter. The concentration of $H^+(aq)$ in the resulting solution is 0.032*M*. What is the ionization constant of cyanoacetic acid?

15.4 In a 0.25*M* solution of benzyl amine, $C_7H_7NH_2$, the concentration of $OH^-(aq)$ is $2.4 \times 10^{-3} M$. What is the ionization constant for this weak base? The reaction is

$$C_7H_7NH_2 + H_2O \rightleftharpoons C_7H_7NH_3^+ + OH^-$$

15.5 Morphine is a weak base, and the aqueous ionization can be indicated as

$$M(aq) + H_2O \rightleftharpoons MH^+(aq) + OH^-(aq)$$

In a 0.0050*M* solution of morphine, the ratio of $OH^-(aq)$ ions to morphine molecules is 1.0/83. What is the value of *K* for the aqueous ionization of morphine?

15.6 (a) What concentration of $HClO_2$ molecules is in equilibrium with 0.030*M* $H^+(aq)$ in a solution prepared from pure chlorous acid? (b) How many moles of $HClO_2$ should be used to prepare one liter of this solution?

15.7 The ionization constant for cacodylic acid, a monobasic, arsenic-containing, organic acid, is 6.4×10^{-7}. What is the concentration of $H^+(aq)$ in a 0.30*M* solution of cacodylic acid?

15.8 (a) What is the concentration of $H^+(aq)$ in a 0.20*M* solution of benzoic acid? (b) What is the percent ionization of $HC_7H_5O_2$ in this solution?

15.9 The ionization constant of propanoic acid, $HC_3H_5O_2$, is 1.3×10^{-5}. (a) What is the concentration of $H^+(aq)$ in a 0.25*M* solution of this acid? (b) What is the percent ionization of $HC_3H_5O_2$ in this solution?

15.10 (a) What are the concentrations of $H^+(aq)$, $OCl^-(aq)$ and $HOCl(aq)$ in a 0.20*M* solution of HOCl? (b) What is the percent ionization of HOCl in this solution?

15.11 What are the concentrations of $C_6H_5NH_3^+(aq)$, $OH^-(aq)$, and $C_6H_5NH_2(aq)$ in a 0.30*M* solution of aniline?

15.12 In an aqueous solution of NH_3, the concentration of $OH^-(aq)$ is 1.8×10^{-3} *M*. What is the concentration of $NH_3(aq)$?

15.13 What is the concentration of $OH^-(aq)$ in a 0.15*M* aqueous solution of hydrazine (N_2H_4)?

15.14 A weak acid, HX, is 0.80% ionized in 0.015*M* solution. What percent of HX is ionized in a 0.10*M* solution?

15.15 A weak acid, HX, is 0.10% ionized in 0.10*M* solution. At what concentration is the acid 1.0% ionized?

15.16 What are the concentrations of $H^+(aq)$, $N_3^-(aq)$, and $HN_3(aq)$ in a solution prepared from 0.23 mol of NaN_3 and 0.10 mol of HCl in a total volume of 1.0 liter?

15.17 What is the concentration of $H^+(aq)$ in a solution prepared from 0.22 mol of sodium formate ($NaCHO_2$) and 0.15 mol of HCl in a total volume of 1.0 liter?

15.18 What are the concentrations of $H^+(aq)$, $C_7H_5O_2^-(aq)$, and $HC_7H_5O(aq)$ in a solution prepared by adding 10 ml of 0.30*M* sodium benzoate, $NaC_7H_5O_2$, to 10 ml of 0.20*M* HCl? Assume that the total volume of the solution is 20 ml.

15.19 What are the concentrations of NH_3(aq), NH_4^+(aq), and OH^-(aq) in a solution prepared from 0.15 mol of NH_4Cl and 0.25 mol of NaOH in a total volume of 1.0 liter?

15.20 What are the concentrations of OH^-(aq), NH_4^+(aq), and NH_3(aq) in a solution prepared by adding 150 ml of 0.45*M* NH_4Cl to 300 ml of 0.30*M* NaOH? Assume that the total volume of the solution is 450 ml.

15.21 What are the concentrations of H^+(aq) and OH^-(aq) in (a) a 0.020*M* solution of HNO_3 and (b) a 0.020*M* solution of $Ba(OH)_2$?

15.22 What *p*H corresponds to each of the following? (a) $[H^+] = 0.35M$, (b) $[OH^-] = 0.15M$, (c) $[OH^-] = 6.0 \times 10^{-6}M$, (d) $[H^+] = 2.5 \times 10^{-8}M$.

15.23 What *p*H corresponds to each of the following: (a) $[H^+] = 7.3 \times 10^{-5}M$, (b) $[H^+] = 0.084M$, (c) $[OH^-] = 3.3 \times 10^{-4}M$, (d) $[OH^-] = 0.042M$.

15.24 What concentration of H^+(aq) corresponds to each of the following? (a) *p*H = 3.33, (b) *p*OH = 3.33, (c) *p*H = 6.78, (d) *p*OH = 11.11.

15.25 What concentration of OH^-(aq) corresponds to each of the following? (a) *p*H = 0.55, (b) *p*OH = 4.32, (c) *p*H = 12.13, (d) *p*OH = 12.34.

15.26 A 0.37*M* solution of a weak acid, HX, has a *p*H of 3.70. What is the ionization constant of the acid?

15.27 How many moles of chlorous acid, $HClO_2$, must be used to prepare 1.00 liter of solution that has a *p*H of 2.60?

15.28 How many moles of benzoic acid, $HC_7H_5O_2$, must be used to prepare 500 ml of solution that has a *p*H of 2.44?

15.29 A 0.23*M* solution of a weak acid, HX, has a *p*H of 2.89. What is the ionization constant of the acid?

15.30 What is the *p*H of a 0.30*M* NH_3 solution?

15.31 Cocaine is a weak base. The aqueous ionization can be indicated as

$$C(aq) + H_2O \rightleftharpoons CH^+(aq) + OH^-(aq)$$

A $5.0 \times 10^{-3}M$ solution of cocaine has a *p*H of 10.04. What is the ionization constant for this substance?

15.32 An indicator, HIn, has an ionization constant of 9.0×10^{-9}. The acid color of the indicator is yellow, and the alkaline color is red. The yellow color is visible when the ratio of yellow form to red form is 30 to 1, and the red color is predominant when the ratio of red form to yellow form is 2 to 1. What is the *p*H range of color change for the indicator?

15.33 In each of the following parts, a given solution affects the indicators listed in the way shown. Refer to Table 15.4, and describe the *p*H of each solution as completely as possible.

(a) Thymol blue turns blue and alizarin yellow turns yellow.
(b) Bromcresol green turns blue and bromthymol blue turns yellow.
(c) Methyl orange turns yellow and litmus turns red.

15.34 A solution is prepared by adding 0.010 mol of sodium formate, $Na(CHO_2)$, to 100 ml of 0.025*M* formic acid, $HCHO_2$. Assume that no volume change occurs. Calculate (a) the *p*H of the solution and (b) the percent ionization of $HCHO_2$.

15.35 A solution is prepared by adding 0.010 mol of sodium nitrite, $NaNO_2$, to 100 ml of 0.035*M* nitrous acid, HNO_2. Assume that the volume of the final solution is 100 ml. Calculate (a) the *p*H of the solution and (b) the percent ionization of HNO_2.

15.36 A solution prepared from 0.028 mol of a weak acid, HX, and 0.0070 mol of NaX diluted to 200 ml has a *p*H of 3.66. What is the ionization constant of HX?

15.37 A solution is prepared from 3.0×10^{-3} mol of a weak acid, HX, and 6.0×10^{-4} mol of NaX diluted to 200 ml. The solution has a *p*H of 4.80. What is the ionization constant of HX?

15.38 A 0.10*M* solution of hydrazine, N_2H_4, containing an unknown concentration of hydrazine hydrochloride, $N_2H_5^+Cl^-$, has a *p*H of 7.15. What is the concentration of hydrazine hydrochloride in the solution?

15.39 A 0.24*M* solution of dimethylamine, $(CH_3)_2NH$, containing an unknown concentration of dimethylamine hydrochloride, $(CH_3)_2NH_2^+Cl^-$, has a *p*H of 10.40. What is the concentration of dimethylamine hydrochloride?

15.40 A solution prepared from 0.060 mol of a weak acid, HX, diluted to 250 ml has a *p*H of 2.89. What is the *p*H of the solution after 0.030 mol of solid NaX is dissolved in it? Assume that no significant volume change occurs when the NaX is dissolved in the solution.

15.41 A 0.050*M* solution of formic acid, $HCHO_2$, containing an unknown concentration of formate ion, CHO_2^-, derived from sodium formate has a *p*H of 3.70. What is the concentration of formate ion in the solution?

15.42 A 1.00 liter solution prepared from 0.049 mol of a weak acid, HX, has a *p*H of 3.55. What is the *p*H of the solution after 0.020 mol of solid NaX is dissolved in it? Assume that the volume of the solution does not change when the NaX is dissolved in it.

15.43 What concentrations should be used to prepare a benzoic acid–benzoate ion buffer with a *p*H of 5.00?

15.44 What concentrations should be used to prepare an ammonia–ammonium ion buffer with a *p*H of 9.50?

15.45 How many moles of sodium hypochlorite, NaOCl, should be added to 200 ml of 0.22*M* hypochlorous acid, HOCl, to produce a buffer with a *p*H of 6.75? Assume that no volume change occurs when the NaOCl is added to the solution.

15.46 (a) What are the concentrations of $H^+(aq)$, $HCO_3^-(aq)$, $CO_3^{2-}(aq)$, and $CO_2(aq)$ in a saturated solution of carbonic acid (0.034*M* in CO_2)? (b) What is the *p*H of the solution?

15.47 Calculate the concentrations of $H^+(aq)$, $H_2AsO_4^-(aq)$, $HAsO_4^{2-}(aq)$, $AsO_4^{3-}(aq)$, and $H_3AsO_4(aq)$ in a 0.30*M* solution of arsenic acid.

15.48 A 0.15*M* solution of HCl is saturated with H_2S. (a) What is the concentration of $S^{2-}(aq)$? (b) What is the concentration of $HS^-(aq)$?

15.49 A solution with a *p*H of 3.30 is saturated with H_2S. What is the concentration of $S^{2-}(aq)$?

15.50 What should the concentration of $H^+(aq)$ be in order to have a sulfide ion concentration of $3.0 \times 10^{-17}M$ when the solution is saturated with H_2S?

15.51 At 25°C, 1.7×10^{-5} mol of $Cd(OH)_2$ dissolves in one liter of water. Calculate the K_{SP} of $Cd(OH)_2$.

15.52 The K_{SP} of $NiCO_3$ is 1.4×10^{-7}. What is the molar solubility of $NiCO_3$? Neglect the hydrolysis of ions.

15.53 At 25°C, 5.2×10^{-6} mol of $Ce(OH)_3$ dissolves in one liter of water. Calculate the K_{SP} of $Ce(OH)_3$.

15.54 The K_{SP} of CaF_2 is 3.9×10^{-11}. What is the molar solubility of CaF_2? Neglect the hydrolysis of ions.

15.55 Which carbonate has the lower molar solubility, Ag_2CO_3 (K_{SP}, 8.2×10^{-12}) or $CuCO_3$ (K_{SP}, 2.5×10^{-10})? Neglect the hydrolysis of ions.

15.56 Which sulfide has the lower molar solubility, Ag_2S (K_{SP}, 5.5×10^{-51}) or CuS (K_{SP}, 8.0×10^{-37})? Neglect the hydrolysis of ions.

15.57 A saturated solution of a slightly soluble hydroxide, $M(OH)_2$, has a *p*H of 9.53. What is the K_{SP} of $M(OH)_2$?

15.58 How many moles of $Ni(OH)_2$ will dissolve in one liter of a solution of NaOH with a *p*H of 12.34? Neglect hydrolysis of ions.

15.59 How many moles of Ag_2CO_3 will dissolve in 150 ml of 0.15*M* Na_2CO_3 solution? Neglect hydrolysis of ions.

15.60 Calculate the final concentrations of $Na^+(aq)$, $C_2O_4^{2-}(aq)$, $Ba^{2+}(aq)$, and $Cl^-(aq)$ in a solution prepared by adding 100 ml of 0.20*M* $Na_2C_2O_4$ to 150 ml of 0.25*M* $BaCl_2$. Neglect the hydrolysis of ions.

15.61 Calculate the final concentrations of $Sr^{2+}(aq)$, $NO_3^-(aq)$, $Na^+(aq)$, and $F^-(aq)$ in a solution prepared by adding 50 ml of 0.30*M* $Sr(NO_3)_2$ to 150 ml of 0.12*M* NaF. Neglect the hydrolysis of ions.

15.62 What concentration of F^- is necessary to start the precipitation of SrF_2 from a saturated solution of $SrSO_4$?

15.63 What concentration of $SO_4^{2-}(aq)$ is necessary to start the precipitation of $BaSO_4$ from a saturated solution of BaF_2?

15.64 What minimum concentration of NH_4^+ is necessary to prevent the formation of a $Fe(OH)_2$ precipitate from a solution that is 0.020*M* in Fe^{2+} and 0.020*M* in NH_3?

15.65 What minimum concentration of NH_4^+ is necessary to prevent formation of $Mn(OH)_2(s)$ from a solution that is 0.030*M* in $Mn^{2+}(aq)$ and 0.030*M* in NH_3?

15.66 A solution is $0.030M$ in Mn^{2+} and $0.025M$ in NH_4^+. What should the concentration of NH_3 be in order to cause $Mn(OH)_2$ to start to precipitate?

15.67 A solution is $0.090M$ in Mg^{2+}(aq) and $0.33M$ in NH_4^+(aq). What minimum concentration of NH_3 will cause $Mg(OH)_2$ to start to precipitate?

15.68 A solution is prepared by mixing 10 ml of $0.50M$ $CaCl_2$ with 10 ml of a solution that is $0.50M$ in NH_3 and $0.050M$ in NH_4^+(aq). Will $Ca(OH)_2$ precipitate?

15.69 A solution that is $0.30M$ in H^+(aq) and $0.15M$ in Ni^{2+} is saturated with H_2S. Should NiS precipitate?

15.70 A solution that is $0.25M$ in H^+(aq) and $0.10M$ in CO^{2+}(aq) is saturated with H_2S. Will CoS precipitate?

15.71 A solution that is $0.50M$ in H^+(aq) and $0.030M$ in Cd^{2+}(aq) is saturated with H_2S. Will CdS precipitate?

15.72 What should the H^+(aq) concentration be in a solution that is $0.25M$ in Co^{2+} to prevent the precipitation of CoS when the solution is saturated with H_2S?

15.73 What is the lowest concentration of H^+(aq) that must be present in a $0.50M$ solution of Zn^{2+}(aq) to prevent the precipitation of ZnS when the solution is saturated with H_2S?

15.74 A solution that is $0.20M$ in H^+(aq) and $0.20M$ in Pb^{2+}(aq) is saturated with H_2S. What is the concentration of Pb^{2+}(aq) after PbS has precipitated? Note that it is necessary to take into account the increase in acidity caused by the precipitation.

15.75 A solution that is $0.10M$ in H^+(aq), $0.30M$ in Cu^{2+}(aq), and $0.30M$ in Fe^{2+}(aq) is saturated with H_2S. Calculate the concentrations of H^+(aq), S^{2-}(aq), Cu^{2+}(aq), and Fe^{2+}(aq). Note that it is necessary to take into account any increase in acidity caused by precipitation.

15.76 What concentration of H^+(aq) should be present in a solution that is $0.20M$ in Ni^{2+}(aq) and $0.20M$ in Cd^{2+}(aq) so that when the solution is saturated with H_2S the maximum amount of CdS will precipitate but no NiS will precipitate at any stage?

15.77 (a) What concentration of H^+(aq) should be present in a solution that is $0.20M$ in Pb^{2+} and $0.20M$ in Zn^{2+} so that upon saturation with H_2S, the maximum amount of PbS precipitates but no ZnS precipitates at any stage? (b) What concentration of Pb^{2+} remains in the solution after the PbS has precipitated? Ignore the hydrolysis of ions, but do not neglect the increase in acidity caused by the precipitation.

15.78 Compare the molar solubilities of (a) AgCl, (b) AgBr, and (c) AgI in $0.50M$ NH_3 solution.

15.79 A $0.010M$ solution of $AgNO_3$ is made $0.50M$ in NH_3, thus forming the $Ag(NH_3)_2^+$ complex ion. (a) Will AgCl precipitate if sufficient NaCl is added to make the solution $0.010M$ in Cl^-?

15.80 What is the pH of a $0.10M$ solution of sodium nitrite ($NaNO_2$)?

15.81 What is the pH of a $0.10M$ solution of aniline hydrochloride ($C_6H_5NH_3^+Cl^-$)?

15.82 What is the pH of a $0.10M$ solution of sodium benzoate ($NaC_7H_5O_2$)?

15.83 What is the pH of a $0.10M$ solution of hydrazine hydrochloride ($N_2H_5^+Cl^-$)?

15.84 A solution of sodium benzoate ($NaC_7H_5O_2$) has a pH of 9.00. What is the concentration of sodium benzoate?

15.85 What concentration of ammonium chloride, NH_4Cl, will produce a solution with a pH of 5.20?

15.86 The pH of a $0.15M$ solution of NaX is 9.77. What is the ionization constant of the weak acid HX?

15.87 The pH of a $0.30M$ solution of NaY is 9.50. What is the ionization constant of the weak acid HY?

15.88 The $Zn(OH)^+$ ion is believed to be the only zinc-containing ion resulting from the hydrolysis of $Zn^{2+}(aq)$ in dilute solutions. What is the pH of a $0.010M$ solution of $Zn(NO_3)_2$? The K_{inst} of $Zn(OH)^+$ is 4.1×10^{-5}.

15.89 The $Cd(OH)^+$ ion is believed to be the only cadmium-containing ion resulting from the hydrolysis of $Cd^{2+}(aq)$ in dilute solutions. What is the concentration of $Cd^{2+}(aq)$ in a solution of $Cd(ClO_4)_2$ that has a pH of 5.48? The K_{inst} of $Cd(OH)^+$ is 6.9×10^{-5}.

15.90 Determine pH values for the titration curve pertaining to the titration of 30.00 ml of $0.100N$ benzoic acid, $HC_7H_5O_2$, with $0.100N$ NaOH (a) after 10.00 ml of NaOH solution has been added, (b) after 30.00 ml of NaOH solution has been added, (c) after 40.00 ml of NaOH solution has been added.

15.91 Determine pH values for the titration curve pertaining to the titration of 25.00 ml of $0.100N$ NH_3 solution with a $0.100N$ solution of HCl (a) after 10.00 ml of the HCl solution has been added, (b) after 25.00 ml of HCl solution has been added, (c) after 35.00 ml of HCl solution has been added.

15.92 In the titration of 25.00 ml of a solution of a weak acid HX with $0.250N$ NaOH, the pH of the solution is 4.50 after 5.00 ml of the NaOH solution have been added. The equivalence point in the titration occurs when 30.40 ml of the NaOH solution have been added. What is the equilibrium constant for the ionization of HX?

15.93 The equivalence point in the titration of 50.00 ml of a solution of acetic acid occurs when 33.00 ml of $0.200N$ NaOH have been added. How many ml of the NaOH solution should be added to 50.00 ml of the acetic acid solution to produce a solution with a pH of 5.04?

Chapter 16

Electrochemistry

All chemical reactions are fundamentally electrical in nature since electrons are involved (in various ways) in all types of chemical bonding. Electrochemistry, however, is primarily the study of oxidation-reduction phenomena.

The relations between chemical change and electrical energy have theoretical as well as practical importance. Chemical reactions can be used to produce electrical energy (in cells that are called either **voltaic** or **galvanic cells**). Electrical energy can be used to bring about chemical transformations (in **electrolytic cells**). In addition, the study of electrochemical processes leads to an understanding, as well as to the systemization, of oxidation-reduction phenomena that take place outside cells. For example, data from electrochemical studies are used to determine the most favorable conditions for a given oxidation-reduction reaction, whether a proposed chemical transformation will occur, and the magnitude and nature of the energy effect that accompanies a given chemical change.

16.1 METALLIC CONDUCTION

An electric current is the flow of electric charge. In metals this charge is carried by electrons, and electrical conduction of this type is called **metallic conduction.** The current results from the application of an electric force supplied by a battery or some other source of electrical energy. A complete circuit is necessary to produce a current.

Metallic crystals may be described in terms of mobile electron clouds permeating relatively fixed lattices of positive metal ions (Section 19.1). When electrons are forced into one end of a metal wire, the impressed electrons displace other electrons of the cloud at the point of entry. The displaced electrons, in turn, assume new positions by pushing neighboring electrons ahead, and this effect is transmitted down the length of the wire until electrons are forced out of the wire at the opposite end. The current source may be regarded as an electron pump, for it serves to force electrons into one end of the circuit and drain them off from the other end. At any position in the wire, electrical neutrality is preserved, since the rate of electrons in equals the rate of electrons out.

The analogy between the flow of electricity and the flow of a liquid is an old one. In earlier times electricity was described in terms of a current of "electric fluid." Conventions of long standing, which may be traced back to Benjamin Franklin (1747) and which were adopted before the electron was identified, ascribe a positive charge to this current. We shall interpret electrical circuits in terms of the movement of electrons. Remember, however, that conventional electric current is arbitrarily described as positive and as flowing in the opposite direction.

Electric current is measured in **amperes.** Quantity of electric charge is measured in **coulombs;** the coulomb is defined as the quantity of electricity carried in one second by a current of one ampere. Therefore,

$$1 \text{ amp} = 1 \text{ coul/s}$$

and

$$1 \text{ coul} = 1 \text{ amp s}$$

The current is forced through the circuit by an electrical potential difference, which is measured in **volts.** It takes one joule of work to move one coulomb from a lower to a higher potential when the potential difference is one volt. One volt, therefore, equals one joule/coulomb, and one volt coulomb is a unit of energy and equals one joule.

$$1 \text{ V} = 1 \text{ J/coul}$$
$$1 \text{ V coul} = 1 \text{ J}$$

The higher the potential difference between two points in a given wire, the more current the wire will carry between those two points. Georg Ohm in

1826 expressed the quantitative relation between potential difference, $\mathscr{E}$, in volts, and current, I, in amperes, as

$$I = \mathscr{E}/R \quad \text{or} \quad \mathscr{E} = IR$$

where the proportionality constant, R, of **Ohm's law** is called the resistance. Resistance is measured in **ohms.** One volt is required to force a current of one ampere through a resistance of one ohm.

Resistance to the flow of electricity in metals is probably caused by the vibration of the metal ions about their lattice positions. These vibrations interfere with the motion of the electrons and retard the current. As the temperature is increased, the thermal motion of the metal ions is increased. Hence, the resistance of metals increases, and the metals become poorer conductors.

16.2 ELECTROLYTIC CONDUCTION

Electrolytic conduction, in which the charge is carried by ions, will not occur unless the ions of the electrolyte are free to move. Electrolytic conduction, therefore, is exhibited principally by molten salts and by aqueous solutions of electrolytes. Furthermore, a sustained current through an electrolytic conductor requires that chemical change accompany the movement of ions.

These principles of electrolytic conduction are best illustrated by reference to an electrolytic cell such as the one diagrammed in Figure 16.1 for the electrolysis of molten NaC1 between inert electrodes.* The current source pumps electrons into the left-hand electrode, which therefore may be con-

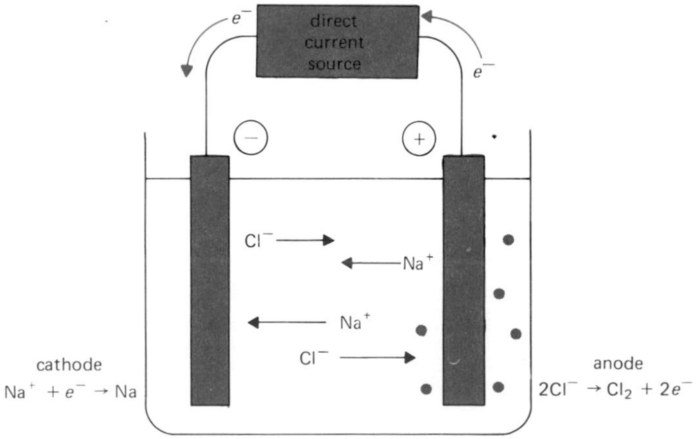

Figure 16.1 Electrolysis of molten sodium chloride.

* Inert electrodes are not involved in electrode reactions.

sidered to be negatively charged. Electrons are drained from the right-hand, positive electrode. In the electric field thus produced, sodium ions (cations) are attracted toward the negative pole (cathode), and chloride ions (anions) are attracted toward the positive pole (anode). Electric charge in electrolytic conduction is carried by cations moving toward the **cathode** and anions moving in the opposite direction, toward the **anode.**

For a complete circuit, electrode reactions must accompany the movement of ions. At the cathode some chemical species (not necessarily the charge carrier) must accept electrons and be reduced. At the anode, electrons must be removed from some chemical species, which as a consequence is oxidized. The conventions relating to the terms anode and cathode are summarized in Table 16.1.

Table 16.1 Electrode conventions

	CATHODE	ANODE
ions attracted	cations	anions
direction of electron movement	into cell	out of cell
half reaction	reduction	oxidation
sign		
electrolysis cell	negative	positive
galvanic cell	positive	negative

In the diagrammed cell, sodium ions are reduced at the cathode

$$Na^+ + e^- \rightarrow Na$$

and chloride ions are oxidized at the anode

$$2Cl^- \rightarrow Cl_2 + 2e^-$$

Proper addition of these two partial equations gives the reaction for the entire cell

$$2NaCl(l) \xrightarrow{\text{electrolysis}} 2Na(l) + Cl_2(g)$$

In the actual operation of the commercial cell used to produce metallic sodium, calcium chloride is added to lower the melting point of sodium chloride, and the cell is operated at a temperature of approximately 600°C. At this temperature sodium metal is a liquid.

We can trace the flow of negative charge through the circuit of Figure 16.1 as follows. Electrons leave the current source and are pumped into the cathode where they are picked up by and reduce sodium ions that have been attracted to this negative electrode. Chloride ions move away from the cathode toward the anode and thus carry negative charge in this direction. At the anode, electrons are removed from the chloride ions, thus oxidizing them to chlorine gas. These electrons are pumped out of the cell by the current source. In this manner the circuit is completed.

Electrolytic conduction, then, rests on the mobility of ions, and anything that inhibits the motion of these ions causes resistance to the current. Factors that influence the electrical conductivity of solutions of electrolytes include interionic attractions, solvation of ions, and viscosity of the solvent. These factors rest on solute–solute attractions, solute–solvent attractions, and solvent–solvent attractions, respectively. The average kinetic energy of the solute ions increases as the temperature is raised, and, therefore, the resistance of electrolytic conductors generally decreases as the temperature is raised (that is, conduction increases). Furthermore, the effect of each of the three previously mentioned factors decreases as the temperature is increased.

At all times the solution is electrically neutral. The total positive charge of all of the cations equals the total negative charge of all of the anions.

16.3 ELECTROLYSIS

The electrolysis of molten sodium chloride serves as a commercial source of sodium metal and chlorine gas. Analogous procedures are used to prepare other very active metals (such as potassium and calcium). When certain aqueous solutions are electrolyzed, however, water is involved in the electrode reactions rather than the ions derived from the solute. Hence, the current-carrying ions are not necessarily discharged at the electrodes.

In the electrolysis of aqueous sodium sulfate, sodium ions move toward the cathode and sulfate ions move toward the anode (Figure 16.2). Both these ions are difficult to discharge. When this electrolysis is conducted between inert electrodes, hydrogen gas is evolved at the cathode, and the solution surrounding the electrode becomes alkaline. Reduction occurs at the cathode, but rather than the reduction of the sodium ion

$$e^- + Na^+ \rightarrow Na$$

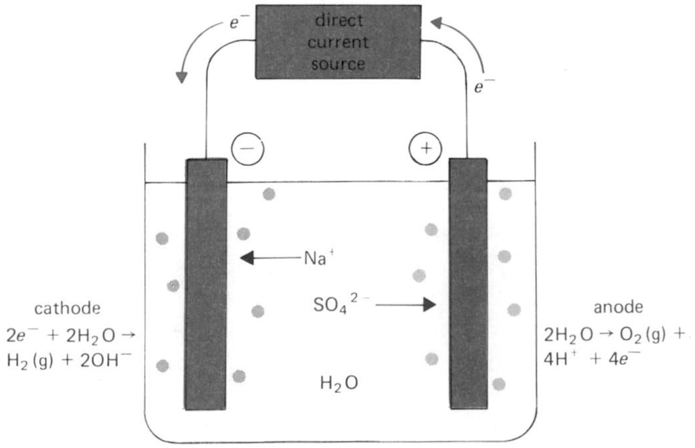

Figure 16.2 Electrolysis of aqueous sodium sulfate.

the *net change* that occurs is the reduction of water

$$2e^- + 2H_2O \rightarrow H_2(g) + 2OH^-$$

Water is an extremely weak electrolyte. Pure water is approximately $2 \times 10^{-7}\%$ ionized at 25°C

$$2H_2O \rightleftharpoons H_3O^+ + OH^-$$

or, more briefly,

$$H_2O \rightleftharpoons H^+ + OH^-$$

The exact mechanism of the cathode reaction in the electrolysis of aqueous Na_2SO_4 is not known. It may be that the hydrogen ions from water are discharged and that the reaction proceeds as follows:

$$H_2O \rightleftharpoons H^+ + OH^-$$
$$2e^- + 2H^+ \rightarrow H_2(g)$$

Multiplication of the first equation by 2 followed by addition of the equations gives the net change

$$2e^- + 2H_2O \rightarrow H_2(g) + 2OH^-$$

In general, water is reduced at the cathode (producing hydrogen gas and hydroxide ions) whenever the cation of the solute is difficult to reduce.

Oxidation occurs at the anode, and in the electrolysis of aqueous Na_2SO_4, the anions (SO_4^{2-}) that migrate toward the anode are difficult to oxidize.

$$2SO_4^{2-} \rightarrow S_2O_8^{2-} + 2e^-$$

Therefore, the oxidation of water occurs preferentially. The mode of this reaction may be

$$H_2O \rightleftharpoons H^+ + OH^-$$
$$4OH^- \rightarrow O_2(g) + 2H_2O + 4e^-$$

Multiplying the first equation by 4 and adding the equations, we get the net change

$$2H_2O \rightarrow O_2(g) + 4H^+ + 4e^-$$

At the anode the evolution of oxygen gas is observed, and the solution surrounding the pole becomes acidic. In general, water is oxidized at the anode (producing oxygen gas and hydrogen ions) whenever the anion of the solute is difficult to oxidize.

The complete reaction for the electrolysis of aqueous Na_2SO_4 may be obtained by adding the cathode and anode reactions

$$\begin{array}{r} 2[2e^- + 2H_2O \rightarrow H_2(g) + 2OH^-] \\ 2H_2O \rightarrow O_2(g) + 4H^+ + 4e^- \\ \hline 6H_2O \rightarrow 2H_2(g) + O_2(g) + 4H^+ + 4OH^- \end{array}$$

If the solution is mixed, the hydrogen and hydroxide ions that are produced neutralize one another, and the net change

$$2H_2O \xrightarrow{\text{electrolysis}} 2H_2(g) + O_2(g)$$

is merely the electrolysis of water. In the course of the electrolysis, the hydrogen ions migrate away from the anode, where they are produced, toward the cathode. In like manner, the hydroxide ions move toward the anode. These ions neutralize one another in the solution between the two electrodes.

The electrolysis of an aqueous solution of NaCl between inert electrodes serves as an example of a process in which the anion of the electrolyte is discharged, but the cation is not.

$$\begin{array}{ll} \text{anode:} & 2Cl^- \rightarrow Cl_2(g) + 2e^- \\ \text{cathode:} & 2e^- + 2H_2O \rightarrow H_2(g) + 2OH^- \\ \hline & 2H_2O + 2Cl^- \rightarrow H_2(g) + Cl_2(g) + 2OH^- \end{array}$$

Since the sodium ion remains unchanged in the solution, the reaction may be indicated

$$2H_2O + 2(Na^+,Cl^-) \xrightarrow{\text{electrolysis}} H_2(g) + Cl_2(g) + 2(Na^+,OH^-)$$

This process is a commercial source of hydrogen gas, chlorine gas, and, by evaporation of the solution left after electrolysis, sodium hydroxide.

In the electrolysis of a solution of $CuSO_4$ between inert electrodes (see right portion of Figure 16.4, given subsequently), the current is carried by the Cu^{2+} and SO_4^{2-} ions. The current-carrying cations are discharged, but the anions are not.

$$\begin{array}{ll} \text{anode:} & 2H_2O \rightarrow O_2(g) + 4H^+ + 4e^- \\ \text{cathode:} & 2e^- + Cu^{2+} \rightarrow Cu(s) \\ \hline & 2Cu^{2+} + 2H_2O \rightarrow O_2(g) + 2Cu(s) + 4H^+ \end{array}$$

It is, of course, possible to have both ions of the solute discharged during the electrolysis of an aqueous solution. An example is the electrolysis of $CuCl_2$ between inert electrodes.

$$\begin{array}{ll} \text{anode:} & 2Cl^- \rightarrow Cl_2(g) + 2e^- \\ \text{cathode:} & 2e^- + Cu^{2+} \rightarrow Cu(s) \\ \hline & Cu^{2+} + 2Cl^- \rightarrow Cu(s) + Cl_2(g) \end{array}$$

It is also possible to have the electrode itself enter into an electrode reaction. If aqueous $CuSO_4$ is electrolyzed between copper electrodes (Figure 16.3), Cu^{2+} ions are reduced at the cathode

$$2e^- + Cu^{2+} \rightarrow Cu$$

but of the *three* possible anode oxidations

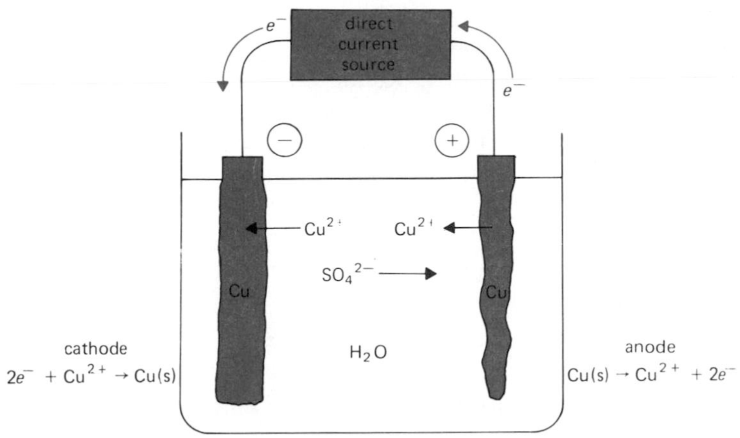

Figure 16.3 Electrolysis of aqueous cupric sulfate between copper electrodes.

$$2SO_4^{2-} \rightarrow 2S_2O_8^{2-} + 2e^-$$
$$2H_2O \rightarrow O_2(g) + 4H^+ + 4e^-$$
$$Cu(s) \rightarrow Cu^{2+} + 2e^-$$

the oxidation of the copper metal of the electrode is observed to occur. Hence, at the anode, copper from the electrode goes into solution as Cu^{2+} ions, and at the cathode, Cu^{2+} ions plate out as Cu(s) on the electrode. This process is used to refine copper. Impure copper is used as the anode of an electrolytic cell, and a solution of $CuSO_4$ is electrolyzed. Pure copper plates out on the cathode. Active electrodes are also used in electroplating processes. In silver plating, silver anodes are employed.

16.4 FARADAY'S LAWS

The quantitative relationships between electricity and chemical change were first described by Michael Faraday. In 1832 he showed that the weight of a chemical substance liberated at an electrode is directly proportional to the amount of current passed through the cell. In 1833 he stated that the weights of different substances produced by a given amount of current are proportional to the equivalent weights of the substances.

Faraday's laws are readily interpreted by reference to the electrolysis of molten sodium chloride. The change at the cathode requires one electron for every sodium ion reduced.

$$Na^+ + e^- \rightarrow Na$$

If Avogadro's number of electrons (1 mol) are consumed at this electrode, 22.9898 g of sodium metal (1 mol) is produced. The corresponding quantity of electricity is called the **faraday** (F) and has been found to equal 96,487 coul (for ordinary problem work this value is customarily rounded off to 96,500 coul). If 2 F of electricity (2 mol of electrons) are used, 2 mol of sodium are produced; 0.5 F would liberate 0.5 mol of sodium. Since the charge on the sodium ion is 1+, the equivalent weight of sodium is the same as the atomic weight.

In the same time that electrons equivalent to 1 F are added to the cathode, that same number of electrons are removed from the anode.

$$2Cl^- \rightarrow Cl_2 + 2e^-$$

The removal of 1 mol of electrons from this anode would result in the discharge of 35.453 g of chloride ion (1 mol), thus producing 0.5 mol of chlorine gas (1 gram equivalent). If 2 F of electricity flow through the cell, 2 mol of chloride ions are discharged.

One faraday, then, will liberate 1 gram equivalent of an element or a compound. Rather than use equivalent weights, however, it is possible to interpret electrode reactions in terms of moles and faradays. Thus, the anode oxidation of the hydroxide ion,

$$4OH^- \rightarrow O_2(g) + 2H_2O + 4e^-$$

may be read: four hydroxide ions produce one oxygen molecule, two water molecules, and four electrons, *or* 4 mol of hydroxide ion produces 1 mol of oxygen and 2 mol of water when 4 F of electricity are passed through the cell.

In Figure 16.4 two electrolytic cells are set up in series. Electricity passes through one cell first and then through the other before returning to the cur-

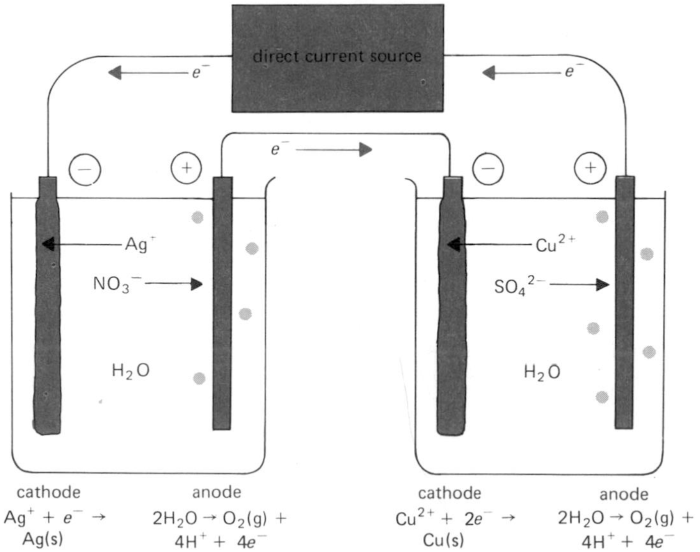

Figure 16.4 Silver coulometer in series with a cell for electrolysis.

rent source. If silver nitrate is electrolyzed in one of the cells, the cathode reaction is

$$Ag^+ + e^- \rightarrow Ag(s)$$

and metallic silver is plated out on the electrode used for the electrolysis. By weighing the electrode before and after the electrolysis, one can determine the quantity of silver plated out and hence the number of coulombs that have passed through the cell. One faraday would plate out 107.868 g of silver. Therefore, 1 coul is equivalent to

$$107.868/96{,}487 = 1.1180 \times 10^{-3} \text{ g of silver}$$

The same number of coulombs pass through both cells in a given time when these cells are arranged in series. Therefore, the number of coulombs used in an electrolysis may be determined through the addition, in series, of this **silver coulometer** to the circuit of the experimental cell.

Example 16.1 The charge on a single electron is 1.6021×10^{-19} coul. Calculate Avogadro's number from the fact that 1 F = 96,487 coul.

Solution

$$? \text{ electrons} = 9.6487 \times 10^4 \text{ coul} \left(\frac{1 \text{ electron}}{1.6021 \times 10^{-19} \text{ coul}}\right)$$

$$= 6.0225 \times 10^{23} \text{ electrons}$$

Example 16.2 In the electrolysis of $CuSO_4$, how much copper is plated out on the cathode by a current of 0.750 amp in 10.0 min?

Solution

The number of faradays used may be calculated as follows:

$$? \text{ F} = 10.0 \text{ min} \left(\frac{60 \text{ s}}{1 \text{ min}}\right)\left(\frac{0.75 \text{ coul}}{1 \text{ s}}\right)\left(\frac{1 \text{ F}}{96{,}500 \text{ coul}}\right)$$

$$= 0.00466 \text{ F}$$

The cathode reaction is $Cu^{2+} + 2e^- \rightarrow Cu(s)$, and therefore 2 F plate out 63.5 g of Cu(s).

$$? \text{ g Cu} = 0.00466 \text{ F}\left(\frac{63.5 \text{ g Cu(s)}}{2\text{F}}\right) = 0.148 \text{ g Cu(s)}$$

Example 16.3 (a) What volume of $O_2(g)$ at STP is liberated at the anode in the electrolysis of $CuSO_4$ described in Example 16.2? (b) If 100 ml of 1.00 M $CuSO_4$ is employed in the cell, what is the H^+ (aq) concentration at the end of the electrolysis? Assume that there is no volume change for the solution during the experiment and that the anode reaction is $2H_2O \rightarrow 4H^+(aq) + O_2(g) + 4e^-$.

Solution (a) Four faradays produce 22.4 liter of $O_2(g)$ at STP.

$$? \text{ liter } O_2(g) = 0.00466 \text{ F}\left(\frac{22.4 \text{ liter } O_2(g)}{4\text{F}}\right)$$

$$= 0.0261 \text{ liter } O_2(g)$$

(b) Four faradays also produce 4 mol of $H^+(aq)$

$$? \text{ mol } H^+(aq) = 0.00466 \text{ F}\left(\frac{1 \text{ mol } H^+(aq)}{1\text{F}}\right)$$

$$= 0.00466 \text{ mol } H^+(aq)$$

The small contribution of $H^+(aq)$ from the ionization of water may be ignored, and we may assume that there are 0.00466 mol $H^+(aq)$ in 100 ml of solution.

$$? \text{ mol } H^+(aq) = 1000 \text{ ml solution } \left(\frac{0.00466 \text{ mol } H^+(aq)}{100 \text{ ml solution}}\right)$$

$$= 0.0466 \text{ mol } H^+(aq)$$

The solution is therefore 0.0466 M in hydrogen ion.

Example 16.4 (a) What mass of copper is plated out in the electrolysis of $CuSO_4$ in the same time that it takes to deposit 1.00 g of Ag in a silver coulometer that is arranged in series with the $CuSO_4$ cell? (b) If a current of 1.00 amp is used, how many minutes are required to plate out this quantity of copper?

Solution (a) From the electrode reactions we see that 2 F deposit 63.5 g Cu and 1 F deposits 107.9 g Ag.

$$? \text{ g Cu} = 1.00 \text{ g Ag}\left(\frac{1\text{F}}{107.9 \text{ g Ag}}\right)\left(\frac{63.5 \text{ g Cu}}{2\text{F}}\right) = 0.294 \text{ g Cu}$$

(b)

$$? \text{ min} = 1.00 \text{ g Ag}\left(\frac{96{,}500 \text{ coul}}{107.9 \text{ g Ag}}\right)\left(\frac{1 \text{ s}}{1 \text{ coul}}\right)\left(\frac{1 \text{ min}}{60 \text{ s}}\right)$$

$$= 14.9 \text{ min}$$

16.5 VOLTAIC CELLS

A cell that is used as a source of electrical energy is called a voltaic cell or a galvanic cell after Alessandro Volta (1800) or Luigi Galvani (1780), who first experimented with the conversion of chemical energy into electrical energy.

The reaction between metallic zinc and copper(II) ions in solution is illustrative of a spontaneous change in which electrons are transferred.

$$Zn(s) + Cu^{2+}(aq) \rightarrow Zn^{2+}(aq) + Cu(s)$$

The exact mechanism by which electron transfer occurs is not known. We may, however, represent the above reaction as a combination of two half reactions.

$$Zn(s) \rightarrow Zn^{2+}(aq) + 2e^-$$
$$2e^- + Cu^{2+}(aq) \rightarrow Cu(s)$$

In a voltaic cell these half reactions are made to occur at different electrodes so that the transfer of electrons takes place through the external electrical circuit rather than directly between zinc metal and copper(II) ions.

The cell diagrammed in Figure 16.5 is designed to make use of this reaction to produce an electric current. The half cell on the left contains a zinc metal electrode and $ZnSO_4$ solution. The half cell on the right consists of a copper metal electrode in a solution of $CuSO_4$. The half cells are separated by a porous partition that prevents the mechanical mixing of the solutions but permits the passage of ions under the influence of the flow of electricity. A cell of this type is called a **Daniell cell.**

When the zinc and copper electrodes are joined by a wire, electrons flow from the zinc electrode to the copper electrode. At the zinc electrode the zinc metal is *oxidized* to zinc ions. This electrode is the anode, and the electrons that are the product of the oxidation *leave* the cell from this pole (Table 16.1). The electrons travel the external circuit to the copper electrode where they are used in the reduction of copper(II) ions to metallic copper. The copper thus produced plates out on the electrode. The copper electrode is the cathode. Here, the electrons *enter* the cell and *reduction* occurs.

Since electrons are produced at the zinc electrode, this anode is designated as the negative pole. Electrons travel from the negative pole to the positive pole in the external circuit of any voltaic cell when the cell is operating. The cathode, where electrons are used in the electrode reaction, is therefore the positive pole. Within the cell the movement of ions completes the electric cir-

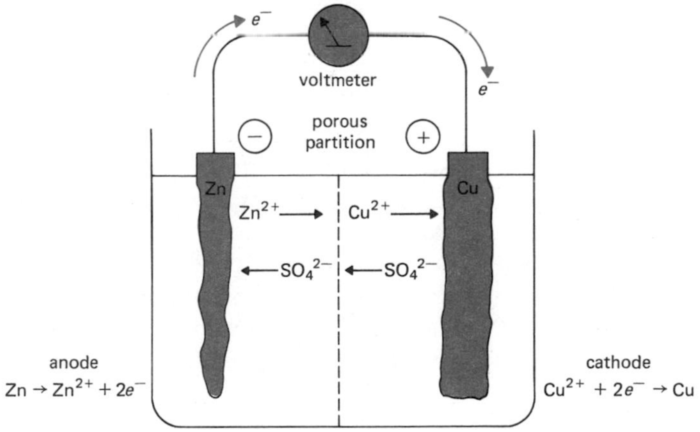

Figure 16.5 The Daniell Cell.

cuit. At first glance, it is surprising that anions, which are negatively charged, should travel toward an anode that is the negative electrode. Conversely, cations, which carry a positive charge, travel toward the cathode, which is the positive pole.

Careful consideration of the electrode reactions provides the answer to this problem. At the anode, zinc ions are being produced and electrons left behind in the metal. At all times the electrical neutrality of the solution is maintained. In the solution surrounding the electrode there must be as much negative charge from anions as there is positive charge from cations. Hence, SO_4^{2-} ions move toward the anode to neutralize the effect of the Zn^{2+} ions that are being produced. At the same time, zinc ions move away from the anode toward the cathode. At the cathode, electrons are being used to reduce Cu^{2+} ions to copper metal. While the Cu^{2+} ions are being discharged, more Cu^{2+} ions move into the region surrounding the cathode to take the place of the ions being removed. If this did not occur, a surplus of SO_4^{2-} ions would build up around the cathode.

The porous partition is added to prevent mechanical mixing of the solutions of the half cells. If Cu^{2+} ions came into contact with the zinc metal electrode, electrons would be transferred directly rather than through the circuit. In the normal operation of the cell, this "short circuit" does not occur because the Cu^{2+} ions move in a direction away from the zinc electrode.

Actually this cell would work if a solution of an electrolyte other than $ZnSO_4$ were used in the anode compartment, and if a metal other than copper were used for the cathode. The substitutes, however, must be chosen so that the electrolyte in the anode compartment does not react with the zinc electrode and the cathode does not react with Cu^{2+} ions.

16.6 ELECTROMOTIVE FORCE

If $1M$ $ZnSO_4$ and $1M$ $CuSO_4$ solutions are employed in the Daniell cell, the cell may be represented by the notation

$$Zn(s)|Zn^{2+}(1M)|Cu^{2+}(1M)|Cu(s)$$

in which the vertical lines represent phase boundaries. By convention, the substance forming the anode is listed first. The other materials of the cell are then listed in the order that one would encounter them leading from the anode to the cathode. The composition of the cathode is given last.

Electric current is produced by a voltaic cell as a result of the **electromotive force** (emf) of the cell, which is measured in volts. The greater the tendency for the cell reaction to occur, the higher the emf of the cell. The emf of a given cell, however, also depends upon the concentrations of the substances used to make the cell.

A **standard emf,** $\mathscr{E}°$, pertains to the electromotive force of a cell, at 25°C, in which all reactants and products are present in their standard states. The standard state of a solid or a liquid is, of course, the pure solid or pure liquid itself. The standard state of a gas or a substance in solution is a defined state of *ideal unit activity*—that is, corrections are applied for deviations from ideality caused by intermolecular and interionic attractions. For our discussion we shall make the assumption that the activity of ions may be represented by their molar concentrations and the activity of gases by their pressures in atmospheres. Hence, according to this approximation, a standard cell would contain ions at $1M$ concentrations and gases at 1 atm pressures. In the cell notations that follow, concentrations will be indicated only if they deviate from standard.

If the emf of a cell is to be used as a reliable measure of the tendency for the cell reaction to occur, the voltage must be the maximum value obtainable for the particular cell under consideration. If there is an appreciable flow of electricity during measurement, the voltage measured, $\mathscr{E}$, will be reduced because of the internal resistance of the cell. In addition, when the cell delivers current, the electrode reactions produce concentration changes that reduce the voltage.

The emf of a cell, therefore, must be measured with no appreciable flow of electricity through the cell. This measurement is accomplished by the use of a potentiometer. The circuit of a potentiometer includes a current source of variable voltage and a means of measuring this voltage. The cell being studied is connected to the potentiometer circuit in such a way that the emf of the cell is opposed by the emf of the potentiometer current source.

If the emf of the cell is larger than that of the potentiometer, electrons will flow in the normal direction for a spontaneously discharging cell of that type. On the other hand, if the emf of the potentiometer current source is larger than that of the cell, electrons will flow in the opposite direction, thus causing the cell reaction to be reversed. When the two emf's are exactly balanced, no electrons flow. This voltage is the **reversible emf** of the cell. The emf of a standard Daniell cell is 1.10 V.

Faraday's laws apply to the cell reactions of voltaic, as well as electrolytic, cells. One precaution must be observed, however. Electricity is generated by the simultaneous oxidation and reduction half reactions that occur at the anode and cathode, respectively. Both must occur if the cell is to deliver current. Two faradays of electricity will be produced, therefore, by the oxidation of 1 mol of zinc at the anode *together with* the reduction of 1 mol of Cu^{2+} ions at the cathode. The partial equations

$$\begin{array}{ll} \text{anode:} & Zn \rightarrow Zn^{2+} + 2e^- \\ \text{cathode:} & 2e^- + Cu^{2+} \rightarrow Cu \end{array}$$

when read in terms of moles, represent the flow of two times Avogadro's number of electrons or the production of 2 F of electricity.

The quantity of electrical energy, in joules, produced by a cell is the product of the quantity of electricity delivered, in coulombs, and the emf of the cell, in volts (see Section 16.1). The electrical energy *produced* by the reaction between 1 mol of zinc metal and 1 mol of copper(II) ions may be calculated as follows:

$$-2(96{,}500 \text{ coul})(1.10 \text{ V}) = -212{,}000 \text{ J} = -212 \text{ kJ}$$

One volt coulomb is a joule.

The $\mathscr{E}°$ used in the preceding calculation is the reversible emf of the standard Daniell cell and hence the maximum voltage for this cell. Therefore, the value secured (−212 kJ) is the maximum work that can be obtained from the operation of this type of cell. The maximum *net* work* that can be obtained from a chemical reaction conducted at a constant temperature and pressure is a measure of the *decrease* in the Gibbs free energy (Section 17.4) of the system. Hence,

$$\Delta G = -nF\mathscr{E}$$

where n is the number of moles of electrons transferred in the reaction (or the number of faradays produced), F is the value of the faraday in appropriate units, and $\mathscr{E}$ is the emf in volts. If F is expressed as 96,487 coul, ΔG is obtained in joules. A change in free energy derived from a standard emf, $\mathscr{E}°$, is given the symbol $\Delta G°$.

The free energy change of a reaction is a measure of the tendency of the reaction to occur. If work must be done on a system to bring about a change, the change is not spontaneous. At constant temperature and pressure a spontaneous change is one from which net work can be obtained. Hence, for any spontaneous reaction the free energy of the system decreases; ΔG is negative. Since $\Delta G = -nF\mathscr{E}$, only if $\mathscr{E}$ is positive will the cell reaction be spontaneous and serve as a source of electrical energy.

16.7 ELECTRODE POTENTIALS

In the same way that a cell reaction may be regarded as the sum of two half reactions, the emf of a cell may be thought of as the sum of two half-cell potentials. However, it is impossible to determine the absolute value of the potential of a single half cell. A relative scale has been established by assigning a value of zero to the voltage of a standard reference half cell and expressing all half-cell potentials relative to this reference electrode.

The reference half cell used is the **standard hydrogen electrode,** which

* Some reactions proceed with an increase in volume, and the system must do work to expand against the atmosphere in order to maintain a constant pressure. The energy for this pressure–volume work is not available for any other purpose; it must be expended in this way if the reaction is to occur at constant pressure. Pressure–volume work is not included in the potentiometric measurement of the electrical work of any cell. Net work (or available work) is work other than pressure–volume work.

consists of hydrogen gas, at 1 atm pressure, bubbling over a platinum electrode (coated with finely divided platinum to increase its surface) that is immersed in an acid solution containing $H^+(aq)$ at unit activity. In Figure 16.6 a standard hydrogen electrode is shown connected by means of a salt bridge to a standard Cu^{2+}/Cu electrode. A **salt bridge** is a tube filled with a concentrated solution of a salt (usually KCl), which conducts the current between the half cells but prevents the mixing of the solutions of the half cells. The cell of Figure 16.6 may be diagrammed as

$$\mathrm{Pt|H_2|H^+||Cu^{2+}|Cu}$$

A double bar indicates a salt bridge. The hydrogen electrode is the anode, the copper electrode is the cathode, the emf of the cell is 0.34 V.

The cell emf is considered to be the sum of the half-cell potential for the oxidation half reaction (which we shall give the symbol $\mathscr{E}_{ox}$) and the half-cell potential for the reduction half reaction (which we shall indicate as $\mathscr{E}$ without subscript). For the cell of Figure 16.6,

$$\begin{array}{llr} \text{anode:} & \mathrm{H_2 \rightarrow 2H^+ + 2}e^- & \mathscr{E}^\circ_{ox} = \ \ 0.00\ \mathrm{V} \\ \text{cathode:} & 2e^- + \mathrm{Cu^{2+} \rightarrow Cu} & \mathscr{E}^\circ = +0.34\ \mathrm{V} \end{array}$$

Since the hydrogen electrode is arbitrarily assigned a potential of zero, the entire cell emf is ascribed to the standard Cu^{2+}/Cu electrode. The value +0.34 V is called the **standard electrode potential** of the Cu^{2+}/Cu electrode. Notice that electrode potentials are given for reduction half reactions.

There are several interpretations of the meaning of the positive sign given to this electrode potential.

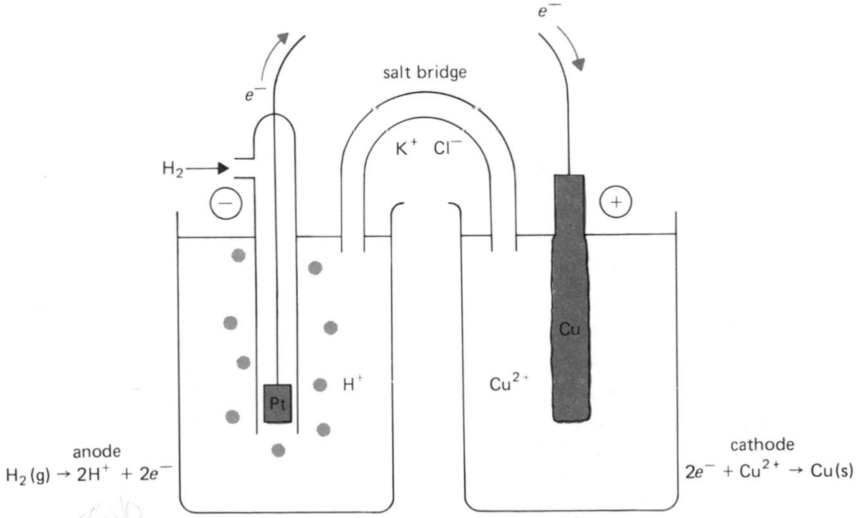

Figure 16.6 Standard hydrogen electrode and a Cu^{2+}/Cu electrode.

1. The copper electrode is the positive electrode (cathode).
2. Copper(II) ions are more active electron acceptors than $H^+(aq)$ ions (greater tendency for reduction).
3. The reduction of Cu^{2+} ions is the spontaneous direction of change for the Cu^{2+}/Cu electrode of this cell.

If a cell is constructed from a standard hydrogen electrode and a standard Zn^{2+}/Zn electrode, the zinc electrode is the anode, and the emf of the cell is 0.76 V. Thus,

$$\text{anode:} \quad Zn \rightarrow Zn^{2+} + 2e^- \qquad \mathscr{E}^\circ_{ox} = +0.76 \text{ V}$$
$$\text{cathode:} \quad 2e^- + 2H^+ \rightarrow H_2 \qquad \mathscr{E}^\circ = 0.00 \text{ V}$$

The value +0.76 V is sometimes called an **oxidation potential,** since it corresponds to an oxidation half reaction. An electrode potential, however, is a **reduction potential.** To obtain the electrode potential of the Zn^{2+}/Zn couple, we must change the sign of the oxidation potential so that the potential corresponds to the reverse half reaction, a reduction.

$$2e^- + Zn^{2+} \rightarrow Zn \qquad \mathscr{E}^\circ = -0.76 \text{ V}$$

The negative sign of this electrode potential may be interpreted in several ways.

1. The zinc electrode is the negative electrode (anode).
2. Zinc ions are less active electron acceptors than $H^+(aq)$ ions (lesser tendency for reduction).
3. Oxidation of Zn (*not* reduction of Zn^{2+}) is the spontaneous direction of change for the Zn^{2+}/Zn electrode of this cell.

It is not necessary to use a cell containing a standard hydrogen electrode to obtain a standard electrode potential. For example, the standard potential of the Ni^{2+}/Ni electrode may be determined from the cell

$$Ni|Ni^{2+}||Cu^{2+}|Cu$$

The emf of this cell is 0.59 V, and the nickel electrode functions as the anode.

$$Ni + Cu^{2+} \rightarrow Ni^{2+} + Cu \qquad \mathscr{E}^\circ = +0.59 \text{ V}$$

The standard electrode potential of the Cu^{2+}/Cu electrode has been determined.

$$2e^- + Cu^{2+} \rightarrow Cu \qquad \mathscr{E}^\circ = +0.34 \text{ V}$$

If we subtract the Cu^{2+}/Cu half reaction from the cell reaction and subtract the half-cell potential from the cell emf, we obtain

$$Ni \rightarrow Ni^{2+} + 2e^- \qquad \mathscr{E}^\circ_{ox} = +0.25 \text{ V}$$

The electrode potential is therefore

$$2e^- + Ni^{2+} \rightarrow Ni \qquad \mathscr{E}^\circ = -0.25 \text{ V}$$

Standard electrode potentials are listed in Table 16.2, and a more complete list is found in the appendix.* The table is constructed with the most positive electrode potential (greatest tendency for reduction) at the bottom. Hence, if a pair of electrodes is combined to make a voltaic cell, the reduction half reaction (cathode) of the cell will be that listed for the electrode that stands lower in the table, and the oxidation half reaction (anode) will be the *reverse* of that shown for the electrode that stands higher in the table.

Table 16.2 Standard electrode potentials at 25°C[a]

HALF REACTION	$\mathscr{E}°$ (VOLTS)
$Li^+ + e^- \rightleftharpoons Li$	−3.045
$K^+ + e^- \rightleftharpoons K$	−2.925
$Ba^{2+} + 2e^- \rightleftharpoons Ba$	−2.906
$Ca^{2+} + 2e^- \rightleftharpoons Ca$	−2.866
$Na^+ + e^- \rightleftharpoons Na$	−2.714
$Mg^{2+} + 2e^- \rightleftharpoons Mg$	−2.363
$Al^{3+} + 3e^- \rightleftharpoons Al$	−1.662
$2H_2O + 2e^- \rightleftharpoons H_2 + 2OH^-$	−0.82806
$Zn^{2+} + 2e^- \rightleftharpoons Zn$	−0.7628
$Cr^{3+} + 3e^- \rightleftharpoons Cr$	−0.744
$Fe^{2+} + 2e^- \rightleftharpoons Fe$	−0.4402
$Cd^{2+} + 2e^- \rightleftharpoons Cd$	−0.4029
$Ni^{2+} + 2e^- \rightleftharpoons Ni$	−0.250
$Sn^{2+} + 2e^- \rightleftharpoons Sn$	−0.136
$Pb^{2+} + 2e^- \rightleftharpoons Pb$	−0.126
$2H^+ + 2e^- \rightleftharpoons H_2$	0
$Cu^{2+} + 2e^- \rightleftharpoons Cu$	+0.337
$Cu^+ + e^- \rightleftharpoons Cu$	+0.521
$I_2 + 2e^- \rightleftharpoons 2I^-$	+0.5355
$Fe^{3+} + e^- \rightleftharpoons Fe^{2+}$	+0.771
$Ag^+ + e^- \rightleftharpoons Ag$	+0.7991
$Br_2 + 2e^- \rightleftharpoons 2Br^-$	+1.0652
$O_2 + 4H^+ + 4e^- \rightleftharpoons 2H_2O$	+1.229
$Cr_2O_7^{2-} + 14H^+ + 6e^- \rightleftharpoons 2Cr^{3+} + 7H_2O$	+1.33
$Cl_2 + 2e^- \rightleftharpoons 2Cl^-$	+1.3595
$MnO_4^- + 8H^+ + 5e^- \rightleftharpoons Mn^{2+} + 4H_2O$	+1.51
$F_2 + 2e^- \rightleftharpoons 2F^-$	+2.87

[a] Data from A. J. de Bethune and N. A. Swendeman Loud, "Table of Electrode Potentials and Temperature Coefficients," pp. 414–424 in *Encyclopedia of Electrochemistry* (C. A. Hampel, editor), Van Nostrand Reinhold, New York, 1964, and from A. J. de Bethune and N. A. Swendeman Loud, *Standard Aqueous Electrode Potentials and Temperature Coefficients,* 19 pp., C. A. Hampel, publisher, Skokie, Illinois, 1964.

* The definition of the standard electrode potential used agrees with that suggested by the International Union of Pure and Applied Chemistry (IUPAC) in 1953. Prior to this time, the common American practice was to tabulate oxidation potentials rather than electrode potentials.

For example, consider a cell constructed from standard Ni^{2+}/Ni and Ag^{+}/Ag electrodes. The table entries for these electrodes are

$$Ni^{2+} + 2e^- \rightleftharpoons Ni \qquad \mathscr{E}^\circ = -0.250 \text{ V}$$
$$Ag^+ + e^- \rightleftharpoons Ag \qquad \mathscr{E}^\circ = +0.799 \text{ V}$$

Of the two ions, the Ag^+ ion shows the greater tendency for reduction. The Ag^+/Ag electrode is therefore the cathode, and the Ni^{2+}/Ni electrode is the anode. The half reaction that takes place at an anode is an oxidation, and the half-cell potential is an oxidation potential.

$$\text{anode:} \qquad Ni \rightarrow Ni^{2+} + 2e^- \qquad \mathscr{E}^\circ_{ox} = +0.250 \text{ V}$$
$$\text{cathode:} \qquad 2e^- + 2Ag^+ \rightarrow 2Ag \qquad \mathscr{E}^\circ = +0.799 \text{ V}$$

The cell reaction and cell emf may be obtained by addition.

$$Ni + 2Ag^+ \rightarrow Ni^{2+} + 2Ag \qquad \mathscr{E}^\circ = +1.049 \text{ V}$$

Notice that the half reaction for the reduction of Ag^+ must be multiplied by 2 before the addition so that the electrons lost and gained in the half reactions will cancel. The $\mathscr{E}^\circ$ for the Ag^+/Ag electrode, however, is *not* multiplied by 2. The magnitude of an electrode potential depends upon the temperature and the concentrations of materials used in the construction of the half cell. These variables are fixed for *standard* electrode potentials. Indication of the stoichiometry of the cell reaction does not imply that a concentration change has been made.

Actually the half reactions implied by the half-cell potentials are

$$\cancel{2H^+} + Ni \rightarrow \cancel{H_2} + Ni^{2+}$$
$$\underline{\cancel{H_2} + 2Ag^+ \rightarrow \cancel{2H^+} + 2Ag}$$
$$Ni + 2Ag^+ \rightarrow Ni^{2+} + 2Ag$$

Notice, however, that in the addition of these half reactions, the H_2 molecules and H^+ ions cancel.

An alternative way to use the table of electrode potentials to calculate a cell emf consists of subtracting the electrode potential of the anode (value nearer the top of the table column) from the electrode potential of the cathode (value nearer the bottom of the table column). This mathematical procedure amounts to the same thing as changing the sign of the electrode potential of the anode (to correspond to an oxidation) and adding the value to the electrode potential of the cathode.

By convention, the *anode* of a cell is indicated on the *left* in a cell notation. Thus, the Cd^{2+}/Cd electrode is the anode of the cell

$$Cd|Cd^{2+}||Ag^+|Ag$$

and the cell reaction indicated is

$$Cd + 2Ag^+ \rightarrow Cd^{2+} + 2Ag$$

The standard emf of the cell is

$$\mathscr{E}° = \mathscr{E}°_{right} - \mathscr{E}°_{left}$$

Substitution of $\mathscr{E}°$ values from the table gives

$$\mathscr{E}° = (+0.799) - (-0.403)$$
$$= +1.202 \text{ V}$$

The positive sign of the cell emf indicates that the cell is properly diagrammed and the reaction proceeds as indicated.

If we had mistakenly diagrammed the cell as

$$Ag|Ag^+||Cd^{2+}|Cd$$

which corresponds to the reaction

$$2Ag + Cd^{2+} \rightarrow 2Ag^+ + Cd$$

we would have obtained the result

$$\mathscr{E}° = \mathscr{E}°_{right} - \mathscr{E}°_{left}$$
$$= (-0.403) - (+0.799)$$
$$= -1.202 \text{ V}$$

The negative sign of the cell emf means that the cell will not function in the manner indicated by the cell notation. Rather, the cell reaction would be the reverse of that shown, and the diagram should be reversed.

Electrode potentials are also useful for the evaluation of oxidation–reduction reactions that take place outside of electrochemical cells. Either of the two ways described for the analysis of a cell may be used to evaluate an oxidation–reduction reaction. The half reactions of Table 16.2 are listed as reductions, from left to right, with the oxidized form of the element, molecule, or ion accepting electrons to produce the reduced form.

$$(\text{oxidized form}) + ne^- \rightleftharpoons (\text{reduced form})$$

Because of the way in which the table is constructed (greatest tendency for reduction at the bottom), any *reduced form,* at unit activity, should be able to reduce the *oxidized form,* at unit activity, of a half reaction listed below it in the table. From the positions in the table of the following half reactions

$$2e^- + I_2 \rightleftharpoons 2I^- \qquad \mathscr{E}° = +0.536 \text{ V}$$
$$2e^- + Cl_2 \rightleftharpoons 2Cl^- \qquad \mathscr{E}° = +1.360 \text{ V}$$

we would correctly predict that the reduced form, I^-, should be able to reduce the oxidized form, Cl_2.

$$Cl_2(g) + 2I^-(aq) \rightarrow 2Cl^-(aq) + I_2(s) \qquad \mathscr{E}° = +0.824$$

The species with the greatest tendency toward oxidation (best reducing agent) is the reduced form of the half reaction at the top of the table (Li in Table

16.2). The oxidized form of the half reaction at the bottom of the table (F_2 in Table 16.2) has the greatest tendency to be reduced (best oxidizing agent).

There are several factors that must be kept in mind when using a table of electrode potentials to predict the course of a chemical reaction. Because $\mathscr{E}$ changes with changes in concentration, many presumably unfavored reactions can be made to occur by altering the concentrations of the reacting species. In addition, some theoretically favored reactions proceed at such a slow rate that they are of no practical consequence.

Correct use of the table also demands that all pertinent half reactions of a given element be considered before making a prediction. On the basis of the half reactions

$$3e^- + Fe^{3+} \rightleftharpoons Fe \qquad \mathscr{E}^\circ = -0.036\ V$$
$$2e^- + 2H^+ \rightleftharpoons H_2 \qquad \mathscr{E}^\circ = 0.000\ V$$

one might predict that the products of the reaction of iron with H^+ would be hydrogen gas and Fe^{3+} ions ($\mathscr{E}^\circ$ for the complete reaction, +0.036 V). The oxidation state iron(II), however, lies between metallic iron and the oxidation state iron(III). Once an iron atom has lost two electrons and becomes an Fe^{2+} ion, further oxidation is opposed, as may be seen from the reverse of the following

$$e^- + Fe^{3+} \rightleftharpoons Fe^{2+} \qquad \mathscr{E}^\circ = +0.771\ V$$

Thus, the reaction yields Fe^{2+} ions only. This fact could have been predicted by an examination of the half reaction

$$2e^- + Fe^{2+} \rightleftharpoons Fe \qquad \mathscr{E}^\circ = -0.440\ V$$

The $\mathscr{E}_{ox}$ for the production of Fe^{2+} ions from the reaction of iron metal and H^+ ions (+0.440 V) is greater than that for the production of Fe^{3+} ions (+0.036V), and hence the former is favored.

We may summarize the electrode potentials for iron and its ions as follows:

$$Fe^{3+} \xrightarrow{+0.771V} Fe^{2+} \xrightarrow{-0.440V} Fe$$
$$\underbrace{Fe^{3+} \longrightarrow Fe}_{-0.036V}$$

The preceding predictions are immediately evident from this diagram, if we remember that oxidation is the reverse of the relation corresponding to an electrode potential.

Occasionally an oxidation state of an element is unstable toward disproportionation. The electrode potentials for copper and its ions may be summarized as follows:

$$Cu^{2+} \xrightarrow{+0.153V} Cu^{+} \xrightarrow{+0.521V} Cu$$
$$\underbrace{Cu^{2+} \longrightarrow Cu}_{+0.337V}$$

From this we see that the Cu^+ ion is not a very stable one. In water, Cu^+ ions disproportionate to copper metal and Cu^{2+} ions.

$$2Cu^+(aq) \rightarrow Cu(s) + Cu^{2+}(aq)$$

The $\mathscr{E}°$ for this reaction is $+0.521 - 0.153 = +0.368$ V. Species instable toward disproportionation may be readily recognized by the fact that the electrode potential for the reduction to the next lower oxidation state is more positive than the electrode potential for the couple with the next higher oxidation state. An inspection of the diagram for iron and its ions shows that the Fe^{2+} ion is stable toward such disproportionation.

16.8 EFFECT OF CONCENTRATION ON CELL POTENTIALS

The potential of a cell in which the constituents are present at concentrations other than standard may be calculated by means of the **Nernst equation,** named for Walther Nernst, who developed it in 1889.

A simple form of the Nernst equation for 25°C is

$$\mathscr{E} = \mathscr{E}° - \frac{0.05916}{n} \log Q$$

where $\mathscr{E}°$ is the standard cell emf, n is the number of moles of electrons transferred in the reaction (or the number of faradays involved), and Q is the reaction quotient. The reaction quotient, Q, is a fraction derived from the activities of the dissolved substances and gases employed in the cell. The activity of each substance is raised to a power equal to the coefficient of that substance in the balanced chemical equation. The *numerator* of Q is the product of the activity terms for the materials on the *right* of the chemical equation. The *denominator* of Q is the product of the activity terms for the substances on the *left* of the chemical equation. Since the activity of a pure solid is assumed to be unity at all times, the activity term for a solid is always equal to 1.

For a hypothetical cell reaction,

$$wW + xX \rightarrow yY + zZ$$

where the lower case letters represent the coefficients of the balanced chemical equation,

$$Q = \frac{(a_Y)^y(a_Z)^z}{(a_W)^w(a_X)^x}$$

For our work we shall make the assumptions that the activity of a substance in solution is given by the molar concentration of the substance and the activity of a gas is equal to the partial pressure of the gas in atmospheres; these assumptions lead to $\mathscr{E}$ values that are approximately correct.

When the activities of all substances are unity (standard states), $Q = 1$, $\log Q = 0$, and $\mathscr{E} = \mathscr{E}°$. The Nernst equation may be used to determine the emf of a cell constructed from nonstandard electrodes or to calculate the electrode potential of a half cell in which all species are not present at unit activity.

Example 16.5 What is the electrode potential of a Zn^{2+}/Zn electrode in which the concentration of Zn^{2+} ions is $0.1M$?

Solution The partial equation

$$2e^- + Zn^{2+} \rightarrow Zn$$

shows 2 electrons gained. If the symbol $[Zn^{2+}]$ is used to designate the molar concentration of Zn^{2+} ions,

$$\mathscr{E} = \mathscr{E}° - \frac{0.0592}{2} \log \left(\frac{1}{[Zn^{2+}]}\right)$$

$\mathscr{E}°$ for the Zn^{2+}/Zn electrode is -0.76 V.

$$\mathscr{E} = -0.76 - \frac{0.0592}{2} \log \left(\frac{1}{0.1}\right)$$

$$\mathscr{E} = -0.76 - 0.0296(1) = -0.79 \text{ V}$$

Example (a) What is the potential for the cell

$$Ni|Ni^{2+}(0.01M)||Cl^-(0.2M)|Cl_2(1 \text{ atm})|Pt$$

(b) What is ΔG for the cell reaction?

Solution (a) Two faradays of electricity are involved in the cell reaction

$$Ni + Cl_2 \rightarrow Ni^{2+} + 2Cl^-$$

and therefore $n = 2$. The standard cell emf is

$$\begin{aligned} \mathscr{E}° &= \mathscr{E}°(Cl_2/Cl^-) - \mathscr{E}°(Ni^{2+}/Ni) \\ &= (+1.36) - (-0.25) = +1.61 \text{ V} \end{aligned}$$

Therefore,

$$\mathscr{E} = \mathscr{E}° - \frac{0.0592}{2} \log \left(\frac{[Cl^-]^2[Ni^{2+}]}{[Cl_2]}\right)$$

$$\mathscr{E} = +1.61 - \frac{0.0592}{2} \log \left(\frac{(0.2)^2(0.01)}{(1)}\right)$$

$$\mathscr{E} = +1.61 - 0.0296 \log (0.0004)$$

$$\mathscr{E} = +1.61 + 0.10 = +1.71 \text{ V}$$

(b)

$$\begin{aligned} \Delta G &= -nF\mathscr{E} \\ \Delta G &= -2(96{,}500 \text{ coul})(1.71 \text{ V}) \\ &= -330{,}000 \text{ J} = -330 \text{kJ} \end{aligned}$$

Example 16.7 What is the $\mathscr{E}$ of the cell

$$\text{Sn}|\text{Sn}^{2+}(1.0M)||\text{Pb}^{2+}(0.0010M)|\text{Pb}$$

Solution The following may be obtained from a table of standard electrode potentials

$$\text{Sn} \rightarrow \text{Sn}^{2+} + 2e^- \qquad \mathscr{E}^\circ_{ox} = +0.136\ \text{V}$$
$$2e^- + \text{Pb}^{2+} \rightarrow \text{Pb} \qquad \mathscr{E}^\circ = -0.126\ \text{V}$$

Thus the reaction in a standard cell is

$$\text{Sn} + \text{Pb}^{2+} \rightarrow \text{Sn}^{2+} + \text{Pb} \qquad \mathscr{E}^\circ = +0.010\ \text{V}$$

For the cell as diagrammed in the problem,

$$\mathscr{E} = \mathscr{E}^\circ - \frac{0.0592}{2} \log\left(\frac{[\text{Sn}^{2+}]}{[\text{Pb}^{2+}]}\right)$$

$$\mathscr{E} = +0.010 - \frac{0.0592}{2} \log\left(\frac{1.0}{0.0010}\right)$$

$$\mathscr{E} = +0.010 \quad 0.0296(3)$$

$$= +0.010 - 0.089 = -0.079\ \text{V}$$

This result means that the cell will not function in the manner implied by the diagram. Instead, it would operate in the reverse order and in a direction opposite to that of the standard cell. The cell is properly diagrammed

$$\text{Pb}|\text{Pb}^{2+}(0.0010M)||\text{Sn}^{2+}(1.0M)|\text{Sn}$$

and the reaction of the cell is

$$\text{Pb} + \text{Sn}^{2+} \rightarrow \text{Pb}^{2+} + \text{Sn} \qquad \mathscr{E} = +0.079\ \text{V}$$

From this example we see that concentration effects can sometimes reverse the direction that a reaction is expected to take.

Notice that the results of the preceding examples are qualitatively in agreement with predictions based on Le Chatelier's principle (Section 14.4). Increasing the concentrations of reactants and decreasing the concentrations of products would be expected to increase the driving force of the reaction and lead to a more positive $\mathscr{E}$. On the other hand, decreasing the concentrations of reactants and increasing the concentrations of products would be expected to retard a reaction and result in a more negative $\mathscr{E}$.

16.9 CONCENTRATION CELLS

Since an electrode potential depends upon the concentration of the ions used in the electrode, a cell may be constructed from two electrodes composed of the same materials but differing in concentration of ions. For example,

$Cu|Cu^{2+}(0.010M)||Cu^{2+}(0.10M)|Cu$

From the partial equation,

$$2e^- + Cu^{2+} \rightleftharpoons Cu \qquad \mathscr{E}° = +0.34 \text{ V}$$

and Le Chatelier's principle, we can predict that *increasing* the concentration of Cu^{2+} ions would drive the reaction to the right and raise the reduction potential, whereas *decreasing* the concentration of Cu^{2+} ions would drive the reaction to the left and raise the oxidation potential (or lower the reduction potential). Hence, of the two electrodes in the cell diagrammed, the left electrode would have the stronger tendency for oxidation and the right would have the stronger tendency for reduction. The "reaction" of the cell is

$$Cu + Cu^{2+}(0.10M) \rightarrow Cu^{2+}(0.010M) + Cu$$

$\mathscr{E}°$ for this cell is zero since the same electrode is involved in each half cell. Then

$$\mathscr{E} = 0.00 - \frac{0.0592}{2} \log\left(\frac{0.010}{0.10}\right)$$

$$\mathscr{E} = -0.0296(-1) = +0.0296 \text{ V}$$

16.10 ELECTRODE POTENTIALS AND ELECTROLYSIS

Electrode potentials are determined under reversible conditions. The measured voltage of a spontaneously discharging voltaic cell will not equal the calculated electromotive force for the cell since in the production of an electric current, the cell is operating in an irreversible manner. The cell emf calculated by means of electrode potentials is the maximum voltage that the cell can develop and is determined by the application of a measured external voltage of a magnitude that will just stop the flow of electrons. If an external voltage infinitesimally greater than that which the cell develops were applied, the cell would run in the reverse direction and electrolysis would occur. Electrode potentials, therefore, give the minimum voltage necessary for electrolysis.

In practice, however, it is found that a higher voltage than calculated must be applied to effect electrolysis. Part of the increased voltage is required to overcome the electrical resistance of the cell, and the remainder is needed to overcome two additional effects. The first is called **concentration polarization** and is caused by the changes in the concentration of the electrolyte around the electrodes as the electrolysis proceeds and ions are removed or produced. The concentration gradient set up within the cell is, in effect, a concentration cell and produces a back emf opposing the applied voltage. Concentration polarization can be reduced by stirring the electrolyte during the course of the electrolysis.

The other effect that makes the voltage required for an electrolysis higher than the reversible emf is called **overvoltage.** Overvoltage is thought to be caused by a slow rate of reaction at the electrodes. Excess applied voltage is required to make the electrolysis proceed at an appreciable rate. Overvoltages for the deposition of metals are low, but those required for the liberation of hydrogen gas or oxygen gas are appreciable, vary with the material used for the electrode, and may amount to a volt or more.

Electrolysis is an important procedure for the quantitative analysis of ores, alloys, and other metal-containing compounds and mixtures. As the voltage is increased in the electrolysis of an aqueous solution, ions or molecules are discharged at the *cathode* in order of decreasing value of electrode potential—the most positive value (nearest to the bottom of the table) first. In addition, the reduction of water is possible. The $\mathscr{E}^\circ$ value for this reduction is -0.828 V, but in neutral aqueous solutions the concentration of OH^- is $10^{-7}\,M$, not $1\,M$. The $\mathscr{E}$ value for neutral aqueous solutions, obtained by means of the Nernst equation, is

$$2e^- + 2H_2O \rightleftharpoons H_2 + 2OH^- \qquad \mathscr{E} = -0.414 \text{ V}$$

One might predict that it would be impossible to reduce any aqueous species for which $\mathscr{E}^\circ$ is more negative than -0.414 V, the reduction potential for water. However, because of the high hydrogen overvoltage, it is possible to plate out, on the cathode, metals that have electrode potentials more negative than the one for the reduction of water—down to, and including, zinc ($\mathscr{E}^\circ = -0.763$ V).

At the *anode* the species present are oxidized in inverse order of electrode potential—most negative value (uppermost on the table) first. The half reactions are the reverse of those listed in the table. In the electrolysis of aqueous solutions, the oxidation of water is possible. For neutral aqueous solutions (concentration of H^+, $10^{-7}\,M$), this oxidation is the reverse of

$$4e^- + 4H^+ + O_2 \rightleftharpoons 2H_2O \qquad \mathscr{E} = +0.815 \text{ V}$$

For the liberation of oxygen, however, high overvoltages are required. Thus, the oxidation of chloride ion to chlorine gas takes place in the electrolysis of aqueous solutions of chlorides.

$$2e^- + Cl_2(g) \rightleftharpoons 2Cl^- \qquad \mathscr{E}^\circ = +1.36 \text{ V}$$

The overvoltage of chlorine is less than the overvoltage of oxygen.

The products of electrolysis vary with the concentration of ions in the solution, since the half-cell emf's are dependent upon concentrations. For example, the electrolysis of *dilute* solutions of chlorides yields oxygen at the anode rather than chlorine. Furthermore, after the primary electrode reactions in which electrons are transferred, secondary reactions may occur. Thus, if chlorine is liberated in an alkaline solution, ClO^- or ClO_3^- ions are formed from the reaction of the chlorine with the hydroxide ions. If chlorine is liberated on a silver anode, AgCl is formed.

Example 16.8 A solution is $1 M$ in each of the following: $AgNO_3$, $Cu(NO_3)_2$, $Fe(NO_3)_2$, and $NaNO_3$. In the electrolysis of this solution, between inert electrodes, the voltage is gradually increased. What would be the order in which the materials would plate out on the cathode, and at what voltage would each be discharged? Assume that the overvoltage of oxygen is 0.600 V, the overvoltage of hydrogen is 0.300 V, and the solution is neutral. Ignore concentration polarization and cell resistance.

Solution The only species at the anode (which is inert) are the nitrate ion and water. Since it is impossible to oxidize the nitrate ion, water will be oxidized:

$$2H_2O \rightarrow O_2(g) + 4H^+ + 4e^- \qquad \mathscr{E}_{OX} = -0.815 \text{ V}$$

If the overvoltage of oxygen is taken into account, the half-cell potential for the oxidation is -1.415 V.

The following reductions should be considered as possible cathode half reactions:

$$e^- + Na^+ \rightarrow Na \qquad \mathscr{E}^\circ = -2.714 \text{ V}$$
$$2e^- + Fe^{2+} \rightarrow Fe \qquad \mathscr{E}^\circ = -0.440 \text{ V}$$
$$2e^- + Cu^{2+} \rightarrow Cu \qquad \mathscr{E}^\circ = +0.337 \text{ V}$$
$$e^- + Ag^+ \rightarrow Ag \qquad \mathscr{E}^\circ = +0.799 \text{ V}$$

In addition, the reduction of water

$$2e^- + 2H_2O \rightarrow H_2(g) + 2OH^- \qquad \mathscr{E} = -0.414 \text{ V}$$

must be considered. If the overvoltage of hydrogen is taken into account, the reduction potential for water is -0.714 V.

The silver will plate out first. The voltage at which this will occur will be equal to the oxidation potential of water (-1.415 V) plus the reduction potential of Ag ($+0.799$ V), or -0.616 V. (The values are negative since the voltage is applied.) After the silver has been removed, the copper will plate out at a voltage of -1.088 V, followed by the iron at a voltage of -1.855 V. The reduction of water (half-cell potential $= -0.714$ V) will occur before the reduction of sodium ions (half-cell potential $= -2.714$ V). Hence, at a voltage of 2.129 V, hydrogen will be evolved at the cathode. Sodium ions can be reduced only in the absence of water, that is, by the electrolysis of molten sodium salts.

16.11 SOME COMMERCIAL VOLTAIC CELLS

Several voltaic cells are of commercial importance. The **dry cell** (Figure 16.7) consists of a zinc metal container (which serves as the anode) that is filled with a moist paste of ammonium chloride and zinc chloride and contains a graphite

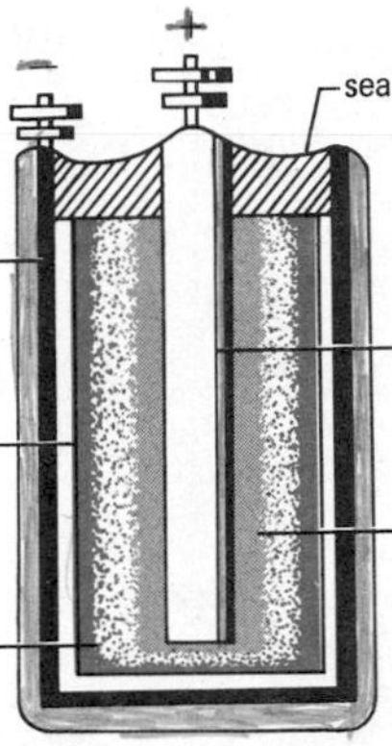

Figure 16.7 The dry cell.

electrode (the cathode) surrounded by manganese dioxide. The electrode reactions are complex, but they may be approximately represented by

anode: $Zn \rightarrow Zn^{2+} + 2e^-$
cathode: $2e^- + 2MnO_2 + 2NH_4^+ \rightarrow Mn_2O_3 + H_2O + 2NH_3$

The dry cell generates a voltage of approximately 1.25 to 1.50 V.

A newer type of dry cell which has found use in small electrical devices (such as hearing aids) consists of a zinc container as the anode, a carbon rod as the cathode, and moist mercury(II) oxide mixed with potassium hydroxide as the electrolyte. A lining of porous paper keeps the electrolyte separated from the zinc anode. The cell has a potential of approximately 1.35 V.

anode: $Zn + 2OH^- \rightarrow Zn(OH)_2 + 2e^-$
cathode: $2e^- + HgO + H_2O \rightarrow Hg + 2OH^-$

The **lead storage cell** consists of a lead anode and a grid of lead packed with lead dioxide as the cathode. The electrolyte is sulfuric acid, and the half-cell reactions are

anode: $Pb(s) + SO_4^{2-} \rightarrow PbSO_4(s) + 2e^-$
cathode: $2e^- + PbO_2(s) + SO_4^{2-} + 4H^+ \rightarrow PbSO_4(s) + 2H_2O$

In practice, the current obtainable from a lead storage cell is increased by constructing the cell from a number of cathode plates joined together and arranged alternately with a number of anode plates, which are also joined together. The potential difference of one cell is approximately 2 V. A storage battery consists of three or six such cells joined in series to produce a 6- or a 12-volt battery.

The electrode reactions of the storage battery can be reversed by the application of an external current source and in this manner the battery can be recharged. Since sulfuric acid is consumed as the storage battery delivers current, the state of charge of the battery can be determined by measuring the density of the battery electrolyte.

The **nickel–cadmium storage cell** has a longer life than the lead storage cell but is more expensive to manufacture.

$$\begin{array}{ll} \text{anode:} & Cd(s) + 2OH^- \rightarrow Cd(OH)_2 + 2e^- \\ \text{cathode:} & 2e^- + NiO_2(s) + 2H_2O \rightarrow Ni(OH)_2 + 2OH^- \end{array}$$

The potential of each cell of a nickel–cadmium battery is approximately 1.4 V, and the battery is rechargeable.

16.12 FUEL CELLS

Electrical cells that are designed to convert the energy from the combustion of fuels such as hydrogen, carbon monoxide, or methane directly into electrical energy are called fuel cells. Since in theory 100% of the free energy released by a combustion (ΔG) should be obtainable from an efficient fuel cell, extensive research into their development is currently being undertaken. Although approximately only 60% to 70% efficiency has been realized as yet, present fuel cells are about twice as efficient as processes in which the heat of combustion is used to generate electricity by mechanical means.

In a typical fuel cell, hydrogen and oxygen are bubbled through porous carbon electrodes into concentrated aqueous sodium hydroxide or potassium hydroxide. Catalysts are incorporated in the electrodes.

$$C|H_2(g)|OH^-|O_2(g)|C$$

The gaseous materials are consumed and are continuously supplied. The electrode reactions are

$$\begin{array}{ll} \text{anode:} & 2H_2(g) + 4OH^- \rightarrow 4H_2O + 4e^- \\ \text{cathode:} & 4e^- + O_2(g) + 2H_2O \rightarrow 4OH^- \end{array}$$

The complete cell reaction is

$$2H_2(g) + O_2(g) \rightarrow 2H_2O(1)$$

The cell is maintained at an elevated temperature, and the water produced by the cell reaction evaporates as it is formed.

Research in the design of fuel cells is currently being directed toward lowering the cost of the processes. Another problem involves the development of electrode catalysts that will cause the half-cell reactions to occur more rapidly.

PROBLEMS

16.1 Given the following: 1 faraday = 96,487 coulombs, 1 volt coulomb = 1 joule, Avogadro's number = 6.0222×10^{23}. Calculate: (a) the charge on a single electron in coulombs, (b) the number of joules equal to one electron volt, (c) the value of the electron volt in kilojoules per mole, (d) the value of the faraday in joules per volt.

16.2 What is the effect of an increase in temperature on the conductivity of metals and on the conductivity of solutions of electrolytes? Discuss the causes of resistance to the current in these two types of conductors.

16.3 Sketch a cell for the electrolysis of $CdCl_2$ between inert electrodes. On the sketch indicate: (a) the signs of the electrodes, (b) the cathode and anode, (c) the directions in which the ions move, (d) the direction in which the electrons move, (e) the electrode reactions.

16.4 Sketch a voltaic cell in which the reaction is

$$Cd + Cl_2 \rightarrow Cd^{2+} + 2Cl^-$$

On the sketch indicate: (a) the signs of the electrodes, (b) the cathode and anode, (c) the directions in which the ions move, (d) the direction in which the electrons move, (e) the electrode reactions, (f) the cell voltage.

16.5 A CO^{2+} solution is electrolyzed using a current of 0.400 amp. How many grams of Co plate out in 30.0 min?

16.6 An acidic solution containing BiO^+ ions is electrolyzed using a current of 0.750 amp. How many grams of Bi metal plate out in 45.0 min? Write the equation for the reaction at the cathode.

16.7 A Cu^{2+} solution is electrolyzed using a current of 2.50 amp. How many grams of Cu plate out in 30.0 min?

16.8 How many minutes will it take to plate out 58.7 g of Ni from a solution of $NiSO_4$ using a current of 0.900 amp?

16.9 How many minutes will it take to plate out 10.0 g of Cu metal from a $CuSO_4$ solution using a current of 3.00 amp?

16.10 (a) If 0.690 g of Ag is deposited on the cathode of a silver coulometer, how many coulombs have passed through the circuit? (b) If the process takes 20.0 min, what was the rate of the current?

16.11 In the Hall process, aluminum is produced by the electrolysis of molten Al_2O_3. The electrode reactions are

anode: $C + 2O^{2-} \rightarrow CO_2 + 4e^-$
cathode: $3e^- + Al^{3+} \rightarrow Al$

In the process the carbon of which the anode is composed is gradually consumed by the anode reaction. What mass is lost by the anode when 1.00 kg of aluminum is prepared?

16.12 How long would it take to produce enough aluminum by the Hall process (see Problem 16.11) to make a case of aluminum soft-drink cans (24 cans) if each can uses 5.00 g of Al, a current of 50,000 amp is employed, and the efficiency of the cell is 90.0%?

16.13 A nuclear power station operates 100,000 amp below capacity for 6 hours each day. A scientist proposes generating hydrogen with this excess current by electrolyzing water so that the hydrogen can be burned during peak consumption hours and produce extra power when it is needed. What volume tank must be constructed to hold the hydrogen (measured at STP)?

16.14 In the electrolysis of anhydrous $MgCl_2$, how many liters of $Cl_2(g)$, measured at STP, are produced in the same time that it takes to produce 1.00 kg of Mg?

16.15 (a) In the electrolysis of aqueous NaCl, what volume of $Cl_2(g)$ is produced in the same time that it takes to liberate 3.00 liter of $H_2(g)$? Assume that the volumes of both gases are measured at STP. (b) The volume of the solution used in the electrolysis is 300 ml. What is the molarity of NaOH after the process described in (a)?

16.16 What is the equivalent weight of a metal if a current of 0.300 amp will cause 0.129 g of the metal to plate out of a solution undergoing electrolysis in 40.0 min?

16.17 If 300 ml of 0.300*M* $NiCl_2$ solution is electrolyzed using a current of 3.00 amp for 40.0 min, what are the final concentrations of Ni^{2+} ions and Cl^- ions? Assume that the volume of the solution does not change during the course of the electrolysis.

16.18 How many hours will it take to plate out all of the copper in 300 ml of a 0.300*M* Cu^{2+} solution using a current of 0.750 amp?

16.19 In the electrolysis of a $CuSO_4$ solution, what volume of O_2 (measured at STP) is liberated at the anode in the time that it takes to plate out 5.08 g of Cu on the cathode?

16.20 What is the molarity of H^+ in the solution after the electrolysis described in Problem 16.19? The final volume of the solution is 300 ml.

16.21 (a) What is $\mathscr{E}^\circ$ for the cell

$$Sn|Sn^{2+}||Ag^+|Ag$$

(b) Write an equation for the cell reaction. (c) Which electrode is positive?

16.22 (a) What is $\mathscr{E}^\circ$ for the cell:

$$Cd|Cd^{2+}||Cu^{2+}|Cu?$$

(b) Write the equation for the cell reaction. (c) Which electrode is positive?

16.23 (a) Give the notation for a cell that utilizes the reaction

$$Cl_2(g) + 2I^-(aq) \rightarrow 2Cl^-(aq) + I_2(s)$$

(b) What is $\mathscr{E}^\circ$ for the cell? (c) Which electrode is the cathode?

16.24 For the cell:

$$Sc|Sc^{3+}||Cu^{2+}|Cu,$$

$\mathscr{E}^\circ$ is +2.414 V. Use the emf of the cell and the $\mathscr{E}^\circ$ for the Cu^{2+}/Cu couple to calculate $\mathscr{E}^\circ$ for the Sc^{3+}/Sc half-reaction.

16.25 For the cell:

$$Mg|Mg^{2+}||Lu^{3+}|Lu$$

$\mathscr{E}^\circ$ is +0.108 V. Use the emf of the cell and the $\mathscr{E}^\circ$ for the Mg^{2+}/Mg

couple to calculate $\mathscr{E}°$ for the Lu^{3+}/Lu half-reaction.

16.26 For the cell

$$Zn|Zn^{2+}||In^{3+}|In$$

$\mathscr{E}°$ is +0.4198 V. Use the emf of the cell and the $\mathscr{E}°$ for the Zn^{2+}/Zn couple to calculate $\mathscr{E}°$ for the In^{3+}/In half-reaction.

16.27 From the table of standard electrode potentials (reactions in acid solution) that appears in the appendix, select a suitable substance for each of the following transformations (assume that all soluble substances are present in 1*M* concentrations): (a) an oxidizing agent capable of oxidizing Ni to Ni^{2+} but not Pb to Pb^{2+}, (b) an oxidizing agent capable of oxidizing H_2S to S but not H_2SO_3 to SO_4^{2-}, (c) a reducing agent capable of reducing Ce^{4+} to Ce^{3+} but not PbO_2 to Pb^{2+}, (d) a reducing agent capable of reducing Cr^{3+} to Cr^{2+} but not Cr^{3+} to Cr.

16.28 Use $\mathscr{E}°$ values to predict whether or not the following skeleton equations represent reactions that will occur in acid solution with all soluble substances present in 1*M* concentrations. All reactants and products other than H^+ and H_2O are shown below. Complete and balance the equation for each reaction that is predicted to occur.

(a) $Au^+ \rightarrow Au + Au^{3+}$
(b) $Fe + Fe^{3+} \rightarrow Fe^{2+}$
(c) $Co^{2+} \rightarrow Co + Co^{3+}$
(d) $Cr_2O_7^{2-} + Cr \rightarrow Cr^{3+}$

16.29 Predict whether or not each of the following reactions will occur spontaneously in acid solution, and write a balanced chemical equation for each reaction that is predicted to occur. Assume that soluble reactants and products are present in 1*M* concentrations. (a) Oxidation of Co^{2+} to Co^{3+} by Br_2 (reduced to Br^-), (b) reduction of Br_2 to Br^- by H_2O_2 (oxidized to O_2), (c) oxidation of Mn^{2+} to MnO_4^- by Ce^{4+} (reduced to Ce^{3+}), (d) oxidation of Cl^- to Cl_2 by NO_3^- (reduced to NO), (e) reduction of Mn^{2+} to Mn by Zn (oxidized to Zn^{2+}).

16.30 Given the following standard electrode potential diagram (acid solution):

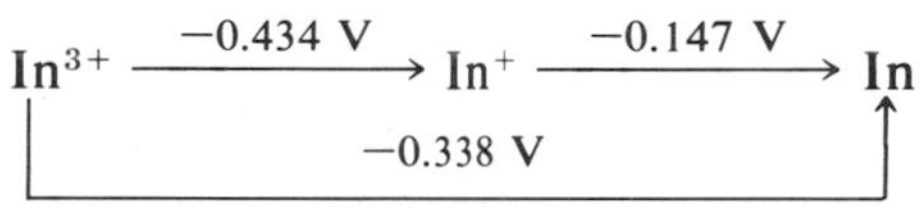

(a) Is the In^+ ion stable toward disproportionation in water solution? (b) Which ion is produced when In metal reacts with $H^+(aq)$? (c) The $\mathscr{E}°$ for Cl_2/Cl^- is +1.36 V. Will In react with Cl_2? What is the product? (d) Write balanced chemical equations for all reactions.

16.31 Given the following standard electrode potential diagram (acid solution):

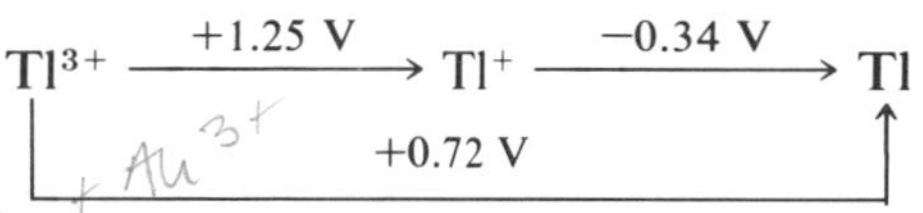

(a) Is the Tl^+ ion stable toward disproportionation in water solution? (b) Which ion is produced when Tl metal reacts with $H^+(aq)$? (c) The $\mathscr{E}°$ for Cl_2/Cl^- is +1.36 V. Will Tl react with Cl_2? What is the product? (d) Write balanced chemical equations for all reactions.

16.32 Given the following standard electrode potential diagram (acid solution):

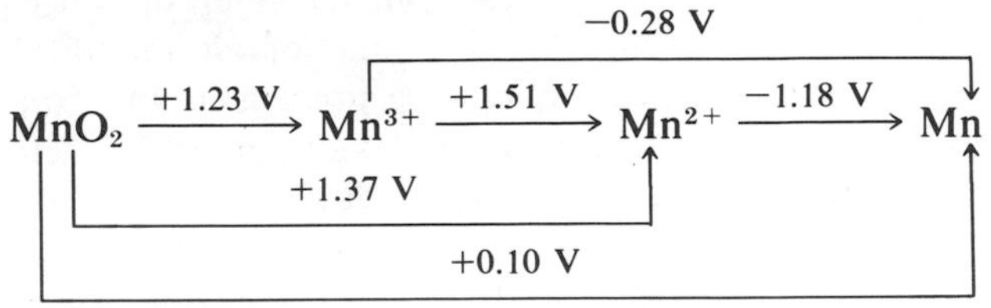

(a) Write a balanced chemical equation for the disproportionation of any species shown that disproportionates. (b) Write the equation for the reaction of Mn metal with $H^+(aq)$.

16.33 The following reactions occur at 25°C with all soluble substances present in 1*M* concentrations:

$$Zn + 2Eu^{3+} \rightarrow Zn^{2+} + 2Eu^{2+}$$
$$Zr + 2Zn^{2+} \rightarrow Zr^{4+} + 2Zn$$
$$4Sc + 3Zr^{4+} \rightarrow 4Sc^{3+} + 3Zr$$

From this information alone, predict whether the following reactions will occur under similar conditions.
(a) $Sc + 3Eu^{3+} \rightarrow Sc^{3+} + 3Eu^{2+}$
(b) $4Eu^{2+} + Zr^{4+} \rightarrow 4Eu^{3+} + Zr$
(c) $2Sc + 3Zn^{2+} \rightarrow 2Sc^{3+} + 3Zn$

16.34 Given the following:

$$PbSO_4 + 2e^- \rightleftharpoons Pb + SO_4^{2-} \qquad \mathscr{E}° = -0.359\ V$$
$$Pb^{2+} + 2e^- \rightleftharpoons Pb \qquad \mathscr{E}° = -0.126\ V$$

(a) Write the notation for a cell utilizing these half-reactions. (b) Write the equation for the cell reaction. (c) Calculate $\mathscr{E}°$ for the cell. (d) Determine $\Delta G°$.

16.35 Given the following:

$$AgI + e^- \rightleftharpoons Ag + I^- \qquad \mathscr{E}° = -0.152\ V$$
$$Ag^+ + e^- \rightleftharpoons Ag \qquad \mathscr{E}° = +0.799\ V$$

(a) Write the notation for a cell utilizing these half-reactions. (b) Write the equation for the cell reaction. (c) Calculate $\mathscr{E}°$ for the cell. (d) Determine $\Delta G°$.

16.36 (a) Diagram a cell for which the cell reaction is

$$H^+ + OH^- \rightarrow H_2O$$

(b) Calculate $\mathscr{E}°$ for the cell and $\Delta G°$ for the reaction.

16.37 (a) Diagram a cell for which the cell reaction in acid solution is

$$2H_2 + O_2 \rightarrow 2H_2O$$

(b) Calculate $\mathscr{E}°$ for the cell and $\Delta G°$ for the reaction.

16.38 (a) For the reaction (at 25°C)

$$Pb(s) + Cl_2(g) \rightarrow PbCl_2(s)$$

$\Delta G^\circ = -314.0$ kJ. Calculate $\mathscr{E}^\circ$ for a cell utilizing this reaction. (b) A half-cell can be prepared in which a Pb electrode is in contact with $PbCl_2(s)$, which in turn is in contact with a solution containing Cl^- ions.

$$2e^- + PbCl_2(s) \rightleftharpoons Pb(s) + 2Cl^-(aq)$$

Determine $\mathscr{E}^\circ$ for this half-reaction by designing a cell for the reaction given in part (a) and by using the cell emf calculated in part (a). (c) Diagram a cell and calculate $\mathscr{E}^\circ$ for the cell using the reaction:

$$Pb^{2+} + 2Cl^- \rightarrow PbCl_2(s)$$

16.39 Why does the measurement of the density of the electrolyte in a lead storage battery give an indication of the state of charge of the battery?

16.40 What is the concentration of Ni^{2+} in the cell:

$$Ni \mid Ni^{2+}(?M) \parallel Cu^{2+}(0.750M) \mid Cu$$

if the emf of the cell is 0.601 V?

16.41 What is the concentration of Cd^{2+} in a cell

$$Cd \mid Cd^{2+}(?M) \parallel Ni^{2+}(2.00M) \mid Ni$$

if the emf of the cell is 0.200 V?

16.42 (a) Calculate the emf of the cell

$$Tl \mid Tl^+(0.0500M) \parallel H^+(2.00M) \mid H_2(0.100 \text{ atm}) \mid Pt$$

(b) Write an equation for the cell reaction. (c) Which electrode is negative?

16.43 (a) Calculate the emf of the cell:

$$Ga \mid Ga^{3+}(0.0300M) \parallel H^+(3.00M) \mid H_2(0.100 \text{ atm}) \mid Pt$$

(b) Write an equation for the cell reaction. (c) Which electrode is negative?

16.44 (a) According to the $\mathscr{E}^\circ$ value, should the following reaction proceed spontaneously at 25°C with all soluble substances present at 1.00*M* concentration?

$$MnO_2(s) + 4H^+(aq) + 2Cl^-(aq) \rightarrow Mn^{2+}(aq) + 2H_2O + Cl_2(g)$$

(b) Would the reaction be spontaneous if the concentrations of H^+ and Cl^- were each increased to 10.0*M*?

16.45 For a half-reaction of the form

$$M^{2+} + 2e^- \rightleftharpoons M$$

what would be the effect on the electrode potential if (a) the concentration of M^{2+} were doubled, (b) the concentration of M^{2+} were cut in half?

16.46 For the half-reaction

$$Cr_2O_7^{2-} + 14H^+ + 6e^- \rightleftharpoons 2Cr^{3+} + 7H_2O$$

$\mathscr{E}°$ is +1.33 V. What would the potential be if the concentration of $H^+(aq)$ were reduced to 0.100*M*?

16.47 Given the following standard cell:

$$Co \mid Co^{2+} \parallel Ni^{2+} \mid Ni$$

(a) Calculate the value of $\mathscr{E}°$ for the cell. (b) When the cell operates, the concentration of Co^{2+} increases and the concentration of Ni^{2+} decreases. What are the concentrations of the ions when $\mathscr{E}$ is zero?

16.48 (a) Determine the emf, at 25°C, of the cell

$$Pt \mid H_2(g,\ 1.00\ atm) \mid H^+(0.0250M) \parallel H^+(5.00M) \mid H_2(g,\ 1.00\ atm) \mid Pt$$

(b) Write an equation for the "cell reaction." (c) Determine $\mathscr{E}$ for the cell if the pressure of $H_2(g)$ is changed to 2.00 atm in the anode half-cell and to 0.100 atm in the cathode half cell.

16.49 (a) Determine the emf of a concentration cell prepared from two Ga^{3+}/Ga half-cells, one in which the concentration of Ga^{3+} is 2.00*M* and a second in which the concentration of Ga^{3+} is 0.300*M*. (b) Write an equation for the "cell reaction." (c) Which electrode is negative?

Chapter 17

Elements of Chemical Thermodynamics

Thermodynamics is the study of the energy changes that accompany physical and chemical changes. An important aspect of the laws of chemical thermodynamics is that they enable us to predict whether a particular chemical reaction is theoretically possible under a given set of conditions. A reaction that has a natural tendency to occur of its own accord is said to be **spontaneous.**

Thermodynamic principles can also be used to determine the extent of a spontaneous reaction—the position of equilibrium. The maximum theoretical yield of the products of a reaction is limited by the equilibrium state of the reaction.

Thermodynamics, however, has nothing to say about the rate or mechanism of a spontaneous reaction. These questions are the concern of chemical kinetics (Chapter 13). Some spontaneous changes occur extremely slowly. The stable form of carbon under ordinary conditions is graphite, not diamond. The change from diamond to graphite, therefore, is thermodynamically spontaneous. This change, however, is so extremely slow that it is not observed at ordinary temperatures and pressures.

17.1 FIRST LAW OF THERMODYNAMICS

Many scientists of the late-eighteenth and early-nineteenth centuries studied the relationship between work and heat. Thermodynamics had its origins in these studies. By the 1840's it became clear that

1. work and heat are both forms of a larger classification called energy
2. one form of energy can be converted into another form
3. energy cannot be created or destroyed

The **first law of thermodynamics** is the law of conservation of energy: energy can be converted from one form into another but it cannot be created or destroyed. In other words, the total energy of the universe is a constant.

In applying thermodynamic concepts, we frequently confine our attention to the changes that occur within definite boundaries. That portion of nature that is included within these boundaries is called a **system.** The remainder is called the **surroundings.** A mixture of chemical compounds, for example, can constitute a system. The container and everything else around the system make up what is called the surroundings.

A system is assumed to have an **internal energy,** E, which includes all possible forms of energy attributable to the system. Important contributions to the internal energy of a system include the attractions and repulsions between the atoms, molecules, ions, and subatomic particles that make up the system and the kinetic energies of all of its parts.

According to the first law of thermodynamics, the internal energy of an *isolated* system is constant. The actual value of E for any system is not known and cannot be calculated. Thermodynamics, however, is concerned only with *changes* in internal energy, and these changes can be measured.

The **state** of a system can be defined by specifying the values of properties such as temperature, pressure, and composition. The internal energy of a system depends upon the state of the system and not upon how the system arrived at that state. Internal energy is therefore called a **state function.** Consider a sample of an ideal gas that occupies a volume of 1 liter at 100 K and 1 atm pressure (state A). At 200 K and 0.5 atm (state B), the sample occupies a volume of 4 liter. According to the first law, the internal energy of the system in state A, E_A, is a constant, as is the internal energy of the system in state B, E_B.

It follows that the difference in the internal energies of the two states, ΔE, is also a constant and is independent of the path taken between state A and state B. It makes no difference whether the gas is heated before the pressure change, whether the heating is done after the pressure change, or, indeed, whether the total change is brought about in several steps,

$$\Delta E = E_B - E_A$$

Suppose that we have a system in an *initial* state and that the internal energy of the system is E_i. If the system absorbs heat from the surroundings, q, the internal energy of the system will now be

$$E_i + q$$

If the system now uses some of its internal energy to do work on the surroundings, w, the internal energy of the system in the *final* state, E_f, will be

$$E_f = E_i + q - w$$
$$E_f - E_i = q - w$$
$$\Delta E = q - w$$

It is important to keep in mind the conventions regarding the signs of these quantities:

q, positive = heat *absorbed* by the system
q, negative = heat *evolved* by the system
w, positive = work done *by* the system
w, negative = work done *on* the system

The values of q and w involved in changing a system from an initial state to a final state depend upon the way in which the change is carried out. The value of $(q - w)$, however, is a constant, equal to ΔE, for the change no matter how it is brought about. If a system undergoes a change in which the internal energy of the system remains constant, the work done by the system equals the heat absorbed by the system.

17.2 ENTHALPY

For ordinary chemical reactions, the work term generally arises as a consequence of pressure–volume changes. The work done against the pressure of the atmosphere if the system expands in the course of the reaction is an example of pressure–volume work. The term PV has the dimensions of work. Pressure, which is force per unit area, may be expressed in newtons per square meter (N/m^2). If volume is expressed in cubic meters (m^3), the product PV is

$$PV = (N/m^2)(m^3) = N \cdot m$$

The newton meter (which is a joule) is a unit of work, since work is defined as force (the newton) times distance (the meter). In like manner, liter atmospheres are units of work. If the pressure is held constant, the work done in expansion from V_A to V_B is

$$w = P(V_B - V_A) = P\Delta V$$

No pressure–volume work can be done by a process carried out at constant volume, and $w = 0$. Thus, at *constant volume* the equation

$$\Delta E = q - w$$

becomes

$$\Delta E = q_V$$

where q_V is the heat absorbed by the system at constant volume.

Processes carried out at constant pressure are far more common in chemistry than those conducted at constant volume. If we restrict our attention to pressure–volume work, the work done in constant pressure processes is $P\Delta V$. Thus, at *constant pressure* the equation

$$\Delta E = q - w$$

becomes

$$\Delta E = q_P - P\Delta V$$

Or, by rearranging,

$$q_P = \Delta E + P\Delta V$$

where q_P is the heat absorbed by the system at constant pressure.

The thermodynamic function **enthalpy,** H, is defined by the equation

$$H = E + PV$$

The change in enthalpy, ΔH, for a reaction conducted at constant pressure is

$$\Delta H = \Delta E + P\Delta V$$

Therefore,

$$q_P = \Delta H$$

The heat absorbed by a reaction conducted at constant pressure is equal to the change in enthalpy. Enthalpy, like internal energy, is a function of the state of the system and is independent of the manner in which the state was achieved. The validity of the law of Hess rests on this fact (see Section 3.5).

When a bomb calorimeter is used to make a calorimetric determination (Section 3.3), the heat effect is measured at constant volume. Ordinarily, reactions are run at constant pressure. The relationship between change in enthalpy and change in internal energy is used to convert heats of reaction at constant volume ($q_V = \Delta E$) to heats of reaction at constant pressure ($q_P = \Delta H$). The conversion is made by considering the change in volume of the system—

the total volume of the reactants subtracted from the total volume of the products. The changes in the volumes of liquids and solids are so small that they are neglected.

For reactions involving gases, however, volume changes may be significant. Let us say that V_A is the total volume of gaseous reactants, V_B is the total volume of gaseous products, n_A is the number of moles of gaseous reactants, n_B is the number of moles of gaseous products, and that the pressure and temperature are constant.

$$PV_A = n_A RT \text{ and } PV_B = n_B RT$$

Thus

$$\begin{aligned} P\Delta V &= PV_B - PV_A \\ &= n_B RT - n_A RT \\ &= (n_B - n_A)RT \\ &= (\Delta n)RT \end{aligned}$$

Since

$$\Delta H = \Delta E + P\Delta V$$
$$\Delta H = \Delta E + (\Delta n)RT$$

where Δn is the number of moles of gaseous products minus the number of moles of gaseous reactants.

In order to solve problems using this equation, we must express the value of R in appropriate units. We have noted that liter atmospheres are units of energy. The value of R (0.082056 liter atm/K mol) may be converted to J/K mol.

$$R = 8.3143 \text{ J/K mol}$$

Various types of calculations involving enthalpy changes are the topic of Chapter 3.

Example 17.1 The heat of combustion at constant volume of $CH_4(g)$ is measured in a bomb calorimeter at 25°C and is found to be −885,389 J/mol. What is the ΔH?

Solution For the reaction

$$CH_4(g) + 2O_2(g) \rightarrow CO_2(g) + 2H_2O(l) \qquad \Delta E = -885{,}389 \text{ J}$$
$$\Delta n = 1 - (2 + 1) = -2$$

Therefore,

$$\begin{aligned} \Delta H &= \Delta E + (\Delta n)RT \\ &= -885{,}389 \text{ J} + (-2 \text{ mol})(8.3143 \text{ J/K mol})(298.2 \text{ K}) \\ &= -885{,}389 \text{ J} - 4959 \text{ J} \\ &= -890{,}348 \text{ J} = -890.348 \text{ kJ} \end{aligned}$$

17.3 SECOND LAW OF THERMODYNAMICS

The first law of thermodynamics puts only one restriction on chemical or physical changes—energy must be conserved. The first law, however, provides no basis for determining whether a proposed change will be spontaneous. The second law of thermodynamics establishes criteria for making this important prediction.

The thermodynamic function **entropy,** S, is central to the second law. Entropy may be interpreted as a measure of the randomness, or disorder, of a system. A highly disordered system is said to have a high entropy. Since a disordered condition is more probable than an ordered one, entropy may be regarded as a probability function. One statement of the **second law of thermodynamics** is: every spontaneous change is accompanied by an increase in entropy.

As an example of a spontaneous change, consider the mixing of two ideal gases. The two gases, which are under the same pressure, are placed in bulbs that are joined by a stopcock (Figure 17.1). When the stopcock is opened, the gases spontaneously mix until each is evenly distributed throughout the entire apparatus. Why did this spontaneous change occur? The first law cannot help us answer this question. Throughout the mixing, the volume, total pressure, and temperature remain constant. Since the gases are ideal, no intermolecular forces exist, and neither the internal energy nor the enthalpy of the system is affected.

This change represents an increase in entropy. The final state is more random and hence more probable than the initial state. The random motion of the gas molecules has produced a more disordered condition. The fact that the gases mix spontaneously is not surprising; one would have predicted it from experience. Indeed, it would be surprising if the reverse were to be observed—a gaseous mixture spontaneously separating into two pure gases, each occupying one of the bulbs.

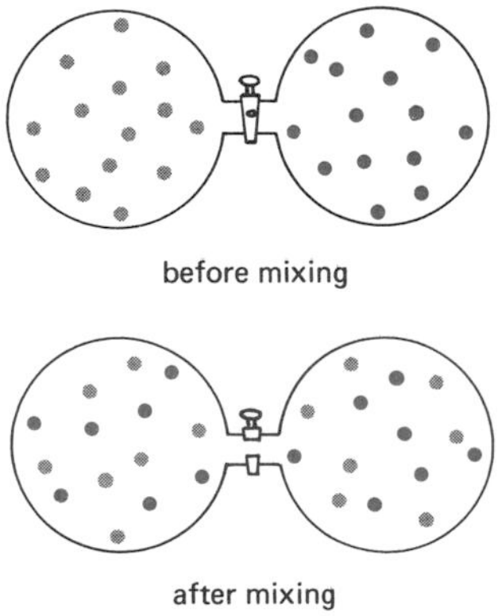

Figure 17.1 Spontaneous mixing of two gases.

For a given substance the solid, crystalline state is the state of lowest entropy (most ordered); the gaseous state is the state of highest entropy (most random); and the liquid state is intermediate between the other two. Hence, when a substance either melts or vaporizes, its entropy increases. The reverse changes, crystallization and condensation, are changes in which the entropy of the substance decreases. Why, then, should a substance spontaneously freeze at temperatures below its melting point, since this change represents a decrease in the entropy of the substance?

All the entropy effects that result from the proposed change must be considered. When two ideal gases mix by the process previously described, there is no exchange of matter or energy between the isolated system in which the change occurs and its surroundings. The only entropy effect is an increase in the entropy of the isolated system itself. Usually, however, a chemical reaction or a physical change is conducted in such a way that the system is not isolated from its surroundings. The total change in entropy is equal to the sum of the change in the entropy of the system (ΔS_{system}) and the change in entropy of the surroundings ($\Delta S_{\text{surroundings}}$).

$$\Delta S_{\text{total}} = \Delta S_{\text{system}} + \Delta S_{\text{surroundings}}$$

When a liquid freezes, the heat of fusion is evolved by the liquid and absorbed by the surroundings. This heat increases the random motion of the surrounding molecules and therefore increases the entropy of the surroundings. The spontaneous freezing of a liquid at a temperature below the melting point occurs, therefore, because the decrease in entropy of the liquid (ΔS_{system}) is more than offset by the increase in entropy of the surroundings ($\Delta S_{\text{surroundings}}$) so that there is a net increase in entropy.

The total change in entropy should always be considered to determine spontaneity. When a substance melts, the entropy increases, but this effect alone does not determine whether or not the transformation is spontaneous. The entropy of the surroundings must also be considered, and spontaneity is indicated only if the total entropy of system and surroundings taken together increases.

The data of Table 17.1 pertain to the freezing of water. The meaning of the units of ΔS values will be discussed in later sections. For the moment, let us be concerned only with the numerical values listed in the table. At $-1°C$ the

Table 17.1 Entropy changes for the transformation $H_2O(l) \rightarrow H_2O(s)$ at 1 atm

TEMPERATURE (°C)	ΔS_{system} (J/K mol)	$\Delta S_{\text{surroundings}}$ (J/K mol)	ΔS_{total} (J/K mol)
+1	−22.13	+22.05	−0.08
0	−21.99	+21.99	0
−1	−21.85	+21.93	+0.08

change is spontaneous; ΔS_{total} is positive. At $+1°C$, however, ΔS_{total} is negative, and freezing is not a spontaneous change. On the other hand, the reverse change, melting, is spontaneous at $+1°C$ (the signs of all ΔS values would be reversed).

At 0°C, the melting point, ΔS_{total} is zero, which means that neither freezing nor melting is spontaneous. At this temperature a water–ice system would be in equilibrium, and no *net* change would be observed. Note, however, that freezing or melting can be made to occur at 0°C by removing or adding heat, but neither change will occur spontaneously.

Thus, the ΔS_{total} of a postulated change may be used as a criterion for whether the change will occur spontaneously. The entropy of the universe is steadily increasing as spontaneous changes occur. Rudolf Clausius summarized the first and second laws of thermodynamics as: "The energy of the universe is constant; the entropy of the universe tends toward a maximum."

Entropy, like internal energy and enthalpy, is a state function. The entropy, or randomness, of a system in a given state is a definite value, and hence, ΔS for a change from one state to another is a definite value depending only on the initial and final states and not on the path between them.

17.4 GIBBS FREE ENERGY

The type of change of primary interest to the chemist is, of course, the chemical reaction. The $\Delta S_{surroundings}$ for a reaction conducted at constant temperature and pressure may be calculated by means of the equation

$$\Delta S_{surroundings} = -\frac{\Delta H}{T}$$

where ΔH is the enthalpy change of the reaction and T is the absolute temperature. The change in the entropy of the surroundings is brought about by the heat transferred into or out of the surroundings because of the enthalpy change of the reaction. Since heat *evolved* by the reaction is *absorbed* by the surroundings (and *vice versa*), the sign of ΔH must be reversed. Hence, the larger the value of $-\Delta H$, the more disorder created in the surroundings and the larger the value of $\Delta S_{surroundings}$.

On the other hand, the change in the entropy of the surroundings is *inversely* proportional to the absolute temperature at which the change takes place. A given quantity of heat added to the surroundings at a low temperature (where the randomness is relatively low initially) will create a larger difference in the disorder of the surroundings than the same quantity of heat added at a high temperature (where the randomness is relatively high to begin with). Entropy is therefore measured in units of J/K.

In the last section we noted that

$$\Delta S_{total} = \Delta S_{system} + \Delta S_{surroundings}$$

If $-\Delta H/T$ is substituted for $\Delta S_{\text{surroundings}}$ and if the symbol ΔS (without a subscript) is used to indicate the entropy change of the system, the following equation is obtained.

$$\Delta S_{\text{total}} = \Delta S - \frac{\Delta H}{T}$$

Multiplication by T gives

$$T\Delta S_{\text{total}} = T\Delta S - \Delta H$$

By reversing the signs of the terms of this equation, we get

$$-T\Delta S_{\text{total}} = \Delta H - T\Delta S$$

We can use this equation to show how a function called the Gibbs free energy determines reaction spontaneity.

Gibbs free energy, G, is defined by the equation

$$G = H - TS$$

For a chemical reaction conducted at constant temperature and pressure,

$$\Delta G = \Delta H - T\Delta S$$

If we compare this equation to

$$-T\Delta S_{\text{total}} = \Delta H - T\Delta S$$

we see that

$$\Delta G = -T\Delta S_{\text{total}}$$

On the basis of this result, we note that

1. *If ΔG is negative, the reaction is spontaneous.* Since ΔS_{total} is greater than zero for a spontaneous change, $T\Delta S_{\text{total}}$ must also be greater than zero and $-T\Delta S_{\text{total}}$ must be less than zero. Therefore, for a spontaneous change at constant temperature and pressure,

   $$\Delta G < 0$$

   and when a spontaneous change occurs, the free energy of the system decreases.

   The negative value of $T\Delta S_{\text{total}}$ is used so that the signs of ΔG values will be like the signs of other energy terms. A *negative* value indicates that energy is *liberated* by the system.
2. *If ΔG is zero, the system is in equilibrium.* For a system in equilibrium, ΔS_{total} is zero and, therefore,

   $$\Delta G = 0$$

3. *If ΔG is positive, the reaction is not spontaneous.* The reverse of the reaction, however, will be spontaneous.

Notice that all the terms in the very important equation

$$\Delta G = \Delta H - T\Delta S$$

pertain to changes in the properties of the *system*. The use of free energy values, therefore, removes the need to consider changes in the surroundings. Notice also that each term in the equation is an energy term. Since ΔS is measured in J/K, the term $T\Delta S$ is expressed in J; ΔH and ΔG are also expressed in J.

Gibbs free energy is named for J. Willard Gibbs, who developed the application of thermodynamic concepts to chemistry. The change in Gibbs free energy, ΔG, combines two factors that influence reaction spontaneity.

$$\Delta G = \Delta H - T\Delta S$$

1. Reactions tend to seek a minimum in energy. Spontaneity is favored if the ΔH value is negative (the system liberates energy). A negative value of ΔH helps to make the ΔG value negative, which indicates a spontaneous reaction.
2. Reactions tend to seek a maximum in randomness. Spontaneity is favored if the value of ΔS is positive (randomness increases). Since ΔS appears in the term $-T\Delta S$, a positive value of ΔS helps to make the value of ΔG negative, which indicates a spontaneous reaction.

The most favorable circumstance for a negative value of ΔG, therefore, is a negative value of ΔH together with a positive value of ΔS. It is possible, however, for a favorable ΔH to outweigh an unfavorable $T\Delta S$ or for a favorable $T\Delta S$ to overshadow an unfavorable ΔH. The value of ΔG depends upon the magnitude of both factors.

Gibbs free energy, like the other thermodynamic functions we have discussed, is a state function. The value of ΔG depends only upon the final and initial states of the system and not upon the path taken between those states.

17.5 STANDARD FREE ENERGIES

A standard free energy change, which is given the symbol ΔG°, is the free energy change for a process at 25°C and 1 atm in which the reactants in their standard states are converted to the products in their standard states. The value of ΔG° for a reaction can be derived from standard free energies of formation in the same way that ΔH° values can be calculated from standard enthalpies of formation.

The **standard free energy of formation** of a compound, ΔG_f°, is defined as the change in standard free energies when 1 mol of the compound is formed from its constituent elements. According to this definition, the standard free energy of formation of any element is zero. The value of ΔG° for a reaction is

Table 17.2 Gibbs free energy of formation at 25°C and 1 atm

COMPOUND	ΔG_f° (kJ/mol)	COMPOUND	ΔG_f° (kJ/mol)
$H_2O(g)$	−228.61	$CO(g)$	−137.28
$H_2O(l)$	−237.19	$CO_2(g)$	−394.38
$HF(g)$	−270.7	$NO(g)$	+86.69
$HCl(g)$	−95.27	$NO_2(g)$	+51.84
$HBr(g)$	−53.22	$NaCl(s)$	−384.05
$HI(g)$	+1.30	$CaO(s)$	−604.2
$H_2S(g)$	−33.0	$Ca(OH)_2(s)$	−896.76
$NH_3(g)$	−16.7	$CaCO_3(s)$	−1128.76
$CH_4(g)$	−50.79	$BaO(s)$	−528.4
$C_2H_6(g)$	−32.89	$BaCO_3(s)$	−1138.9
$C_2H_4(g)$	+68.12	$Al_2O_3(s)$	−1576.41
$C_2H_2(g)$	+209.20	$Fe_2O_3(s)$	−741.0
$C_6H_6(l)$	+129.66	$AgCl(s)$	−109.70
$SO_2(g)$	−300.37	$ZnO(s)$	−318.19

equal to the sum of the standard free energies of formation of the products minus the sum of the standard free energies of formation of the reactants. Some standard free energies of formation are given in Table 17.2.

Example 17.2 Use ΔG_f° values from Table 17.2 to calculate ΔG° for the transformation

$$2NO(g) + O_2(g) \rightarrow 2NO_2(g)$$

Solution

$$\begin{aligned}\Delta G^\circ &= 2\Delta G_f^\circ(NO_2) - 2\Delta G_f^\circ(NO)\\ &= 2(+51.84 \text{ kJ}) - 2(+86.69 \text{ kJ})\\ &= -69.70 \text{ kJ}\end{aligned}$$

17.6 ABSOLUTE ENTROPIES

The addition of heat to a substance results in an increase in molecular randomness. Hence, the entropy of a substance increases as the temperature increases. Conversely, cooling a substance makes it more ordered and decreases its entropy. At absolute zero the entropy of a perfect crystalline substance may be taken as zero. This statement is sometimes called the **third law of thermodynamics** and was first formulated by Walther Nernst in 1906. The entropy of an imperfect crystal, a glass, or a solid solution is not zero at 0 K.

On the basis of the third law, absolute entropies can be calculated from heat capacity data. The **standard absolute entropy** of a substance, S°, is the entropy of the substance in its standard state at 25°C and 1 atm. Some $S°$ values are given in Table 17.3. The $\Delta S°$ value for a reaction is equal to the sum of the absolute entropies of the products minus the sum of the absolute entropies of the reactants. Note that the absolute entropy of an element is *not* equal to zero and that the absolute entropy of a compound is *not* the entropy change when the compound is formed from its constituent elements.

Example 17.3 (a) Use absolute entropies from Table 17.3 to calculate the standard entropy change, $\Delta S°$, for the formation of 1 mol of HgO(s) from its elements. (b) The standard enthalpy of formation, ΔH_f°, of HgO(s) is −90.7 kJ/mol. Calculate the standard free energy of formation, ΔG_f°, of HgO(s).

Solution

(a)

$$\mathrm{Hg(l)} + \tfrac{1}{2}\mathrm{O_2(g)} \rightarrow \mathrm{HgO(s)}$$

$$\begin{aligned}\Delta S° &= S°(\mathrm{HgO}) - [S°(\mathrm{Hg}) + \tfrac{1}{2}S°(\mathrm{O_2})] \\ &= (72.0\ \mathrm{J/K}) - [(77.4\ \mathrm{J/K}) + \tfrac{1}{2}(205.0\ \mathrm{J/K})] \\ &= -107.9\ \mathrm{J/K}\end{aligned}$$

(b)

$$\begin{aligned}\Delta G° &= \Delta H° - T\Delta S° \\ &= -90{,}700\ \mathrm{J} - (298\ \mathrm{K})(-107.9\ \mathrm{J/K}) \\ &= -58{,}500\ \mathrm{J} = -58.5\ \mathrm{kJ}\end{aligned}$$

17.7 FREE ENERGY AND CHEMICAL EQUILIBRIUM

Since ΔG is zero for a system in equilibrium, free energy changes can be used to evaluate the equilibrium state of a chemical reaction. The free energy of a substance in a state other than its standard state, G, is related to its standard free energy, $G°$, by the equation

$$G = G° + 2.30RT \log a$$

where R is the ideal gas constant (8.3143 J/K mol), T is the absolute temperature, and a is the activity of the substance (its effective concentration).

This equation can be used to derive an expression for the free energy change that accompanies a chemical reaction. The free energy change of the general reaction

$$w\mathrm{W} + x\mathrm{X} \rightleftharpoons y\mathrm{Y} + z\mathrm{Z}$$

is given by the equation

$$\Delta G = \Delta G° + 2.303RT \log\left(\frac{(a_\mathrm{Y})^y(a_\mathrm{Z})^z}{(a_\mathrm{W})^w(a_\mathrm{X})^x}\right)$$

Table 17.3 Absolute entropy at 25°C and 1 atm

SUBSTANCE	$S°$ (J/ K mol)	SUBSTANCE	$S°$ (J/ K mol)
$H_2(g)$	130.6	$HBr(g)$	198.5
$F_2(g)$	203.3	$HI(g)$	206.3
$Cl_2(g)$	223.0	$H_2S(g)$	205.6
$Br_2(l)$	152.3	$NH_3(g)$	192.5
$I_2(s)$	116.7	$CH_4(g)$	186.2
$O_2(g)$	205.03	$C_2H_6(g)$	229.5
S(rhombic)	31.9	$C_2H_4(g)$	219.5
$N_2(g)$	191.5	$C_2H_2(g)$	200.8
C(graphite)	5.69	$SO_2(g)$	248.5
Li(s)	28.0	CO(g)	197.9
Na(s)	51.0	$CO_2(g)$	213.6
Ca(s)	41.6	NO(g)	210.6
Al(s)	28.3	$NO_2(g)$	240.5
Ag(s)	42.72	NaCl(s)	72.38
Fe(s)	27.2	CaO(s)	39.8
Zn(s)	41.6	$Ca(OH)_2(s)$	76.1
Hg(l)	77.4	$CaCO_3(s)$	92.9
La(s)	57.3	$Al_2O_3(s)$	51.00
$H_2O(g)$	188.7	$Fe_2O_3(s)$	90.0
$H_2O(l)$	69.96	HgO(s)	72.0
HF(g)	173.5	AgCl(s)	96.11
HCl(g)	186.7	ZnO(s)	43.9

The fraction contains the activities of the substances participating in the reaction with each activity raised to a power equal to the coefficient in the balanced chemical equation. The logarithmic term of the equation corrects $\Delta G°$ to the more general condition in which the activities of the substances are not each equal to one.

At equilibrium, $\Delta G = 0$ and, therefore,

$$0 = \Delta G° + 2.303RT \log\left(\frac{(a_{\mathrm{Y}})^y(a_{\mathrm{Z}})^z}{(a_{\mathrm{W}})^w(a_{\mathrm{X}})^x}\right)$$

Since the system is in equilibrium, the activities that appear in this equation are equilibrium activities, and the fraction is the thermodynamic equilibrium constant, K. Therefore,

$$\Delta G° = -2.303RT \log K$$

In using equilibrium constants, we have made the assumption that the activities of substances are equal to their concentrations. This approximation is satisfactory if the concentrations are low.

The ideal gas constant, R, is 8.3143 J/K mol. For an equilibrium system at 25°C (which is 298.15 K), therefore,

$$\begin{aligned}\Delta G^\circ &= -2.303RT \log K\\ &= -2.303(8.3143 \text{ J/K mol})(298.15 \text{ K}) \log K\\ &= -(5709 \text{ J/mol}) \log K\\ &= -(5.709 \text{ kJ/mol}) \log K\end{aligned}$$

Example 17.4 Calculate K_p for the following reaction at 25°C.

$$2SO_2(g) + O_2(g) \rightleftharpoons 2SO_3(g)$$

For $SO_2(g)$, ΔG_f° is -300.4 kJ/mol. For $SO_3(g)$, ΔG_f° is -370.4 kJ/mol.

Solution The standard free energy change for the reaction is

$$\begin{aligned}\Delta G^\circ &= 2\Delta G_f^\circ(SO_3) - 2\Delta G_f^\circ(SO_2)\\ &= 2 \text{ mol}(-370.4 \text{ kJ/mol}) - 2 \text{ mol}(-300.4 \text{ kJ/mol})\\ &= -140.0 \text{ kJ}\end{aligned}$$

We use this value to find K_p. Since

$$\Delta G^\circ = -(5.709 \text{ kJ/mol}) \log K$$

$$\log K = \frac{\Delta G^\circ}{-5.709 \text{ kJ/mol}}$$

$$= \frac{-140.0 \text{ kJ/mol}}{-5.709 \text{ kJ/mol}} = 24.52$$

Therefore,

$$K = \text{antilog } 24.52 = 3.3 \times 10^{24}$$

The type of K obtained by this method depends upon the definition of the standard states used in the determination of ΔG° values. For the reactions of gases, K_p's are obtained since the standard state of a gas is defined as the gas at a partial pressure of 1 atm.

For the reaction

$$N_2O_4(g) \rightleftharpoons 2NO_2(g)$$
$$\Delta G^\circ = +5.40 \text{ kJ}$$

Since ΔG° is positive, we may predict that N_2O_4 in its standard state at 25°C and 1 atm would not dissociate into NO_2 at all and that the reverse reaction (the formation of N_2O_4 from NO_2) would go to completion. Both these predictions are incorrect.

If we use the value of ΔG° to calculate the equilibrium constant for this system at 25°C, we find

$$K_p = 0.113$$

At equilibrium, therefore, some of the N_2O_4 is dissociated.

The free energy of a system in which this reaction occurs at 25°C and 1 atm is plotted against the fraction of N_2O_4 dissociated in Figure 17.2. Point A represents the standard free energy of 1 mol of N_2O_4, point B represents the standard free energy of 2 mol of NO_2, and the intervening points on this curve represent the free energies of mixtures of N_2O_4 and NO_2. Absolute values of free energies are not known, and no scale is indicated on the vertical axis of the diagram. Differences in free energies, however, can be calculated so that the shape of the curve is accurately represented.

The free energy curve exhibits a minimum at the equilibrium point, E, where 16.6% of the N_2O_4 is dissociated. The difference between the standard free energy of 2 mol of NO_2 (point B) and the standard free energy of 1 mol of N_2O_4 (point A) is $\Delta G°$ for the reaction (+5.40 kJ) and is indicated on the figure. However, ΔG for the preparation of the equilibrium mixture (point E) from 1 mol of N_2O_4 (point A) is −0.84 kJ, which indicates that N_2O_4 will spontaneously dissociate until equilibrium is reached.

The figure shows that equilibrium can be approached from either direction. Thus, $\Delta G = -6.23$ kJ for the preparation of the equilibrium mixture (point E) from 2 mol of pure NO_2 (point B). The negative values of ΔG for both changes (from A to E and from B to E) indicate that both changes are spontaneous.

We must be careful, therefore, when we interpret the meaning of $\Delta G°$ values in relation to reaction spontaneity. Values of K corresponding to various $\Delta G°$ values are listed in Table 17.4. A large *negative* value of $\Delta G°$ means that K for the reaction is a large *positive* value, and therefore the reaction from left

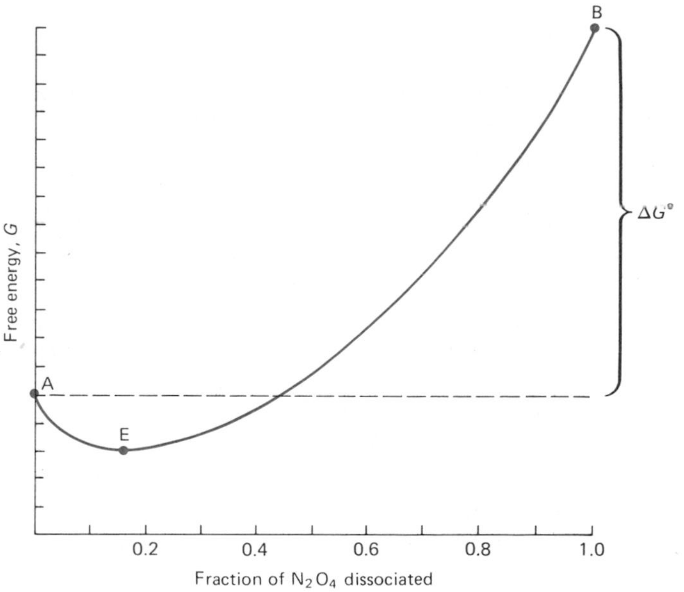

Figure 17.2 Free energy of a system that contains the equivalent of 1 mol of N_2O_4 as the reaction $N_2D_4(g) \rightleftharpoons NO_2(g)$ occurs (25°C and 1 atm).

Table 17.4 Values of K corresponding to ΔG° values according to the equation $\Delta G^\circ = -2.303RT \log K$

ΔG° (kJ)	K
−200	1.1×10^{35}
−100	3.3×10^{17}
−50	5.7×10^{8}
−25	2.4×10^{4}
−5	7.5
0	1.0
+5	0.13
+25	4.2×10^{-5}
+50	1.7×10^{-9}
+100	3.0×10^{-18}
+200	9.3×10^{-36}

to right will go virtually to completion. On the other hand, if ΔG° is a large positive value, K will be extremely small, thus indicating that the reverse reaction, from right to left, will go virtually to completion. Only if the value of ΔG° is neither very large nor very small (see Table 17.4) will the value of K indicate a situation in which the reaction will not go essentially to completion in one direction or the other.

17.8 ELECTROCHEMICAL MEASUREMENT OF ΔG° AND ΔS°

Any spontaneous reaction has the capacity to do work. If work must be done on a system to bring about a chemical change, that change is not spontaneous. For a spontaneous reaction conducted at constant temperature and pressure, the decrease in Gibbs free energy is a measure of the maximum *useful* work that the system can do. The word *useful* is used to indicate that ΔG does not include pressure–volume work.

Some reactions are accompanied by an increase or decrease in volume (because gases are produced or used). If such reactions are run at constant pressure, work is done by the system to push back the atmosphere (in the case of an expanding volume) or work is done on the system by the surroundings (if the volume of the system contracts as the reaction proceeds). Such pressure–volume work must be done if the reaction is to take place at constant pressure, and, therefore, work of this type is not available for any other purpose. The change in free energy measures the maximum amount of work, other than pressure–volume work, that can be derived from a reaction under conditions of constant pressure. Since only a spontaneous reaction has the capacity to do work, ΔG is a measure of spontaneity.

One way to obtain useful work from a reaction is to make it produce electrical work by having it serve as the cell reaction of a voltaic cell. We can use the value of the reversible emf of the cell (which can be derived from standard electrode potentials) to calculate the maximum useful work that the cell can theoretically produce (Section 16.6). The maximum quantity of electrical work that can be obtained from the reaction, which is the *decrease* in Gibbs free energy for the reaction, can be calculated from

$$\Delta G^\circ = -nF\mathscr{E}^\circ$$

where n is the number of moles of electrons transferred in the reaction (or the number of faradays of charge), F is the value of the faraday, and $\mathscr{E}^\circ$ is the reversible standard emf of the cell in volts. If 96,487 coul is used as the value of the faraday, the answer is expressed in volt–coulombs, which are joules.

Example 17.5 (a) Use standard electrode potentials to determine the value of ΔG° for the following reaction at 25°C.

$$2Ag(s) + Cl_2(g) \rightarrow 2AgCl(s)$$

(b) The standard enthalpy of formation, ΔH_f°, of AgCl(s) is −127.0 kJ/mol. What is ΔS° for the reaction?

Solution (a) The reversible emf of the cell is determined from the standard electrode potentials.

anode:	$2Ag(s) + 2Cl^-(aq) \rightarrow 2AgCl(s) + 2e^-$	$\mathscr{E}^\circ_{ox} = -0.222$ V
cathode:	$2e^- + Cl_2(g) \rightarrow 2Cl^-(aq)$	$\mathscr{E}^\circ = +1.359$ V
cell reaction:	$2Ag(s) + Cl_2(g) \rightarrow 2AgCl(s)$	$\mathscr{E}^\circ = +1.137$ V

Since 2F of charge are transferred,

$$\begin{aligned}\Delta G^\circ &= -nF\mathscr{E}^\circ \\ &= -2(96{,}487 \text{ coul})(+1.137 \text{ V}) \\ &= -219{,}400 \text{ J} = -219.4 \text{ kJ}\end{aligned}$$

(b) For the reaction as written,

$$\begin{aligned}\Delta H^\circ &= 2\Delta H_f^\circ(\text{AgCl}) \\ &= 2(-127.0 \text{ kJ}) = -254.0 \text{ kJ}\end{aligned}$$

Therefore,

$$\begin{aligned}\Delta G^\circ &= \Delta H^\circ - T\Delta S^\circ \\ -219.4 \text{ kJ} &= -254.0 \text{ kJ} - T\Delta S^\circ \\ T\Delta S^\circ &= -34.6 \text{ kJ} \\ \Delta S^\circ &= -34{,}600 \text{ J}/298 \text{ K} = -116 \text{ J/K}\end{aligned}$$

Example 17.6 (a) What is ΔG° (in kJ) for the reaction

$$Cl_2(g) + 2I^-(aq) \rightarrow 2Cl^-(aq) + I_2(s)$$

For I^-(aq), $\Delta G_f^\circ = -51.7$ kJ/mol; for Cl^-(aq), $\Delta G_f^\circ = -131.2$ kJ/mole. (b) Use the value of ΔG° to calculate $\mathscr{E}^\circ$ for a cell utilizing this reaction as the cell reaction.

Solution

(a)

$$\begin{aligned}\Delta G^\circ &= 2\Delta G_f^\circ(Cl^-) - 2\Delta G_f^\circ(I^-)\\ &= 2(-131.2 \text{ kJ}) - 2(-51.7 \text{ kJ})\\ &= -159.0 \text{ kJ}\end{aligned}$$

(b)

$$\Delta G^\circ = -n + \mathscr{E}^\circ$$

$$\mathscr{E}^\circ = \frac{\Delta G^\circ}{-nF}$$

$$\mathscr{E}^\circ = \frac{-159{,}000 \text{ J}}{-2(96{,}500 \text{ coul})} = +0.824 \text{ J/coul}$$

$$\mathscr{E}^\circ = +0.824 \text{ V}$$

Note that 1 J = 1 V coul, therefore, 1 J/coul = 1 V.
The value of $\mathscr{E}^\circ$ derived from standard electrode potentials is +0.82 V.

Example 17.7 Using electrochemical data, calculate K for the reaction

$$Fe^{2+}(aq) + Ag^+(aq) \rightleftharpoons Fe^{3+}(aq) + Ag(s)$$

Solution The half reactions are

$$\begin{aligned} Fe^{2+} &\rightarrow Fe^{3+} + e^- \qquad & \mathscr{E}^\circ_{ox} &= -0.771 \text{ V}\\ e^- + Ag^+ &\rightarrow Ag(s) & \mathscr{E}^\circ &= +0.799 \text{ V}\end{aligned}$$

Therefore, $\mathscr{E}^\circ = +0.028$ V and 1 F of charge is transferred. The standard change in free energy is

$$\begin{aligned}\Delta G^\circ &= -nF\mathscr{E}^\circ\\ &= -1(96{,}500 \text{ coul})(+0.028 \text{ V})\\ &= -2700 \text{ J} = -2.7 \text{ kJ}\end{aligned}$$

The equilibrium constant may be calculated from the equation

$$\Delta G^\circ = -(5.71 \text{ kJ/mol}) \log K$$

$$\log K = \frac{-2.70 \text{ kJ/mol}}{-5.71 \text{ kJ/mol}} = 0.473$$

$$K = 3.0$$

Since the standard states and activities of solutes in aqueous solutions are defined in terms of ideal concentrations, K obtained in this calculation pertains to activities so defined.

KELVIN AND THE AGE OF THE EARTH

"When you can measure what you are talking about and express it in numbers, you know something about it . . . ," said William Thomson, Lord Kelvin, in 1891. To be sure, the more mathematical science has become, the more rapidly it has advanced. Few would disagree with Kelvin's statement. The descriptive, nonmathematical aspects of science, however, are also of importance—particularly to a young developing field. At least one time, Kelvin's condescending attitude toward descriptive science led him into error.

"British popular geology at the present time is in direct opposition to the principles of Natural Philosophy," said Kelvin in 1866. At the time, physicists and geologists were engaged in a controversy about the age of the earth—an argument that lasted from roughly 1860 to 1900.

Geologists estimated that the earth was from 200 to 400 million years old. They based their estimates on the thickness of rock strata and the length of time that they thought it would take for such strata to form.

Kelvin, over a period of years, finally arrived at an estimate of about 20 million years for the age of the earth. He based his estimate on the rate of cooling of the earth and the rate at which the sun is dissipating its heat to warm the earth. By using thermodynamic principles, Kelvin demonstrated mathematically that the earth could not be as old as the geologists thought it was.

When Darwin developed his theory of evolution, he relied on the estimates of the geologists. If the earth was only as old as Kelvin later said it was, Darwin's theory of natural selection was wrong. To Kelvin, who was given to making pronouncements, the age of the earth demonstrated the "utter futility" of Darwin's theory.

At the time of the controversy, nothing was known about atomic energy, which is now believed to be the source of the sun's heat. Since Kelvin thought that the sun was an incandescent mass that is rapidly radiating its energy, his estimate of the earth's age is far too low. His work, however, was entirely correct in terms of the science of his age. Today, the earth is believed to be from 4,000 to 5,000 million years old. As it turned out, the geologists were nearer to being right than Kelvin.

PROBLEMS

17.1 The combustion of 1.000 g of ethyl alcohol, $C_2H_5OH(l)$, in a bomb calorimeter (at constant volume) evolves 29.62 kJ of heat at 25°C. The products of the combustion are $CO_2(g)$ and $H_2O(l)$. The molecular weight of ethyl alcohol is 46.06. (a) What is $\Delta E°$ for the combustion of 1 mol of ethyl alcohol? (b) Write the chemical equation for the reaction. What is $\Delta H°$ for the combustion? (c) Use values from Table 3.1 to calculate the enthalpy of formation of ethyl alcohol.

17.2 The combustion of 1.000 g of thiourea, $CS(NH_2)_2(s)$, in a bomb calorimeter (at constant volume) evolves 15.37 kJ of heat at 25°C. The products of the combustion are $CO_2(g)$, $SO_2(g)$, $N_2(g)$, and $H_2O(l)$. The molecular weight of thiourea is 76.12. (a) What is $\Delta E°$ for the combustion of 1 mol of thiourea? (b) Write the chemical equation for the reaction. What is $\Delta H°$ for the combustion? (c) Use values from Table 3.1 to calculate the enthalpy of formation of thiourea.

17.3 The combustion of 1.000 g of cyclohexane, $C_6H_{12}(l)$, in a bomb calorimeter (at constant volume) evolves 46.48 kJ of heat. The products of the combustion are $CO_2(g)$ and $H_2O(l)$. The molecular weight of cyclohexane is 84.16. (a) What is $\Delta E°$ for the combustion of 1 mol of cyclohexane? (b) Write the chemical equation for the reaction. What is $\Delta H°$ for the combustion? (c) Use values from Table 3.1 to calculate the enthalpy of formation of cyclohexane.

17.4 Calculate $\Delta E°$ for the combustion of octane, $C_8H_{18}(l)$, to $CO_2(g)$ and $H_2O(l)$ at 25°C. The value of $\Delta H°$ for this reaction is −5470.71 kJ/mol of octane.

17.5 Calculate $\Delta E°$ for the combustion of ethylene, $C_2H_4(g)$, to $CO_2(g)$ and $H_2O(l)$ at 25°C. The value of $\Delta H°$ for this reaction is −1410.8 kJ/mol of ethylene.

17.6 For the reaction

$$3NO_2(g) + H_2O(l) \rightarrow 2HNO_3(l) + NO(g)$$

$\Delta H°$ is −71.53 kJ. What is $\Delta E°$?

17.7 For the reaction

$$Ca_3P_2(s) + 6H_2O(l) \rightarrow 3Ca(OH)_2(s) + 2PH_3(g)$$

$\Delta H°$ is −721.70 kJ. What is $\Delta E°$?

17.8 For the reaction

$$CaNCN(s) + 3H_2O(l) \rightarrow CaCO_3(s) + 2NH_3(g)$$

$\Delta E°$ is −261.75 kJ. Use values from Table 3.1 to calculate the enthalpy of formation of CaNCN(s).

17.9 For the reaction

$$NH_4NO_3(s) \rightarrow N_2O(g) + 2H_2O(l)$$

$\Delta E°$ is −127.49 kJ. Use values from Table 3.1 to calculate the enthalpy of formation of $N_2O(g)$.

17.10 Use the data given in Table 17.2 to calculate $\Delta G°$ for the following proposed reactions

$$SO_2(g) + H_2(g) \rightarrow H_2S(g) + O_2(g)$$

$$SO_2(g) + 3H_2(g) \rightarrow H_2S(g) + 2H_2O(l)$$

What would you predict the products of the reaction of $H_2S(g)$ and $H_2(g)$ at 25°C to be?

17.11 A chemist claims that the following reaction occurs at 25°C.

$$SF_6(g) + 8HI(g) \rightarrow H_2S(g) + 6HF(g) + 4I_2(g)$$

Is it theoretically possible? Use values from Table 17.2 plus the fact that ΔG_f° for $SF_6(g)$ is −992 kJ/mol to calculate ΔG° for the reaction.

17.12 Will both $BF_3(g)$ and $BCl_3(l)$ hydrolyze at 25°C according to the following equation (where X is F or Cl)?

$$BX_3 + 3H_2O(l) \rightarrow H_3BO_3(aq) + 3HX(aq)$$

The pertinent ΔG_f° values are: $H_3BO_3(aq)$, −963.32 kJ/mol; HF(aq), −276.48 kJ/mol; HCl(aq), −131.17 kJ/mol; $H_2O(l)$, −237.19 kJ/mol; $BF_3(g)$, −1093.28 kJ/mol; $BCl_3(l)$, −379.07 kJ/mol.

17.13 For oxygen difluoride, $OF_2(g)$, ΔG_f° is +40.6 kJ/mol. (a) Is the preparation of $OF_2(g)$ from its elements at 25°C a spontaneous reaction? (b) For ozone, $O_3(g)$, ΔG_f° is +163.43 kJ/mol. Is it theoretically possible to prepare $OF_2(g)$ at 25°C by the reaction

$$3F_2(g) + O_3(g) \rightarrow 3OF_2(g)$$

17.14 The signs of the standard free energies of formation of all the oxides of nitrogen are positive. What implications does this fact have regarding (a) the preparation of these oxides from the elements at 25°C, and (b) the products of the combustion of a nitrogen-containing compound in oxygen?

17.15 For the reaction

$$HCOOH(l) \rightarrow CO_2(g) + H_2O(l)$$

ΔH° is +15.79 kJ and ΔS° is +215.27 J/K. Calculate ΔG°. Is the decomposition of formic acid, HCOOH(l), spontaneous at 25°C?

17.16 For the reaction

$$2CHCl_3(l) + O_2(g) \rightarrow 2COCl_2(g) + 2HCl(g)$$

ΔH° is −366 kJ and ΔS° is +340 J/K. Calculate ΔG°. Is the formation of the poisonous gas phosgene, $COCl_2$, from chloroform, $CHCl_3$, and oxygen a spontaneous reaction at 25°C?

17.17 For the reaction

$$C_2H_4(g) + H_2(g) \rightarrow C_2H_6(g)$$

ΔH° is −136.98 kJ. (a) Use values from Table 17.2 to calculate ΔG° for the reaction. (b) Use ΔG° and ΔH° to calculate ΔS° for the reaction. (c) Calculate the value of ΔS° for the reaction by using S° values from Table 17.3.

17.18 The standard enthalpy of formation, ΔH_f°, *of* $CS_2(l)$ is +87.9 kJ/mol. The absolute entropy, S°, of C(graphite) is 5.69 J/K mol, of S(rhombic) is 31.9 J/K mol, and of $CS_2(l)$ is 151.0 J/K mol. Calculate the standard free energy of formation, ΔG_f°, of $CS_2(l)$.

17.19 For the reaction

$$NH_3(g) + HCl(g) \rightarrow NH_4Cl(s)$$

ΔH° is −176.9 kJ. (a) The absolute entropy, S°, of $NH_4Cl(s)$ is 94.56 J/K mol. Use this value and others from Table 17.3 to find ΔS° for the reaction. (b) What is ΔG° for the reaction at 25°C?

17.20 What is the standard free energy of formation, ΔG_f°, of $PH_3(g)$? For $PH_3(g)$: ΔH_f° is +9.25 kJ/mol, and S° is 210.0 J/K mol. For the standard state of $P_4(s)$, S° is 177.6 J/K mol, and for $H_2(g)$, S° is 130.5 J/K mol.

17.21 For the reaction

$$CO(g) + 2H_2(g) \rightarrow CH_3OH(l)$$

(a) Calculate ΔH° using values from Table 3.1. (b) The absolute entropy, S°, of $CH_3OH(l)$ is 126.78 J/K mol. Use this fact and values from Table 17.3 to calculate ΔS° for the reaction. (c) Use ΔH° and ΔS° to calculate ΔG° for the reaction. (d) Use the value of ΔG° for the reaction and the value of ΔG_f° for CO(g) from Table 17.2 to calculate the standard free energy of formation of $CH_3OH(l)$.

17.22 Use the reaction

$$SO_2(g) + Cl_2(g) \rightarrow SO_2Cl_2(l)$$

to find the absolute entropy, S°, of $SO_2Cl_2(l)$. For $SO_2Cl_2(l)$, ΔH_f° is −389.1 kJ/mol, and ΔG_f° is −313.8 kJ/mol. Thermodynamic values for $SO_2(g)$ and $Cl_2(g)$ are found in Tables 3.1, 17.2, and 17.3.

17.23 For $HgBr_2(s)$, ΔH_f° is −169.5 kJ/mol and ΔG_f° is −162.3 kJ/mol. The absolute entropies of Hg(l) and $Br_2(l)$ are given in Table 17.3. What is S° for $HgBr_2(s)$?

17.24 Graphite is the standard state of carbon, and S° for graphite is 5.694 J/K mol. For diamond, ΔH_f° is +1.895 kJ/mol, and ΔG_f° is +2.866 kJ/mol. What is the absolute entropy, S°, of diamond? Which form of carbon is the more ordered?

17.25 The combustion of cyanogen, $C_2N_2(g)$, in oxygen forms $CO_2(g)$ and $N_2(g)$ and produces an extremely hot flame. For $C_2N_2(g)$, ΔG_f° is +296.27 kJ/mol and S° is 242.09 J/K mol. Use these data together with values from Tables 17.2 and 17.3 to calculate ΔH° for the reaction.

17.26 The value of ΔG_f° for $ZnCO_3(s)$ is −731.36 kJ/mol. Use this fact plus values from Table 17.2 to determine K_p for the reaction

$$ZnCO_3(s) \rightleftharpoons ZnO(s) + CO_2(g)$$

17.27 For urea, $CO(NH_2)_2(s)$, ΔG_f° is −197.15 kJ/mol. Use this value together with values from Table 17.2 to determine K_p for the reaction

$$CO_2(g) + 2NH_3(g) \rightleftharpoons H_2O(g) + CO(NH_2)_2(s)$$

17.28 For $SO_3(g)$, ΔG_f° is −370.37 kJ/mol. Use this value together with values from Table 17.2 to determine K_p for the reaction

$$SO_2(g) + NO_2(g) \rightleftharpoons SO_3(g) + NO(g)$$

17.29 For the reaction

$$Br_2(l) + Cl_2(g) \rightleftharpoons 2BrCl(g)$$

ΔG° is −1.80 kJ/mol. What is K_p for the reaction?

17.30 Use values given in Table 17.2 to calculate K_p for the equilibrium

$$CaCO_3(s) \rightleftharpoons CaO(s) + CO_2(g)$$

17.31 Use values given in Table 17.2 to calculate K_p for the equilibrium

$$BaO(s) + CO_2(g) \rightleftharpoons BaCO_3(s)$$

17.32 Given that the value of ΔG_f° for $Br_2(g)$ is +3.14 kJ/mol, what is the vapor pressure of Br_2(l) at 25°C? For Br_2(l), $\Delta G_f^\circ = 0.0$.

17.33 At 25°C, $K_p = 0.108$ for the reaction

$$NH_4HS(s) \rightleftharpoons NH_3(g) + H_2S(g)$$

Values of ΔG_f° for NH_3(g) and H_2S(g) are given in Table 17.2. What is the value of ΔG_f° for NH_4HS(s)?

17.34 For the equilibrium

$$PCl_5(g) \rightleftharpoons PCl_3(g) + Cl_2(g)$$

at 25°C, $K_p = 1.8 \times 10^{-7}$. What is ΔG° for the reaction?

17.35 When the substances of the equation

$$I_2(s) + Br_2(l) \rightleftharpoons 2IBr(g)$$

are at equilibrium at 25°C, the partial pressure of IBr(g) is 0.214 atm. The S° of IBr(g) is 258.6 J/K mol. Other absolute entropies may be found in Table 17.3. Calculate a value for the enthalpy change, ΔH, of the reaction.

17.36 (a) Use electrode potentials (Appendix E) to calculate the emf of a standard cell utilizing the reaction

$$Cl_2(g) + 2Br^-(aq) \rightarrow 2Cl^-(aq) + Br_2(l)$$

(b) What is ΔG° for the reaction? (c) If ΔH° for the reaction is −93.09 kJ, what is ΔS°?

17.37 (a) Use electrode potentials (Appendix E) to calculate the emf of a standard cell utilizing the reaction

$$2H_2O_2(aq) \rightarrow O_2(g) + 2H_2O$$

(b) What is ΔG° for the reaction? (c) The ΔH_f° of H_2O(l) is −285.85 kJ/mol, and the ΔH_f° of H_2O_2(aq) is −191.13 kJ/mol. What is ΔH° for the reaction? (d) What is ΔS° for the reaction?

17.38 (a) Use electrode potentials (Appendix E) to calculate the emf of a standard cell utilizing the reaction

$$I_2(s) + H_2S(aq) \rightarrow S(\text{rhombic}) + 2H^+(aq) + 2I^-(aq)$$

(b) What is ΔG° for the reaction? (c) The absolute entropy of H_2S(aq) is 122.2 J/K mol, that of I^-(aq) is 26.14 cal/K mol [or 109.37 J/K mol, and that of H^+(aq) is 0.0 J/K mol; other values can be found in Table 17.3. What is ΔS° for the reaction? (d) What is ΔH° for the reaction?

17.39 For the half-reaction

$$XeF_2(aq) + 2H^+(aq) + 2e^- \rightarrow Xe(g) + 2HF(aq)$$

$\mathscr{E}^\circ$ has been reported to equal +2.64 V. (a) Calculate ΔG° for the reaction

$$XeF_2(aq) + H_2(g) \rightarrow Xe(g) + 2HF(aq)$$

(b) The free energy of formation, ΔG_f°, of HF(aq) is 276.48 kJ/mol. What is ΔG_f° of XeF_2(aq)?

17.40 Given the following electrode potentials

$$H_3BO_3(s) + 3H^+(aq) + 3e^- \rightarrow B(s) + 3H_2O \qquad \mathscr{E}^\circ = -0.869 \text{ V}$$

$$4H^+(aq) + O_2(g) + 4e^- \rightarrow 2H_2O \qquad \mathscr{E}^\circ = +1.229 \text{ V}$$

Calculate the value of ΔG_f° for H_3BO_3(s). For H_2O(l), ΔG_f° is −237.19 kJ/mol.

17.41 Perbromates have only recently been prepared. For the half-reaction,

$$BrO_4^-(aq) + 8H^+(aq) + 7e^- \rightarrow \tfrac{1}{2}Br_2(l) + 4H_2O$$

$\mathscr{E}°$ is reported to be +1.59 V. (a) Calculate ΔG_f° for $BrO_4^-(aq)$. For $H_2O(l)$, ΔG_f° is −237.19 kJ/mol, and for $H^+(aq)$, ΔG_f° is 0.0. (b) Calculate $\mathscr{E}°$ for the half-reaction

$$BrO_4^-(aq) + 2H^+(aq) + 2e^- \rightarrow BrO_3^-(aq) + H_2O$$

The value of ΔG_f° for $BrO_3^-(aq)$ is +21.71 kJ/mol. (c) What oxidizing agent, listed in Appendix E, *should* be capable of oxidizing $BrO_3^-(aq)$ to $BrO_4^-(aq)$?

17.42 Consider the standard cell

$$Cu(s)|CuCl(s)|Cl^-(aq)|Cl_2(g)\,|\,Pt(s)$$

(a) Write the equation for the cell reaction. (b) Calculate $\mathscr{E}°$ for the half reaction

$$e^- + CuCl(s) \rightarrow Cu(s) + Cl^-(aq)$$

The free energy of formation of CuCl(s) is −117.99 kJ/mol, and $\mathscr{E}°$ for the Cl_2/Cl^- electrode is +1.360 V.

17.43 The following half-reactions take place in a standard cell

$$Te(s) + 2H_2O \rightarrow TeO_2(s) + 4H^+(aq) + 4e^- \qquad \mathscr{E}°_{ox} = ?$$

$$4e^- + 4H^+(aq) + O_2(g) \rightarrow 2H_2O \qquad \mathscr{E}° = +1.229 \text{ V}$$

(a) Write the equation for the cell reaction. (b) The free energy of formation of $TeO_2(s)$ is −270.29 kJ/mol. What is $\mathscr{E}°$ for the TeO_2/Te electrode?

17.44 For the reaction

$$2Li(s) + 2H^+(aq) \rightarrow 2Li^+(aq) + H_2(g)$$

$\Delta H°$ is −556.9 kJ. The value of $S°$ for $Li^+(aq)$ is 14.2 J/K mol, and that of $H^+(aq)$ is 0.0. Other values can be found in Table 17.3. Calculate (a) $\Delta S°$ and (b) ΔG for the reaction. (c) Determine $\mathscr{E}°$ for the half-reaction

$$Li^+(aq) + e^- \rightarrow Li(s)$$

17.45 For the reaction

$$2La(s) + 6H^+(aq) \rightarrow 2La^{3+}(aq) + 3H_2(g)$$

$\Delta H°$ is −1474.4 kJ. The value of $S°$ for $La^{3+}(aq)$ is −163.2 J/K mol, and that for $H^+(aq)$ is 0.0; other values can be found in Table 17.3. Calculate (a) $\Delta S°$ and (b) $\Delta G°$ for the reaction. (c) Determine $\mathscr{E}°$ for the half-reaction

$$La^{3+}(aq) + 3e^- \rightarrow La(s)$$

17.46 Use electrode potentials (Appendix E) to calculate the equilibrium constant for the reaction

$$Ni(s) + Sn^{2+}(aq) \rightleftharpoons Ni^{2+}(aq) + Sn(s)$$

17.47 Use electrode potentials (given in Appendix E) to calculate the equilibrium constant at 25°C for the reaction

$$3Au^+(aq) \rightleftharpoons 2Au(s) + Au^{3+}(aq)$$

17.48 Use standard electrode potentials (given in Appendix E) to determine the equilibrium constant at 25°C for the reaction

$$Cl_2(g) + H_2O \rightleftharpoons H^+(aq) + Cl^-(aq) + HOCl(aq)$$

17.49 Given the following

$$2H_2O + 2e^- \rightarrow H_2(g) + 2OH^-(aq) \quad \mathscr{E}° = -0.828 \text{ V}$$

calculate the value of the water constant, K_w.

$$H_2O \rightleftharpoons H^+(aq) + OH^-(aq)$$

17.50 Use standard electrode potentials (given in Appendix E) to calculate the equilibrium constant at 25°C for the reaction

$$4H^+(aq) + 4Br^-(aq) + O_2(g) \rightleftharpoons 2Br_2(l) + 2H_2O$$

17.51 Calculate $\mathscr{E}°$ for the half-reaction

$$PbBr_2(s) + 2e^- \rightarrow Pb(s) + 2Br^-(aq)$$

from the following data:

$$PbBr_2(s) \rightleftharpoons Pb^{2+}(aq) + 2Br^-(aq) \quad K = 4.60 \times 10^{-6}$$

$$Pb^{2+}(aq) + 2e^- \rightarrow Pb(s) \quad \mathscr{E}° = -0.126 \text{ V}$$

Chapter 18

The Nonmetals

In his book *Traité Elémentaire de Chimie* (*Elementary Treatise on Chemistry*) published in 1789, Lavoisier proposed the modern definition of a chemical element and correctly identified 23 elements. During the next forty years, 31 elements were discovered—principally as the result of the analysis of minerals. By the end of 1830, 54 elements were known.

The search for new elements was given fresh stimulation in 1859 by Bunsen and Kirchhoff, who introduced the spectroscope as a means of identifying elements. As a result, approximately 60 elements were known by the time that Mendeleev and Lothar Meyer published their versions of the periodic table (in 1869 and 1870, respectively). Mendeleev left blanks for undiscovered elements in his table, and he predicted the properties of these unknown substances. Several elements were discovered as the result of the efforts of chemists to fill in these blanks.

The existence of the noble gases was unforeseen by Mendeleev. After the fortuitous discovery of argon by Rayleigh in 1892, five of the noble gases were isolated and identified in the years 1892–98. By 1925, 88 naturally-occurring elements were known. Elements that do not occur in nature (or occur in extremely small amounts) have been made by nuclear processes. Work of this type (from 1940 to the present) has expanded the number of known elements to its present value, 106.

18.1 THE GROUP VII A ELEMENTS

The elements of group VII A, fluorine, chlorine, bromine, iodine, and astatine, are called the **halogens.** The name "halogen" is derived from Greek and means "salt former." These elements, with the exception of astatine, occur extensively in nature in the form of halide salts. Astatine probably occurs in nature, in extremely small amounts, as a short-lived intermediate of natural radioactive decay processes.

Each halogen atom has one electron less than the noble gas that follows it in the periodic classification. There is a marked tendency on the part of a halogen atom to attain a noble-gas configuration by the formation of a uninegative ion or a single covalent bond. Positive oxidation states exist for all the elements except fluorine.

Under ordinary conditions the halogens exist as diatomic molecules with a single covalent bond joining the atoms. The molecules are held together in the solid and liquid state by London forces. Of all the halogen molecules, I_2 is the largest, has the most electrons, and is the most polarizable. It is not surprising, therefore, that the intermolecular attractions between I_2 molecules are the strongest and that iodine has the highest melting point and boiling point. At ordinary temperatures and pressures, I_2 is a solid, Br_2 is a liquid, and Cl_2 and F_2 are gases.

Each halogen is the most reactive nonmetal of its period, and fluorine is the most reactive of all the nonmetals. Fluorine has the highest electronegativity of any element and is one of the strongest oxidizing agents known. The electronegativity of the halogens decreases in the order $F > Cl > Br > I$, and the oxidizing power of the halogens decreases in the same order.

The relative oxidizing ability of the halogens may be observed in displacement reactions. Thus, fluorine can displace chlorine, bromine, and iodine from their salts; chlorine can displace bromine and iodine from their salts; and bromine can displace iodine from iodides.

$$F_2(g) + 2NaCl(s) \rightarrow 2NaF(s) + Cl_2(g)$$
$$Cl_2(g) + 2Br^-(aq) \rightarrow 2Cl^-(aq) + Br_2(l)$$
$$Br_2(l) + 2I^-(aq) \rightarrow 2Br^-(aq) + I_2(s)$$

Since fluorine actively oxidizes water (producing O_2), displacement reactions involving F_2 cannot be run in water solution.

Fluorine must be prepared by an electrochemical process because no suitable chemical agent is sufficiently powerful to oxidize fluoride ion to fluorine. Furthermore, since the oxidation of water is easier to accomplish than the oxidation of the fluoride ion, the electrolysis must be carried out under anhydrous conditions. In actual practice, a solution of potassium fluoride in anhydrous hydrogen fluoride is electrolyzed. Pure HF does not conduct electric current. The KF reacts with HF to produce ions (K^+ and HF_2^-) that can act as charge carriers. The HF_2^- ion is formed from an F^- ion hydrogen-bonded to an HF molecule ($F—H\cdots F^-$). This hydrogen bond is so strong that the H atom is exactly midway between the two F atoms. The hydrogen fluoride used in the electrolysis is commercially derived from fluorospar, CaF_2.

$$2KHF_2(l) \xrightarrow[\text{heat}]{\text{electrolysis}} H_2(g) + F_2(g) + 2KF(l)$$

The principal industrial source of chlorine is the electrolysis of aqueous sodium chloride, from which sodium hydroxide and hydrogen are also products.

$$2Na^+(aq) + 2Cl^-(aq) + 2H_2O \xrightarrow{\text{electrolysis}} H_2(g) + Cl_2(g) + 2Na^+(aq) + 2OH^-(aq)$$

Chlorine is also obtained, as a by-product, from the industrial processes in which the reactive metals Na, Ca, and Mg are prepared. In each of these processes an anhydrous molten chloride is electrolyzed and Cl_2 gas is produced at the anode. For example,

$$2NaCl(l) \xrightarrow{\text{electrolysis}} 2Na(l) + Cl_2(g)$$

Bromine is commercially prepared by the oxidation of the bromide ion of salt brines or sea water with chlorine as the oxidizing agent.

$$Cl_2(g) + 2Br^-(aq) \rightarrow 2Cl^-(aq) + Br_2(l)$$

The liberated Br_2 is removed from the solution by a stream of air and, in subsequent steps, collected from the air and purified.

The principal commercial source of iodine is the iodide ion found in oil-well brines. Free iodine is obtained by chlorine displacement.

$$Cl_2(g) + 2I^-(aq) \rightarrow 2Cl^-(aq) + I_2(s)$$

With the exception of fluorine (which must be prepared electrochemically), the free halogens are usually prepared in the laboratory by the action of oxidizing agents on aqueous solutions of the hydrogen halides or on solutions containing the sodium halides and sulfuric acid. In actual practice, $KMnO_4$, $K_2Cr_2O_7$, PbO_2, and MnO_2 are frequently used to prepare the free halogens from halide ions.

Table 18.1 Some reactions of the halogens ($X_2 = F_2, Cl_2, Br_2$, or I_2)

GENERAL REACTION	REMARKS
$nX_2 + 2M \rightarrow 2MX_n$	F_2, Cl_2 with practically all metals; Br_2, I_2 with all except noble metals
$X_2 + H_2 \rightarrow 2HX$	
$3X_2 + 2P \rightarrow 2PX_3$	with excess P; similar reactions with As, Sb, and Bi
$5X_2 + 2P \rightarrow 2PX_5$	with excess X_2, but not with I_2; SbF_5, $SbCl_5$, AsF_5, $AsCl_5$, and BiF_5 may be similarly prepared
$X_2 + 2S \rightarrow S_2X_2$	with Cl_2, Br_2
$X_2 + H_2O \rightarrow H^+ + X^- + HOX$	not with F_2
$2X_2 + 2H_2O \rightarrow 4H^+ + 4X^- + O_2$	F_2 rapidly; Cl_2, Br_2 slowly in sunlight
$X_2 + H_2S \rightarrow 2HX + S$	
$X_2 + CO \rightarrow COX_2$	Cl_2, Br_2
$X_2 + SO_2 \rightarrow SO_2X_2$	F_2, Cl_2
$X_2 + 2X'^- \rightarrow X_2' + 2X^-$	$F_2 > Cl_2 > Br_2 > I_2$
$X_2 + X_2' \rightarrow 2XX'$	formation of the interhalogen compounds (all except IF)

$$MnO_2(s) + 4H^+(aq) + 2Cl^-(aq) \rightarrow Mn^{2+}(aq) + Cl_2(g) + 2H_2O$$
$$2MnO_4^-(aq) + 16H^+(aq) + 10Br^-(aq) \rightarrow 2Mn^{2+}(aq) + 5Br_2(l) + 8H_2O$$
$$Cr_2O_7^{2-}(aq) + 14H^+(aq) + 6I^-(aq) \rightarrow 2Cr^{3+}(aq) + 3I_2(s) + 7H_2O$$

Some of the reactions of the halogens are summarized in Table 18.1.

18.2 COMPOUNDS OF THE HALOGENS

Each of the hydrogen halides may be prepared by the direct reaction of hydrogen with the corresponding free halogen. The vigor of the reaction decreases markedly from fluorine to iodine.

Both HF and HCl are made industrially by the action of warm concentrated sulfuric acid on the corresponding naturally occurring halide, CaF_2 and NaCl.

$$CaF_2(s) + H_2SO_4(l) \rightarrow CaSO_4(s) + 2HF(g)$$
$$NaCl(s) + H_2SO_4(l) \rightarrow NaHSO_4(s) + HCl(g)$$

All hydrogen halides are colorless gases at room temperature. Sulfuric acid, on the other hand, is a high-boiling liquid. Thus, the foregoing reactions are examples of a general method for the preparation of a volatile acid from its salts by means of a nonvolatile acid. At higher temperatures (ca. 500°C), further reaction occurs between $NaHSO_4$ and NaCl.

$$NaCl(s) + NaHSO_4(l) \rightarrow HCl(g) + Na_2SO_4(s)$$

Hydrogen bromide and hydrogen iodide cannot be made by the action of concentrated sulfuric acid on bromides and iodides because hot, concentrated sulfuric acid oxidizes these anions to the free halogens. The bromide and iodide ions are easier to oxidize than the fluoride and chloride ions.

$$2NaBr(s) + 2H_2SO_4(l) \rightarrow Br_2(g) + SO_2(g) + Na_2SO_4(s) + 2H_2O(g)$$

Since the iodide ion is a stronger reducing agent (more easily oxidized) than the bromide ion, S and H_2S, as well as SO_2, are obtained as reduction products from the reaction of NaI with hot concentrated sulfuric acid.

Pure HBr or HI can be obtained by the action of phosphoric acid on NaBr or NaI. Phosphoric acid is an essentially nonvolatile acid and is a poor oxidizing agent.

$$NaBr(s) + H_3PO_4(l) \rightarrow HBr(g) + NaH_2PO_4(s)$$
$$NaI(s) + H_3PO_4(l) \rightarrow HI(g) + NaH_2PO_4(s)$$

The hydrogen halides may also be prepared by the reaction of water on the appropriate phosphorus trihalide.

$$PX_3 + 3H_2O \rightarrow 3HX(g) + H_3PO_3(aq)$$

Hydrogen fluoride molecules associate with each other through hydrogen bonding. The vapor consists of aggregates up to $(HF)_6$ at temperatures near the boiling point (19.4°C) but is less highly associated at higher temperatures. Gaseous HCl, HBr, and HI consist of single molecules. Liquid HF and solid HF are more highly hydrogen bonded than gaseous HF, and the boiling point and melting point of HF are abnormally high in comparison to those of the other hydrogen halides.

All hydrogen halides are very soluble in water. Water solutions are called hydrohalic acids. For example, aqueous HI is called hydroiodic acid. The H—F bond is stronger than any other H—X bond; HF is a weak acid in water solution, whereas HCl, HBr, and HI are completely dissociated.

$$HF(aq) \rightleftharpoons H^+(aq) + F^-(aq)$$

The F^- ions from this dissociation are largely associated with HF molecules.

$$F^-(aq) + HF(aq) \rightleftharpoons HF_2^-(aq)$$

Concentrated HF solutions are more strongly ionic than dilute solutions and contain high concentrations of ions of the type HF_2^-, $H_2F_3^-$, and higher.

Hydrofluoric acid reacts with silica, SiO_2, and glass, which is made from silica.

$$SiO_2(s) + 6HF(aq) \rightarrow 2H^+(aq) + SiF_6^{2-}(aq) + 2H_2O$$

When warmed, the reaction is

$$SiO_2(g) + 4HF(aq) \rightarrow SiF_4(g) + 2H_2O(g)$$

For this reason, hydrofluoric acid must be stored in wax or plastic containers instead of glass bottles.

Table 18.2 Oxyhalogen acids

OXIDATION STATE OF THE HALOGEN	FORMULA OF ACID				NAME OF ACID	NAME OF ANION DERIVED FROM ACID
1+	HOF	HOCl	HOBr	HOI	hypohalous acid	hypohalite ion
3+	—	$HClO_2$	—	—	halous acid	halite ion
5+	—	$HClO_3^a$	$HBrO_3^a$	HIO_3^a	halic acid	halate ion
7+	—	$HClO_4^a$	$HBrO_4^a$	HIO_4, $H_4I_2O_9$, H_5IO_6	perhalic acid	perhalate ion

[a] Strong acids.

The fluorosilicate ion, SiF_6^{2-}, is an example of a large group of complex ions formed by the halide ions. Halo complexes are formed by most metals (with the notable exceptions of the group I A, group II A, and lanthanide metals) and with some nonmetals (for example, BF_4^-). The formulas of these complex ions are most commonly of the types $(MX_4)^{n-4}$ and $(MX_6)^{n-6}$, where n is the oxidation number of the central atom of the complex. Metal halides can be prepared by direct interaction of the elements, by the reactions of the hydrogen halides with hydroxides or oxides, and by the reactions of the hydrogen halides with carbonates.

$$K_2CO_3(s) + 2HF(l) \rightarrow 2KF(s) + CO_2(g) + H_2O$$

The halides are discussed in Section 8.10.

The oxyacids of the halogens are listed in Table 18.2. The only oxyacid of fluorine that has been prepared is hypofluorous acid, HOF, a thermally unstable compound. The acids of chlorine and their salts are the most important of these compounds.

Lewis formulas for the oxychlorine acids are shown in Figure 18.1. Removal of H^+ from each of these structures gives the electronic formula of the corresponding anion.

18.3 THE GROUP VI A ELEMENTS

Group VI A includes oxygen, sulfur, selenium, tellurium, and polonium. Polonium is a metal. Since all the isotopes of Po are intensely radioactive, the study of the chemistry of this element is difficult.

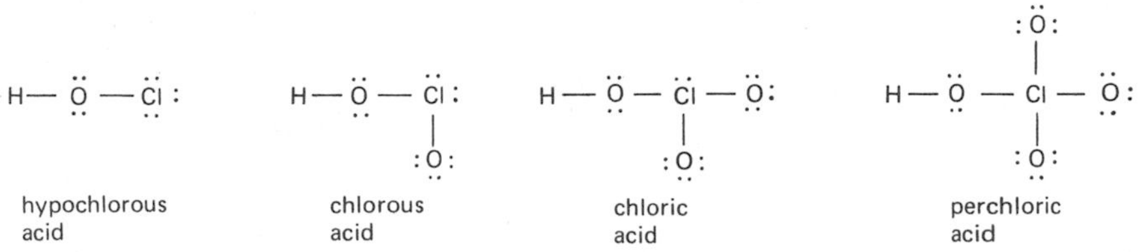

Figure 18.1 Lewis formulas for the oxychlorine acids.

Each element of the group has six electrons in its valence level. The nonmetals, therefore, can attain noble-gas configurations by forming 2− ions (for example, O^{2-} and S^{2-}) or two electron-pair covalent bonds (for example, H_2O and H_2S).

Sulfur, selenium, and tellurium exist in positive oxidation states in compounds in which they are combined with more electronegative elements (such as oxygen and the halogens). Oxygen is considered to have a positive oxidation number only in the few compounds that it forms with fluorine. For sulfur, selenium, and tellurium, the oxidation states of 4+ and 6+ are particularly important.

The existence of an element in more than one form in the same physical state is called **allotropy,** and the forms are called **allotropes.** All the members of group VI A exist in more than one allotropic form. Oxygen occurs in a triatomic form called ozone, O_3, in addition to the common diatomic form, O_2. The O_2 molecule is paramagnetic (Section 7.8, Figure 7.19). The O_3 molecule is diamagnetic and angular.

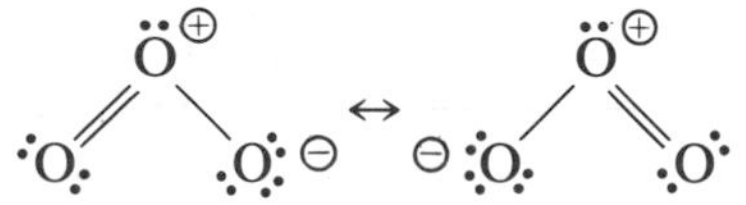

Ozone is produced by passing a silent electric discharge through O_2 gas. The reaction is endothermic.

$$\tfrac{3}{2}\,O_2(g) \rightarrow O_3(g) \qquad \Delta H = +142 \text{ kJ}$$

Ozone is a blue gas that is highly reactive. It is explosive at temperatures above 300°C or in the presence of substances that catalyze its decomposition. Ozone will react with many substances at temperatures that are not high enough to produce reaction with O_2.

The most important allotropes of sulfur belong to the orthorhombic and monoclinic crystal systems (Section 9.12, Figure 9.13). The crystals of both allotropes are built from S_8 molecules. These molecules are in the form of puckered, eight-membered rings of S atoms in which the S atoms are bonded to each other by single covalent bonds and the S—S—S bond angle is about 105° (Figure 18.2). In chemical equations, therefore, elementary sulfur should be indicated by the formula S_8. The usual practice, however, is to designate sulfur by the symbol S. This practice leads to less complicated equations that are nevertheless stoichiometrically valid.

If sulfur is heated to approximately 200°C and then poured into cold water, a red–brown rubbery mass called plastic sulfur is obtained. It is assumed that plastic sulfur consists mainly of long chains of sulfur atoms. X-ray analysis of plastic sulfur has shown that it has the molecular structure characteristic of fibers. At room temperature, plastic sulfur, which is a supercooled liquid, slowly crystallizes and the S_8 rings re-form.

The stable modification of selenium at room temperature is a gray, hexagonal form, the crystals of which are constructed of zigzag chains of selenium

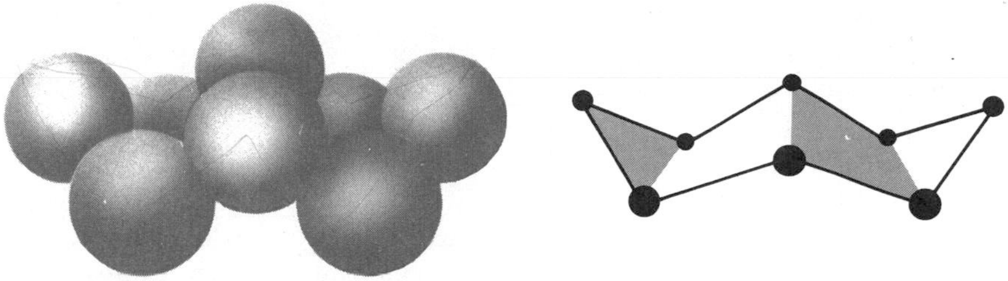

Figure 18.2 Structure of the S_8 molecule.

atoms. It is this form that is used in photoelectric cells. The normally low electrical conductivity of hexagonal selenium is increased about 200 times by exposure of the element to light. There are two monoclinic forms of selenium, both red, that are made up of Se_8 rings that are similar to S_8 rings. In addition, amorphous forms of selenium have been described.

The common form of tellurium consists of silver-white, hexagonal crystals built from zigzag chains of tellurium atoms. A black, amorphous form of tellurium exists.

Oxygen is the most abundant element in the earth's crust. The earth's atmosphere is about 23.2% oxygen by mass, or 21.0% oxygen by volume. Water is approximately 89% oxygen by mass. Oxygen occurs in most minerals. Common types include oxides, carbonates, silicates, and sulfates.

Most of the oxygen produced industrially is obtained from the liquefaction of air (Section 18.5). Small amounts of pure oxygen are produced commercially by the electrolysis of water.

$$2H_2O \xrightarrow{\text{electrolysis}} 2H_2(g) + O_2(g)$$

Oxygen may be prepared by the thermal decomposition of certain oxygen-containing compounds. The oxides of metals of low reactivity (for example, Hg, Ag, and Au) decompose on heating to give oxygen and the free metals.

$$2HgO(s) \rightarrow 2Hg(l) + O_2(g)$$

Oxygen and the oxide ion are produced when the peroxide ion, O_2^{2-}, is heated.

$$2\left[:\ddot{\underset{..}{O}}:\ddot{\underset{..}{O}}:\right]^{2-} \rightarrow 2\left[:\ddot{\underset{..}{O}}:\right]^{2-} + O_2$$

Thus,

$$2Na_2O_2(s) \rightarrow 2Na_2O(s) + O_2(g)$$
$$2BaO_2(s) \rightarrow 2BaO(s) + O_2(g)$$

Potassium chlorate decomposes upon heating to yield oxygen. A catalyst (MnO_2) is usually used so that the decomposition occurs at a lower temperature than would otherwise be required.

$$2KClO_3(s) \rightarrow 2KCl(s) + 3O_2(g)$$

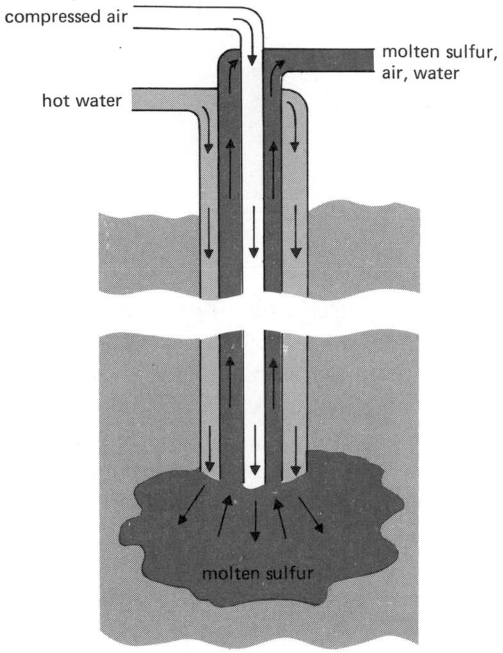

Figure 18.3 The Frasch Process.

Sulfur is obtained from large underground beds of the free element by the **Frasch process** (Figure 18.3). The sulfur is melted underground by water that is heated to approximately 170°C under pressure and forced down to the deposits. A froth of sulfur, air, and water is forced to the surface by hot, compressed air. The sulfur obtained in this way is about 99.5% pure.

The principal commercial source of both selenium and tellurium is the anode sludge obtained from the electrolytic refining of copper (Section 19.5).

18.4 COMPOUNDS OF THE GROUP VI A ELEMENTS

Oxygen reacts directly with all elements except the noble gases, the halogens, and a few of the less-reactive metals (such as Ag and Au). Oxides of most of these elements are known, but they must be made indirectly. The oxides are discussed in Section 8.9.

Most metals react with S, Se, and Te.

$$Hg + S \rightarrow HgS$$
$$Zn + Se \rightarrow ZnSe$$
$$2Al + 3Te \rightarrow Al_2Te_3$$

The hydrogen compounds are readily obtained by the action of dilute acid on sulfides, selenides, and tellurides. For example,

$$FeS(s) + 2H^+(aq) \rightarrow Fe^{2+}(aq) + H_2S(g)$$

The reaction of thioacetamide with water is a convenient laboratory source of H_2S.

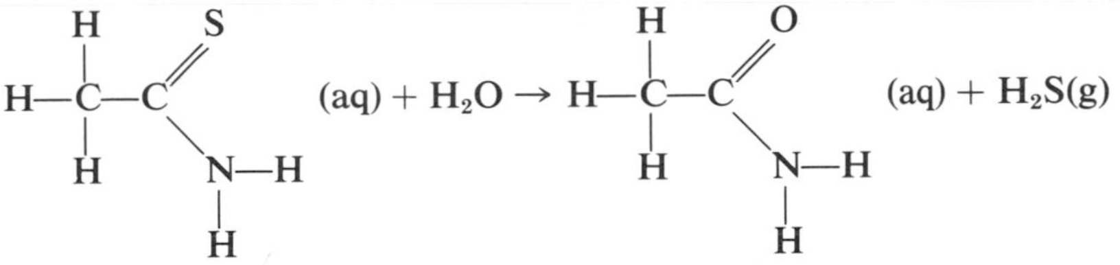

thioacetamide acetamide

Hydrogen sulfide, hydrogen selenide, and hydrogen telluride are colorless, unpleasant-smelling, highly poisonous gases. They are composed of angular molecules, similar to the water molecule.

Hydrogen sulfide reacts with oxygen to yield water and either SO_2 or free S depending upon the amount of oxygen employed. The combustion of H_2Se or H_2Te in oxygen produces water and either Se or Te.

The compounds H_2S, H_2Se, and H_2Te are all moderately soluble in water and are weak diprotic acids (see Section 8.8). The aqueous H_2S system is discussed in Section 15.7.

The sulfides of the group I A and group II A metals are water soluble. The sulfides of the remaining metals are either very slightly soluble or decompose in the presence of water to form slightly soluble hydroxides.

$$Al_2S_3(s) + 6H_2O \rightarrow 2Al(OH)_3(s) + 3H_2S(g)$$

Precipitation of the slightly soluble sulfides, under varying conditions, is extensively used in analytical procedures for the separation and identification of cations in solution (Section 15.9).

The important compounds in which sulfur appears in a 4+ oxidation state are sulfur dioxide (SO_2), sulfurous acid (H_2SO_3), and the salts of sulfurous acid – the sulfites. Both selenium and tellurium form compounds analogous to those of sulfur.

Sulfur dioxide is commercially obtained by burning sulfur

$$S(s) + O_2(g) \rightarrow SO_2(g)$$

or by roasting sulfide ores (such as ZnS, PbS, Cu_2S, and FeS_2) in air (Section 19.3)

$$2ZnS(s) + 3O_2(g) \rightarrow 2ZnO(s) + 2SO_2(g)$$

Sulfur dioxide is a colorless gas. It has a sharp, irritating odor and is poisonous. The molecules of SO_2 are angular, and the structure of the compound may be represented as a resonance hybrid (Section 7.5).

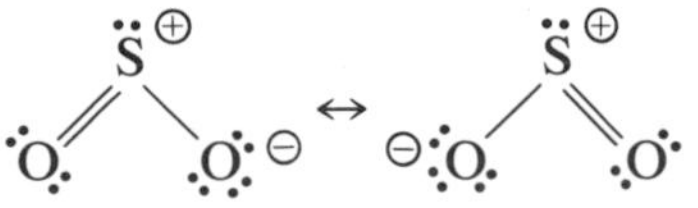

Sulfur dioxide is moderately soluble in water, producing solutions of sulfurous acid, H_2SO_3. The acid is not very stable, and pure H_2SO_3 cannot be isolated. Sulfurous acid is a weak, diprotic acid. The acid, therefore, forms two series of salts: normal salts (for example, Na_2SO_3, sodium sulfite) and acid salts (for example, $NaHSO_3$, sodium bisulfite or sodium hydrogen sulfite).

Sulfites are often prepared by bubbling SO_2 gas through a solution of a hydroxide.

$$2OH^- + SO_2(g) \rightarrow SO_3^{2-}(aq) + H_2O$$

If the addition is continued, acid sulfites are produced.

$$H_2O + SO_3^{2-}(aq) + SO_2(g) \rightarrow 2HSO_3^-(aq)$$

The salts of sulfurous acid can be isolated from solution.

Since sulfurous acid is unstable, the addition of an acid to a sulfite or to an acid sulfite liberates SO_2 gas.

$$SO_3^{2-}(aq) + 2H^+(aq) \rightarrow SO_2(g) + H_2O$$

The dioxides of selenium and tellurium may be prepared by direct combination of the elements with oxygen. Both SeO_2 and TeO_2 are white solids. Selenous acid, H_2SeO_3, is formed when the very soluble SeO_2 is dissolved in water. It is a weak, diprotic acid and may be obtained in pure form by the evaporation of a solution of SeO_2. Tellurium dioxide is only slightly soluble in water, and pure H_2TeO_3 has never been prepared.

Sulfur trioxide is produced when sulfur dioxide reacts with atmospheric oxygen. Since the reaction is very slow at ordinary temperatures, the commercial preparation is conducted at elevated temperatures (400°C to 700°C) and in the presence of a catalyst (such as V_2O_5 or Pt).

$$2SO_2(g) + O_2(g) \rightarrow 2SO_3(g)$$

Sulfur trioxide is a volatile material (boiling point, 44.8°C). In the gas phase it consists of single molecules that are planar triangular in form with O—S—O bond angles of 120°. The electronic structure of the molecule may be represented as a resonance hybrid:

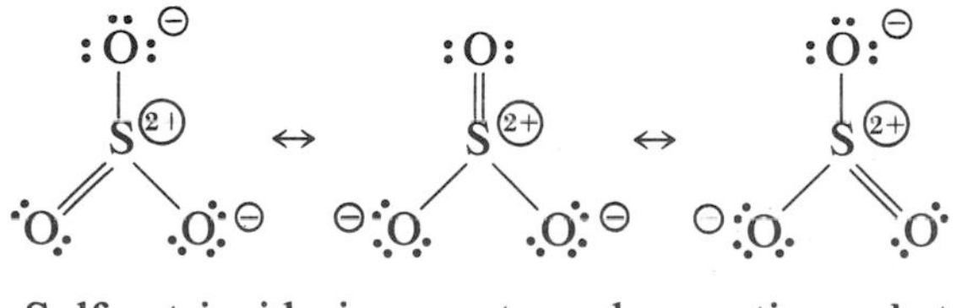

Sulfur trioxide is an extremely reactive substance and a strong oxidizing agent. It is the anhydride of sulfuric acid, H_2SO_4, and reacts vigorously with water to produce the acid and with metallic oxides to produce sulfates (Section 8.9).

Sulfuric acid is an important industrial chemical. Most sulfuric acid is made by the **contact process** in which SO_2 is catalytically oxidized to SO_3 in the manner previously described. The SO_3 vapor is bubbled through H_2SO_4 and pyrosulfuric acid ($H_2S_2O_7$) is formed.

$$SO_3(g) + H_2SO_4(l) \rightarrow H_2S_2O_7(l)$$

Water is then added to the pyrosulfuric acid to make sulfuric acid of the desired concentration.

$$H_2S_2O_7(l) + H_2O \rightarrow 2H_2SO_4(l)$$

This procedure, in which pyrosulfuric acid is formed, is easier to control than the direct reaction of SO_3 with water.

The **lead chamber process** is an older method for the manufacture of sulfuric acid. Sulfur dioxide, oxygen, water vapor, and oxides of nitrogen are mixed in a lead-lined chamber. The reactions are complex and are not completely understood, but it is recognized that nitrosylsulfuric acid, $HOSO_2(ONO)$, is formed as an intermediate and that this compound then reacts with water to form sulfuric acid. We may summarize the process by the simplified equations

$$2SO_2(g) + NO(g) + NO_2(g) + O_2(g) + H_2O(g) \rightarrow 2HOSO_2(ONO)(s)$$
$$2HOSO_2(ONO)(s) + H_2O(g) \rightarrow 2H_2SO_4(l) + NO(g) + NO_2(g)$$

The oxides of nitrogen are regenerated and are reused. The sulfuric acid produced by the lead chamber process has a concentration of only 60% to 78% and is less pure than that made by the contact process.

Sulfuric acid is a colorless, oily liquid that freezes at 10.4°C and begins to boil at approximately 290°C with decomposition into water and sulfur trioxide. The electronic structure of sulfuric acid may be represented as

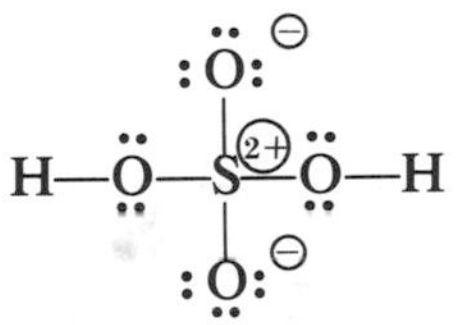

When concentrated sulfuric acid is added to water, a great deal of heat is evolved. This heat effect may occur because of the formation of the hydronium ion, H_3O^+. Sulfuric acid has a strong affinity for water and forms a series of hydrates (such as $H_2SO_4 \cdot H_2O$, $H_2SO_4 \cdot 2H_2O$, and $H_2SO_4 \cdot 4H_2O$). Thus, sulfuric acid is used as a drying agent. Gases that do not react with H_2SO_4 may be dried by being bubbled through the acid.

In aqueous solution, sulfuric acid ionizes in two steps.

$$H_2SO_4 \rightarrow H^+(aq) + HSO_4^-(aq)$$
$$HSO_4^-(aq) \rightleftharpoons H^+(aq) + SO_4^{2-}(aq)$$

Sulfuric acid is a strong electrolyte as far as the first dissociation is concerned. The second dissociation, however, is not complete. The acid forms two series of salts: normal salts (such as sodium sulfate, Na_2SO_4) and acid salts (such as sodium bisulfate, $NaHSO_4$).

Since sulfuric acid has a relatively high boiling point (or decomposition temperature), it is used to secure more volatile acids from their salts. This use is illustrated by the preparations of HF and HCl (Section 18.2), as well as by the preparation of HNO_3 (Section 18.8).

The oxidizing ability of hot, concentrated H_2SO_4 on bromides and iodides has already been noted (Section 18.2). This reagent will also oxidize many nonmetals.

$$C(s) + 2H_2SO_4(l) \rightarrow CO_2(g) + 2SO_2(g) + 2H_2O(g)$$

Most metals are oxidized by hot, concentrated sulfuric acid. Copper, which will not displace hydrogen from aqueous acids, is oxidized by hot, concentrated sulfuric acid (although hydrogen gas is not a product of the reaction).

$$Cu(s) + 2H_2SO_4(l) \rightarrow CuSO_4(s) + SO_2(g) + 2H_2O(g)$$

Selenic acid, H_2SeO_4, is prepared by the oxidization of selenous acid, H_2SeO_3. Selenium trioxide, the acid anhydride of selenic acid, is not very stable and decomposes to SeO_2 and O_2 on warming. Low yields of SeO_3 mixed with SeO_2 are produced when an electric discharge is passed through selenium vapor and oxygen.

Selenic acid is very similar to sulfuric acid. In aqueous solution the first dissociation of H_2SeO_4 is complete and the second dissociation is not. The acid forms normal and acid salts. Like the sulfate ion, the selenate ion is tetrahedral.

Telluric acid is prepared by the action of vigorous oxidizing agents on elementary tellurium. Unlike H_2SO_4 and H_2SeO_4, the formula of telluric acid is H_6TeO_6, which may be regarded as a hydrated form of the nonexistent H_2TeO_4. Telluric acid is not very similar to sulfuric acid, and H_6TeO_6 functions in aqueous solution as a very weak diprotic acid. If H_6TeO_6 is heated to approximately 350°C, water is driven off and solid tellurium trioxide results. This compound is not very water soluble but reacts with alkalies to produce tellurates.

There are other acids of sulfur in which S is in a 6+ oxidation state. Pyrosulfuric acid (also called disulfuric acid)

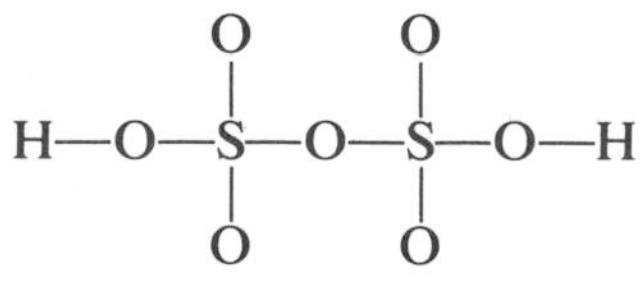

has been mentioned as the product of the reaction of SO_3 with H_2SO_4 in a 1:1 molar ratio. The pyrosulfate ion has been shown to have a structure in which two SO_4 tetrahedra are joined by an O atom common to both tetrahedra: $[O_3SOSO_3]^{2-}$. Pyrosulfuric acid is a stronger oxidizing agent and a stronger dehydrating agent than sulfuric acid.

A peroxy acid is an acid that contains a peroxide group (—O—O—) somewhere in the molecule. Two peroxy acids of sulfur exist: peroxymonosulfuric acid (H_2SO_5) and peroxydisulfuric acid ($H_2S_2O_8$).

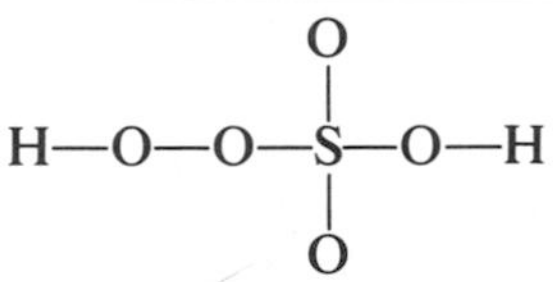

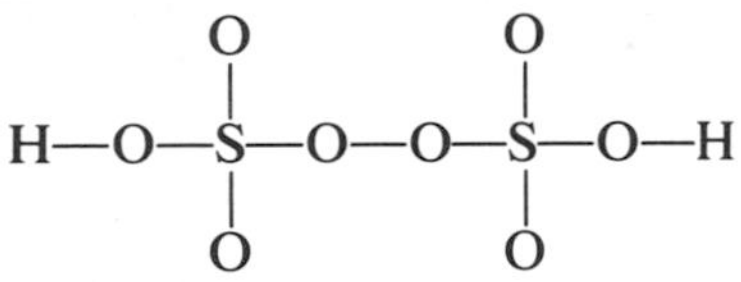

The structure of the peroxydisulfate ion consists of two complete SO_4 tetrahedra joined by an O—O bond.

There is another significant group of sulfur-containing anions. The most important representative of the group is the thiosulfate ion, $S_2O_3^{2-}$.

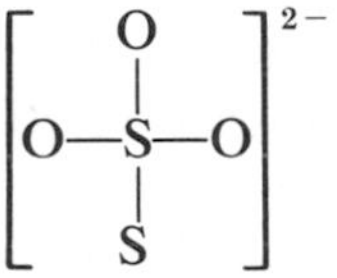

This tetrahedral ion may be regarded as a sulfate ion in which one oxygen atom has been replaced by a sulfur atom. In fact, the prefix "thio-" is used to name any species that may be considered to be derived from another compound by replacing an oxygen atom by a sulfur atom. The prefix is added to the base name of the unaltered compound. Thus, CNO^- is the cyanate ion and CNS^- is the thiocyanate ion; $CO(NH_2)_2$ is urea and $CS(NH_2)_2$ is thiourea.

Thiosulfates may be prepared by the reaction of sulfur with sulfites in aqueous solution.

$$SO_3^{2-}(aq) + S(s) \rightarrow S_2O_3^{2-}(aq)$$

The corresponding acid does not exist. Upon acidification, thiosulfates decompose to elementary sulfur and SO_2 gas.

$$S_2O_3^{2-}(aq) + 2H^+(aq) \rightarrow S(s) + SO_2(g) + H_2O$$

18.5 THE ATMOSPHERE

Air is a mixture. The composition varies with altitude and to a lesser extent with location. The analysis of air is made after water and solid particles (such as dust and spores) have been removed. Some of the components of air are listed in Table 18.3. The percentages given (by volume) are for clean, dry air at sea level.

The amount of water vapor in air varies widely. In the tropics, it may be as high as 5% and in colder climates, it may be less than 0.01%. The **relative hu-**

Table 18.3 Composition of dry air

SUBSTANCE	PERCENT BY VOLUME	SUBSTANCE	PERCENT BY VOLUME
N_2	78.00	CH_4	2×10^{-4}
O_2	20.95	Kr	1×10^{-4}
Ar	0.93	N_2O	5×10^{-5}
CO_2	0.03	H_2	5×10^{-5}
Ne	0.0018	Xe	8×10^{-6}
He	0.0005	O_3	1×10^{-6}

midity is a measure of the quantity of water vapor that a sample of air holds in comparison to the amount it would hold if it were saturated. In other words,

$$\text{relative humidity} = \left(\frac{p_{H_2O}}{p^{\circ}_{H_2O}}\right) 100$$

where p_{H_2O} is the partial pressure of water vapor in the air and $p^{\circ}_{H_2O}$ is the equilibrium vapor pressure of water at the temperature of the air (Table 4.4, Section 4.10). Suppose, for example, that the partial pressure of water vapor in a sample of air at 20°C is 0.0135 atm. The vapor pressure of water at 20°C is 0.0231 atm. Therefore,

$$\text{relative humidity} = \left(\frac{0.0135 \text{ atm}}{0.0231 \text{ atm}}\right) 100 = 58.4\%$$

Since the vapor pressure of water at 25°C is 0.0313 atm, the relative humidity of this sample of air if it were warmed to 25°C would be

$$\text{relative humidity} = \left(\frac{0.0135 \text{ atm}}{0.0313 \text{ atm}}\right) 100 = 43.1\%$$

Nitrogen is a constituent element of all plant and animal protein. The cells of living systems cannot assimilate the nitrogen of the air to use in the synthesis of proteins. However, nitrogen of the air is converted by several natural **nitrogen-fixation processes** into compounds that can be used by plants. During storms, lightning flashes cause some nitrogen and oxygen of the air to form nitrogen oxide

$$N_2(g) + O_2(g) \rightarrow 2NO(g)$$

Nitrogen dioxide is produced by the reaction of NO with additional O_2 from the air.

$$2NO(g) + O_2(g) \rightarrow 2NO_2(g)$$

The NO_2 reacts with water to form nitric acid

$$3NO_2(g) + H_2O(l) \rightarrow 2HNO_3(l) + NO(g)$$

The nitric acid is washed to the earth where it forms nitrates in the soil, which can be used by plants as nutrients.

Certain soil bacteria, as well as nitrogen-fixing bacteria in the root nodules of leguminous plants (such as peas, beans, and alfalfa) fix atmospheric nitrogen into compounds that plants can assimilate. The Haber process for the manufacture of ammonia from nitrogen and hydrogen

$$N_2(g) + 3H_2(g) \rightleftharpoons 2NH_3(g)$$

is the principal commercial nitrogen-fixation process. A large proportion of the ammonia produced by the Haber process is used to make fertilizers.

These fixation processes are the first step in the **nitrogen cycle** of nature. Plants use the fixed nitrogen to make plant protein, which is eaten by animals. Animals use plant protein to make animal protein. The decay of animal protein, plant protein, and the waste products of animal metabolism produces N_2 which is returned to the atmosphere, thus completing the cycle.

There is also an **oxygen–carbon dioxide cycle** in nature. Oxygen is removed from the air and CO_2 is added to the air by the respiration of animals, the combustion of fuels, and the decay of organic matter. The reverse occurs in photosynthesis (Section 20.6). In the formation of carbohydrates by plants, CO_2 is removed from the air and O_2 is released to the air. The geochemical equilibrium between CO_2, H_2O, and limestone, $CaCO_3$, also serves to control the level of CO_2 in the air.

$$CaCO_3(s) + CO_2(g) + H_2O \rightleftharpoons Ca^{2+}(aq) + 2HCO_3^-(aq)$$

At the present time, the rate at which CO_2 is being added to the atmosphere exceeds the rate at which it is being removed by other processes. The concentration of CO_2 in the atmosphere, therefore, is steadily increasing. The concentration today is about 10% higher than it was a century ago.

The principal source of nitrogen, oxygen, and some of the noble gases is the fractional distillation of liquid air. Air is liquified by compression and cooling. When the air is allowed to warm, nitrogen (boiling point, −196°C) boils away from the oxygen (boiling point, −183°C). The noble gases are obtained from the nitrogen and oxygen fractions by repeated distillation and other separation techniques.

Oxygen is used in metallurgical processes—for example, in the blast furnace used for the reduction of iron ore and the basic-oxygen process for the refining of pig iron. Oxygen is also used in the oxyacetylene and oxyhydrogen welding torches. A small amount is used for medicinal purposes.

Nitrogen is used in the manufacture of ammonia and other nitrogen-containing compounds. Nitrogen is comparatively unreactive, and it is used when an inert atmosphere is desired. It is used, for example, to pack oxidizable foods (such as coffee) in cans. Since N_2 does not support combustion, liquid N_2 is a safer coolant to use than liquid air.

All the noble gases occur in the atmosphere, and Ne, Ar, Kr, and Xe are by-products of the fractionation of liquid air. Certain natural gas deposits (located principally in Kansas) contain a higher percentage of helium than is found in air. These deposits constitute the major commercial source of helium.

Argon is used to fill electric light bulbs. The gas does not react with the hot filament but rather conducts heat away from it, thus prolonging its life. Argon is also used as an inert atmosphere in welding and high-temperature metallurgical processes. The gas protects the hot metals from air oxidation. Helium is used in lighter-than-air craft (its density is about 14% of the density of air) and in low temperature work (it has the lowest boiling point of any known substance). Neon signs are made from discharge tubes containing neon gas at a low pressure. Compounds of the noble gases are discussed in Section 8.10.

18.6 AIR POLLUTION

Modern technological civilization is introducing foreign substances into the atmosphere at an ever-increasing rate. The principal air pollutants in terms of quantities present are the following.

1. *Carbon monoxide* is produced by the incomplete combustion of fuels. The automobile's internal-combustion engine is the principal source of this pollutant. The mass of CO produced by this source is about equal to half the mass of gasoline consumed.

   Carbon monoxide is toxic because it combines with the hemoglobin of the blood and prevents the hemoglobin from carrying oxygen to the body tissues (Section 20.1). In other ways, however, CO is not very reactive. Reaction with O_2 of the air to form CO_2 does occur, but very slowly.
2. *Oxides of sulfur* (SO_2 and SO_3) result from the combustion of coal, metallurgical processes, and petroleum combustion and refining. The major source is the combustion of coal (which contains from 0.5% to 3.0% S) in the generation of electricity. The roasting of sulfide ores (Section 19.3) is also a significant source of SO_2 pollution.

   $$2PbS(s) + 3O_2(g) \rightarrow 2PbO(s) + 2SO_2(g)$$

   Sulfur dioxide from these sources is slowly oxidized to SO_3 by O_2 in the air. The SO_3 forms sulfuric acid with atmospheric water, and SO_2 forms sulfurous acid. These substances are, of course, extremely corrosive.

   The oxides of sulfur are in some ways the most serious air pollutants. They are toxic, cause respiratory ailments, and are a serious health threat. They damage plant life, corrode metals, and erode marble and limestone. Ancient monuments (such as the Parthenon in Athens) which have stood for centuries are crumbling because of the pollution of modern civilization.

3. *Oxides of nitrogen* (NO and NO_2) are produced from the N_2 and O_2 of the air at the high temperatures characteristic of some combustions. Significant amounts of NO are formed in the combustions carried out in automobile engines and in electric generating plants. Nitrogen dioxide is formed in the air by the oxidation of NO.

   Small amounts of NO and NO_2 ordinarily occur in air and form a minor part of the nitrogen cycle. Nitrogen dioxide is considerably more toxic than NO. The gases, however, usually occur at relatively low concentrations so that the *direct* effect of these pollutants is not serious. The significance of these oxides lies in the role that they play in the formation of other, more serious pollutants (see Hydrocarbons).

4. *Hydrocarbons* are compounds that contain carbon and hydrogen. They are found in petroleum, natural gas, and coal. The compounds are released into the atmosphere by evaporation, petroleum refining, and incomplete combustion of fuels. The unburned hydrocarbons in automobile exhaust constitute a major source of this type of contamination.

   A few hydrocarbons are carcinogenic (cancer-producing). The principal danger associated with hydrocarbon pollution, however, lies in the pollutants that are produced from hydrocarbons in the air. Nitrogen dioxide decomposes in sulight to give O atoms.

$$NO_2(g) \rightarrow NO(g) + O(g)$$

   These O atoms react with O_2 to produce ozone, O_3.

$$O(g) + O_2(g) \rightarrow O_3(g)$$

   Ozone is highly reactive and reacts with some hydrocarbons to produce oxygen-containing organic compounds. Since the process is initiated by sunlight, the products are sometimes called **photochemical pollutants.**

   These substances are toxic and very irritating to the eyes, skin, and respiratory tract. They cause extensive crop damage and the deterioration of materials. They constitute what is called photochemical smog.

5. *Small particles* suspended in air are another important contributor to air pollution. The particles may consist of liquids or solids and vary in size from about 0.01 μm to 100 μm (the micrometer, μm, previously called the micron, is 10^{-4} cm). A principal source is the smoke from the combustion of coal. Industrial processes (for example, the manufacture of cement) also introduce particulate matter into the air. Automobile exhaust contains suspended particles.

   Particles reduce visibility and pose a threat to health. They cause lung damage and may be toxic. Carcinogenic substances may be contained in soot particles. Particles in automobile exhaust contain lead, a toxic substance which comes from the lead compounds that are added to gasoline to improve its engine performance.

18.7 THE GROUP V A ELEMENTS

Within any group of the periodic classification, metallic character increases (and nonmetallic character decreases) with increasing atomic number, atomic weight, and atomic size. This trend is particularly striking in group V A. Nitrogen and phosphorus are generally regarded as nonmetals, arsenic and antimony as semimetals or metalloids, and bismuth as a metal.

For the group as a whole, the 3−, 3+, and 5+ oxidation states are most common. The importance and stability of the 5+ and 3− states decline from the lighter to the heavier elements. Nitrogen, however, appears in every oxidation state from 3− to 5+. Nitrogen also has a tendency toward the formation of multiple bonds (for example, in the cyanide ion, $C{\equiv}N^-$).

Nitrogen is produced commercially by the fractional distillation of liquid air. Very pure nitrogen is conveniently prepared by heating an aqueous solution saturated with ammonium chloride and sodium nitrite.

$$NH_4^+(aq) + NO_2^-(aq) \rightarrow N_2(g) + 2H_2O$$

or by heating either sodium azide or barium azide,

$$2NaN_3(s) \rightarrow 2Na(l) + 3N_2(g)$$

Free nitrogen is surprisingly unreactive, in part because of the great strength of the bonding in the N_2 molecule.

$$:N{\equiv}N:$$

Phosphorus is industrially prepared by heating a mixture of phosphate rock, sand, and coke in an electric furnace.

$$2Ca_3(PO_4)_2(s) + 6SiO_2(s) \rightarrow 6CaSiO_3(l) + P_4O_{10}(g)$$
$$P_4O_{10}(g) + 10C(s) \rightarrow P_4(g) + 10CO(g)$$

The calcium silicate is withdrawn as a molten slag from the bottom of the furnace, and the product gases are passed through water, which condenses the phosphorus vapor into a white solid.

Arsenic, antimony, and bismuth are obtained by carbon reduction of their oxides at elevated temperatures.

$$As_4O_6(s) + 6C(s) \rightarrow As_4(g) + 6CO(g)$$

An important industrial source of the oxides is the flue dust obtained from the processes used in the production of certain metals, notably copper and lead. In addition, the oxides are obtained by roasting the sulfide ores of the elements in air. For example,

$$2Sb_2S_3(s) + 6O_2(g) \rightarrow Sb_4O_6(g) + 6SO_2(g)$$

Phosphorus, arsenic, and antimony occur in allotropic modifications. There are three important forms of phosphorus: white, red, and black. White phos-

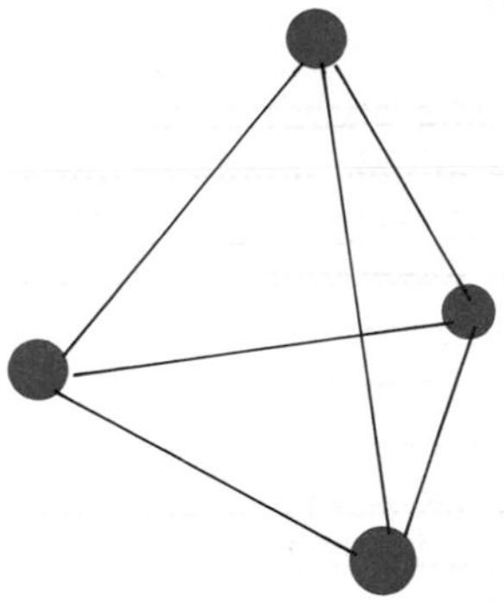

Figure 18.4 Structure of the P_4 molecule.

phorus, a waxy solid, is obtained by condensing phosphorus vapor. Crystals of white phosphorus are formed from P_4 molecules (Figure 18.4). White phosphorus is the most reactive form of the element and is stored under water to protect it from atmospheric oxygen.

Red phosphorus may be prepared by heating white phosphorus to about 250°C in the absence of air. It is a polymeric material in which many phosphorus atoms are joined in a network, but the details of the structure of red phosphorus are not known. Red phosphorus is considerably less reactive than the white variety. It does not react with oxygen at room temperature.

Black phosphorus, a less common allotrope, is made by subjecting the element to very high pressures. Crystalline black phosphorus consists of layers of phosphorus atoms covalently joined into a network (Figure 18.5). Black phosphorus is the least reactive form of the element.

Arsenic and antimony exist in soft, yellow, nonmetallic modifications which are thought to be formed from tetrahedral As_4 and Sb_4 molecules analogous to the P_4 molecules of white phosphorus. These yellow forms may be obtained by the rapid condensation of vapors. They are unstable and are readily converted into stable, gray, metallic modifications.

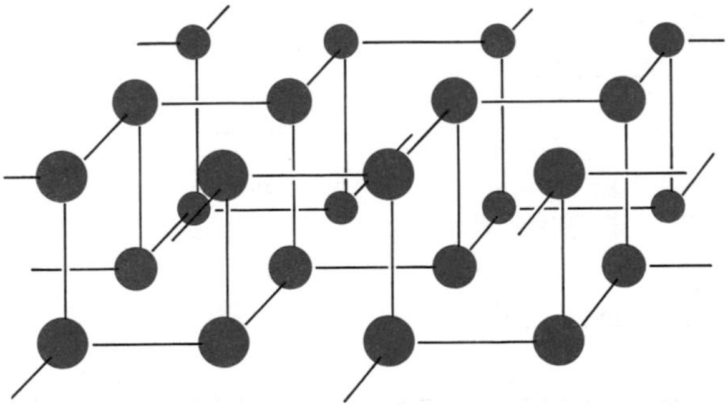

Figure 18.5 Structure of a layer of the black phosphorus crystal.

18.8 COMPOUNDS OF THE GROUP V A ELEMENTS

Elementary nitrogen reacts with a number of metals at high temperatures. It forms ionic compounds that contain the N^{3-} ion with lithium, the group II A metals, cadmium, and zinc. These **ionic nitrides** are high-melting, white, crystalline solids. They readily react with water to yield ammonia and hydroxides.

$$Li_3N(s) + 3H_2O \rightarrow 3Li^+(aq) + 3OH^-(aq) + NH_3(g)$$
$$Ca_3N_2(s) + 6H_2O \rightarrow 3Ca^{2+}(aq) + 6OH^-(aq) + 2NH_3(g)$$

Interstitial nitrides may be made at elevated temperatures from many transition metals, which are in powdered form, and nitrogen or ammonia. These substances, such as VN, Fe_4N, W_2N, and TiN, contain nitrogen atoms in the interstices of the metal lattices. They frequently deviate slightly from exact stoichiometry. Interstitial nitrides have extremely high melting points, resemble metals in appearance, are good electrical conductors, are very hard, and are chemically inert.

Covalent nitrides include such compounds as S_4N_4, P_3N_5, Si_3N_4, Sn_3N_4, BN, and AlN. Some of these compounds are molecular in form. Others are materials in which a large number of the constituent atoms are covalently joined into a crystal lattice.

Many metals unite with white phosphorus directly to form phosphides. These compounds have not been so well characterized as the nitrides. The group II A metals form phosphides of formula M_3P_2. Lithium forms Li_3P, and sodium forms Na_3P. In addition, the I A metals form compounds with formulas corresponding to M_2P_5. All these compounds readily hydrolyze to form phosphine, PH_3. For example,

$$Ca_3P_2(s) + 6H_2O \rightarrow 3Ca^{2+}(aq) + 6OH^-(aq) + 2PH_3(g)$$

The III A metals and many transition metals also form phosphides.

The reactions of metals with arsenic, antimony, and, to a lesser extent, bismuth yield arsenides, stibnides, and bismuthides. These compounds become progressively more difficult to prepare as the atomic number of the group V A element increases. The group V A elements all form hydrogen compounds, the most important of which is ammonia, NH_3. Large quantities of ammonia are commercially prepared by the direct union of the elements (**Haber process**).

$$N_2(g) + 3H_2(g) \rightleftharpoons 2NH_3(g)$$

Ammonia is the only hydrogen compound of the V A elements that can be prepared directly. The reaction is conducted under high pressures (from 100 to 1000 atm), at 400°C to 550°C, and in the presence of a catalyst. One catalyst, so employed, consists of finely divided iron and Fe_3O_4 containing small amounts of K_2O and Al_2O_3.

In the laboratory, ammonia is conveniently prepared by heating an ammonium salt with a strong alkali [such as NaOH or $Ca(OH)_2$].

$$NH_4^+(aq) + OH^-(aq) \rightarrow NH_3(g) + H_2O$$

Aqueous solutions of ammonia are alkaline.

$$NH_3(aq) + H_2O \rightleftharpoons NH_4^+(aq) + OH^-(aq)$$

In solution or as a dry gas, ammonia reacts with acids to produce ammonium salts.

$$NH_3(g) + HCl(g) \rightarrow NH_4Cl(s)$$

Hydrazine, N_2H_4, may be considered to be derived from NH_3 by the replacement of an H atom by a —NH_2 group.

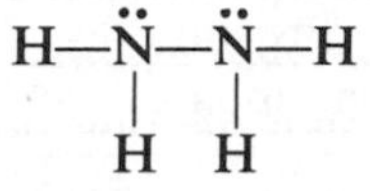

The compound, which is a liquid, may be prepared by oxidizing NH_3 with NaOCl. Hydrazine is less basic than NH_3 but does form cations in which one proton or two protons are bonded to the free electron pairs of the molecule. For example,

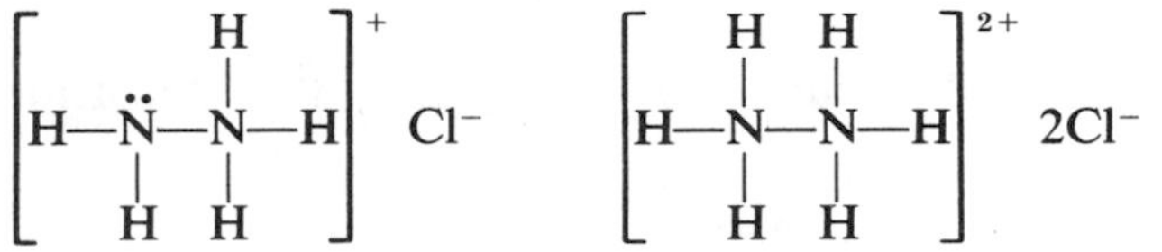

Hydrazine is a strong reducing agent and has found some use in rocket fuels.

Hydroxylamine, NH_2OH, is another compound which, like hydrazine, may be considered to be derived from the NH_3 molecule.

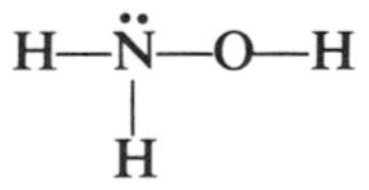

Like hydrazine, hydroxylamine is a weaker base than NH_3, but salts containing the $[NH_3OH]^+$ ion may be prepared.

Phosphine (PH_3) is a very poisonous, colorless gas that can be prepared by the hydrolysis of phosphides.

$$Ca_3P_2(s) + 6H_2O \rightarrow 3Ca(OH)_2(s) + 2PH_3(g)$$

Phosphine is much less basic than ammonia. Phosphonium compounds, such as $PH_4I(s)$ which can be made from dry $PH_3(g)$ and HI(g), are unstable. They decompose at relatively low temperatures, or in aqueous solution, to yield the component gases.

Arsine (AsH_3) and stibine (SbH_3) are extremely poisonous gases that may be produced by the hydrolysis of arsenides and stibnides (for example, Na_3As and Zn_3Sb_2). The stability of the hydrogen compounds declines in the series from NH_3 to SbH_3. Arsine and stibine are decomposed by warming. Arsine and stibine have no basic properties and do not form salts with acids.

The most important halides of the V A elements are the trihalides (for example, NF_3) and the pentahalides (for example, PF_5). All four binary trihalides of each of the V A elements have been made, but the tribromide and triiodide of nitrogen can be isolated only in the form of ammoniates ($NBr_3 \cdot 6NH_3$ and $NI_3 \cdot xNH_3$). Nitrogen trifluoride is a very stable colorless gas, whereas the other trihalides of nitrogen are explosively unstable.

The pentahalide series is not so complete as the trihalide series. Since N has no *d* orbitals in its valence level, N can form no more than four covalent bonds. Pentahalides of N, therefore, do not exist. All pentahalides of phosphorus are known with the exception of the pentaiodide. Presumably, there is not sufficient room around a phosphorus atom to accommodate five large iodine atoms. In addition, AsF_5, SbF_5, BiF_5, and $SbCl_5$ have been prepared.

The pentahalides may be prepared by direct reaction of the elements using an excess of the halogen and by the reaction of the halogen with the trihalide.

$$PCl_3(l) + Cl_2(g) \rightleftharpoons PCl_5(s)$$

The preceding reaction is reversible. In the gas phase the pentahalides dissociate in varying degrees.

The phosphorus pentahalides undergo hydrolysis in two steps; for example,

$$PCl_5(s) + H_2O \rightarrow POCl_3(l) + 2H^+(aq) + 2Cl^-(aq)$$
$$POCl_3(l) + 3H_2O \rightarrow H_3PO_4(aq) + 3H^+(aq) + 3Cl^-(aq)$$

The phosphoryl halides, POX_3 (X = F, Cl, or Br), can be prepared by the hydrolysis of the appropriate pentahalide in a limited amount of water or by the reaction of the trihalide with oxygen.

$$2PCl_3(l) + O_2(g) \rightarrow 2POCl_3(l)$$

Oxides are known for every oxidation state of nitrogen from 1+ to 5+ (see Table 18.4). The most important oxides are NO and NO_2. Nitrogen oxide (NO) may be prepared by the direct combination of the elements at high temperatures. Commercially, however, the most important source of NO is the catalytic oxidation of ammonia from the Haber process.

Nitrogen is formed when ammonia is burned in pure oxygen.

$$4NH_3(g) + 3O_2(g) \rightarrow 2N_2(g) + 6H_2O(g)$$

When a mixture of ammonia and air, however, is passed over platinum gauze at 1000°C, NO is produced.

$$4NH_3(g) + 5O_2(g) \rightarrow 4NO(g) + 6H_2O(g)$$

Nitrogen dioxide is produced by the reaction of NO with O_2 at room temperature. This oxide is used to make nitric acid in the **Ostwald process.** Nitrogen

Table 18.4 Nitrogen and its oxides

OXIDATION NUMBER	FORMULA	PREPARATION	
5+	N_2O_5	$4HNO_3(g) + P_4O_{10}(s) \rightarrow 4HPO_3(s) + 2N_2O_5(g)$	(heat)
4+	NO_2	$2NO(g) + O_2(g) \rightarrow 2NO_2(g)$	
	N_2O_4	$2NO_2(g) \rightleftharpoons N_2O_4(l)$	(cool)
3+	N_2O_3	$NO(g) + NO_2(g) \rightleftharpoons N_2O_3(l)$	(cool)
2+	NO	$4NH_3(g) + 5O_2(g) \rightarrow 4NO(g) + 6H_2O(g)$	(heat, catalyst)
		$N_2(g) + O_2(g) \rightleftharpoons 2NO(g)$	(high temperatures)
1+	N_2O	$NH_4NO_3(s) \rightarrow N_2O(g) + 2H_2O(g)$	(heat)
0	N_2	$NH_4^+(aq) + NO_2^-(aq) \rightarrow N_2(g) + 2H_2O$	(heat)

dioxide gas together with excess oxygen, is passed into a tower where it reacts with warm water.

$$3NO_2(g) + H_2O \rightarrow 2H^+(aq) + 2NO_3^-(aq) + NO(g)$$

The excess oxygen converts the NO into NO_2. This NO_2 then reacts with water as before. In this cyclic manner the nitric oxide is eventually completely converted into nitric acid. The product of the Ostwald process is about 70% HNO_3 and is known as concentrated nitric acid. Solutions which are more concentrated may be prepared from it by distillation.

Pure nitric acid is a colorless liquid that boils at 83°C. It may be prepared in the laboratory by heating sodium nitrate with concentrated sulfuric acid.

$$NaNO_3(s) + H_2SO_4(l) \rightarrow NaHSO_4(s) + HNO_3(g)$$

This preparation (from Chilean saltpeter) is a minor commercial source of the acid. Nitric acid is a strong acid. Most salts of nitric acid, which are called nitrates, are very soluble in water.

Nitric acid is a powerful oxidizing agent. It oxidizes most nonmetals (generally to oxides or oxyacids of their highest oxidation state) and all metals with the exception of a few of the noble metals.

In nitric acid oxidations, hydrogen is almost never obtained. Instead, a variety of nitrogen-containing compounds, in which nitrogen is in a lower oxidation state, is produced. The product to which HNO_3 is reduced depends upon the concentration of the acid, the temperature, and the nature of the material being oxidized. Generally, a mixture of products is obtained, but the principal product in many cases is NO when dilute HNO_3 is employed and NO_2 when concentrated HNO_3 is used.

Dilute:

$$3Cu(s) + 8H^+(aq) + 2NO_3^-(aq) \rightarrow 3Cu^{2+}(aq) + 2NO(g) + 4H_2O$$

Concentrated:

$$Cu(s) + 4H^+(aq) + 2NO_3^-(aq) \rightarrow Cu^{2+}(aq) + 2NO_2(g) + 2H_2O$$

Pure nitrous acid, HNO_2, has never been isolated, but a solution of this acid may be prepared by adding a strong acid to a cold aqueous solution of a nitrite.

$$H^+(aq) + NO_2^-(aq) \rightarrow HNO_2(aq)$$

The acid is weak. It decomposes on warming.

$$3HNO_2(aq) \xrightarrow{\text{heat}} H^+(aq) + NO_3^-(aq) + H_2O + 2NO(g)$$

The nitrites of the I A metals are formed when the nitrates are heated. They may also be prepared by heating the nitrate with a reducing agent, such as lead, iron, or coke.

$$NaNO_3(l) + C(s) \rightarrow NaNO_2(l) + CO(g)$$

When white phosphorus is burned in a limited supply of air, the chief product is phosphorus(III) oxide (also called phosphor*ous* oxide), P_4O_6. It is a colorless material that melts at 23.8°C. The combustion of phosphorus in excess oxygen yields phosphorus(V) oxide (also called phosphor*ic* oxide), P_4O_{10}, a white powder that sublimes at approximately 360°C. The structures of these molecules are based on P_4 tetrahedra (Figure 18.6).

Phosphorus(V) oxide has a great affinity for water and is a very effective drying agent. Many different phosphoric acids may be prepared by the addition of water to P_4O_{10}. The most important acid of phosphorus in the 5+ state, orthophosphoric acid (or merely phosphoric acid) results from the complete hydration of the oxide.

$$P_4O_{10} + 6H_2O \rightarrow 4H_3PO_4$$

Phosphoric acid is obtained commercially by this means or by treating phosphate rock with sulfuric acid.

$$Ca_3(PO_4)_2(s) + 3H_2SO_4(l) \rightarrow 2H_3PO_4(l) + 3CaSO_4(s)$$

The compound is a colorless solid but is generally sold as an 85% solution. Phosphoric acid is a weak, triprotic acid. Three types of salts, therefore, can be obtained from H_3PO_4 (for example, Na_3PO_4, Na_2HPO_4, and NaH_2PO_4). The product of a given neutralization depends upon the stoichiometric ratio of H_3PO_4 to base.

Phosphates are important ingredients of commercial fertilizers. Phosphate rock is too insoluble in water to be used for this purpose. The more soluble calcium dihydrogen phosphate, however, may be obtained by treatment of phosphate rock with an acid.

$$Ca_3(PO_4)_2(s) + 2H_2SO_4(l) \rightarrow Ca(H_2PO_4)_2(s) + 2CaSO_4(s)$$

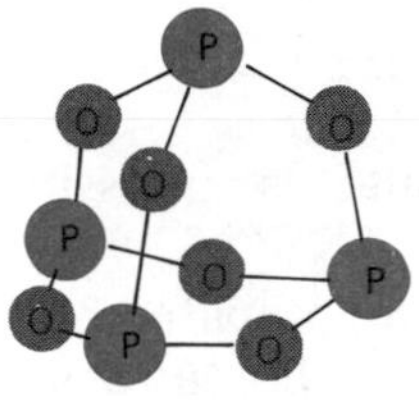

P_4O_6

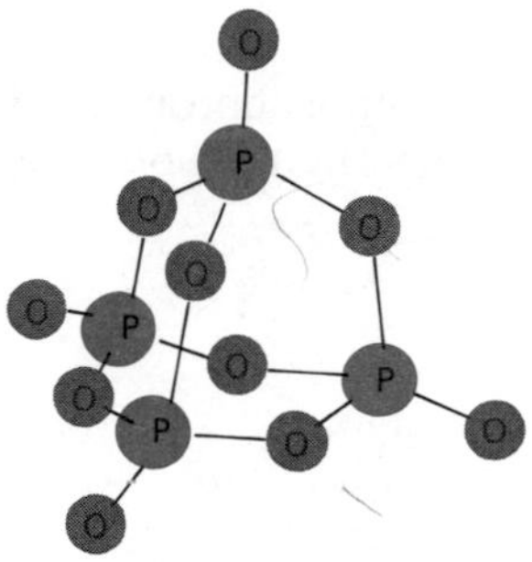

P_4O_{10}

Figure 18.6 Structures of P_4O_6 and P_4O_{10}.

The mixture of $Ca(H_2PO_4)_2$ and $CaSO_4$ is called **superphosphate fertilizer.** A higher yield of the dihydrogen phosphate is obtained if phosphoric acid is employed in the reaction instead of sulfuric acid.

$$Ca_3(PO_4)_2(s) + 4H_3PO_4(l) \rightarrow 3Ca(H_2PO_4)_2(s)$$

A number of condensed phosphoric acids (that have more than one P atom per molecule) are known. The condensed phosphoric acids may be obtained by the controlled addition of water to P_4O_{10}. For example,

$$P_4O_{10}(s) + 4H_2O \rightarrow \underset{\text{diphosphoric acid}}{2H_4P_2O_7(s)}$$

$$P_4O_{10}(s) + 2H_2O \xrightarrow{0°C} \underset{\text{tetrametaphosphoric acid}}{H_4P_4O_{12}(s)}$$

On standing in water, all condensed phosphoric acids revert to H_3PO_4.

Phosphor*ous* acid, H_3PO_3, may be prepared by adding P_4O_6 to cold water,

$$P_4O_6(s) + 6H_2O \rightarrow 4H_3PO_3(aq)$$

or by the hydrolysis of PCl_3 (Section 18.2). Condensed phosphorous acids also exist. Even though the H_3PO_3 molecule contains three hydrogen atoms, phosphorous acid is a weak, *diprotic* acid and is probably better formulated as $H_2(HPO_3)$.

$$H_2(HPO_3)(aq) \rightleftharpoons H^+(aq) + H(HPO_3)^-(aq)$$
$$H(HPO_3)^-(aq) \rightleftharpoons H^+(aq) + HPO_3^{2-}(aq)$$

The sodium salts NaH_2PO_3 and Na_2HPO_3 are known, but it is impossible to prepare Na_3PO_3.

Solutions of salts of hypophosphorous acid, H_3PO_2, are prepared by boiling white phosphorus with solutions of alkalies. Hypophosphorous acid is a weak monoprotic acid; hence the formula of the compound is sometimes written $H(H_2PO_2)$.

$$H(H_2PO_2)(aq) \rightleftharpoons H^+(aq) + H_2PO_2^-(aq)$$

The numbers of protons released by orthophosphoric, phosphorous, and hypophosphorous acids may be explained by the electronic structures of these compounds.

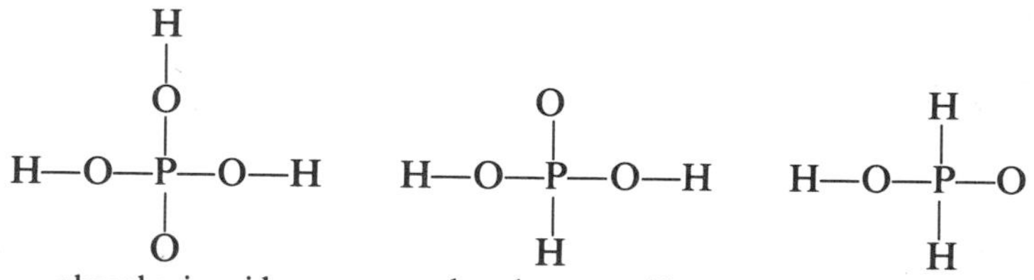

phosphoric acid   phosphorous acid   hypophosphorous acid

Only the hydrogen atoms bonded to oxygen atoms are acidic. Those bonded to phosphorus atoms do not dissociate as $H^+(aq)$ in water.

When arsenic or antimony is heated in air, As_4O_6 or Sb_4O_6 is formed. Aqueous solutions of As_4O_6 are acidic and thought to contain arsenious acid (H_3AsO_3). Antimony (III) oxide is insoluble in water. Both As_4O_6 and Sb_4O_6 dissolve in aqueous alkalies to form arsenites (salts of AsO_3^{3-}) and antimonites (salts of SbO_2^-).

The structures of the 5+ oxides of arsenic and antimony are not known, and the empirical formulas As_2O_5 and Sb_2O_5 are used. Orthoarsenic acid (H_3AsO_4, a weak, triprotic acid) forms when As_2O_5 is dissolved in water. Arsenates (salts of AsO_4^{3-}) and antimonates (salts of $Sb(OH)_6^-$) can be prepared by dissolving As_2O_5 or Sb_2O_5 in aqueous alkalies.

18.9 CARBON AND SILICON

Carbon and silicon are the nonmetallic elements of group IV A. The compounds of carbon are more numerous than the compounds of any other element except hydrogen. In fact, approximately ten compounds that contain carbon are known for every compound that does not. The chemistry of the hydrocarbons and their derivatives is the subject of organic chemistry (Chapter 21). The most important way that carbon differs from all other elements is the pronounced ability of carbon to catenate (see Section 8.8).

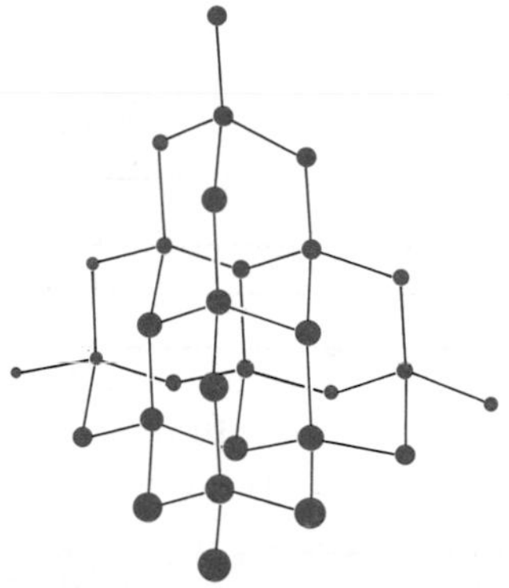

Figure 18.7 Arrangement of atoms in a diamond crystal.

Carbon constitutes approximately 0.03% of the earth's crust. In addition, the atmosphere contains 0.03% CO_2 by volume and carbon is also a constituent of all plant and animal matter. The allotropes of carbon, diamond and graphite, as well as impure forms of the element such as coal (which also contains combined carbon), occur in nature. In combined form the element occurs in hydrocarbons which are found in natural gas and petroleum, in the atmosphere as CO_2, and in carbonate minerals such as limestone ($CaCO_3$), dolomite ($CaCO_3 \cdot MgCO_3$), siderite ($FeCO_3$), and witherite ($BaCO_3$).

In diamond each carbon atom is bonded through sp^3 hybrid orbitals to four other carbon atoms arranged tetrahedrally (Figure 18.7). Strong bonds hold this network crystal together. The diamond is extremely hard, high melting, stable, and a nonconductor of electricity.

Graphite is a soft, black solid with a slight metallic luster. The graphite crystal is composed of layers formed from hexagonal rings of carbon atoms (Figure 18.8). The layers are held together by relatively weak London forces.

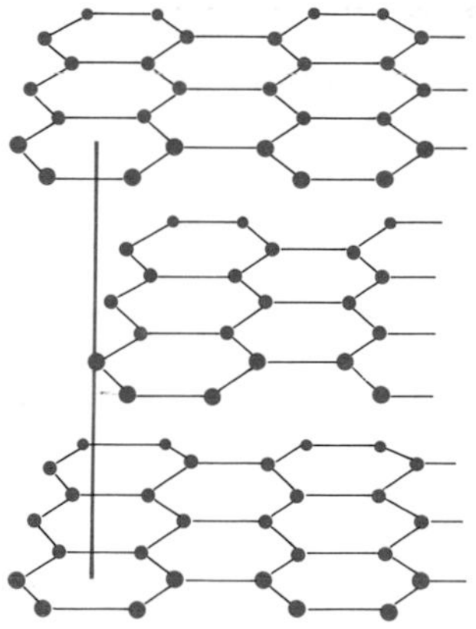

Figure 18.8 Arrangement of atoms in a graphite crystal.

Since it is easy for the layers to slide over one another, graphite is soft and has a slippery feel. It is less dense than diamond.

The nature of the bonding in the layers of the graphite crystal accounts for some of the properties of this substance. Each carbon atom is bonded to three other carbon atoms, and all the bonds are perfectly equivalent. Graphite may be represented as a resonance hybrid (Figure 18.9) in which each bond is a $1\frac{1}{3}$ bond.

Each C atom in graphite forms σ bonds with three other C atoms through the use of sp^2 hybrid orbitals. Therefore, the structure is planar with the three σ bonds of each atom directed to the corners of equilateral triangles. This bonding accounts for three of the four valence electrons of each C atom. The fourth, a p electron, is not involved in σ bond formation (Figure 18.10a).

In graphite *each* C atom has an additional p electron, and the resonance forms depict the possibility of forming conventional double bonds in three ways. Since each p orbital overlaps with more than one other p orbital, an extended π bonding system that encompasses the entire structure forms (Figure 18.10b). The electrons in this π bonding system are not localized between two atoms but are free to move throughout the entire layer. Hence, graphite has a metallic luster and is an electrical conductor. The conductivity is fairly large in a direction parallel to the layers but is small in a direction perpendicular to the planes of the crystal.

Silicon, which constitutes approximately 28% of the earth's crust, is the second most abundant element (oxygen is first). The element does not occur free in nature. It is found as silicon dioxide (sometimes called silica) and in an enormous variety of silicate minerals.

Silicon is prepared by the reduction of silicon dioxide by coke at high temperatures in an electric furnace.

$$SiO_2(l) + 2C(s) \rightarrow Si(l) + 2CO(g)$$

If a larger quantity of carbon is employed, silicon carbide (which is called Carborundum), SiC, is produced rather than silicon. The only known modification of silicon has a structure similar to diamond. Crystalline silicon is a gray, lustrous solid that is a semiconductor (Section 19.1).

A large number of carbides and silicides are known. The carbides have been much more extensively studied than the silicides. Carbides may be made by heating the appropriate metal or its oxide with carbon, carbon monoxide, or a hydrocarbon. Carbides may be classified as **saltlike, interstitial,** or **covalent.**

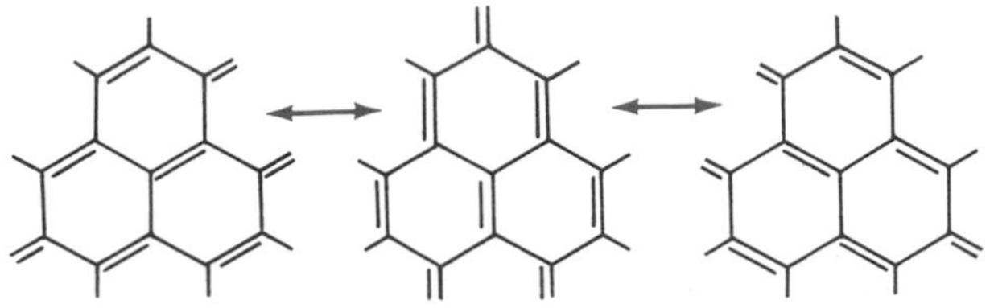

Figure 18.9 Resonance forms for a fragment of a graphite layer.

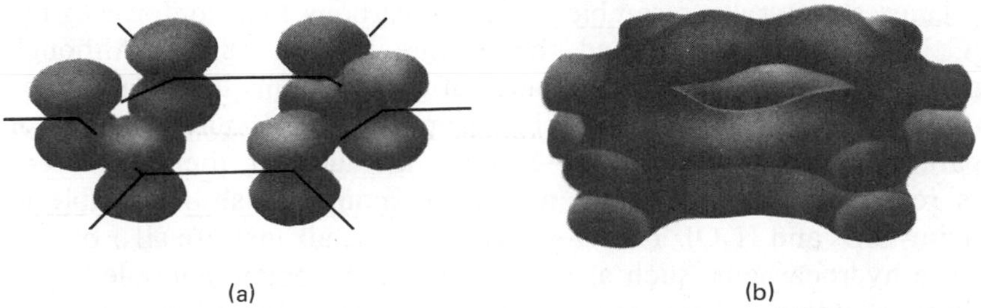

Figure 18.10 Schematic representation of the formation of the multicenter bonding system of a graphite fragment (σ bonds are shown as solid lines).

The saltlike carbides may be divided into two types. The I A and II A metals, as well as Cu^+, Ag^+, Au^+, Zn^{2+}, and Cd^{2+}, form carbides that are believed to contain the acetylide ion, C_2^{2-}, which has the structure

$$[:C{\equiv}C:]^{2-}$$

Upon hydrolysis, acetylides yield acetylene, C_2H_2, H—C≡C—H

$$CaC_2(s) + 2H_2O \rightarrow Ca(OH)_2(aq) + C_2H_2(g)$$

Beryllium carbide, Be_2C, and aluminum carbide, Al_4C_3, contain the methanide ion, C^{4-}, and yield methane, CH_4, upon hydrolysis.

$$Al_4C_3(s) + 12H_2O \rightarrow 4Al(OH)_3(s) + 3CH_4(g)$$

Interstitial carbides are formed by transition elements and consist of metallic lattices with carbon atoms in the interstices. These materials, such as TiC, TaC, W_2C, VC, and Mo_2C, are in general very hard, high melting, and chemically inert. They resemble metals in appearance and electrical conductivity.

Silicon carbide, produced by the reaction of SiO_2 and C, is a covalent carbide. It is hard, stable, and chemically inert. The silicon carbide crystal consists of a three-dimensional network formed from tetrahedra of alternating carbon and silicon atoms. Boron carbide, B_4C, which is even harder than silicon carbide, is made by the reduction of B_2O_3 by carbon in an electric furnace.

Silicon dissolves in almost all molten metals, and in many of these instances, definite compounds, called silicides, are produced. The compounds Mg_2Si, $CaSi_2$, Li_3Si, FeSi, Fe_2Si, Ni_2Si, NiSi, VSi_2, and $CoSi_3$ are among those known. Although probably none of the silicides are truly ionic, certain of them hydrolyze to produce hydrogen-silicon compounds, called **silicon hydrides** or **silanes.**

The silanes are compounds of general formula Si_nH_{2n+2}. Compounds where n equals 1 to 6 are well characterized. The hydrolysis of Mg_2Si by dilute acids gives a mixture of silanes. For example,

$$Mg_2Si(s) + 4H^+(aq) \rightarrow 2Mg^{2+}(aq) + SiH_4(g)$$

The silanes structurally resemble the hydrocarbons that conform to the formula C_nH_{2n+2}, which are called the **alkanes** (Figure 18.11). Although there is presumably no limit to the number of carbon atoms that can join together to form alkanes, the number of silanes appears to be limited because of the comparative weakness of the Si—Si bond. In addition, the alkanes are much less reactive than the silanes, which are spontaneously flammable in air (producing SiO_2 and H_2O). The C—C bonds of the alkanes are all σ bonds, but there are hydrocarbons, such as acetylene, which contain multiple bonds between carbon atoms. Silicon hydrides that contain multiple bonds are unknown.

The principal oxides of carbon are carbon monoxide and carbon dioxide. Carbon monoxide is formed by the combustion of carbon in a limited supply of oxygen. It is also produced by the reaction of CO_2 and C at high temperatures. It is used as a reducing agent in metallurgical processes. At high temperatures, it reacts with many metal oxides to yield the free metal and CO_2 (see Section 19.4).

$$FeO(s) + CO(g) \rightarrow Fe(l) + CO_2(g)$$

The role of CO as an air pollutant is discussed in Section 18.6.

Carbon monoxide burns in air to form carbon dioxide.

$$2CO(g) + O_2(g) \rightarrow 2CO_2(g) \qquad \Delta H = -283 \text{ kJ}$$

Since this reaction is highly exothermic, CO is used as a fuel. Carbon dioxide is also produced by the complete combustion of carbon or compounds of carbon. It may be prepared by the reaction of carbonates or hydrogen carbonates with acids,

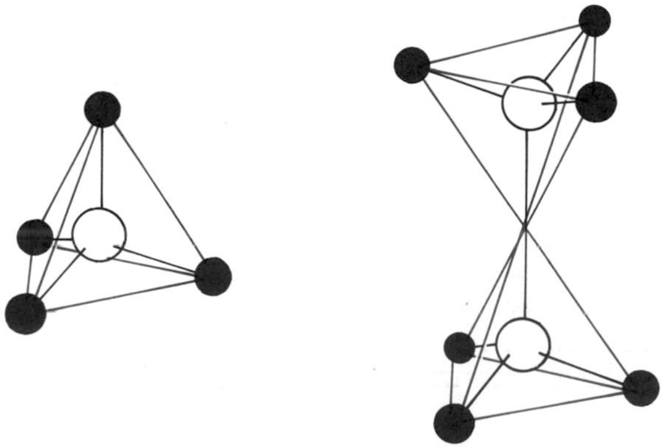

(a) (b)

Figure 18.11 Arrangement of atoms in (a) CH_4 and SiH_4 and (b) C_2H_6. Open circles, C or Si; dark circles, H.

$$CaCO_3(s) + 2H^+(aq) \rightarrow Ca^{2+}(aq) + CO_2(g) + H_2O$$

and by heating carbonates,

$$CaCO_3(s) \rightarrow CaO(s) + CO_2(g)$$

Carbon dioxide is the acid anhydride of carbonic acid, a weak, diprotic acid (see Section 15.7). It forms two series of salts: the normal carbonates, such as Na_2CO_3 and $CaCO_3$, and the hydrogen carbonates, such as $NaHCO_3$ and $Ca(HCO_3)_2$.

Sodium carbonate is an important industrial chemical, and more than 90% of the compound is produced commercially by the **Solvay process.** Ammonia and carbon dioxide, under pressure, are passed into a saturated solution of sodium chloride. The ammonia and carbon dioxide react.

$$NH_3(g) + CO_2(g) + H_2O \rightleftharpoons NH_4^+(aq) + HCO_3^-(aq)$$

Sodium bicarbonate precipitates from the solution because of the high concentration of sodium ion, and ammonium chloride is left behind.

$$Na^+(aq) + HCO_3^-(aq) \rightarrow NaHCO_3(s)$$

The sodium bicarbonate is removed by filtration and washed; heating decomposes it into the normal carbonate.

$$2NaHCO_3(s) \rightarrow Na_2CO_3(s) + CO_2(g) + H_2O(g)$$

The ammonia is recovered, for reuse, from the original solution by heating this solution with quicklime, CaO.

$$2NH_4^+(aq) + CaO(s) \rightarrow 2NH_3(g) + Ca^{2+}(aq) + H_2O$$

The calcium oxide is obtained from limestone, $CaCO_3$, by heating. The CO_2 produced as a by-product of the preparation of CaO, as well as that produced by heating $NaHCO_3$, is used in the initial step of the process.

The only well-characterized oxide of silicon is SiO_2. In contrast to the oxides of carbon, which are volatile molecular species held together by London forces in the solid state, SiO_2 forms very stable, nonvolatile, three-dimensional network crystals (melting point, ~ 1700°C). One of the three crystal modifications of SiO_2 has a lattice that may be considered to be derived from the diamond lattice with silicon atoms replacing carbon atoms and an oxygen atom midway between each pair of silicon atoms.

The chemistry of silicon is dominated by compounds that contain the Si—O linkage. Silicon dioxide is the product of the reaction of the elements. It is produced by reaction of the spontaneously flammable silicon hydrides with air. Hydrous SiO_2 is the product of the hydrolysis of many silicon compounds. Silicon dioxide occurs in several forms in nature. Among them are sand, flint, agate, jasper, onyx, and quartz.

Silicon dioxide is an acidic oxide. No acids of silicon, however, have ever been isolated. The oxide does not react directly with water. Acidification of a water solution of a soluble silicate yields only hydrous SiO_2. Silicates may be made by heating metal oxides or metal carbonates with SiO_2. Certain silicates of the I A metals (with a molar ratio of silicon dioxide to metal oxide of not more than 2 to 1) are water soluble. A large number of silicates, of various types, occur in nature.

Glass is a mixture of silicates made by fusing SiO_2 with metal oxides and carbonates. Common soda-lime glass is made from Na_2CO_3, $CaCO_3$, and SiO_2. Special glasses may be made by the addition of other acidic and basic oxides (such as Al_2O_3, B_2O_3, PbO, and K_2O). Cement is a complex aluminosilicate mixture made from limestone ($CaCO_3$) and clay ($H_4Al_2Si_2O_9$).

Carbon disulfide, CS_2, is a volatile liquid prepared by heating carbon and sulfur together in an electric furnace. The electronic structure of the molecule is analogous to CO_2. Silicon disulfide, SiS_2, prepared by the reaction of the elements at high temperatures, is a colorless, crystalline material that hydrolyzes to SiO_2 and H_2S. Solid SiS_2 has a chainlike structure

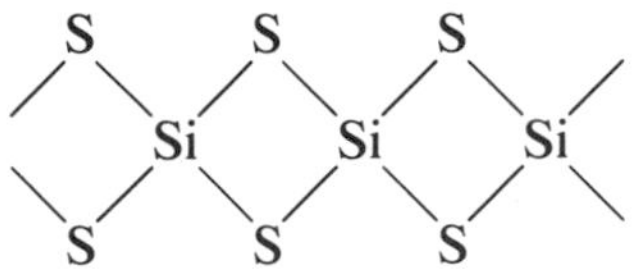

Carbon tetrachloride is made by heating chlorine with carbon disulfide

$$CS_2(g) + 3Cl_2(g) \rightarrow CCl_4(g) + S_2Cl_2(g)$$

Carbon tetrachloride, which is a liquid under ordinary conditions, is a good solvent for many nonpolar materials. Carbon tetrafluoride, CF_4, which may be obtained by the fluorination of almost any carbon-containing compound, is a very stable gas. Carbon tetrabromide, CBr_4, and carbon tetraiodide, CI_4, are unstable solids.

The silicon tetrahalides, SiX_4, are formed by the reactions of the elements; SiF_4 is a gas; $SiCl_4$ and $SiBr_4$ are liquids; and SiI_4 is a solid. Unlike the tetrahalides of carbon, the silicon tetrahalides readily hydrolyze

$$SiX_4 + (n + 2)H_2O \rightarrow SiO_2 \cdot nH_2O(s) + 4H^+(aq) + 4X^-(aq)$$

The difference in chemical reactivity of the tetrahalides of carbon and silicon is attributed to the availability of unoccupied *d* orbitals in silicon which make it possible for a water molecule to attach itself and thus facilitate hydrolysis. Catenated halides of carbon and silicon are known (for example, C_2Cl_6 and Si_2Cl_6).

18.10 BORON

Boron is the only group III A element that is a nonmetal. The principal ores of boron, which does not occur free in nature, are borates (such as borax,

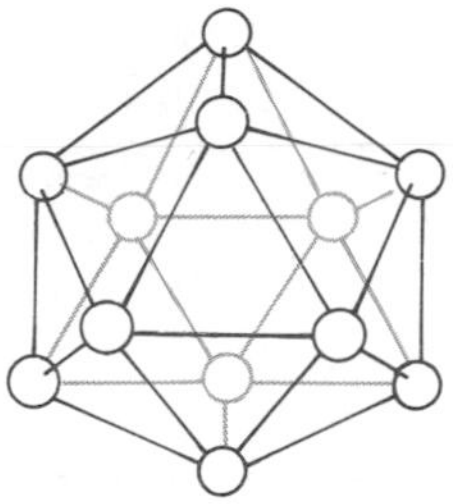

Figure 18.12 Arrangement of atoms in a B_{12} icosahedron.

$Na_2B_4O_7 \cdot 10H_2O$). Pure, crystalline boron may be obtained by reducing BBr_3 by hydrogen on a hot tungsten filament (at about 1500°C).

$$2BBr_3(g) + 3H_2(g) \rightarrow 2B(s) + 6HBr(g)$$

The crystals, which condense on the filament, are black with a metallic luster.

The structures of three types of boron crystals have been determined, and at least one other allotrope is known to exist. Groups of 12 B atoms arranged in regular icosahedra (solid figures, each with 20 equilateral-triangular faces) occur in each of these crystalline modifications (Figure 18.12). Each B atom participates not only in the bonding of its icosahedron but also in bonding that icosahedron to others. Ordinary electron-pair bonds cannot account for the large number of boron–boron interactions of the structure since a boron atom has only three electrons and four orbitals in its valence level.

The electrons used in the bonding of elemental boron, therefore, are extensively delocalized over many B atoms. The properties of crystalline boron are those of a covalent network crystal similar to diamond. Boron is extremely hard, brittle, and has a high melting point (about 2250°C).

At very high temperatures (about 2000°C) boron reacts with many metals to form borides. These substances are very hard, are chemically stable, and have metallic conductivity.

Boron reacts with ammonia or nitrogen at elevated temperatures to produce boron nitride, BN. This substance is isoelectronic with carbon and has a crystal structure similar to graphite but with alternating boron and nitrogen atoms. At very high temperatures and pressures this modification of BN is converted to another form that has a diamond-type lattice and is harder than diamond.

At high temperatures, boron reacts with the halogens to yield the trihalides. BF_3 and BCl_3 are gases, BBr_3 is a liquid, and BI_3 is a solid. The molecules of the trihalides are triangular planar. Presumably, sp^2 hybrid orbitals of boron are used for bonding.

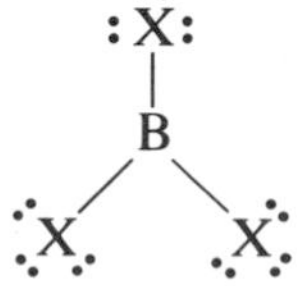

Since the boron atom does not have an octet of electrons in each of these molecules, the trihalides react as electron acceptors (Lewis acids, see Section 12.6).

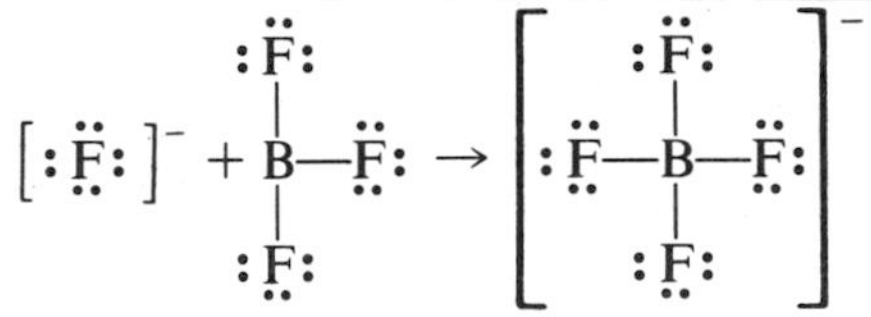

$$NH_3 + BF_3 \rightarrow H_3N\!:\!BF_3$$
$$H_2O + BF_3 \rightarrow H_2O\!:\!BF_3$$

The fluoborate ion, BF_4^-, is tetrahedral and is bonded by sp^3 hybrid bonds.

Many borates occur in nature. The acidification of an aqueous solution of any borate precipitates orthoboric acid (also called boric acid), H_3BO_3 or $B(OH)_3$, as soft, white crystals. This acid may also be prepared by the complete hydration of B_2O_3. Boric acid and the borates are toxic.

Boric acid is a very weak, *monobasic* acid in solution. Its ionization in fairly dilute solution may be represented as

$$B(OH)_3(aq) + H_2O \rightleftharpoons H^+(aq) + B(OH)_4^-(aq)$$

or

$$(H_2O)B(OH)_3(aq) \rightleftharpoons H^+(aq) + B(OH)_4^-(aq)$$

The $B(OH)_4^-$ ion is tetrahedral. In more concentrated solutions, polymeric anions, such as $B_3O_3(OH)_4^-$, exist.

When boric acid is heated to about 175°C, metaboric acid, HBO_2, results. The meta- acid reverts to the more highly hydrated ortho- form in water. Although salts of many hypothetical boric acids are known, these are the only two acids that exist in the pure condition.

Metaboric acid may be dehydrated by heating into boric oxide, B_2O_3. The oxide may also be prepared by heating boron in air.

A number of hydrides of boron are known. These electron-deficient compounds are discussed in Section 8.8.

18.11 HYDROGEN

The most important naturally occurring compound of hydrogen is water, which is the principal source of the element. Hydrogen also occurs in the hydrocarbons found in coal, natural gas, and petroleum. A few minerals, such as clay and certain hydrates, contain hydrogen. Compounds of hydrogen (with carbon, oxygen, and other elements) constitute the principal part of all plant and animal matter.

The very reactive metals (Ca, Sr, Ba, and the group I A metals) react with water at room temperature to produce hydrogen and solutions of hydroxides.

$$2Na(s) + 2H_2O \rightarrow 2Na^+(aq) + 2OH^-(aq) + H_2(g)$$
$$Ca(s) + 2H_2O \rightarrow Ca^{2+}(aq) + 2OH^-(aq) + H_2(g)$$

At high temperatures some additional metals, which have slightly lower reactivity, will displace hydrogen from water (steam). In these cases, metal oxides (the anhydrides of hydroxides) are obtained instead of metal hydroxides.

$$Mg(s) + H_2O(g) \rightarrow MgO(s) + H_2(g)$$

The reaction of iron and steam serves as a commercial source of hydrogen.

$$3Fe(s) + 4H_2O(g) \rightarrow Fe_3O_4(s) + 4H_2(g)$$

A still longer list of metals (including Mg, Al, Mn, Zn, Cr, Fe, Cd, Co, Ni, Sn, and Pb) will react with aqueous solutions of acids to produce hydrogen. This method is a convenient laboratory preparation, but it is too expensive to be of commercial importance. The very reactive metals react too violently to be useful.

$$Zn(s) + 2H^+(aq) \rightarrow Zn^{2+}(aq) + H_2(g)$$
$$Fe(s) + 2H^+(aq) \rightarrow Fe^{2+}(aq) + H_2(g)$$
$$2Al(s) + 6H^+(aq) \rightarrow 2Al^{3+}(aq) + 3H_2(g)$$

The anions of certain acids (for example, nitric acid) also react with the metals. These acids cannot be used to prepare pure hydrogen.

Coke and steam react at high temperatures (1000°C) to produce a gaseous mixture known as water gas.

$$C(s) + H_2O(g) \rightarrow CO(g) + H_2(g)$$

This mixture is used as a fuel. Since it is difficult to separate hydrogen from carbon monoxide, the water gas is subjected to a second step if hydrogen alone is desired. Water gas, mixed with additional steam, is passed over a catalyst at a temperature of about 500°C. This step converts the carbon monoxide into carbon dioxide.

$$CO(g) + H_2O(g) \rightarrow CO_2(g) + H_2(g)$$

The carbon dioxide and hydrogen are separated by taking advantage of the solubility of carbon dioxide in cold water under pressure.

Hydrocarbons react with steam in the presence of suitable catalysts to give mixtures of carbon monoxide and hydrogen.

$$CH_4(g) + H_2O(g) \rightarrow CO(g) + 3H_2(g)$$

The production of hydrogen and water gas from these reactions is industrially important.

Hydrogen is obtained from the catalytic decomposition of hydrocarbons at high temperatures.

$$CH_4(g) \rightarrow C(s) + 2H_2(g)$$

In gasoline refining, petroleum hydrocarbons are "cracked" into compounds of lower molecular weight. Hydrogen is a by-product.

Very pure, but relatively expensive, hydrogen is commercially prepared by the electrolysis of water.

$$2H_2O \xrightarrow{\text{electrolysis}} 2H_2(g) + O_2(g)$$

Hydrogen is a by-product (as is chlorine) in the industrial preparation of sodium hydroxide by the electrolysis of concentrated aqueous solutions of sodium chloride.

$$2Na^+(aq) + 2Cl^-(aq) + 2H_2O \xrightarrow{\text{electrolysis}} Na^+(aq) + 2OH^-(aq) + H_2(g) + Cl_2(g)$$

Certain metals and nonmetals displace hydrogen from alkaline solutions.

$$Zn(s) + 2OH^-(aq) + 2H_2O \rightarrow H_2(g) + \underset{\text{zincate ion}}{Zn(OH)_4^{2-}(aq)}$$

$$2Al(s) + 2OH^-(aq) + 6H_2O \rightarrow 3H_2(g) + \underset{\text{aluminate ion}}{2Al(OH)_4^-(aq)}$$

$$Si(s) + 2OH^-(aq) + H_2O \rightarrow 2H_2(g) + \underset{\text{silicate ion}}{SiO_3^{2-}(aq)}$$

The reactions of hydrogen are discussed in Section 8.8.

PROBLEMS

18.1 Write chemical equations to show how to prepare the following: (a) F_2 from CaF_2, (b) Cl_2 from aqueous NaCl, (c) Cl_2 from molten NaCl, (d) Br_2 from sea water, (e) I_2 from oil well brines.

18.2 Write chemical equations to show how $Cl_2(g)$ may be prepared from $Cl^-(aq)$ by use of (a) $MnO_2(s)$, (b) $PbO_2(s)$, (c) $MnO_4^-(aq)$, (d) $Cr_2O_7^{2-}(aq)$. Analogous equations may be written for reactions in which bromine and iodine are prepared from bromide ions and iodide ions. Fluorine, however, cannot be prepared from fluoride ions by use of these oxidizing agents. Why not?

18.3 Write chemical equations for the reactions of Cl_2 with (a) H_2, (b) Zn, (c) P, (d) S, (e) H_2S, (f) CO, (g) SO_2, (h) $I^-(aq)$, (i) cold H_2O.

18.4 Write equations for the reaction of HF with (a) SiO_2, (b) Na_2CO_3, (c) KF, (d) CaO.

18.5 Write equations to show why concentrated HF(aq) is more strongly ionic than dilute HF(aq).

18.6 Since HCl(g) can be prepared from NaCl and H_2SO_4, why is it that the reaction of NaBr and H_2SO_4 cannot be used to prepare HBr(g)?

18.7 Describe the allotropes of oxygen and sulfur. Draw the Lewis structures for the molecules involved.

18.8 Describe the Frasch process for mining sulfur.

18.9 Compare the contact and lead-chamber processes for the manufacture of sulfuric acid. Which process yields the purer product?

18.10 Discuss the reactions of sulfuric acid.

18.11 What are the oxidation numbers of sulfur in the thiosulfate ion ($S_2O_3^{2-}$) and in the peroxydisulfuric acid molecule ($H_2S_2O_8$)?

18.12 Write equations for the reactions of SO_2 with (a) O_2 (Pt catalyst), (b) H_2O, (c) OH^-(aq), (d) SO_3^{2-}(aq) and H_2O.

18.13 Draw Lewis structures and describe the geometric structure of (a) SO_3^{2-}, (b) SO_4^{2-}, (c) $S_2O_3^{2-}$, (d) H_2SO_5, (e) $H_2S_2O_7$, (f) SO_3.

18.14 Write equations for the reactions of water with (a) $CH_3C(NH_2)S$, (b) Al_2S_3, (c) SeO_2, (d) $H_2S_2O_7$.

18.15 For each of the following acids, an oxide exists that reacts with water to produce the acid. What are these acid anhydrides? (a) H_2SO_3, (b) H_2SeO_4, (c) H_6TeO_6, (d) HOCl, (e) $HClO_4$, (f) HIO_3, (g) H_5IO_6.

18.16 Write equations for the reaction of HCl(aq) with (a) Na_2SO_3, (b) Na_2S, (c) $Na_2S_2O_3$, (d) ZnO.

18.17 Write a sequence of equations representing reactions that could be used to synthesize each of the following and that start with elementary S. (a) H_2S (not by direct union of the elements), (b) H_2SO_3, (c) $Na_2S_2O_3$, (d) $NaHSO_4$, (e) $H_2S_2O_7$.

18.18 Explain why (a) H_6SO_6 does not exist but H_6TeO_6 does, (b) HCl_2^- is not known but HF_2^- is, (c) OF_4 cannot be prepared but SF_4 can be.

18.19 What is the difference in meaning between the prefixes *per-* and *peroxy-* as applied in the naming of acids?

18.20 Write equations for the preparation of oxygen from (a) HgO, (b) H_2O, (c) BaO_2, (d) $KClO_3$.

18.21 What is the relative humidity of air at 22°C if the partial pressure of water vapor in the air is 0.0170 atm? What is the relative humidity if the air is warmed to 29°C? Vapor pressures of water are given in Table 4.4.

18.22 On a hot summer day, the temperature is 34°C and the relative humidity if 90.0%. What is the partial pressure of water vapor in the air?

18.23 Which steps of the nitrogen cycle in nature increase the concentration of N_2 in the air? Which decrease the concentration?

18.24 Which steps of the oxygen–carbon dioxide cycle of nature increase the concentration of CO_2 in the air? Which decrease the concentration?

18.25 What substances are commercially obtained from liquid air?

18.26 What are the five principal air pollutants?

18.27 What is photochemical smog? How is it thought to form?

18.28 Why is it advantageous for a farmer to rotate the crops sown in a given field?

18.29 In 1892 Lord Rayleigh observed that the density of nitrogen prepared from air (by the removal of H_2O, CO_2, and O_2) was different from the density of nitrogen prepared from NH_4Cl and $NaNO_2$. On the basis of this observation, Rayleigh and William Ramsey isolated argon (1894). Assume that air is 1.00% Ar, 78.00% N_2, and 21.00% O_2 (by volume). How should the densities of the two "nitrogen" samples compare?

18.30 The normal boiling point of the compound ethylene diamine, $H_2NCH_2CH_2NH_2$, is 117°C and that of propyl amine, $CH_3CH_2CH_2NH_2$, is 49°C. The molecules, however, are similar in size and molecular weight. What reason can you give for the difference in boiling point?

18.31 Why is nitrogen a relatively unreactive element?

18.32 Discuss the molecular structures and reactivities of the three allotropic forms of phosphorus.

18.33 What are the characteristics of the three types of nitrides?

18.34 Discuss the preparation, structure, and properties of ammonia. Why is the boiling point of NH_3 high in comparison to the boiling points of PH_3, AsH_3, and SbH_3?

18.35 Write equations for the following reactions of ammonia.

(a) $NH_3(aq) + Ag^+(aq) \rightarrow$

(b) $NH_3(aq) + H^+(aq) \rightarrow$

(c) $NH_3(aq) + H_2O + CO_2(aq) \rightarrow$

(d) $NH_3(g) + O_2(g) \xrightarrow{\text{heat}}$

(e) $NH_3(g) + O_2(g) \xrightarrow[\text{heat}]{\text{Pt}}$

(f) $NH_3(g) + HCl(g) \rightarrow$

(g) $NH_3(g) + V(s) \xrightarrow{\text{heat}}$

18.36 Write a series of equations for the preparation of nitric acid starting with N_2 as the source of nitrogen.

18.37 Write an equation for the reaction of HNO_3 with (a) Cu, (b) P_4O_{10}, (c) NH_3, (d) $Ca(OH)_2$.

18.38 List the oxides of nitrogen. Write a chemical equation for the preparation of each of them.

18.39 State the products of the thermal decompositions of (a) NH_4NO_3, (b) NH_4NO_2, (c) $NaNO_3$, (d) NaN_3, (e) $NaHCO_3$, (f) $CaCO_3$.

18.40 Write equations for the following oxidation–reduction reactions: (a) the reaction of NH_3 with OCl^- that produces N_2H_4 and Cl^-; (b) the reaction of NO_2 with H_2 in HCl solution that produces $[NH_3OH]Cl$; (c) the electrolysis of NH_4F in anhydrous HF that produces NF_3 and H_2.

18.41 Draw electronic structures for orthophosphoric, phosphorus, and hypophosphorous acids. Tell how these structures explain the number of $H^+(aq)$ dissociated per mole when each of these compounds is dissolved in water.

18.42 Write an equation for the reaction of each of the following with water. (a) Li_3N, (b) Ca_3P_2, (c) PCl_3, (d) PCl_5, (e) $H_4P_2O_7$, (f) NO_2, (g) CaC_2, (h) Be_2C, (i) $SiCl_4$, (j) BF_3.

18.43 Determine the oxidation number of the central atom and the acid anhydride of each of the following.

(a) $H_2N_2O_2$, (b) HNO_2, (c) HNO_3, (d) $H_4P_2O_7$, (e) $H_3P_3O_9$, (f) H_3PO_3, (g) H_3AsO_4, (h) H_2CO_3, (i) $B(OH)_3$.

18.44 Write equations for the reactions of aqueous NaOH with the following. (a) NH_4Cl, (b) P_4, (c) As_4O_6, (d) Si, (e) Zn.

18.45 Write equations for the reactions of HCl with (a) H_2NNH_2, (b) H_2NOH, (c) $NaNO_2$, (d) $CaCO_3$, (e) $Na_2S_2O_3$.

18.46 Explain the difference that CF_4 and SiF_4 show in their reactivity toward water.

18.47 Write an equation for the reaction that occurs when each of the following is heated with carbon. (a) P_4O_{10}, (b) As_4O_6, (c) CaO, (d) $NaNO_3$, (e) SiO_2.

18.48 Describe the crystal structure of (a) diamond, (b) graphite, (c) SiC, (d) Si, (e) SiO_2.

18.49 Describe the shape of the following molecules. (a) P_4, (b) P_4O_6, (c) P_4O_{10}.

18.50 Describe the commercial processes for the manufacture of (a) P_4, (b) NH_3, (c) H_3PO_4, (d) HNO_3, (e) superphosphate fertilizer, (f) SiC, (g) water gas.

18.51 Write an equation to show how orthoboric acid, $B(OH)_3$, ionizes in water.

18.52 Write chemical equations for the preparation of hydrogen from (a) Na and H_2O, (b) Mg and H_2O, (c) Fe and H_2O, (d) C and H_2O, (e) Zn and $H^+(aq)$, (f) Zn and $OH^-(aq)$.

Chapter 19

Metals and Metallurgy

Several physical and chemical properties that are typical of metals are used to define this classification of elements—although the extent of each of these properties varies widely from metal to metal. Metals have superior electrical and thermal conductivities, characteristic luster, and the ability to be deformed under stress without cleaving. The tendency of these elements toward the formation of cations through electron loss and their formation of basic oxides are among the chemical characteristics of metals.

More than three-quarters of all known elements are metals. The stepped line that appears in the periodic table marks an approximate division between metals and nonmetals—the nonmetals appear in the upper right corner of the table. The division, however, is slightly arbitrary. Elements that appear close to the line have intermediate properties.

19.1 THE METALLIC BOND

Metals have comparatively low ionization energies and electronegativities. Consequently, the outer electrons of metal atoms are relatively loosely held. In a metallic crystal, positive ions (which consist of metal atoms minus the outer electrons) occupy positions in the crystal lattice. The outer electrons move freely throughout this lattice and bind the crystal together.

The **band theory** describes this bonding in terms of molecular orbitals that extend over the entire crystal. A general idea of the band model for the lithium crystal can be developed in the following way. In the Li_2 molecule, the $2s$ orbital of one Li atom overlaps the $2s$ orbital of another, and $\sigma\ 2s$ and σ^*2s molecular orbitals are formed (see Section 7.8). In the ground state of Li_2, only the $\sigma 2s$ molecular orbital is occupied. Nevertheless, the combination of two atomic orbitals leads to the formation of two molecular orbitals even if one of them is not occupied.

The $2s$ orbitals of two separate Li atoms have the same energy and are said to be **degenerate** (see the right portion of the curves in Figure 19.1). When the two Li atoms are brought together, the degeneracy is split (move

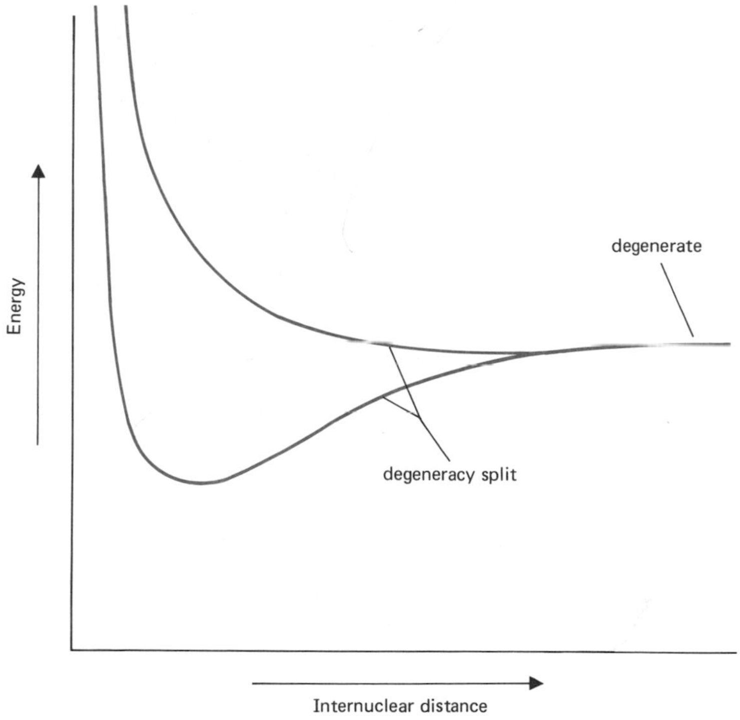

Figure 19.1 Energy of the $2s$ orbitals of two lithium atoms as a function of internuclear distance.

toward the left in Figure 19.1). The $\sigma 2s$ orbital (lower curve of Figure 19.1) has a lower energy than the $\sigma^* 2s$ orbital (upper curve of Figure 19.1).

If *seven* Li atoms are brought together, *seven* molecular orbitals result from the overlap of the *seven* 2*s* orbitals [Figure 19.2 (a)]. These molecular orbitals taken together make up what is called a **band.** In Figure 19.2(a), the internuclear distance R_0 corresponds to the normal distance between Li nuclei in the Li crystal. Figure 19.2(b) illustrates the relative spacing in terms of energy of the seven molecular orbits within the band at R_0.

If a large number, *N*, of lithium atoms are brought together, *N* molecular orbitals will be formed. These molecular orbitals, or energy levels, will be crowded together into a band. The levels will be very closely spaced in terms of energy and for all practical purposes will form a continuous energy band. Even at low temperatures, the electrons have enough energy to move from level to level within a band.

Each level of a band may accommodate two electrons of opposed spin. If *N* atoms contribute *N* levels to a band, the band is capable of holding 2*N* electrons. Since *N* lithium atoms have a total of *N* valence electrons, it would appear that the band should be one-half occupied. The band derived from the 2*p* orbitals of lithium must also be considered, however.

In an isolated lithium atom the three 2*p* orbitals are only slightly higher in energy than the 2*s* orbital. The band derived from the 2*p* orbitals fans out and

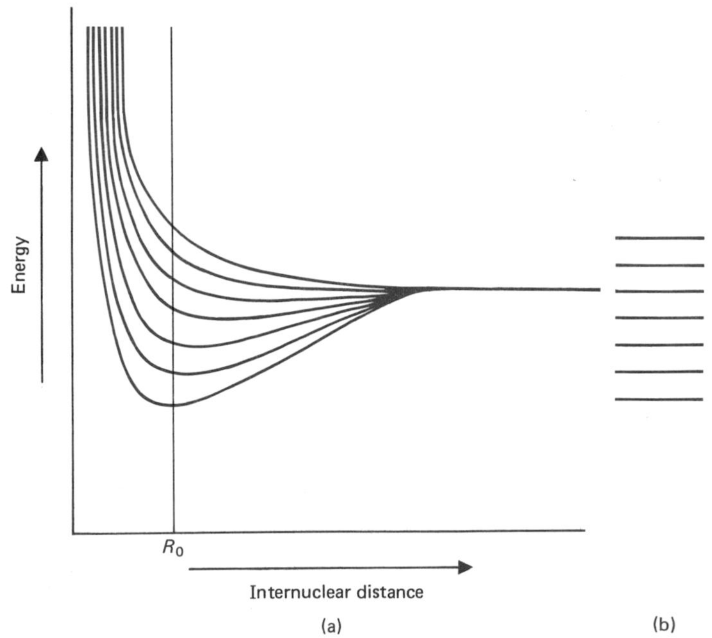

Figure 19.2 (a) Energy of 2*s* orbitals versus internuclear distance for a collection of seven atoms. (b) Energy-level diagram corresponding to the internuclear distance R_0.

overlaps the band from the $2s$ orbitals (Figure 19.3). At the normal equilibrium internuclear distance, R_0, the two bands may be considered to form one. Since each atom contributes three $2p$ orbitals and one $2s$ orbital, the band resulting from the combination of N atoms would consist of $4N$ levels that are capable of holding $8N$ electrons. The lithium band, therefore, is only one-eighth filled.

If this band overlap did not occur, the valence band of beryllium (electronic configuration, $1s^2 2s^2$) would be full, and beryllium would not conduct electricity. Beryllium is a metallic conductor because the band resulting from the overlap of the $2s$ and $2p$ orbitals is only one-quarter filled.

In a metallic crystal, therefore, the valence electrons are held by all the atoms of the crystal and are highly mobile. The valence band, which may be only partially filled, is overlapped by an unfilled conduction band. Transference of an electron to a higher level within a band requires the addition of very little energy since the levels are close together. Thus the valence electrons of a metal can move to higher levels by absorbing light of a wide range of wavelengths. When these electrons fall to lower energy levels, light is radiated. The lustrous appearance of metals is caused by these electron transitions.

The freely moving electrons of metallic crystals account for the high

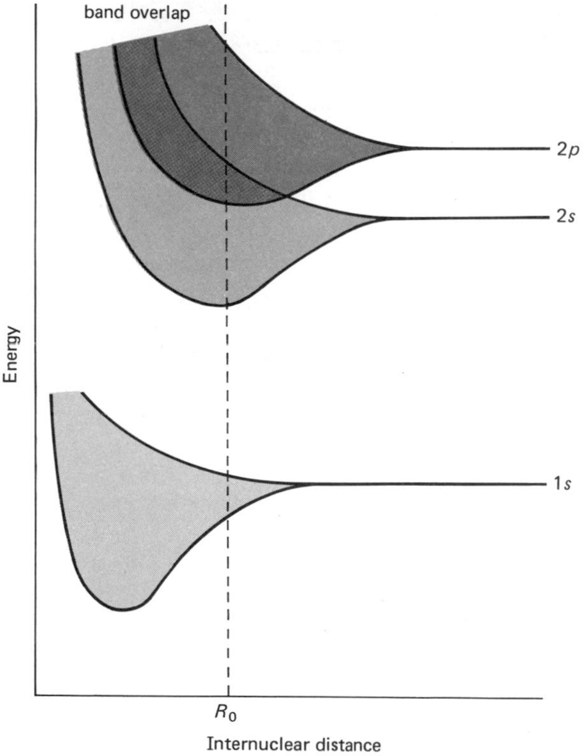

Figure 19.3 Overlay of $2s$ and $2p$ bands.

thermal and electrical conductivities of metals. Valence electrons of a metal absorb heat as kinetic energy and transfer it rapidly to all parts of the metal since their motion is relatively unrestricted.

Energy-level diagrams for **conductors, insulators,** and **intrinsic semiconductors** are shown in Figure 19.4. In the diagram for a conductor (such as Li), the valence band (from the 2*s* orbitals) is overlapped by an empty conduction band (from the 2*p* orbitals). These two bands are separated from an empty, upper conduction band (derived from orbitals of the third shell) by a **forbidden energy zone.** Electrical conduction takes place by means of the motion of electrons within the lower conduction band. There is no need to supply the energy required to bridge the forbidden zone and use the upper conduction band.

The diagram for the insulator, however, shows a completely filled valence band that is widely separated from a conduction band by a forbidden energy zone. Electron motion, and hence electrical conductivity, is only possible if energy is provided to promote electrons across the comparatively large forbidden zone to the conduction band. Normally such promotion does not occur, and hence the conductivities of insulators are extremely low.

An intrinsic semiconductor is a material that has a low electrical conductivity that is intermediate between that of a conductor and an insulator. This conductivity increases markedly with increasing temperature. For a semiconductor, the forbidden zone is sufficiently narrow that electrons can be promoted from the valence band to the conduction band by heat (see Figure 19.4). The vacancies left by the removal of electrons from the valence band permit the electrons remaining in the valence band to move under the influence of an electric field. Conduction takes place through the motion of electrons in the valence, as well as in the conduction, band.

Pure silicon and germanium function as intrinsic semiconductors. Both these elements crystallize in the diamond-type network lattice in which each atom is bonded to four other atoms (Figure 9.12). At room temperature the conductivity of either silicon or germanium is extremely low, since the electrons are fixed by the bonding of the crystal. At higher temperatures, however,

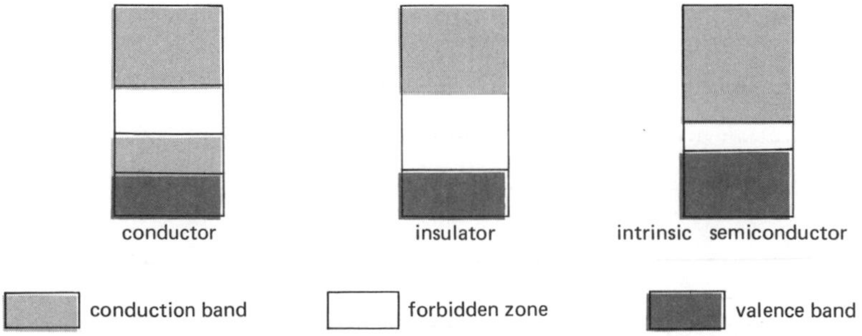

Figure 19.4 Energy-level band diagrams for three types of solids.

the crystal bonding begins to break down. Electrons are freed and are able to move through the structure, thereby causing the conductivity to increase.

The addition of small traces of certain impurities to either silicon or germanium enhances the conductivity of these materials and produces what are called **extrinsic** (or **impurity**) **semiconductors.** Each Si atom has four valence electrons that are used in the bonding of the network lattice. A boron atom, however, has only three valence electrons. Hence, a B atom that assumes a position of a Si atom in the crystal can form only three of the four bonds required for a perfect lattice, and an electron vacancy (or hole) is introduced. An electron from a nearby bond can move into this vacancy, thus completing the four bonds on the B atom but, at the same time, leaving a vacancy at the original site of the electron. In this way electrons move through the structure, and the holes move in a direction opposite to that of the conduction electrons.

This type of extrinsic semiconductor, in which holes enable electron motion, is called a ***p*-type** semiconductor. The *p* stands for positive. The term is somewhat misleading, however, since the crystal, which is formed entirely from neutral atoms, is electrically neutral. The structure is electron deficient only with regard to the covalent bonding requirements of the lattice. It never has an excess of positive charge.

In addition to boron, other III A elements (Al, Ga, or In) can be added in small amounts to pure silicon or germanium to produce *p*-type semiconductors. Impurities that owe their effect to their ability to accept electrons by means of the holes they produce in the crystal lattice are known as **acceptor impurities.**

Traces of group V A elements (P, As, Sb, or Bi) when added to silicon or germanium produce a second type of extrinsic semiconductor known as an ***n*-type.** The *n* stands for negative. In this instance, each impurity atom has five valence electrons—one more than required by the bonding of the host crystal. The extra electron can move through the structure and function as a conduction electron. Impurities of this type are known as **donor impurities,** since they provide conduction electrons. The *n*-type semiconductor is negative only in the peculiar sense that more electrons are present than are required by the bonding scheme of the crystal. The substance is electrically neutral.

Both *n*- and *p*-type semiconductors are used in transistors, photocells, and solar batteries.

19.2 NATURAL OCCURRENCE OF METALS

An **ore** is a naturally occurring material from which one or more metals can be profitably extracted. The principal types of ores and examples of each are listed in Table 19.1. A few of the less reactive metals occur in nature in elementary form. For many of these metals, **native ores** constitute the most important source. The greatest tonnage of metals is derived from oxides—either oxide ores or metal oxides that are produced by the roasting of carbonate or sulfide ores.

Table 19.1 Occurrence of metals

TYPE OF ORE	EXAMPLES
native metals	Cu, Ag, Au, As, Sb, Bi, Pd, Pt
oxides	Al_2O_3, Fe_2O_3, Fe_3O_4, SnO_2, MnO_2, TiO_2, $FeO{\cdot}Cr_2O_3$, $FeO{\cdot}WO_3$, Cu_2O, ZnO
carbonates	$CaCO_3$, $CaCO_3{\cdot}MgCO_3$, $MgCO_3$, $FeCO_3$, $PbCO_3$, $BaCO_3$, $SrCO_3$, $ZnCO_3$, $MnCO_3$, $CuCO_3{\cdot}Cu(OH)_2$, $2CuCO_3{\cdot}Cu(OH)_2$
sulfides	Ag_2S, Cu_2S, CuS, PbS, ZnS, HgS, FeS·CuS, FeS_2, Sb_2S_3, Bi_2S, MoS_2, NiS, CdS
halides	NaCl, KCl, AgCl, $KCl{\cdot}MgCl_2{\cdot}6H_2O$, NaCl and $MgCl_2$ in sea water
sulfates	$BaSO_4$, $SrSO_4$, $PbSO_4$, $CaSO_4{\cdot}2H_2O$, $CuSO_4{\cdot}2Cu(OH)_2$
silicates	$Be_3AlSi_6O_{18}$, $ZrSiO_4$, $Sc_2Si_2O_7$, ($NiSiO_3$, $MgSiO_3$)[a]
phosphates	[$CePO_4$, $LaPO_4$, $NdPO_4$, $PrPO_4$, $Th_3(PO_4)_4$][a], $LiF{\cdot}AlPO_4$

[a] Occur in a single mineral but not in fixed proportions.

Silicate minerals are abundant in nature. The extraction of metals from silicates is difficult, however, and the cost of such processes may be prohibitive. Consequently, only the less common metals are commercially derived from silicate ores. Phosphate minerals are, in general, rare and occur in low concentrations.

A number of metals occur as impurities in the ores of other metals so that both metals are derived from the same commercial operation. Cadmium metal, for example, is obtained as a by-product in the production of zinc.

Ores, as mined, generally contain variable amounts of unwanted materials (such as silica, clay, and granite), which are called **gangue.** The concentration of the desired metal must be sufficiently high to make its extraction chemically feasible and economically competitive. Ores of low concentration are worked only if they can be processed comparatively easily and inexpensively or if the metal product is scarce and valuable. The required concentration varies greatly from metal to metal. For aluminum or iron it should be 30% or more; for copper it may be 1% or less.

19.3 METALLURGY: PRELIMINARY TREATMENT OF ORES

Metallury is the science of extracting metals from their ores and preparing them for use. Metallurgical processes may be conveniently divided into three principal operations:

1. **preliminary treatment,** in which the desired component of the ore is concentrated, specific impurities removed, and/or the mineral is put into a suitable form for subsequent treatment
2. **reduction,** in which the metal compound is reduced to the free metal

3. **refining,** in which the metal is purified and, in some cases, substances added to give desired properties to the final product

Since the problems encountered in each step vary from metal to metal, many different metallurgical procedures exist.

The processing of many ores requires, as a first step, that most of the gangue be removed. Such **concentration** procedures, which are usually carried out on ores that have been crushed and ground, may be based on physical or chemical properties. **Physical separations** are based on differences between the physical properties of the mineral and the gangue. Thus, through washing with water the particles of rocky impurities may often be separated from the heavier mineral particles. This separation may be accomplished by shaking the crushed ore in a stream of water on an inclined table. The heavier mineral particles settle to the bottom and are collected.

Flotation is a method of concentration applied to many ores, especially those of copper, lead, and zinc. The finely crushed ore is mixed with a suitable oil and water in large tanks. The mineral particles are wetted by the oil, whereas the gangue particles are wetted by the water. Agitation of the mixture with air produces a froth which contains the oil and mineral particles. This froth floats on the top of the water and is skimmed off.

Magnetic separation is employed to separate Fe_3O_4 (the mineral magnetite) from gangue. The ore is crushed, and electromagnets are used to attract the particles of Fe_3O_4, which is ferromagnetic.

The removal of free metals from native ores may be considered to be a type of concentration. Certain ores, such as native bismuth and native copper, are heated to a temperature just above the melting point of the metal, and the liquid metal is poured away from the gangue. The process is called **liquation.**

Mercury will dissolve silver and gold to form what are called **amalgams.** Hence, the native ores of silver and gold are treated with mercury, the resulting liquid amalgam collected, and the free silver or gold recovered from the amalgam by distilling the mercury away.

The chemical properties of minerals are the basis of **chemical separations.** An important example is the **Bayer method** of obtaining pure aluminum oxide (sometimes called alumina) from the ore bauxite. Since aluminum hydroxide is amphoteric, the alumina is dissolved away from the ore impurities (principally iron(III) oxide and silicates) by treatment of the crushed ore with hot sodium hydroxide solution.

$$Al_2O_3(s) + 2OH^-(aq) + 3H_2O \rightarrow 2Al(OH)_4^-(aq)$$

The solution of sodium aluminate is filtered and cooled. Its *p*H is adjusted downward by dilution and/or neutralization with carbon dioxide so that aluminum hydroxide precipitates (Section 15.11). Pure aluminum oxide, ready for reduction, is obtained by heating the aluminum hydroxide.

$$2Al(OH)_3(s) \rightarrow Al_2O_3(s) + 3H_2O(g)$$

The first step in the extraction of magnesium from sea water is a chemical concentration. The Mg^{2+} in sea water (approximately 0.13%) is precipitated as magnesium hydroxide by treatment of the sea water with a slurry of calcium hydroxide. The calcium hydroxide is obtained from CaO that is produced by roasting oyster shells ($CaCO_3$).

$$CaCO_3(s) \xrightarrow{\text{heat}} CaO(s) + CO_2(g)$$
$$CaO(s) + H_2O \rightarrow Ca(OH)_2(s)$$
$$Mg^{2+}(aq) + Ca(OH)_2(s) \rightarrow Mg(OH)_2(s) + Ca^{2+}(aq)$$

The magnesium hydroxide is converted to magnesium chloride by reaction with hydrochloric acid, and the magnesium chloride solution is evaporated to produce the dry $MgCl_2$ necessary for electrolytic reduction.

The metal component of some ores may be obtained by **leaching.** Thus, low-grade carbonate and oxide ores of copper may be leached with dilute sulfuric acid

$$CuO(s) + 2H^+(aq) \rightarrow Cu^{2+}(aq) + H_2O$$
$$CuCO_3(s) + 2H^+(aq) \rightarrow Cu^{2+}(aq) + CO_2(g) + H_2O$$

and the resulting copper sulfate solutions may be directly subjected to electrolysis.

After concentration many ores are **roasted** in air. The sulfide ores of the less-reactive metals are directly reduced to the free metal by heating (see Section 19.4). The majority of the sulfide ores, as well as the carbonate ores, however, are converted into oxides by roasting.

$$2ZnS(s) + 3O_2(g) \rightarrow 2ZnO(s) + 2SO_2(g)$$
$$PbCO_3(s) \rightarrow PbO(s) + CO_2(g)$$

The free metals are more readily obtained from oxides than from sulfides or carbonates.

19.4 METALLURGY: REDUCTION

By far the largest quantity of metals, as well as the largest number of metals, are produced by **smelting** operations—high-temperature reduction processes in which the metal is usually secured in a molten state. In most of these processes a **flux** (such as limestone, $CaCO_3$) is used to remove the gangue that remains after ore concentration. The flux forms a **slag** with silica and silicate impurities. For limestone and silica the simplified equations are

$$CaCO_3(s) \rightarrow CaO(s) + CO_2(g)$$
$$CaO(s) + SiO_2(s) \rightarrow CaSiO_3(l)$$

The slag, which is a liquid at the smelting temperatures, generally floats on top of the molten metal and is readily separated from the metal.

The reducing agent employed for a given smelting operation is the least expensive material that is capable of yielding a product of the required purity. For the ores of the metals of low reactivity—for example, the sulfide ores of mercury, copper, and lead—no chemical agent is required. Mercury is produced by roasting cinnabar, HgS, in air.

$$HgS(s) + O_2(g) \rightarrow Hg(g) + SO_2(g)$$

The mercury vapor is condensed in a receiver and requires no further purification.

Iron, zinc, tin, cadmium, antimony, nickel, cobalt, molybdenum, lead, and other metals are produced by **carbon reduction** of oxides, which are obtained as ores or as the products of roasting operations.

The reactions that occur in a high-temperature carbon reduction are complex. In most cases the reduction is effected principally by carbon monoxide, not carbon. Since both the mineral and coke are solids that are not readily melted, contact between them is poor and direct reaction slow.

$$MO(s) + C(s) \rightarrow M(l) + CO(g)$$

(M stands for a metal.) A gas and a solid, however, make better contact and react more readily.

$$2C(s) + O_2(g) \rightarrow 2CO(g)$$
$$MO(s) + CO(g) \rightarrow M(l) + CO_2(g)$$

The carbon dioxide that is produced is converted into carbon monoxide by reaction with coke.

$$C(s) + CO_2(g) \rightarrow 2CO(g)$$

The most important commercial metal, iron, is produced by carbon reduction in a **blast furnace** designed to operate continuously (Figure 19.5). The ore, coke, and a limestone flux are charged into the top of the furnace and heated air, which is sometimes enriched with oxygen, is blown in at the bottom. The incoming air reacts with carbon to form carbon monoxide and to liberate considerable amounts of heat. At this point the temperature of the furnace is the highest (about 1500°C).

The iron ore (usually Fe_2O_3) is reduced in stages depending upon the temperature. Near the top of the furnace, where the temperature is lowest, Fe_3O_4 is the reduction product.

$$3Fe_2O_3(s) + CO(g) \rightarrow 2Fe_3O_4(s) + CO_2(g)$$

The descending Fe_3O_4 is reduced to FeO in a lower, hotter zone.

$$Fe_3O_4(s) + CO(g) \rightarrow 3FeO(s) + CO_2(g)$$

In the hottest zone, reduction to metallic iron occurs.

$$FeO(s) + CO(g) \rightarrow Fe(l) + CO_2(g)$$

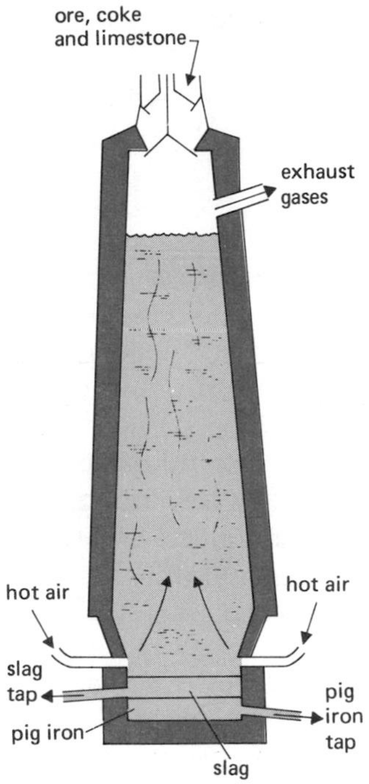

Figure 19.5 Diagram of a blast furnace (schematic).

The molten iron collects in the bottom of the furnace. Molten slag, principally calcium silicate which is produced from the flux and the gangue, also collects on the bottom. The slag floats on top of the molten iron, thus protecting the metal from oxidation by the incoming air. The slag and iron are periodically drawn off. The hot exhaust gases are used to heat incoming air.

The impure iron produced by the blast furnace **(pig iron)** contains up to 4% carbon, up to 2% silicon, some phosphorus, and a trace of sulfur. In the manufacture of steel (Section 19.5), these impurities are removed or their concentrations are adjusted. The presence of some carbon in steel is desirable, and the amount of carbon in the final product is readily controlled in the refining process.

The carbon reduction of some oxides (for example, those of chromium, manganese, and tungsten) yields products that contain an appreciable quantity of carbon. Since carbon impurities are usually difficult to remove, carbon reduction is not a satisfactory method for preparing these metals in a pure form. In such instances, as well as those in which carbon is incapable of effecting a reduction, reactive metals (such as Na, Mg, and Al) may be used as reducing agents. Processes that use metals as reducing agents are more expensive than those that employ carbon because the cost of the production of the metal used as a reducing agent must be taken into account.

The reduction of an oxide by aluminum is called a **Goldschmidt process** or a **thermite process.**

$$Cr_2O_3(s) + 2Al(s) \rightarrow 2Cr(l) + Al_2O_3(l)$$
$$3MnO_2(s) + 4Al(s) \rightarrow 3Mn(l) + 2Al_2O_3(l)$$
$$3BaO(s) + 2Al(s) \rightarrow 3Ba(l) + Al_2O_3(l)$$

The reactions are highly exothermic, and molten metals are produced. The reaction of Fe_2O_3 and aluminum is used at times for the production of molten iron in welding operations and as the basis of incendiary (thermite) bombs. Other oxides commercially reduced by metals include: UO_3 (by Al or Ca), V_2O_5 (by Al), Ta_2O_5 (by Na), MoO_3 (by Al), ThO_2 (by Ca), and WO_3 (by Al).

The **Kroll process** involves the reduction of metal halides (such as $TiCl_4$, $ZrCl_4$, UF_4, and $LaCl_3$) by magnesium, sodium, or calcium. In the production of titanium, titanium tetrachloride is prepared by the reaction of titanium dioxide, carbon, and chlorine. The chloride is a liquid and is readily purified by distillation. Purified $TiCl_4$ is passed into molten magnesium or sodium at approximately 700°C under an atmosphere of argon or helium (which prevents oxidation of the product).

$$TiCl_4(g) + 2Mg(l) \rightarrow Ti(s) + 2MgCl_2(l)$$

Hydrogen reduction is used for the preparation of some metals (at high temperatures) when carbon reduction yields an unsatisfactory product. Examples are

$$GeO_2(s) + 2H_2(g) \rightarrow Ge(s) + 2H_2O(g)$$
$$WO_3(s) + 3H_2(g) \rightarrow W(s) + 3H_2O(g)$$
$$MoO_3(s) + 3H_2(g) \rightarrow Mo(s) + 3H_2O(g)$$

The metals are produced as powders by this method. Germanium is melted and cast into ingots, but tungsten (wolfram) and molybdenum have high melting points. They are heated and pounded into compact form. Hydrogen cannot be used to reduce the oxides of some metals because of the formation of undesirable hydrides.

The most reactive metals are prepared by **electrolytic reduction** of molten compounds. Sodium and magnesium (as well as the other I A and II A metals) are prepared by electrolysis of the fused chlorides. Sodium chloride is electrolyzed in a **Downs cell** (Figure 19.6), which is designed to keep the products separate from one another so that they do not react.

anode: $2Cl^- \rightarrow Cl_2(g) + 2e^-$
cathode: $e^- + Na^+ \rightarrow Na(l)$

Purified Al_2O_3 is electrolyzed in the **Hall process** for the production of aluminum (Figure 19.7). Alumina dissolved in cryolite, Na_3AlF_6, produces a solution that will conduct electric current. The carbon lining of the cell serves as the cathode, and carbon anodes are used. The primary reaction at the anode

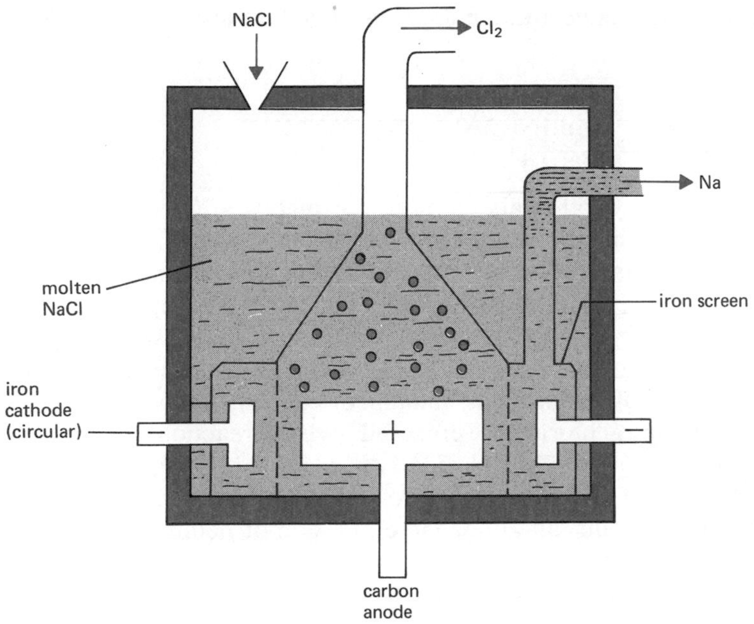

Figure 19.6 Schematic diagram of the Downs cell for the production of sodium and chlorine.

may be considered to be the discharge of oxygen from oxide ions. The oxygen, however, reacts with the carbon anodes so that the cell reactions are

$$\text{anode:} \quad C(s) + 2O^{2-} \rightarrow CO_2(g) + 4e^-$$
$$\text{cathode:} \quad 3e^- + Al^{3+} \rightarrow Al(l)$$

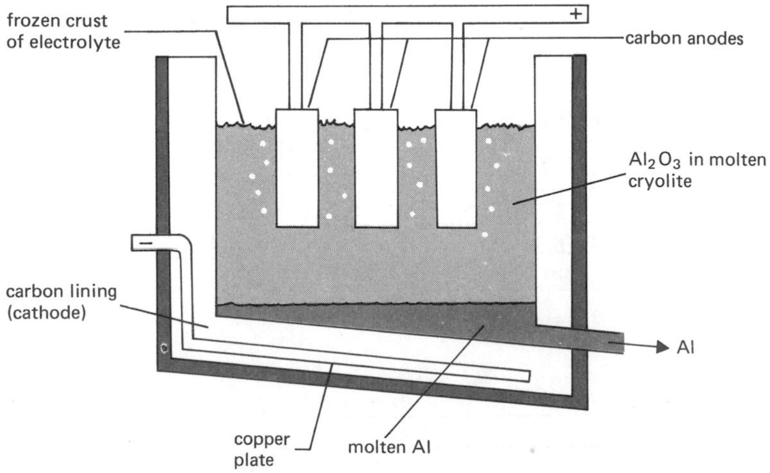

Figure 19.7 Schematic diagram of an electrolytic cell for the production of aluminum.

The molten aluminum collects at the bottom of the cell and is periodically withdrawn.

Some metals are prepared by the electrolysis of aqueous solutions of their salts. Very pure zinc is produced by the electrolysis of zinc sulfate solutions. The zinc sulfate is obtained by treating zinc oxide (obtained from zinc sulfide ores by roasting) with sulfuric acid.

anode: $2H_2O \rightarrow O_2(g) + 4H^+(aq) + 4e^-$
cathode: $2e^- + Zn^{2+}(aq) \rightarrow Zn(s)$

Aqueous solutions of salts of copper, cadmium, chromium, cobalt, gallium, indium, manganese, and thorium are electrolyzed to produce the corresponding metals. Electrolytically prepared metals generally require no further refining.

19.5 METALLURGY: REFINING

Most of the metals obtained from reduction operations require refining to rid them of objectionable impurities. Refining processes vary widely from metal to metal, and for a given metal the method employed may vary with the proposed use of the final product. Along with the removal of material that impart undesirable properties to the metal, the refining step may include the addition of substances to give the product desired characteristics. Some refining processes are designed to recover valuable metal impurities, such as gold, silver, and platinum.

Crude tin, lead, and bismuth are purified by **liquation.** Ingots of the impure metals are placed at the top of a sloping hearth that is maintained at a temperature slightly above the melting point of the metal. The metal melts and flows down the inclined hearth into a well, leaving behind the solid impurities.

Some low-boiling metals, such as zinc and mercury, are purified by **distillation.**

The **Parkes process** for refining lead relies upon the selective dissolution of silver in molten zinc. A small amount of zinc (1% to 2%) is added to molten lead that contains silver as an impurity. Silver is much more soluble in zinc than in lead. Lead and zinc are insoluble in each other. Hence, most of the silver dissolves in the zinc, which comes to the top of the molten lead. The zinc layer solidifies first upon cooling and is removed. The silver is obtained by remelting the zinc layer and distilling away the zinc, which is collected and used over again.

In the **Mond process,** impure nickel is treated with carbon monoxide at 60°C to 80°C. The nickel reacts to form gaseous nickel carbonyl, leaving solid impurities, including cobalt, behind.

$$Ni(s) + 4CO(g) \rightarrow Ni(CO)_4(g)$$

Pure nickel is recovered from the carbonyl by heating the gas to temperatures of from 180°C to 200°C.

$$Ni(CO)_4(g) \rightarrow Ni(s) + 4CO(g)$$

The carbon monoxide is recirculated.

The **Van Arkel process** is also based upon the thermal decomposition of a metal compound. The method, which is used for the purification of titanium, hafnium, and zirconium, involves the decomposition of a metal iodide on a hot metal filament. For example, impure zirconium is heated with a limited amount of iodine in an evacuated glass apparatus.

$$Zr(s) + 2I_2(g) \rightarrow ZrI_4(g)$$

The gaseous zirconium tetraiodide is decomposed upon contact with a hot filament, and pure zirconium metal is deposited upon the filament.

$$ZrI_4(g) \rightarrow Zr(s) + 2I_2(g)$$

The regenerated iodine reacts with more zirconium. The process is very expensive and is employed for the preparation of limited amounts of very pure metals for special uses.

Another process capable of producing metals of very high purity is **zone refining.** A circular heater is fitted around a rod of an impure metal, such as germanium (Figure 19.8). The heater, which is slowly moved down the rod, melts a band of the metal. As the heater moves along, pure metal recrystallizes out of the melt, and impurities are swept along in the molten zone to one end of the rod, which is subsequently discarded. More than one pass of the heater may be made on the same rod.

Electrolytic refining is an important and widely used method of purification. Many metals, including copper, tin, lead, gold, zinc, chromium, and nickel, are refined electrolytically. Plates of the impure metal are used as anodes, and the electrolyte is a solution of a salt of the metal. The pure metal plates out on the cathode. For copper, copper sulfate is employed as the electrolyte, and the electrode reactions are

$$\begin{aligned} \text{anode:} \quad & Cu(s) \rightarrow Cu^{2+}(aq) + 2e^- \\ \text{cathode:} \quad & 2e^- + Cu^{2+}(aq) \rightarrow Cu(s) \end{aligned}$$

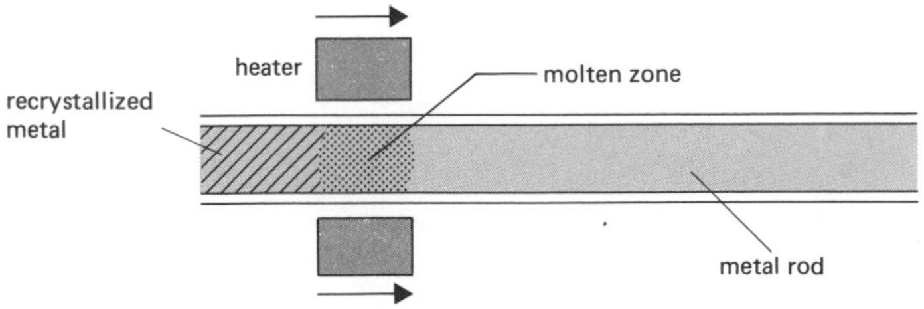

Figure 19.8 Schematic diagram of zone refining.

The more reactive metals in the crude copper anode, such as iron, are oxidized and pass into solution where they remain. They are not reduced at the cathode. The less reactive metals, such as silver, gold, and platinum, are not oxidized. As the copper anode dissolves away, they fall to the bottom of the cell from where they are recovered as a valuable "anode sludge."

Several methods are used to refine pig iron into steel. In these processes the impurities (C from the coke used in the reduction, and such substances as Si, P, and S from the ore) are oxidized and volatilize out of the iron or go into a slag.

The **Bessemer process** (1855) was the first method developed capable of producing large quantities of steel. Molten pig iron from the blast furnace is poured into an egg-shaped converter, and a blast of air is blown through the melt to oxidize the impurities. The process is rapid (10 to 15 min), but the quality of the product is variable.

In the **open hearth process,** pig iron, scrap iron, and iron(III) oxide are heated in a shallow hearth lined with CaO or MgO. A blast of burning fuel and hot air is directed on the molten iron. The impurities are oxidized by excess air and the iron oxide. The CO escapes and burns, and the oxides of Si, P, and S combine with the oxides of the hearth lining to form a slag. Since it takes about 8 to 10 hours to produce a batch of steel, the quality of the product is easily controlled. Alloying metals (such as Mn, Cr, Ni, W, Mo, and V) may be added before the charge is poured.

The **basic oxygen process** is rapid (20 to 50 min) and gives a high-quality product. Molten pig iron is placed in a converter that is lined with basic oxides. The impurities are oxidized by pure oxygen, which is blown onto the surface of the molten metal through a water-cooled lance.

EARTHS

Earth is one of the four terrestial elements of the ancient Greeks (earth, air, fire, and water). For centuries earth was regarded as an element and various substances were identified as forms of earth. Gradually, the term *earth* came to be applied to substances that we now know to be oxides that are difficult to reduce. Since the oxides of the group II A elements, CaO, SrO, and BaO, give alkaline solutions when dissolved in water, they became known as the *alkaline earths.* Magnesium oxide, MgO, is not soluble in water, but it too is usually called an alkaline earth.

Lavoisier included the alkaline earths in his list of elements (1789). He noted, however, that "It is quite probable that the four alkaline earths . . . contain oxygen. . . . These substances may be nothing but oxidized metals. . . ." He thought that oxygen "had more affinity" for the metals than for carbon and that, therefore, the earths could not be reduced by carbon to give the free metals. In 1808, Humphry Davy prepared all four of the free metals, Mg, Ca, Sr, and Ba, by electrolytic reduction.

In the past, most of the earths were identified by an old system of naming. The ending *-a* was used as a substitute for the ending of the name of an element to indicate the oxide of the element. The names *alumina* (for Al_2O_3), *magnesia* (for MgO), and *silica* (for SiO_2) are examples of this practice.

The lanthanides (elements 58 to 71) were at one time known as the *rare-earth elements* because they were first isolated in the form of their 3+ oxides, and they were thought to be rare. Promethium (number 61) probably occurs in nature only in trace amounts. With this exception, however, the lanthanides are not rare elements. They are more abundant, for example, than Hg, Cd, or Se. The name *lanthanide* is also subject to criticism since the ending *-ide* is customarily applied to binary compounds. *Lanthanon* has been suggested as a substitute for *lanthanide.*

The lanthanides constitute the series of inner-transition elements that follows lanthanum in the periodic table. These elements together with yttrium, which is a member of the same group as lanthanum but in the period before, occur together in nature and are very similar chemically. The oxides of the elements, the earths, are difficult to separate from one another.

Lanthanum (number 57) and elements 58 through 64 were eventually isolated from a mixture of oxides originally thought to be a pure oxide and named *ceria.* Yttrium (number 39) and elements 65 through 71 were isolated from another mixture of oxides named *yttria.*

Yttria was obtained in 1794 by Johan Gadolin from a mineral found in Ytterby, Sweden, and was named for this Swedish village. In the course of over one hundred years, the earth yttria was shown to consist of eight different earths. Time and again, an earth thought to be a single oxide was shown to consist of several oxides. In this way, four of the eight earths from yttria, and consequently four elements, came to be named for the village of Ytterby. The elements are: yttrium (39), terbium (65), erbium (68), and ytterbium (70).

PROBLEMS

19.1 How does the band theory of metallic bonding account for the electrical conductivity, thermal conductivity, and luster of metals?

19.2 Explain the difference between an *n*-type and a *p*-type semiconductor.

19.3 Use energy level diagrams to explain the differences among a conductor, an insulator, and a semiconductor.

19.4 What are the principal types of ores? Give an example of each type.

19.5 Explain why it is not surprising that the following ores are found in nature: (a) native ores of Pt, Ag, and Au; (b) sulfate ores of Ba, Sr, and Pb; (c) Na^+ and Mg^{2+} in sea water; (d) sulfide ores of Pb, Bi, and Ni.

19.6 What types of ores are roasted? Write a chemical equation for the roasting of an ore of each type. Why is this procedure used?

19.7 Give chemical equations for the reactions that occur in the blast furnace for the production of pig iron.

19.8 Describe, with equations, the steps used in the production of aluminum from bauxite.

19.9 Use chemical equations to show how magnesium metal is obtained from sea water.

19.10 What is the function of a flux, such as limestone, in a smelting operation?

19.11 What processes are used to refine pig iron into steel? How do they differ?

19.12 Identify the following: (a) ore, (b) gangue, (c) amalgam, (d) slag.

19.13 Describe the following metallurgical processes. Use chemical equations where possible: (a) froth flotation, (b) Parkes process, (c) Van Arkel process, (d) Kroll process, (e) liquation, (f) zone refining.

19.14 Describe, with chemical equations, how copper is obtained from low-grade $CuCO_3$ ores by leaching followed by electrolytic reduction.

19.15 Why is carbon reduction not used to obtain some metals from their oxide ores?

19.16 Write equations for the thermite reduction of (a) UO_3 (by Al), (b) V_2O_5 (by Al), (c) Ta_2O_5 (by Na), (d) ThO_2 (by Ca), (e) WO_3 (by Al).

19.17 How is sodium produced commercially?

19.18 Write equations to show how lead is obtained by roasting PbS followed by carbon reduction.

19.19 Write chemical equations for the Mond process, which is used to purify nickel.

19.20 Explain, using chemical equations, how impure copper can be refined electrolytically. What is the "anode sludge"?

Chapter 20

Complex Compounds

Chemists of the late-nineteenth century had difficulty in understanding how "molecular compounds" or "compounds of higher order" are bonded. The formation of a compound such as $CoCl_3 \cdot 6NH_3$ was baffling—particularly in this case since simple $CoCl_3$ does not exist. In 1893 Alfred Werner proposed a theory to account for compounds of this type. Werner wrote the formula of the cobalt compound as $[Co(NH_3)_6]Cl_3$. He assumed that the six ammonia molecules are symmetrically "coordinated" to the central cobalt atom by "subsidiary valencies" of cobalt while the "principal valencies" of cobalt are satisfied by the chloride ions. Werner spent over twenty years preparing and studying coordination compounds and perfecting and proving his theory. Although modern work has amplified his theory, it has required little modification.

Many practical applications have been derived from the study of complex compounds. Advances have resulted in such fields as metallurgy, analytical chemistry, biochemistry, water purification, textile dyeing, electrochemistry, and bacteriology. In addition, the study of these compounds has enlarged our understanding of chemical bonding, certain physical properties (for example, spectral and magnetic properties), minerals (many minerals are complex compounds), and metabolic processes (both heme of blood and chlorophyll of plants are complex compounds).

20.1 STRUCTURE

A complex ion or complex compound consists of a central metal cation to which several anions and/or molecules (called **ligands**) are bonded. With few exceptions, *free* ligands have at least one electron pair that is not engaged in bonding.

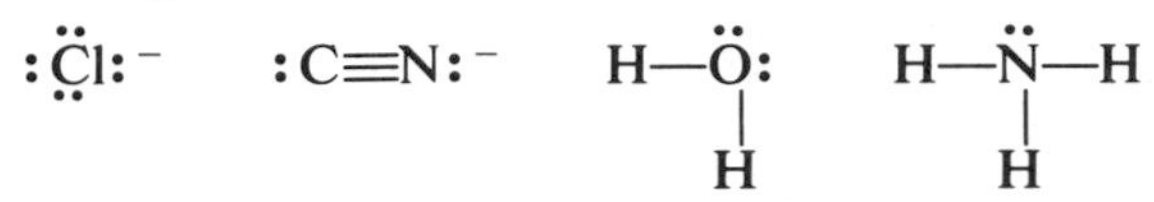

These electron pairs may be considered to be donated to the electron-deficient metal ions in the formation of complexes. Ligands, therefore, are substances that are capable of acting as Lewis bases. The bonding of complexes, however, shows a wide variation in character—from strongly covalent to predominantly ionic (Section 20.5).

The ligands are said to be coordinated around the central cation in a **first coordination sphere.** In the formulas of complex compounds, such as $K_3[Fe(CN)_6]$ and $[Cu(NH_3)_4]Cl_2$, the first coordination sphere is indicated by square brackets. The ligands are disposed about the central ion in a regular geometric manner (Figure 20.1). The number of atoms *directly* bonded to the central metal ion, or the number of coordination positions, is called the **coordination number** of the central ion.

The charge of a complex is the sum of the charges of the constituent parts. Complexes may be cations, anions, or neutral molecules. In each of the complexes of platinum(IV)—$[Pt(NH_3)_5Cl]^{3+}$, $[Pt(NH_3)_2Cl_4]^0$, and $[PtCl_6]^{2-}$—the platinum contributes 4+, each chlorine contributes 1−, and the coordinated ammonia molecules do not contribute to the charge of the complex.

An interesting series of platinum(IV) complexes appears in Table 20.1. The list is headed by the chloride of the $[Pt(NH_3)_6]^{4+}$ ion, and in each subsequent entry an ammonia molecule of the coordination sphere is replaced by a chloride ion. The coordinated ammonia molecules and chloride ions are tightly held and do not dissociate in water solution. Those chloride ions of the compound that are not coordinated to the platinum are ionizable. Aqueous solutions of the last three compounds of the table do not precipitate silver chloride upon the addition of silver nitrate, whereas solutions of the first four precipitate AgCl in amounts proportional to 4/4, 3/4, 2/4, and 1/4, respectively, of their total chlorine content. In each case, the total number of ions per formula unit derived from conductance data agrees with the formula listed in Table 20.1.

Table 20.1 Some platinum(IV) complex compounds

	MOLAR CONDUCTANCE 0.001*N* SOLUTION[a] (25°C)	NUMBER OF IONS PER FORMULA UNIT	NUMBER OF Cl^- IONS PER FORMULA UNIT
$[Pt(NH_3)_6]Cl_4$	523	5	4
$[Pt(NH_3)_5Cl]Cl_3$	404	4	3
$[Pt(NH_3)_4Cl_2]Cl_2$	228	3	2
$[Pt(NH_3)_3Cl_3]Cl$	97	2	1
$[Pt(NH_3)_2Cl_4]$	0	0	0
$K[Pt(NH_3)Cl_5]$	108	2	0
$K_2[PtCl_6]$	256	3	0

[a] cm^2/ohm mol.

In general, the most stable complexes are formed by metal ions that have a high positive charge and a small ionic radius. The transition elements and the metals immediately following (notably the III A and IV A metals) have a marked tendency to form complexes. Few complexes are known for the lanthanides and the I A and II A metals (with the exception of beryllium). The bonding of transition-metal complexes involves the *d* orbitals of the central metal atom.

Complexes containing metal ions with coordination numbers ranging from two to nine are known. The majority of complexes, however, are two-, four-, and sixfold coordinate (Figure 20.1), and the coordination number six is by far the most common.

Six-coordinate complexes are octahedral. The regular octahedral arrangement of atoms is frequently represented

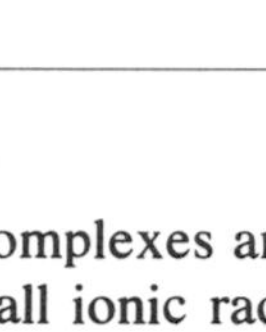

This representation is a convenient way to create a three-dimensional illusion. No difference between the bonds of the vertical axis and the other bonds should be inferred from this drawing. *All* the bonds are equivalent. Tetragonal geometry is a distorted form of the octahedral in which the bond distances along one of the axes are longer or shorter than the remaining bonds.

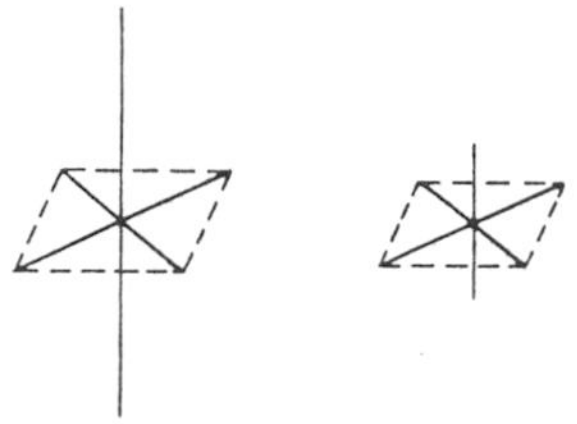

(a) $[Ag(NH_3)_2]^+$

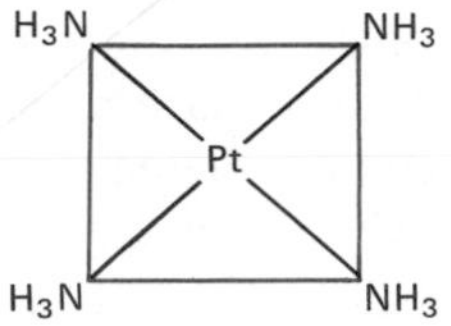

(b) $[Pt(NH_3)_4]^{2+}$

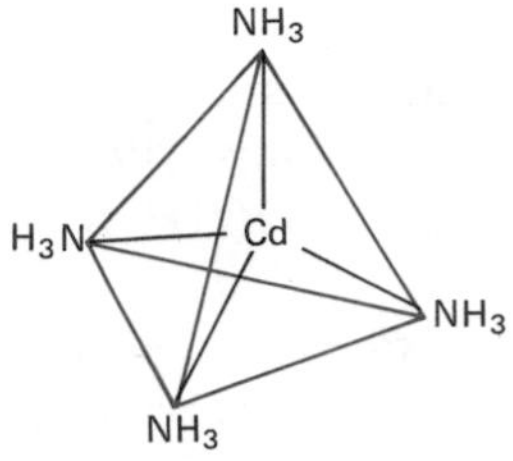

(c) $[Cd(NH_3)_4]^{2+}$

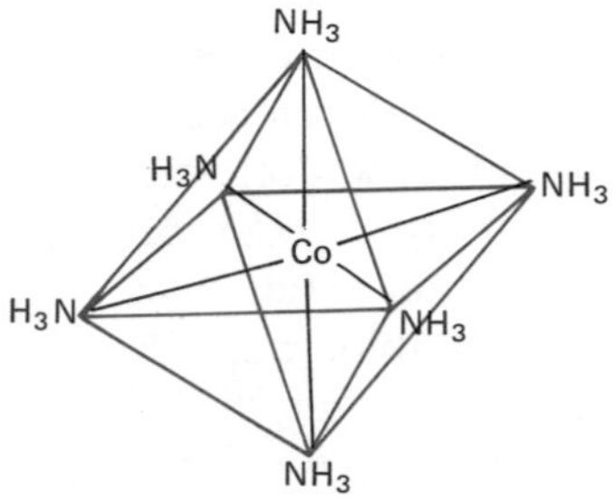

(d) $[Co(NH_3)_6]^{3+}$

Figure 20.1 Common configurations of complex ions: (a) linear, (b) square planar, (c) tetrahedral, (d) octahedral.

Four-coordinate complexes are known in tetrahedral and square-planar configurations. The square-planar configuration is the usual one for Pt^{II}, Pd^{II}, and Au^{III} and is assumed by many complexes of Ni^{II} and Cu^{II}. Examples include $[Pt(NH_3)_4]^{2+}$, $[PdCl_4]^{2-}$, $[AuCl_4]^-$, $[Ni(CN)_4]^{2-}$, and $[Cu(NH_3)_4]^{2+}$. In some instances there is evidence that square complexes may be tetragonal forms with two groups, along a vertical axis, located at greater distances from the central ion than the ligands of the plane. Thus, $[Cu(NH_3)_4]^{2+}$ in water solution may have two water molecules coordinated in the manner shown in Figure 20.2. Hence, square-planar and octahedral geometries may be considered to merge.

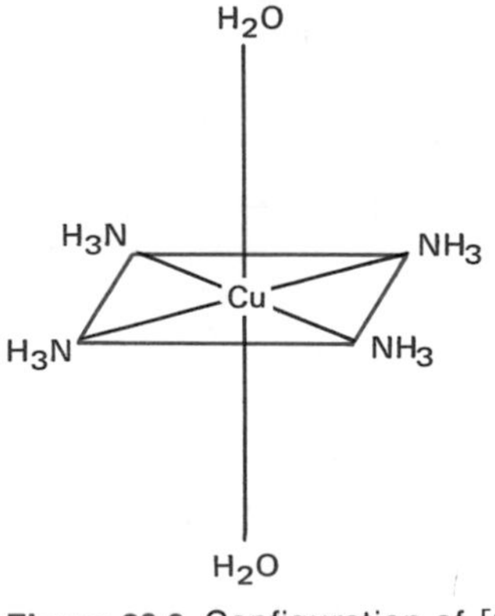

Figure 20.2 Configuration of $[Cu(NH_3)_4(H_2O)_2]^{2+}$.

For four-coordinate complexes the tetrahedral configuration is encountered more frequently than the square-planar and is particularly common for complexes of the nontransition elements. Certain complexes of Cu^{I}, Ag^{I}, Au^{I}, Be^{II}, Zn^{II}, Cd^{II}, Hg^{II}, Al^{III}, Ga^{III}, In^{III}, Fe^{III}, Co^{II}, and Ni^{0} are tetrahedral. Examples include $[Cu(CN)_4]^{3-}$, $[BeF_4]^{2-}$, $[AlF_4]^{-}$, $[FeCl_4]^{-}$, $[Cd(CN)_4]^{2-}$, $[ZnCl_4]^{2-}$, and $[Ni(CO)_4]^{0}$. The oxyanions of certain transition metals (such as VO_4^{3-}, CrO_4^{2-}, FeO_4^{2-}, and MnO_4^{-}) are tetrahedral, resembling the tetrahedral oxyanions of nonmetals (such as SiO_4^{4-}, PO_4^{3-}, AsO_4^{3-}, SO_4^{2-}, and ClO_4^{-}). Although a number of tetrahedral complexes of transition elements are known, the majority of complexes of these elements are octahedral.

Linear, two-coordinate complexes are not so common as the other forms previously mentioned. However, well-characterized complexes of this type are known for Cu^{I}, Ag^{I}, and Hg^{II}; examples are $[CuCl_2]^{-}$, $[Ag(NH_3)_2]^{+}$, $[Au(CN)_2]^{-}$, $[Hg(NH_3)_2]^{2+}$, and $[Hg(CN)_2]^{0}$.

In general, each metal exhibits more than one coordination number and geometry in its complexes. Although all known complexes of Co^{III} are octahedral, most cations form more than one type of complex. For example, Al^{III} forms tetrahedral and octahedral complexes, Cu^{I} forms linear and tetrahedral complexes, and Ni^{II} forms square-planar, tetrahedral, and octahedral complexes.

Complexes of the transition elements are frequently highly colored. Examples of complexes that exhibit a wide range of colors are $[Co(NH_3)_6]^{3+}$ (yellow), $[Co(NH_3)_5(H_2O)]^{3+}$ (pink), $[Co(NH_3)_5Cl]^{2+}$ (violet), $[Co(H_2O)_6]^{3+}$ (purple), and $[Co(NH_3)_4Cl_2]^{+}$ (a violet form and a green form—see Section 19.4).

The ligands we have discussed thus far have been capable of forming only one bond with the central ion. They are referred to as **unidentate** (from Latin, meaning *one toothed*) ligands. Certain ligands are capable of occupying more than one coordination position of a metal ion. Ligands that coordinate through two bonds from different parts of the molecule or anion are called **bidentate.** Examples are

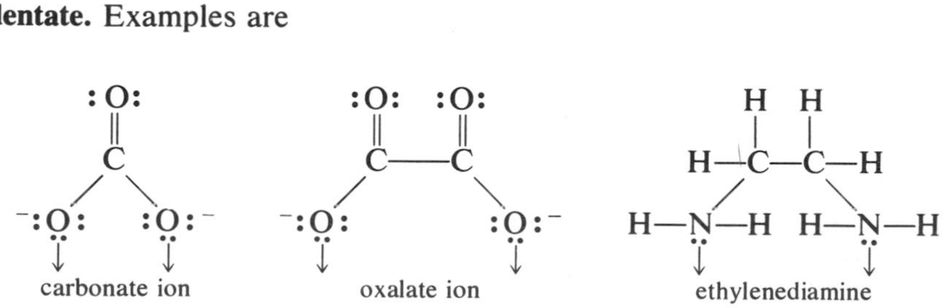

carbonate ion oxalate ion ethylenediamine

The carbonate and oxalate ions each coordinate through two oxygen atoms. The ethylenediamine molecule (abbreviated *en*) coordinates through both nitrogen atoms. These positions are marked with arrows. Bidentate ligands form rings with the central metal ions.

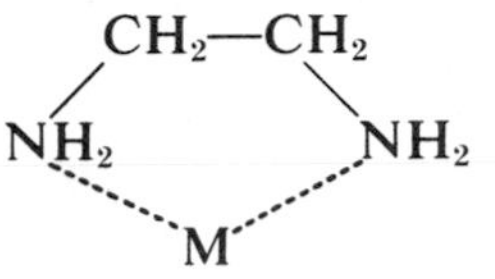

The resulting metal complexes are called **chelates** (from Greek, meaning *claw*). The formation of five- or six-membered rings is generally favored.

Multidentate ligands have been prepared that can coordinate at 2, 3, 4, 5, or 6 positions. In general, chelates are more stable than complexes containing only unidentate ligands. Thus, the sexadentate complexing agent ethylenediaminetetraacetate ion (EDTA)

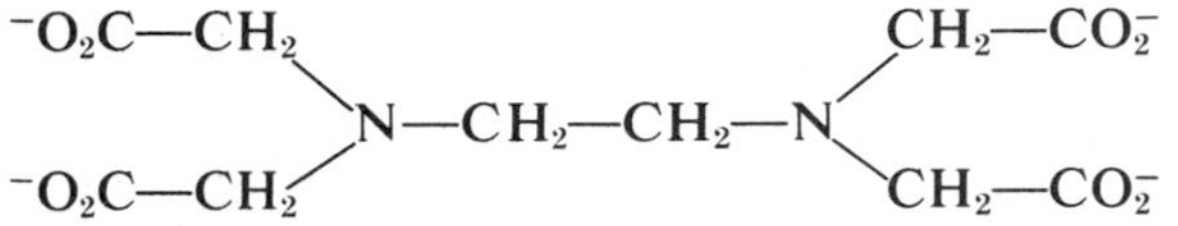

is capable of forming a very stable complex with the calcium ion—an ion with one of the least tendencies toward the formation of complexes (Figure 20.3).

Heme of hemoglobin is a chelate of Fe^{2+}, and chlorophyll is a chelate of Mg^{2+}. In both these substances the metal atom is coordinated to a quadridentate ligand, which may be considered to be derived from a porphin structure (Figure 20.4) by the substitution of various groups for the H atoms of the porphin. The substituted porphins are called **porphyrins.**

Coordination around the iron atom of heme is octahedral. Four of the

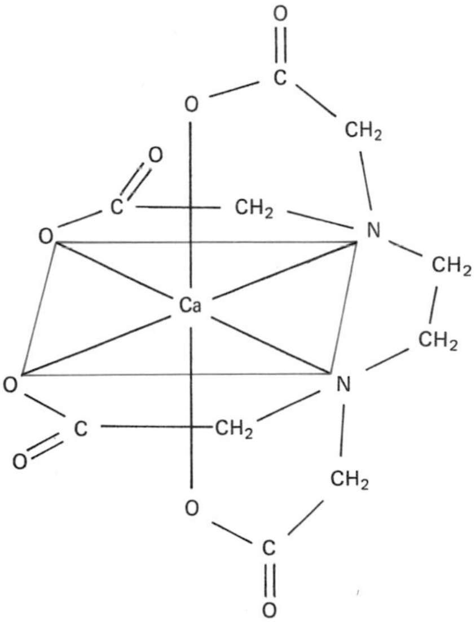

Figure 20.3 Ethylenediaminetetraacetate complex of Ca^{2+} ($[Ca(EDTA)]^{2-}$).

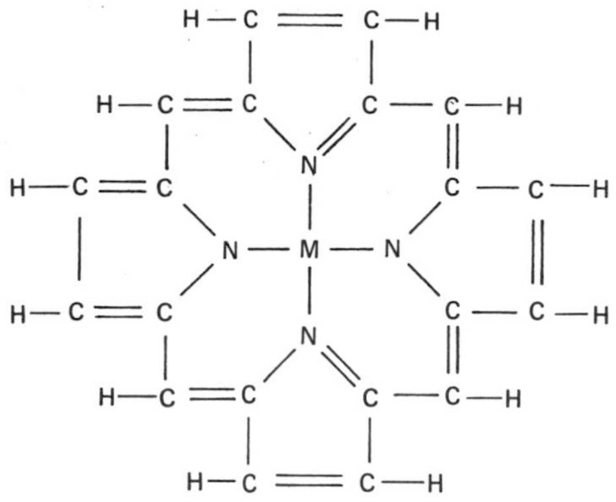

Figure 20.4 Metal porphin.

coordination positions are utilized in the formation of the heme (which is essentially planar), the fifth is used to bond the heme to a protein molecule (globin), and the sixth is used to coordinate either H_2O (hemoglobin) or O_2 (oxyhemoglobin). Coordination about this sixth position is reversible

$$\text{hemoglobin} + O_2 \rightleftharpoons \text{oxyhemoglobin} + H_2O$$

and dependent upon the pressure of O_2. Hemoglobin picks up O_2 in the lungs and releases it in the body tissues, where it is used for the oxidation of food. Hemoglobin reacts with carbon monoxide to form a complex that is more stable than oxyhemoglobin, and therefore CO is toxic.

Chlorophyll, the green pigment of plants, serves as a catalyst for the process of photosynthesis in which CO_2 and H_2O are converted into a carbohydrate (glucose) and O_2. The energy for photosynthesis comes from sunlight, and the chlorophyll molecule initiates this process by absorbing a quantum of light.

20.2 NOMENCLATURE

Since thousands of complexes are known and the number is constantly expanding, a system of nomenclature has been adopted for these compounds. The following paragraphs summarize the important rules of the system. They are adequate for naming the simple and frequently encountered complexes.

1. If the complex compound is a salt, the cation is named first whether or not it is the complex ion.
2. The constituents of the complex are named in the following order: anions, neutral molecules, central metal ion.
3. Anionic ligands are given *-o* endings. Examples are: OH^-, hydroxo; O^{2-}, oxo; S^{2-}, thio; Cl^-, chloro; F^-, fluoro; CO_3^{2-}, carbonato; CN^-, cyano; CNO^-, cyanato; $C_2O_4^{2-}$, oxalato; NO_3^-, nitrato; NO_2^-, nitro; SO_4^{2-}, sulfato; and $S_2O_3^{2-}$, thiosulfato.

4. The names of neutral ligands are not changed. Exceptions to this rule are: H_2O, aquo; NH_3, ammine; CO, carbonyl; and NO, nitrosyl.
5. The number of ligands of a particular type is indicated by a prefix: di-, tri-, tetra-, penta-, and hexa- (for two to six). For complicated ligands (such as ethylenediamine), the prefixes bis-, tris-, and tetrakis- (two to four) are employed.
6. The oxidation number of the central ion is indicated by a Roman numeral, which is set off by parentheses and placed after the name of the complex.
7. If the complex is an anion, the ending *-ate* is employed. If the complex is a cation or a neutral molecule, the name is not changed.

Examples of the rules of nomenclature are

$[Ag(NH_3)_2]Cl$	Diamminesilver(I) chloride
$[Co(NH_3)_3Cl_3]$	Trichlorotriamminecobalt(III)
$K_4[Fe(CN)_6]$	Potassium hexacyanoferrate(II)
$[Ni(CO)_4]$	Tetracarbonylnickel(0)
$[Cu(en)_2]SO_4$	Bis(ethylenediamine)copper(II) sulfate
$[Pt(NH_3)_4][PtCl_6]$	Tetraammineplatinum(II) hexachloroplatinate(IV)
$[Co(NH_3)_4(H_2O)Cl]Cl_2$	Chloroaquotetraamminecobalt(III) chloride

Common names are frequently employed when they are clearly more convenient than the systematic name (for example, ferrocyanide rather than hexacyanoferrate(II) for $[Fe(CN)_6]^{4-}$) or when the structure of the complex is not certain (for example, the aluminate ion).

20.3 ISOMERISM

Two compounds with the same molecular formula but different arrangements of atoms are called **isomers.** Such compounds differ in their chemical and physical properties. **Structural isomerism** is displayed by compounds that have different ligands within their coordination spheres. Several types of structural isomers may be identified.

The following pair of compounds of Co^{III} serve as an example of **ionization isomers.**

(a) $[Co(NH_3)_5(SO_4)]Br$ red   (b) $[Co(NH_3)_5Br]SO_4$ violet

Conductance data show that both compounds dissociate into two ions in aqueous solution. In the first compound, the SO_4^{2-} ion is a part of the coordination sphere, and the Br^- ion is ionizable. An aqueous solution of compound (a) gives an immediate precipitate of AgBr upon the addition of $AgNO_3$, but since the SO_4^{2-} ion is not free, no precipitate forms upon the addition of

$BaCl_2$. For compound (b), the reverse is true. An aqueous solution of this compound gives a precipitate of $BaSO_4$ but not AgBr, since the SO_4^{2-} is ionizable and the Br^- is coordinated. Note that SO_4^{2-} functions as a unidentate ligand and that the charge on the complex ion of compound (a) is 1+ and that of compound (b) is 2+. There are numerous additional examples of ionization isomers, such as,

$$[Pt(NH_3)_4Cl_2]Br_2 \qquad [Pt(NH_3)_4Br_2]Cl_2$$

Hydrate isomerism is analogous to ionization isomerism and is probably best illustrated by the following series of compounds that have the formula $CrCl_3{\cdot}6H_2O$.

(a) $[Cr(H_2O)_6]Cl_3$
violet

(b) $[Cr(H_2O)_5Cl]Cl_2{\cdot}H_2O$
green

(c) $[Cr(H_2O)_4Cl_2]Cl{\cdot}2H_2O$
green

In a mole of each of these compounds there are six moles of water. However, in compound (a), six water molecules are coordinated; in compound (b), five; and in compound (c), four. The uncoordinated water molecules occupy separate positions in the crystals and are readily lost when compounds (b) and (c) are exposed to desiccants. The coordinated water, however, is not so easily removed. Further evidence for the structures of the compounds is afforded by conductance data (the compounds are composed of four, three, and two ions, respectively) and the quantity of AgCl precipitated (the compounds have three, two, and one ionizable chloride ions, respectively).

Another example of hydrate isomerism is given by the following pair of compounds.

$$[Co(NH_3)_4(H_2O)Cl]Cl_2 \qquad [Co(NH_3)_4Cl_2]Cl{\cdot}H_2O$$

Coordination isomers may exist in compounds that have two or more centers of coordination. Isomers arise through the exchange of ligands between these coordination centers. In simple examples which involve only two complex ions per compound, the coordinated metal ions may be the same

$$[Cr(NH_3)_6][Cr(NCS)_6] \qquad [Cr(NH_3)_4(NCS)_2][Cr(NH_3)_2(NCS)_4]$$

or different

$$[Cu(NH_3)_4][PtCl_4] \qquad [Pt(NH_3)_4][CuCl_4]$$

and the oxidation state of the metals may vary

$[Pt(NH_3)_4][PtCl_6]$ tetraammineplatinum(II)- hexachloroplatinate(IV)	$[Pt(NH_3)_4Cl_2][PtCl_4]$ dichlorotetraammineplatinum(IV)- tetrachloroplatinate(II)

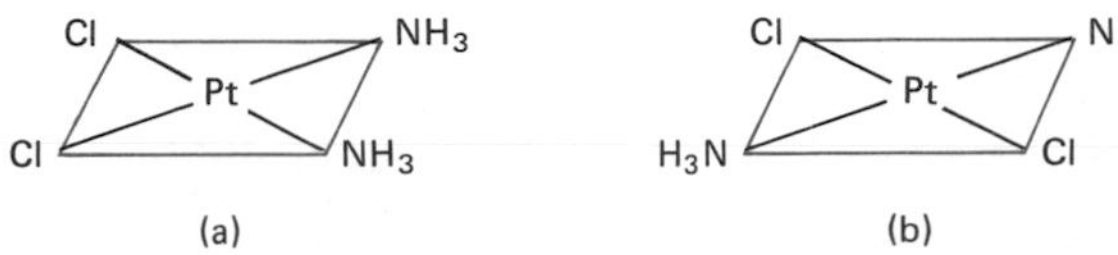

Figure 20.5 (a) *Cis* and (b) *trans* isomers of dichlorodiammineplatinum(II).

Linkage isomerism is a rare but interesting type. It arises when ligands are capable of coordinating in two ways. The nitrite ion, NO_2^-, for example, can coordinate through an oxygen atom (—ONO, nitrito compounds) or through the nitrogen atom (—NO_2, nitro compounds).

$[Co(NH_3)_5(NO_2)]Cl_2$	$[Co(NH_3)_5(ONO)]Cl_2$
(yellow)	(red)
nitropentaamminecobalt(III)-chloride	nitritopentaamminecobalt(III)-chloride

Theoretically, many ligands might be capable of forming linkage isomers: CN^- might coordinate either through the C atom or the N atom; SCN^-, through N or S; CO, through C or O. Few authentic cases of linkage isomerism are known.

Stereoisomerism is a second general classification of isomers. Compounds are stereoisomers when they both contain the same ligands in their coordination spheres but differ in the way that these ligands are arranged in space. One type of stereoisomerism is **geometric, or *cis-trans,* isomerism.**

An example of geometric isomerism is afforded by the *cis* and *trans* isomers of the square planar dichlorodiammineplatinum(II); see Figure 20.5. In the *cis* isomer the chlorine atoms are situated on adjacent corners of the square (along an edge), whereas in the *trans* isomer they occupy opposite corners (along the diagonal).

Since all the ligands of a tetrahedral complex have the same relationship to one another, *cis-trans* isomerism does not exist for this geometry. Many geometric isomers are known for octahedral complexes, however. There are two isomers of the dichlorotetraamminecobalt(III) ion: a violet *cis* form and a green *trans* form (Figure 20.6).

A second type of stereoisomerism is **optical isomerism.** Some molecules and ions can exist in two forms that are not superimposable and that bear the

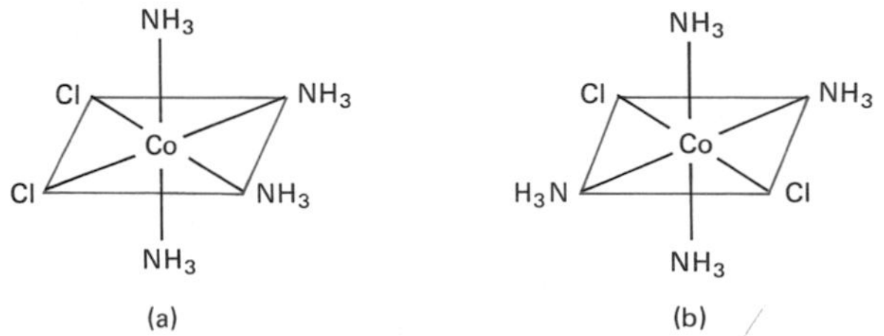

Figure 20.6 (a) *Cis* and (b) *trans* isomers of the dichlorotetraamminecobalt(III) ion.

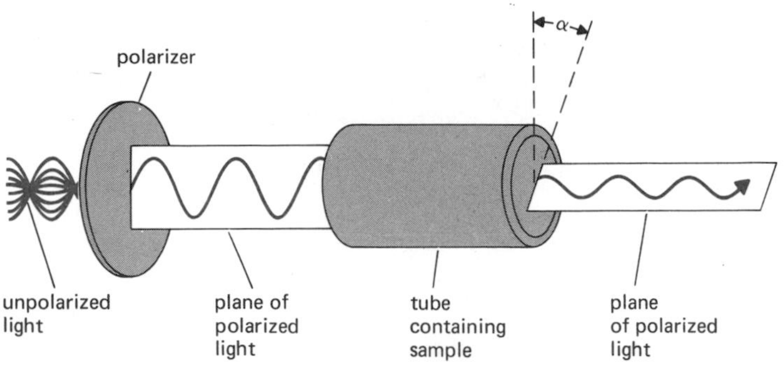

Figure 20.7 Rotation of the plane-polarized light. The plane is rotated by an angle after passage through a solution of an optically active isomer. In this case, rotation is to the right and the same is said to be dextrorotatory.

same relationship that a right hand bears to a left hand. That the hands are not superimposable is readily demonstrated by attempting to put a left-handed glove on a right hand. Such molecules and ions are spoken of as being **dissymmetric,** and the forms are called **enantiomorphs** (from Greek, meaning *opposite forms*) or **mirror images** (since the one may be considered to be a mirror reflection of the other).

Enantiomorphs have identical physical properties except for their effects on plane-polarized light. Light that has been passed through a polarizer consists of waves that vibrate in a single plane. One enantiomorph (the **dextro** form), whether pure or in solution, will rotate the plane of the plane-polarized light to the right; the other (the **levo** form) will rotate the plane an equal extent to the left (see Figure 20.7).

For this reason enantiomorphs are called optical isomers or **optical antipodes.** An equimolar mixture of enantiomorphs, called a **racemic modification,** has no effect on plane-polarized light since it contains an equal number of *dextro*rotary and *levo*rotary forms.

The tris(ethylenediamine)cobalt(III) ion exists in enantiomorphic forms. Examination of the diagrams of Figure 20.8 confirms that the enantiomorphs are not superimposable and that the ions are dissymmetric. Note that bidentate chelating agents can span only *cis* positions.

Both types of stereoisomerism—geometric and optical—are illustrated by the isomers of the dichlorobis(ethylenediamine)cobalt(III) ion (Figure 20.9). The *trans* configuration of this ion is optically inactive, and a mirror image

Figure 20.8 Dextro and levo forms of the tris(ethylenediamine)cobalt(III) ion.

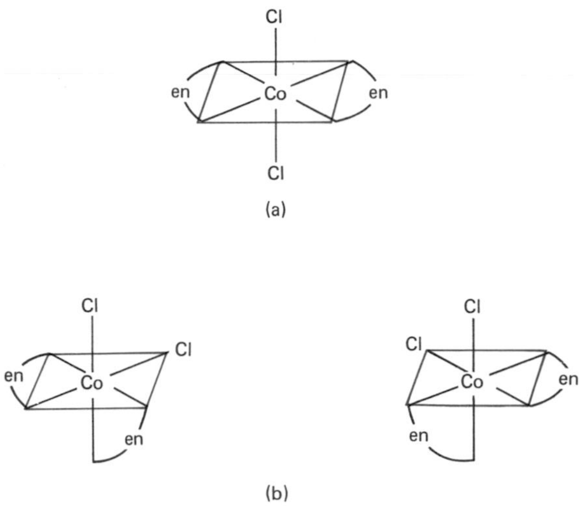

Figure 20.9 Isomers of the dichlorobis(ethylenediamine)cobalt(III) ion. (a) *trans* isomer, (b) optical isomers of the *cis* form.

would be identical to the original. The *cis* arrangement, however, is dissymmetric and exists in *dextro* and *levo* forms. The *trans* modification is said to be a *diastereoisomer* of either the *dextro* or *levo cis* modification.

20.4 THE BONDING IN COMPLEXES

There are several approaches to the theoretical treatment of the bonding in complexes. The **valence bond theory** describes the bonding in terms of hybridized orbitals of the central metal ion. The metal ion is assumed to have a number of unoccupied orbitals available for complex formation equal to its coordination number. Each ligand donates a pair of electrons toward the formation of a coordinate covalent bond, and a bond arises from the overlap of a vacant hybrid orbital of the metal atom and a filled orbital of a ligand. The bonds usually have an appreciable degree of polarity. The common types of hybridization (Section 7.6) are

sp	linear
sp^3	tetrahedral
dsp^2	square planar
d^2sp^3	octahedral

The valence bond diagrams of some simple complexes together with the electronic configurations of the central metal ions are shown in Figure 20.10. Hund's rule applies to the electrons occupying the nonbonding *d* orbitals, and, accordingly, some of these electrons are unpaired. In each of the complexes shown, the number of unpaired electrons determined by magnetic susceptibility measurement agrees with the structure diagrammed. Thus, $[Co(en)_3]^{3+}$ and $[Ni(CN)_4]^{2-}$ are diamagnetic, and the others are paramagnetic. In addition,

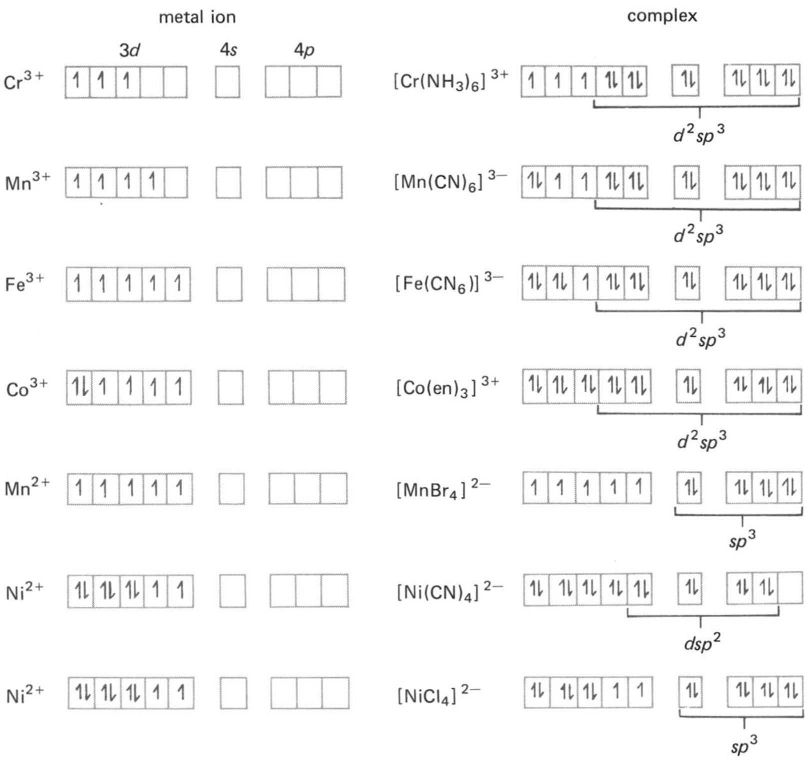

Figure 20.10 Valence-bond diagrams for some simple complexes.

the experimentally determined geometry of each complex agrees with that predicted on the basis of the assumed type of hybridization.

There are some defects in this attractively simple picture. From the diagrams of Figure 20.10 one might suppose that all of the octahedral complexes of Fe^{III} would have one unpaired electron (as in $[Fe(CN)_6]^{3-}$) and that all of the octahedral complexes of CO^{III} would be diamagnetic (like $[Co(en)_3]^{3+}$). Such is not the case. The magnetic moments of $[FeF_6]^{3-}$ and $[CoF_6]^{3-}$ indicate the existence of five and four unpaired electrons, respectively, in these complexes. These numbers of unpaired electrons are the same as those found in the free Fe^{3+} and Co^{3+} ions. At one time, the bonding in this type of complex was thought to be ionic with the *d* orbitals of the central ion undisturbed by complex formation. According to a later explanation, the 4*d*, rather than the 3*d*, orbitals are assumed to be used for the octahedral hybridization. Hence, **inner** (d^2sp^3) and **outer** (sp^3d^2) **complexes** are postulated (Figure 20.11). Since it is impossible to clear two inner *d* orbitals of the Ni^{2+} ion for inner d^2sp^3 hybridization, all octahedral Ni^{II} complexes must be assumed to be of the outer type.

A further complication arises in the case of $[Co(NO_2)_6]^{4-}$—an octahedral complex with one unpaired electron. In order to fit this complex into the d^2sp^3

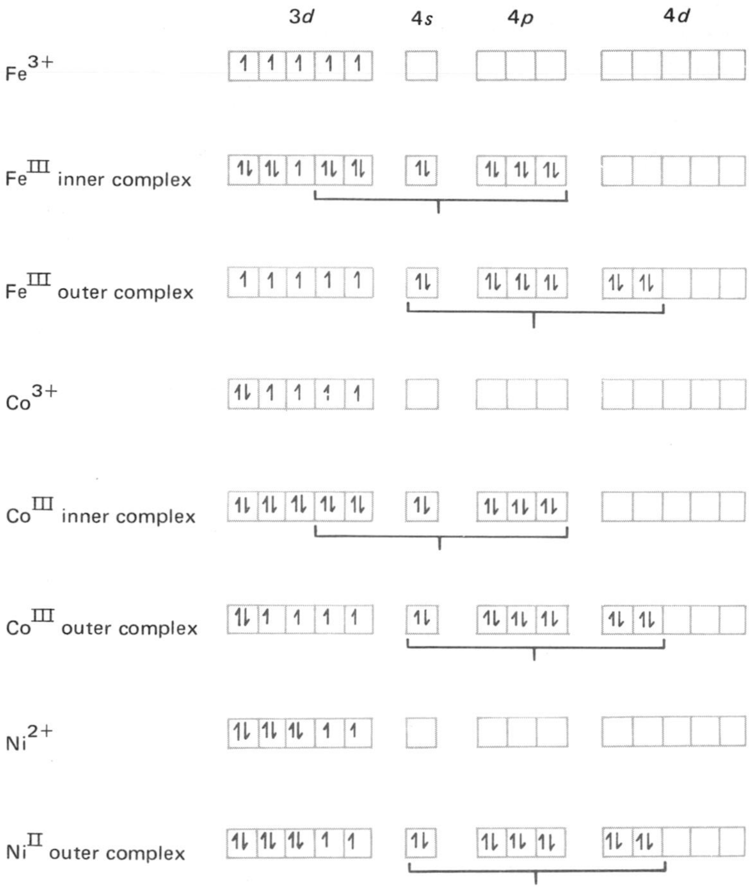

Figure 20.11 Inner and outer complexes of Fe^{III}, Co^{III}, and Ni^{II}.

category, it is necessary to postulate the promotion of an electron into the 4*d* sublevel.

	3*d*	4*s*	4*p*	4*d*
Co^{2+}	↑↓ ↑↓ ↑ ↑ ↑	_	_ _ _	_ _ _ _ _

	3*d*	d^2sp^3	4*d*
$[Co(NO_2)_6]^{4-}$	↑↓ ↑↓ ↑↓	↑↓ ↑↓ ↑↓ ↑↓ ↑↓ ↑↓	↑ _ _ _ _

It is also necessary to assume such an unlikely promotion for all of the square complexes of Cu^{II}.

	3*d*	4*s*	4*p*
Cu^{2+}	↑↓ ↑↓ ↑↓ ↑↓ ↑	_	_ _ _

	3*d*	dsp^2	4*p*
$[CuCl_4]^{2-}$	↑↓ ↑↓ ↑↓ ↑↓	↑↓ ↑↓ ↑↓ ↑↓	↑

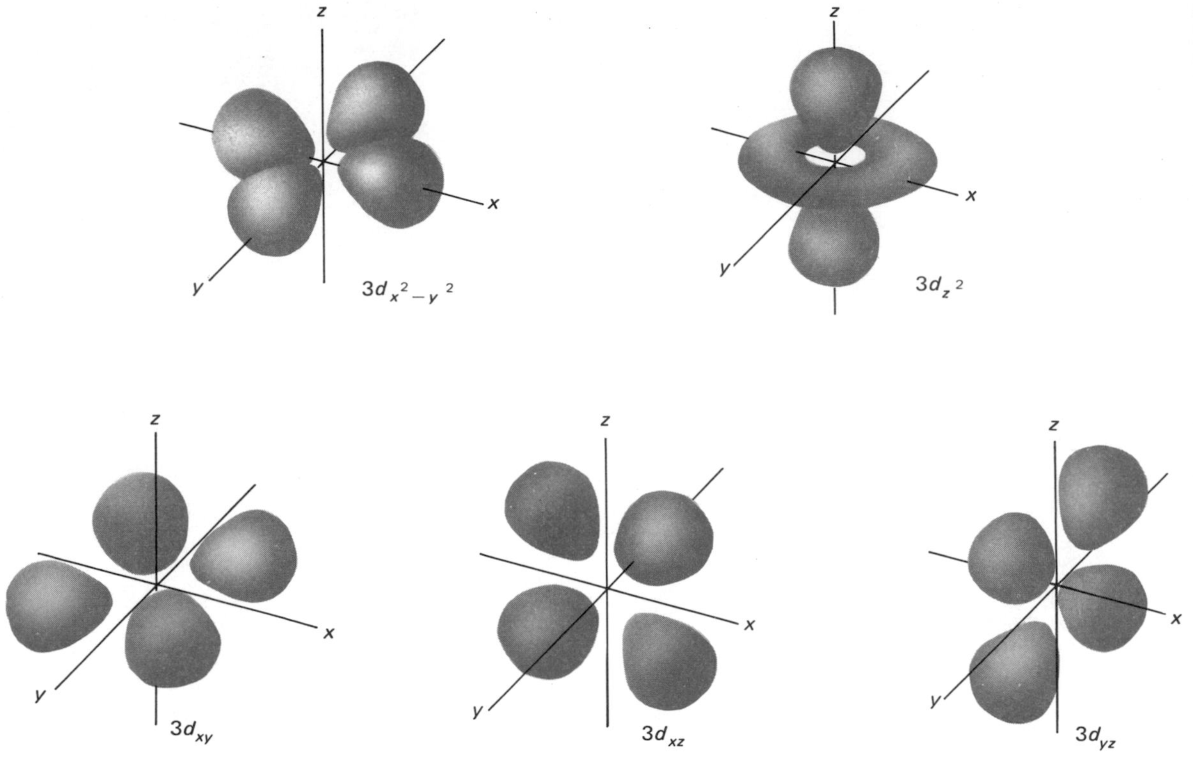

Figure 20.12 Boundary surface diagrams for the *d* orbitals.

The fundamental defect of the valence bond theory is that it fails to take into account the existence of the antibonding molecular orbitals that are produced, along with bonding molecular orbitals, in the formation of the complexes. Many of the experimental observations that are difficult to explain on the basis of the valence bond approach are readily understood in terms of electrons occupying antibonding orbitals.

The **crystal field theory** and its outgrowth, the **ligand field theory,** center on the *d* orbitals of the metal ion. In a free transition-metal ion all five of the *d* orbitals have equal energies—they are **degenerate.** All the *d* orbitals are not equivalent, however, when the metal ion engages in complex formation; the degeneracy is split.

Consider the relation of the *d* orbitals (Figure 20.12) to the arrangement of the ligands of an octahedral complex (Figure 20.13). The d_{z^2} and $d_{x^2-y^2}$ orbitals have lobes that point toward ligands, whereas the lobes of the d_{xy}, d_{xz}, and d_{yz} orbitals lie between ligands. Thus, in the complex two sets of *d* orbitals exist. The d_{xy}, d_{xz}, and d_{yz} orbitals (or t_{2g} orbitals) are equivalent to each other, and the d_{z^2} and $d_{x^2-y^2}$ orbitals (or e_g orbitals) are equivalent to each other and different from the first three. The symbols t_{2g} and e_g are applied to threefold degenerate and twofold degenerate sets of orbitals, respectively.

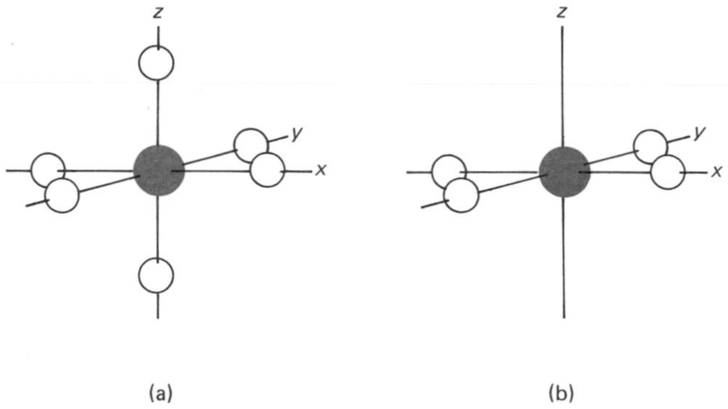

Figure 20.13 Arrangement of ligands of (a) octahedral and (b) square planar complexes in relation to sets of Cartesian coordinates.

It is not immediately obvious that the d_{z^2} orbital is perfectly equivalent to the $d_{x^2-y^2}$ orbital. The d_{z^2} orbital may be regarded as a combination, in equal parts, of two hypothetical orbitals, $d_{z^2-y^2}$ and $d_{z^2-x^2}$, which have shapes exactly like that of the $d_{x^2-y^2}$ orbital (see Figure 20.14). Since the number of d orbitals is limited to five, the $d_{z^2-y^2}$ and $d_{z^2-x^2}$ orbitals have no independent existence.

In the crystal field theory the assumption is made that an electrostatic field surrounding the central metal ion is produced by the negative ends of the dipolar molecules or by the anions that function as ligands. A metal-ion electron in a d orbital that has lobes directed toward ligands has a higher energy (owing to electrostatic repulsion) than an electron in an orbital with lobes that point between ligands. In octahedral complexes the orbitals of the e_g group, therefore, have higher energies than those of the t_{2g} group. The difference between the energies of the t_{2g} and e_g orbitals in an octahedral complex is given the symbol Δ_o.

In a square-planar complex (see Figure 20.13), the d orbitals exhibit four

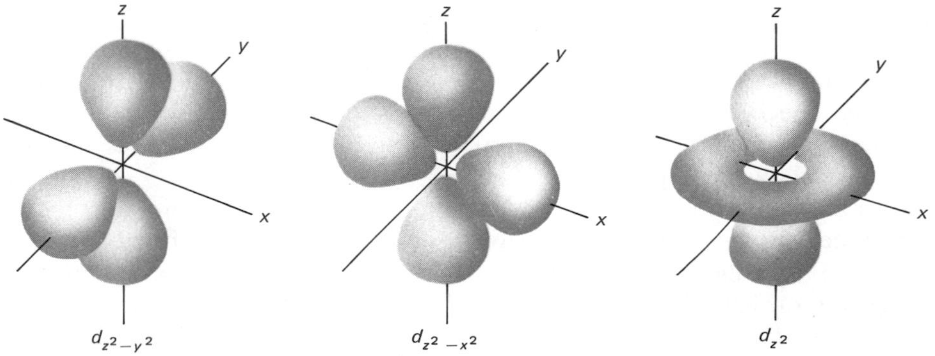

Figure 20.14 Diagram showing that the d_{z^2} orbitals may be considered to be a combination of the $d_{z^2-y^2}$ and $d_{z^2-x^2}$ orbitals.

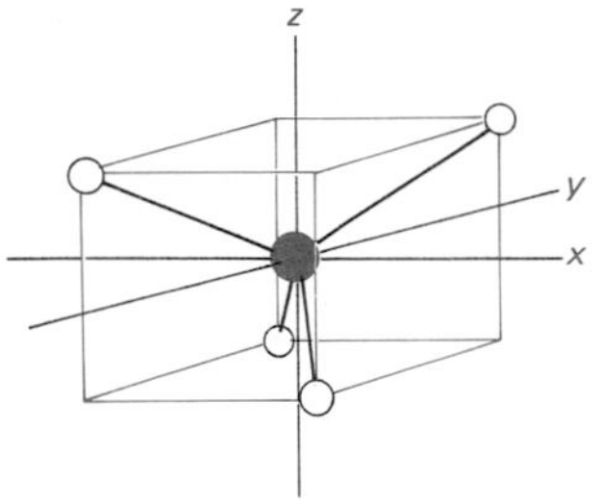

Figure 20.15 The relation of tetrahedrally arranged ligands to a set of Cartesian coordinates.

different relationships. The lobes of the $d_{x^2-y^2}$ orbital point toward ligands, and this orbital has the highest energy. The lobes of the d_{xy} orbital lie between orbitals but are coplanar with them, and hence this orbital is next highest in energy. The lobes of the d_{z^2} orbital point out of the plane of the complex, but the belt around the center of the orbital (which contains about a third of the electron density) lies in the plane. The d_{z^2} orbital, therefore, is next highest in energy. The d_{xz} and d_{yz} orbitals, which are degenerate, are least affected by the electrostatic field of the ligands since their lobes point out of the plane of the complex. These orbitals are lowest in energy.

The order of splitting of the *d* orbitals in a tetrahedral complex may be derived from an examination of Figure 20.15, which shows the relation of tetrahedrally arranged ligands to a system of Cartesian coordinates and a hypothetical cube. The lobes of the d_{xy}, d_{xz}, and d_{yz} orbitals point toward cube edges, and the lobes of the d_{z^2} and $d_{x^2-y^2}$ orbitals point toward the centers of cube faces. Notice that the distance from the center of a cube face to a ligand is farther than the distance from the center of a cube edge to a ligand. The order of orbital energies, therefore, is the reverse of that for octahedral coordination, and the threefold degenerate set, t_{2g}, is of higher energy than the twofold degenerate set, e_g. The difference in energy is given the symbol Δ_t.

The splitting of *d*-orbital energies in tetrahedral, octahedral, and square-planar complexes is summarized in Figure 20.16. The crystal field theory fails to take into account the mixing of metal and ligand orbitals and the covalent character of the bonding in complexes, and it is ultimately unsatisfactory. The ligand field theory offers a more realistic explanation, which is derived from molecular orbital theory. In both theories the orders of splitting are those shown in Figure 20.16.

In an octahedral complex the $3d_{z^2}$ and $3d_{x^2-y^2}$ orbitals along with the 4*s* and three 4*p* orbitals are assumed to overlap the six ligand σ orbitals with the attendant formation of six bonding molecular orbitals and six antibonding molecular orbitals (Section 7.8). The d_{xy}, d_{xz}, and d_{yz} orbitals (the t_{2g} set), which do not overlap the σ orbitals of the ligands, are essentially nonbonding. (The t_{2g} set can be used in π bonding, however.)

A bonding molecular orbital concentrates electron density between the

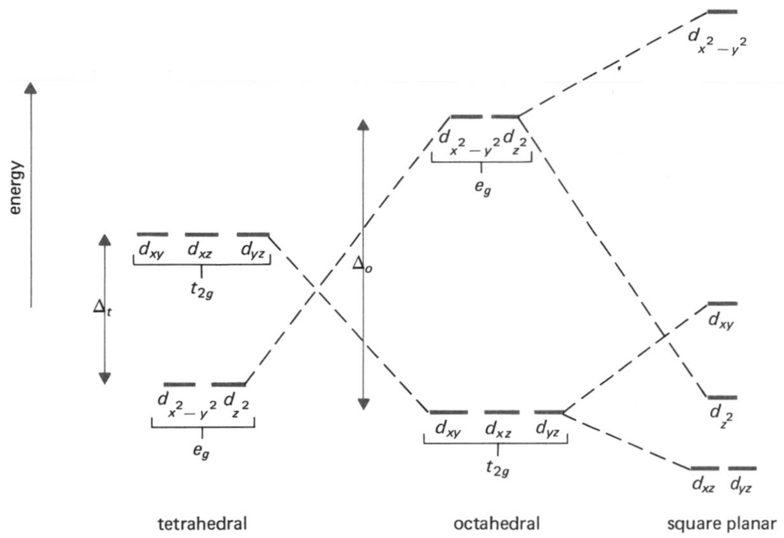

Figure 20.16 The splitting of *d*-orbital energies by ligand fields of three different geometries.

atoms and is of relatively low energy in comparison to an antibonding molecular orbital, which has a low electron density between the atoms and acts as a disruptive force. A molecular orbital energy level diagram for an octahedral complex with no π bonding is given in Figure 20.17.

Whenever two atomic orbitals of different energies combine, the character

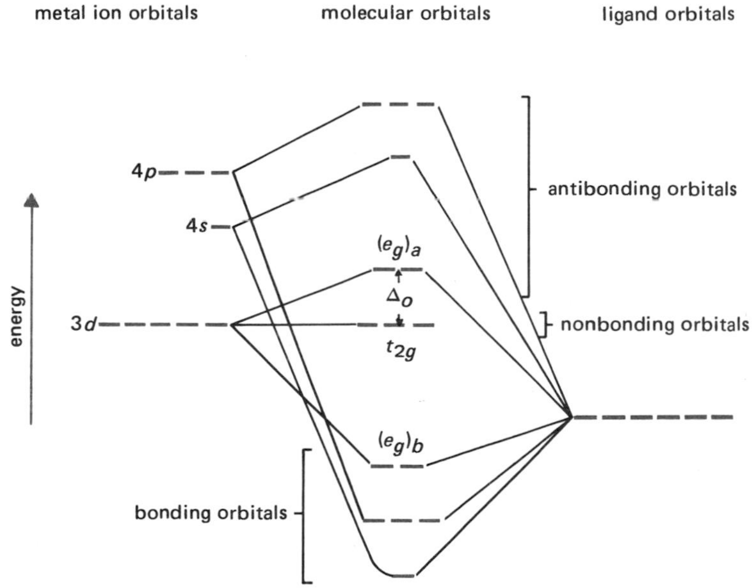

Figure 20.17 Diagram indicating the formation of molecular orbitals for an octahedral complex with no π binding.

of the resulting bonding molecular orbital is predominantly that of the atomic orbital of lower energy, and the antibonding molecular orbital has mainly the character of the higher energy atomic orbital. In an octahedral complex the bonding orbitals have predominantly the character of ligand orbitals. The antibonding orbitals resemble metal orbitals more than ligand orbitals. The t_{2g} set, which are nonbonding, may be considered as purely metal orbitals.

In an octahedral complex, the six electron pairs from the ligands completely occupy the bonding molecular orbitals. The *d* electrons of the central metal ion are accommodated in the t_{2g} nonbonding orbitals and the $(e_g)_a$ antibonding orbitals. The difference between the energies of these sets is Δ_o. The four remaining antibonding orbitals are never occupied in the ground states of any known complex.

The conclusion reached by this treatment is much the same as that postulated by the crystal field theory. In an octahedral complex the degeneracy of the metal *d* orbitals may be considered to be split into a threefold degenerate set, t_{2g}, and a higher energy, metal-like, twofold degenerate set, which may be labeled $(e_g)_a$ or simply e_g.

The molecular orbital treatment may be applied to tetrahedral and square-planar complexes, but the applications are more complicated. The conclusions reached are in essential agreement with the splittings diagrammed in Figure 20.16.

For the octahedral complexes of a given metal ion, the magnitude of Δ_o is different for each set of ligands, and the electronic configurations of many complexes depend upon the size of Δ_o.

For complexes of transition-element ions with one, two, or three *d* electrons (which are referred to as d^1, d^2, or d^3 ions), the orbital occupancy is certain and is independent of the magnitude of Δ_o. The electrons enter the lower energy t_{2g} orbitals singly with their spins parallel. For d^4, d^5, d^6, and d^7 ions, a choice of two configurations is possible (see Figure 20.18).

In the case of an octahedral complex of a d^4 ion, the fourth electron can singly occupy a higher energy e_g orbital or it can enter a t_{2g} orbital and pair with an electron already present. The former configuration has four unpaired electrons and is called the **high-spin state.** The latter configuration, which has two unpaired electrons, is called the **low-spin state.** Which configuration is assumed depends upon which is energetically more favorable.

If Δ_o is small, the electron may be promoted to the e_g level where it occupies an orbital singly. However, if Δ_o is large, the electron may be forced to pair with an electron in a t_{2g} orbital even though this pairing requires the expenditure of a pairing energy, *P*, to overcome the inter-electronic repulsion. Thus, the high-spin configuration results when

$$\Delta_o < P$$

and the low-spin configuration results when

$$\Delta_o > P$$

The value of *P* depends upon the metal ion; Δ_o is different for each complex.

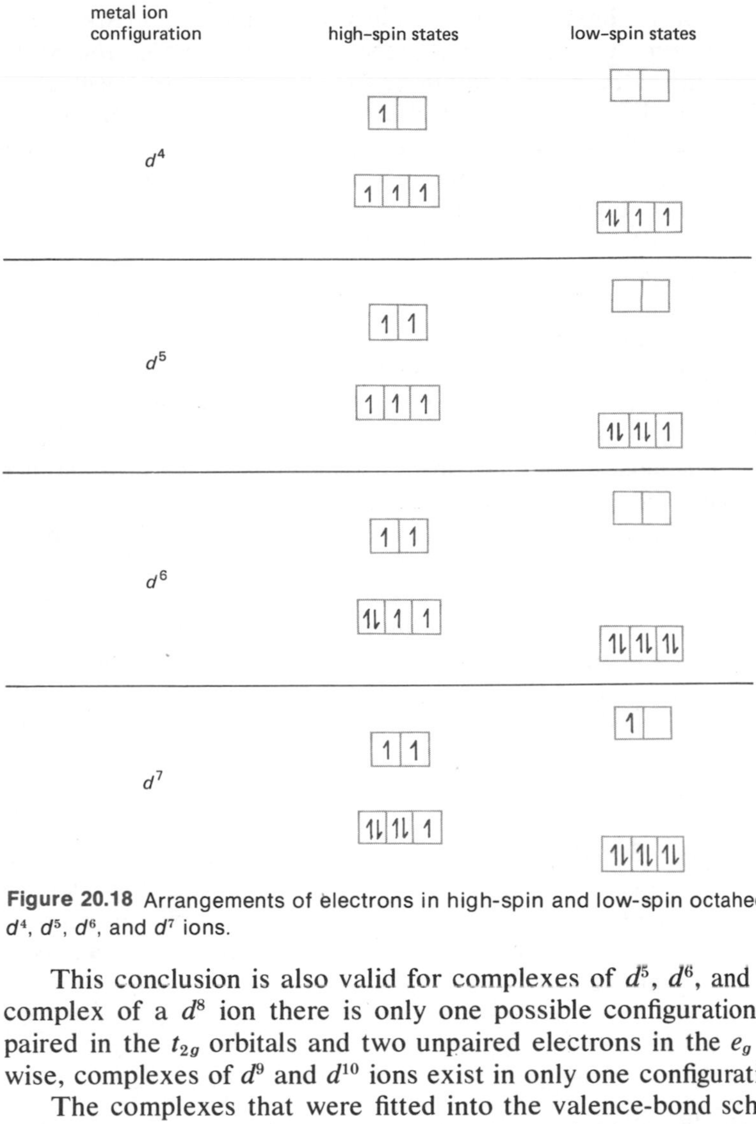

Figure 20.18 Arrangements of electrons in high-spin and low-spin octahedral complexes of d^4, d^5, d^6, and d^7 ions.

This conclusion is also valid for complexes of d^5, d^6, and d^7 ions. For a complex of a d^8 ion there is only one possible configuration: six electrons paired in the t_{2g} orbitals and two unpaired electrons in the e_g orbitals. Likewise, complexes of d^9 and d^{10} ions exist in only one configuration.

The complexes that were fitted into the valence-bond scheme only with difficulty are readily classified by the ligand field theory. Hence, the inner complex $[Fe(CN)_6]^{3-}$ with one unpaired electron is a low-spin state of this d^5 ion, and the outer complex $[FeF_6]^{3-}$ with five unpaired electrons has the high-spin configuration of a d^5 ion. The octahedral complexes of Ni^{II}, all of which must be assumed to be of the outer valence-bond type, have the only configuration possible for a complex of a d^8 ion: six electrons in the t_{2g} orbitals and two unpaired electrons in the e_g orbitals. The complex $[Co(NO_2)_6]^{4-}$, for which it was necessary to postulate the promotion of an electron to a $4d$ atomic orbital in the valence-bond treatment, is a low-spin d^7 complex according to the ligand field theory.

Values of Δ_o can be obtained from studies of the absorption spectra of complexes. In the complex $[Ti(H_2O)_6]^{3+}$, there is only one electron to be accommodated in either the t_{2g} or e_g orbitals. In the ground state this electron occupies a t_{2g} orbital. Excitation of the electron to an e_g orbital is possible when the energy required for this transition, Δ_o, is supplied. The absorption of light by the complex can bring about such excitations. The wavelength of light absorbed most strongly by the $[Ti(H_2O)_6]^{3+}$ ion is approximately 490 nm, which corresponds to a Δ_o of about 243 kJ/mol. The single absorption band of this complex spreads out over a considerable portion of the visible spectrum. Most of the red and violet light, however, is not absorbed, and this causes the red-violet color of the complex.

The interpretation of the absorption spectra of complexes with more than one *d* electron is considerably more complicated, since more than two arrangements of *d* electrons are possible. In general, for a given metal ion the replacement of one set of ligands by another causes a change in the energy difference between the t_{2g} and e_g orbitals, Δ_o, which gives rise to different light-absorption properties. In many instances a striking color change is observed when the ligands of a complex are replaced by other ligands.

Ligands may be arranged in a **spectrochemical series** according to the magnitude of Δ_o they bring about. From the experimental study of the spectra of many complexes, it has been found that the order is generally the same for the complexes of all of the transition elements in their common oxidation states with only occasional inversions of order between ligands that stand near to one another on the list. The order of some common ligands is

$$I^- < Br^- < Cl^- < F^- < OH^- < C_2O_4^{2-} < H_2O < NH_3 < en < NO_2^- < CN^-$$

The values of Δ_o induced by the halide ions are generally low, and complexes of these ligands usually have high-spin configurations. The cyanide ion, which stands at the opposite end of the series from the halide ions, induces the largest *d*-orbital splittings of any ligand listed. Cyano complexes generally have low-spin configurations.

A given ligand, however, does not always produce complexes of the same spin type. Thus, the hexaammine complex of Fe^{II} has a high-spin configuration, whereas the hexaammine complex of Co^{III} (which is isoelectronic with Fe^{II}) has a low-spin configuration.

For each metal ion there is a point in the series that corresponds to the change from ligands that produce high-spin complexes to ligands that form low-spin complexes. For example, Co^{II} forms high-spin complexes with NH_3 and ethylenediamine, but the NO_2^- and CN^- complexes of Co^{II} have low-spin configurations. The actual position in the series at which this change from high- to low-spin complex formation occurs depends upon the electron-pairing energy, *P*, for the metal ion, as well as the values of Δ_o for the complexes under consideration.

THE ARRANGEMENT OF ATOMS IN SPACE

About 1810, John Dalton had a set of molecular models made. The set consisted of "a number of equal balls about an inch in diameter" into which holes were drilled (either 12, 8, 6, 5, 4, 3, 2, or 1). The "holes had pins connecting other balls." The amazing thing is that Dalton, without any knowledge of chemical bonding, had the holes arranged in exactly the way that they are today. He wrote, in part,

> The 6 are an equidistant number of atoms, 90° asunder, 2 at the poles, and 4 at the equator. The 5 are a *symmetrical number,* 2 at the poles, and 3 at the equator 120° asunder. The 4 is a split of the 8, a regular number, and equidistant. The 3 are around the equator and 120° asunder. The 2 are at two opposite poles.

The chemical world was not ready for molecular models and stereochemistry, and Dalton's ideas had little effect and were soon forgotten. The importance of the spatial arrangement of atoms in molecules was not realized until more than sixty years later.

In 1874, Joseph Le Bel and Jacobus van't Hoff independently proposed that the four bonds of a carbon atom in an organic molecule are arranged in a tetrahedral manner. The proposal, which was made to explain the existence of optical isomers, proved to be extremely important to the development of organic chemistry.

The concept did not gain immediate acceptance. Hermann Kolbe, an older, well-established chemist, severely criticized van't Hoff's theory, which appeared in a German translation under the title of *The Arrangement of Atoms in Space.* Kolbe wrote

> Anyone . . . may read, if he is able to, the book . . . on *The Arrangement of Atoms in Space,* which has recently appeared and which overflows with fantastic foolishness. . . .
>
> A Dr. J. H. van't Hoff, of the Veterinary School at Utrecht, has no liking, it seems, for exact chemical investigation. He has considered it more convenient to mount Pegasus (apparently borrowed from the Veterinary School) and to proclaim . . . how the atoms appear to him to be arranged in space, when he is on the chemical Mount Parnassus which he has reached by bold flight.
>
> It is indicative of the present day . . . that . . . unknown chemists . . . judge with such assurance the most important problems of chemistry, which may never be solved. . . .

PROBLEMS

20.1 Define the following terms: (a) coordination number, (b) ligand, (c) chelate, (d) enantiomorph, (e) inert complex, (f) outer complex, (g) low-spin state.

20.2 State the oxidation number of the central atom in each of the following complexes:
(a) $[Co(NO_2)_6]^{3-}$
(b) $[Au(CN)_4]^-$
(c) $[V(CO)_6]$
(d) $[Co(NH_3)_4Br_2]^+$
(e) $[Co(en)Cl_4]^{2-}$

20.3 Write formulas for (a) potassium pentachloroaquorhodate(III), (b) sulfatotetraamminecobalt(III) nitrate, (c) sodium dioxotetracyanorhenate(V), (d) diamminebis-(ethylenediamine)-cobalt(II) chloride, (e) potassium tetracyanonickelate(0), (f) potassium hexacyanonickelate(II), (g) tetraamminecopper(II) hexachlorochromate-(III).

20.4 Write formulas for:
(a) hexacarbonylvanadium(0)
(b) zinc hexachloroplatinate(IV)
(c) chloronitrotetraamminecobalt(III) sulfate
(d) sodium dithiosulfatoargentate(I)
(e) tetraammineplatinum(II) trichloroammineplatinate(II)
(f) tetachlorodiammineplatinum(IV)
(g) potassium hexabromoaurate(III)
(h) hexaamminenickel(II) hexanitrocobaltate(III)

20.5 Name the following compounds:
(a) $K_4[Ni(CN)_4]$
(b) $K_2[Ni(CN)_4]$
(c) $(NH_4)_2[Fe(H_2O)Cl_5]$
(d) $[Cu(NH_3)_4][PtCl_4]$
(e) $[Ir(NH_3)_5(ONO)]Cl_2$
(f) $[Co(NH_3)_6]_2[Ni(CN)_4]_3$

20.6 Name the following compounds:
(a) $[Co(NO)(CO)_3]$
(b) $[Pt(NH_3)_2Cl_4]$
(c) $K_4[Pt(CN)_4]$
(d) $[Co(NH_3)_6]_4[Co(NO_2)_6]_3$
(e) $Na[Au(CN)_2]$
(f) $[Co(en)_2(SCN)Cl]Cl$

20.7 Addition of a solution of potassium hexacyanoferrate(II) to Fe^{3+}(aq) yields a precipitate called Prussian blue. Addition of a solution of potassium hexacyanoferrate(III) to Fe^{2+}(aq) also yields a blue precipitate (Turnbull's blue). The blue precipitates are now known to be identical and are called potassium iron(III) hexacyanoferrate(II). Write the formula for (a) Prussian blue, (b) iron(III) hexacyanoferrate(III), a brown precipitate, (c) copper(II) hexacyanoferrate(II), a purple precipitate, (d) potassium iron(II) hexacyanoferrate(II), a white precipitate.

20.8 Two compounds have the same empirical formula: $Co(NH_3)_3(H_2O)_2ClBr_2$. One mole of compound A readily loses one mole of water in a desiccator, whereas compound B does not lose any water under the same conditions. An aqueous solution of A has a conductivity equivalent to that of a compound with two ions per formula unit. The conductivity of an aqueous solution of B corresponds to that of a compound with three ions per formula unit. When $AgNO_3$ is added to a solution of compound

A, one mole of AgBr is precipitated per mole of A. A solution of compound B yields two moles of AgBr per mole of B. (a) Write the formulas of A and B. (b) What type of isomers are A and B?

20.9 Write a formula for an example of (a) an ionization isomer of $[Co(NH_3)_5(NO_3)]SO_4$, (b) a linkage isomer of $[Co(en)_2(NO_2)_2]$, (c) a coordination isomer of $[Pt(NH_3)_4][PtCl_4]$, (d) a hydrate isomer of $[Co(en)_2(H_2O)_2]Br_3$.

20.10 Write a formula for an example of (a) an ionization isomer of $[Pt(NH_3)_4(OH)_2]SO_4$, (b) a linkage isomer of $[Pd(dipy)(SCN)_2]$ (dipy is a bidentate ligand—2,2′-dipyridine), (c) a hydrate isomer of $[Co(NH_3)_4(H_2O)Cl]Cl_2$, and (d) a coordination isomer of $[Cr(NH_3)_4(C_2O_4)][Cr(NH_3)_2(C_2O_4)]$.

20.11 Identify all the possible isomers (stereoisomers as well as coordination isomers) of the compound $[Pt(NH_3)_4][PtCl_6]$.

20.12 Diagram the structures of all of the possible stereoisomers of each of the following octahedral complexes. Classify the structures as geometric or optical isomers.

(a) $[Cr(NH_3)_2(NCS)_4]^-$
(b) $[Co(NH_3)_3(NO_2)_3]$
(c) $[Co(en)(NH_3)_2Cl_2]^+$
(d) $[Co(en)Cl_4]^-$
(e) $[Co(en)_2ClBr]^+$
(f) $[Cr(NH_3)_2(C_2O_4)_2]^-$
(g) $[Cr(C_2O_4)_3]^{3-}$

20.13 Diagram all the stereoisomers of each of the following molecules. In the formulas, py stands for pyridine, C_5H_5N, a unidentate ligand.

(a) $[Pt(NH_3)(py)ClBr]$ (square planar)
(b) $[Pt(NH_3)(py)Cl_2]$ (square planar)
(c) $[Pt(py)_2Cl_2]$ (square planar)
(d) $[Co(py)_2Cl_2]$ (tetrahedral)

20.14 How can dipole-moment measurement distinguish between the *cis*- and *trans*-isomers of the square planar $[Pt(NH_3)_2Cl_2]$?

20.15 Diagram the four stereoisomers of $[Pt(en)(NO_2)_2Cl_2]$.

20.16 Draw valence bond diagrams for the bonding in the following, diamagnetic ions. What geometric configuration does each have? (a) $AuCl_2^-$, (b) $PdCl_4^{2-}$, (c) $ZnCl_4^{2-}$, (d) $BeCl_4^{2-}$, (e) $RhCl_6^{3-}$.

20.17 The complex $[Ni(CN)_4]^{2-}$ is square planar, and the complex $[NiCl_4]^{2-}$ is tetrahedral. Draw valence bond diagrams for these complexes, and predict the number of unpaired electrons in each.

20.18 The octahedral complexes $[Fe(CN)_6]^{3-}$ and $[FeF_6]^{3-}$ have one and five unpaired electrons, respectively. Offer an explanation for these observations based on (a) the valence bond theory, (b) the ligand field theory.

20.19 (a) Contrast the valence bond and ligand field explanations for the electronic structures of the two types of octahedral Co^{2+} complexes (one with three unpaired electrons and another with only one unpaired electron). (b) Why is the valence bond description of the complex with one unpaired electron unsatisfactory?

20.20 For Fe^{2+} the electron-pairing energy, *P*, is approximately 210 kJ/mol. Approximate values of Δ_O for the complexes $[Fe(H_2O)_6]^{2+}$ and $[Fe(CN)_4]^{4-}$ are 120 kJ/mol and 390 kJ/mol, respectively. (a) Do these complexes have high-spin or low-spin configurations? (b) Draw a *d*-orbital splitting diagram for each.

20.21 For $[Mn(H_2O)_6]^{3+}$, which is a high-spin complex, Δ_O is approximately 250 kJ/mol. For $[Mn(CN)_6]^{3-}$, which is a low-spin complex, Δ_O is approximately 460 kJ/mol. (a) What conclusion can you reach in regard to the magnitude of the electron-pairing energy, *P*, for Mn^{3+}? (b) Would you expect the complex $[Mn(C_2O_4)_3]^{3-}$ to be a high-spin or a low-spin complex?

20.22 Draw *d*-orbital splitting diagrams for $[Co(NH_3)_6]^{2+}$ and $[Co(NH_3)_6]^{3+}$. The Δ_O values for these two complexes are approximately 120 kJ/mol and 270 kJ/mol, respectively. The pairing energy of Co^{2+} is approximately 270 kJ/mol and of Co^{3+} is approximately 210 kJ/mol.

20.23 Draw *d*-orbital splitting diagrams for the high-spin and low-spin octahedral complexes of (a) Zn^{2+}, (b) Cr^{2+}, (c) Ni^{2+}, (d) Ni^{3+}, (e) Mn^{2+}, (f) Fe^{3+}, (g) V^{2+}, (h) V^{3+}, (i) V^{4+}, (j) V^{5+}.

20.24 Draw *d*-orbital splitting diagrams for (a) the octahedral complexes of Cr^{3+}, (b) the octahedral complexes of Ni^{2+}, (c) the square planar complexes of Ni^{2+} (all of which are diamagnetic), (d) the square planar complexes of Co^{2+} (all of which have one unpaired electron), (e) the tetrahedral complexes of Co^{2+}, (f) the octahedral complexes of Ru^{2+} (all of which are diamagnetic), (g) the octahedral complexes of Ir^{4+} (all of which have one unpaired electron).

20.25 The $[Ti(H_2O)_6]^{3+}$ complex is red-violet. What change in color would be expected if the ligands of this complex were replaced by ligands that induce a larger Δ_O? Note that the color of the complex corresponds to light transmitted, not absorbed.

20.26 Given the following data, calculate the instability constant, K_{inst}, of the $Al(OH)_4^-$ ion.

$$3e^- + Al(OH)_4^- \rightleftharpoons Al + 4OH^- \qquad \mathscr{E}° = -2.330 \text{ V}$$

$$3e^- + Al^{3+} \rightleftharpoons Al \qquad \mathscr{E}° = -1.662 \text{ V}$$

Chapter 21

Organic Chemistry

The name organic chemistry derives from the early concept that substances of plant or animal origin (organic substances) were different from those of mineral origin (inorganic substances). In the middle of the nineteenth century, however, the idea that organic substances could only be synthesized by living organisms was gradually discounted. Not only has a large number of natural products been synthesized in the laboratory, but also countless related materials have been made that do not occur in nature. All these compounds contain carbon. Over a million carbon compounds are known.

Because it represents a convenient division of chemistry, the term organic chemistry is retained, but it is now commonly defined as the chemistry of carbon and its compounds. Since some carbon compounds (such as carbonates, carbides, and cyanides) are traditionally classed as inorganic compounds, organic chemistry is probably better defined as the chemistry of the hydrocarbons (compounds containing only carbon and hydrogen) and their derivatives.

Carbon forms an unusually large number of compounds because of its exceptional ability to catenate (Section 8.8). In addition, the carbon atom can form four, very stable, single-covalent bonds. It also has the ability to form multiple bonds with other carbon atoms or with atoms of other elements.

21.1 THE ALKANES

In the first four sections of this chapter various types of hydrocarbons are characterized. The reactions of the hydrocarbons is the topic of Section 21.5. The simplest hydrocarbon is methane, CH_4. This molecule contains four equivalent carbon–hydrogen bonds arranged tetrahedrally (Figure 21.1). The bonding may be considered to arise through the use of sp^3 hybrid orbitals of the carbon atom. Each of the bonds of methane is of the same length (109.5 pm), and each of the H—C—H bond angles is 109° 28′ (the so-called tetrahedral angle). The commonly employed structural formula of methane,

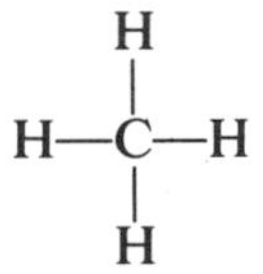

does not accurately represent the tetrahedral arrangement of the molecule.

The **alkanes** are hydrocarbons in which all of the carbon–carbon bonds are single bonds. They may be considered to be derived from methane by the successive addition of —CH_2— units. Such a series of compounds is said to be **homologous.** Thus, the formula of the second member of the family, ethane

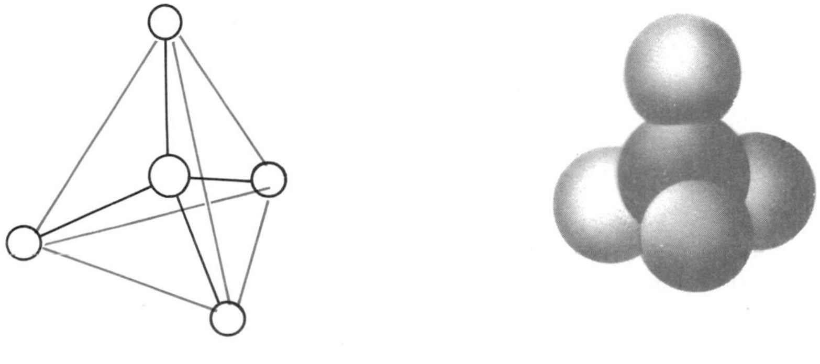

(a) (b)

Figure 21.1 Representations of the structure of methane, CH_4. The bonds in diagram (a) are of exaggerated length in comparison to the atomic sizes.

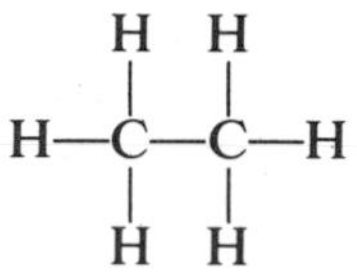

may be formally derived by the introduction of a $—CH_2—$ unit between the carbon and a hydrogen of methane. This molecule is also represented by the formula $CH_3—CH_3$.

The alkanes conform to the general formula C_nH_{2n+2}, where n is the number of carbon atoms in the compound. A few subsequent members of the family, and their names, are

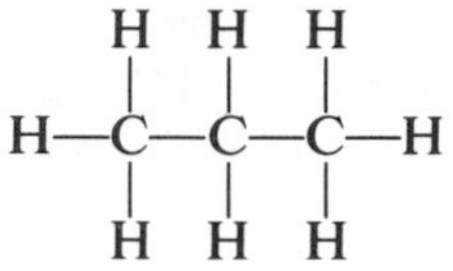

$CH_3CH_2CH_3$
propane

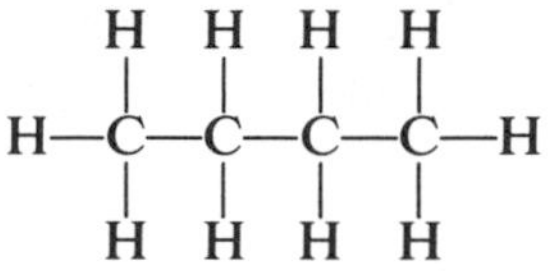

$CH_3CH_2CH_2CH_3$
butane

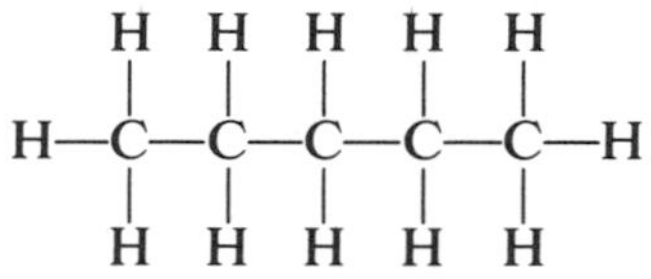

$CH_3CH_2CH_2CH_2CH_3$
pentane

These compounds are spoken of as **straight-chain** compounds even though the carbon chains are far from linear (Figure 21.2). All the bond angles are approximately tetrahedral, and axial rotation of any carbon atom around a single bond of the chain is possible.

There is only one compound for each of the formulas CH_4, C_2H_6, and C_3H_8. There are two compounds, however, with the formula C_4H_{10}: the compound with a straight chain and one with a **branched chain.**

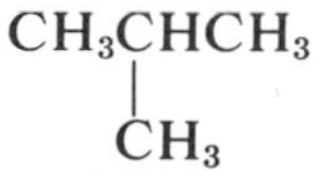

methylpropane

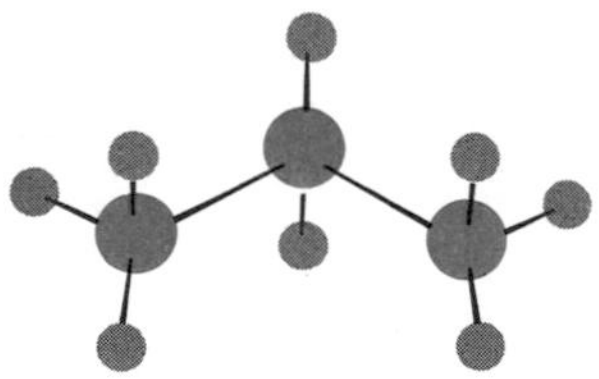

Figure 21.2 Representation of the structure of propane, $CH_3CH_2CH_3$.

The two compounds of formula C_4H_{10} are **structural isomers** (compounds with the same molecular formula but different structural formulas). They have different properties.

There are three structural isomers of C_5H_{12}: pentane (the straight-chain compound),

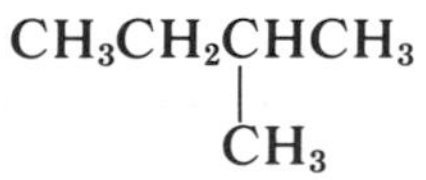

methylbutane

and

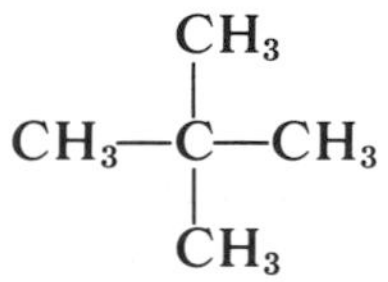

dimethylpropane

In the alkane series the number of possible structural isomers increases rapidly. There are five structural isomers for C_6H_{14}, nine for C_7H_{16}, and 75 for $C_{10}H_{22}$. It has been calculated that more than 4×10^9 isomers are possible for $C_{30}H_{62}$.

A branched-chain compound may be named in terms of the longest straight chain in the molecule. The side chains are named as alkyl radicals, and their positions on the straight chain are usually indicated by a numbering system. Alkyl radicals are fragments of alkane molecules from which a hydrogen atom has been removed. Their names are derived from the name of the parent alkane with the ending changed to *-yl*. A list of common alkyl radicals appears in Table 21.1.

Table 21.1 Simple alkyl radicals

FORMULA	NAME
CH_3—	methyl
CH_3CH_2—	ethyl
$CH_3CH_2CH_2$—	*normal*-propyl or *n*-propyl
CH_3CH— | CH_3	isopropyl
$CH_3CH_2CH_2CH_2$—	*normal*-butyl or *n*-butyl
CH_3CHCH_2— | CH_3	isobutyl
CH_3CH_2CH— | CH_3	*secondary*-butyl or *sec*-butyl
CH_3 | CH_3C— | CH_3	*tertiary*-butyl or *tert*-butyl

For example, the longest straight chain in the molecule

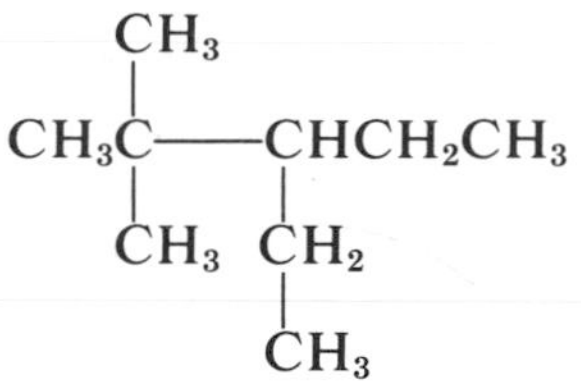

consists of five carbon atoms, and the compound is named as a derivative of pentane. The pentane chain is numbered starting at the end that will give the side chains the lowest numbers.

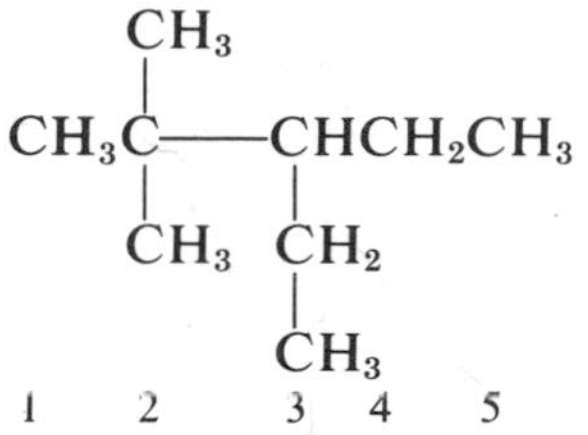

In the name a number is used to indicate the position of each substituent radical. Thus, the name of the compound is

2,2-dimethyl-3-ethylpentane

No numbers appear in the names of our earlier examples of structural isomers. The name methylpropane requires no number to indicate the position of the methyl radical with regard to the propane chain since there is only one possible position for the methyl group (number 2). If the methyl group were placed on either terminal carbon atom (number 1 or number 3), the compound would be the straight-chain isomer, butane. In like manner, the names

Table 21.2 Physical properties of some straight-chain alkanes

COMPOUND	FORMULA	MELTING POINT (°C)	BOILING POINT (°C)
methane	CH_4	−183	−162
ethane	C_2H_6	−172	−88
propane	C_3H_8	−187	−42
butane	C_4H_{10}	−135	1
pentane	C_5H_{12}	−131	36
hexane	C_6H_{14}	−94	69
heptane	C_7H_{16}	−91	99
octane	C_8H_{18}	−57	126
nonane	C_9H_{20}	−54	151
decane	$C_{10}H_{22}$	−30	174
hexadecane	$C_{16}H_{34}$	20	288
heptadecane	$C_{17}H_{36}$	23	303

methylbutane and dimethylpropane do not require the use of numbers. There is only one possible structure corresponding to each name.

Names for some of the higher straight-chain homologs may be obtained from Table 21.2, which lists the melting points and boiling points of some of the straight-chain alkanes. The melting point and boiling point increase as the length of the chain (and hence the molecular weight) of the hydrocarbon increases. The first four compounds are gases under ordinary conditions. Higher homologs are liquids (C_5H_{12} to $C_{15}H_{32}$) and solids (from $C_{16}H_{34}$ on).

Aliphatic hydrocarbons are open-chain structures; in addition, **cyclic hydrocarbons** are known. The **cycloalkanes** are ring structures that contain only single carbon–carbon bonds. They have the general formula C_nH_{2n}. Examples are,

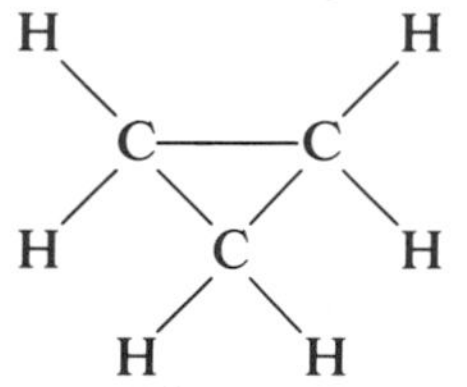

cyclopropane

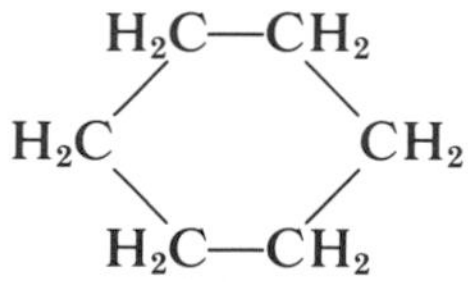

cyclohexane

The cyclohexane ring is not planar but is puckered in such a way that the C—C—C bond angles are 109° 28′, the tetrahedral angle. Three- and four-membered rings are strained because the C—C—C bond angles deviate significantly from the tetrahedral angle. The strain is greatest in cyclopropane, in which the C atoms form an equilateral triangle and the bond angles are 60°. Thus, the cyclopropane ring is readily broken by H_2 (Ni catalyst) to give propane. Cyclobutane reacts with H_2 in an analogous manner (to give butane), but a higher temperature is required.

Petroleum is the principle source of alkanes and cycloalkanes. In the refining of petroleum the crude oil is first separated into fractions by distillation (Table 21.3). Some higher-boiling fractions are subjected to **cracking**

Table 21.3 Petroleum fractions

FRACTION	BOILING RANGE (°C)	CARBON CONTENT OF COMPOUNDS	USE
gas	below 20	C_1–C_4	fuel
petroleum ether	20–90	C_5–C_7	solvent
gasoline	35–220	C_5–C_{12}	motor fuel
kerosene	200–315	C_{12}–C_{16}	jet fuel
fuel oil	250–375	C_{15}–C_{18}	diesel fuel
lubricating oils, greases	350 up	C_{16}–C_{20}	lubrication
paraffin wax	50–60 (m.p.)	C_{20}–C_{30}	candles
asphalt	viscous liquid		paving
residue	solid		fuel

processes in which large molecules are broken down into smaller molecules, thus increasing the yield of the most valuable fraction, gasoline. In addition, small molecules are converted into larger ones in what are called **alkylation processes.** Gases that are by-products of these processes, together with the petroleum fractions and substances obtained from the fractions, comprise important starting materials for the manufacture of many chemical products. More than 2000 such products (such as rubber, detergents, plastics, and textiles) are derived from the substances obtained from petroleum and natural gas.

21.2 THE ALKENES

The **alkenes,** which are also called **olefins,** are aliphatic hydrocarbons that have a carbon–carbon double bond somewhere in their molecular structure. The first member of the series is ethene (also called ethylene),

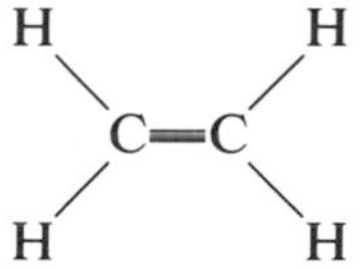

Each carbon atom of ethene uses sp^2 hybrid orbitals to form three sigma bonds: one with the other C atom and two with H atoms. All six atoms lie in the same plane, and all bond angles are approximately 120°. In addition, each carbon atom has an electron in a *p* orbital that is not engaged in the formation of sp^2 σ bonds. These two *p* orbitals are coplanar and perpendicular to the plane of the molecule. They overlap to form a π bonding orbital with regions of charge density above and below the plane of the molecule (Section 7.9, Figure 7.22).

The *p* orbitals can overlap and form a π bonding orbital only when they lie in the same plane. Consequently, free rotation around the carbon–carbon double bond is not possible without breaking the π bond. The carbon–carbon bond distance in ethene (133 pm) is shorter than that in ethane (154 pm) because of the double bond. The π bond is not so strong as the σ bond. The carbon–carbon bond energy in ethane is approximately 340 kJ/mol, and the bond energy of the carbon–carbon double bond in ethene is about 610 kJ/mol.

Alkanes are called **saturated** compounds because all the valence electrons of the carbon atoms are engaged in single bond formation, and no more hydrogen atoms or atoms of other elements can be accommodated by the carbon atoms of the chain. **Unsaturated** compounds, such as the alkenes, can undergo **addition reactions** (Section 21.5) because of the availability of the π electrons of the double bond. In one such reaction hydrogen adds to ethene to form ethane.

$$CH_2{=}CH_2 + H_2 \rightarrow CH_3{-}CH_3$$

Alkenes conform to the general formula C_nH_{2n}. The name of an alkene is derived from the name of the corresponding alkane by changing the ending from *-ane* to *-ene*. A number is used, when necessary, to indicate the position of the double bond. The second member of the series is propene.

$$CH_2{=}CH{-}CH_3$$

New types of isomerism arise in the alkene series. The type of structural isomerism displayed by the alkanes (for example, butane and methylpropane) is known as **chain isomerism.** In addition to chain isomerism, **position isomerism** occurs in the olefin series. There is only one straight-chain butane. There are, however, two straight-chain butenes, and they differ in the **position** of the double bond.

$$\underset{\text{1-butene}}{CH_2{=}CH{-}CH_2{-}CH_3} \qquad \underset{\text{2-butene}}{CH_3{-}CH{=}CH{-}CH_3}$$

In the first compound the double bond is located between carbon atoms number 1 and number 2, and the lower number (1) is used to indicate its position. In like manner, the lower number is used to indicate the position of the double bond (which occurs between carbon atoms number 2 and number 3) in the name of the second compound. In each case the chain is numbered from the end that gives the lowest possible number for the name. The name of a branched-chain alkene is derived from that of the longest straight chain that contains the double bond.

The alkenes also exhibit **geometric** (or **cis-trans**) **isomerism,** which is a type of **stereoisomerism.** Stereoisomers have the same structural formula but differ in the arrangement of the atoms in space. An example is provided by the isomers of 2-butene. Because of the restricted rotation around the double bond, one isomer exists with both methyl groups on the same side of the double bond (the *cis* isomer), and another isomer exists with the methyl groups on opposite sides of the double bond (the *trans* isomer). All carbon atoms of the molecule lie in the same plane.

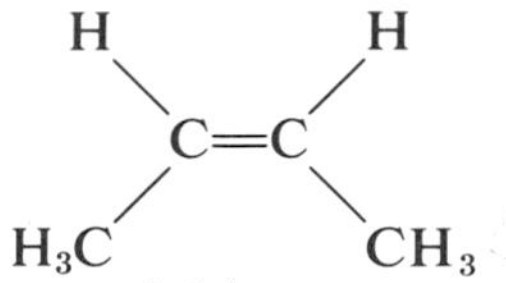

cis-2-butene
boiling point, 1°C

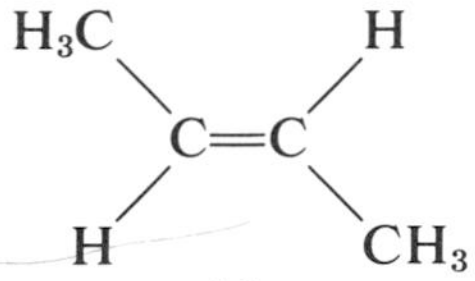

trans-2-butene
boiling point, 2.5°C

The properties of the 2-butenes do not differ greatly. Other *cis-trans* isomers exhibit wide variation in their physical properties.

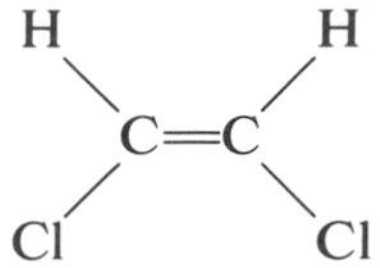

cis-1,2-dichloroethene
boiling point, 60.1°C

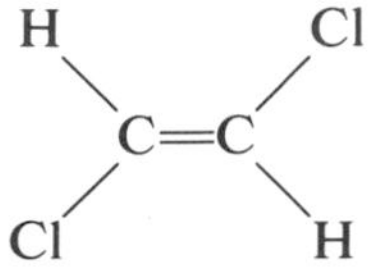

trans-1,2-dichloroethene
boiling point, 48.4°C

The physical properties of the alkenes are very similar to the alkanes. The C_2H_4, C_3H_6, and C_4H_8 compounds are gases under ordinary conditions. The C_5H_{10} to $C_{18}H_{36}$ compounds are liquids, and the higher alkenes are solids. All the compounds are only slightly soluble in water.

Molecules can contain more than one double bond. For example,

$$CH_2{=}CH{-}CH{=}CH_2$$

1,3-butadiene

which is used to produce a type of synthetic rubber (Section 21.12). Cycloalkenes are known, for example,

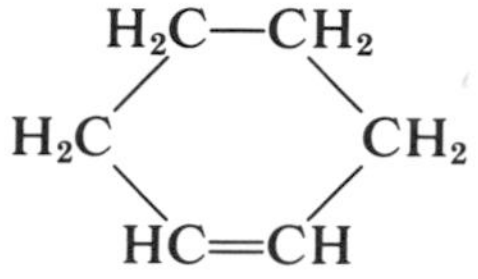

cyclohexene

21.3 THE ALKYNES

Molecules that contain carbon–carbon triple bonds are known as alkynes. These unsaturated hydrocarbons have the general formula C_nH_{2n-2} and constitute the **alkyne, or acetylene, series.** The first member of the series is ethyne, or acetylene.

$$H{-}C{\equiv}C{-}H$$

In acetylene each carbon atom uses *sp* hybrid orbitals to form a σ bond with a hydrogen atom and a σ bond with the other carbon atom. Consequently, the four atoms of the molecule lie in a straight line. In addition to the carbon–carbon σ bond, two π bonds are formed between the carbon atoms by the overlap of *p* orbitals (Section 7.9, Figure 7.23). The resultant carbon–carbon bond distance (121 pm) is shorter than the carbon–carbon double-bond distance (133 pm). The π bonds are weaker than the C—C σ bond. The bond energy of the triple bond is approximately 830 kJ/mol (compared to the single bond energy of 340 kJ/mol). Alkynes readily undergo addition reactions across the triple bond because of the availability of the four electrons of the π bonds.

Alkynes are named in a manner analogous to that employed in naming alkenes. The ending *-yne* is used in place of *-ene*. Their physical properties are similar to those of the other aliphatic hydrocarbons. Both chain and position isomerism occur in the alkyne series, but *cis-trans* isomerism is not possible because of the linear geometry of the group X—C≡C—X, where X is a carbon atom or an atom of another element.

Notice that a branched-chain isomer of C_4H_6 is impossible. Only two isomers of C_4H_6 exist

$$HC{\equiv}CCH_2CH_3 \qquad CH_3C{\equiv}CCH_3$$

1-butyne   2-butyne

Both branched-chain and straight-chain isomers of C_5H_8 are known.

$HC\equiv CCH_2CH_2CH_3$	$CH_3C\equiv CCH_2CH_3$	$HC\equiv CCHCH_3$ | CH_3
1-pentyne	2-pentyne	methyl-1-butyne

21.4 AROMATIC HYDROCARBONS

Aromatic hydrocarbons are compounds that have molecular structures based on that of benzene, C_6H_6. The six carbon atoms of benzene are arranged in a ring from which the hydrogen atoms are radially bonded (Figure 21.3). The entire structure is planar, and all of the bond angles are 120°. The carbon–carbon bond distance is 139 pm, which is between the single-bond distance of 154 pm and the double-bond distance of 133 pm.

The electronic structure of benzene may be represented as a resonance hybrid

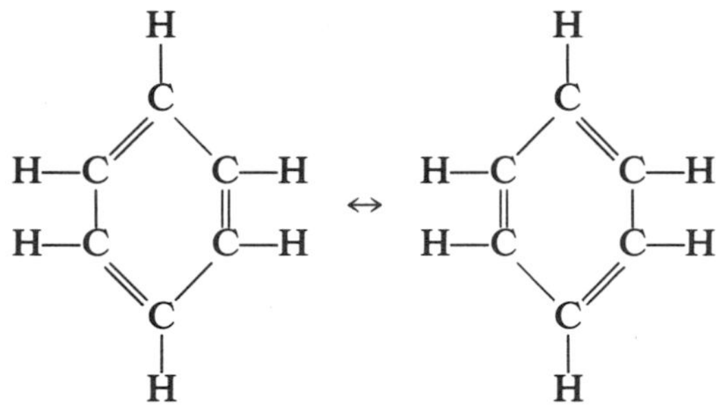

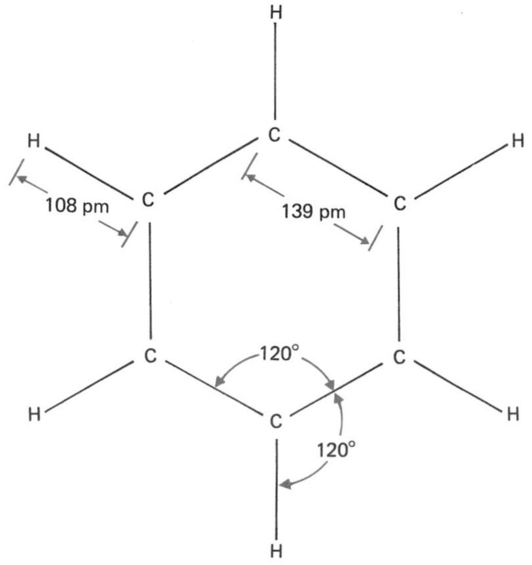

Figure 21.3 Geometry of the benzene molecule.

This representation correctly shows that all of the carbon–carbon bonds are equivalent and that each one is intermediate between a single bond and a double bond.

Each carbon atom of the ring uses sp^2 hybrid orbitals to form σ bonds with two adjacent carbon atoms and with a hydrogen atom. Therefore, the resulting framework of the molecule is planar with bond angles of 120° (Figure 21.3). A $2p$ electron of each carbon atom is not employed in the formation of the sp^2-hybrid, σ bonds. The axes of these p orbitals are perpendicular to the plane of the molecule, and each p orbital can overlap two adjacent p orbitals. The result is the formation of a system of π orbitals with regions of charge density above and below the plane of the ring (Figure 21.4).

The electron pairs of the σ bonds of the benzene molecule (as well as the electron pair of an ordinary π bond) are localized with respect to the two nuclei that they serve to bond. The six electrons that occupy the π bonding system (one contributed by each carbon atom) are delocalized. On the average, the π electrons tend to be distributed evenly about the benzene ring, which explains the bond length and equivalence of all the C to C bonds. Since delocalization minimizes electron–electron repulsion, the structure, which is reminiscent of that of graphite (Section 18.9), is very stable and is responsible for the properties that are typical of aromatic compounds. The benzene ring does not readily undergo addition reactions in the way that alkenes and alkynes do.

Notice that the resonance and molecular-orbital pictures of benzene are

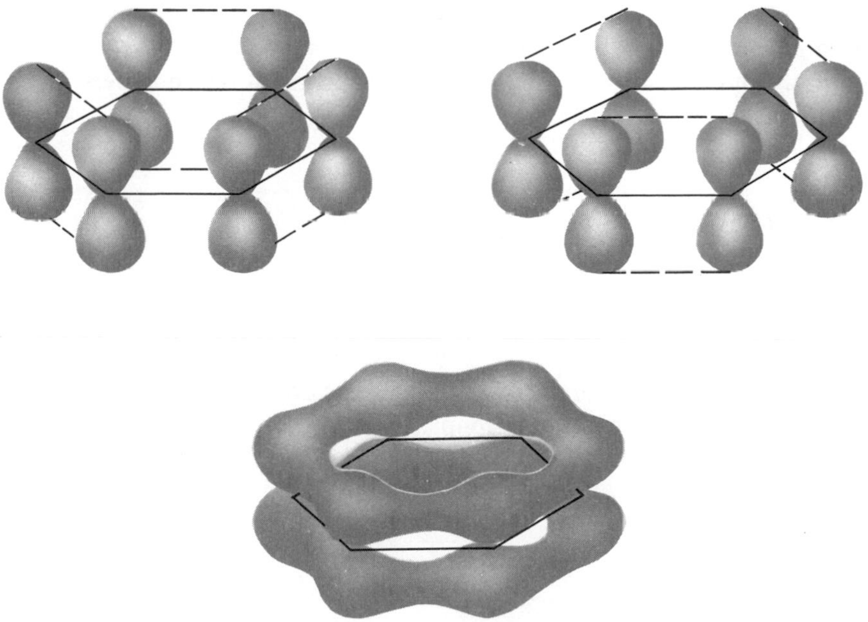

Figure 21.4 The bonding system of benzene; σ bonds are indicated by the solid lines.

qualitatively equivalent. If the overlap of the *p* orbitals is imagined to occur only between orbitals of adjacent carbon atoms in such a way that traditional electron-pair bonds between only two atoms result, the two resonance forms are derived (Figure 21.4). According to the resonance theory, neither of these forms is correct. The true structure is a hybrid of both of them.

The benzene ring is usually represented as

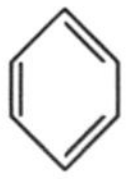
or

The carbon and hydrogen atoms of the ring are not shown. In formulas for benzene derivatives, such as

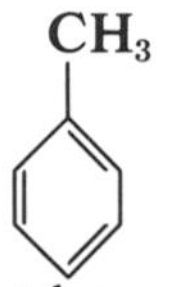

toluene
methylbenzene

and

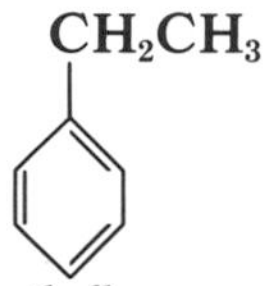

ethylbenzene

it is understood that the substituents replace hydrogen atoms of the ring. The radical formed from benzene by the removal of a hydrogen atom

is called a **phenyl radical.**

There are three position isomers of any disubstituted benzene derivative. For example,

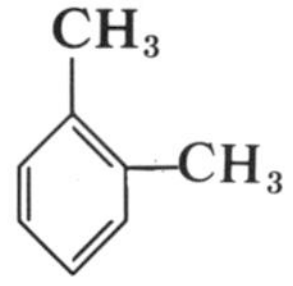

o-xylene

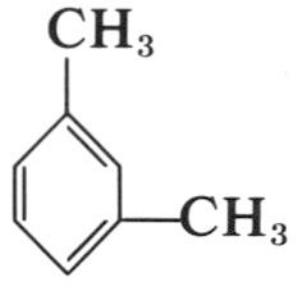

m-xylene

p-xylene

The prefixes *ortho-* (*o-*), *meta-* (*m-*), and *para-* (*p-*) are used to designate the positions of the two substituent groups. Numbers are used when more than two substituents are present on the ring.

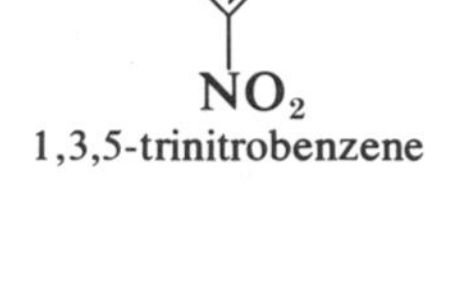

1,3,5-trinitrobenzene

Compounds are known in which several rings are fused together. For example, naphthalene, $C_{10}H_8$, is a planar molecule that is a resonance hybrid,

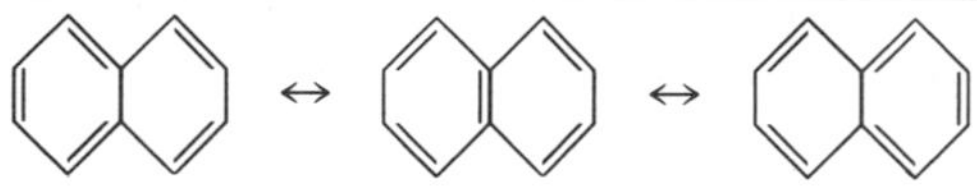

There are no hydrogen atoms bonded to the carbon atoms through which the rings are fused.

Coal is a major source of aromatic compounds. In the preparation of coke for industrial use (principally for the reduction of iron ore), bituminous coal is heated out of contact with air in retorts. Coal tar, a by-product of this process, is a black, viscous material from which various aromatic compounds may be separated. Aromatic compounds are also obtained from processes that utilize the alkanes and cycloalkanes derived from petroleum.

21.5 REACTIONS OF THE HYDROCARBONS

Combustion of any hydrocarbon in excess oxygen yields carbon dioxide and water.

$$CH_4(g) + 2O_2(g) \rightarrow CO_2(g) + 2H_2O(g)$$

The reactions are highly exothermic, a fact which accounts for the use of hydrocarbons as fuels. Carbon compounds exist in oxidation states intermediate between the hydrocarbons and carbon dioxide. These compounds are discussed in sections that follow. They are not generally prepared by direct reaction with elementary oxygen.

The replacement of a hydrogen atom of a hydrocarbon molecule by another atom or group of atoms is called a **substitution reaction.** The alkanes react with chlorine or bromine in the presence of sunlight or ultraviolet light by means of a free-radical chain mechanism (Section 13.8). The chain is initiated by the light-induced dissociation of a chlorine molecule into atoms.

$$Cl_2 \rightarrow 2Cl\cdot$$

Propagation of the chain occurs by the reactions

$$Cl\cdot + CH_4 \rightarrow CH_3\cdot + HCl$$
$$CH_3\cdot + Cl_2 \rightarrow CH_3Cl + Cl\cdot$$

The chain is terminated by the reactions

$$CH_3\cdot + Cl\cdot \rightarrow CH_3Cl$$
$$Cl\cdot + Cl\cdot \rightarrow Cl_2$$
$$CH_3\cdot + CH_3\cdot \rightarrow CH_3CH_3$$

The reactions that terminate the chain occur at the walls of the container. The energy liberated by the formation of the new bond is transferred to the walls of the container. In a collision between radicals in the gas phase, the united fragments fly apart almost immediately because a third body is not present to carry away the energy released by bond formation. Hence, chain-terminating reactions do not occur with great frequency, and the initial reactants are rapidly consumed.

The overall reaction for the preceding mechanism is

$$CH_4 + Cl_2 \rightarrow CH_3Cl + HCl$$

In addition to chloromethane, CH_3Cl, the reaction produces dichloromethane, CH_2Cl_2, trichloromethane (chloroform), $CHCl_3$, and tetrachloromethane (carbon tetrachloride), CCl_4. In terms of the chain mechanism, these compounds arise through collisions of chlorine atoms with already chlorinated methane molecules. For example,

$$Cl\cdot + CH_3Cl \rightarrow CH_2Cl\cdot + HCl$$
$$CH_2Cl\cdot + Cl_2 \rightarrow CH_2Cl_2 + Cl\cdot$$

Complex mixtures of mono- and polysubstituted isomers are obtained as products of the chlorination of higher alkanes.

Addition reactions are characteristic of unsaturated hydrocarbons. Examples follow:

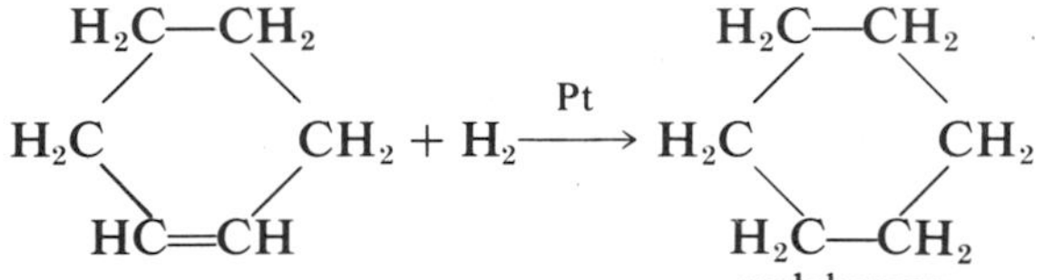

cyclohexane

$$CH_2{=}CH_2 + Br_2 \longrightarrow CH_2Br{-}CH_2Br$$
1,2-dibromoethane

$$HC{\equiv}CH + HCl \longrightarrow CH_2{=}CHCl$$
1-chloroethene

$$CH_2{=}CHCl + HCl \longrightarrow CH_3{-}CHCl_2$$
1,1-dichloroethane

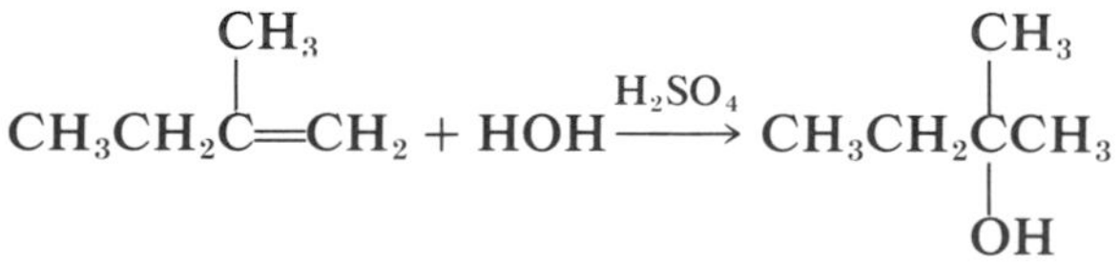

2-methyl-2-butanol

2-bromopropane

When unsymmetrical molecules (such as HBr) add to the double bond of an olefin, two isomeric compounds are sometimes produced. Thus, in the last reaction of the previous series, two compounds are obtained.

2-bromopropane 1-bromopropane

Over 90% of the product, however, is the 2-bromo-isomer. The principal product of such an addition is that in which the hydrogen atom of the addend is bonded to the carbon that originally had the larger number of hydrogen atoms (**Markovnikov's rule).**

The initial step of the mechanism of the addition of HBr is thought to be an electrophilic attack by a proton (a Lewis acid) on the π electrons of the double bond (a Lewis base). The product of this step is a positive ion called a **carbonium ion.** For propene two such ions are possible.

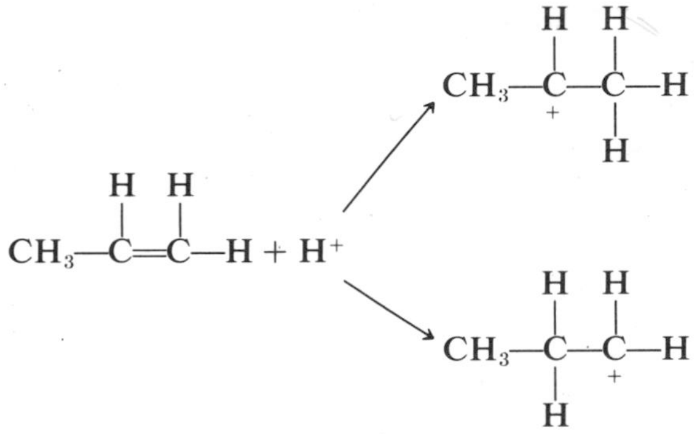

The π electrons are used to bond the incoming proton to one of the carbon atoms of the double bond which leaves the other carbon atom with a positive charge. The stability of carbonium ions is known to decrease in the order

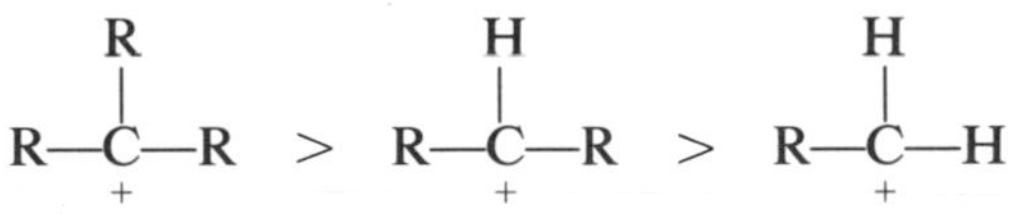

where R is an alkyl radical. Consequently, in the reaction of propene with HBr, the carbonium ion produced in larger quantity is the one shown on top.

Carbonium ions are highly reactive Lewis acids. They exist only transiently and rapidly combine with bromide ion (a Lewis base).

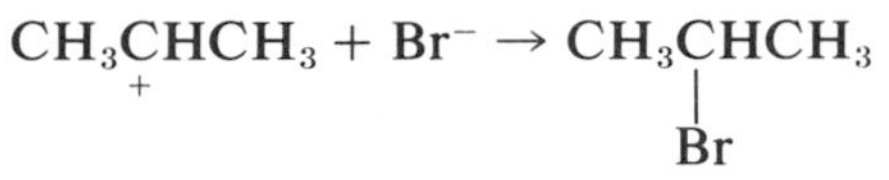

$$CH_3CH_2CH_2^+ + Br^- \rightarrow CH_3CH_2CH_2Br$$

Thus, the relative stabilities of the carbonium ions are responsible for the fact that 2-bromopropane is the principal product of the reaction.

In contrast to unsaturated aliphatic hydrocarbons, the principal reactions of benzene are substitutions, not additions.

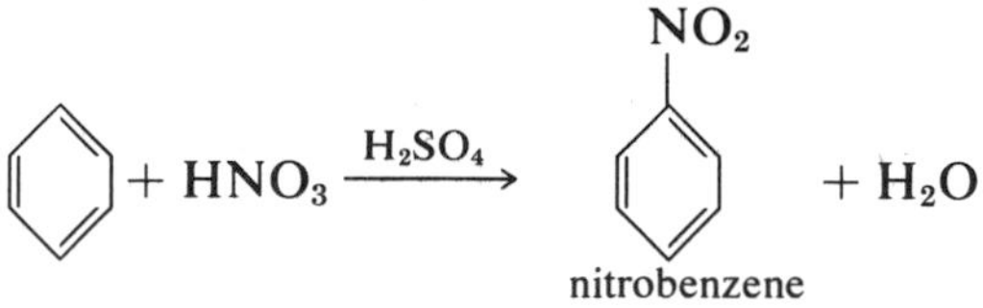

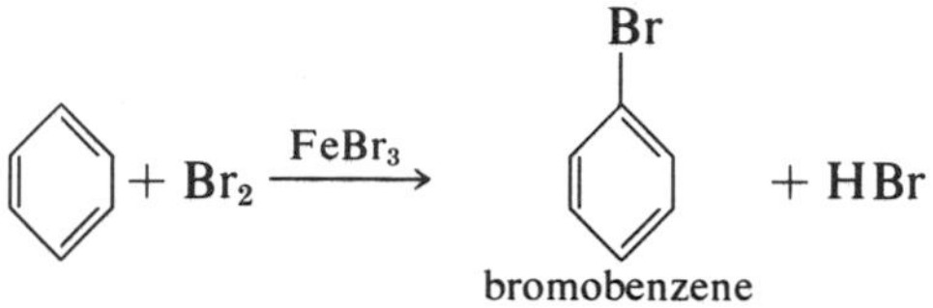

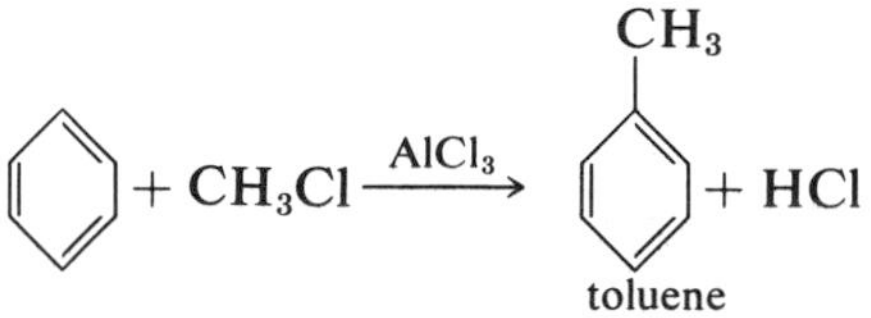

The last reaction is the Friedel–Crafts synthesis for the preparation of aromatic hydrocarbons.

The catalysts, which are indicated over the arrows in the preceding equations, produce powerful Lewis acids (NO_2^+, Br^+, and CH_3^+) as follows:

$$HNO_3 + 2H_2SO_4 \rightarrow NO_2^+ + H_3O^+ + 2HSO_4^-$$
$$Br_2 + FeBr_3 \rightarrow Br^+ + FeBr_4^-$$
$$CH_3Cl + AlCl_3 \rightarrow CH_3^+ + AlCl_4^-$$

These cationic Lewis acids, which are short-lived reaction intermediates, are electrophilic, and we shall indicate them as E^+ in the mechanism that follows.

The benzene ring is attacked by the electrophilic group, E^+, which bonds to a carbon atom by means of a pair of π electrons and creates an activated complex with a positive charge.

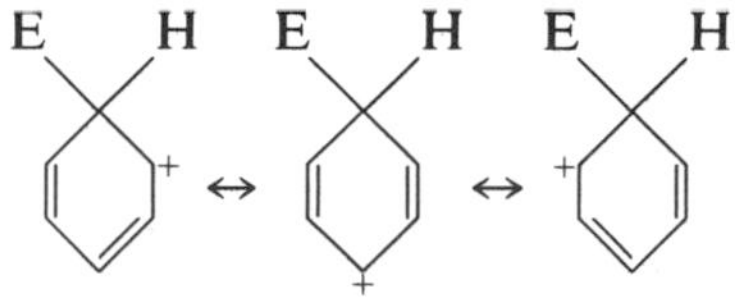

Notice that the carbon atom to which E is bonded holds two groups in the complex. The remaining five carbon atoms hold only their customary hydrogen atoms. The complex is a resonance hybrid, and the charge is delocalized, thus providing a degree of stability. The π bonding system of benzene is more stable, however, and is restored by the loss of a proton.

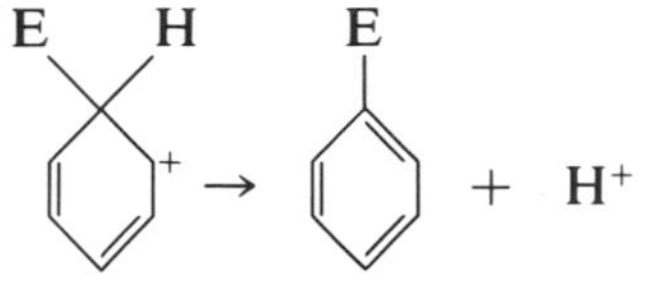

The proton is lost to the anion produced in the reaction that generated the cationic electrophile.

$$H^+ + HSO_4^- \rightarrow H_2SO_4$$
$$H^+ + FeBr_4^- \rightarrow HBr + FeBr_3$$
$$H^+ + AlCl_4^- \rightarrow HCl + AlCl_3$$

21.6 ALCOHOLS AND ETHERS

The alcohols may be considered as derivatives of hydrocarbons in which a hydroxyl group, OH, replaces a hydrogen. The hydroxyl group is one of a number of **functional groups,** which are groups of atoms that give organic compounds bearing them characteristic chemical and physical properties.

The alcohols are named as derivatives of a hydrocarbon consisting of the longest straight chain that contains the OH group. The ending of the parent hydrocarbon is changed from *-ane* to *-anol,* and numbers are used to indicate the positions of substituents. The lowest possible number is assigned to the OH group. Alcohols may be classified as *primary, secondary,* or *tertiary,* according to the number of alkyl radicals on the carbon atom holding the OH. Thus,

1-butanol	3-methyl-2-butanol	2-methyl-2-butanol
$CH_3CH_2CH_2CH_2OH$	$CH_3—CH(CH_3)—CH(OH)CH_3$	$CH_3C(CH_3)(OH)CH_2CH_3$
a primary alcohol	a secondary alcohol	a tertiary alcohol

Alcohols associate through hydrogen bonding,

```
R          R          R
|          |          |
O—H · · · O—H · · · O—H
```

For this reason, the melting points and boiling points of the alcohols are higher than those of alkanes of corresponding molecular weight (Table 21.4). The lower alcohols are miscible with water in all proportions because of intermolecular hydrogen bonding.

```
H          R          H
|          |          |
O—H · · · O—H · · · O—H
```

Table 21.4 Physical properties of some alcohols

NAME	FORMULA	MELTING POINT (°C)	BOILING POINT (°C)	SOLUBILITY IN WATER (g/100 g H_2O, 20°C)
methanol	CH_3OH	−98	65	miscible
ethanol	CH_3CH_2OH	−115	78	miscible
1-propanol	$CH_3(CH_2)_2OH$	−127	97	miscible
1-butanol	$CH_3(CH_2)_3OH$	−90	117	7.9
1-pentanol	$CH_3(CH_2)_4OH$	−79	138	2.4
1-hexanol	$CH_3(CH_2)_5OH$	−52	157	0.6

However, as the size of the alkyl radical increases, the alcohols become more like alkanes in their physical properties, and the higher alcohols are only slightly soluble in water.

Alcohols, like water, are amphiprotic substances. They can act as Brønsted bases (proton acceptors) with very strong acids.

$$H_2SO_4 + ROH \rightleftharpoons ROH_2^+ + HSO_4^-$$

With very strong bases they are Brønsted acids (proton donors).

$$ROH + H^- \rightarrow OR^- + H_2$$

Alcohols, however, are more weakly acidic and basic than water. Consequently, the conjugate acids (ROH_2^+) and conjugate bases (OR^-) are stronger than H_3O^+ and OH^-. The anion OR^-, known as an alkoxide ion, results from the reaction of an alcohol with a reactive metal (compare the reaction of water with sodium).

$$2CH_3CH_2OH + 2Na \rightarrow 2CH_3CH_2ONa + H_2$$

The first member of the series, methanol (or methyl alcohol), is known as wood alcohol because it can be obtained by the destructive distillation of wood. The principal commercial source of this alcohol is the catalytic hydrogenation of carbon monoxide.

$$CO + 2H_2 \rightarrow CH_3OH$$

The alcohol of alcoholic beverages is ethanol, or ethyl alcohol. It is prepared by the fermentation of starches or sugars. Ethanol is also commercially prepared from ethene. The indirect addition of water to olefins by means of sulfuric acid is a general method of preparing alcohols. The synthesis of ethanol proceeds by the following steps.

$$CH_2{=}CH_2 + H_2SO_4 \rightarrow CH_3CH_2OSO_2OH$$
$$CH_3CH_2OSO_2OH + H_2O \rightarrow CH_3CH_2OH + H_2SO_4$$

Alcohols can be prepared from alkyl halides by displacement, or nucleophilic substitution, reactions.

$$RX + OH^- \rightarrow ROH + X^-$$

This is an important classification of organic reactions. It includes the reactions of the alkyl halides, as well as the alcohols themselves, with a wide variety of nucleophilic substances. The reactions are usually reversible and are run under conditions that favor the formation of the desired compound. An excess of the nucleophilic reagent may be employed or the product may be removed from the reaction mixture as it forms (for example, by distillation).

Primary and secondary halides or alcohols generally react by what is called an S_N2 mechanism, which can be illustrated by the reaction of hydroxide ion with methyl bromide. The hydroxide ion, a Lewis base, attacks the carbon atom and displaces the bromide ion (also a Lewis base), which takes along the electron pair with which it had been bonded.

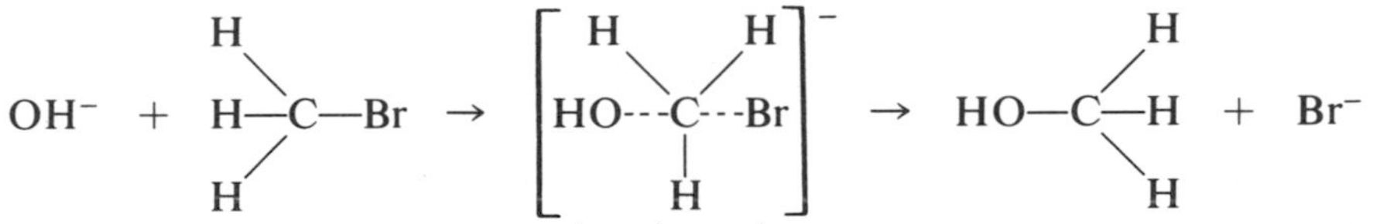

activated complex

The attack of the OH^- ion takes place on the side of the carbon atom that is opposite the bromine atom. As the OH^- ion approaches, it begins to form a covalent bond, and the bromide ion begins to break away. The activated complex of the reaction is the form in which both nucleophilic groups are partially bonded to the carbon atom. The reaction causes inversion of the geometric arrangement of the groups attached to the carbon atom, a process that is customarily compared to the inversion of an umbrella in a high wind. The mechanism is given the designation S_N2 because it is the substitution of one nucleophilic group for another, and the rate-determining step (the formation of the activated complex) is bimolecular. The nucleophilic substitution reaction of HBr with CH_3OH proceeds through the formation of $CH_3OH_2^+$ and the displacement of H_2O from this ion by Br^-.

In the case of tertiary alkyl halides, the three alkyl groups bonded to the carbon atom bearing the halogen inhibit the rearward approach of the OH^- ion, and it is thought that tertiary alkyl halides undergo displacement reactions by a different mechanism. The rate-determining step is thought to be the formation of a carbonium ion,

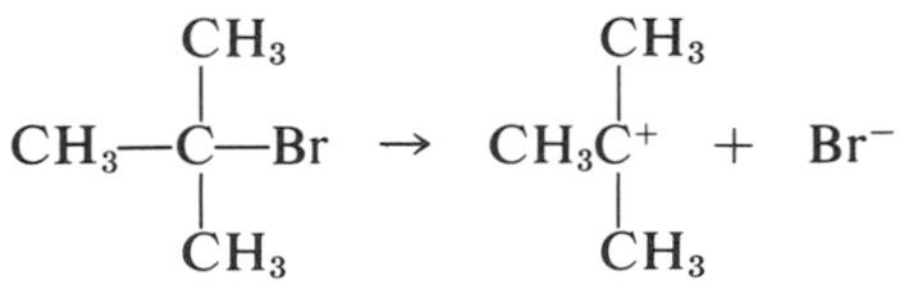

The central carbon atom of the carbonium ion has only six electrons in its valence level and displays sp^2 hybridization. The carbonium ion is therefore

planar. It is a powerful Lewis acid and rapidly adds hydroxide ion (a Lewis base).

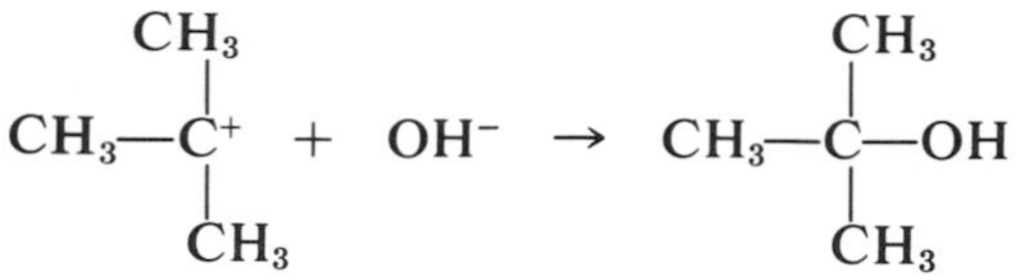

This mechanism is given the designation S_N1 because it is a nucleophilic substitution in which the rate-determining step is unimolecular. The S_N1 mechanism is favored when the reaction is run in polar solvents (such as water), which aid the first-step ionization of the halide.

Tertiary alcohols also undergo S_N1 reactions. The steps of a typical substitution involving HBr are

$$(CH_3)_3COH + H^+ \rightleftharpoons (CH_3)_3COH_2^+ \qquad \text{(rapid, equilibrium)}$$
$$(CH_3)_3COH_2^+ \rightarrow (CH_3)_3C^+ + H_2O \qquad \text{(rate determining)}$$
$$(CH_3)_3C^+ + Br^- \rightarrow (CH_3)_3CBr$$

In all these nucleophilic substitution reactions, appreciable quantities of olefins are formed along with the substitution products. This is particularly true of the reactions of the tertiary compounds. Olefin formation is thought to proceed by the elimination of a proton by a carbonium ion.

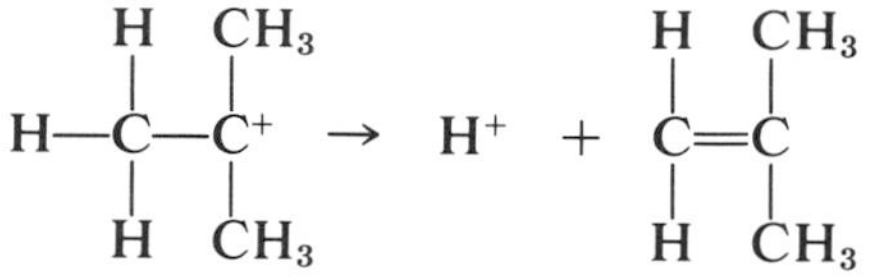

Tertiary carbonium ions are formed more readily than any other type (Section 21.5). Elimination reactions of primary and secondary alcohols are thought to occur by a mechanism that does not involve the formation of carbonium ions.

The elimination of one water molecule from one molecule of an alcohol produces an olefin.

$$CH_3CH_2OH \xrightarrow[\text{above } 150°C]{H_2SO_4} CH_2{=}CH_2 + H_2O$$

Under milder conditions only one molecule of water is removed from two molecules of alcohol, and an ether is produced.

$$CH_3CH_2OH + HOCH_2CH_3 \xrightarrow[130\text{–}140°C]{H_2SO_4} \underset{\text{diethyl ether}}{CH_3CH_2OCH_2CH_3} + H_2O$$

Ethers, ROR, may be regarded as derivatives of water, HOH, in which both hydrogen atoms are replaced by alkyl groups, R. The alkyl groups may be alike or different. Ethers are also produced by nucleophilic substitution reactions involving alkoxide ions and alkyl halides.

$$CH_3CH_2O^- + CH_3CH_2CH_2Br \rightarrow \underset{\text{ethyl propyl ether}}{CH_3CH_2OCH_2CH_2CH_3} + Br^-$$

Unlike alcohols, ethers do not associate by hydrogen bonding. Therefore, the boiling points of ethers are much lower than those of alcohols of corresponding molecular weight, and the ethers are much less soluble in water.

CH_3CH_2OH	CH_3OCH_3
ethanol	dimethyl ether
boiling point, 78°C	boiling point, −24°C

Notice that ethanol and dimethyl ether are isomeric. The type of isomerism displayed by this pair of compounds is known as **functional group isomerism.** The ethers are less reactive than the alcohols.

Polyhydroxy alcohols contain more than one OH group. Examples include 1,2-ethanediol (ethylene glycol),

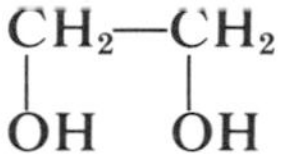

and 1,2,3-propanetriol (glycerol), which is derived from fats,

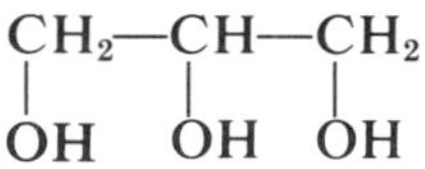

Unlike aliphatic alcohols, which do not dissociate in water solution, the aromatic alcohols are weak acids in water.

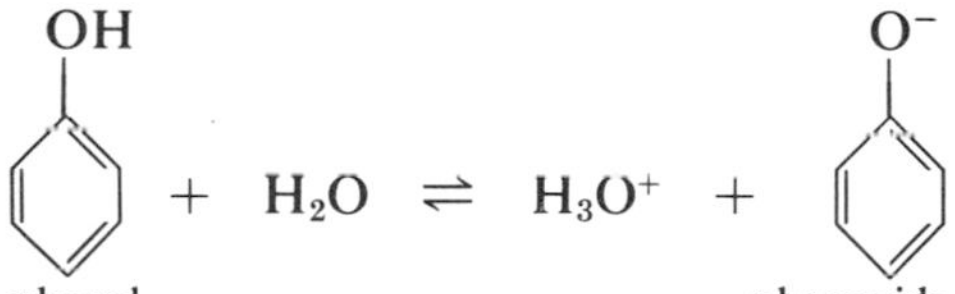

phenol phenoxide ion

The acidity of phenol (carbolic acid) is attributed to the stability of the phenoxide ion in which the charge is delocalized by the π bonding system of the aromatic ring.

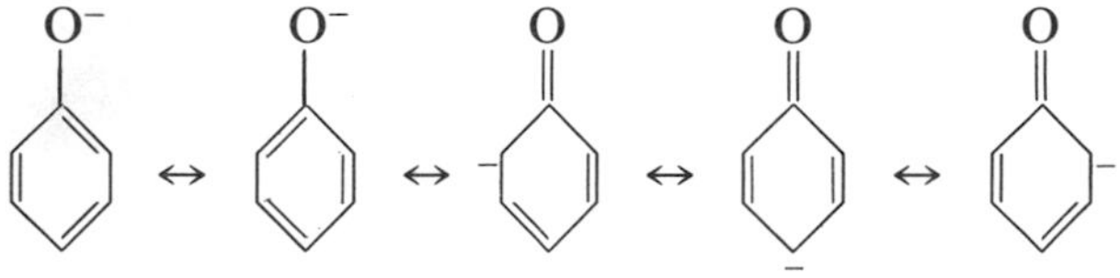

Aromatic halides do not readily undergo displacement reactions. Sodium phenoxide is commercially obtained from chlorobenzene by reaction with NaOH at elevated temperatures and under high pressure.

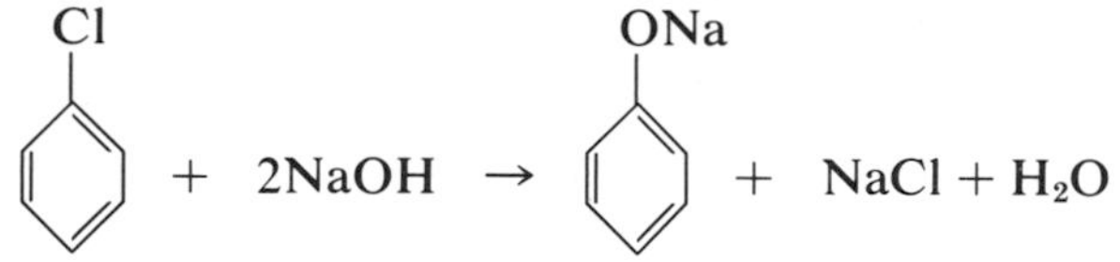

Phenol is derived by acidification of the phenoxide.

21.7 CARBONYL COMPOUNDS

A carbon atom can form a double bond with an oxygen atom to produce what is called a **carbonyl group**

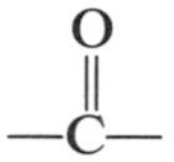

The double bond, like the carbon–carbon double bond, consists of a σ bond and a π bond between the bonded atoms. The carbon atom of the carbonyl group and the atoms bonded to it are coplanar and form bond angles of 120°, a geometry that is typical of sp^2 hybridized species. The carbonyl double bond differs from the olefinic double bond in that it is markedly polar (oxygen is more electronegative than carbon).

If the carbonyl group is bonded to one alkyl group and one hydrogen or to two hydrogens the compound is an **aldehyde.**

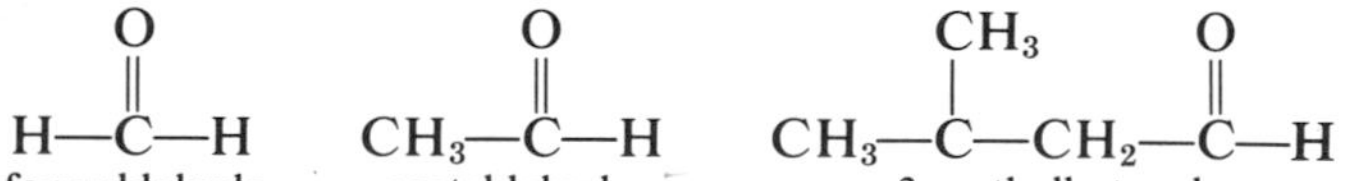

formaldehyde methanal; acetaldehyde ethanal; 3-methylbutanal

Systematic nomenclature of aldehydes employs the ending *-al.* Substituent groups are given numbers based on the assignment of the number 1 (understood) to the carbonyl carbon.

In a **ketone** the carbonyl group is bonded to two alkyl groups, which may be alike or different.

acetone propanone; methyl ethyl ketone butanone; 5-methyl-3-hexanone

The ending *-one* is used to designate a ketone. Numbers are employed to indicate the positions of substituents on the longest continuous chain containing the carbonyl group, which is assigned the lowest possible number.

Aldehydes may be prepared by the mild oxidation of primary alcohols.

An oxidizing agent such as potassium dichromate in dilute sulfuric acid may be used.

$$3RCH_2OH + Cr_2O_7^{2-} + 8H^+ \rightarrow 3R\overset{\overset{\Large O}{\|}}{C}{-}H + 2Cr^{3+} + 7H_2O$$

Equations for a reaction such as this are usually written with the oxidizing agent indicated over the arrow and only the organic reactant and product shown. Thus,

$$CH_3CH_2OH \xrightarrow{Cr_2O_7^{2-}/H^+} CH_3\overset{\overset{\Large O}{\|}}{C}{-}H$$

When an aldehyde is prepared by the oxidation of a primary alcohol, provision must be made to prevent the aldehyde from being destroyed by further oxidation, since aldehydes are easily oxidized to carboxylic acids (Section 21.8). Many aldehydes may be distilled out of the reaction mixture as they are formed.

Some aldehydes are made commercially by reacting alcohol vapors with air over a copper catalyst at elevated temperatures.

$$2CH_3OH + O_2 \xrightarrow{Cu} 2H{-}\overset{\overset{\Large O}{\|}}{C}{-}H + 2H_2O$$

The oxidation of alcohols may also be conducted by passing the hot alcohol vapor over a heated copper catalyst in the absence of oxygen. This process results in the removal of a molecule of hydrogen from an alcohol molecule and is called a **dehydrogenation.** It avoids the danger of secondary oxidation of the aldehyde product.

$$CH_3CH_2CH_2OH \xrightarrow{Cu} CH_3CH_2\overset{\overset{\Large O}{\|}}{C}{-}H + H_2$$

The oxidation of secondary alcohols produces ketones.

$$CH_3CH_2\overset{\overset{\large OH}{|}}{C}HCH_3 \xrightarrow{Cr_2O_7^{2-}/H^+} CH_3CH_2\overset{\overset{\Large O}{\|}}{C}CH_3$$

Acetone (propanone) may be prepared by the oxidation of 2-propanol with oxygen or by the dehydrogenation of the alcohol.

$$CH_3\overset{\overset{\large OH}{|}}{C}HCH_3 \xrightarrow{Cu} CH_3\overset{\overset{\Large O}{\|}}{C}CH_3 + H_2$$

The oxidation of tertiary alcohols results in the destruction of the carbon skeleton of the molecule.

The double bond of the carbonyl group readily undergoes many addition reactions. Reduction of aldehydes or ketones with hydrogen in the presence of a nickel or platinum catalyst yields primary or secondary alcohols.

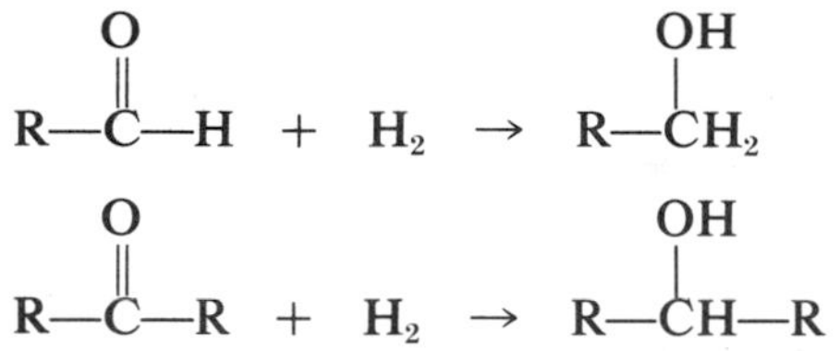

When an unsymmetrical reagent (such as HCN) adds to the double bond, the positive part of the addend (the hydrogen) adds to the oxygen, and the negative part of the addend (the cyanide group) bonds to the carbon. This mode of addition reflects the polarity of a carbonyl double bond. The oxygen is negative with respect to the carbon.

a cyanohydrin

The proposed mechanism for additions of this type consists of the electrophilic attack of the anionic Lewis base

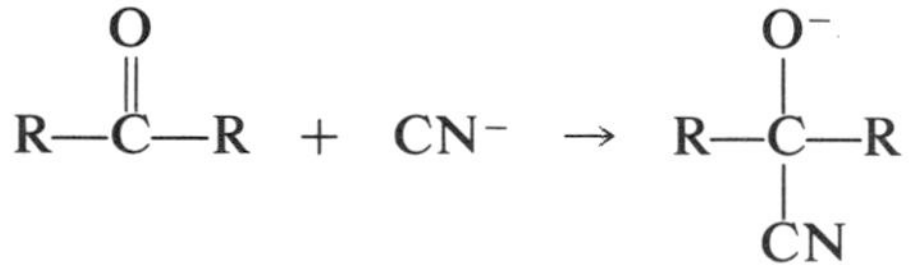

followed by combination with a proton

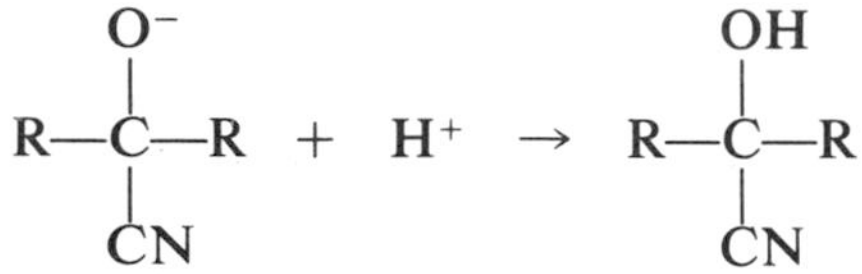

An important addition reaction involves organomagnesium compounds known as **Grignard reagents.** Alkyl halides react with magnesium metal in dry diethyl ether

$$R{-}X + Mg \rightarrow R{-}Mg{-}X$$

These reagents are usually given the formula RMgX. The materials, however, possess a high degree of ionic character and may be mixtures of magnesium dialkyls (MgR_2) and magnesium halides (MgX_2). Water must be excluded from a Grignard reaction because these reagents are easily hydrolyzed.

$$R{-}Mg{-}X + HOH \rightarrow R{-}H + Mg(OH)X$$

The formula Mg(OH)X stands for an equimolar mixture of magnesium hydroxide and magnesium halide.

A Grignard reagent adds to an aldehyde as follows

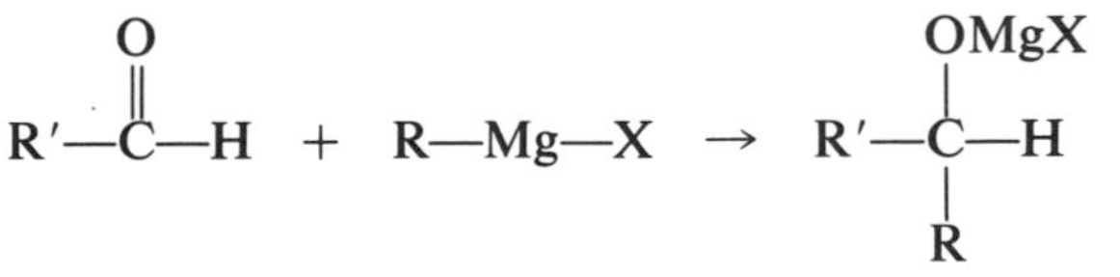

where the alkyl groups R and R′ may be alike or different. The addition compound is readily decomposed by water or by dilute acid.

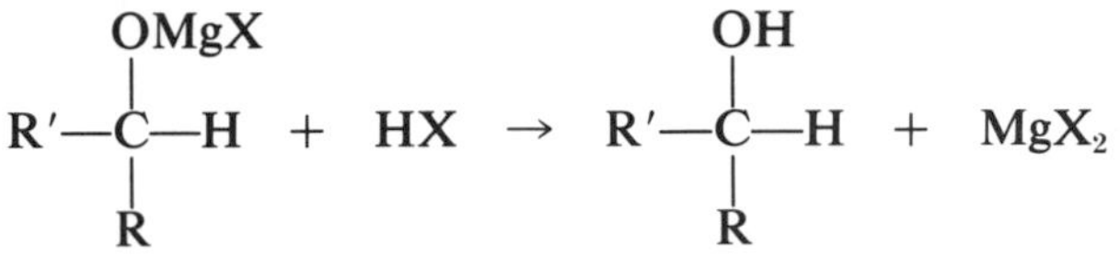

Thus, the ultimate product of the reaction of a Grignard reagent and an aldehyde is a secondary alcohol.

Hydrolysis of an addition compound formed by a Grignard reagent and a ketone yields a tertiary alcohol.

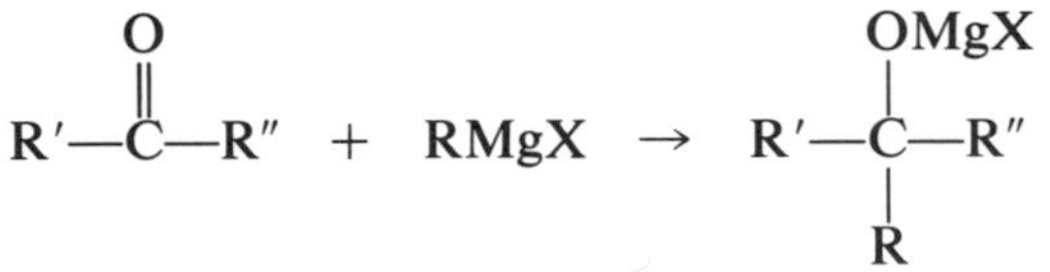

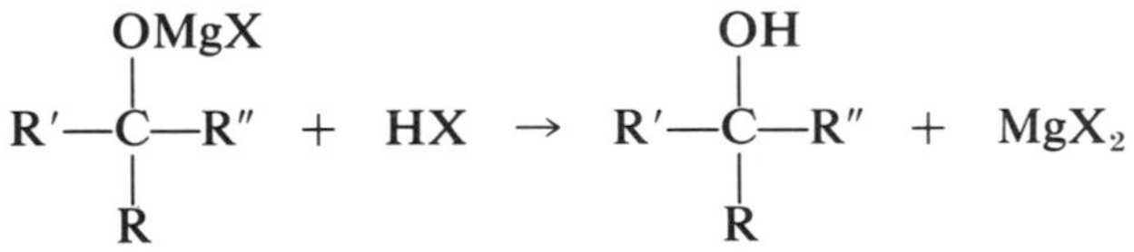

The alkyl groups may be alike or different.

In these reactions of Grignard reagents, new carbon–carbon bonds are formed, and the reactions are frequently useful as steps in organic syntheses. The alcohols produced may be oxidized to carbonyl compounds, subjected to displacement reactions, or dehydrated to olefins. Thus, a series of reactions may be employed to synthesize a desired compound from compounds of lower molecular weight.

21.8 CARBOXYLIC ACIDS AND ESTERS

Oxidation of the aldehyde group yields a **carboxyl group**

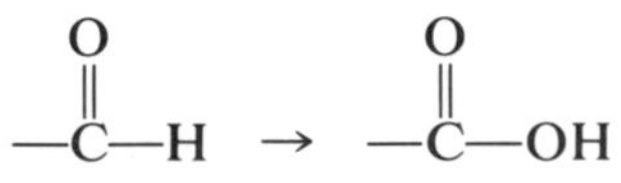

Compounds containing the —COOH group are weak acids **(carboxylic acids).** Ionization constants for some of these acids are listed in Table 21.5.

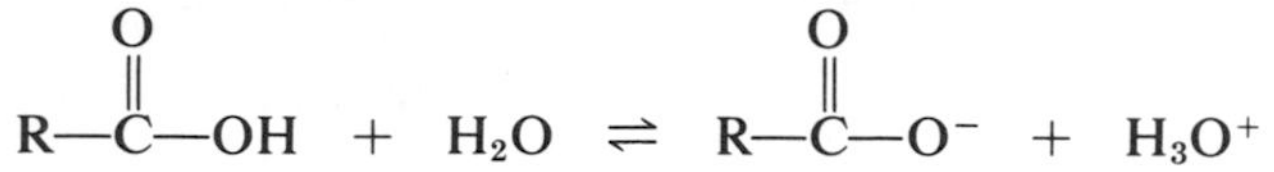

The charge of the carboxylate anion is delocalized.

The acids are associated through hydrogen bonding. Lower members of the series form dimers in the vapor state.

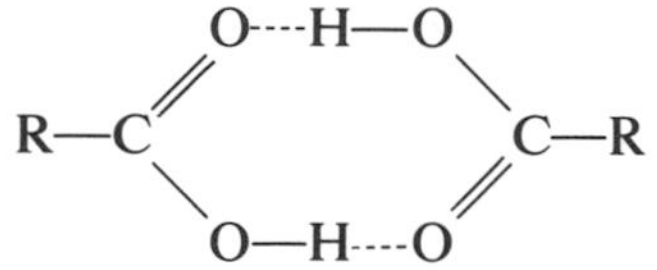

According to systematic nomenclature, the name of a carboxylic acid is derived from the parent hydrocarbon by elision of the final *-e,* addition of the ending *-oic,* and addition of the separate word *acid.* The numbers used to designate the positions of substituents on the carbon chain are derived by numbering the chain starting with the carbon of the carboxyl group.

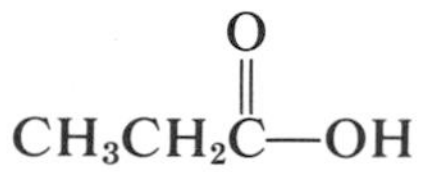
propanoic acid

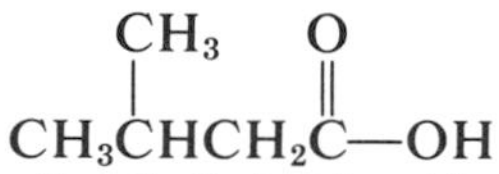
3-methylbutanoic acid

Carboxylic acids may be prepared by the oxidation of primary alcohols or aldehydes. Potassium dichromate or potassium permanganate are frequently employed as the oxidizing agent.

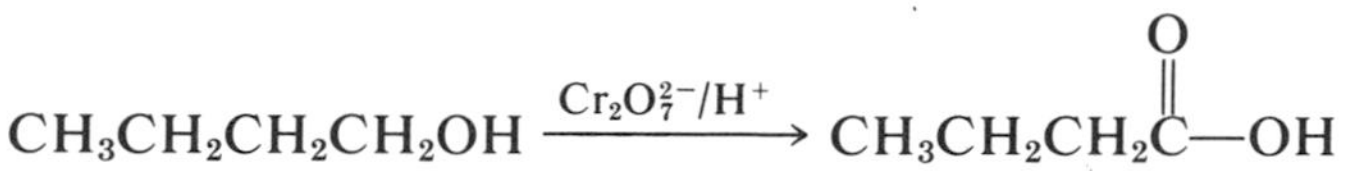

Table 21.5 Properties of some carboxylic acids

ACID	FORMULA	MELTING POINT (°C)	BOILING POINT (°C)	IONIZATION CONSTANT AT 25°C
methanoic (formic)	$HCOOH$	8	101	1.8×10^{-4}
ethanoic (acetic)	CH_3COOH	17	118	1.8×10^{-5}
propanoic (propionic)	CH_3CH_2COOH	−22	141	1.4×10^{-5}
butanoic (butyric)	$CH_3(CH_2)_2COOH$	−8	164	1.5×10^{-5}
pentanoic (valeric)	$CH_3(CH_2)_3COOH$	−35	187	1.6×10^{-5}
benzoic	C_6H_5COOH	122	249	6.0×10^{-5}

Acetic acid is commercially prepared by the air oxidation of acetaldehyde.

$$2CH_3\overset{\overset{O}{\|}}{C}{-}H + O_2 \xrightarrow{Mn(C_2H_3O_2)_2} 2CH_3\overset{\overset{O}{\|}}{C}{-}OH$$

Vigorous oxidation of an alkyl-substituted aromatic compound converts the side chain to a carbonyl group and thus yields benzoic acid.

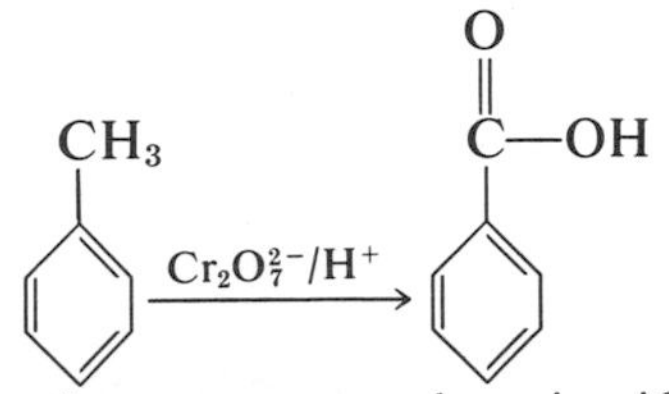

toluene benzoic acid

These oxidations do not touch the aromatic ring, an illustration of the stability of this structure.

The oxidation of an unsaturated hydrocarbon yields a variety of products, including carboxylic acids, depending upon conditions. Dilute aqueous permanganate at room temperature oxidizes olefins to glycols.

$$R{-}CH{=}CH{-}R \xrightarrow{MnO_4^-} R{-}\underset{\underset{OH}{|}}{CH}{-}\underset{\underset{OH}{|}}{CH}{-}R$$

Under more vigorous conditions (more concentrated solutions and heating), the carbon chain is cleaved at the double bond and carboxylic acids are produced.

$$R{-}CH{=}CH{-}R \xrightarrow{MnO_4^-} R{-}\overset{\overset{O}{\|}}{C}{-}OH + HO{-}\overset{\overset{O}{\|}}{C}{-}R$$

Alkynes also may be cleaved to yield acids.

$$R{-}C{\equiv}C{-}R \xrightarrow{MnO_4^-} R{-}\overset{\overset{O}{\|}}{C}{-}OH + HO{-}\overset{\overset{O}{\|}}{C}{-}R$$

Oxidation of some olefins yields ketones.

$$R{-}\underset{\underset{R}{|}}{C}{=}\underset{\underset{R}{|}}{C}{-}R \xrightarrow{MnO_4^-} R{-}\underset{\underset{R}{|}}{C}{=}O + O{=}\underset{\underset{R}{|}}{C}{-}R$$

Salts of carboxylic acids may be produced by the alkaline hydrolysis of alkyl cyanides. The cyanides are the product of nucleophilic substitution reactions of the CN^- ion with alkyl halides.

$$CH_3CH_2Br + CN^- \rightarrow CH_3CH_2CN + Br^-$$

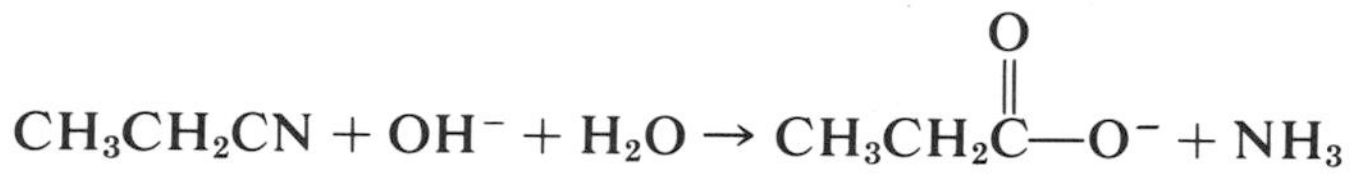

Some compounds contain more than one carboxyl group. Examples include

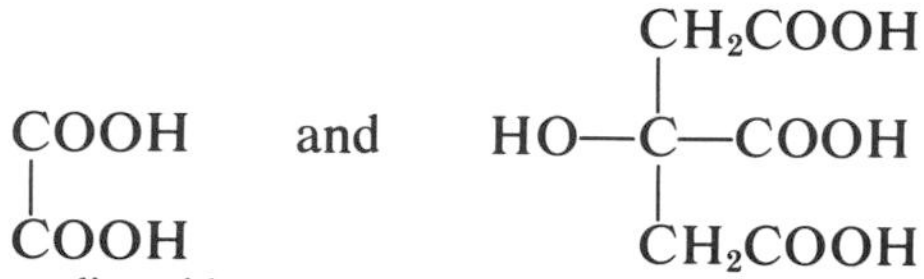

oxalic acid citric acid

Carboxylic acids react with alcohols to produce compounds known as esters.

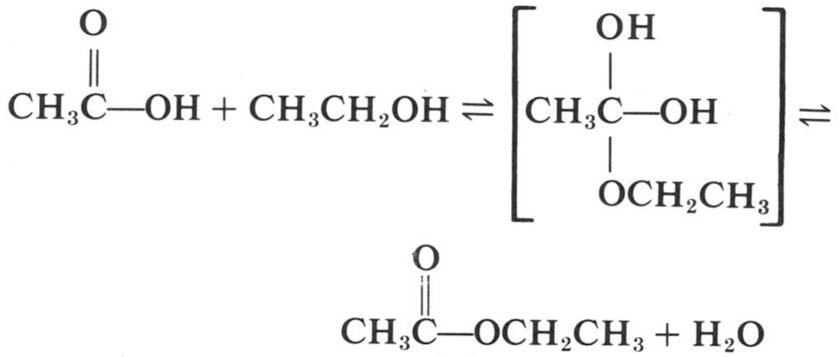

ethyl acetate

Esterification reactions are reversible and proceed to equilibrium. They may be forced in either direction by the appropriate choice of conditions. The name of an ester reflects the alcohol and acid from which it is derived, the ending *-ate* being employed with the base of the name of the acid to give the second portion of the name of the ester.

The alkaline hydrolysis of an ester produces an alcohol and a salt of a carboxylic acid.

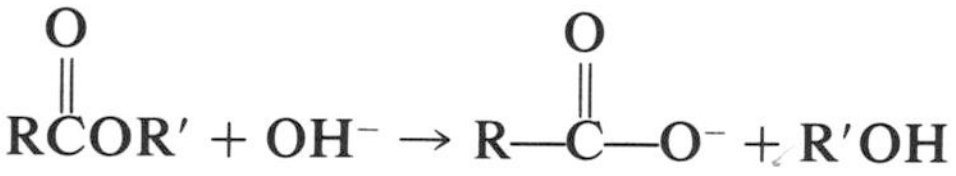

21.9 AMINES AND AMIDES

The **amines** may be considered as derivatives of ammonia with one, two, or three hydrogen atoms replaced by alkyl groups.

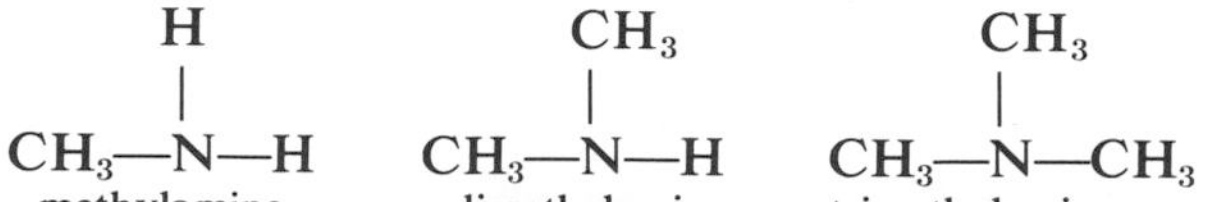

methylamine, a primary amine — dimethylamine, a secondary amine — trimethylamine, a tertiary amine

The amines resemble ammonia in that they are weak bases.

$$CH_3NH_2 + H_2O \rightleftharpoons CH_3NH_3^+ + OH^-$$
$$(CH_3)_2NH + H_2O \rightleftharpoons (CH_3)_2NH_2^+ + OH^-$$
$$(CH_3)_3N + H_2O \rightleftharpoons (CH_3)_3NH^+ + OH^-$$

Ionization constants for some amines are listed in Table 21.6. The amines form salts with acids

$$CH_3NH_2 + HCl \rightarrow CH_3NH_3^+ + Cl^-$$

which may be decomposed by hydroxides

$$CH_3NH_3^+ + OH^- \rightarrow CH_3NH_2 + H_2O$$

The last reaction is the reverse of the ionization of methylamine in water. It is forced to yield the free amine by employment of an excess of OH^- or by warming.

Amines may be prepared by reacting alkyl halides with ammonia. These reactions are nucleophilic substitutions.

$$NH_3 + CH_3CH_2Br \rightarrow CH_3CH_2NH_3^+ + Br^-$$

Treatment of the amine salt with hydroxide ion yields the free amine. Mixtures of primary, secondary, and tertiary amines are produced by this type of reaction.

Primary amines may be produced by the catalytic hydrogenation of alkyl cyanides.

$$CH_3CH_2C{\equiv}N + 2H_2 \rightarrow CH_3CH_2CH_2{-}NH_2$$

Aniline is produced by the reduction of nitrobenzene. Iron and steam, with a trace of hydrochloric acid, serve as the reducing agent.

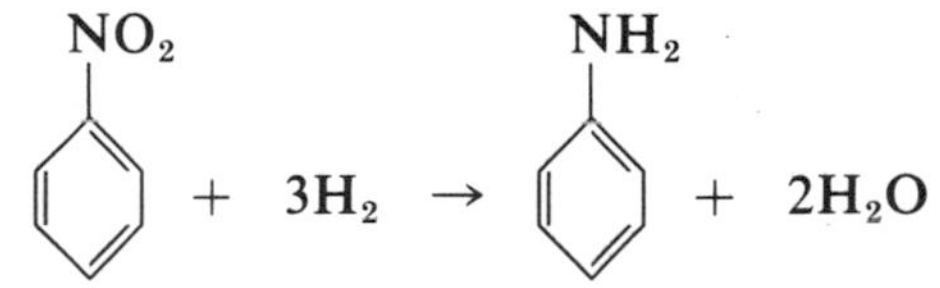

Table 21.6 Properties of some amines

AMINE	FORMULA	MELTING POINT (°C)	BOILING POINT (°C)	IONIZATION CONSTANT AT 25°C
methylamine	CH_3NH_2	−93	−7	5×10^{-4}
dimethylamine	$(CH_3)_2NH$	−96	7	7.4×10^{-4}
trimethylamine	$(CH_3)_3N$	−124	4	7.4×10^{-5}
ethylamine	$CH_3CH_2NH_2$	−81	17	5.6×10^{-4}
propylamine	$CH_3CH_2CH_2NH_2$	−83	49	4.7×10^{-4}
aniline	$C_6H_5NH_2$	−6	184	4.6×10^{-10}

Some compounds that contain more than one amino group are important. Ethylenediamine (1,2-diaminoethane, $H_2N—CH_2—CH_2—NH_2$) is a useful chelating agent (Section 20.1), and hexamethylenediamine (1,6-diaminohexane, $H_2N(CH_2)_6NH_2$) is used in the manufacture of nylon.

Amides may be produced from ammonia and carboxylic acids. The preparation proceeds through the formation of ammonium salts which eliminate water upon heating.

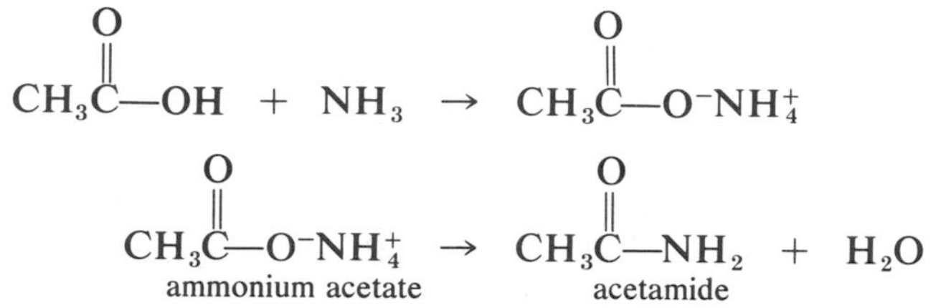

Amides may also be prepared by the nucleophilic substitution reaction of ammonia on an ester.

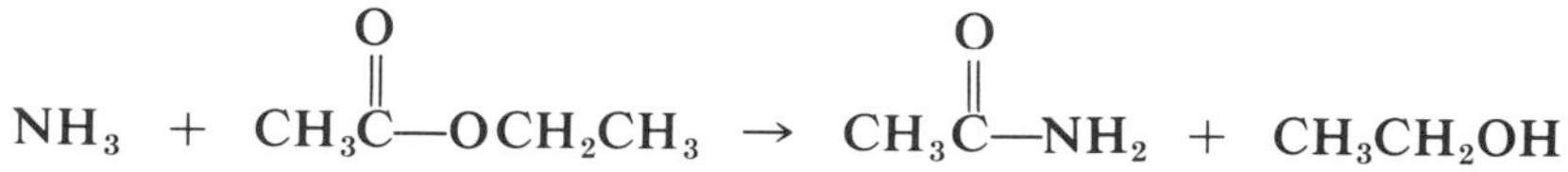

Primary and secondary amines react with acids in a similar manner to produce substituted amides, compounds in which one or two alkyl radicals replace hydrogen atoms of the $—NH_2$ group.

21.10 POLYMERS

Starch, cellulose, and proteins are examples of natural **polymers**—molecules of high molecular weight that are formed from simpler molecules, called **monomers.** Important polymers, or **macromolecules,** that do not occur in nature have been synthesized. Most of these are linear, or chain-type, structures although some are cross-linked.

Many polymers are formed from compounds that contain carbon–carbon double bonds by a process that is called **addition polymerization.** For example, ethene (ethylene) polymerizes upon heating (100°C–400°C) under high pressure (1000 atm). The product is called polyethylene.

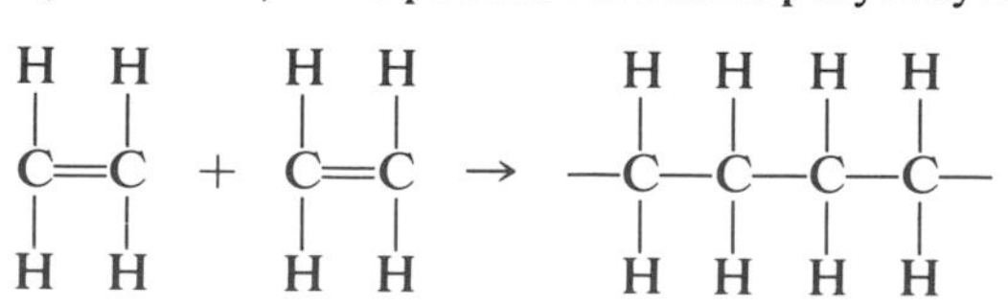

The preceding equation shows the combination of only two molecules of ethene. The actual polymerization process continues by successive additions of $CH_2=CH_2$ molecules until chains of hundreds or thousands of

—CH_2—CH_2— units are produced. The product, polyethylene, is a tough, waxy solid.

Important polymers have been made from ethene derivatives. Orlon and acrilan are polymers of acrylonitrile, $CH_2{=}CH{-}C{\equiv}N$. The polymers have the structure

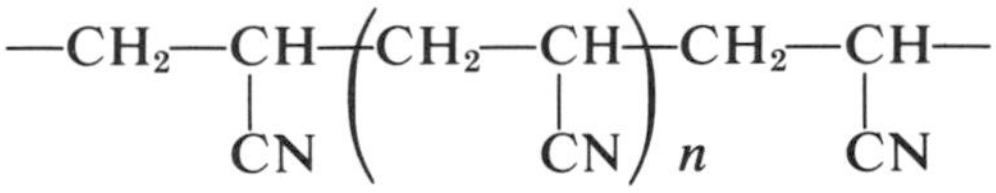

Teflon is a polymer of tetrafluoroethene, $CF_2{=}CF_2$. Vinyl chloride, $CH_2{=}CHCl$, and vinyl acetate, $CH_2{=}CH{-}O{-}\overset{\overset{\displaystyle O}{\|}}{C}{-}CH_3$, are important monomers in the preparation of vinyl plastics. Lucite, or Plexiglas, is a polymer of methyl methacrylate, $CH_2{=}C(CH_3)COOCH_3$.

Natural rubber is a polymer of the diolefin 2-methyl-1,3-butadiene, or isoprene.

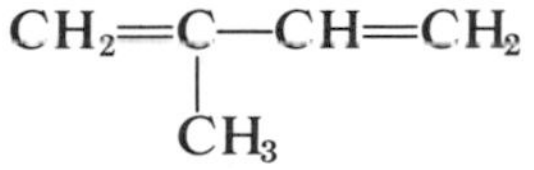

Such a system of alternating double and single bonds is called a **conjugated system.** Rubber consists of thousands of these units joined in a chain.

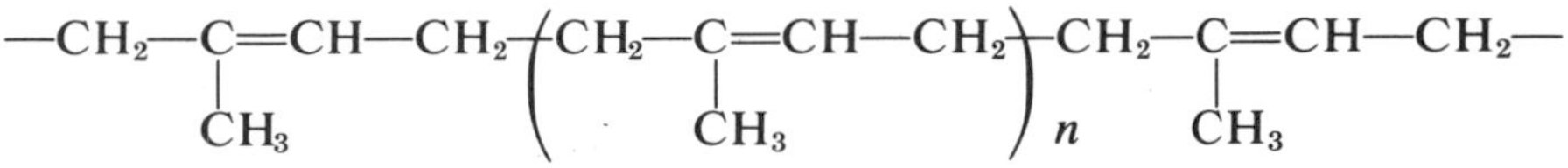

Notice that the polymer contains double bonds. Crude rubber is vulcanized by heating with sulfur. It is thought that the sulfur atoms add to some of the double bonds, thereby linking adjacent chains into a complex network. This process adds strength to the final product.

A number of types of synthetic rubber have been made utilizing such monomers as 1,3-butadiene ($CH_2{=}CH{-}CH{=}CH_2$), 2-chloro-1,3-butadiene (chloroprene, $CH_2{=}C(Cl){-}CH{=}CH_2$), and 2,3-dimethyl-1,3-butadiene ($CH_2{=}C(CH_3){-}C(CH_3){=}CH_2$) alone or in combination. The product formed by the polymerization of two different monomers is called a **copolymer.** The molecular structures and properties of such materials depend upon the proportions of the two monomers employed. Buna S rubber is a copolymer of 1,3-butadiene and styrene ($C_6H_5CH{=}CH_2$). A section of the chain of this material has the structure

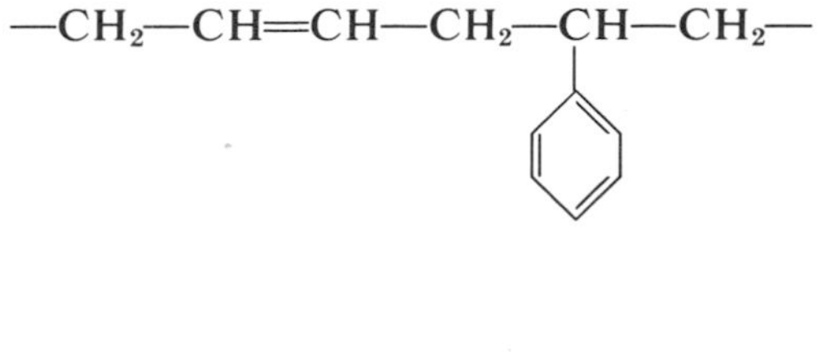

An addition polymerization is thought to proceed by either a free-radical or a carbonium-ion chain mechanism. The free-radical type may be initiated by small amounts of hydrogen peroxide or organic peroxides, which are readily split into free radicals (indicated by Z· in the equations that follow).

$$Z_2 \rightarrow 2Z\cdot$$

A free radical from an initiator combines with one of the π electrons of a molecule of the monomer to form a new free radical.

$$Z\cdot + CH_2{=}CH_2 \rightarrow Z{-}CH_2{-}CH_2\cdot$$

Chain propagation occurs by the combination of this free radical with another molecule of the monomer

$$Z{-}CH_2{-}CH_2\cdot + CH_2{=}CH_2 \rightarrow Z{-}CH_2{-}CH_2{-}CH_2{-}CH_2\cdot$$

and the molecule grows by repeated additions. Chain termination is caused by the combination of two free radicals or by other means, such as the elimination of a hydrogen atom.

A carbonium-ion chain polymerization is initiated by a Lewis acid, such as $AlCl_3$ or BF_3. In the equations that follow, the proton is used to indicate the Lewis acid, or electrophilic, initiator, although stronger Lewis acids are more frequently employed. Combination of the initiator with a molecule of the monomer produces a carbonium ion.

$$H^+ + CH_2{=}CH_2 \rightarrow CH_3{-}CH_2^+$$

Chain propagation occurs through the combination of the carbonium ion with both π electrons of a molecule of the monomer.

$$CH_3{-}CH_2^+ + CH_2{=}CH_2 \rightarrow CH_3{-}CH_2{-}CH_2{-}CH_2^+$$

Chain termination may occur by the elimination of a proton from a long-chain carbonium ion.

$$CH_3{-}CH_2(CH_2{-}CH_2)_nCH_2{-}CH_2^+ \rightarrow$$
$$CH_3{-}CH_2(CH_2{-}CH_2)_nCH{=}CH_2 + H^+$$

Condensation polymerization occurs between molecules of monomers by the elimination of a small molecule, usually water. Proteins and polysaccharides are condensation polymers. Nylon, like the proteins, is a polyamide. It is formed by the condensation of a diamine [such as hexamethylenediamine, $H_2N(CH_2)_6NH_2$] and a dicarboxylic acid [such as adipic acid, $HOOC(CH_2)_4$-COOH]. The polymerization is accompanied by the loss of water, and the chain has the structure

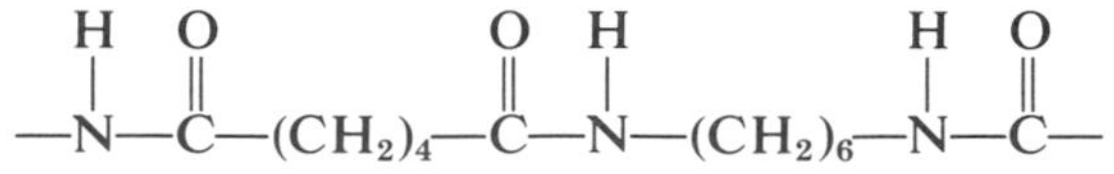

Dacron is a polyester formed by the elimination of water between ethylene glycol ($HOCH_2CH_2OH$) and a dicarboxylic acid (such as terephthalic acid, *p*-$HOOCC_6H_4COOH$).

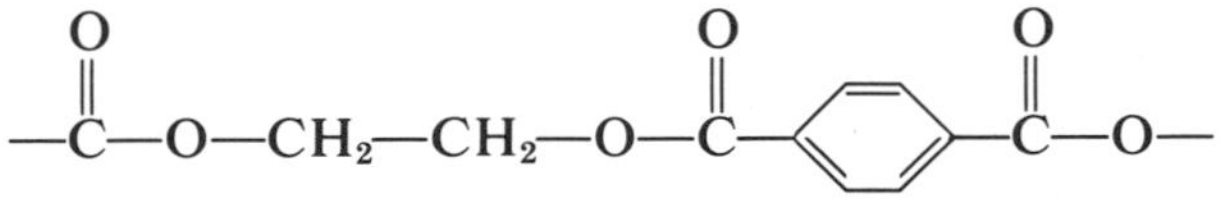

Bakelite is a cross-linked polymer formed from phenol and formaldehyde by the elimination of water. Notice that the formaldehyde units condense in the *ortho* and *para* positions of phenol.

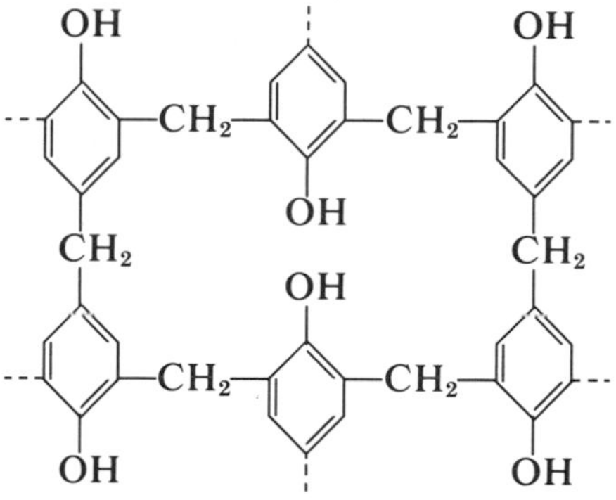

THE TIME OR THE PERSON?

According to the "great-man theory of history," the progress of science is largely the result of concepts discovered by a relatively small number of individuals. On the other hand, there is a theory that proposes that a concept emerges when there is an accumulation of work to support it and a need for it. According to this theory, if one individual does not make the discovery, then another will, provided the time is right.

In the history of science, there are many examples of concepts proposed, not accepted, forgotten, and revived or rediscovered at a later time. Presumably the time was not right for Avogadro's hypothesis when Avogadro proposed it in 1811; it was not accepted until after 1860. Mendel's work on genetics, published in 1866, was ignored for 34 years. Huygens' wave theory of light, proposed in 1678, was revived in 1801 by Young.

There are also many examples of discoveries made independently by two people at the same time: the discovery of oxygen by Priestley and Scheele, the development of calculus by Newton and Leibnitz, the formulation of the periodic table by Lothar Meyer and Mendeleev, the description of the tetrahedral nature of the carbon atom by Le Bel and van't Hoff.

We should not depreciate the work of great scientists. Neither should we forget Newton's statement: "If I have seen farther [than others], it is because I have stood upon the shoulders of giants."

PROBLEMS

21.1 Write structural formulas for the following compounds: (a) 2.4-dimethyl-3-ethylpentane, (b) 2,4-dimethyl-1,4-pentadiene, (c) 3-isopropyl-1-hexyne, (d) 2,2,3-trichlorobutane, (e) 2-methyl-2-butanol, (f) methylcyclopropane.

21.2 Write structural formulas for (a) butanal, (b) 2-butanone, (c) methyl ethyl ketone, (d) methyl ethyl amine, (e) methyl ethyl ether, (f) methyl ethanoate.

21.3 Write structural formulas for (a) *p*-xylene, (b) 2,3,6-trinitrotoluene, (c) naphthalene, (d) sodium benzoate, (e) *o*-bromophenol, (f) diphenyl amine.

21.4 Name each of the following compounds.

(a)

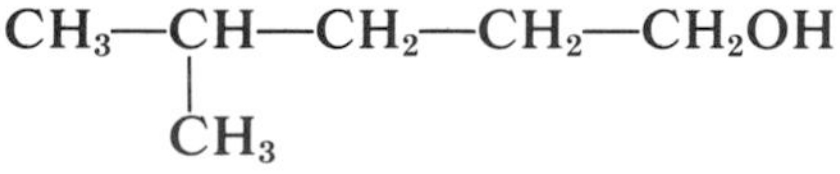

(b) 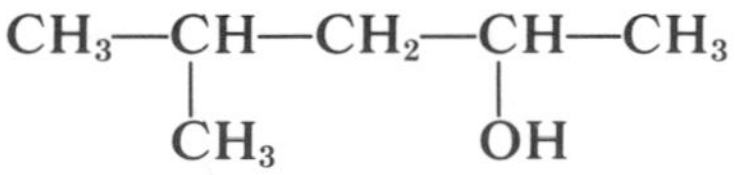

(c) 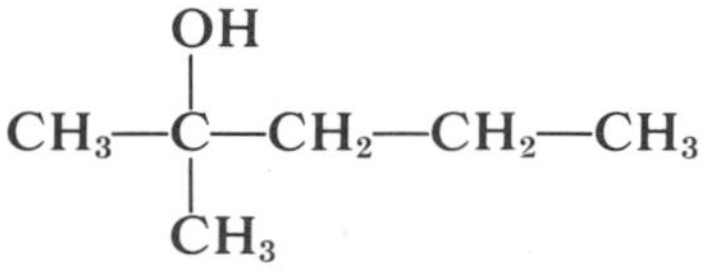

(d) 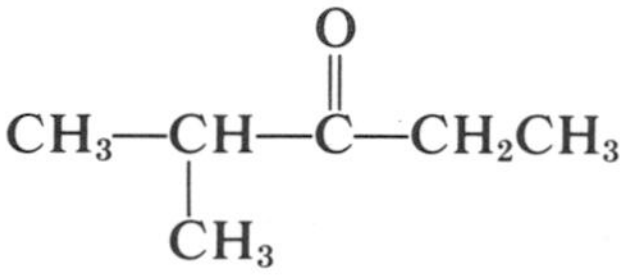

(e) 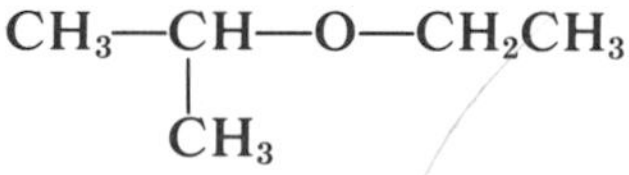

(f)

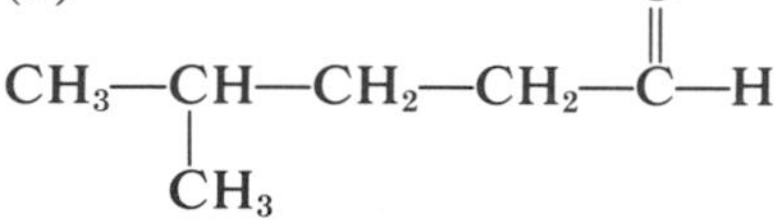

21.5 Name the following compounds.

(a) 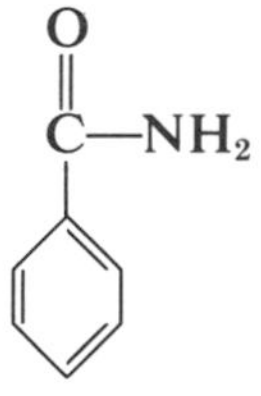

(b) 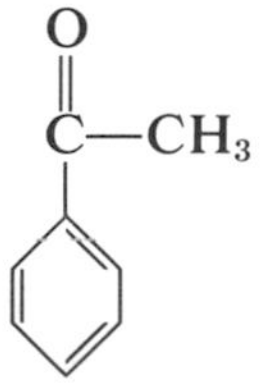

(c) 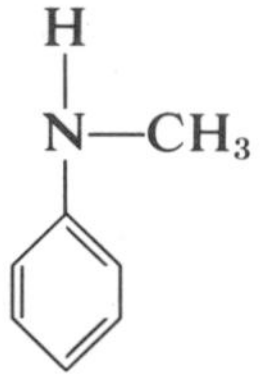

(d)

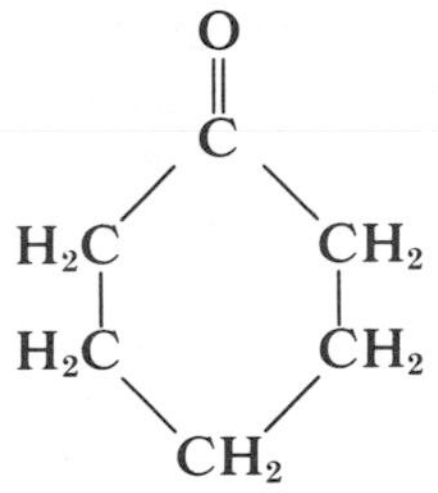

(e)

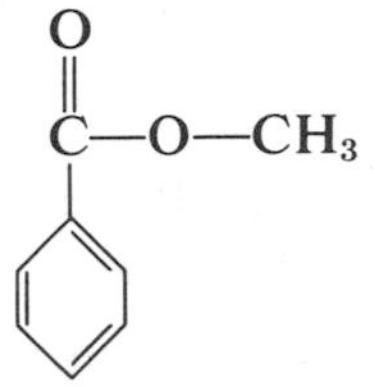

(f)

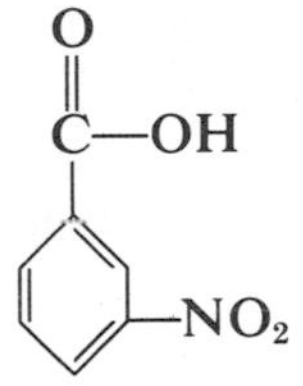

21.6 What is incorrect about each of the following names? (a) 3-pentene, (b) 1,2-dimethylpropane, (c) 2-methyl-2-butyne, (d) 2-methyl-3-butyne, (e) 3,3-dimethyl-2-butene, (f) 2-ethylpentane.

21.7 The general formula for the alkanes is C_nH_{2n+2}. What are the general formulas for the (a) alkenes, (b) alkynes, (c) alkadienes, (d) cycloalkanes, (e) alcohols?

21.8 Write structural formulas for all the mono-, di-, and tri-substituted chloro derivatives of propane.

21.9 Write structural formulas for all the isomers that have the formula C_4H_8.

21.10 Write structural formulas for all the isomers that have the formula $C_4H_{10}O$.

21.11 Write structural formulas for all the isomers that have the formula $C_4H_8Cl_2$.

21.12 Give an example of an isomer of (a) butane, (b) methyl methanoate (methyl formate), (c) dimethyl ether, (d) dimethyl amine, (e) acetone, (f) aminoacetic acid.

21.13 When the terms primary, secondary, and tertiary are applied to amines, their meanings are not the same as when they are applied to alcohols. Explain the difference.

21.14 Give examples of each of the following types of isomers: (a) chain, (b) position, (c) functional group, (d) geometric, (e) optical.

21.15 Write structural formulas for all the isomers of (a) bromophenol, (b) dibromophenol, (c) tribromobenzene.

21.16 *Ortho-*, *meta-*, and *para*-isomers can be identified by determining the number of compounds that can be derived from each isomer when a third substituent is introduced into the ring (Körner's method). If one nitro group is introduced into the benzene ring of each of the following compounds, how many isomeric products would be secured from each? (a) *o*-dibromobenzene, (b) *m*-dibromobenzene, (c) *p*-dibromobenzene.

21.17 Write structural formulas for the ten dichloro substitution isomers of naphthalene.

21.18 *Cis-* and *trans*-isomers are known for certain alkenes. Why do not the alkynes display this type of isomerism?

21.19 For which of the following compounds are *cis-* and *trans-* isomers possible? (a) 2,3-dimethyl-2-butene, (b) 3-methyl-2-pentene, (c) 2-methyl-2-pentene, (d) $CH(COOH){=}CH(COOH)$.

21.20 Define the following: (a) olefin, (b) homologous series, (c) cracking, (d) conjugated system, (e) cyanohydrin.

21.21 Why do alcohols have higher boiling points than hydrocarbons of corresponding molecular weight?

21.22 What is the difference between (a) an amine and an amide, (b) an ester and an ether, (c) an aliphatic hydrocarbon and an aromatic hydrocarbon, (d) a saturated hydrocarbon and an unsaturated hydrocarbon.

21.23 What is the product of the reaction of 1-bromobutane with each of the following? (a) Mg, (b) NaCN, (c) NH_3, (d) CH_3NH_2, (e) OH^-(aq), (f) $NaOCH_2CH_3$.

21.24 What is the product of the reaction of 2-propanol with each of the following? (a) Na, (b) H_2SO_4 (high temperature), (c) H_2SO_4 (moderate temperature), (d) propanoic acid, (e) $K_2Cr_2O_7$ in acid solution.

21.25 What are the products when the following compounds are oxidized under vigorous conditions? (a) 2-hexene, (b) 2-methyl-2-butene, (c) 3-hexyne, (d) 1-propanol, (e) 2-propanol, (f) propanol, (g) *p*-nitrotoluene. What are the products when the following compounds are oxidized using mild conditions? (a) 2-hexene, (b) 1-propanol.

21.26 What are the products when the following compounds are reacted with hydrogen in the presence of a suitable catalyst? (a) pentanal, (b) 2-pentanone, (c) 2-pentene, (d) 2-pentyne, (e) ethyl cyanide, (f) *p*-nitrotoluene.

21.27 What is (are) the product(s) of the reaction of bromine with each of the following? (a) methane, (b) 2-butene, (c) 2-butyne, (d) benzene (in the presence of $FeBr_3$).

21.28 State the formulas of all products formed by the reaction of HBr with each of the following. If more than one compound is formed, tell which is produced in the greatest quantity.

(a) CH_3CH_2OH
(b) CH_3CH_2MgBr
(c) $(CH_3CH_2)_2C{=}CH_2$
(d) $CH_3C{\equiv}CH$
(e) $CH_3CH_2COO^-Na^+$
(f) $(CH_3CH_2)_2NH$

21.29 State the compound formed when each of the following is treated with aqueous NaOH.

(a) $(CH_3)_2CHCOOH$
(b) $(CH_3)_2CHBr$
(c) $CH_3CH_2CH_2CN$
(d) $CH_3CH_2CH_2COOCH_2CH_3$
(e) $CH_3CH_2NH_3^+Cl^-$

21.30 Write equations to show the reactions of NH_3 with (a) 2-bromopropane, (b) propanoic acid, (c) methyl propanoate.

21.31 What is the final product (obtained by hydrolysis of the initial product) of the reaction of ethyl magnesium bromide with (a) methanal, (b) butanal, (c) butanone.

21.32 What Grignard reagent and what carbonyl compound should be used to prepare the following alcohols? (a) 3-methyl-2-butanol, (b) 2-methyl-2-butanol, (c) 2-methyl-1-butanol.

21.33 Write equations to show how the following can be made by means of addition reactions: (a) 1,2-dibromoethane from ethene, (b) 1,1-dibromoethane from acetylene, (c) cyclohexanol from cyclohexene.

21.34 Write equations to show how ethanol may be converted into the following substances. More than one step may be required for a given preparation. (a) ethene, (b) ethanal, (c) diethyl ether, (d) ethane, (e) acetic acid, (f) bromoethane, (g) ethyl cyanide, (h) 2-hydroxypropanoic acid.

21.35 Using 1-propanol and 2-propanol as the only organic starting materials, write a series of equations to show how to prepare: (a) 3-hexanol, (b) 2,3-dimethyl-2-butanol, (c) 2-methyl-2-pentanol, (d) 2-methyl-3-pentanol. Use a Grignard reaction as one of the steps in each preparation.

21.36 Show by a series of reactions how to convert 1-propanol into 2-propanol.

21.37 Describe with examples: (a) addition polymers, (b) copolymers, (c) condensation polymers.

21.38 (a) What is Markovnikov's rule? (b) How does the mechanism of the addition of HBr to an alkene explain why such an addition follows Markovnikov's rule?

21.39 Compare and contrast the mechanism of a substitution reaction of an alkane with the mechanism of a substitution reaction of an aromatic hydrocarbon.

21.40 Interpret the following in terms of the Lewis theory: (a) an S_N1 reaction of an alkyl halide, (b) an S_N2 reaction of an alcohol, (c) the addition of HBr to an olefin, (d) the addition of HCN to a ketone.

21.41 Describe the function of $AlCl_3$ in (a) a Friedel-Crafts reaction, (b) a polymerization.

21.42 Arrange the following compounds in decreasing order of strength as Brønsted acids: CH_3CH_2OH, $CH_3CH_2OH_2^+$, H_2O, H_3O^+, CH_3COOH, $CH_3CH_2NH_2$.

21.43 Arrange the following compounds in decreasing order of strength as Brønsted bases: CH_3CH_2OH, $CH_3CH_2O^-$, OH^-, CH_3COO^-, $CH_3CH_2NH_2$.

21.44 Draw the resonance structures of (a) benzene, (b) naphthalene, (c) the acetate ion.

21.45 (a) Why is phenol a stronger acid than methanol? (b) Why is aniline a weaker base than methylamine?

Chapter 22

Biochemistry

Biochemistry is the science that is concerned with the composition and structure of substances that occur in living systems and the chemical reactions that take place in these systems. Impressive progress has been made in this branch of chemistry over the last thirty years. It is currently an area of intense research activity. In this chapter, we will survey some aspects of the field.

The principal component of living material is *water*. It constitutes from about 60% to 90% of all plants and animals. The quantity of inorganic substances present is comparatively small—usually less than 4%; the amount is largest in organisms that have a skeleton. The remainder consists of organic compounds, most of which have complicated structures.

The four most abundant elements of the compounds found in living material are oxygen, carbon, hydrogen, and nitrogen. These four elements account for approximately 96% of the mass of the human body.

22.1 PROTEINS

Proteins are an essential constituent of all living cells—both plant and animal. These substances, therefore, are of prime importance. The word *protein* is derived from a Greek word that means *first.* Thousands of different proteins are found in the human body. They are important components of muscles, tendons, hair, bones, body organs, and skin. The biochemical catalysts called enzymes, hemoglobin of blood, hormones, and antibodies are all proteins.

Proteins are macromolecules that have molecular weights ranging from about 6000 to over 1,000,000. They are formed from simpler compounds called **α-amino acids.** An α-amino acid is a carboxylic acid that has an amino group, —NH_2, bonded to the C atom next to the carboxyl group, —COOH. The designation α (alpha) denotes the position of the amino group. The C atom adjacent to the carboxyl group is called the α-carbon atom. The general formula of these compounds is

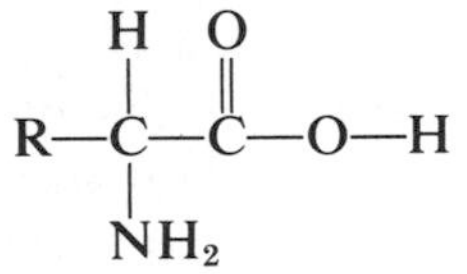

The —NH_2 group is basic and will react with acids to form —NH_3^+. At low *p*H's, therefore, an amino acid exists as a cation, which will move toward the negative pole in an electric field.

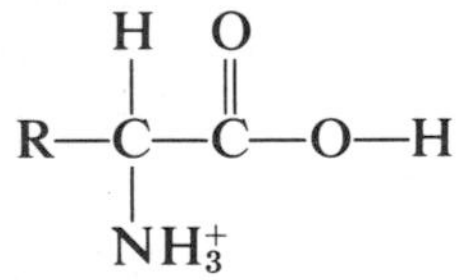

The —COOH group is acidic and forms —COO^- with bases. At high *p*H's, therefore, amino acids are anions and move toward the positive pole in an electric field.

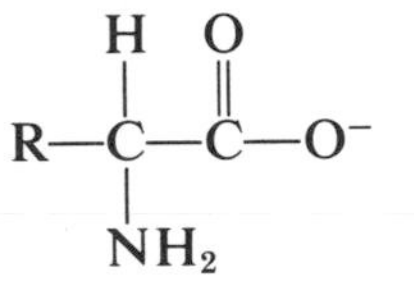

Internal neutralization, in which the proton from the —COOH group moves to the —NH_2 group, produces a **zwitterion.**

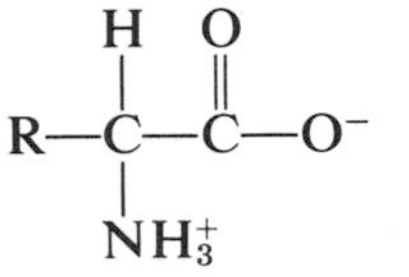

For a given compound, the acid and base strengths, which depend upon the molecular structure, are usually not equal. The **isoelectric point** is the *p*H at

which the internal neutralization is complete. At the isoelectric point, all the amino acid molecules are in the form of zwitterions, which are attracted equally by the positive and the negative poles of an electric field and move toward neither.

The amino acids provide examples of a type of stereoisomerism called **optical isomerism** (see Section 20.4). The simplest optical isomers are compounds that contain asymmetric carbon atoms—carbon atoms to which four different kinds of groups are bonded (Figure 22.1). The mirror image (called an **enantiomorph**) of a molecule of this type is not superimposable on the original molecule.

The structures shown in Figure 22.1 are mirror reflections of one another. The dashed line represents the reflection plane. Start with any group (1, 2, 3, or 4) of the structure on the left and trace the shortest distance to the reflection plane. Continue to the right of the reflection plane for the same distance, and the same group will be found in the structure on the right.

A study of this figure reveals that the two structures cannot be superimposed. If groups 1 and 3 of the structures are superimposed, group 4 of one structure coincides with group 2 (not 4) of the other. No matter how the two structures are turned, it is impossible to make all four groups of one structure coincide with the same four groups, respectively, of the other structure. The two structures are isomers.

The physical properties of optical isomers are identical except for their effects on plane-polarized light. The isomers are said to be optically active because they rotate the plane of plane-polarized light in either a clockwise or a counterclockwise direction. The isomer that rotates the plane to the right (clockwise) is said to be **dextrorotatory** (+). The other isomer rotates the plane an equal amount to the left and is said to be **levorotatory** (−).

With the exception of glycine (for which R = H), the α-carbon atoms of all α-amino acids have four different kinds of groups bonded to them and are, therefore, asymmetric. Consequently, optical isomers exist for the α-amino acids. The structures of the isomers are usually diagrammed

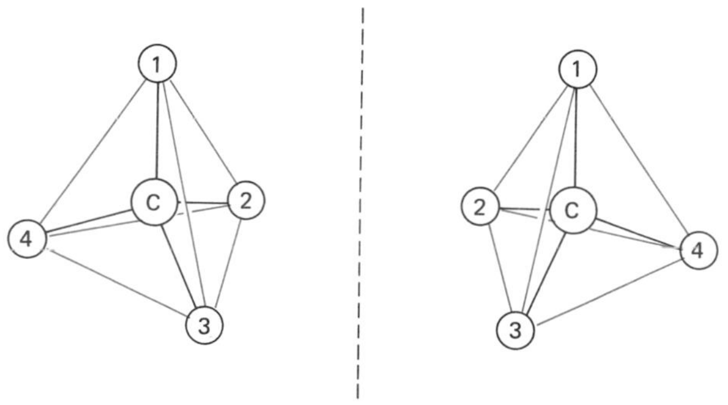

Figure 22.1 Optical isomers of a compound containing an asymmetric carbon atom. (Dashes represent reflection plane.)

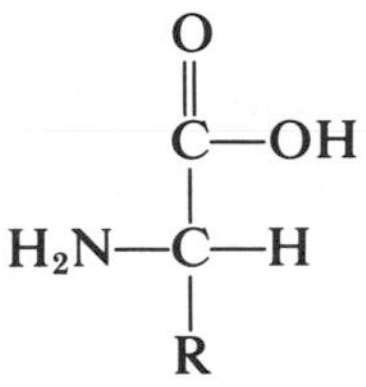

L-amino acid

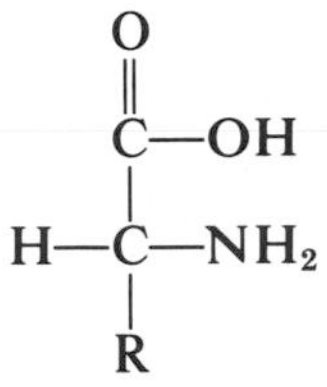

D-amino acid

The diagrams represent structures in which the asymmetric carbon atoms are in the plane of the paper, the —COOH and —R groups are behind the plane of the paper, and the —H and $—NH_2$ groups are in front of the plane of the paper. With such an arrangement of groups, the L-form is the one in which the amino group is to the left of the chain, and the D-form is the one in which the amino group is to the right. The structures can be related to the diagrams of Figure 22.1. The L-form is the one on the left in the figure if group 1 = —COOH, 2 = —R, 3 = —H, and 4 = $—NH_2$.

All the amino acids derived from proteins are L-amino acids. Only the L-forms can be utilized in the metabolic processes of the body. The sign of rotation should not be confused with the symbols D- or L- that are used to designate the configuration of the asymmetric carbon atom. Thus, the L-isomer of alanine (2-aminopropanoic acid) is dextrorotatory (+).

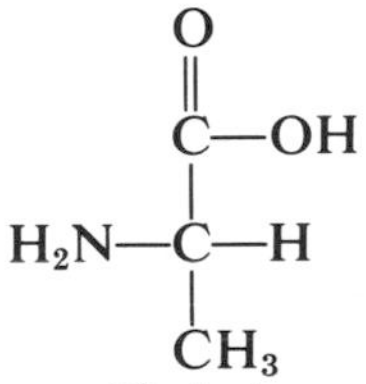

L-(+)-alanine

In protein molecules, many amino acid units are linked together to form long chains. When two amino acids combine, the amino group of one molecule joins, with the elimination of water, to the carboxyl group of the second molecule. A **peptide linkage,** 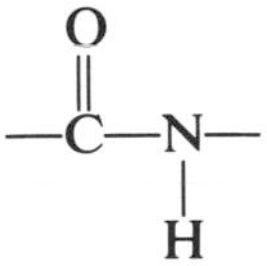 is formed.

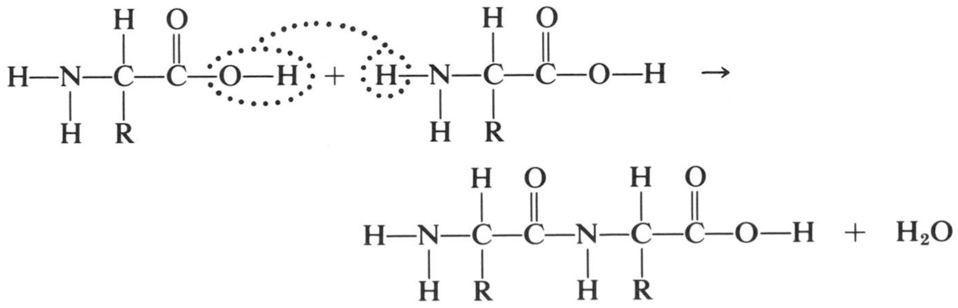

The product, called a **dipeptide,** has a free amino group at one end and a free carboxyl group at the other. It can condense with other amino acids to form long chains called **polypeptides.** Proteins are polypeptides. The hydrolysis of a polypeptide is the reverse of the condensation process and produces amino acids.

The radicals, R, in the preceding formulas vary considerably in their structures. Some of them are simple hydrocarbons, some contain rings (including aromatic rings), some contain additional $—NH_2$ or —COOH groups, and some incorporate —SH or —OH groups in their structures (see Figure 22.2). Twenty different α-amino acids are regularly found in proteins. A few others are occasionally found as minor constituents.

Proteins contain from about 50 to over 10,000 amino acid residues in a chain. The average molecular weight of an amino acid residue is about 110 to 120. A protein with a molecular weight of 300,000, therefore, contains roughly 300,000/120 or 2500 amino acid residues.

The amino acids in a given protein occur in a definite order. The same type of amino acid may be used many times in the polypeptide chain. Since a protein may contain hundreds or thousands of amino acid residues, a tremendous number of arrangements are possible even though only about 20 different amino acids are used.

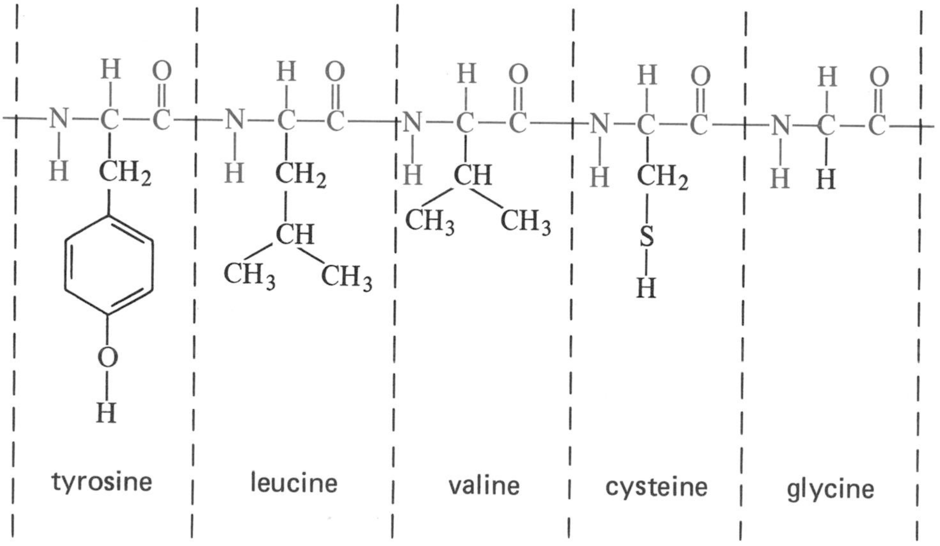

Figure 22.2 Fragment of bovine insulin protein chain. The polypeptide chain is shown in color. The R radicals are shown in black. The names of the amino acids that make up the chains appear at the bottom.

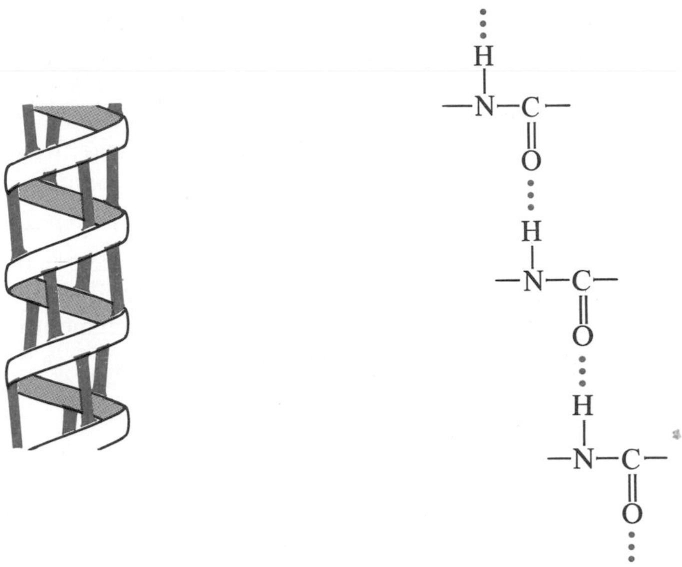

(a) (b)

Figure 22.3 (a) Schematic representation for an α-helix. The colored lines are hydrogen bonds. (b) Hydrogen bonds between coils of an α-heix.

The sequence of amino acid residues in a protein is called its **primary structure.** The primary structure of bovine insulin was determined in 1945–52 by Frederick Sanger. A fragment of the structure is shown in Figure 22.2. The complete structure contains 51 amino acid residues in two chains – one of 21 amino acid residues and the other of 30.

The spatial arrangement of the polypeptide chain of a protein is called the **secondary structure** of the protein. The most important arrangement is the **alpha helix.** In an α-helix, the chain is coiled in a form that resembles a spring (Figure 22.3). There are about 3.6 amino acid residues in each turn. The helix is held together by hydrogen bonds between the loops of the coil. The hydrogen bonds are formed by the H atom of the N—H of one peptide group with the O atom of the C═O group of another peptide group (see Figure 22.3). Each peptide group is joined, by hydrogen bonding, to the third peptide group before it and the third peptide group following it.

Another type of secondary structure is the **pleated sheet** arrangement. Chains are held together in the form of pleated sheets by hydrogen bonds between adjacent chains. This arrangement is not so common as the α-helix arrangement.

The **tertiary structure** of a protein is the arrangement of the secondary structures brought about by interactions between the R groups of the amino acid residues. The structure of a globular protein (so called because these proteins are usually approximately spherical in shape) is shown schematically in Figure 22.4.

The interactions that hold the coiled segments of the chain into a compact tertiary form do not involve the peptide linkages of the chain. Instead, these interactions take place between substituents of the R groups of the amino acid residues.

1. Some R radicals contain a —COOH group over and above the one involved in peptide formation. Other R radicals contain a second $-NH_2$ group. These two functional groups—one attached to one segment of the coil, the other attached to another segment—can react.

$$-\overset{\overset{\displaystyle O}{\|}}{C}-O-H + H_2N- \rightarrow -\overset{\overset{\displaystyle O}{\|}}{C}-O^- + {}^+H_3N-$$

2. By mild oxidation, a disulfide linkage can form between two —SH groups attached to different places on the coil.

$$-S-H + H-S- \rightarrow -S-S- + 2H^+ + 2e^-$$

3. Substituent groups can hydrogen-bond to one another.

$$\begin{matrix} & & O- \\ & & | \\ -O\cdot\cdot\cdot\cdot & & H \\ | & & \\ H & & \end{matrix}$$

These interactions occur in places where the coils overlap.

The primary, secondary, and tertiary structures describe the arrangement

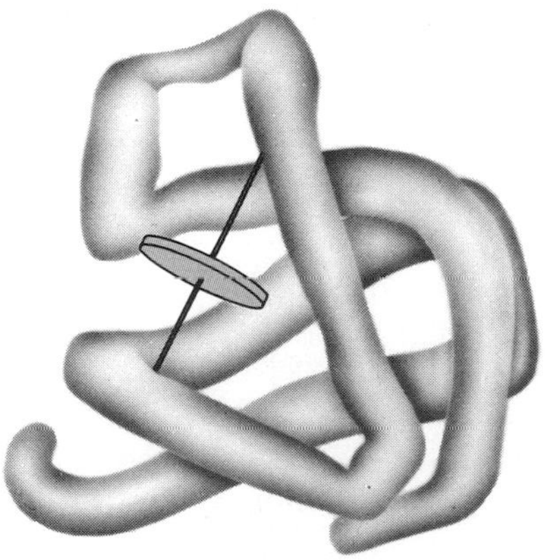

Figure 22.4 Tertiary structure of a globular protein. The colored disc is an iron-containing heme group.

of a single polypeptide chain. Some proteins contain more than one chain. The number of chains and the way in which they are arranged in the complete protein make up what is called the **quaternary structure** of the protein. For example, hemoglobin, the oxygen carrier in blood, is formed from four chains of two different types. Two of the chains each contain 140 amino acid residues. Each of the other two contains 146 amino acid residues. Each of the four chains has an iron-containing heme group (Section 20.4) and has a tertiary structure similar to that shown in Figure 22.4.

22.2 CARBOHYDRATES

Sugars, starches, and cellulose belong to a group of organic compounds called **carbohydrates.** The name was derived from the fact that most (but not all) compounds of this type have the general formula $C_x(H_2O)_y$, and might, therefore, appear to be hydrates of carbon. Carbohydrates, however, do not contain water. They are hydroxy aldehydes, hydroxy ketones, or substances derived from them. Carbohydrates are the principal substances of which plants are composed. They are an important food for animals—they serve as a source of energy and provide carbon chains for compounds that are synthesized by living organisms.

The simplest carbohydrates are the **monosaccharides,** or simple sugars. The most important monosaccharide is *glucose*. It is the most abundant sugar found in nature, and in animals it is a normal constituent of blood. The two optical isomers of glucose are

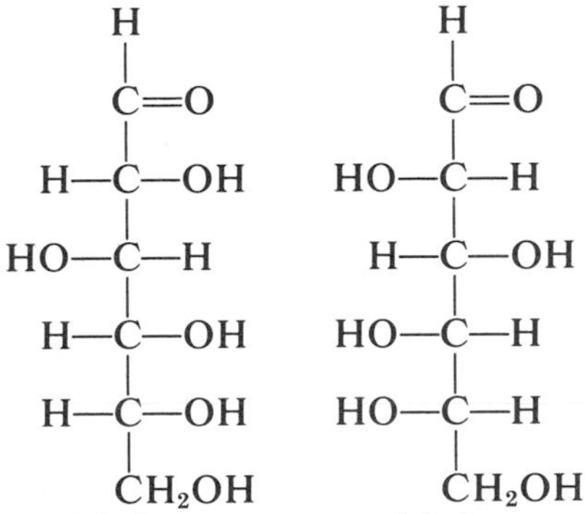

D-(+)-glucose L-(−)-glucose

There are four asymmetric carbon atoms in the molecule. The carbon atoms at each end of the molecule do not hold four different groups and are not asymmetric. Sixteen optical isomers (eight D,L pairs) of this formula exist. Only the preceding pair are called glucose.

D-glucose forms a cyclic molecule by an addition reaction involving the carbonyl group and a hydroxy group.

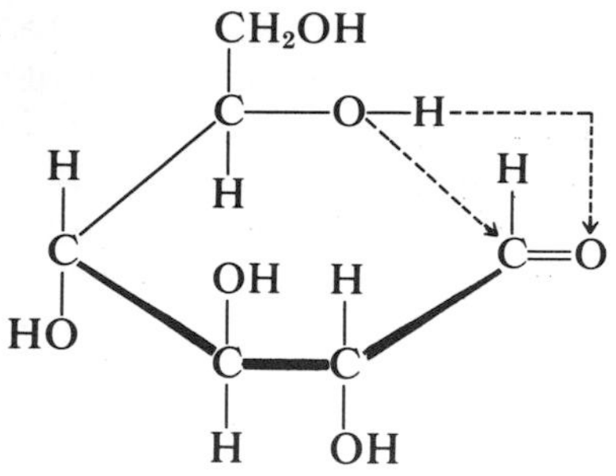

Ring formation produces a new asymmetric center, and two ring isomers of D-glucose exist which differ in the orientation of the new OH group.

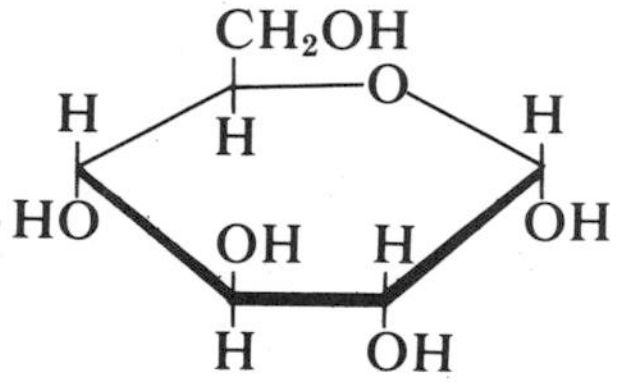

α-D-glucose

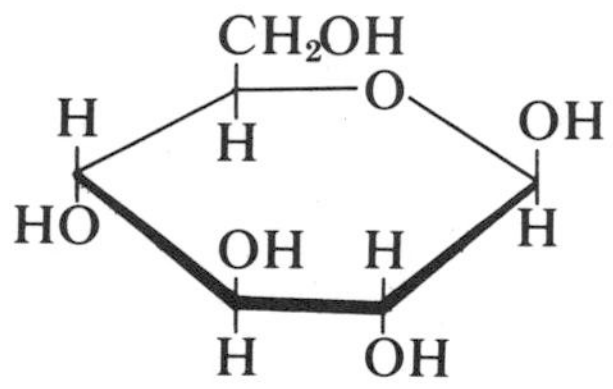

β-D-glucose

In these diagrams the carbon atoms of the ring are not shown. The rings are actually puckered, not planar. Notice that in the α form the OH group of the extreme right-hand carbon atom is on the same side of the ring as the OH group of the adjacent carbon atom. These OH groups may be said to be *cis* to each other. In aqueous solution, α and β forms of D-glucose exist in equilibrium, together with a low concentration of the open-chain form.

Fructose is a hydroxy ketone.

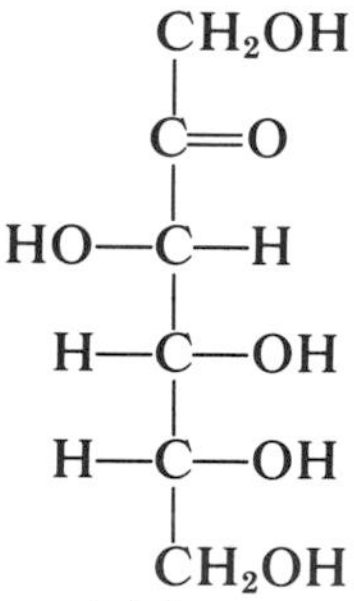

D-(−)-fructose

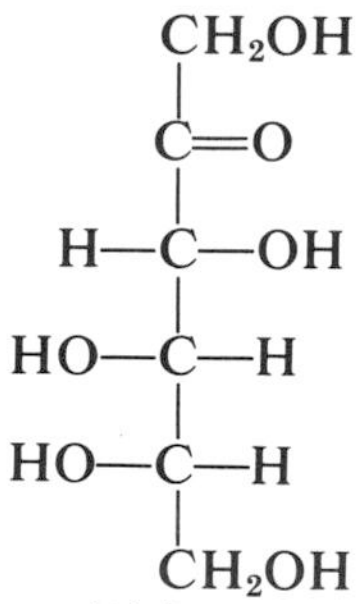

L-(+)-fructose

There are three asymmetric carbon atoms in this molecule and eight stereoisomers (or four D,L pairs) of this formula are known. Cyclic forms of D-fructose are known with five- or six-membered rings.

Disaccharides are composed of two monosaccharide units. *Sucrose,* which is cane sugar, is a disaccharide. Acid hydrolysis of sucrose yields the two simple sugars D-glucose and D-fructose. In the sucrose molecule an α

D-glucose unit and the β five-membered ring form of a D-fructose unit are joined through two OH groups by the elimination of a molecule of water.

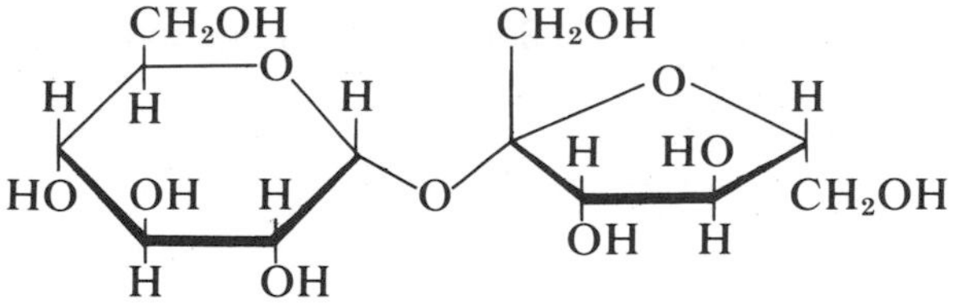

Cellulose and starch are **polysaccharides** that yield only D-glucose upon hydrolysis. It is estimated that the number of D-glucose units in the molecular structures of these substances may be as high as several thousand. The D-glucose units of cellulose are linked in long chains in β combination.

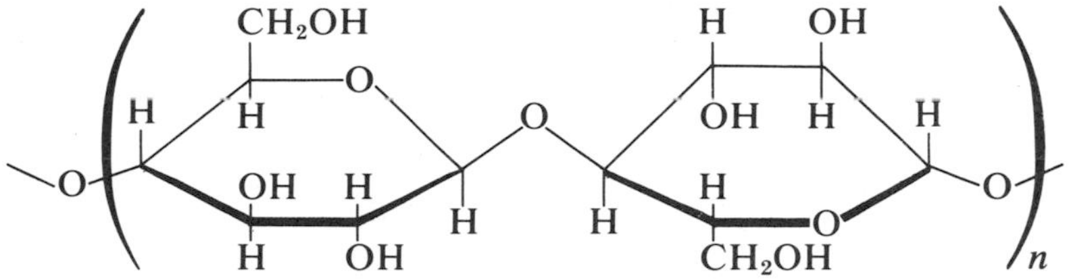

The D-glucose units of starch are linked in a different manner. There are occasional cross links between long chains of D-glucose units arranged in α combination.

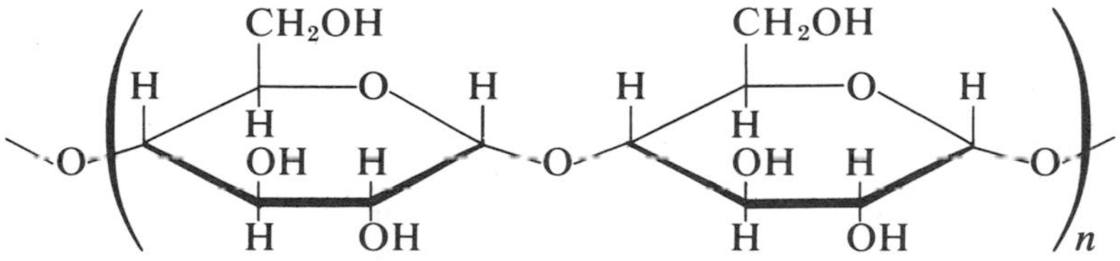

Starch is an important food material, but cellulose cannot be digested by man —an important distinction brought about by difference in the way that the D-glucose units are linked together.

22.3 FATS AND OILS

Fats and oils belong to a larger classification called **lipids.** Lipids are substances that can be dissolved away from biological material by solvents that are nonpolar or slightly polar (such as hydrocarbons, carbon tetrachloride, and diethyl ether). Since the classification is based on solubility, not structure, a wide variety of compounds are lipids. Fats, oils, some vitamins and hormones, and certain constituents of the cell walls are examples of some of the substances classified as lipids.

Fats and oils are **esters** of fatty acids and glycerol. The **fatty acids** are long straight-chain carboxylic acids (Table 22.1). Some of them are saturated and some contain one or more double bonds. Almost all the fatty acids isolated from natural sources contain an even number of carbon atoms.

Glycerol is a trihydroxy alcohol: 1,2,3-propanetriol.

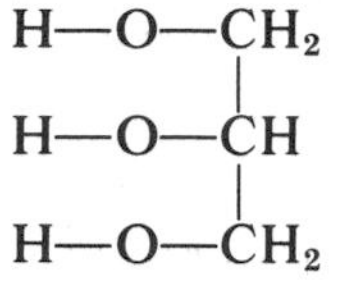

The esters formed from 1 mol of glycerol and 3 mol of fatty acid, $R{-}C({=}O){-}O{-}H$, are called **triglycerides.**

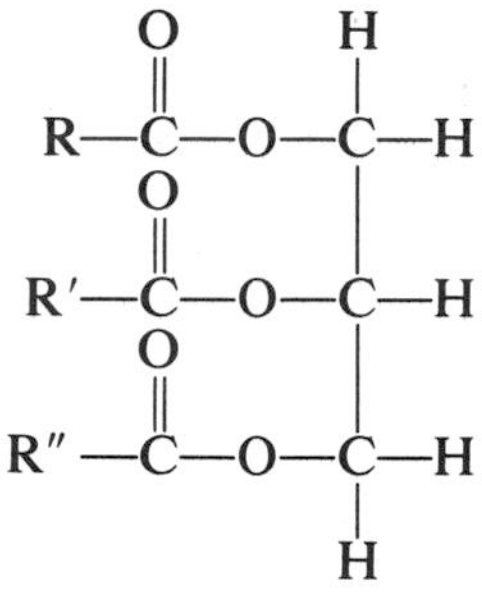

Table 22.1 Some common fatty acids

SATURATED ACIDS		
ACID	FORMULA	OCCURRENCE
lauric	$CH_3(CH_2)_{10}COOH$	coconut oil, palm oil, animal fat
myristic	$CH_3(CH_2)_{12}COOH$	coconut oil, palm oil, animal fat
palmitic	$CH_3(CH_2)_{14}COOH$	animal fat, cottonseed oil, palm oil
stearic	$CH_3(CH_2)_{16}COOH$	animal fat

UNSATURATED ACIDS			
ACID	FORMULA	POSITIONS OF DOUBLE BONDS*	OCCURRENCE
oleic	$C_{17}H_{33}COOH$	9	corn oil, cottonseed oil, olive oil
linoleic	$C_{17}H_{31}COOH$	9, 12	cottonseed oil, corn oil, linseed oil
linolenic	$C_{17}H_{29}COOH$	9, 12, 15	linseed oil
arachidonic	$C_{19}H_{31}COOH$	5, 8, 11, 14	sardine oil, corn oil, animal fat

* All these acids have straight chains. The carbon atom of the COOH group is number 1. The number given in the table indicates the first carbon of the double bond. Number 9, for example, indicates a double bond between carbon atoms 9 and 10.

A natural fat or oil is a mixture of different triglycerides. In fact, a single triglyceride molecule may be formed from three different fatty acids. For this reason, the radicals are indicated in the preceding formula as R, R′, and R″. Palmitic and oleic acids are present in most vegetable and animal fats and oils (Table 22.1).

Fats are mixtures of triglycerides that are solids at room temperature. Oils are mixtures that are liquids at room temperature. Oils contain a higher percentage of the glyceryl esters of unsaturated fatty acids than fats do. Vegetable oils (such as corn oil and coconut oil) are changed into solids (vegetable shortening and margarine) by reaction with hydrogen. The hydrogen adds to some of the double bonds of the unsaturated triglycerides. In practice, the process is continued until a solid of the desired consistency is obtained. Not all of the double bonds are saturated with hydrogen.

When fats and oils are heated with aqueous solutions of bases, glycerol and the salts of the fatty acids are obtained.

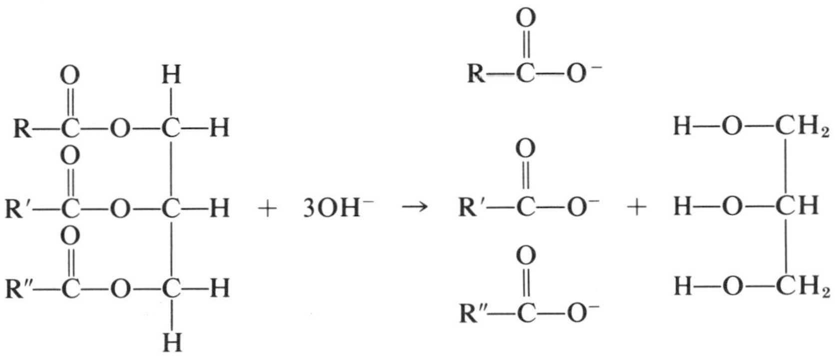

The process is called **saponification,** which means *soap making*. Soaps are salts of fatty acids. The type of soap obtained depends, of course, upon the base used in the saponification. Sodium soaps, $RCOO^- Na^+$, are the most common.

Fats and oils are used as fuels by the body, and carbon dioxide and water are the products of the oxidation. Fats are also stored as a reserve source of energy. Carbohydrates may be converted into fat and stored also. Fats and oils are also used by the body in the syntheses of constituents of tissues.

22.4 NUCLEIC ACIDS

Nucleic acids were first isolated from cell nuclei in the 1860's, and the name was derived from the name of their source. Many different proteins must be constantly made by an organism to enable it to grow, function, renew itself, and reproduce itself. The nucleic acids direct the syntheses of these proteins. The nucleic acids store, use, and pass on to the next generation the pattern for the reproduction of the organism.

Nucleic acids are high molecular weight, long-chain polymers. Each unit of the chain is formed from three components.

1. a phosphoric acid molecule
2. a five-carbon sugar molecule
3. a molecule of a nitrogen-containing base

A schematic diagram of a fragment of a nucleic acid chain that contains only three units (called **nucleotides**) is shown in Figure 22.5. Actual nucleic acid molecules contain from about 80 to 100 million nucleotides.

There are two types of nucleic acids: **deoxyribonucleic acids** (called **DNA**) and **ribonucleic acids** (called **RNA**). The principal difference between the two is in the sugar molecule employed. DNA contains deoxyribose and RNA contains ribose.

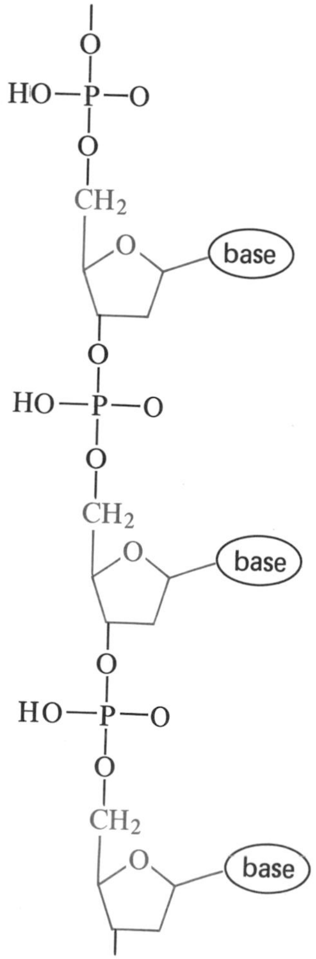

Figure 22.5 Schematic diagram of the primary structure of a nucleic acid. Three nucleotide units are shown.

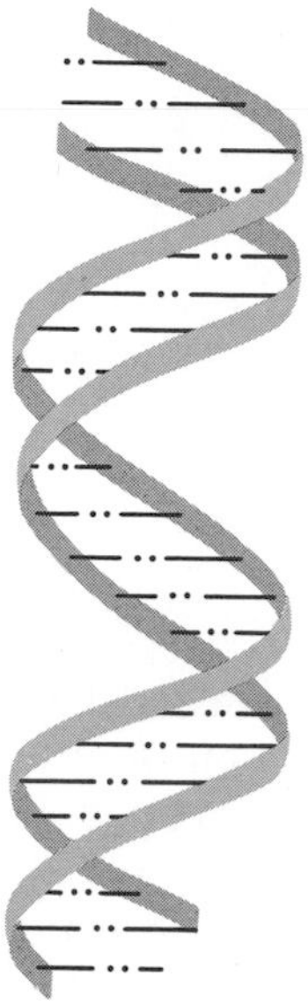

Figure 22.6 Schematic representation of the double helix. Dots represent hydrogen bonding between complementary bases of the two DNA strands.

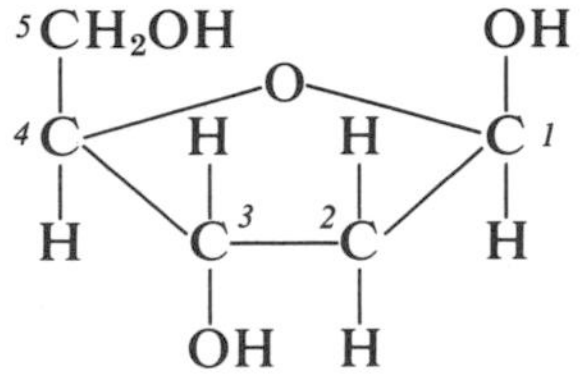

β-D-deoxyribose

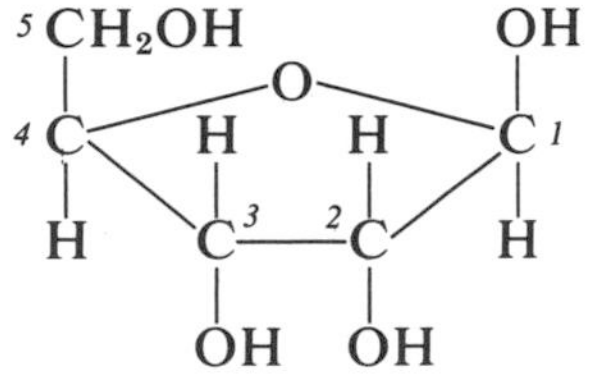

β-D-ribose

The prefix *deoxy-* indicates removal of an oxygen atom. If an atom of oxygen is removed from the number *2* carbon atom of ribose, the structure of deoxyribose is obtained.

In the chain of a nucleic acid, carbon atom number *3* of one sugar molecule and carbon atom number *5* of the next sugar molecule are joined by ester linkages to a molecule of phosphoric acid (see Figure 22.5). One of four different nitrogen base radicals replaces the OH group of carbon atom number *1* of each sugar molecule.

The secondary structure of DNA is a **double helix** (Figure 22.6). Two chains of DNA are coiled together in such a way that the bases are in the interior of the coils. The structure is held together by hydrogen bonds between the bases of one chain and the bases of the other chain.

The structures of the four different base radicals found in DNA are shown in Figure 22.7. The formulas are drawn in pairs with hydrogen bonds indicated between the pairs. Notice that **adenine** (abbreviated **A**) and **thymine** (abbreviated **T**) complement each other. The positions of the atoms make it possi-

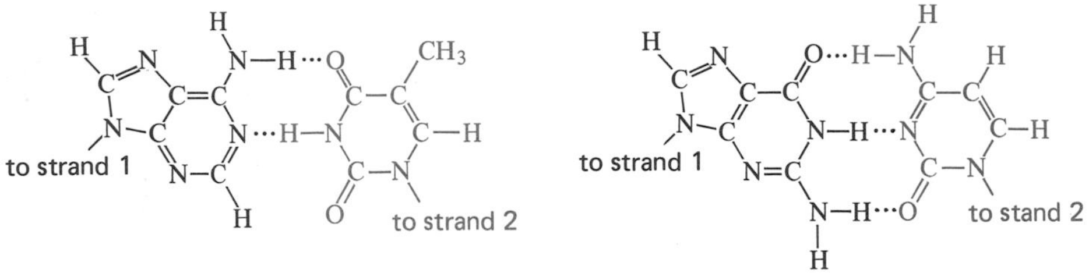

Figure 22.7 The four base radicals found in DNA. The formulas are drawn in complementary pairs.

ble for two strong hydrogen bonds to form between A on one chain and T on the other chain of the double helix. **Guanine (G)** and **cytosine (C)** complement each other in the same way. Three strong hydrogen bonds are formed between this pair of bases.

In any sample of DNA there is the same amount of A as there is T and there is the same amount of G as there is C. The bases attached to one DNA chain complement the bases attached to the other DNA chain. If there is an A on chain 1, T is found on chain 2 opposite it. If T is found on chain 1, A will be opposite it on chain 2. The same type of base pairing occurs between G and C. Notice that the two hydrogen-bonded pairs shown in Figure 22.7 are about the same length. Consequently, they hold the two chains of the double helix a uniform distance apart.

When a cell divides, the two chains of a DNA double helix come apart. Each chain serves as a template, or pattern, for the synthesis of a new, complementary chain. Two identical double helices result from the process—each one containing one of the chains of the original double helix (see Figure 22.8).

Nucleotides from the surrounding solution form the new chains. The base of a nucleotide pairs with the complementary base of the DNA chain through the formation of hydrogen bonds. The nucleotides line up, therefore, in a way that is dictated by the order in which the bases occur in the DNA chain. The new chains formed from the nucleotides are complementary to the original DNA chains.

The order in which the bases occur in a DNA molecule constitutes the information used for the synthesis of proteins. Two types of RNA are used for these processes: messenger RNA and transfer RNA. RNA employs the same bases that DNA does with the exception of thymine. In RNA, **uracil (U)** is used in place of thymine (T). The structure of uracil is the same as that of thymine except that a H atom replaces the —CH_3 group. U, therefore, complements A in the same way that T does (see Figure 22.7).

An RNA chain is synthesized from nucleotides in the nucleus of the cell in such a way that the order of bases in the RNA chain complements the order of a section of the DNA chain. The DNA chain, therefore, acts as a template.

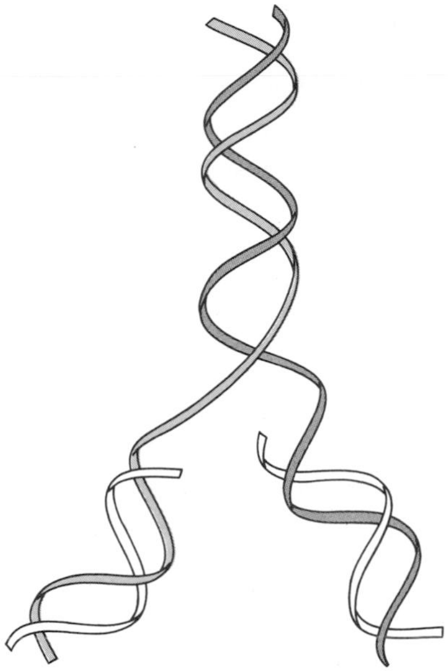

Figure 22.8 Replication of DNA double helix. The strands unwind and each serves as a template for the synthesis of a new complementary strand.

Suppose, for example, that the order of bases of a small portion of the DNA chain is

C—C—A—A—C—G

The order of bases in the RNA chain that is made by using this part of the DNA chain as a pattern is

G—G—U—U—G—C

since G complements C and U complements A.

The chain formed in this way is called **messenger RNA.** A messenger RNA chain may contain about 500 bases in an order dictated by a *section* of a DNA chain, which contains altogether more than 100 million bases. The information in the section of the DNA chain (the order of bases) is said to be **transcribed** by the messenger RNA. This information is the message that the messenger carries from the cell nucleus to the ribosomes, where the protein is made.

The order of bases directs the order in which different amino acids are assembled in the formation of the protein. Each amino acid of the protein is specified by a set of three base groups (a base triplet) of the messenger RNA chain. The sequence

G—G—U—U—G—C

pertains to two amino acids. The code G—G—U specifies the amino acid glycine. The code U—G—C specifies the amino acid cysteine. The portion of the messenger RNA chain that contains the three bases of a code is called a **codon.** The preceding sequence described two codons.

A second type of RNA, called **transfer RNA,** is used to decipher the code and build the protein. Transfer RNA chains are the smallest nucleic acid chains. They consist of approximately 80 nucleotides and may contain a few bases other than the customary four.

A particular transfer RNA molecule bonds with a specific amino acid and carries it to the messenger RNA where it is inserted into the protein being made at the position directed by the code of the messenger RNA. The transfer RNA is said to **translate** the code. Each transfer RNA contains a base triplet (called an **anticodon**) that complements the triplet of a codon of messenger RNA. The anticodon of a particular transfer RNA corresponds to the specific amino acid that it carries.

Suppose, for example, that the codon of messenger RNA specifies the code GGU. The anticodon CCA that corresponds to this codon is found on transfer RNA that carries the amino acid glycine.

The synthesis of a protein is thought to occur in steps (Figure 22.9). Transfer RNA carrying an amino acid becomes attached to the messenger RNA by codon–anticodon pairing. A second transfer RNA becomes attached to the next codon of the messenger RNA chain. A peptide linkage is formed between the two amino acids of the two transfer RNA's. The first transfer RNA is then released and leaves behind its amino acid now bound into a dipeptide. A third transfer RNA, of a type specified by the next codon, becomes attached to the messenger RNA. As soon as a peptide bond connects the dipeptide to the amino acid of the third transfer RNA, the second transfer RNA is released. The process stops when a codon that does not correspond to any amino acid is encountered. The synthesis of more than one protein is directed by a single messenger RNA.

Specific enzymes catalyze every step of the entire process including the formation of the messenger RNA and the bonding of an amino acid to the transfer RNA. Formation of the polypeptides is rapid. It is estimated that hundreds of peptide bonds are formed in 1 min.

Any change in the DNA of the cell will result in a change in the mes-

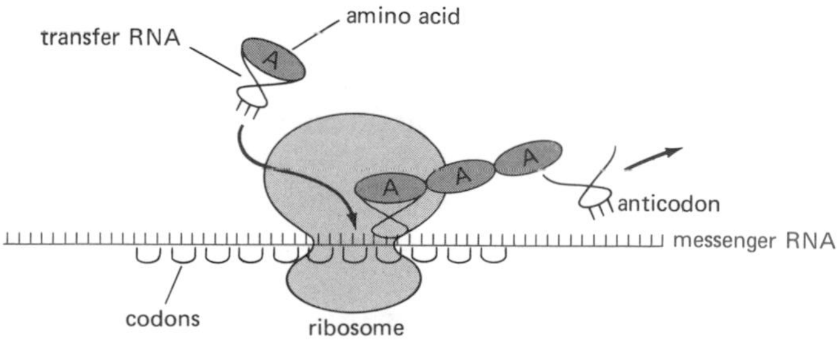

Figure 22.9 Synthesis of a protein by transfer RNA on messenger RNA.

senger RNA produced from it and, consequently, an alteration in the resulting protein. Such changes are called **mutations.** The changes may prove to be beneficial, harmful, or trivial. They are passed on from generation to generation. Mutations are responsible for genetic diseases, such as Tay-Sachs disease, sickle-cell anemia, hemophilia, and Huntington's chorea. They also explain the mechanism of evolution.

22.5 ENZYMES

Enzymes are biological catalysts. The term is derived from a Greek word that means *in leaven* since the action of yeast in fermentation was the first enzymatic action identified. Enzymes are involved in every reaction that occurs in a cell.

A cell contains over 1000 enzymes. A large number is required because enzymes are **specific** in their action. Some are so specific that they catalyze only one reaction of one particular compound. Enzymes are known that will hydrolyze only one definite type of peptide bond—one that involves a particular kind of amino acid residue. The enzymes that are least specific are probably those that hydrolyze most ester linkages.

Some enzymes are **stereochemically specific.** Such an enzyme might act on only the L form of an amino acid and not the D form, or an enzyme might be active only toward β linkages between glucose units of a carbohydrate and not α linkages.

Enzymes are amazingly efficient. The **turnover number** of an enzyme is the number of moles of reactant transformed in 1 min by the action of the enzyme. Turnover numbers vary from about 10,000 mol/min to 5,000,000 mol/min. Since these values are so high, only a small amount of each enzyme is required by a cell.

The action of enzymes permits the cell reactions to occur rapidly at the relatively low temperature of the body and at *p*H's near to 7. When comparable reactions are run in the laboratory without the aid of enzymes, they require higher temperatures and the addition of acids or bases.

The **substrate** of an enzyme is the compound upon which it acts. If E is used to designate the enzyme, S the substrate, and P the product of the reaction, the overall process can be indicated as follows.

$$E + S \rightleftharpoons ES \rightarrow E + P$$

ES, formed from the enzyme and the substrate, may be considered to be the activated complex for the reaction. The catalytic effect is brought about by a significant lowering of the energy of activation for the reaction (see Section 13.4).

Enzymes are surface catalysts. They are globular proteins and are much larger than the substrate upon which they act. Reaction occurs at an **active site** on the surface of the enzyme. The substrate is held to the active site by hydrogen bonding, ionic attraction, or some other type of chemical interaction.

The contour of an active site fits the shape of the substrate exactly, or more probably, the active site changes its conformation to fit the substrate molecule after it is attached. In the process, bonds within the substrate are stretched and weakened. In this way, the reaction is facilitated. After the products of the reaction leave, the active site is free to bind another substrate molecule.

Enzymes are specific in their action because they function by means of active sites that are tailored to fit a specific substrate or type of substrate. The stereochemical specificity of some enzymes is a result of the conformation of their active sites—the L form of a molecule might fit and the D form might not.

Even though the reaction occurs on a relatively small part of the surface—the active site—the rest of the enzyme molecule is important. It probably helps to form the active site into the required conformation and to hold it that way until after the reaction has occurred.

Most enzymes are **conjugated proteins,** that is, they contain nonprotein parts together with protein parts. If the nonprotein part is not easily removed, it is called a **prosthetic group.** If it is easily removed, it is called a **coenzyme.** Neither the coenzyme nor the protein part from which it is removed is active when the two are separated. Prosthetic groups and coenzymes are important because they help to form the active sites of enzymes. Many coenzymes are vitamins or compounds derived from vitamins. Some enzymes require the presence of metal ions (such as Co^{2+}, Fe^{2+}, Zn^{2+}, Cu^{2+}, and Mg^{2+}) to function.

The action of enzymes can be inhibited, or retarded, in several ways. An **inhibitor** is a substance that combines with an enzyme at the active site and prevents the enzyme from functioning. In **competitive inhibition,** the enzyme, E, and the inhibitor, I, combine reversibly.

$$E + I \rightleftharpoons EI$$

The combination of the enzyme and substrate is also reversible.

$$E + S \rightleftharpoons ES$$

The concentrations of S and I determine the amounts of ES and EI formed. Since the desired reaction depends upon the formation of ES, the reaction can be slowed down or stopped altogether if the concentration of I is relatively high.

The product of a reaction may serve as an inhibitor. **Product inhibition** is important since it serves to control the extent of reactions that occur in a cell. As the concentration of the product of some reactions increases, product inhibition gradually slows the reaction down and prevents an undesirable accumulation of product in the cell.

If a process occurs in a series of steps,

$$A \rightarrow B \rightarrow C \rightarrow D$$

control of the entire sequence can be achieved if a product of the last step (D) inhibits a previous step (say, A → B). Such inhibition is called **feedback inhibition.** It too controls the amounts of materials produced in a cell.

If an inhibitor combines with an enzyme in an irreversible manner, the effect is called **noncompetitive inhibition.**

$$E + I \rightarrow EI$$

The substrate cannot displace this type of inhibitor from the active site. Many poisons work by inhibiting enzyme reactions that are necessary for life processes.

Some drugs are competitive inhibitors. Antibiotics probably work by inhibiting enzyme reactions that are required by microorganisms. Sulfanilamide is believed to act as an inhibitor of an enzyme for which the substrate is *p*-aminobenzoic acid.

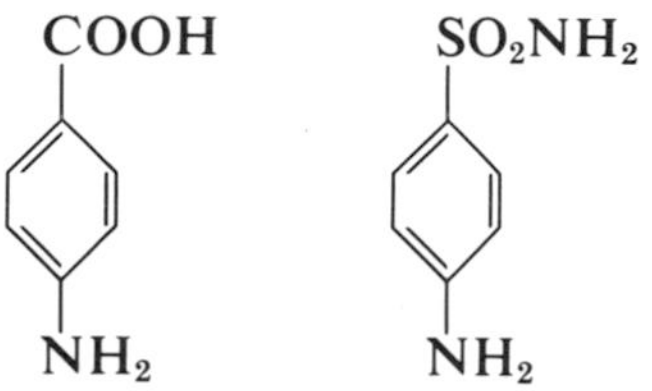

p-aminobenzoic acid sulfanilamide

Notice the similarity between the structures of the two compounds, which compete for the active site. Certain microorganisms utilize the enzymatic reaction of *p*-aminobenzoic acid. Inhibition of the reaction by sulfanilamide causes the death of these microorganisms.

Many insecticides and herbicides are effective because they are enzymatic inhibitors.

22.6 METABOLISM

Metabolism is a general term that refers to all the reactions that occur in a living organism. Metabolic reactions are divided into two groups.

1. **Catabolic processes** are those in which substances are broken down into simpler substances. These reactions provide the energy required by the organism. They also produce simple compounds from which larger ones are made.
2. **Anabolic processes** are those in which large molecules are synthesized from smaller ones. These processes require energy and produce substances needed for the reproduction and survival of the organism.

The primary source of energy for all life is the light from the sun. In **photosynthesis,** the chlorophyll of plants absorbs sunlight, which is used as a source of energy to bring about the manufacture of carbohydrates from CO_2 and H_2O. In this way, energy of the sun is stored as chemical energy in the plants. Humans obtain the energy required to sustain life from their food, which comes from plants or from animals that have been fed on plants.

A large amount of energy is liberated by the oxidation of glucose and other substances derived from food. If 1 mol of glucose is oxidized to CO_2 and H_2O, about 2870 kJ of energy is liberated. In the body, the oxidation is conducted gradually, in a controlled manner, through a sequence of many steps. Each step is catalyzed by an enzyme and involves a comparatively small energy effect. Enzymes cannot function at temperatures much higher than body temperature.

A part of the energy from the oxidation of glucose is dissipated as heat, which maintains body temperature. About 44% of the total energy available, however, is captured and stored. The energy is stored in high-energy molecules. The most important molecule of this type is **adenosine triphosphate,** or **ATP.**

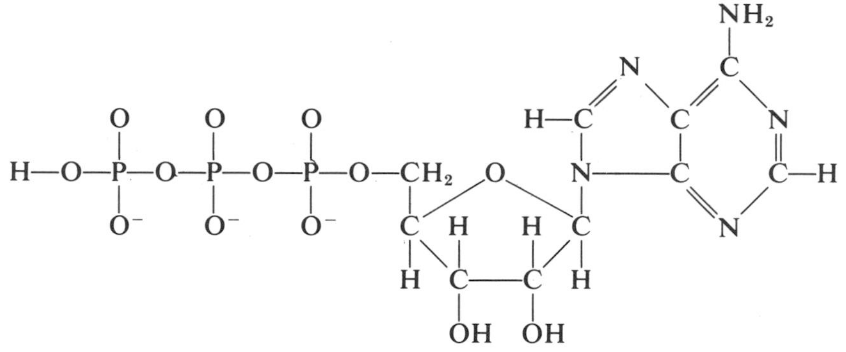

ATP is prepared from **adenosine diphosphate, ADP.** The structures of the two compounds are the same except that ATP has three phosphate groups and ADP has two.

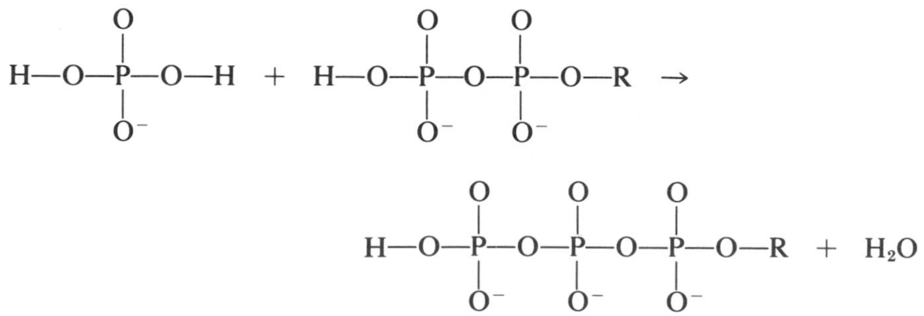

In this equation, R is used to indicate the organic part of the molecules.

The free energy change for the preparation of ATP from ADP is about +33 kJ/mol—which means that energy is required and the reaction will not occur by itself. In the cell, the reaction is **coupled** with another reaction that liberates more energy than the preparation of ATP requires. The overall change in free energy for the coupled reactions is negative. The following figures are typical.

cell reaction	$\Delta G = -50$ kJ	
preparation of ATP	$\Delta G = +33$ kJ	(stored)
coupled reactions	$\Delta G = -17$ kJ	(lost as heat)

The energy stored in ATP is released by the reverse of the preceding reaction. When ATP loses a phosphate group, ADP is produced and the free energy change is about −33 kJ. The energy released is used, by means of reaction coupling, to drive reactions in which components of the organism are synthesized. All reactions of this type (anabolic reactions) require energy. The energy of ATP hydrolysis is also used to produce muscle contractions, but the mechanism of the process is not known.

The oxidation of glucose in the body takes place in many steps. Several of the steps lead to the production of one or more molecules of ATP. In all, the oxidation of 1 mol of glucose produces 38 mol of ATP. The process, therefore, stores 38×33 kJ = 1254 kJ of energy. Since the total energy liberated by the oxidation of 1 mol of glucose is 2870 kJ,

$$(1254 \text{ kJ}/2870 \text{ kJ})100 = 44\%$$

of the energy liberated is stored.

PROBLEMS

22.1 Alanine is 2-aminopropanoic acid. Write the structure of the polypeptide formed from three molecules of alanine.

22.2 What is the difference in meaning between the symbols D and (+) as applied to amino acids?

22.3 Write a chemical equation for the reaction of valine, 2-amino-3-methylbutanoic acid, (a) with $H^+(aq)$ in water solution, (b) with $OH^-(aq)$ in water solution, (c) in which a zwitterion is prepared.

22.4 Do optical isomers of glycine (2-aminoacetic acid) exist? Why or why not?

22.5 What is the isoelectric point of an aqueous solution of an amino acid?

22.6 What is meant by the primary, secondary, tertiary, and quaternary structures of a protein?

22.7 What types of interactions hold a protein into its secondary structure? tertiary structure?

22.8 Approximately how many amino acid residues are present in a protein with a molecular weight of 1,000,000?

22.9 Consider two different amino acids, which we will call A and B. If we prepare a dipeptide from these amino acids, how many arrangements

(AA, AB, etc.) are possible? If we prepare a polypeptide containing three amino acid residues, how many arrangements of these two amino acids (AAA, AAB, etc.) are possible? If x is the number of different amino acids used, and y is the number of amino acid residues in a polypeptide, what is the formula for the number of possible polypeptides?

22.10 What is a carbohydrate?

22.11 What is a monsaccharide, a disaccharide, and a polysaccharide?

22.12 Glyceraldehyde is 2,3-dihydroxypropanal. Draw the structure of the compound. How many asymmetric carbon atoms does the molecule contain? How many optical isomers of glyceraldehyde should exist?

22.13 The number of optical isomers of a compound is 2^n, where n is the number of asymmetric carbon atoms in the compound. How many optical isomers of 2,3,4-trihydroxybutanal should exist? How many optical isomers of 1,3,4-trihydroxy-2-butanone should exist?

22.14 Write an equation to show how a soap can be made from a trigylceride.

22.15 (a) Draw the structure of the triglyceride made from 3 mol of linolenic acid (see Table 22.1) and 1 mol of glycerol. (b) Write a chemical equation for the hydrogenation of this compound.

22.16 (a) Draw the structure of the trigylceride made from 3 mol of lauric acid (see Table 22.1) and 1 mol of glycerol. (b) Write a chemical equation for the combustion of this compound in oxygen.

22.17 What is the chemical difference between a fat and an oil?

22.18 If oleic acid, linoleic acid, and linolenic acid were hydrogenated completely, what saturated fatty acid would result? See Table 22.1.

22.19 What is the difference between an α-helix and a double helix?

22.20 Are the two chains of a DNA double helix identical? Explain.

22.21 Describe the process by which a DNA double helix is duplicated in cell division.

22.22 What is a mutation?

22.23 Assume that the average molecular weight of an RNA nucleotide is 340, and the average molecular weight of an amino acid residue found in a protein is 120. What must the minimum molecular weight of a messenger RNA molecule be if it is used to synthesize a protein that has a molecular weight of 90,000? The molecular weight is a minimum since some messenger RNA molecules are used to synthesize more than one protein.

22.24 The anticodon of a transfer RNA molecule that carries the amino acid serine is A-G-A. (a) What is the corresponding codon of messenger RNA? (b) What sequence of bases on the DNA chain (strand 1) would lead to the production of this codon on a messenger RNA molecule? (c) What sequence of bases on strand 2 of the DNA double helix would occur opposite the portion described in part (b)?

22.25 What is the turnover number of an enzyme?

22.26 Compare the action of competitive and noncompetitive inhibitors.

22.27 Describe how product inhibition and feedback inhibition work. Why are they important to the operation of the cell?

22.28 By what mechanism are enzymes thought to function? Why are they specific in their action?

22.29 What is the difference between catabolic processes and anabolic processes?

22.30 What are coupled reactions?

22.31 What purpose do high energy molecules serve? How do they function?

Chapter 23

Nuclear Chemistry

Ordinary chemical reactions involve only the extranuclear electrons. In such reactions the nucleus is important only insofar as it influences the electrons. However, matter does undergo important transformations that involve the nucleus directly. The study of nuclear transformations has enlarged our understanding of the nature of matter and the courses of many chemical and biological processes. Many technological applications have resulted from these investigations.

When Wilhelm Röntgen discovered X rays in 1895, he noticed that these invisible rays expose photographic plates and cause fluorescent salts to glow. Henri Becquerel in 1896 investigated the hypothesis that salts of this type emit invisible radiation independent of any external stimulus. Becquerel found that a double sulfate of potassium and uranium (that he happened to have on hand) emits radiation capable of exposing a photographic plate that is well protected from light. He subsequently identified uranium as the source of the "radioactive" rays. Following Becquerel's discovery, other radioactive elements were identified (notably by Marie and Pierre Curie). The nature of the rays was explained principally by the work of Ernest Rutherford. Radioactive rays originate from transformations that take place within the nucleus.

23.1 THE NUCLEUS

The nuclei of atoms are thought to contain protons and neutrons, particles that are collectively called **nucleons.** The number of protons in a specific atomic nucleus corresponds to the atomic number (or nuclear charge), Z, and the total number of nucleons is given by the mass number, A. Thus, the number of neutrons equals $(A-Z)$. This information is indicated on the chemical symbol of a given nuclide by appending the atomic number as a subscript and the mass number as a superscript. The symbols for the two naturally occurring isotopes of lithium are

$${}^{6}_{3}\mathrm{Li} \qquad \text{and} \qquad {}^{7}_{3}\mathrm{Li}$$

Much is known about the structure of the nucleus, but much more remains to be learned. Determination of the radii of a large number of atomic nuclei shows that the radius of a given nucleus, r, is directly related to the cube root of its mass number.

$$r = (1.3 \times 10^{-13}\ \mathrm{cm})A^{1/3}$$

The volume of a sphere is $\frac{4}{3}\pi r^3$, and if we assume a spherical nucleus, it follows that nuclear volume varies directly with mass number. In other words, the mass of a nucleus determines its volume. Nuclear density is therefore approximately constant for all atomic nuclei. This density is about 2.44×10^{14} g/cm^3, an amazingly high value.

The density of a liquid is constant and independent of the size of any drop considered. Since all nuclei have approximately the same density, a fluid-droplet model of the nucleus has been proposed. The nature of the cohesive forces holding the nucleons into a nuclear fluid is far from being completely understood.

It is clear, however, that powerful cohesive forces between nucleons exist and that they effectively overcome the electrostatic forces of repulsion between the protons of the nucleus. The attractive forces between nucleons are charge independent, that is, proton–proton, proton–neutron, and neutron–neutron attractions are identical. The range of the force is believed to be on the order of 2×10^{-13} cm, the diameter of a nucleon.

In 1935 Hidekei Yukawa postulated that neutrons and protons are bound together by the very rapid exchange of a nuclear particle, which was identified as a π meson, or pion. Three types of pions are known to exist: positive, π^+, negative, π^-, and neutral, π^0. According to the exchange theory, a neutron, n^0_A, is converted into a proton, p^+_A, by the emission of a negative pion. The emitted π^- is accepted by a proton, p_B^+, which is converted into a neutron, n_B^0.

$$n_A^0 \rightarrow p_A^+ + \pi^-$$
$$\pi^- + p_B^+ \rightarrow n_B^0$$

If we consider the deuterium nucleus, which consists of one proton and one neutron, this exchange then reverses. The neutron n_B^0 becomes a proton, p_B^+, and the proton p_A^+ becomes a neutron, n_A^0. The exchange is extremely rapid. Approximately 10^{24} transfers occur in 1 s.

The same type of interaction can be postulated for the positive pion,

$$p_A^+ \rightarrow n_A^0 + \pi^+$$
$$\pi^+ + n_B^0 \rightarrow p_B^+$$

or the neutral pion,

$$p_A^+ \rightarrow p_A^+ + \pi^0$$
$$\pi^0 + n_B^0 \rightarrow n_B^0$$

The exchange of a neutral pion can also be used to account for the existence of neutron–neutron and proton–proton forces.

In Figure 23.1 the number of neutrons is plotted against the number of protons for the naturally occurring, nonradioactive nuclei. The points, which represent stable combinations of protons and neutrons, lie in what may be called a **zone of stability.** Nuclei that have compositions represented by points that lie outside of this zone spontaneously undergo radioactive transformations that tend to bring their compositions into or closer to this zone (Section 23.2).

The stable nuclei of the lighter elements contain approximately equal numbers of neutrons and protons, a neutron/proton ratio of 1. The heavier nuclei contain more neutrons than protons. With increasing atomic number more and more protons are packed into a tiny nucleus, and the electrostatic forces of repulsion increase sharply. A larger and larger excess of neutrons is required to diminish the effect of these repulsion forces, and the neutron/proton ratio increases with increasing atomic number until the ratio is approximately 1.5 at the end of the curve of Figure 23.1. There appears to be an upper limit to the number of protons that can be packed into a nucleus, no matter how many neutrons are present. The largest stable nucleus is $^{209}_{83}Bi$. Nuclei that are larger than this exist, but all of them are radioactive.

Most naturally occurring stable nuclides have an even number of protons and an even number of neutrons. Only five ($^{2}_{1}H$, $^{6}_{3}Li$, $^{10}_{5}B$, $^{14}_{7}N$, and $^{180}_{73}Ta$) have an odd number of protons and an odd number of neutrons (Table 23.1). For each odd atomic number there are never more than two stable nuclides, whereas for an even atomic number, as many as ten stable nuclides may occur. The two elements of atomic number less than 83 that have never been proven to be naturally existing ($_{43}Tc$ and $_{61}Pm$) have odd atomic numbers.

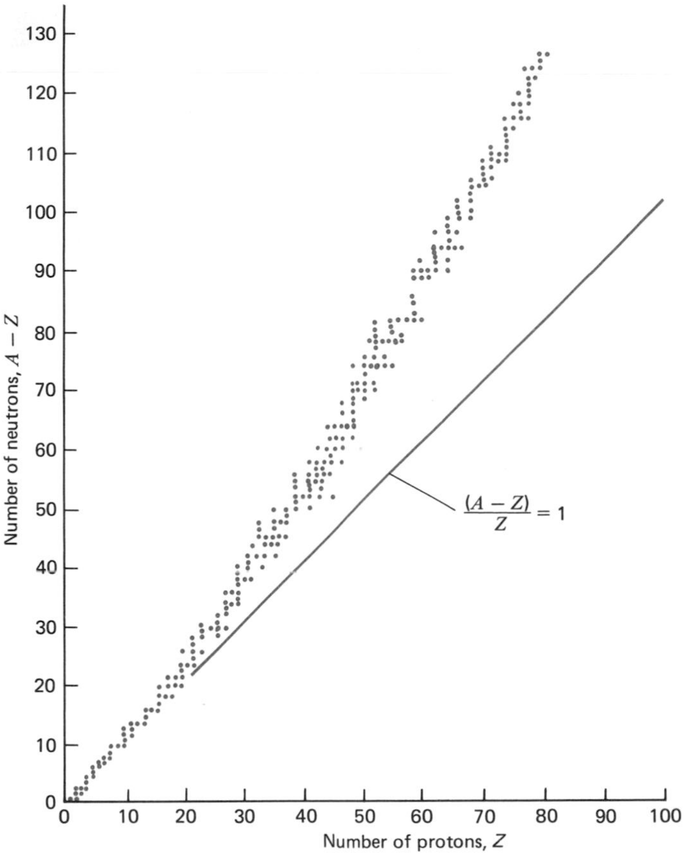

Figure 23.1 Neutron/proton ratio.

Empirical observations such as these suggest that there is a periodicity in nuclear structure similar to the periodicity of atomic structure. A nuclear shell model, as yet incompletely developed, has been suggested.

Periodic variations are observed in many nuclear properties. Certain nuclides have relatively high binding energies (which are indicative of comparatively great stability; Section 23.6) compared to nuclides of close atomic

Table 23.1 Distribution of naturally occurring stable nuclides

PROTONS	NEUTRONS	NUMBER OF NUCLIDES
even	even	157 } 209
even	odd	52 }
odd	even	50 } 55
odd	odd	5 }

number or mass number. Exceptional nuclear stability is also shown by certain nuclides that have a poor ability to capture neutrons (Section 23.5). Comparison of data from these and other studies of nuclear properties indicates that unusual nuclear stability is associated with nuclides having either a number of protons or a number of neutrons equal to a **magic number:** 2, 8, 20, 28, 50, 82, and 126. It is thought that the magic numbers indicate closed nuclear shells in the same way that the atomic numbers of the noble gases, 2, 10, 18, 36, 54, and 86, indicate stable electronic configurations. In general, elements that have an atomic number equal to a magic number have a larger number of stable isotopes than neighboring elements. The nuclides with a magic number of protons as well as a magic number of neutrons, ${}^{4}_{2}He$, ${}^{16}_{8}O$, ${}^{40}_{20}Ca$, and ${}^{208}_{82}Pb$, have notably high stabilities in terms of neutron capture (Section 23.5), binding energy (Section 23.6), and relative abundance.

23.2 RADIOACTIVITY

Unstable nuclei spontaneously undergo certain changes that result in the attainment of more stable nuclear compositions. Some unstable nuclides are naturally occurring, others are man-made. Certain synthetic radioactive nuclides undergo some types of radioactive decay that have not been observed for any naturally occurring unstable nuclide.

Alpha emission consists of the ejection of α particles, which have an atomic number of 2 and a mass number of 4 and may be considered to be ${}^{4}_{2}He$ nuclei. Both synthetic and natural nuclides undergo α decay, which is common only for nuclides of mass number greater than 209 and atomic number greater than 82. Nuclides such as these have too many protons for stability. Points indicating the composition of these nuclides fall outside the plot of Figure 23.1 to the upper right, beyond the zone of stability. The emission of α particles reduces the number of protons by two and the number of neutrons by two and adjusts the composition of the nucleus downward, closer to the stability zone.

An example of α decay is

$${}^{210}_{84}Po \rightarrow {}^{206}_{82}Pb + {}^{4}_{2}He$$

Notice that the equation indicates the conservation of mass number (superscripts) and atomic number (subscripts). Equations such as these are written to indicate nuclear changes only. The extranuclear electrons are customarily ignored. An α particle is emitted as a ${}^{4}_{2}He$ nucleus without electrons and with a 2+ charge. Subsequent to its emission, however, the particle attracts electrons from other atoms (which become cations), and the α particle becomes a neutral atom of ${}^{4}_{2}He$. The ${}^{206}_{82}Pb$ is left therefore with a surplus of two electrons and a 2− charge. These excess electrons are rapidly lost to surrounding cations.

The energy released in this process is readily calculated from the difference in mass between the reactant (the parent nucleus) and products (the daughter nucleus and the α particle). The masses of neutral *atoms*, rather than the masses of *nuclei,* are generally recorded, but this causes no trouble in the calculation. If the mass of the $^{210}_{84}Po$ *atom* is used, the mass of 84 electrons is included. For the products the mass of the $^{206}_{82}Pb$ *atom* includes the mass of 82 electrons, and the mass of the $^{4}_{2}He$ *atom* includes the mass of two electrons. Thus, the masses of the extranuclear electrons cancel when atomic masses are employed for the calculation.

$$(\text{mass } ^{210}_{84}\text{Po}) - (\text{mass } ^{206}_{82}\text{Pb} + \text{mass } ^{4}_{2}\text{He})$$
$$209.9829 \text{ u} - (205.9745 \text{ u} + 4.0026 \text{ u}) = 0.0058 \text{ u}$$

The energy equivalent of this mass difference may be calculated by means of Einstein's equation, $E = mc^2$. The energy unit customarily employed is the MeV (megaelectron volts, which is a million electron volts). An electron volt (eV) is the energy acquired by an electron when it is accelerated through a potential difference of 1 V. The energy equivalent of 1 u is 931.478 MeV, and therefore the energy released by the α decay of $^{210}_{84}Po$ is

$$0.0058 \text{ u} \times 931 \text{ MeV/u} = 5.4 \text{ MeV}$$

If the daughter nucleus in an α-decay process is left in the ground state, the kinetic energy of the emitted α particle accounts for most of the energy released, and the kinetic energy of the recoiling daughter nucleus accounts for the remainder. If all the daughter nuclei are left in the ground state, all the α particles emitted have the same energy. In some α-decay processes, however, several energy groups of α particles are emitted. In these cases the α particles of highest energy correspond to daughter nuclei in the ground state, and the α particles of lower energies originate from nuclei left in excited states. A nucleus in an excited state subsequently emits energy in the form of γ radiation to reach the ground state. The sum of the kinetic energy of the α particle, the recoil energy of the daughter nucleus, and the energy of the γ radiation equals the decay energy.

Gamma radiation is electromagnetic radiation of very short wavelength. Its emission is caused by energy changes within the nucleus. Its emission alone does not cause changes in the mass number or in the atomic number of the nucleus. At times, nuclides are produced in excited states by nuclear reactions (Section 23.5), and such nuclides revert to their ground states by the emission of the excess energy in the form of γ radiation.

$$\underset{\text{excited state}}{[^{125}_{52}\text{Te}]^*} \rightarrow \underset{\text{ground state}}{^{125}_{52}\text{Te}} + \gamma$$

The γ rays emitted by a specific nucleus have a definite energy value or set of energy values because they correspond to transitions between discrete energy levels of the nucleus. Thus, an emission spectrum of γ radiation is analogous to the line spectrum that results from transitions of electrons between energy levels in an excited atom.

Gamma radiation frequently accompanies all other types of radioactive decay. The following α-decay process is an example.

$$^{240}_{94}Pu \rightarrow [^{236}_{92}U]^* + ^{4}_{2}He$$
$$[^{236}_{92}U]^* \rightarrow ^{236}_{92}U + \gamma$$

In cases such as this, deductions can be made concerning the energy levels of the daughter nucleus. The emission of a 5.16 MeV α particle results directly in the production of $^{236}_{92}U$ in the ground state. No γ radiation accompanies such α particles. For the process outlined in the set of equations, however, a 5.12 MeV α particle is emitted along with a 0.04 MeV γ ray. It is assumed that the energy of the γ ray corresponds to the energy difference between the ground state and the first excited state of $^{236}_{92}U$. Alpha particles of other energies are also emitted, and the complete analysis of these energies results in a more detailed picture of the energy levels of the $^{236}_{92}U$ nucleus.

Beta emission is observed for nuclides that have too high a neutron/proton ratio for stability. Points representing such nuclides lie to the left of the zone of stability of Figure 23.1. The β particle is an electron, indicated by the symbol $_{-1}^{0}e$, which may be considered to result from the transformation of a nuclear neutron into a nuclear proton. Electrons, as such, do not exist in the nucleus. The net effect of β emission is that the number of neutrons is decreased by 1 and the number of protons is increased by 1. Thus, the neutron/proton ratio is decreased. The mass number does not change

Beta decay is a very common mode of radioactive disintegration and is observed for both natural and synthetic nuclides. Examples include

$$^{186}_{73}Ta \rightarrow ^{186}_{74}W + _{-1}^{0}e$$
$$^{82}_{35}Br \rightarrow ^{82}_{36}Kr + _{-1}^{0}e$$
$$^{27}_{12}Mg \rightarrow ^{27}_{13}Al + _{-1}^{0}e$$
$$^{14}_{6}C \rightarrow ^{14}_{7}N + _{-1}^{0}e$$

Notice that the sums of the subscripts and superscripts on the right side of the equation equal the subscript and superscript of the parent nucleus on the left.

The last equation represents the mode of decay of the radioactive carbon isotope that is present in small amount in the atmosphere. The energy released by this process may be calculated by using atomic masses instead of nuclear masses. If we add six orbital electrons to both sides of the equation, we have the equivalent of the *atomic* mass of $^{14}_{6}C$ (which includes the mass of six electrons) on the left and, taking into account the electron ejected as a β particle, the equivalent of the *atomic* mass $^{14}_{7}N$ (which includes the mass of seven electrons) on the right. Therefore,

$$(\text{mass } ^{14}_{6}C) - (\text{mass } ^{14}_{7}N)$$
$$14.00324 \text{ u} - 14.00307 \text{ u} = 0.00017 \text{ u} = 0.16 \text{ MeV}$$

In β decay the recoil energy of the daughter nucleus is negligible since the ejected electron has a small mass. One would expect that the decay energy would be taken up by the kinetic energy of the β particle and that all the β particles emitted would have energies corresponding to this value. Instead, however, a continuous spectrum of β-particle energies is observed with the highest energy almost equal to the decay energy. It is postulated that when a β particle of less than maximum energy is ejected, a **neutrino,** ν, that carries off the excess energy is emitted at the same time.

$$^{14}_{6}\text{C} \rightarrow {}^{14}_{7}\text{N} + {}^{0}_{-1}e + \nu$$

The neutrino is assumed to be a chargeless particle of a vanishingly small mass.

Unstable nuclides that have neutron/proton ratios below those required for stability (points below the zone of stability of Figure 23.1) do not occur in nature. Many such artificial nuclides are known, however. Two types of radioactive processes are observed that increase the neutron/proton ratio of this type of nuclide: positron emission and electron capture.

Positron emission, or β^+ emission, consists of the ejection of a positive electron, which is called a positron and indicated by the symbol ${}^{0}_{1}e$, from the nucleus. A positron has the same mass as an electron but an opposite charge; it arises from the conversion of a nuclear proton into a neutron. Positron emission results in a decrease of *one* in the number of protons and an increase of *one* in the number of neutrons; no change in mass number occurs. Hence, positron emission raises the numerical value of the neutron/proton ratio.

Examples of this mode of radioactive decay include

$$^{122}_{53}\text{I} \rightarrow {}^{122}_{52}\text{Te} + {}^{0}_{1}e$$
$$^{38}_{19}\text{K} \rightarrow {}^{38}_{18}\text{Ar} + {}^{0}_{1}e$$
$$^{23}_{12}\text{Mg} \rightarrow {}^{23}_{11}\text{Na} + {}^{0}_{1}e$$
$$^{15}_{8}\text{O} \rightarrow {}^{15}_{7}\text{N} + {}^{0}_{1}e$$

We can use atomic masses to calculate the energy released by the process described by the last equation. If we add eight orbital electrons to both sides of the equation, the equivalent of the atomic mass of ${}^{15}_{8}\text{O}$ (eight electrons) would be indicated on the left. The equivalent of the atomic mass of ${}^{15}_{7}\text{N}$ (seven electrons) plus the mass of one orbital electron (making up the eight added) plus the mass of the ejected positron would be indicated on the right. Thus, the mass difference is

$$(\text{mass } {}^{15}_{8}\text{O}) - (\text{mass } {}^{15}_{7}\text{N} + \text{mass } {}^{0}_{-1}e + \text{mass } {}^{0}_{1}e)$$

Since the mass of a positron is identical to that of an electron,

$$(\text{mass } {}^{15}_{8}\text{O}) - [\text{mass } {}^{15}_{7}\text{N} + 2(\text{mass } {}^{0}_{-1}e)]$$
$$15.00308 \text{ u} - [15.00011 \text{ u} + 2(0.00055 \text{ u})]$$
$$15.00308 \text{ u} - 15.00121 \text{ u} = 0.00187 \text{ u} = 1.74 \text{ MeV}$$

For spontaneous positron emission the atomic mass of the parent must exceed the atomic mass of the daughter by at least 0.00110 u, the mass of two electrons.

In positron emission a spectrum of β^+ energies is observed similar to that observed for β emission, and the simultaneous ejection of neutrinos is postulated.

$$^{15}_{8}\text{O} \rightarrow {}^{15}_{7}\text{N} + {}^{0}_{1}e + \nu$$

The positron was the first antiparticle to be observed. It is similar to the electron in all respects except charge. When a positron and an electron collide, they annihilate each other, and γ radiation equivalent to the masses of the two particles is produced. It is believed that antiparticles exist for all particles except the neutral pion, π^0, and the photon. The negative proton (antiproton) has been detected. The annihilation of proton with antiproton usually produces several high-energy pions. Since the antineutron, like the neutron, has no charge, its detection rests on annihilation phenomena.

Electron capture (ec), sometimes called *K* capture, is another process through which the neutron/proton ratio of an unstable, proton-rich nuclide may be increased. Unlike positron emission, electron capture can occur when the mass difference between parent and daughter does not exceed 0.00110 u. In this process the nucleus captures an orbital electron from the *K* or *L* shell, and the captured electron converts a nuclear proton into a neutron. The transformation results in a daughter nuclide with *one* less proton and *one* more neutron than the parent. Consequently, the atomic number of the daughter is *one* less than that of the parent, and the mass number does not change. Examples are

$$^{0}_{-1}e + {}^{197}_{80}\text{Hg} \xrightarrow{\text{ec}} {}^{197}_{79}\text{Au}$$

$$^{0}_{-1}e + {}^{106}_{47}\text{Ag} \xrightarrow{\text{ec}} {}^{106}_{46}\text{Pd}$$

$$^{0}_{-1}e + {}^{37}_{18}\text{Ar} \xrightarrow{\text{ec}} {}^{37}_{17}\text{Cl}$$

$$^{0}_{-1}e + {}^{7}_{4}\text{Be} \xrightarrow{\text{ec}} {}^{7}_{3}\text{Li}$$

$$^{0}_{-1}e + {}^{55}_{26}\text{Fe} \xrightarrow{\text{ec}} {}^{55}_{25}\text{Mn}$$

The energy released can be calculated directly from the atomic masses of the parent and daughter. Thus, for the last example,

$$(\text{mass } {}^{55}_{26}\text{Fe}) - (\text{mass } {}^{55}_{25}\text{Mn})$$
$$54.93830 \text{ u} - 54.93805 \text{ u} = 0.00025 \text{ u} = 0.23 \text{ MeV}$$

The recoil energy of the daughter nucleus is negligible, and if the daughter nucleus is left in the ground state, all the available energy is carried away by a neutrino that is ejected in the process. In addition, electron capture is accompanied by the production of X rays. The capture of an orbital electron leaves a vacancy in the *K* or *L* shell, and when an outer electron falls into this vacancy, X-ray emission follows (Section 5.5).

23.3 RATE OF RADIOACTIVE DECAY

Many techniques are employed to study the emissions of radioactive substances. Radiations from these materials affect photographic film in the same way that ordinary light does. Photographic techniques for the qualitative and quantitative detection of radiation are employed, but they are not very accurate nor are they suitable for rapid analysis.

The energy of emissions from radioactive sources is absorbed by some materials (for example, zinc sulfide) and transformed into radiant energy of visible wavelength. The zinc sulfide is said to **fluoresce,** and a little flash of light may be observed from the impact of each particle from the radioactive source. This property has been put to use in an instrument known as a **scintillation counter.** The window of a sensitive photoelectric tube is coated with ZnS, and the flash of light emitted by the ZnS when it is struck by a particle causes a pulse of electric current to pass through the photoelectric tube. These signals are amplified and made to operate various kinds of counting devices.

The **Wilson cloud chamber** enables the path of ionizing radiation to be seen. The chamber contains air saturated with water vapor. By the movement of a piston, the air in the chamber is suddenly expanded and cooled. This causes droplets of water to condense on the ions that are formed by the particles as they move through the vapor, and makes the paths of the particles visible. Photographs of these cloud tracks may be made and studied. Such photographs provide information on the length of the paths, collisions undergone by the particles, the speed of the particles, and the effects of external forces on the behavior of the particles.

The essential features of a **Geiger-Müller counter** are diagrammed in Figure 23.2. The radiation enters the tube through a thin window. As a particle or a γ ray traverses the tube, which contains argon gas, it knocks electrons off the argon atoms in its path and forms Ar^+ ions. A potential of about 1000 to 1200 V is applied between the electrodes of the tube, and the electrons and

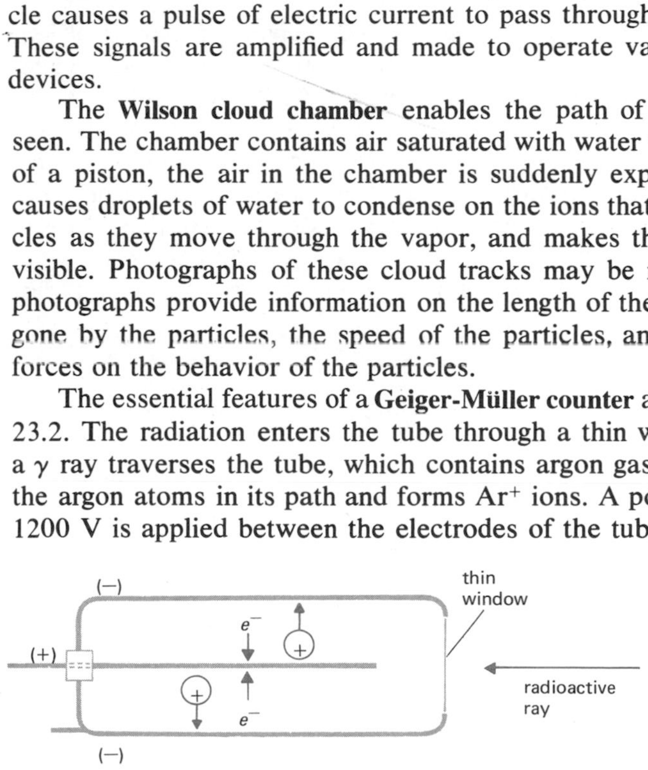

Figure 23.2 Essential features of a Geiger-Müller counter tube.

Ar^+ ions cause a pulse of electric current to flow through the circuit. This pulse is amplified and causes a clicker to sound or an automatic counting device to operate.

The rates of decay of all radioactive substances have been found to be first order (Section 13.6) and to be independent of temperature. This lack of temperature dependence implies that the activation energy of any radioactive-decay process is zero. The rate of decay, therefore, depends only upon the amount of radioactive material present. If we let N equal the number of atoms of radioactive material, ΔN is the number of atoms that disintegrate in a time interval, Δt,

$$-\frac{\Delta N}{\Delta t} = kN$$

where k is the rate constant. The rate expression is negative because it represents the disappearance of the radioactive substance.

Rearrangement of the rate expression gives

$$-\frac{\Delta N}{N} = k\Delta t$$

which states that the fraction lost ($-\Delta N/N$) in a given time interval (Δt) is directly proportional to the length of the interval. The time required for half of the sample to decay depends upon a time interval known as the **half-life,** $t_{1/2}$ which is a constant.

The curve of Figure 23.3, which shows number of radioactive atoms versus time, is typical of first-order processes. Let N_0 equal number of radioactive atoms present at the start. After a single half-life period has elapsed, one-half of the original number of atoms remain ($\frac{1}{2}N_0$). This number is reduced by half (to $\frac{1}{4}N_0$) by the time another half-life has passed. Each radio-

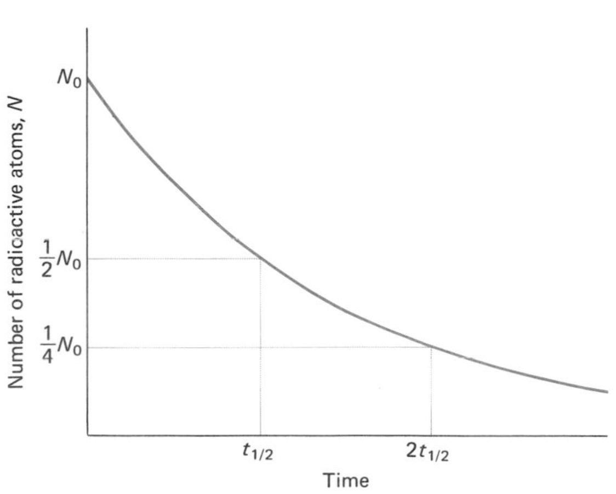

Figure 23.3 Curve showing the rate of a radioactive decay.

active isotope has a characteristic half-life, and these vary widely. For example, $^{5}_{3}Li$ has a half-life estimated to be 10^{-21} s, and $^{238}_{92}U$ has a half-life of 4.51×10^9 years.

The relationship between the number of radioactive atoms present at the start (N_0), and the number (N) present at time t is

$$\log\left(\frac{N_0}{N}\right) = \frac{kt}{2.303}$$

After a half-life period the number of radioactive atoms left is equal to $\frac{1}{2} N_0$.

$$N = \tfrac{1}{2} N_0$$

Therefore,

$$\frac{N_0}{N} = 2$$

Substitution gives

$$\log\left(\frac{N_0}{N}\right) = \frac{kt}{2.303}$$

$$\log 2 = \frac{kt_{1/2}}{2.303}$$

$$t_{1/2} = \frac{2.303 \log 2}{k}$$

$$t_{1/2} = \frac{0.693}{k}$$

According to the rate law, a radioactive substance never completely disappears. The rate law describes the behavior of a sample containing a large number of atoms. When the number of atoms in the sample declines to a very small value, the rate law is no longer followed exactly. How can a fraction of an atom disintegrate?

Example 23.1 The isotope $^{60}_{27}Co$ has a half-life of 5.27 years. What amount of a 0.0100 g sample of $^{60}_{27}Co$ remains after 1.00 year?
Solution The rate constant for this disintegration is

$$k = \frac{0.693}{t_{1/2}}$$

$$= \frac{0.693}{5.27 \text{ years}} = 0.132/\text{year}$$

The fraction remaining undecomposed at the end of 1.00 year may be found in the following way.

$$\log\left(\frac{N_0}{N}\right) = \frac{kt}{2.30}$$

$$= \frac{(0.132/\text{year})(1.00\ \text{year})}{2.30} = 0.0573$$

$$\log\left(\frac{N}{N_0}\right) = -0.0573$$

$$\frac{N}{N_0} = 0.876$$

Therefore, the amount remaining after 1.00 year is

$$0.876 \times 0.0100\ \text{g} = 0.00876\ \text{g}$$

The radioactive isotope $^{14}_{6}C$ is produced in the atmosphere by the action of cosmic-ray neutrons on $^{14}_{7}N$.

$$^{14}_{7}N + ^{1}_{0}n \rightarrow ^{14}_{6}C + ^{1}_{1}H$$

The $^{14}_{6}C$ is oxidized to CO_2, and this radioactive CO_2 mixes with nonradioactive CO_2. The radiocarbon disappears through radioactive decay, but it is also constantly being made. The steady state is reached when the proportion is one atom of $^{14}_{6}C$ to 10^{12} ordinary carbon atoms, which represents 15.3 ± 0.1 disintegrations per minute per gram of carbon.

The CO_2 of the atmosphere is absorbed by plants through the process of photosynthesis, and the ratio of $^{14}_{6}C$ to ordinary carbon in plant materials that are alive and growing is the same ratio as that in the atmosphere. When the plant dies, however, the amount of $^{14}_{6}C$ diminishes through radioactive decay and is not replenished by the assimilation of atmospheric CO_2 by the plant.

The half-life of $^{14}_{6}C$, a β emitter, is 5770 years. The age of a wooden object can be determined by comparing the radiocarbon activity of the object with that of growing trees. This method of **radiocarbon dating** has been applied to many archeological finds and objects of historical interest. By this means the Dead Sea scrolls were determined to be 1917 ± 200 years old.

Example 23.2 A sample of carbon from a wooden artifact is found to give 7.00 $^{14}_{6}C$ counts per minute per gram of carbon. What is the approximate age of the artifact?

Solution The half-life of $^{14}_{6}C$ is 5770 years. Therefore,

$$k = \frac{0.693}{t_{1/2}} = \frac{0.693}{5770\ \text{years}} = 1.20 \times 10^{-4}/\text{year}$$

The $^{14}_{6}C$ from wood recently cut down decays at the rate of 15.3 disintegrations per minute per gram of carbon. Therefore,

$$\log\left(\frac{N_0}{N}\right) = \frac{kt}{2.30}$$

$$\log\left(\frac{15.3 \text{ disintegrations/min}}{7.00 \text{ disintegrations/min}}\right) = \frac{(1.20 \times 10^{-4}/\text{year})t}{2.30}$$

$$t = \frac{2.30 \log 2.19}{1.20 \times 10^{-4}/\text{year}}$$

$$= 6520 \text{ years}$$

23.4 RADIOACTIVE DISINTEGRATION SERIES

The examples of α, β^-, and β^+ emission and of electron capture given in Section 23.2 are one-step processes that lead to stable nuclides. Frequently, however, the daughter nucleus produced by a radioactive process is itself radioactive. The repetition of this situation creates a chain, or series, of disintegration processes involving many radioactive nuclides and leading ultimately to the production of a stable nuclide.

Three such **disintegration series,** which involve only α and β emission, occur in nature: the $^{232}_{90}Th$, $^{238}_{92}U$, and $^{235}_{92}U$ series. The $^{238}_{92}U$ series, which leads finally to the stable nuclide $^{206}_{82}Pb$, is diagrammed in Figure 23.4. In several places in the series branching occurs, and the series proceeds by two different routes. The branches, however, always rejoin at a later point, and commonly one branch is preferred over the other (note the percentage figures in Figure 23.4). By any given route the $^{238}_{92}U$ series consists of 14 steps—8 involving α decay and 6 involving β decay.

The radioactive nuclides that occur in nature are those that have very long half-lives or those that are constantly being produced by the disintegration of other nuclides. This last type of naturally occurring radioactive nuclide eventually exists in a steady state at which time the amount of the nuclide remains essentially constant because the material is being produced at the same rate that it is decomposing.

The natural disintegration series serve as the basis of a method of geological dating. A sample of rock may be analyzed for its $^{206}_{82}Pb$ and $^{238}_{92}U$ content, and the length of time to produce this ratio of lead to uranium may be calculated from the decay constants of the series. Results from studies such as these place the age of some rocks at from 3,000 to 3,500 million years. The age of the earth is estimated to be 4,000 to 5,000 million years.

Decay series, some more elaborate than others, are known for artificial nuclides. Positron emission and K capture are observed in some of these series as well as α and β emission. Following are three examples of simple

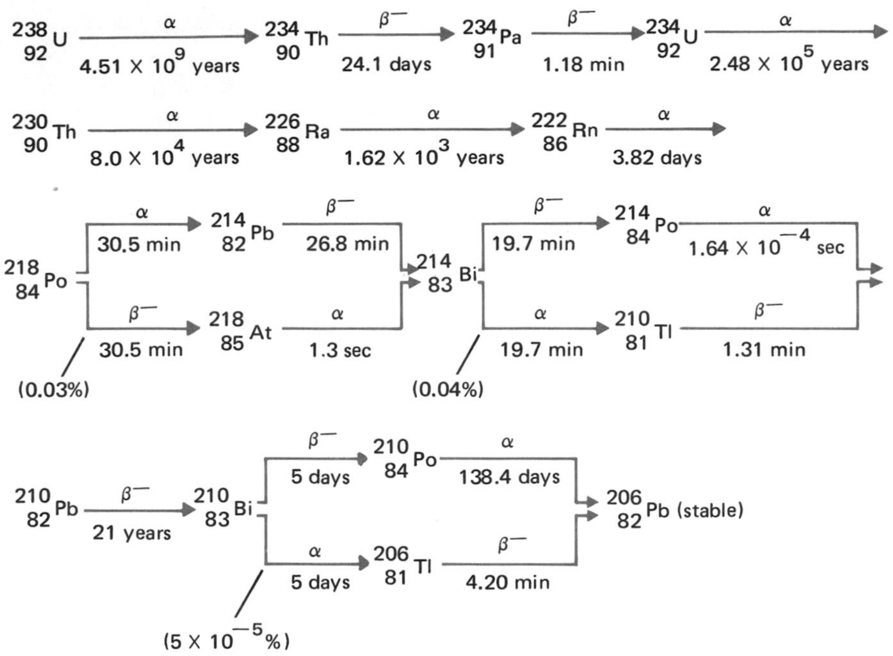

Figure 23.4 Disintegration series of $^{238}_{92}U$. Half-lives of isotopes are indicated.

two-step disintegration series of artificial nuclides. The half-lives of the nuclides are indicated.

$$^{20}_{8}O \xrightarrow[14\ s]{\beta^-} {}^{20}_{9}F \xrightarrow[11\ s]{\beta^+} {}^{20}_{10}Ne$$

$$^{30}_{16}S \xrightarrow[1.4\ s]{\beta^+} {}^{30}_{15}P \xrightarrow[2.6\ min]{\beta^+} {}^{30}_{14}Si$$

$$^{76}_{36}Kr \xrightarrow[10\ hours]{ec} {}^{76}_{35}Br \xrightarrow[16.5\ hours]{\beta^+ \text{ or ec}} {}^{76}_{34}Se$$

23.5 NUCLEAR REACTIONS

In 1915 Ernest Rutherford reported that the following transformation occurs when α particles from $^{214}_{84}Po$ are passed through nitrogen.

$$^{14}_{7}N + {}^{4}_{2}He \rightarrow {}^{17}_{8}O + {}^{1}_{1}H$$

This was the first artificial transmutation of one element into another to be reported. In the years following, thousands of such nuclear transformations have been studied. It is assumed that the projectile (in this case, an α particle) forms a compound nucleus with the target ($^{14}_{7}N$) and that the compound nucleus very rapidly ejects a subsidiary particle ($^{1}_{1}H$) to form the product nucleus ($^{17}_{8}O$).

Other projectiles, such as neutrons, deuterons (^2_1H), protons, and ions of low atomic number, are used in addition to α particles.

Particle–particle reactions are usually classified according to the type of projectile employed and the subsidiary particle ejected. Thus, the preceding reaction is called an (α, p) reaction, and the complete transformation is indicated by the notation $^{14}_{7}N\ (\alpha, p)\ ^{17}_{8}O$. Examples of several of the more common types of nuclear transformations are listed in Table 23.2.

The first artificial, radioactive nuclide produced was made by the (α, n) reaction

$$^{27}_{13}Al + ^{4}_{2}He \rightarrow ^{30}_{15}P + ^{1}_{0}n$$

The product, $^{30}_{15}P$, decays by positron emission,

$$^{30}_{15}P \rightarrow ^{30}_{14}Si + ^{0}_{1}e$$

Except for the nature of the product there is no difference between nuclear reactions that produce stable nuclides and those that yield radioactive nuclides.

Projectile particles that bear a positive charge are repelled by target nuclei. This is particularly true of the heavier nuclei, which have high charges. Consequently, only a small number of nuclear transformations can be brought about by the positive particles emitted by radioactive sources. Various particle accelerators are used to give protons, deuterons, α particles, and other cationic projectiles sufficiently high kinetic energies to overcome the electrostatic repulsions of the target nuclei. The **cyclotron** (Figure 23.5) is one such instrument.

The ion source is located between two hollow D-shaped plates (D_1 and D_2), called *dees,* that are separated by a gap. The dees are enclosed in an evacuated chamber located between the poles of a powerful electromagnet (not shown in the figure). A high-frequency generator keeps the dees op-

Table 23.2 Examples of nuclear reactions

TYPE	REACTION	RADIOACTIVITY OF PRODUCT NUCLIDE
(α, n)	$^{75}_{33}As + ^{4}_{2}He \rightarrow ^{78}_{35}Br + ^{1}_{0}n$	β^+
(α, p)	$^{106}_{46}Pd + ^{4}_{2}He \rightarrow ^{109}_{47}Ag + ^{1}_{1}H$	stable
(p, n)	$^{7}_{3}Li + ^{1}_{1}H \rightarrow ^{7}_{4}Be + ^{1}_{0}n$	ec
(p, γ)	$^{14}_{7}N + ^{1}_{1}H \rightarrow ^{15}_{8}O + \gamma$	β^+
(p, α)	$^{9}_{4}Be + ^{1}_{1}H \rightarrow ^{6}_{3}Li + ^{4}_{2}He$	stable
(d, p)	$^{31}_{15}P + ^{2}_{1}H \rightarrow ^{32}_{15}P + ^{1}_{1}H$	β^-
(d, n)	$^{209}_{83}Bi + ^{2}_{1}H \rightarrow ^{210}_{84}Po + ^{1}_{0}n$	α
(n, γ)	$^{59}_{27}Co + ^{1}_{0}n \rightarrow ^{60}_{27}Co + \gamma$	β^-
(n, p)	$^{45}_{21}Sc + ^{1}_{0}n \rightarrow ^{45}_{20}Ca + ^{1}_{1}H$	β^-
(n, α)	$^{27}_{13}Al + ^{1}_{0}n \rightarrow ^{24}_{11}Na + ^{4}_{2}He$	β^-

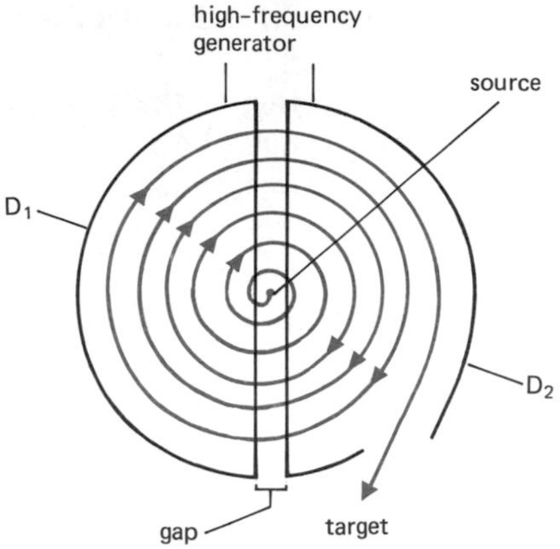

Figure 23.5 Path of a particle in a cyclotron.

positely charged. Under the influence of the magnetic and electrical fields the ions move from the source in a circular path. Each time they reach the gap between the dees, the polarity of the dees is reversed. Thus, the positively charged particles are pushed out of a positive dee and attracted into a negative dee. Each time they traverse the gap, therefore, they are accelerated. Because of this, the particles travel an ever-increasing spiral path. Eventually they penetrate a window in the instrument and, moving at extremely high speed, impinge on a target.

The **linear accelerator** (Figure 23.6) operates in much the same way except that no magnetic field is employed. The particles are accelerated through a series of tubes enclosed in an evacuated chamber. A positive ion from the source is attracted into tube 1, which is negatively charged. At this time, the odd-numbered tubes have negative charges and the even-numbered tubes have positive charges. As the particle emerges from tube 1, the charges of the tubes are reversed so that the even-numbered tubes are now negatively charged. The particle is repelled out of tube 1 (now positive) and attracted into tube 2 (now negative). As a result, it is accelerated.

Each time the particle leaves one tube to enter another, the charges of the tubes are reversed. Since the polarity of the tubes is reversed at a constant time interval and since the speed of the particle increases constantly, each tube must be longer than the preceding one. The accelerated particles leave the last tube at high speed and strike the target.

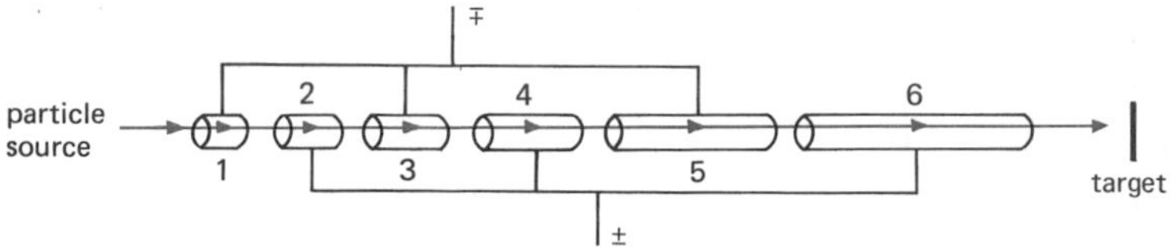

Figure 23.6 Schematic representation of a linear accelerator.

Neutrons are particularly important projectiles because they bear no charge and therefore are not repelled by the positive charge of the target nuclei. A mixture of beryllium and an α emitter (such as $^{222}_{86}Rn$) is a convenient neutron source.

$$^{9}_{4}Be + ^{4}_{2}He \rightarrow ^{12}_{6}C + ^{1}_{0}n$$

This reaction was used by James Chadwick in his experiments that characterized the neutron (1932). The bombardment of beryllium by accelerated deuterons from a cyclotron is a more intense source of neutrons.

$$^{9}_{4}Be + ^{2}_{1}H \rightarrow ^{10}_{5}B + ^{1}_{0}n$$

A very important source of neutrons is the nuclear reactor (Section 23.6).

Neutrons from nuclear reactions are known as **fast neutrons.** They cause reactions in which a subsidiary particle is ejected [such as (n, α) and (n, p) reactions]. **Slow neutrons,** or **thermal neutrons,** are produced when the neutrons derived from a nuclear reaction are passed through a moderator (such as carbon, paraffin, hydrogen, deuterium, or oxygen). Through collisions with the nuclei of the moderator, the kinetic energies of the neutrons are decreased to values approximating those of ordinary gas molecules. Bombardments using slow neutrons bring about (n, γ) reactions, which are also called **neutron-capture reactions** since no subsidiary particle is ejected.

$$^{34}_{16}S + ^{1}_{0}n \rightarrow ^{35}_{16}S + \gamma$$

Isotopes of practically every element have been prepared by this type of reaction.

Nuclear reactions have been used to prepare isotopes belonging to elements that do not exist in nature or that exist in extremely minute concentrations. Thus, isotopes of technetium and astatine have been prepared by the reactions

$$^{96}_{42}Mo + ^{2}_{1}H \rightarrow ^{97}_{43}Tc + ^{1}_{0}n$$
$$^{209}_{83}Bi + ^{4}_{2}He \rightarrow ^{211}_{85}At + 2^{1}_{0}n$$

The elements following uranium in the periodic classification are called the **transuranium elements.** None of these elements are naturally occurring, but many have been made by nuclear reactions. Some of these reactions use targets of artificial nuclides, and in these cases the final products are therefore the result of syntheses consisting of several steps. Examples of these preparations are

$$^{238}_{92}U + ^{1}_{0}n \rightarrow ^{239}_{92}U + \gamma$$
$$^{239}_{92}U \rightarrow ^{239}_{93}Np + ^{0}_{-1}e$$
$$^{239}_{93}Np \rightarrow ^{239}_{94}Pu + ^{0}_{-1}e$$
$$^{239}_{94}Pu + ^{2}_{1}H \rightarrow ^{240}_{95}Am + ^{1}_{0}n$$
$$^{239}_{94}Pu + ^{4}_{2}He \rightarrow ^{242}_{96}Cm + ^{1}_{0}n$$

In addition to the common projectiles, ions of elements of low atomic number are used in some bombardment reactions.

$$^{238}_{92}U + ^{12}_{6}C \rightarrow ^{244}_{98}Cf + 6\ ^{1}_{0}n$$

$$^{238}_{92}U + ^{14}_{7}N \rightarrow ^{246}_{99}Es + 6\ ^{1}_{0}n$$

$$^{238}_{92}U + ^{16}_{8}O \rightarrow ^{250}_{100}Fm + 4\ ^{1}_{0}n$$

$$^{252}_{98}Cf + ^{10}_{5}B \rightarrow ^{257}_{103}Lr + 5\ ^{1}_{0}n$$

23.6 NUCLEAR FISSION AND FUSION

Nuclear fission reactions are more famous and infamous than particle-particle or particle-capture transformations. In a **fission** process a heavy nucleus is split into nuclei of lighter elements and several neutrons. The fission of a given nuclide results in more than one set of products. Two possible reactions for the slow neutron induced fission of $^{235}_{92}U$ are

$$^{235}_{92}U + ^{1}_{0}n \rightarrow ^{93}_{36}Kr + ^{140}_{56}Ba + 3\ ^{1}_{0}n$$

$$^{235}_{92}U + ^{1}_{0}n \rightarrow ^{90}_{38}Sr + ^{144}_{54}Xe + 2\ ^{1}_{0}n$$

Heavy nuclei have much larger neutron/proton ratios than nuclei of moderate mass (Figure 23.1). The neutrons released in the fission process lower the neutron/proton ratios of the product nuclei. Nevertheless, these nuclei are generally radioactive, and further adjustment of the neutron/proton ratios occurs, usually through β emission. Certain fissions can be induced by protons, deuterons, or α particles, but the most important are those that are brought about by neutrons.

The **binding energy** of a nucleus may be considered to be the energy *required* to pull the nucleons of the nucleus apart or the energy *released* by the hypothetical formation of the nucleus by the condensation of the individual nucleons. The binding energy of a nucleus may be calculated from the difference between the sum of the masses of the constituent nucleons and the mass of the corresponding nucleus. For $^{35}_{17}Cl$ this mass difference is 0.320 u, and therefore

$$0.320 \text{ u} \times 931 \text{ MeV/u} = 298 \text{ MeV}$$

The magnitude of the binding energy of a given nucleus indicates the stability of that nucleus toward radioactive decay. For the purpose of comparison, the values are usually given in terms of binding energy per nucleon, and the largest values are characteristic of the most stable nuclei. The binding energy per nucleon in the case of $^{35}_{17}Cl$ is

$$\frac{298 \text{ MeV}}{35 \text{ nucleons}} = 8.5 \text{ MeV/nucleon}$$

In Figure 23.7 mass number is plotted against binding energy per nucleon for the nuclides. Inspection of the curve shows that nuclides of intermediate mass have larger values of binding energy per nucleon than the heavier nuclides. Thus, the fission of $^{235}_{92}U$ produces lighter nuclei with higher binding energy/nucleon values, and energy is liberated. The sum of the masses of the products of a fission reaction is less than the sum of the masses of the reactants. A typical fission of a single $^{235}_{92}U$ nucleus releases approximately 200 MeV.

The fission of each $^{235}_{92}U$ nucleus is induced by a single neutron and produces several neutrons. In the overall process an average of approximately 2.5 neutrons are produced per fission. If each neutron causes the fission of another $^{235}_{92}U$ nucleus, an explosively rapid chain reaction ensues. Using a reproduction factor of 2 for simplicity, the first fission would cause 2 first-generation fissions. Each of these would cause 2 fissions—a total of 4 second-generation fissions. These would be followed by 8, 16, 32, . . . , fissions in succeeding generations. In the *n*th generation, 2^n fissions would occur. Since each fission is extremely rapid, an explosion results. If 200 MeV is released by each fission, a tremendous amount of energy is liberated.

In a small amount of $^{235}_{92}U$ undergoing fission, many of the neutrons produced are lost from the surface of the mass before they can bring about nuclear reactions. If the size of the fissionable material exceeds a certain **critical mass,** however, the neutrons are captured before they can leave the

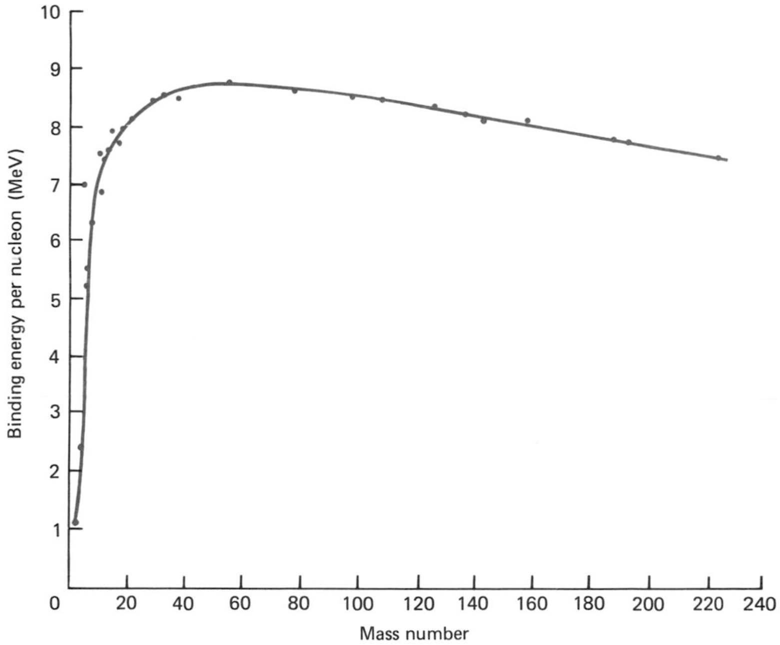

Figure 23.7 Binding energy per nucleon versus mass number.

mass and an explosive chain reaction results. An atomic bomb is detonated by bringing together two pieces of fissionable material, each of subcritical size, together into one piece of supercritical size. The reaction is started by a stray neutron.

A number of nuclei, such as $^{238}_{92}U$, $^{231}_{91}Pa$, $^{237}_{93}Np$, and $^{232}_{90}Th$, undergo fission with fast, but not with slow, neutrons. Slow neutrons, however, can induce fission in $^{233}_{92}U$, $^{235}_{92}U$, and $^{239}_{94}Pu$. The latter two isotopes are those commonly employed as nuclear fuels. The isotope $^{235}_{92}U$, which constitutes only about 0.7% of natural uranium, was employed in the construction of the first atomic bomb. The separation of $^{235}_{92}U$ from the principal uranium isotope, $^{238}_{92}U$, was carried out in many ways, the most successful being the separation of gaseous $^{235}UF_6$ from $^{238}UF_6$ by thermal diffusion through porous barriers (Section 4.11). The isotope $^{239}_{94}Pu$ does not occur in nature but may be prepared from $^{238}_{92}U$ by the series of nuclear reactions given in Section 23.5.

The controlled fission of nuclear fuels in a **nuclear reactor** serves as a source of energy or as a source of neutrons and γ radiation for scientific research and the production of artificial isotopes. In a nuclear reactor, cylinders containing at least a critical mass of nuclear fuel ($^{235}_{92}U$, $^{233}_{92}U$, $^{239}_{94}Pu$, natural U, or natural U enriched with $^{235}_{92}U$) are surrounded by a moderator (graphite, water, or heavy water). The moderator serves to slow down the neutrons produced by the fission, thus increasing the possibility of their capture.

The neutron reproduction factor must be maintained at a value close to 1. If it falls below this value, the reaction will eventually stop. If it increases much above this value, the reaction may become explosively violent. Control rods of cadmium, boron steel, or other materials, which may be inserted into the reactor to any desired depth, serve to capture excess neutrons and control the rate of the reaction.

$$^{113}_{48}Cd + ^{1}_{0}n \rightarrow ^{114}_{48}Cd + \gamma$$
$$^{10}_{5}B + ^{1}_{0}n \rightarrow ^{11}_{5}B + \gamma$$

The $^{114}_{48}Cd$ and $^{11}_{5}B$ isotopes are not radioactive. Provision is made for the insertion of research samples into the reactor.

Early reactors that used natural uranium as a fuel and graphite blocks as a moderator were large—cubes on the order of 7m on an edge. The use of enriched fuels and moderators other than graphite has reduced the size required considerably. Breeder reactors produce a fissionable isotope (for example, $^{239}_{94}Pu$ from $^{238}_{92}U$) within the reactor and thus provide for the maintenance of a supply of nuclear fuel. Nuclear reactors that are designed as sources of energy employ a circulating coolant to remove the heat from the reactor to the outside where it is used in power production. All nuclear reactors must be heavily shielded.

Nuclear **fusion** is a process in which very light nuclei are fused into a heavier nucleus. The curve of Figure 23.7 shows that such processes should liberate more energy than fission processes. The hydrogen bomb is based on nuclear fusion. Extremely high activation energies are required for fusion reactions, and in actual practice, a fission bomb is used to supply the high temperatures required. The energy of the sun is believed to be derived from the conversion of hydrogen nuclei into helium by nuclear fusion. Reactions such as the following are postulated.

$$ {}^{1}_{1}\mathrm{H} + {}^{1}_{1}\mathrm{H} \rightarrow {}^{2}_{1}\mathrm{H} + {}^{0}_{1}e $$

$$ {}^{2}_{1}\mathrm{H} + {}^{1}_{1}\mathrm{H} \rightarrow {}^{3}_{2}\mathrm{He} + \gamma $$

$$ {}^{3}_{2}\mathrm{He} + {}^{3}_{2}\mathrm{He} \rightarrow {}^{4}_{2}\mathrm{H} + 2\,{}^{1}_{1}\mathrm{H} $$

The liquid drop model of the nucleus interprets fusion and fission in terms of the forces of repulsion and attraction between the nucleons in the nucleus. The size of the nucleus is critical in determining the total effect of these two opposing forces. The nucleons on the surface of the nucleus are subjected to fewer forces of attraction than the nucleons that are surrounded on all sides by other nucleons in the center of the nucleus. Therefore, the total effect of the attractive forces is increased when the surface area is decreased. One large drop has a smaller surface area than several small drops comprising the same amount of nuclear fluid.

On the other hand, the electrostatic forces of repulsion between protons tend to disrupt a large drop into smaller ones. Whether the process of fusion or fission occurs depends upon the balance between these two effects, which do not vary with the size of the drop in a parallel manner. Since the surface area of a sphere is $4\pi r^2$, the increase or decrease in the energy of attraction brought about by a change in surface area is proportional to the square of the radius, r^2 (or $A^{2/3}$). The coulombic forces of repulsion, however, are proportional to the square of the nuclear charge divided by the radius, Z^2/r (or $Z^2/A^{1/3}$). For light nuclei the increase in the effect of the attractive forces brought about by the reduction in nuclear surface area more than offsets the increase in the effect of the forces of repulsion, and fusion occurs. For heavy nuclei the reverse applies. Fission occurs since it leads to a reduction in the effect of the forces of repulsion that is of more significance than the accompanying reduction in the effect of the forces of attraction.

23.7 USES OF ISOTOPES

A large number of uses have been developed for the isotopes that are the products of the processes run in nuclear reactors. Thickness gauges have been developed in which a radioactive source is placed on one side of the material to be tested (cigarettes, metal plates, and so forth) and a counting device on the other. The amount of radiation reaching the counter is a measure of the thickness of the material.

The effectiveness of lubricating oils is measured in an engine constructed from metal into which radioactive isotopes have been incorporated. After the engine has been run for a fixed time period, the oil is withdrawn and tested for the presence of radioactive particles accumulated through engine wear.

When a single pipeline is used to transfer more than one petroleum derivative, a small amount of a radioactive isotope is placed in the last portion of one substance to signal its end and the start of another. The radiation may be used to activate an automatic valve system so that the liquids are diverted into different tanks.

The use of radiocarbon in determining the age of materials of plant origin has been mentioned. Photosynthesis has been studied by tracing the absorption of CO_2 containing ${}^{14}_{6}C$ by plants. In this way much has been learned about the conversion of CO_2 into sugars, starches, and cellulose.

Radioactive isotopes have found many uses in medical research, therapy, and diagnosis. The radiations from ${}^{60}_{27}Co$, a β and γ emitter, are used in cancer therapy. The isotope ${}^{131}_{53}I$ is used in the diagnosis and treatment of thyroid disorders since this gland concentrates ingested iodine.

Radioactive tracers have found wide use in chemical studies. The following structure of the thiosulfate ion

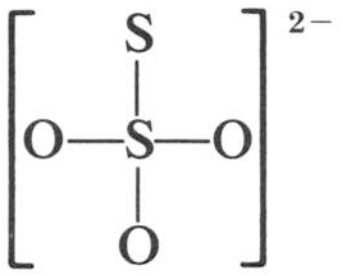

is indicated by studies using ${}^{35}_{16}S$. The thiosulfate ion is prepared by heating a sulfite with the radioactive sulfur.

$$^{35}S(s) + SO_3^{2-}(aq) \rightarrow [^{35}SSO_3]^{2-}(aq)$$

Upon the acidification of this thiosulfate solution, the radioactive sulfur is quantitatively precipitated. None is found in the resulting SO_2 gas.

$$[^{35}SSO_3]^{2-}(aq) + 2H^+(aq) \rightarrow {}^{35}S(s) + SO_2(g) + H_2O$$

In addition, upon the decomposition of silver thiosulfate derived from this ion, all the radioactive sulfur ends in the silver sulfide.

$$Ag_2[^{35}SSO_3](s) + H_2O \xrightarrow{\text{heat}} Ag_2{}^{35}S(s) + SO_4^{2-}(aq) + 2H^+(aq)$$

These results indicate that the two sulfur atoms of the thiosulfate ion are not equivalent and that the proposed structure is likely.

Radioactive isotopes have been used to study reaction rates and mechanisms as well as the action of catalysts. By using radioactive tracers it becomes possible to follow the progress of tagged molecules and atoms through a chemical reaction.

For example, the mechanism of the reaction between the sulfite ion and the chlorate ion

$$SO_3^{2-} + ClO_3^- \rightarrow SO_4^{2-} + ClO_2^-$$

has been shown to proceed by the exchange of an oxygen atom. The SO_3^{2-} ion used contained ordinary oxygen atoms, but the ClO_3^- ion was prepared with $^{18}_{8}O$, which is not radioactive but which may be detected by a mass spectrograph. The SO_4^{2-} ion produced by the reaction contained $^{18}_{8}O$ in an amount that would indicate that one $^{18}_{8}O$ atom had been added to each SO_3^{2-} ion.

The mechanism proposed, therefore, is a Lewis nucleophilic displacement of ClO_2^- by SO_3^{2-} on one of the oxygen atoms of the ClO_3^- ion. The $^{18}_{8}O$ atoms are marked with asterisks.

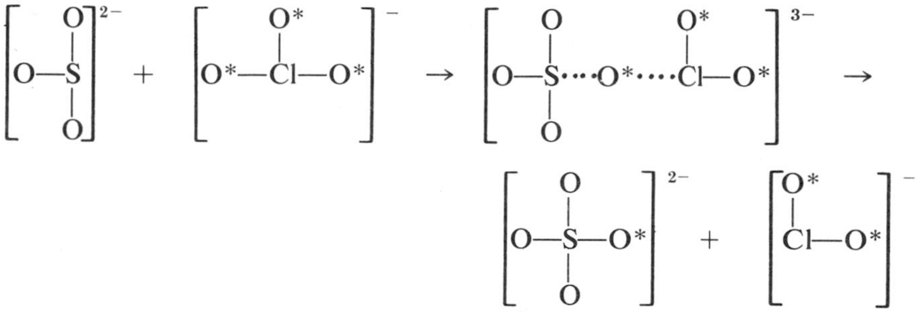

Activation analysis is the determination of the quantity of an element in a sample by bombarding the sample with suitable nuclear projectiles and measuring the intensity of the radioactivity induced in the element being investigated. The induced activity is not influenced by the chemical bonding of the element. Neutrons are the most frequently employed projectiles. Techniques have been developed for the determination of more than 50 elements. This method is particularly valuable for the determination of elements present in extremely low concentrations.

Determinations of very small vapor pressures and very low solubilities may be made conveniently by using tagged materials. For example, the solubility of water in benzene may be determined by using water that has been enriched with the radioactive $^{3}_{1}H$ and measuring the activity of the water-saturated benzene.

WERE THE ALCHEMISTS FAKERS?

In 1941, the dream of the alchemists was realized—an isotope of mercury was transmuted into an isotope of gold. An extremely small amount of a radioactive gold isotope was produced by a nuclear reaction. This expensive process, however, hardly qualifies as a significant source of gold.

The alchemists sought to transmute appreciable amounts of base metals, such as iron and lead, into the noble metals silver and gold by *ordinary chemical means*. This aim today seems fanciful, and there is a tendency to regard the alchemists as either fools or charlatans.

Yet in the minds of these early investigators, transmutation was a real possibility. Cinnabar is an attractive, red crystalline mineral consisting of mercury(II) sulfide. Upon heating in air, cinnabar decomposes. Sulfur dioxide gas is given off, and silvery droplets of mercury remain. In the light of a spectacular transformation such as this, who would rule out the possibility of changing lead into gold?

Other evidence from alchemical laboratories reinforced the belief that transmutation is possible. The addition of a small amount of copper to the white metal zinc produces a yellow alloy, brass, that appears similar to gold. Few alchemists believed that brass is identical to gold, but the study of alloys encouraged them to continue their search.

If an iron sword is dipped into an aqueous solution of copper sulfate, a displacement reaction occurs.

$$Fe(s) + Cu^{2+}(aq) \rightarrow Fe^{2+}(aq) + Cu(s)$$

Some of the iron goes into solution as Fe^{2+} ions, and a chemically equivalent amount of Cu^{2+} ions plates out on the sword as copper metal. The demonstration is dramatic—the iron sword is apparently converted into a copper sword.

Galena (which is PbS) and iron pyrites (which is FeS_2) are minerals that have metallic lusters. Galena is silvery, and iron pyrites is a yellow mineral that looks somewhat like gold and is called *fool's gold.* The alchemists subjected each of these minerals to long involved processes. At the conclusion, all that remained in a crucible that originally contained a sample of galena was a small amount of pure silver. In a similar way, a sample of iron pyrites gave a small amount of pure gold. Silver and gold, however, occur as impurities in these minerals. All that the alchemists had accomplished was the isolation of these impurities. To them, however, the procedures were genuine transmutations.

Most alchemists were serious and honest people who labored within a faulty theoretical framework. Some alchemists, however, particularly in the later years of alchemy, were undoubtedly charlatans.

In one fraudulent "transmutation," a hollow iron rod was filled with powdered gold and the open end plugged with wax. When the rod was used to stir a hot crucible of liquid material (with suitable incantations), the wax melted and the gold was discharged into the crucible. Chaucer (1345–1400) describes a cheating alchemist in his *Canon's Yeoman's Tale,* and the hero of Ben Jonson's play *The Alchemist* (1610) is a swindler.

Nevertheless, until the seventeenth century, most people believed that it might be possible to transmute base metals into gold. In 1689, Robert Boyle was active in securing the repeal of an English law that had been enacted almost 300 years earlier and that prohibited the manufacture of silver and gold.

PROBLEMS

23.1 Briefly describe the exchange theory of nuclear bonding.

23.2 Define the following terms: (a) nuclide, (b) nucleon, (c) critical mass, (d) thermal neutron, (e) fission, (f) fusion, (g) transuranium element.

23.3 Write equations for the following examples of radioactive decay: (a) alpha emission by $^{218}_{85}At$, (b) beta emission by $^{198}_{79}Au$, (c) positron emission by $^{25}_{13}Al$, (d) electron capture by $^{108}_{47}Ag$.

23.4 Write equations for the following examples of radioactive decay: (a) alpha emission by $^{181}_{78}Pt$, (b) beta emission by $^{25}_{11}Na$, (c) positron emission by $^{62}_{29}Cu$, (d) electron capture by $^{37}_{18}Ar$.

23.5 Write equations for the following examples of radioactive decay: (a) alpha emission by $^{221}_{87}Fr$, (b) beta emission by $^{66}_{29}Cu$, (c) positron emission by $^{18}_{9}F$, (d) electron capture by $^{133}_{56}Ba$.

23.6 Write equations for the following examples of radioactive decay: (a) alpha emission by $^{227}_{91}Pa$, (b) beta emission by $^{114}_{47}Ag$, (c) positron emission by $^{39}_{20}Ca$, (d) electron capture by $^{55}_{26}Fe$.

23.7 The nuclide $^{192}_{78}Pt$ decays to $^{188}_{76}Os$ by alpha emission. The mass of $^{192}_{78}Pt$ is 191.9614 u, and the mass of $^{188}_{76}Os$ is 187.9560 u. Calculate the energy released in this process.

23.8 The nuclide $^{213}_{85}At$ decays to $^{209}_{83}Bi$ by alpha emission. The mass of $^{213}_{85}At$ is 212.9931 u, and the mass of $^{209}_{83}Bi$ is 208.9804 u. Calculate the energy released in this process.

23.9 When $^{221}_{87}Fr$ decays, 6.34 MeV and 6.12 MeV α particles are emitted. What is the energy of and the source of the γ radiation accompanying this alpha emission?

23.10 When $^{222}_{88}Ra$ decays, 6.56 MeV and 6.23 MeV α particles are emitted. What is the energy of and the source of the γ radiation accompanying this α emission?

23.11 The disintegration energy is 7.03 MeV for the β decay of $^{20}_{9}F$ to $^{20}_{10}Ne$ (mass, 19.99244 u). What is the mass of $^{20}_{9}F$?

23.12 The disintegration energy is 11.5 MeV for the β decay of $^{11}_{4}Be$ to $^{11}_{5}B$ (mass, 11.00931 u). What is the mass of $^{11}_{4}Be$?

23.13 The nuclide $^{21}_{11}Na$ (mass, 20.99883 u) decays to $^{21}_{10}Ne$ (mass, 20.99395 u) by positron emission. What is the energy released by this radioactive decay process?

23.14 The nuclide $^{18}_{10}Ne$ (mass, 18.00572 u) decays to $^{18}_{9}F$ (mass, 18.00095 u) by positron emission. What is the energy released by this radioactive decay process?

23.15 The energy released by the decay of $^{31}_{16}S$ by positron emission is 5.40 MeV. The nuclide produced, $^{31}_{15}P$, has a mass of 30.97376 u. What is the mass of $^{31}_{16}S$?

23.16 The energy released by the decay of $^{27}_{14}Si$ by positron emission is 3.80 MeV. The nuclide produced, $^{27}_{13}Al$, has a mass of 26.98154 u. What is the mass of $^{27}_{14}Si$?

23.17 The nuclide $^{37}_{18}Ar$ (mass, 36.96678 u) decays to $^{37}_{17}Cl$ (mass, 36.96590 u). Is it energetically possible for the process to occur by positron emission, or must it occur by electron capture?

23.18 Is it energetically possible for $^{41}_{20}Ca$ (mass, 40.96227 u) to decay to $^{41}_{19}K$ (mass, 40.96183 u) by positron emission, or must the process occur by electron capture?

23.19 The disintegration energy for the decay of $^{51}_{24}Cr$ by electron capture is 0.75 MeV, and the mass of the nuclide produced, $^{51}_{23}V$, is 50.9440 u. What is the mass of $^{51}_{24}Cr$?

23.20 Why is electron capture accompanied by the production of X rays?

23.21 The nuclide $^{76}_{35}Br$ has a half-life of 16.5 hours. How much of a 0.0100 g sample remains at the end of 1.00 day?

23.22 The nuclide $^{198}_{79}Au$ has a half-life of 64.8 hours. How much of a 0.0100 g sample remains at the end of 1.00 day?

23.23 The half-life of $^{35}_{16}S$ is 86.7 days. How long will it take for 90.0% of a sample to disappear?

23.24 The half-life of $^{112}_{47}Ag$ is 3.20 hours. How long will it take for 75.0% of a sample to disappear?

23.25 The rate of decay of $^{18}_{9}F$ is such that 10.0% of the original quantity remains after 369 min. (a) What is the rate constant for this radioactive disintegration? (b) What is the half-life of $^{18}_{9}F$?

23.26 The nuclide $^{22}_{11}Na$ decays by positron emission. The rate is such that 76.6% of the original quantity remains after 1.00 year. (a) What is the rate constant? (b) What is the half-life of $^{22}_{11}Na$?

23.27 A sample of a radioactive material initially gives 2500 counts/min and 15 min later gives 2400 counts/min. What is the half-life of the nuclide?

23.28 The carbon from the heartwood of a giant sequoia tree gives 10.8 $^{14}_{6}C$ counts per minute per gram of carbon, whereas the wood from the outer portion of the tree gives 15.3

$^{14}_{6}C$ counts per minute per gram of carbon. How old is the tree?

23.29 Starting with $^{154}_{68}Er$, the successive steps in an artificial decay chain are: α, β^+, ec, α, α. What are the members of the chain?

23.30 One of the naturally occurring decay series is that of the nuclide $^{232}_{90}Th$. The particles successively emitted in one route are: α, β, β, α, α, α, β, α, β, α. Determine in order the members of the chain.

23.31 It is believed that a $^{237}_{93}Np$ disintegration series existed in nature at one time. The members of the series (with the exception of the stable nuclide that ends the series), however, have virtually disappeared through radioactive decay in the time since the earth was created. The particles successively emitted in the $^{237}_{93}Np$ series are: α, β, α, α, β, α, α, α, β, α, β. Determine in order the members of the chain.

23.32 The half-life of $^{237}_{93}Np$ is 2.1×10^6 years, and the age of earth is approximately 4.5×10^9 years. What fraction of the quantity of $^{237}_{93}Np$ that was formed when the earth was created is now present?

23.33 Write equations for the following induced nuclear reactions: (a) $^{82}_{35}Br(n, \gamma)$, (b) $^{10}_{5}B(n, \alpha)$, (c) $^{35}_{17}Cl(n, p)$, (d) $^{7}_{3}Li(p, n)$, (e) $^{130}_{52}Te(d, 2n)$, (f) $^{43}_{20}Ca(\alpha, p)$, (g) $^{237}_{93}Np(\alpha, n)$, (h) $^{238}_{92}U$ ($^{22}_{10}Ne$, 4n).

23.34 Write equations for the following induced nuclear reactions: (a) $^{10}_{5}B(\alpha, n)$, (b) $^{6}_{3}Li(n, \alpha)$, (c) $^{7}_{3}Li$ (d, n), (d) $^{12}_{6}C(p, \gamma)$, (e) $^{96}_{42}Mo(p, n)$, (f) $^{75}_{33}As$ (d, p), (g) $^{45}_{21}Sc$ (α, p), (h) $^{51}_{23}V$ (d, 2n).

23.35 Some transuranium nuclides and the types of induced nuclear reactions that have been used to prepare them are listed. In each case what isotope was used as the starting material? (a) (d, 6n) $^{231}_{93}Np$, (b) (d, 3n) $^{243}_{97}Bk$, (c) (α, $^{3}_{1}H$) $^{250}_{99}Es$, (d) ($^{12}_{6}C$, 4n) $^{245}_{99}Es$, (e) ($^{16}_{8}O$, 5n) $^{249}_{100}Fm$, (f) (α, 2n) $^{255}_{101}Md$, (g) ($^{22}_{10}Ne$, 4n) $^{256}_{102}No$, (h) ($^{10}_{5}B$, 3n) $^{257}_{103}Lr$.

23.36 An isotope of element 105 was prepared by bombarding a $^{249}_{98}Cf$ target with $^{15}_{7}N$ nuclei. Four neutrons were emitted when a $^{249}_{98}Cf$ nucleus absorbed a $^{15}_{7}N$ nucleus. Write an equation for the transformation.

23.37 Calculate the binding energy per nucleon of $^{235}_{92}U$ (mass, 325.0439 u). The masses of the proton, neutron, and electron are 1.007277 u, 1.008665 u, and 0.0005486 u, respectively.

23.38 (a) Calculate the energy released by the fission:

$$^{235}_{92}U + ^{1}_{0}n \rightarrow ^{94}_{38}Sr + ^{139}_{54}Xe + 3^{1}_{0}n$$

The atomic masses are: $^{235}_{92}U$, 235.0439 u; $^{94}_{38}Sr$, 93.9154 u; $^{139}_{54}Xe$, 138.9179 u. The mass of the neutron is 1.0087 u. (b) What percent of the total mass of the starting materials is converted into energy?

23.39 (a) Calculate the energy released by the fusion:

$$^{1}_{1}H + ^{2}_{1}H \rightarrow ^{3}_{2}He$$

The atomic masses are: $^{1}_{1}H$, 1.0078 u; $^{2}_{1}H$, 2.0141 u; $^{3}_{2}He$, 3.0160 u. (b)

What percent of the total mass of the starting materials is converted into energy?

23.40 Each of the following nuclides represents one of the products of a neutron-induced fission of $^{235}_{92}U$. In each case another nuclide and three neutrons are also produced. Identify the other nuclides. (a) $^{148}_{58}Ce$, (b) $^{95}_{37}Rb$, (c) $^{134}_{52}Te$.

Appendix A

INTERNATIONAL SYSTEM OF UNITS (SI)

SI BASE UNITS

MEASUREMENT	UNIT	SYMBOL
length	meter	m
mass	kilogram	kg
time	second	s
electric current	ampere	amp[b]
temperature[a]	kelvin	K
amount of substance	mole	mol
luminous intensity	candela	cd

[a] Temperature may also be expressed in degrees Celsius (symbol °C).
[b] The SI symbol for ampere is A.

SI SUPPLEMENTARY UNITS

	SI UNIT	
MEASUREMENT	NAME	SYMBOL
plane angle	radian	rad
solid angle	steradian	sr

SI PREFIXES

	FACTOR	PREFIX	SYMBOL
10^{12}	1 000 000 000 000	tera-	T-
10^{9}	1 000 000 000	giga-	G-
10^{6}	1 000 000	mega-	M-
10^{3}	1 000	kilo-	k-
10^{2}	100	hecto-	h-
10^{1}	10	deka-	da-
10^{-1}	0.1	deci-	d-
10^{-2}	0.01	centi-	c-
10^{-3}	0.001	milli-	m-
10^{-6}	0.000 001	micro-	μ-

SI PREFIXES (*cont.*)

	FACTOR	PREFIX	SYMBOL
10^{-9}	0.000 000 001	nano-	n-
10^{-12}	0.000 000 000 001	pico-	p-
10^{-15}	0.000 000 000 000 001	femto-	f-
10^{-18}	0.000 000 000 000 000 001	atto-	a-

Appendix B

VALUES OF SOME CONSTANTS AND CONVERSION FACTORS

PHYSICAL CONSTANTS

CONSTANT	SYMBOL	VALUE
Avogadro's number	N	6.0222×10^{23}/mol
Bohr radius	a_0	5.2917×10^{-9} cm
coulomb	coul[a]	amp·s
electron rest mass	m	9.1096×10^{-28} g 5.4859×10^{-4} u
electronic charge, unit charge	e	1.6022×10^{-19} coul 4.8033×10^{-10} $g^{1/2}cm^{3/2}/s$ (or esu)
Faraday	F	9.6487×10^{4} coul/mol
gas constant	R	8.2056×10^{-2} liter·atm/K·mol 8.3143 J/K·mol
joule	J	N·m = kg·m^2/s^2 = V·coul
molar volume, ideal gas at STP	–	22.4136 liter
neutron rest mass	–	1.6749×10^{-24} g 1.008665 u
newton	N	kg·m/s^2
pascal	Pa	N/m^2 = kg/s^2·m
Planck's constant	h	6.6262×10^{-27} erg·s 6.6262×10^{-34} J·s
proton rest mass	–	1.6726×10^{-24} g 1.007277 u
speed of light in vacuum	c	2.9979×10^{10} cm/s
volt	V	J/coul = kg·m^2/amp·s^2

[a] The SI symbol for coulomb is C.

CONVERSION FACTORS

UNIT	ABBREVIATION	DEFINITION OR EQUIVALENT
ångstrom	Å	10^{-8} cm = 10^{-10} m = 0.1 nm = 100 pm
atmosphere	atm	760 torr (or mm of mercury) 1.01325×10^5 N/m^2 (or Pa)
calorie	cal	4.1840 J
curie	Ci	3.7×10^{10}/s
electron volt	eV	1.6022×10^{-19} J
electrostatic unit	esu	3.33564×10^{-10} coul
erg	erg	10^{-7} J
joule	J	10^7 erg
Kelvin temperature scale	K	°K = °C + 273.15° Triple point of water (0.01°C) is 273.16K Freezing point of water (0°C) is 273.15K
torr (or mm of mercury)	torr	1.31579×10^{-3} atm
unified atomic mass unit	u	1.6605×10^{-24}g 931.48 MeV

Appendix C

NOTES ON MATHEMATICAL OPERATIONS

EXPONENTS

An **exponent** is a superscript added to a **base.** It indicates a mathematical operation that is to be performed on the base. In the expression a^n the exponent is n and the base is a. The following types of exponents are frequently encountered.

1. *Exponent is a positive integer.* In the expression a^n the exponent n is the number of times that the base a is to be taken as a factor in the expansion. Therefore, $(n - 1)$ is the number of times the base is to be multiplied by itself. Hence,

$$a^4 = a \times a \times a \times a$$

2. *Exponent is a negative integer.* The expression a^{-n} is the reciprocal of a^n. For example,

$$a^{-4} = \frac{1}{a^4} = \frac{1}{a \times a \times a \times a}$$

3. *Exponent is a fraction of the type* $1/n$. The value of n is the index of a root of the base. Thus,

$$a^{1/2} = \sqrt{a}$$

$$a^{1/3} = \sqrt[3]{a}$$

4. *Exponent is a fraction of the type* m/n. This exponent indicates two operations (those of parts 1 and 3). Hence, $a^{m/n}$ is $\sqrt[n]{a^m}$, and

$$a^{3/2} = \sqrt{a^3} = \sqrt{a \times a \times a}$$

5. *Exponent is zero.* Provided the base is not zero, the value of the expression is unity. Thus,

$$a^0 = 1 \qquad (a \neq 0)$$

Some properties of exponents are summarized in the following equations.

1. $a^m a^n = a^{m+n}$ Thus, $a^4 a^2 = a^6$
2. $(a^m)^n = a^{mn}$ Thus, $(a^4)^2 = a^8$
3. $(ab)^n = a^n b^n$ Thus, $(ab)^3 = a^3 b^3$

4. $\frac{a^m}{a^n} = a^{m-n}$ Thus, $\frac{a^5}{a^2} = a^3$; $\frac{a^2}{a^5} = a^{-3} = \frac{1}{a^3}$

5. $\frac{a^n}{a^n} = 1$ Thus, $\frac{a^3}{a^3} = 1$

SCIENTIFIC NOTATION

Scientific notation is used to simplify the handling of very large or very small numbers. When scientific notation is employed, a value is expressed in the form

$$a \times 10^n$$

where a is a number with one digit to the left of the decimal point, and the exponent n is a positive or negative integer. A value can be converted into this form by moving the decimal point until there is one digit to the left of it. For each place the decimal point is moved to the *left*, n is *increased* by one. For each place the decimal point is moved to the *right*, n is *decreased* by one. For example

$$1{,}904 = 1.904 \times 10^3$$
$$0.000053 = 5.3 \times 10^{-5}$$

1. When numbers expressed in scientific notation are *multiplied*, the exponents of 10 are added.

$$\begin{aligned}(3.0 \times 10^5)(4.0 \times 10^{-2}) &= (3.0 \times 4.0)(10^5 \times 10^{-2}) \\ &= 12 \times 10^3 \\ &= 1.2 \times 10^4\end{aligned}$$

2. In *division*, the powers of 10 are subtracted.

$$\begin{aligned}\frac{6.89 \times 10^{-7}}{3.36 \times 10^3} &= \left(\frac{6.89}{3.36}\right)\left(\frac{10^{-7}}{10^3}\right) \\ &= 2.05 \times 10^{-10}\end{aligned}$$

3. To take a *square root*, we divide the exponent of 10 by 2. If necessary, the number is rewritten so that the exponent of 10 is perfectly divisible by 2.

$$\begin{aligned}(2.21 \times 10^{-7})^{1/2} &= (22.1 \times 10^{-8})^{1/2} \\ &= 4.70 \times 10^{-4}\end{aligned}$$

4. When a *cube root* is taken, the exponent is divided by 3.

$$\begin{aligned}(1.86 \times 10^8)^{1/3} &= (186 \times 10^6)^{1/3} \\ &= 5.71 \times 10^2\end{aligned}$$

5. When numbers expressed in scientific notation are *added* or *subtracted,* the powers of 10 must be the same.

$$(6.25 \times 10^3) + (3.0 \times 10^2) = (6.25 \times 10^3) + (0.30 \times 10^3)$$
$$= 6.55 \times 10^3$$

LOGARITHMS

The **logarithm** of a number is the power to which a base must be raised in order to secure the number. Common logarithms (abbreviated log) employ the base 10. If

$$a = 10^n$$
$$\log a = n$$

and, therefore,

$$\log 1000 = \log 10^3 = 3$$
$$\log 0.01 = \log 10^{-2} = -2$$

The logarithm of a number that is merely 10 raised to a power is the exponent of 10. The logarithm of a number such as 3.540, however, cannot be determined by inspection. The logarithms of numbers from 1 to 10 can be obtained from the table of logarithms found in Appendix D. Decimal points are omitted from this table. Each of the numbers listed is assumed to have a decimal point following the first digit. Each of the logarithms should have a decimal point preceding the value listed. The logarithm of 3.540 is 0.5490.

The logarithm of a number greater than 10 or less than 1 can be obtained in the following way. The number is expressed in scientific notation. For example, consider

$$3.540 \times 10^{12}$$

Since logarithms are exponents and since $a^m a^n = a^{m+n}$, the logarithm of this value is obtained by adding the logarithm of 3.540 to the logarithm of 10^{12}. Hence,

$$\log (3.540 \times 10^{12}) = \log 3.540 + \log 10^{12}$$
$$= 0.5490 + 12$$
$$= 12.5490$$

Another example is

$$\log (2.00 \times 10^{-5}) = \log 2.00 + \log 10^{-5}$$
$$= 0.310 + (-5)$$
$$= -4.699$$

Notice that the number of digits that follow the decimal point in the recorded logarithm is equal to the number of significant figures in the original value.

Sometimes it is necessary to find an **antilogarithm,** the number that corresponds to a given logarithm. In such instances the procedure used to find a logarithm is reversed. The given logarithm is written in two parts: a decimal fraction (called a **mantissa**) and a positive or negative whole number (called a **characteristic**). For example,

$$\text{antilog}\ (3.740) = \text{antilog}\ (0.740 + 3)$$

The antilogarithm of the mantissa (0.740) is obtained by finding the number corresponding to this logarithm in the table (it is 5.50), and the antilogarithm of the characteristic (3) is merely 10^3. Therefore,

$$\text{antilog}\ (3.740) = 5.50 \times 10^3$$

or

$$3.740 = \log\ (5.50 \times 10^3)$$

All the mantissas recorded in a table of logarithms are *positive*. This fact must be taken into account when an antilogarithm of a negative number is found. For example, to take the antilogarithm of −3.158, we must write the number in such a way that the mantissa is positive. Thus,

$$\begin{aligned}\text{antilog}\ (-3.158) &= \text{antilog}\ (0.842 - 4)\\ &= 6.95 \times 10^{-4}\end{aligned}$$

or

$$-3.158 = \log\ (6.95 \times 10^{-4})$$

Since logarithms are exponents, mathematical operations involving logarithms follow the rules for the use of exponents. When each of the following operations is performed, the logarithms of the values involved are found, the logarithms are treated as indicated, and the antilogarithm of the result is secured as the final answer.

1. *Multiplication.* $\log\ (ab) = \log a + \log b$
2. *Division.* $\log\ (a/b) = \log a - \log b$
3. *Extraction of a root.* $\log\ (a^{1/n}) = \frac{1}{n} \log a$
4. *Raising to a power.* $\log\ (a^n) = n \log a$

Logarithms that employ the base 10 are called **common logarithms.**

QUADRATIC EQUATIONS

An algebraic equation in the form

$$ax^2 + bx + c = 0$$

is called a **quadratic equation** in one variable. An equation of this type has two solutions given by the **quadratic formula**

$$x = \frac{-b \pm \sqrt{b^2 - 4ac}}{2a}$$

When the quadratic formula is used to find the answer to a chemical problem, two solutions are obtained. One of them, however, must be discarded because it represents a physical impossibility.

Assume that x in the equation

$$x^2 + 0.50\,x - 0.15 = 0$$

is the number of moles of a gas that dissociate under a given set of conditions. The values of the coefficients are: $a = 1$, $b = 0.50$, and $c = -0.15$; and

$$x = \frac{-0.50 \pm \sqrt{(0.50)^2 - 4(1)(-0.15)}}{2(1)}$$

$$x = \frac{-0.50 \pm 0.92}{2}$$

$$x = +0.21, -0.71$$

The value -0.71 is discarded because a negative amount of a substance is physically impossible.

Appendix D

LOGARITHMS

	0	1	2	3	4	5	6	7	8	9
10	0000	0043	0086	0128	0170	0212	0253	0294	0334	0374
11	0414	0453	0492	0531	0569	0607	0645	0682	0719	0755
12	0792	0828	0864	0899	0934	0969	1004	1038	1072	1106
13	1139	1173	1206	1239	1271	1303	1335	1367	1399	1430
14	1461	1492	1523	1553	1584	1614	1644	1673	1703	1732
15	1761	1790	1818	1847	1875	1903	1931	1959	1987	2014
16	2041	2068	2095	2122	2148	2175	2201	2227	2253	2279
17	2304	2330	2355	2380	2405	2430	2455	2480	2504	2529
18	2553	2577	2601	2625	2648	2672	2695	2718	2742	2765
19	2788	2810	2833	2856	2878	2900	2923	2945	2967	2989
20	3010	3032	3054	3075	3096	3118	3139	3160	3181	3201
21	3222	3243	3263	3284	3304	3324	3345	3365	3385	3404
22	3424	3444	3464	3483	3502	3522	3541	3560	3579	3598
23	3617	3636	3655	3674	3692	3711	3729	3747	3766	3784
24	3802	3820	3838	3856	3874	3892	3909	3927	3945	3962
25	3979	3997	4014	4031	4048	4065	4082	4099	4116	4133
26	4150	4166	4183	4200	4216	4232	4249	4265	4281	4298
27	4314	4330	4346	4362	4378	4393	4409	4425	4440	4456
28	4472	4487	4502	4518	4533	4548	4564	4579	4594	4609
29	4624	4639	4654	4669	4683	4698	4713	4728	4742	4757
30	4771	4786	4800	4814	4829	4843	4857	4871	4886	4900
31	4914	4928	4942	4955	4969	4983	4997	5011	5024	5038
32	5051	5065	5079	5092	5105	5119	5132	5145	5159	5172
33	5185	5198	5211	5224	5237	5250	5263	5276	5289	5302
34	5315	5328	5340	5353	5366	5378	5391	5403	5416	5428
35	5441	5453	5465	5478	5490	5502	5514	5527	5539	5551
36	5563	5575	5587	5599	5611	5623	5635	5647	5658	5670
37	5682	5694	5705	5717	5729	5740	5752	5763	5775	5786
38	5798	5809	5821	5832	5843	5855	5866	5877	5888	5899
39	5911	5922	5933	5944	5955	5966	5977	5988	5999	6010
40	6021	6031	6042	6053	6064	6075	6085	6096	6107	6117
41	6128	6138	6149	6160	6170	6180	6191	6201	6212	6222
42	6232	6243	6253	6263	6274	6284	6294	6304	6314	6325
43	6335	6345	6355	6365	6375	6385	6395	6405	6415	6425
44	6435	6444	6454	6464	6474	6484	6493	6503	6513	6522
45	6532	6542	6551	6561	6571	6580	6590	6599	6609	6618
46	6628	6637	6646	6656	6665	6675	6684	6693	6702	6712
47	6721	6730	6739	6749	6758	6767	6776	6785	6794	6803
48	6812	6821	6830	6839	6848	6857	6866	6875	6884	6893
49	6902	6911	6920	6928	6937	6946	6955	6964	6972	6981

	0	1	2	3	4	5	6	7	8	9
50	6990	6998	7007	7016	7024	7033	7042	7050	7059	7067
51	7076	7084	7093	7101	7110	7118	7126	7135	7143	7152
52	7160	7168	7177	7185	7193	7202	7210	7218	7226	7235
53	7243	7251	7259	7267	7275	7284	7292	7300	7308	7316
54	7324	7332	7340	7348	7356	7364	7372	7380	7388	7396
55	7404	7412	7419	7427	7435	7443	7451	7459	7466	7474
56	7482	7490	7497	7505	7513	7520	7528	7536	7543	7551
57	7559	7566	7574	7582	7589	7597	7604	7612	7619	7627
58	7634	7642	7649	7657	7664	7672	7679	7686	7694	7701
59	7709	7716	7723	7731	7738	7745	7752	7760	7767	7774
60	7782	7789	7796	7803	7810	7818	7825	7832	7839	7846
61	7853	7860	7868	7875	7882	7889	7896	7903	7910	7917
62	7924	7931	7938	7945	7952	7959	7966	7973	7980	7987
63	7993	8000	8007	8014	8021	8028	8035	8041	8048	8055
64	8062	8069	8075	8082	8089	8096	8102	8109	8116	8122
65	8129	8136	8142	8149	8156	8162	8169	8176	8182	8189
66	8195	8202	8209	8215	8222	8228	8235	8241	8248	8254
67	8261	8267	8274	8280	8287	8293	8299	8306	8312	8319
68	8325	8331	8338	8344	8351	8357	8363	8370	8376	8382
69	8388	8395	8401	8407	8414	8420	8426	8432	8439	8445
70	8451	8457	8463	8470	8476	8482	8488	8494	8500	8506
71	8513	8519	8525	8531	8537	8543	8549	8555	8561	8567
72	8573	8579	8585	8591	8597	8603	8609	8615	8621	8627
73	8633	8639	8645	8651	8657	8663	8669	8675	8681	8686
74	8692	8698	8704	8710	8716	8722	8727	8733	8739	8745
75	8751	8756	8762	8768	8774	8779	8785	8791	8797	8802
76	8808	8814	8820	8825	8831	8837	8842	8848	8854	8859
77	8865	8871	8876	8882	8887	8893	8899	8904	8910	8915
78	8921	8927	8932	8938	8943	8949	8954	8960	8965	8971
79	8976	8982	8987	8993	8998	9004	9009	9015	9020	9025
80	9031	9036	9042	9047	9053	9058	9063	9069	9074	9079
81	9085	9090	9096	9101	9106	9112	9117	9122	9128	9133
82	9138	9143	9149	9154	9159	9165	9170	9175	9180	9186
83	9191	9196	9201	9206	9212	9217	9222	9227	9232	9238
84	9243	9248	9253	9258	9263	9269	9274	9279	9284	9289
85	9294	9299	9304	9309	9315	9320	9325	9330	9335	9340
86	9345	9350	9355	9360	9365	9370	9375	9380	9385	9390
87	9395	9400	9405	9410	9415	9420	9425	9430	9435	9440
88	9445	9450	9455	9460	9465	9469	9474	9479	9484	9489
89	9494	9499	9504	9509	9513	9518	9523	9528	9533	9538
90	9542	9547	9552	9557	9562	9566	9571	9576	9581	9586
91	9590	9595	9600	9605	9609	9614	9619	9624	9628	9633
92	9638	9643	9647	9652	9657	9661	9666	9671	9675	9680
93	9685	9689	9694	9699	9703	9708	9713	9717	9722	9727
94	9731	9736	9741	9745	9750	9754	9759	9763	9768	9773
95	9777	9782	9786	9791	9795	9800	9805	9809	9814	9818
96	9823	9827	9832	9836	9841	9845	9850	9854	9859	9863
97	9868	9872	9877	9881	9886	9890	9894	9899	9903	9908
98	9912	9917	9921	9926	9930	9934	9939	9943	9948	9952
99	9956	9961	9965	9969	9974	9978	9983	9987	9991	9996

Appendix E

STANDARD ELECTRODE POTENTIALS AT 25°C[a]

ACID SOLUTION

Half Reaction	$\mathscr{E}°$ *(volts)*
$Li^+ + e^- \rightleftharpoons Li$	−3.045
$K^+ + e^- \rightleftharpoons K$	−2.925
$Rb^+ + e^- \rightleftharpoons Rb$	−2.925
$Cs^+ + e^- \rightleftharpoons Cs$	−2.923
$Ra^{2+} + 2e^- \rightleftharpoons Ra$	−2.916
$Ba^{2+} + 2e^- \rightleftharpoons Ba$	−2.906
$Sr^{2+} + 2e^- \rightleftharpoons Sr$	−2.888
$Ca^{2+} + 2e^- \rightleftharpoons Ca$	−2.866
$Na^+ + e^- \rightleftharpoons Na$	−2.714
$Ce^{3+} + 3e^- \rightleftharpoons Ce$	−2.483
$Mg^{2+} + 2e^- \rightleftharpoons Mg$	−2.363
$Be^{2+} + 2e^- \rightleftharpoons Be$	−1.847
$Al^{3+} + 3e^- \rightleftharpoons Al$	−1.662
$Mn^{2+} + 2e^- \rightleftharpoons Mn$	−1.180
$Zn^{2+} + 2e^- \rightleftharpoons Zn$	−0.7628
$Cr^{3+} + 3e^- \rightleftharpoons Cr$	−0.744
$Ga^{3+} + 3e^- \rightleftharpoons Ga$	−0.529
$Fe^{2+} + 2e^- \rightleftharpoons Fe$	−0.4402
$Cr^{3+} + e^- \rightleftharpoons Cr^{2+}$	−0.408

[a] Data from A. J. de Bethune and N. A. Swendeman Loud, "Table of Electrode Potentials and Temperature Coefficients," pp. 414–424 in *Encyclopedia of Electrochemistry* (C. A. Hampel, editor), Van Nostrand Reinhold, New York, 1964, and from A. J. de Bethune and N. A. Swendeman Loud, *Standard Aqueous Electrode Potentials and Temperature Coefficients,* 19 pp., C. A. Hampel, publisher, Skokie, Illinois, 1964.

$Cd^{2+} + 2e^- \rightleftharpoons Cd$	−0.4029
$PbSO_4 + 2e^- \rightleftharpoons Pb + SO_4^{2-}$	−0.3588
$Tl^+ + e^- \rightleftharpoons Tl$	−0.3363
$Co^{2+} + 2e^- \rightleftharpoons Co$	−0.277
$H_3PO_4 + 2H^+ + 2e^- \rightleftharpoons H_3PO_3 + H_2O$	−0.276
$Ni^{2+} + 2e^- \rightleftharpoons Ni$	−0.250
$Sn^{2+} + 2e^- \rightleftharpoons Sn$	−0.136
$Pb^{2+} + 2e^- \rightleftharpoons Pb$	−0.126
$2H^+ + 2e^- \rightleftharpoons H_2$	0.0000
$S + 2H^+ + 2e^- \rightleftharpoons H_2S$	+0.142
$Sn^{4+} + 2e^- \rightleftharpoons Sn^{2+}$	+0.15
$SO_4^{2-} + 4H^+ + 2e^- \rightleftharpoons H_2SO_3 + H_2O$	+0.172
$AgCl + e^- \rightleftharpoons Ag + Cl^-$	+0.2222
$Cu^{2+} + 2e^- \rightleftharpoons Cu$	+0.337
$H_2SO_3 + 4H^+ + 4e^- \rightleftharpoons S + 3H_2O$	+0.450
$Cu^+ + e^- \rightleftharpoons Cu$	+0.521
$I_2 + 2e^- \rightleftharpoons 2I^-$	+0.5355
$MnO_4^- + e^- \rightleftharpoons MnO_4^{2-}$	+0.564
$O_2 + 2H^+ + 2e^- \rightleftharpoons H_2O_2$	+0.6824
$Fe^{3+} + e^- \rightleftharpoons Fe^{2+}$	+0.771
$Hg_2^{2+} + 2e^- \rightleftharpoons 2Hg$	+0.788
$Ag^+ + e^- \rightleftharpoons Ag$	+0.7991
$2NO_3^- + 4H^+ + 2e^- \rightleftharpoons N_2O_4 + 2H_2O$	+0.803
$Hg^{2+} + 2e^- \rightleftharpoons Hg$	+0.854
$2Hg^{2+} + 2e^- \rightleftharpoons Hg_2^{2+}$	+0.920
$NO_3^- + 4H^+ + 3e^- \rightleftharpoons NO + 2H_2O$	+0.96
$Br_2 + 2e^- \rightleftharpoons 2Br^-$	+1.0652
$O_2 + 4H^+ + 4e^- \rightleftharpoons 2H_2O$	+1.229
$MnO_2 + 4H^+ + 2e^- \rightleftharpoons Mn^{2+} + 2H_2O$	+1.23
$Tl^{3+} + 2e^- \rightleftharpoons Tl^+$	+1.25
$Cr_2O_7^{2-} + 14H^+ + 6e^- \rightleftharpoons 2Cr^{3+} + 7H_2O$	+1.33
$Cl_2 + 2e^- \rightleftharpoons 2Cl^-$	+1.3595
$Au^{3+} + 2e^- \rightleftharpoons Au^+$	+1.402
$PbO_2 + 4H^+ + 2e^- \rightleftharpoons Pb^{2+} + 2H_2O$	+1.455
$Au^{3+} + 3e^- \rightleftharpoons Au$	+1.498
$Mn^{3+} + e^- \rightleftharpoons Mn^{2+}$	+1.51
$MnO_4^- + 8H^+ + 5e^- \rightleftharpoons Mn^{2+} + 4H_2O$	+1.51
$Ce^{4+} + e^- \rightleftharpoons Ce^{3+}$	+1.61
$2HOCl + 2H^+ + 2e^- \rightleftharpoons Cl_2 + 2H_2O$	+1.63
$PbO_2 + SO_4^{2-} + 4H^+ + 2e^- \rightleftharpoons PbSO_4 + 2H_2O$	+1.682
$Au^+ + e^- \rightleftharpoons Au$	+1.691
$MnO_4^- + 4H^+ + 3e^- \rightleftharpoons MnO_2 + 2H_2O$	+1.695
$H_2O_2 + 2H^+ + 2e^- \rightleftharpoons 2H_2O$	+1.776
$Co^{3+} + e^- \rightleftharpoons Co^{2+}$	+1.808

$S_2O_8^{2-} + 2e^- \rightleftharpoons 2SO_4^{2-}$	+2.01
$O_3 + 2H^+ + 2e^- \rightleftharpoons O_2 + H_2O$	+2.07
$F_2 + 2e^- \rightleftharpoons 2F^-$	+2.87

ALKALINE SOLUTION

$Al(OH)_4^- + 3e^- \rightleftharpoons Al + 4OH^-$	−2.33
$Zn(OH)_4^{2-} + 2e^- \rightleftharpoons Zn + 4OH^-$	−1.215
$Fe(OH)_2 + 2e^- \rightleftharpoons Fe + 2OH^-$	−0.877
$2H_2O + 2e^- \rightleftharpoons H_2 + 2OH^-$	−0.82806
$Cd(OH)_2 + 2e^- \rightleftharpoons Cd + 2OH^-$	−0.809
$S + 2e^- \rightleftharpoons S^{2-}$	−0.447
$CrO_4^{2-} + 4H_2O + 3e^- \rightleftharpoons Cr(OH)_3 + 5OH^-$	−0.13
$NO_3^- + H_2O + 2e^- \rightleftharpoons NO_2^- + 2OH^-$	+0.01
$O_2 + 2H_2O + 4e^- \rightleftharpoons 4OH^-$	+0.401
$NiO_2 + 2H_2O + 2e^- \rightleftharpoons Ni(OH)_2 + 2OH^-$	+0.490
$HO_2^- + H_2O + 2e^- \rightleftharpoons 3OH^-$	+0.878

Appendix F

EQUILIBRIUM CONSTANTS AT 25°C

IONIZATION CONSTANTS

MONOPROTIC ACIDS

acetic	$HC_2H_3O_2 \rightleftharpoons H^+ + C_2H_3O_2^-$	1.8×10^{-5}
benzoic	$HC_7H_5O_2 \rightleftharpoons H^+ + C_7H_5O_2^-$	6.0×10^{-5}
chlorous	$HClO_2 \rightleftharpoons H^+ + ClO_2^-$	1.1×10^{-2}
cyanic	$HOCN \rightleftharpoons H^+ + OCN^-$	1.2×10^{-4}
formic	$HCHO_2 \rightleftharpoons H^+ + CHO_2^-$	1.8×10^{-4}
hydrazoic	$HN_3 \rightleftharpoons H^+ + N_3^-$	1.9×10^{-5}
hydrocyanic	$HCN \rightleftharpoons H^+ + CN^-$	4.0×10^{-10}
hydrofluoric	$HF \rightleftharpoons H^+ + F^-$	6.7×10^{-4}
hypobromous	$HOBr \rightleftharpoons H^+ + OBr^-$	2.1×10^{-9}
hypochlorous	$HOCl \rightleftharpoons H^+ + OCl^-$	3.2×10^{-8}
nitrous	$HNO_2 \rightleftharpoons H^+ + NO_2^-$	4.5×10^{-4}

POLYPROTIC ACIDS

arsenic	$H_3AsO_4 \rightleftharpoons H^+ + H_2AsO_4^-$	$K_1 = 2.5 \times 10^{-4}$
	$H_2AsO_4^- \rightleftharpoons H^+ + HAsO_4^{2-}$	$K_2 = 5.6 \times 10^{-8}$
	$HAsO_4^{2-} \rightleftharpoons H^+ + AsO_4^{3-}$	$K_3 = 3 \times 10^{-13}$
carbonic	$CO_2 + H_2O \rightleftharpoons H^+ + HCO_3^-$	$K_1 = 4.2 \times 10^{-7}$
	$HCO_3^- \rightleftharpoons H^+ + CO_3^{2-}$	$K_2 = 4.8 \times 10^{-11}$
hydrosulfuric	$H_2S \rightleftharpoons H^+ + HS^-$	$K_1 = 1.1 \times 10^{-7}$
	$HS^- \rightleftharpoons H^+ + S^{2-}$	$K_2 = 1.0 \times 10^{-14}$
oxalic	$H_2C_2O_4 \rightleftharpoons H^+ + HC_2O_4^-$	$K_1 = 5.9 \times 10^{-2}$
	$HC_2O_4^- \rightleftharpoons H^+ + C_2O_4^{2-}$	$K_2 = 6.4 \times 10^{-5}$
phosphoric	$H_3PO_4 \rightleftharpoons H^+ + H_2PO_4^-$	$K_1 = 7.5 \times 10^{-3}$
	$H_2PO_4^- \rightleftharpoons H^+ + HPO_4^{2-}$	$K_2 = 6.2 \times 10^{-8}$
	$HPO_4^{2-} \rightleftharpoons H^+ + PO_4^{3-}$	$K_3 = 1 \times 10^{-12}$
phosphorous	$H_3PO_3 \rightleftharpoons H^+ + H_2PO_3^-$	$K_1 = 1.6 \times 10^{-2}$
(diprotic)	$H_2PO_3^- \rightleftharpoons H^+ + H_2PO_3^{2-}$	$K_2 = 7 \times 10^{-7}$
sulfuric	$H_2SO_4 \rightarrow H^+ + HSO_4^-$	strong
	$HSO_4^- \rightleftharpoons H^+ + SO_4^{2-}$	$K_2 = 1.3 \times 10^{-2}$
sulfurous	$SO_2 + H_2O \rightleftharpoons H^+ + HSO_3^-$	$K_1 = 1.3 \times 10^{-2}$
	$HSO_3^- \rightleftharpoons H^+ + SO_3^{2-}$	$K_2 = 5.6 \times 10^{-8}$

BASES

ammonia	$NH_3 + H_2O \rightleftharpoons NH_4^+ + OH^-$	1.8×10^{-5}
aniline	$C_6H_5NH_2 + H_2O \rightleftharpoons C_6H_5NH_3^+ + OH^-$	4.6×10^{-10}
dimethylamine	$(CH_3)_2NH + H_2O \rightleftharpoons (CH_3)_2NH_2^+ + OH^-$	7.4×10^{-4}
hydrazine	$N_2H_4 + H_2O \rightleftharpoons N_2H_5^+ + OH^-$	9.8×10^{-7}
methylamine	$CH_3NH_2 + H_2O \rightleftharpoons CH_3NH_3^+ + OH^-$	5.0×10^{-4}
pyridine	$C_5H_5N + H_2O \rightleftharpoons C_5H_5NH^+ + OH^-$	1.5×10^{-9}
trimethylamine	$(CH_3)_3N + H_2O \rightleftharpoons (CH_3)_3NH^+ + OH^-$	7.4×10^{-5}

SOLUBILITY PRODUCTS

BROMIDES

$PbBr_2$	4.6×10^{-6}
Hg_2Br_2	1.3×10^{-22}
$AgBr$	5.0×10^{-13}

CARBONATES

$BaCO_3$	1.6×10^{-9}
$CdCO_3$	5.2×10^{-12}
$CaCO_3$	4.7×10^{-9}
$CuCO_3$	2.5×10^{-10}
$FeCO_3$	2.1×10^{-11}
$PbCO_3$	1.5×10^{-15}
$MgCO_3$	1×10^{-15}
$MnCO_3$	8.8×10^{-11}
Hg_2CO_3	9.0×10^{-17}
$NiCO_3$	1.4×10^{-7}
Ag_2CO_3	8.2×10^{-12}
$SrCO_3$	7×10^{-10}
$ZnCO_3$	2×10^{-10}

CHLORIDES

$PbCl_2$	1.6×10^{-5}
Hg_2Cl_2	1.1×10^{-18}
$AgCl$	1.7×10^{-10}

CHROMATES

$BaCrO_4$	8.5×10^{-11}
$PbCrO_4$	2×10^{-16}
Hg_2CrO_4	2×10^{-9}
Ag_2CrO_4	1.9×10^{-12}
$SrCrO_4$	3.6×10^{-5}

FLUORIDES

BaF_2	2.4×10^{-5}
CaF_2	3.9×10^{-11}
PbF_2	4×10^{-8}
MgF_2	8×10^{-8}
SrF_2	7.9×10^{-10}

HYDROXIDES

$Al(OH)_3$	5×10^{-33}
$Ba(OH)_2$	5.0×10^{-3}
$Cd(OH)_2$	2.0×10^{-14}
$Ca(OH)_2$	1.3×10^{-6}
$Cr(OH)_3$	6.7×10^{-31}
$Co(OH)_2$	2.5×10^{-16}
$Co(OH)_3$	2.5×10^{-43}
$Cu(OH)_2$	1.6×10^{-19}
$Fe(OH)_2$	1.8×10^{-15}
$Fe(OH)_3$	6×10^{-38}
$Pb(OH)_2$	4.2×10^{-15}
$Mg(OH)_2$	8.9×10^{-12}
$Mn(OH)_2$	2×10^{-13}
$Hg(OH)_2$ (HgO)	3×10^{-26}
$Ni(OH)_2$	1.6×10^{-16}
$AgOH$ (Ag_2O)	2.0×10^{-8}
$Sr(OH)_2$	3.2×10^{-4}
$Sn(OH)_2$	3×10^{-27}
$Zn(OH)_2$	4.5×10^{-17}

IODIDES

PbI_2	8.3×10^{-9}
Hg_2I_2	4.5×10^{-29}
AgI	8.5×10^{-17}

OXALATES

BaC_2O_4	1.5×10^{-8}
CaC_2O_4	1.3×10^{-9}
PbC_2O_4	8.3×10^{-12}
MgC_2O_4	8.6×10^{-5}
$Ag_2C_2O_4$	1.1×10^{-11}
SrC_2O_4	5.6×10^{-8}

PHOSPHATES

$Ba_3(PO_4)_2$	6×10^{-39}
$Ca_3(PO_4)_2$	1.3×10^{-32}
$Pb_3(PO_4)_2$	1×10^{-54}
Ag_3PO_4	1.8×10^{-18}
$Sr_3(PO_4)_2$	1×10^{-31}

SULFATES

$BaSO_4$	1.5×10^{-9}
$CaSO_4$	2.4×10^{-5}
$PbSO_4$	1.3×10^{-8}
Ag_2SO_4	1.2×10^{-5}
$SrSO_4$	7.6×10^{-7}

SULFIDES

Bi_2S_3	1.6×10^{-72}
CdS	1.0×10^{-28}
CoS	5×10^{-22}
CuS	8×10^{-37}
FeS	4×10^{-19}
PbS	7×10^{-29}
MnS	7×10^{-16}
HgS	1.6×10^{-54}
NiS	3×10^{-21}
Ag_2S	5.5×10^{-51}
SnS	1×10^{-26}
ZnS	2.5×10^{-22}

MISCELLANEOUS

$NaHCO_3$	1.2×10^{-3}
$KClO_4$	8.9×10^{-3}
$K_2[PtCl_6]$	1.4×10^{-6}
$AgC_2H_3O_2$	2.3×10^{-3}
AgCN	1.6×10^{-14}
AgCNS	1.0×10^{-12}

Appendix G

ANSWERS TO SELECTED NUMERICAL PROBLEMS

CHAPTER 1

1.6 (a) 1×10^5 cm, **(b)** 1×10^{-6} kg, **(c)** 1×10^7 ns, **(d)** 1×10^{-16} Tm, **1.9** 1.667 km, **1.14 (a)** 328.6, **(b)** 131.00, **(c)** 731.9, **(d)** 4.8×10^{-6}, **(e)** 7.8, **(f)** 1.77×10^{16}, **1.19** 80 km/hr, **1.21** 9×10^{-1} t, 1.10 short tons

CHAPTER 2

2.1 (a) 3.239×10^{-22} g, **(b)** 3.087×10^{22} atoms, **2.5** SO_3 highest, Cl_2O lowest, **2.12** $C_6H_{10}NOS$, **2.13** $C_{12}H_{20}N_2O_2S_2$, **2.14** 15% Au, 85% Pd, **2.18** $FeTiO_3$, **2.25** 69.9% Fe, **2.33** 24.7 g BeF_2, **2.36** 84.7% Fe_2O_3, **2.41** 624 kg CaNCN, **2.46** 32.4 g $CsXeF_7$, **2.49** 4.49 g P_2I_4, **2.52** 0.32 mg $KMnO_4$

CHAPTER 3

3.2 37.0°C, **3.6** 2.51 kJ, **3.10** 0.138 J/g°C, **3.15 (a)** 48.18°C, **(b)** −5960 kJ/mol, **3.19** 30.25°C, **3.21** 2.500 kJ, **3.24** 2200 kJ/mol, **3.34** −721.7 kJ, **3.38** −156.6 kJ/mol, **3.45** −44.0 kJ/mol

CHAPTER 4

4.1 (a) 9.00 liter, **(b)** 30 liter, **(c)** 0.9000 liter, **4.5 (a)** 273 K, **(b)** 373 K, **(c)** 210 K, **(d)** 820 K, **4.7** 473°C, **4.9 (a)** 4.10 atm, **(b)** −91°C, **(c)** 91°C, **4.15** 0.204 mol, **4.17** 5.03 liter, **4.22** 1.33 g/liter, **4.24** 60.1 g/mol, **4.26** 15.0 liter $CH_4(g)$, 22.5 liter $O_2(g)$, 15.0 liter $NH_3(g)$ required, 45.0 liter $H_2O(g)$ produced, **4.37** 4.61 liter $O_2(g)$, **4.39** 42.8 liter, **4.41** 5.21 g $La_2(C_2)_3$, **4.49** 1.40 liter $C_2H_2(g)$, **4.51** 140 ml, **4.56** 2.55 g/liter, **4.61 (a)** 22.4 atm, **(b)** 21.8 atm

CHAPTER 5

5.5 $1.48 \times 10^{-12}\%$, **5.6** 0.225 cm, **5.8** 20% ^{10}B and 80% ^{11}B, **5.12** 24.3 u, **5.17** $^{22}Ne^+$, $^{21}Ne^+$, $^{20}Ne^+$, $^{22}Ne^{2+}$, $^{20}Ne^{2+}$, **5.18** 4.29×10^{14}/s, 2.84×10^{-19} J; 7.50×10^{14}/s, 4.97×10^{-19} J, **5.20** $Z = 43$, Tc, **5.22** 121 nm, 91.2 nm, **5.24** $n = 5 \rightarrow n\ 2$

CHAPTER 6

6.9 13.6 eV, **6.13** −1005.1 kJ/mol, **6.15** −294.6 kJ/mol

CHAPTER 7

7.9 43.4%

CHAPTER 8

8.15 29.4 g H_2SO_4, **8.18** 12.0 ml KOH sol'n., **8.20** **(a)** 1/3, **(b)** 1/2, **(c)** 1/3, **(d)** 1, **8.21** 30.3 ml, **8.25** 73.5 g, **8.30** 65.0% $H_2C_2O_4$,**8.32** 73.9% Na_2CO_3

CHAPTER 9

9.16 +18 kJ/mol, **9.18** 93.2 g/mol, **9.20** C_6H_6, **9.37** 3.66 g/cm³, **9.41** 195 g/mol, **9.43** 556 pm, **9.49** 171 pm, **9.51** 135 pm, **9.54** 22° 9′, 49° 2′, **9.56** **(a)** 571 pm, **(b)** 143 pm, **9.58** 333 pm

CHAPTER 10

10.4 −785 kJ/mol, **10.6** −47 kJ/mol, **10.11** 4.81×10^{-2} mol CO_2, 2.12 g CO_2, **10.13** **(a)** N_2, 0.800 atm, O_2, 0.200 atm, **(b)** $X_{N_2} = 1.49 \times 10^{-5}$, $X_{O_2} = 0.80 \times 10^{-5}$, **(c)** −0.0024°C, 10.16 $X_{C_2H_5OH} = 0.143$, $X_{H_2O} = 0.857$, **10.18** **(a)** 8.90 *M* HBr, **(b)** 11.4 *m* HBr, **10.20** 85.7 ml conc $HC_2H_3O_2$, **10.27** 0.642 *M* $AgNO_3$, 0.654 *m* $AgNO_3$, **10.31** 18.00 *N* H_3PO_4, 6.00 *N* HCl, 12.00 *N* H_2SO_4, 0.265 atm, **10.38** 1.12 atm, **10.41** 70 g/mol, **10.44** 500 g $C_2H_4(OH)_2$, **10.48** 114 g/mol, **10.49** 78.6°C, −10.43°C, **10.50** 186 g/mol, **10.52** 180 g/mol, **10.58** 2.67

CHAPTER 11

11.18 **(a)** 1/8, **(b)** 1/4, **(c)** 1/2, **(d)** 1/8, **11.20** **(a)** 33.38 g I_2O_5, 14.0 g CO, **(b)** 0.419 g CO, **11.24** **(a)** 31.0 g $H_4C_2(OH)_2$, 96.0 g HIO_4, **(b)** 2.48 g $H_4C_2(OH)_2$, **11.27** **(a)** 17.0 g H_2S, 126.9 g I_2, **(b)** 0.407 g H_2S, **(c)** 51.3% Bi_2S_3, **11.30** **(a)** 0.340% Fe, 4 atoms Fe, **11.32** 23.9% N_2H_4

CHAPTER 13

13.8 **(a)** 0.0080 mol/liter·s, **(b)** 0.040/s, **(c)** 0.20 liter/mol·s, **13.10** **(a)** rate of formation of C $=k[B]^2$, **(b)** 0.03 liter/mol·s, **13.12** 139 kJ/mol, **13.14** 4.7×10^{-3} liter/mol·s, **13.16** 688 K

CHAPTER 14

14.7 61, **14.8** 0.942 atm², **14.9** **(a)** $[PCl_3] = [Cl_2] = 0.050$ mol/liter, $[PCl_5] = 0.060$ mol/liter, **(b)** 0.042 mol/liter, **14.12** 7.6×10^{-4} mol/liter, **14.14** 8.2×10^{-5} mol C_2H_5OH/liter, **14.16** 0.068 mol BrCl/liter, **14.8** **(a)** 1.35×10^{-3} mol/liter, **(b)** 0.0127 mol/liter, **(c)** 21.3%, **14.21** **(a)** 2.19×10^{-10} mol/liter, **(b)** 6.70×10^{-9} atm, **14.24** 3.35×10^{-3} mol/liter, **14.28** **(a)** 0.50 atm, **(b)** $p_{NO_2} = 0.78$ atm, $p_{N_2O_4} = 1.22$ atm

CHAPTER 15

15.2 3.3×10^{-2}, **15.6** **(a)** 0.082 *M*, **(b)** 0.112 mol $HClO_2$, **15.9** **(a)** 1.8×10^{-3} *M*, **(b)** 0.72%, **15.12** 0.18 *M*, **15.15** 1.0×10^{-3} *M*, **15.16** $[H^+] = 1.5 \times 10^{-5}$ *M*, $[N_3^-] = 0.13$ *M*, $[HN_3] = 0.10$ *M*, **15.19** $[NH_4^+] = 2.7 \times 10^{-5}$ *M*, $[OH^-] = 0.10$ *M*, $[NH_3] = 0.15$ *M*, **15.23** **(a)** 4.14, **(b)** 1.08, **(c)** 3.48, **(d)** 10.52, **15.25** **(a)** 3.6×10^{-14} *M*, **(b)** 4.8×10^{-5} *M*, **(c)** 1.3×10^{-2} *M*, **(d)** 4.6×10^{-13} *M*, **15.31** 2.5×10^{-6}, **15.32** 6.57 to 8.35, **15.35** **(a)** 3.80, **(b)** 0.46%, **15.41** 4.5×10^{-2} *M*, **15.44** $[NH_4^+]/[NH_3] = 0.56$, **15.47** $[H^+] = [H_2AsO_4^-] = 3.9 \times 10^{-3}$ *M*, $[HAsO_4^{2-}] = 5.6 \times 10^{-8}$ *M*, $[AsO_4^{3-}] = 2.0 \times 10^{-18}$ *M*, $[H_3AsO_4] = 0.29$ *M*, **15.51** 2.0×10^{-14}, **15.54** 2.1×10^{-4} *M*, **15.58** 3.3×10^{-13} mol, **15.67** 0.18 *M*, **15.73** 0.47 *M*, **15.76** 0.086 *M*, **15.80** 8.18, **15.86** 4.3×10^{-7}, **15.88** 5.82, **15.89** 0.078 *M*

CHAPTER 16

16.6 1.46 g Bi, **16.8** 3570 min, **16.14** 922 liter Cl_2, **16.18** 6.43 hr, **16.21** **(a)** +0.935 V, **(b)** $Sn + Ag^+ \rightarrow Sn^{2+} + 2Ag$, **(c)** the Ag electrode, **16.25** −2.255 V, **16.36** **(a)** $Pt \mid H_2(g) \mid OH^- \parallel H^+ \mid H_2(g) \mid Pt$, **(b)** +0.828 V, −79.9 kJ, **16.40** 0.25 *M*, **16.46** +1.19 V

CHAPTER 17

17.1 **(a)** −1364.3 kJ/mol, **(b)** −1366.8 kJ, **(c)** −277.9 kJ/mol, **17.4** −5459.55 kJ, **17.8** −184.79 kJ/mol, **17.15** −48.41 kJ, **17.17** **(a)** −101.01 kJ, **(b)** −120.6 J/K, **(c)** −120.6 J/K, **17.19** **(a)** −284.6 J/K mol, **(b)** −92.0 kJ, **17.22** 208 J/K mol, **17.25** −1095.02 kJ, **17.26** 5.12×10^{-4}, **17.32** 0.282 atm, **17.34** +38.5 kJ, **17.36** **(a)** +0.2943 V, **(b)** −56.79 kJ, **(c)** −121.7 J/K, **17.39** **(a)** −509.5 kJ, **(b)** −43.5 kJ/mol, **17.42** **(a)** $2Cu(s) + Cl_2(g) \rightarrow 2CuCl(s)$ **(b)** +0.137 V, **17.46** 7.1×10^3, **17.51** −0.284 V

CHAPTER 18

18.21 65.1%, 43.0% **18.29** pure N_2, 1.250 g/liter; N_2 from air, 1257 g/liter

CHAPTER 20

20.26 1.3×10^{-34}

CHAPTER 22

22.8 8333, **22.9** 4, 8, x^y, **22.23** 7.65×10^5

CHAPTER 23

23.7 2.6 MeV, **23.11** 19,99999 u, **23.24** 6.40 hr, **23.28** 2890 yr, **23.37** 7.59 MeV/nucleon

Index

$Co(NO_2^{1-})_6^{3-}$